Springer Series on
SIGNALS AND COMMUNICATION TECHNOLOGY

Signals and Communication Technology

DVB
The Family of International Standards
for Digital Video Broadcasting, 2nd ed.
U. Reimers
ISBN 3-540-43545-X

Digital Interactive TV and Metadata
Future Broadcast Multimedia
A. Lugmayr, S. Niiranen, and S. Kalli
ISBN 0-387-20843-7

Adaptive Antenna Arrays
Trends and Applications
S. Chandran (Ed.)
ISBN 3-540-20199-8

**Digital Signal Processing
with Field Programmable Gate Arrays**
U. Meyer-Baese
ISBN 3-540-21119-5

**Neuro-Fuzzy and Fuzzy-Neural Applications
in Telecommunications**
P. Stavroulakis (Ed.)
ISBN 3-540-40759-6

SDMA for Multipath Wireless Channels
Limiting Characteristics and Stochastic Models
I.P. Kovalyov
ISBN 3-540-40225-X

Digital Television
A Practical Guide for Engineers
W. Fischer
ISBN 3-540-01155-2

Multimedia Communication Technology
Representation, Transmission
and Identification of Multimedia Signals
J.R. Ohm
ISBN 3-540-01249-4

Information Measures
Information and its Description in Science
and Engineering
C. Arndt
ISBN 3-540-40855-X

Processing of SAR Data
Fundamentals, Signal Processing,
Interferometry
A. Hein
ISBN 3-540-05043-4

Chaos-Based Digital Communication Systems
Operating Principles, Analysis Methods,
and Performance Evaluation
F.C.M. Lau and C.K. Tse
ISBN 3-540-00602-8

Adaptive Signal Processing
Applications to Real-World Problems
J. Benesty and Y. Huang (Eds.)
ISBN 3-540-00051-8

**Multimedia Information Retrieval
and Management**
Technological Fundamentals
and Applications
D. Feng, W.C. Siu, and H.J. Zhang (Eds.)
ISBN 3-540-00244-8

Structured Cable Systems
A.B. Semenov, S.K. Strizhakov,
and I.R. Suncheley
ISBN 3-540-43000-8

UMTS
The Physical Layer of the Universal Mobile
Telecommunications System
A. Springer and R. Weigel
ISBN 3-540-42162-9

Advanced Theory of Signal Detection
Weak Signal Detection
in Generalized Observations
I. Song, J. Bae, and S.Y. Kim
ISBN 3-540-43064-4

Wireless Internet Access over GSM and UMTS
M. Taferner and E. Bonek
ISBN 3-540-42551-9

Gabriel Vasilescu

Electronic Noise and Interfering Signals

Principles and Applications

With 385 Figures and 58 Tables

Springer

Dr. Gabriel Vasilescu
Université Pierre & Marie Curie
Laboratoire des Instruments et Systèmes d'Ile de France (LISIF)
4, Place Jussieu, BC 252
75252 Paris Cedex 05
France

ISBN 3-540-40741-3 **Springer Berlin Heidelberg New York**

Library of Congress Control Number: 2004112251

This work is subject to copyright. All rights are reserved, whether the whole or part of the material is concerned, specifically the rights of translation, reprinting, reuse of illustrations, recitation, broadcasting, reproduction on microfilm or in other ways, and storage in data banks. Duplication of this publication or parts thereof is permitted only under the provisions of the German Copyright Law of September 9, 1965, in its current version, and permission for use must always be obtained from Springer-Verlag. Violations are liable to prosecution under German Copyright Law.

Springer is a part of Springer Science+Business Media

springeronline.com

© Springer-Verlag Berlin Heidelberg 2005
Originally published in French by Dunod Éditeur in 1999.
Printed in Germany

The use of general descriptive names, registered names, trademarks, etc. in this publication does not imply, even in the absence of a specific statement, that such names are exempt from the relevant protective laws and regulations and therefore free for general use.

Typesetting and final processing by PTP-Berlin Protago-TeX-Production GmbH, Germany
Cover-Design: design & production GmbH, Heidelberg
Printed on acid-free paper   62/3020/Yu - 5 4 3 2 1 0

In memory of those magnificent and charming people, amateur radio operators in the late 1920s and early 1930s, who enthusiastically contributed, in a quite anonymous but effective way, to the progress of radio communication systems and related techniques.

This book is dedicated especially to the operator of short-wave station YR5VI, which was my father.

Preface

This book addresses the noise theory of electronic circuits, with emphasis on low-noise design principles. Various noise sources are presented, together with their associated mechanisms and models, and with methods for calculating their contribution to the overall circuit noise. Intrinsic noise and interfering signals having different causes, they require separate treatment.

Intrinsic noise arises within an electronic circuit itself, and is merely a consequence of all noise sources associated with the physical operation of any electronic device. Some of these sources can be controlled through careful processing during fabrication, but most of them are fundamental and not dependent on the technology. Globally, their cumulative effects represent the ultimate limit on a circuit's ability to process weak signals.

On the other hand, interference is the result of circuit contamination by extraneous signals emitted by external sources. Several means to protect a circuit against perturbing electromagnetic fields are available, but the best protection is still reducing (if possible) the emission of spurious signals.

This book is organized in three parts.

Part I covers intrinsic noise generated inside the most common devices. An important group of noise signals is random in nature, and since they are not always well-understood, Chap. 2 provides the underlying principles. Chapter 3 is dedicated to the study of the most frequently encountered noise sources and their mechanisms. Since each resistor has two noise sources associated with it, and each transistor at least three, it is obvious that at the circuit level, noise sources seem to be distributed throughout. When trying to evaluate the resulting noise using circuit theory concepts, this is not easy even for small circuits. For large circuits, it becomes all but impossible. As a practical remedy, the effects of all internal noise sources are often lumped into two equivalent generators, situated either at the input or the output of a noiseless, but otherwise identical, circuit. This technique is described in Chap. 5. Its merit is to separate noise sources from the electrical circuit, which makes calculation easier. The definitions of various noise parameters are given in Chap. 4, preceded by the background needed to introduce them. Chapter 6 is concerned with the computation of the noise parameters of the circuit (regarded as a two-port), either by hand (with the method of Hillbrand and Russer), or by simulation with NOF (which is a general computer program for noise analysis of RF and microwave circuits, proposed by the author). Finally, Chap. 7 details noise models of the most commonly encountered circuit devices (resistors, capacitors, diodes, various transistor types, operational amplifiers, etc.), with emphasis on the models found in SPICE.

Part II aims for a comprehensive study of extrinsic noise (interfering signals), which is in part the subject of books on electromagnetic compatibility. The present book, however, is mainly concerned with noise originating from nearby electronic circuits, rather that from lightning discharges or military applications.

Traditionally, so called extrinsic noise arises from unintentionally coupled signals emitted either by remote noise sources (sky noise, galactic noise, distant radio or TV transmitters, radars, etc.) or local noise sources (mains supply lines or nearby perturbing circuits). One may ask why we must worry about this kind of noise? The reason is that when a circuit intended to process weak signals fails to operate correctly, from the user's point of view it doesn't much matter whether the noise contaminating the useful signal arises from the circuit itself or is picked up from the outside. In both cases the noise will obscure the information content of the signal, which, in the worst scenarios, cannot be recovered. Therefore, it is not reasonable to strive exclusively to suppress the intrinsic noise of a circuit, lest for want of protection against interfering signals, circuit operation is compromised!

An overview of all extrinsic noise sources is given in Chap. 8. Chapter 9 is dedicated to classical methods of eliminating interference (grounding, balancing and filtering, shielding, etc.). Chapter 10 deals with various methods of controlling the emission of conducted or radiated interfering signals from electronic systems, Chap. 11 presents interconnection modeling and crosstalk, and Chap. 12 suggests methods of enhancing the immunity of victim circuits to interfering signals.

Part III is concerned with practical applications. Chapter 13 exposes the general principles of low-noise circuit design. Chapter 14 provides a collection of case studies, including some simple techniques to measure the most significant noise quantities with the standard equipment found in any electronics laboratory. Chapter 15 is a collection of case studies related to sensors and their associated amplifiers, mostly encountered in industrial applications, while Chap. 16 covers case studies specific to communication systems. Chapter 17 is devoted to computer-aided noise analysis (with PSPICE and NOF), while Chap. 18 is dedicated to case studies covering various topics, such as protection against spurious signals, grounding, filtering, crosstalk estimation between adjacent interconnects, contact protection and so on.

Overall, more than 100 case studies and applications are proposed, and all are provided with detailed numerical solutions.

This book, and particularly the pool of case studies, is intended to be particularly useful to people concerned with noise and low-noise design: solid-state analog circuit designers, electronics engineers designing with integrated circuits or discrete devices, researchers in the solid-state circuits area, materials scientists and physicists, and electrical engineers. It should be of general interest to students of electronics enrolled in undergraduate or postgraduate programs, to industrial maintenance personnel, and to workers in fields such as noise metrology, low-noise design, and communication systems.

This book requires no advanced mathematical training, since Chap. 2 reviews the fundamental tools before using them later on. However, a knowledge of basic calculus, elementary circuit theory, and general principles of electronic circuits is assumed.

Acknowledgments

This textbook is based on the first French edition of "Bruits et signaux parasites" published by Dunod, France in 1999. It is evident that translating, updating and adapting it into the English language for publication by Springer, Heidelberg, is a task that cannot be undertaken alone.

I would therefore like to thank all those who have given advice and closely participated in the preparation of this edition. I should like to express my gratitude to Maryvonne Vitry (Dunod, France) for her useful advice and effort in presenting the French edition at the Frankfurt Book Fair. Both Maryvonne and Caroline Davis (Springer, Heidelberg) were the enchanting fairies that gently built bridges to ensure the success of this project.

My thanks go equally to Dr. Dieter Merkle, Engineering Editor, who wisely supervised this project and took the right decisions which enabled the author to focus on up-dating the book. Special thanks to Petra Jantzen, Dr. Merkle's assistant, for her skills in rounding-off corners and finding the right words.

I am also indebted to Dr. Mark Damashek for his valuable comments and careful revision of the English manuscript, and to Ian Mulvany for his final touches to the text. Last but not least, I should like to thank my student Sylvain Feruglio for helping me in drawing Figs. 2.4, 2.6, 18.32 and to PTP-Berlin Protago-Tex-Production GmbH, Germany, for their high-quality work of preparing and layouting figures and overall text.

Paris, August 2004 *Gabriel Vasilescu*

Contents

1 Introduction 1
   1.1 Noise Definitions 1
      1.1.1 Traditional Definitions 1
      1.1.2 Definitions Pertinent to Particular Fields 3
   1.2 Overview 4
      1.2.1 Intrinsic Noise 4
      1.2.2 Extrinsic Noise 5
   1.3 Comparing Intrinsic and Extrinsic Noise 5
      1.3.1 General Properties 5
      1.3.2 Bandwidth 6
      1.3.3 Amplitude 6
      1.3.4 Noise Calculation 6
      1.3.5 Low-Noise Design 7
      1.3.6 Noise Immunity 7
   1.4 Why Is Noise a Concern? 8
      1.4.1 Noise in Communication Systems 9
      1.4.2 Noise in Industrial Applications 10
      1.4.3 Benefits of Noise 11
      1.4.4 Conclusion 12

Part I Intrinsic Noise

2 Fundamental Concepts 15
   2.1 Introduction 15
   2.2 Fluctuations: Signal Theory Approach 16
      2.2.1 Average Value 17
      2.2.2 Mean Square Value 18
      2.2.3 Form Factor 20
      2.2.4 Peak Factor 20
      2.2.5 Correlation 20

Contents

- 2.3 Fluctuations: Probabilistic Approach ... 23
  - 2.3.1 One Random Variable ... 23
  - 2.3.2 Two Random Variables ... 29
  - 2.3.3 Energy and Power Spectra ... 33
  - 2.3.4 Fourier Analysis of Fluctuations ... 35
  - 2.3.5 Correlation Matrix ... 39

3 Physical Noise Sources ... 45
- 3.1 Introduction ... 45
- 3.2 Noise Sources ... 46
  - 3.2.1 Thermal Noise ... 46
  - 3.2.2 Diffusion Noise ... 51
  - 3.2.3 Shot Noise ... 52
  - 3.2.4 Quantum Noise ... 56
  - 3.2.5 Generation-Recombination Noise (G-R Noise) ... 58
  - 3.2.6 1/f Noise (Flicker Noise or Excess Noise) ... 60
  - 3.2.7 Popcorn (Burst) Noise ... 64
  - 3.2.8 Avalanche Noise ... 65
- 3.3 Conclusion ... 67

4 Noise Parameters ... 69
- 4.1 Introduction ... 69
- 4.2 Definitions Concerning Electrical Power and Bandwidth ... 70
  - 4.2.1 Normalized Power ... 70
  - 4.2.2 Various Definitions of Power ... 72
  - 4.2.3 Available Power and Available Power Gain ... 74
  - 4.2.4 Exchangeable Power and Exchangeable Power Gain ... 77
  - 4.2.5 Various Gain Definitions ... 79
  - 4.2.6 Noise Bandwidth ... 80
- 4.3 Noise Parameters of Linear One-Ports (Spot Values) ... 83
  - 4.3.1 Equivalent Noise Resistance ... 83
  - 4.3.2 Equivalent Noise Current ... 85
  - 4.3.3 Noise Temperature ... 85
  - 4.3.4 Noise Ratio ... 87
  - 4.3.5 Noise Equivalent Power (NEP) ... 88
  - 4.3.6 Signal-to-Noise Ratio (S/N Ratio) ... 88
- 4.4 Noise Parameters of Linear One-Ports (Average Values) ... 89
- 4.5 Noise Parameters of Linear Two-Ports (Spot Values) ... 89
  - 4.5.1 Equivalent Input Noise ... 89
  - 4.5.2 Signal-to-Noise Ratio ... 91
  - 4.5.3 Input Noise Temperature ... 92
  - 4.5.4 Operating Noise Temperature ... 96
  - 4.5.5 Effective Noise Temperature ... 97
  - 4.5.6 Noise Factor (Noise Figure) ... 97
  - 4.5.7 Operating Noise Factor ... 103

		4.5.8	Extended Noise Factor 104
		4.5.9	Noise Measure 105
	4.6	Average Values of Two-Port Noise Parameters 107
	4.7	Noise of a Linear Multiport 108
		4.7.1	Output Noise Power 108
		4.7.2	Input Noise Temperature 109
		4.7.3	Operating Noise Temperature 110
	4.8	Conclusion 110

5 Noise Analysis of Linear Circuits 113
  5.1 Introduction 113
  5.2 Noise Models of One-Port Circuits 114
       5.2.1 One-Port at Uniform Temperature................... 115
       5.2.2 One-Port at Different Temperatures 119
       5.2.3 Conclusion 121
  5.3 Time Domain Analysis of Noisy Two-Ports 121
       5.3.1 Background 121
       5.3.2 Noisy Two-Port 124
       5.3.3 Model of Rothe and Dahlke........................ 125
       5.3.4 Basic Relationships Among Noise Parameters 129
  5.4 Correlation Matrices 134
       5.4.1 Linear Two-Port Circuits........................... 134
       5.4.2 Linear Multiport Circuits........................... 140
  5.5 Conclusion 142

6 Frequency Domain Noise Analysis of Linear Multiports... 143
  6.1 Introduction 143
  6.2 Method of Hillbrand and Russer.......................... 144
       6.2.1 Description 144
       6.2.2 Noise Parameters of an Attenuating Pad 146
       6.2.3 Collection of Elementary Passive Circuits............ 149
  6.3 Noise Analysis of Linear Multiport Circuits 152
  6.4 Algorithm 156
  6.5 Example 158
       6.5.1 FET Microwave Amplifier 158
       6.5.2 Concluding Remark 161

7 Noise Models of Electronic Devices 163
  7.1 Resistor Noise.. 163
  7.2 Capacitor Noise 169
  7.3 Inductor Noise 171
  7.4 Noise in Junction Diodes 172
       7.4.1 Ideal PN Junctions 172
       7.4.2 Forward-Biased Diodes............................. 174
       7.4.3 Reverse-Biased Diodes 176

Contents

- 7.4.4 PSPICE Model 177
- 7.5 Battery Noise 177
- 7.6 Noise in Bipolar Transistors 178
  - 7.6.1 Preliminary Remarks 179
  - 7.6.2 Physical Aspects of Noise 180
  - 7.6.3 Nielsen's Model 184
  - 7.6.4 Hawkins's Model 186
  - 7.6.5 Model of Motchenbacher and Fitchen 188
  - 7.6.6 Fukui Model 191
  - 7.6.7 Model of Voinigescu et al. 195
  - 7.6.8 PSPICE Model 196
  - 7.6.9 Conclusion 198
  - 7.6.10 Noise in Heterojunction Bipolar Transistors (HBTs) .. 199
- 7.7 Noise of Junction Field Effect Transistors (JFETs) 202
  - 7.7.1 Background 202
  - 7.7.2 Noise Mechanisms 206
  - 7.7.3 Van der Ziel Model 208
  - 7.7.4 Robinson Model 209
  - 7.7.5 Ambrozy Model 210
  - 7.7.6 Bruncke Model 210
  - 7.7.7 PSPICE Model 212
  - 7.7.8 Conclusion 212
- 7.8 Noise in MOS Transistors 213
  - 7.8.1 Introduction 213
  - 7.8.2 PSPICE Model 214
  - 7.8.3 Model of Nicollini, Pancini, and Pernici 216
  - 7.8.4 Model of Wang, Hellums, and Sodini 217
  - 7.8.5 Lee's Scalable Model 218
  - 7.8.6 Conclusion 219
- 7.9 Noise of MESFET Transistors 219
  - 7.9.1 Introduction 219
  - 7.9.2 Physical Aspects 220
  - 7.9.3 Bächtold Model 222
  - 7.9.4 Model of Pucel, Haus, and Statz 223
  - 7.9.5 Fukui Model 226
  - 7.9.6 Podell Model 228
  - 7.9.7 Heinrich Model 230
  - 7.9.8 Model of Escotte and Mollier 233
  - 7.9.9 HSPICE Model 234
  - 7.9.10 Conclusion 235
- 7.10 Noise of HEMT Transistors 236
  - 7.10.1 Introduction 236
  - 7.10.2 Cappy Model 239
  - 7.10.3 Pospieszalski Model 240
  - 7.10.4 Model of Hickson, Gardner, and Paul 243

7.10.5 Model of Klepser, Bergamaschi, Schefer, Diskus, Patrick, and Bächtold 245
7.10.6 Conclusion 247
7.11 Noise in Operational Amplifiers 249

Part II Interfering Signals

8 External Noise 255
   8.1 External Noise Sources 256
      8.1.1 Natural Noise Sources 256
      8.1.2 Man-Made Noise Sources 260
   8.2 Glossary of Terms 265
   8.3 Interference Problem 267
   8.4 Coupling Paths 268
      8.4.1 Methods of Noise Coupling 269
      8.4.2 Coupling Modes 273

9 Interference Reduction Methods 279
   9.1 Electromagnetic Shielding 279
      9.1.1 Background 279
      9.1.2 Circuit Shielding 286
      9.1.3 Estimating Shielding Effectiveness 289
      9.1.4 Performance Degradation 290
      9.1.5 Conclusion 295
   9.2 Filtering and Balancing 295
      9.2.1 Filtering Interfering Signals 296
      9.2.2 Balanced Circuits 299
   9.3 Grounding and Bonding 301
      9.3.1 Equipment Grounding 301
      9.3.2 Noise Related to Grounding 306
      9.3.3 Miscellaneous 308
   9.4 Proper Use of Cables 309

10 Methods of Reducing Emission of Interfering Signals 315
   10.1 Disturbances Associated with Mains Distribution 316
   10.2 Noise Arising from DC Power Supplies 320
      10.2.1 DC Power Supplies with Full-Wave Rectifiers 320
      10.2.2 Voltage Regulators 322
      10.2.3 Ungrounded (Floating) Power Supplies 324
      10.2.4 Switching-Mode Power Supplies 325
      10.2.5 Ripple Filtering 326
   10.3 Noise Generated by Mechanical Contact Switching 327
      10.3.1 Gas Discharge 328
      10.3.2 Arc Discharge 328

10.4 Noise Emitted by Digital Circuits 332
    10.4.1 Introduction 332
    10.4.2 Inductive Noise 334
    10.4.3 Noise Related to Clock Radiation 336
    10.4.4 Reflections Due to Mismatch on the Lines 338
    10.4.5 Abrupt Demand for DC Supply Current 339
10.5 Transformer Noise 340
    10.5.1 Noise Sources 340
    10.5.2 Solutions to Reduce Noise Coupling 341
10.6 Noise Due to Electrostatic Discharge 343
    10.6.1 Accumulation of Electrostatic Charge 344
    10.6.2 Discharge Phase 347
    10.6.3 Prevention and Control 349

11 Interconnection Modeling and Crosstalk 351
11.1 Introduction 351
11.2 Interconnect Modeling 352
    11.2.1 Interconnect Resistance 353
    11.2.2 Mutual Capacitance and Mutual Inductance 354
    11.2.3 Capacitance Estimation 358
    11.2.4 Inductance Estimation 361
    11.2.5 Modeling a Multiconductor Line 364
    11.2.6 Modeling a Trace on a Printed Circuit Board 366
11.3 Crosstalk 370
    11.3.1 Basic Concepts 370
    11.3.2 Crosstalk Due to Dominant Capacitive Coupling 371
    11.3.3 Crosstalk Due to Dominant Inductive Coupling 373
    11.3.4 Crosstalk Due to Electromagnetic Coupling 375
11.4 Interconnect Optimization 380
    11.4.1 Layout and Printed Circuit Board 380
    11.4.2 Managing PCB Optimization 381
    11.4.3 Coupling Effects in VLSI Design 385

12 Methods of Increasing Immunity to Interfering Signals ... 389
12.1 Balancing 389
12.2 Filtering 393
    12.2.1 Decoupling Filters 393
    12.2.2 Filtering of Wires and Cables 396
12.3 Grounding 398
12.4 Practical Advice on Reducing Noise and Interference
    at the Circuit Level 402
    12.4.1 Interference Control 403
    12.4.2 Guidelines for Circuit Design 404
12.5 Increasing System Immunity to Interference:
    Bluetooth Approach 408

Part III Case Studies

13 Low-Noise Circuit Design 413
13.1 Introduction 413
13.2 Low-Noise Design Techniques for Low-Frequency Circuits 415
13.2.1 Rules of Low-Noise Design 415
13.2.2 Noise Performance of Amplifiers 417
13.2.3 Noise Matching with a Coupling Transformer 424
13.2.4 Noise Matching by Paralleling Input Devices 426
13.2.5 Selection of Active Devices 428
13.2.6 Feedback 430
13.2.7 Application 1: Sensor and Its Preamplifier 434
13.2.8 Application 2: Dolby Noise Reduction System 437
13.3 Low-Noise Design Techniques for Microwave Circuits 439

14 Noise Performance Measurement 443
14.1 Noise Sources 443
14.1.1 Introduction 443
14.1.2 Case Studies 444
14.2 Noise Power Measurement 451
14.2.1 Introduction 451
14.2.2 Case Studies 451
14.3 Two-Port Noise Performance Measurement 457
14.3.1 Introduction 457
14.3.2 Case Studies 457
14.4 Miscellaneous 468
14.4.1 Passive Circuits 468
14.4.2 Impedances at Unequal Temperatures 476
14.4.3 Low-Frequency Amplifier 482

15 Noise in Sensing Circuits 485
15.1 Preamplifiers 485
15.1.1 Underlying Principles 485
15.1.2 Case Studies 487
15.2 Sensing Circuits 504
15.2.1 Underlying Principles 504
15.2.2 Case Studies 505
15.3 Circuits with Operational Amplifiers 515

16 Noise in Communication Systems 535
16.1 Attenuators 535
16.1.1 Underlying Principles 535
16.1.2 Case Studies 535
16.2 Multistage Amplifiers 542

XVIII   Contents

    16.3 Low-Noise Input Stages 549
    16.4 Receivers 553
        16.4.1 Background 553
        16.4.2 Case Studies 554
    16.5 Space Communication Systems 567
        16.5.1 Background 567
        16.5.2 Case Studies 572

17 Computer-Aided Noise Analysis 583
    17.1 Noise Simulation with PSPICE 584
        17.1.1 SPICE – An Overview 584
        17.1.2 Noise Analysis 592
        17.1.3 Simulation Techniques 592
        17.1.4 Case Studies 597
    17.2 Noise Simulation with NOF 621
        17.2.1 NOF – An Overview 621
        17.2.2 Case Studies 623

18 Protection Against Interfering Signals 629
    18.1 Techniques to Reduce Interference 629
        18.1.1 Shielding 629
        18.1.2 Filtering 640
        18.1.3 Grounding 648
    18.2 Interconnect Modeling 654
        18.2.1 Evaluation of Stray Elements Associated
             with Interconnects 654
        18.2.2 Crosstalk 663
    18.3 Interfering Signals 669
        18.3.1 Transducers and Associated Circuits 669
        18.3.2 Logic Circuits 677
        18.3.3 Contact Protection 684

References 689

Appendix 703

Index 705

1

Introduction

> "The chess board is the world; the pieces are the phenomena of the universe; the rules of the game are what we call the laws of Nature. The player on the other side is hidden from us. We know that his play is always fair, just and patient. But also we know, to our cost, that he never overlooks a mistake, or makes the smallest allowance for ignorance."
>
> (T.H. Huxley)

1.1 Noise Definitions

Actually, *noise* is a term that is overused and sometimes misused. Many definitions have been proposed, referring either to the classical term or to terms emerging from nonconventional fields.

1.1.1 Traditional Definitions

In the author's view, the most noteworthy definitions of noise (relating to electronic devices and circuit theory) are the following:

- The IEEE Standard Dictionary of Electrical and Electronics Terms [1] defines noise (as a general term) as *unwanted disturbances superposed upon a useful signal that tend to obscure its information content*. This definition is versatile, since it applies both to intrinsic and extrinsic noise (*intrinsic noise* is the noise generated inside a system, while the term *extrinsic noise* designates noise originating elsewhere).
  A second definition, also presented in [1], states that noise is *an undesired disturbance within the useful frequency band*, with the stipulation that *undesired disturbances within the useful frequency band produced by other*

services may be called interference. Although this formulation is equivalent to the previous ones, it has the merit of introducing a definition for *interference.*

- According to the Oxford English Reference Dictionary [2], *noise* is defined as *irregular fluctuations accompanying a transmitted signal but not relevant to it.* Note that this definition is quite general and applies to both intrinsic and extrinsic noise. It is as powerful as the IEEE definition.
- As stated in the Modern Dictionary of Electronics [3], *noise* is defined as *any unwanted electrical disturbance or spurious signal which modifies the transmitting, indicating, or recording of desired data.* It is easy to see that this definition (apparently close to that proposed by IEEE) makes provision for intrinsic and extrinsic noise, and the meaning is illuminated by explicitly indicating ways in which noise can corrupt information content during processing. However, in my opinion, insisting on the fact that noise *modifies* the transmitted, displayed, or recorded data restricts the generality of the term (as a matter of fact, noise exists even if no damage to the processed data is observed).
- The New Penguin Dictionary of Science [4] proposes the following definition of the term noise: *any undesirable disturbance, usually random, in an electronic, electrical or communication system which interferes with its intended operation.* The merit of this definition is its versatility, because it covers both intrinsic and extrinsic noise; nevertheless, as with the previous definition, the presence of interference is a sufficient but not necessary condition for noise to exist.

From a historic point of view, people working in the field of analog circuits were the first to be concerned with the noise problem. Since 1955, Van der Ziel noted that from a conceptual standpoint, the term "spontaneous fluctuation" is more appropriate than "noise".

Later, Rheinfelder [5], followed by Hartmann [6], proposed definitions similar to that given by IEEE. They pointed out that "unwanted signal" means a signal which *does not convey useful information.*

In 1966, Chenette elaborated on the definition on noise by insisting on the fact that the "unwanted disturbances" must be described in terms of their *statistical properties.*

Another apparently equivalent term is often encountered in the literature, namely *electrical noise.* I suspect that the term "electrical" was coupled to "noise" to avoid confusion with "acoustic noise" or other forms of noise. Curiously, as we shall see, the effect was to over-restrict the meaning.

According to [1], *electrical noise is the unwanted electrical energy other than crosstalk present in a transmission system.* Note that in [3] a very similar definition is proposed. Compared with the definition of *noise*, it is obvious that the content of this term is restricted, since only transmission systems are considered and crosstalk is excluded.

1.1.2 Definitions Pertinent to Particular Fields

- In the field of digital circuits (which is quite a different universe, since digital circuit designers think exclusively in terms of the time domain) **noise** is defined as *any deviation of a signal from its nominal value in those subintervals of time when it should otherwise be stable* [7]. As a matter of fact, this deviation from the nominal value is called *fluctuation* in analog circuit theory, and with this remark the proposed definition recovers the classical definition of noise. However, here the *deviation* may result from such causes as variations in the DC power supply, shifting values of parameters due to processing, temperature variation, ageing, incident radiation, etc., which are not commonly taken into account in analog circuits.
- Another special term encountered in logic circuits is **inductive noise**. This concept is associated with voltage transients that appear across an inductor when the current in it varies abruptly (as in switching circuits). Due to unwanted parasitic coupling, the voltage transients are transferred to nearby circuits and perturb them. Note that this kind of noise is not random, since its waveform can often be predicted.
- In [8], **static noise** is introduced with respect to an evaluation node in a CMOS digital integrated circuit. An evaluation node is any node that is used to carry information between the logic gates of the circuit. In this context, *static noise is any deviation from the nominal supply or ground voltages at evaluation nodes which should otherwise represent stable logic one or zero*.
- The designers of digital circuits associated with active pixel sensors (APS) have introduced the so-called **temporal noise** [9,10]. One special feature of this field is that logic circuits are dynamic circuits (i.e., their topology, as well as the values of their elements, are time-dependent, because transistors continually switch between saturation and cutoff). Furthermore, in most APS applications, more time is spent in transients than in stable states. It is clear that noise must be considered in a different way, because there is no such thing as steady-state operation. According to [9], *temporal noise refers to the time-dependent fluctuations that are of fundamental origin, unlike fixed-pattern noise. Circuit-oriented temporal noise originating from substrate coupling or poor power-supply rejection is not included*. Note that in the proposed definition, "time-dependent fluctuations" does not refer to their amplitude (which of course is time-dependent!) but rather to the fact that their characteristics are related to the continual modification of MOS transistor quiescent point (which can explore many regions of the i-v characteristic, from weak inversion to saturation). Note also that in contrast to integrated circuit theory, where noise originating from substrate coupling is considered to be very important, here it is neglected.

- The term **input noise** introduced in [11] refers to *a pulse/glitch that appears at the inputs of dynamic gates and discharges the dynamic node*. Balamurugan and Shanbhag also defined another term, namely **deep-submicron noise**, to be *any phenomenon that causes the voltage at a non switching node to deviate from its nominal value*.
- **Simultaneous switching noise** (SSN), also known as *Delta-I noise* or *ground bounce*, has its origin in chip-package interface power distribution parasitics. It becomes crucial in CMOS devices scaled down into the submicron region, operating at frequencies over 60 MHz. According to [12], it is *the voltage bounce generated by the simultaneous switching of output buffers due to fast-changing currents across the parasitic inductance of V_{DD} and ground lines of integrated circuits*. Note that it is merely inductive noise as applied to integrated circuit parasitics.

In conclusion, as noise represents a limiting factor in almost all practical applications emerging from a variety of fields, it is very difficult to propose a general all-purpose definition. This explains why there are so many proposed definitions, which are not forcefully mutually consistent.

1.2 Overview

1.2.1 Intrinsic Noise

This term refers to the noise generated *inside* an investigated device or circuit. In linear systems the physical origin of noise is the *discrete nature of charge carriers*. Consequently, the number of carriers in some specific plane fluctuates in time. These fluctuations are both universal and unavoidable.

A typical example is thermal noise, originating from the random motion of free electrons inside a piece of conductive material. This leads to temporary agglomeration of electrons at one or the other end, and consequently a fluctuating voltage appears between the ends. As no permanent accumulation of charge at one end may exist, it is obvious that the mean value of the fluctuation must be zero. However, its root mean square (rms) value is not zero, and is given by

$$V_n = \sqrt{4kTR\,\Delta f} \qquad (1.1)$$

where k is Boltzmann's constant, T the temperature expressed in kelvins, R the resistance of the material, and Δf the bandwidth of the equipment employed to measure this noise.

Other kinds of intrinsic noise exist: shot noise (due to random passage of charge carriers across a barrier of potential), flicker or 1/f noise (whose origins are not yet understood, but it apparently depends on lattice defects), diffusion noise (related to carrier velocity fluctuations induced by collisions), and so on.

One of the most important features of intrinsic noise is its *randomness*. This means that we are unable to predict the amplitude of a fluctuating voltage or current. Hence, we are forced to adopt a *statistical description*.

Another feature is that the amplitude of intrinsic noise is very low (seldom does it exceed a fraction of a millivolt, and usually it is well below $1\,\mu V$). As for the frequency spectrum, many noise mechanisms yield *white noise* (i.e., the noise power is equally distributed over all engineering frequencies). One notable exception is flicker noise, which has a 1/f spectrum (the power increases with decreasing frequency), called sometimes *pink noise*.

1.2.2 Extrinsic Noise

The sources of extrinsic noise are situated outside the investigated circuit, which merely acts as a receiving antenna; for this reason this kind of noise is also called *extraneous signals*, or *spurious signals*, or *perturbations*. According to its possible origins, two main categories exist:

1) *Environmental perturbations*, such as sky noise (which includes strong broadband noise sources like the Sun and the Milky Way), atmospheric noise (caused mainly by lightning discharges in thunderstorms), man-made noise (due to electric motors, arc welding, power lines, neon signs, electrostatic discharge, electrical power equipment, radio and TV broadcast services, motors, switches, spark plugs in ignition systems, household appliances, cellular telephones, mobile radios, etc.). All industrial perturbations are characterized by relatively high amplitudes, and a spectrum that cuts off before reaching visible wavelengths. Many are regular in form and periodic.
2) Signals which are useful in one circuit, but unfortunately pass via parasitic coupling into nearby circuits, where they are undesired and therefore act as perturbations. Generally this phenomenon is called *crosstalk between systems* (or *crosstalk noise*). As a general rule, the user discovers the interference (i.e., the undesirable effect of spurious signals) during operation and not before! Sometimes, such coupling can be reduced by modifying the relative position of various cables or equipment in a rack.

1.3 Comparing Intrinsic and Extrinsic Noise

The comparison can be carried out from different points of view. Each will be considered separately and discussed in detail.

1.3.1 General Properties

Intrinsic noise is essentially random in nature and requires a statistical description; the extrinsic noise is usually of a deterministic nature and consequently is less amenable to statistical treatment. In many practical situations, the frequency of interfering signals is known in advance (for instance, signals on assigned frequencies, their harmonics and subharmonics). Often the perturbation contaminating a circuit results from the superposition of several deterministic signals.

1.3.2 Bandwidth

Intrinsic noise shows up at all engineering frequencies, rising towards the low end of the spectrum due to inherent flicker noise (which is as universal as thermal noise).

Once intrinsic noise is present in a circuit, nothing can be done to filter it out, because its spectrum extends to very high frequencies, usually beyond the frequency of any actual useful signal.

In contrast, interfering signals have a much more restricted spectrum; this allows us either to reject them (by inserting appropriate filters) or to shift the band of the useful signal outside their spectrum.

1.3.3 Amplitude

As a general rule, the intrinsic noise level is rather low; for instance, the thermal noise generated by a $100\,k\Omega$ resistor at the ambient temperature ($T = 300\,K$) and in an elementary bandwidth of $1\,Hz$ is close to $40\,nV$ (rms). To compare with parasitic signals, let us recall Ohm's law: "one microampere through a $4\,m\Omega$ resistance yields $4\,nV$", exactly the thermal noise voltage generated by a $1\,k\Omega$ resistor, in a $1\,Hz$ bandwidth, at $290\,K$. This means that a leakage current flowing through a wire (or trace on a printed circuit board) as insignificant as it may seem, *does contribute* to the circuit noise.

In contrast, very often interfering signals emerging from nearby industrial installations reach vulnerable circuits at relatively high amplitudes, and can sometimes exceed the amplitude of useful signals. As a general rule, the amplitude of man-made noise decreases with increasing frequency, but this varies considerably with location.

1.3.4 Noise Calculation

It is much easier to compute the level of intrinsic noise than that of extrinsic noise. For the former, two possibilities exist, either to hand calculate the noise parameters (a lengthy procedure resulting in cumbersome expressions) or to use existing software packages. In the case of extrinsic noise, we are dealing with electromagnetic disturbances, described by Maxwell's equations. The problem is that the latter are differential equations with respect to time and three spatial variables. An accurate solution requires complicated boundary conditions, which are easily expressed in the time domain but almost impossible to formulate in the spatial domain (where they depend on the relative locations of various systems). This explains why for real-world situations solving Maxwell's equations is not practical, and to determine the man-made noise at a given site, it is necessary to make noise *measurements* (which vary according to time of day, season and direction).

1.3.5 Low-Noise Design

Intrinsic noise must be taken into account early in the design process. Besides the specified electrical performance, a high signal-to-noise ratio constitutes one of the basic objectives of any design. With this in mind, the designer makes appropriate choices concerning circuit topology, input stage, type of active device (bipolar or field-effect transistors), bias, feedback, etc. Once this process is complete, almost nothing can be done to reduce the resulting intrinsic noise. Therefore, leaving the intrinsic noise problem to be solved at the end can lead to disaster, and an expensive redesign process.

The extrinsic noise is treated in the layout and prototype stages. Various techniques can be used to prevent interference, such as proper positioning of traces on the printed circuit board, appropriate grounding, filtering and balancing, shielding of the most sensitive parts, using shielded cables, etc. From an economic point of view, it should not be forgotten that *limiting the emission of perturbations is more cost-effective than protecting the victims.*

1.3.6 Noise Immunity

Generally, noise corrupts the information content of a useful signal. However, estimates of the "damage" are closely related to the function performed by the affected circuit.

For instance, many analog circuits are intended to amplify weak signals, and are consequently more sensitive to intrinsic noise than any digital circuit, where the amplitudes of logic waveforms are relatively high. In many analog applications, the useful weak signals are either delivered by an antenna or generated by sensors. As their amplitudes can be as low as a fraction of millivolt (or even a fraction of microvolt), they are highly susceptible to noise, which can easily obscure their information content. This imposes a strong constraint on the electronic equipment processing them, which requires high-gain, low-noise front-end amplifiers.

In contrast, logic signals have considerable higher, rapidly varying amplitudes, and there is no need to amplify them before processing. The minimum noise margin of TTL circuits is 0.4 V (and even higher for a classic CMOS family), and logic circuits are therefore generally less sensitive to intrinsic noise or small-amplitude disturbances.

1.4 Why Is Noise a Concern?

To briefly answer this question, *noise is a concern because it sets a fundamental limit on the normal operation of all electronic circuits and systems.* Traditionally, it is stated that noise is important whenever we are dealing with weak signals. That's true, but perhaps a more appropriate statement would be that noise is important whenever the amplitudes of the processed signals are similar to those of the existing noise. Clearly, if we refer exclusively to the intrinsic noise, only very weak signals risk loosing their information content in a noisy environment. The information carried by signals with high amplitudes is not corrupted. Suppose now that the extrinsic noise is also considered. Since its amplitudes are considerably greater than those of the intrinsic noise, even strong signals are not immune to noise.

Therefore, *what really matters is not the signal level, but the signal-to-noise (S/N) ratio.* Low S/N ratios indicate vulnerability to noise, while high S/N ratios indicate immunity to noise.

The following is a non-exhaustive set of statements showing how noise can affect the operation of electronic systems:

- The minimum usable signal in analog circuit applications is limited by externally and internally generated noise. *The noise level sets a lower bound on the magnitude of a signal that can be amplified.*
- *The noise level determines the upper bound on the gain of an amplifier,* since if the gain is excessive, the amplifier will saturate.
- In analog circuits, a *large dynamic range means that a large swing and low noise are both required.*
- *Noise sets the limit on the minimum detectable signal* in detectors and receivers.
- Noise is responsible for errors in measuring the level of a weak analog signal. Consequently, *noise establishes the ultimate limit of measurement sensitivity.*
- *Noise can give rise to errors in measured signal phase.* This can affect the accuracy of distance estimation in systems using pulse-echo techniques (like radars, sonars, medical imaging systems, etc.), where an error in phase translates into an error in position.
- *Binary data stored in memories can be corrupted by noise,* a "0" accidentally being transformed into a "1" or vice versa.
- Although a priori logic circuits are more immune to noise, the following trends in deep submicrometer technology, in conjunction with increases in operating frequencies, require that noise be considered a key issue in the design process (as important as timing or power):
  - supply voltages are lower and lower, this being a popular low-power strategy (so smaller noise margins result);
  - continuous device down-scaling;
  - scaling of threshold voltages to maintain drive under low-power conditions;

- increasing interconnect density, giving rise to higher coupling capacitance, which becomes more important than the sum of the area capacitance and the fringing capacitance of a wire. In the future, it is expected that the role of coupling capacitance will be even more dominant as sizes shrink. Crosstalk noise analysis will become a critical factor in circuit design.
- faster clock rates, leading to more power supply noise.

In order to illustrate harmful noise effects, let us take a detailed look at two practical applications, one concerning communications systems and the other emerging from the field of industrial electronics.

1.4.1 Noise in Communication Systems

Noise limits the performance of any communication system. Consider the simplified block diagram of an analog communication system, shown in Fig. 1.1.

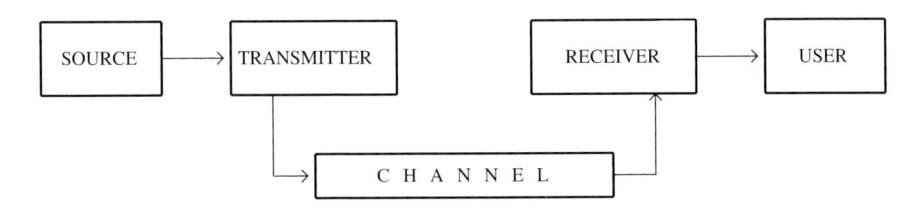

Fig. 1.1. Simplified structure of an analog communication system

The role of the transmitter is to encode the signals emitted from the source, in order to preserve their information integrity during propagation through the channel (this encoding operation is also called "modulation").

The channel conveys signals from one point to another. In its simplest form, it is a free-space path linking the transmitter's antenna and the antenna of the receiver. Other types of channel can exist, such as optical fibers, cables, or a combination of wireless and wireline channels.

The receiver's function is to extract useful information from the signal and convert it into a form suitable for use. The first task to be performed is amplification, since the incoming signal is severely attenuated during propagation. Information is then extracted by demodulation.

Assuming that the source is ideal (no noise added), the signal reaching the channel is contaminated solely by the noise of the transmitter.

During propagation through the channel, the signal undergoes further degradation. This often results from interference or linear and nonlinear distortion, but it can also include other effects such as signal fading, multipath transmission or filtering. At the same time, since the link path can exceed 10000 km (for example, in satellite communication systems), the signals will be drastically attenuated. Consequently, the signal-to-noise (S/N) ratio is drastically lowered and weak signals reaching the receiving station must first of all be amplified. As a general rule, a preamplifier is used at the front-end, immediately following the receiver's antenna. To avoid false alarm (i.e., the system indicates the presence of a signal when in reality it is merely noise), the preamplifier must be designed for high gain and minimum noise. Note that this implies not only reduced intrinsic noise, but also efficient protection against interfering signals.

Typical requirements on the minimum S/N ratio for various applications are +15 dB for radar after integration, +18 dB for a cellular telephone system, and +30 dB for broadcast communication systems for entertainment [13].

Therefore, in any communication system, noise is doubly important:

- It establishes the *minimum signal level* which can be processed, and therefore the *maximum range* of the communication system.
- It determines the *channel capacity* (i.e, the highest rate at which information can be transmitted with zero probability of error), which strongly depends on the S/N ratio at its output. Shannon proved that an ideal channel has a data-carrying capacity proportional to the system bandwidth multiplied by the logarithm of the ratio (S+N)/N, where S is the average signal power and N the noise power.

1.4.2 Noise in Industrial Applications

In instrumentation, measurement and control systems, a typical application implying weak signals involves a transducer (sensor) which delivers an electric signal proportional to some non-electric quantity (humidity, pressure, flow, speed, acceleration, temperature, etc.). The first task is to amplify this weak signal without overwhelming it by noise. A low-noise amplifier is required, preceded by a coupling network which matches the sensor impedance to the amplifier input (Fig. 1.2).

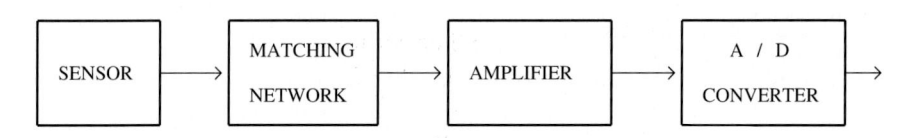

SENSOR → MATCHING NETWORK → AMPLIFIER → A / D CONVERTER →

Fig. 1.2. Simplified block diagram of a sensor and its associated circuits

At the amplifier output, an analog-to-digital converter transforms the analog signals into binary data, which can be easily stored, retrieved, processed, and displayed.

Any sensor produces an electrical signal which is not ideal, i.e., it is inherently affected by distortions. According to [14], there are two categories of distortions:

1) *Systematic distortions*, related to the sensor's linearity, transfer function, dynamic characteristics, etc. Since they result from the sensor design, fabrication, calibration, etc. they are known in advance, drift slowly in time, and are specified in the data sheets. The designer can provide solutions to compensate for them.
2) *Stochastic distortions*, which are quite irregular, change abruptly in time, and are unpredictable. These can be described only in terms of their statistical properties, and are due to the internal noise of the sensor. The S/N ratio of the system shown in Fig. 1.2 is mainly degraded by the sensor noise, the noise of the coupling network, and that of the first stage of the amplifier.

To improve accuracy in measuring the non-electrical quantity of interest, the designer adopts an analog-to-digital converter with good digital resolution. In this case, the value of the least significant bit (LSB) necessarily decreases. For instance, the LSB of a 16-bit system with a 5-V full scale reading is as low as 76 µV. Supposing that the total noise affecting the useful signal is around 180 µV, the LSB will be rather more relevant to the noise than to the signal content! Clearly, *it makes no sense to reduce the value of the LSB below the noise floor.*

Therefore, noise is a resolution-limiting factor in all electronic systems which monitor weak signals delivered by sensors.

1.4.3 Benefits of Noise

Up to this point, we have stressed only the negative effects of noise on the circuit performances. However, it is acknowledged that random noise is beneficial in several situations:

- It helps start oscillations in oscillator circuits, where the small fluctuation at the amplifier input play the role of the signal which must satisfy the Barkhausen criterion (according to which the input variation is reinforced and signal regeneration will occur provided that the gain around the loop is positive and higher than unity). Perhaps white noise is more helpful for this purpose, since its broad spectrum certainly includes the frequency where the circuit satisfies this condition.
- When noise is added to the training data set of a neural network it improves the network capability of generalization [15] (i.e. the ability of a particular neural network to fit real data outside its training set).
- In certain nonlinear systems (including biological sensory systems and electronic circuits), adding the right amount of noise *improves* the performances of some physical devices. Recently, a phenomenon called *stochastic resonance (SR)* was observed in physical nonlinear systems. For instance, by adding the right amount of noise to a bistable system it is possible to reach the SR regime, and consequently help the system to switch even when the amplitude of the command is under the threshold [16].

It has been reported that a number of electronic circuits exhibit stochastic resonance, especially when the nonlinear system does not modulate any component of the input noise on the output signal.

The stochastic resonance seems to be particularly attractive in signal processing when the useful (weak) signals are embedded in noise. This phenomenon could be successfully employed to enhance the S/N ratio at the receiver output [17].

1.4.4 Conclusion

Perhaps the best way to finish this overview of noise is to quote from Sheingold [18]:

> "Like diseases, noise is never eliminated, just prevented, cured, or endured, depending on its nature, seriousness, and the cost/difficulty of treating".

Part I

Intrinsic Noise

2
Fundamental Concepts

> "As far as the laws of mathematics refer to reality, they are not certain; and as far as they are certain, they do not refer to reality"
>
> (A. Einstein)

2.1 Introduction

Electrical noise theory has the reputation of being arcane and rather complicated. There are several reasons for this:

- To understand the physics of noise, as well as noise modeling and computation, background information is needed from various fields like signal theory, circuit theory, quantum mechanics, wave propagation, probability, and statistical analysis.

  As a matter of fact, the phenomena which are pertinent to noise generation are usually investigated with statistical methods and/or thermodynamic concepts. To describe random processes, we use the concepts of probability, average, and correlation. Then, we must take into account that circuits process noise exactly like any useful signal; therefore, we use the Fourier series and integral approach and power or energy spectra, as well as the concept of transfer function.

  If we look at the location of noise sources in any electronic circuit, we realize that they are distributed everywhere. For computational convenience and supposing that the circuit can be regarded as a two-port, we prefer to lump the noise into two equivalent voltage or current generators, conveniently located at the circuit terminals. This approach involves the basic principles of circuit theory.

  In microwave circuits it is common practice to model fluctuations using the noise-wave approach, which offers a unified treatment for signal and noise. However, in this case we need the S-parameters concept and the ability to represent noise by elements of Hilbert space.

As soon as the response of a system excited by random signals has been found, the next step is to process the results with the methods of statistical analysis, in order to derive significant quantities.

In conclusion, it is obvious that knowledge from various fields is required, and this constitutes one of the chief difficulties of noise theory.

- In looking at various papers and textbooks that treat noise theory, it is obvious that many noise parameters have been introduced over the last 40 years, perhaps more than necessary. These parameters arise from different fields, and each textbook shows a preference for some of them. Worse, still in recent decades we tolerated a proliferation of definitions covering the same noise quantities, emerging from various contributors.

  Little effort has been made to provide a unified and coherent set of definitions, and as a result multiple definitions for the same parameter can actually be found. Sometimes they are equivalent, under special conditions not always specified. Finally, people have difficulty correctly interpret the noise performance of given equipment or selecting the best among several candidates, knowing that various manufacturers do not necessarily employ the same definitions.

- Every electrical engineer has been basically trained to solve linear circuits by using Kirchhoff's laws and voltage (or current) superposition. As a matter of fact, this approach has structured our reasoning so strongly that we are conditioned to always think in terms of it. However, in noise theory it is of little help. Indeed, you must resist "instinctually" thinking in terms of Kirchhoff's laws, since for uncorrelated noise signals, only powers can be added, not voltages or currents! This stems from the fact that we are dealing with random signals instead of deterministic signals, and largely explains the feeling that noise computation is not straightforward.

The purpose of this chapter is to recall some basic concepts from various fields which are important to noise theory. An effort is made to introduce them as simply as possible by focusing on their physical meaning, since the mathematical proofs of given theorems are beyond the scope of this book.

2.2 Fluctuations: Signal Theory Approach

■ **Fluctuation.** The term *fluctuation* pertains to *a physical quantity which exhibits a small irregular variation around its mean value, but not at a steady rate*. Van der Ziel was the first to show preference for "fluctuation" over other existing terms like "random signal," "random variation," "electrical disturbance," "noise signal," or "stochastic signal." In fact, the term "fluctuation" will be preferred in this book. However, in order to avoid repetition, "random signal" or "noise signal" will sometimes be adopted.

■ **Comment.** The signal theory approach has more physical meaning than the probabilistic approach, since the basic concepts can be introduced by means of sinusoidal signals and then extended to fluctuations. Generally, this approach is preferred whenever the final purpose is to characterize fluctuations by measurement and not by computation.

■ **Signal.** In a broad sense, *a signal is a phenomenon used to transmit or convey information.* From a circuit theory point of view, *a signal is the form energy takes in order to propagate through a circuit.* It follows that signals are related to the time history of some quantity (current or voltage).

■ **Classification.** Signals are either *deterministic* (if they can be modeled as completely specified functions of time) or *stochastic* (taking on random values in time). A deterministic signal is often described by its peak value, peak-to-peak value, mean value, and/or root-mean-square value. Since at each instant the amplitude of a random signal has a different (but non-repetitive) value, its evolution is not predictable. Because of its very nature, the peak value is no longer a characteristic of the random signal, but instead we use the probability density function (PDF). This quantity offers a statistical description of the random signal in terms of its amplitude probability distribution over time.

From another point of view, signals can be grouped into two categories:

1) *Periodic* signals. A signal is said to be periodic if for all t and a constant T, called the *period*

$$v(t) = v(t + T) \tag{2.1}$$

This means that the signal repeats itself after any interval of T seconds.

2) *Aperiodic* signals, if no value of T satisfies (2.1). This is particularly the case of transients, cosmic radio noise, and random signals.

A partial description of any random signal [20] requires the specification of the following quantities: a) mean (average) value, b) mean square value, c) form factor, and d) peak factor. For the reader's convenience, each one will next be introduced by means of periodic signals.

2.2.1 Average Value

■ **Definition.** *The average value of a signal is the ratio of: (1) the area under its waveform over one period, to (2) the period.*

■ **Computation.** Consider a periodic signal i(t); in this case the average value I_o is given by

$$I_o = \frac{1}{T} \int_0^T i(t) \, dt \tag{2.2}$$

18   2 Fundamental Concepts

where T is the period. If i(t) is aperiodic (or random), then T is identified with the time interval during which the signal is observed. Note that in this case the average may depend on the monitoring time (which must be as long as possible).

■ **Example.** For the periodic signal depicted in Fig. 2.1, the average value computed according to the definition is

$$E_o = \frac{1}{(2\text{ms})}\Big((4\text{V})(1\text{ms}) + (-1\text{V})(1\text{ms})\Big) = 1.5 \text{ V}$$

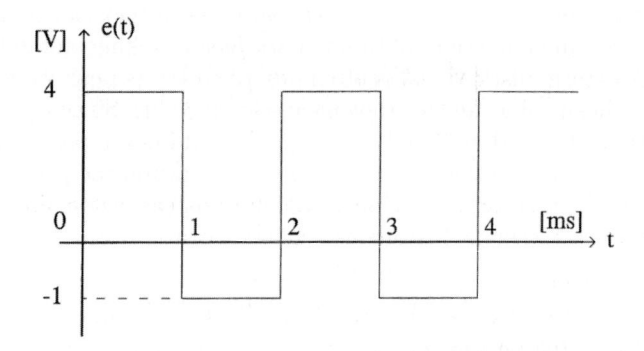

Fig. 2.1. A sample of a periodic signal

■ **Remark.** There are many fluctuations with average value equal to zero. This explains why the following quantities are required.

2.2.2 Mean Square Value

■ **Notation.** Consider a harmonic signal $v = V \cos\omega t$. Its root-mean-square value is denoted by V_{rms}, while its mean square value is $\overline{v^2}$. Of course: $\overline{v^2} = V_{\text{rms}}^2$.

■ **Root-Mean-Square (rms) Value.** This is a measurable quantity, which is traditionally introduced by means of the heat dissipated in a resistor. By definition, *the root-mean-square value (or the effective value) of an alternating voltage (or current) is equal to the value of a DC voltage (or current) which would dissipate the same amount of heat as the AC signal, in the same resistor, during the same time interval (usually one second).*

Applying this definition to the previous harmonic signal

$$\underbrace{\frac{1}{R}\frac{1}{T}V^2\int_0^T \cos^2\omega t \, dt}_{\text{AC signal power}} = \underbrace{\frac{1}{R}V_{\text{dc}}^2}_{\text{DC signal power}} \qquad (2.3)$$

2.2 Fluctuations: Signal Theory Approach

it follows that
$$V_{rms} \equiv V_{dc} = V/\sqrt{2} \tag{2.4}$$
where V is the amplitude of the harmonic signal.

Finally, the concept of root-mean-square value is useful because it allows the computation of the dissipated power, regardless of the particular waveform of the signal. This is particularly interesting for stochastic (random) signals.

■ **Mean Square Value.** By definition, this is *the average of the squared value of the signal.* Therefore

$$\boxed{\overline{v^2} = (V_{rms})^2 = \frac{1}{T}\int_0^T v^2 \, dt} \tag{2.5}$$

■ **Computation.** In order to obtain the mean square value of any periodic signal, the following steps have to be considered:

1. Take the square of the original waveform.
2. Evaluate the area under this waveform during one period.
3. Divide the area by the period.

■ **Example.** For the signal in Fig. 2.1, the mean square value is

$$\overline{e^2} = \frac{(4V)^2(1ms) + (-1V)^2(1ms)}{(2ms)} = 8.5 \text{ V}^2$$

Consequently, the root mean square value is $E_{rms} = \sqrt{8.5 \text{ V}^2} = 2.91$ V. This means that a DC voltage of 2.91 V applied to a resistor R yields the same heating effect as the AC signal depicted in Fig. 2.1.

■ **Remark.** It is obvious that the power developed in a resistor R by a voltage e(t) may be written as

$$\boxed{P = E_{rms}^2/R = \overline{e(t)^2} \, / \, R} \tag{2.6}$$

Expression (2.6) suggests that *the mean square value is numerically equal to the power dissipated per unit resistance* (also called the *normalized power*).

It is important to remember that the mean square value corresponds to the normalized power, since *for many random signals the average value is zero, and the first significant quantity is the mean square value.*

2.2.3 Form Factor

■ **Definition.** *The form factor of a signal is the ratio of: (1) the mean square value to (2) the average value of the signal.*

■ **Example.** For the waveform shown in Fig. 2.1, the form factor is $8.5/1.5 = 5.67$.

■ **Note.** For a noise signal whose average value is zero, the form factor cannot be defined.

2.2.4 Peak Factor

■ **Definition.** The peak factor of a waveform is taken to be *the ratio of: (1) the peak value to (2) the rms value of the signal.*

■ **Note.** If the waveform is alternating there are two peak values (one positive, the other negative). It is common practice to consider the peak value to be the one with the greater magnitude.

■ **Example.** Once again, consider the waveform of Fig. 2.1, where the greatest peak value is taken to be 4 V; therefore, the peak factor is $4/2.91 = 1.37$.

2.2.5 Correlation

■ **Purpose.** In this section we focus on the concept of correlation, which is fundamental to noise theory. In fact, there are two possible ways to introduce this concept: 1) by means of probability theory, which provides an accurate but less intuitive approach; 2) by means of signal theory, with emphasis on the physical meaning. In this paragraph the later is preferred, while the former will be introduced in Sect. 2.3.

■ **Definition.** According to [20], *two waveforms are said to be coherent if they exhibit similar behavior, except for their amplitudes, which may be different.*

■ **Discussion.** Two coherent signals are said to be *completely correlated*. In this case

- the mathematical functions describing them are identical;
- the phase difference is zero;
- their amplitudes are not necessarily equal.

If the phase difference of the signals is not zero but rather small (the remaining conditions being fulfilled), then the signals are said to be *partially correlated*.

However, when the phase difference becomes important, the signals are no longer considered partially correlated, since the mathematical function may be completely altered (for instance, a phase difference of $\pi/2$ transforms a sine into a cosine).

■ **Harmonic Signals.** In order to illustrate this concept, consider two harmonic signals v_a and v_b with the same frequency but different amplitudes. Their phase difference corresponds to a small angle Φ; hence, they are partially correlated:

$$v_a = A \sin \omega t \qquad (2.7a)$$

$$v_b = B \sin(\omega t + \phi) \qquad (2.7b)$$

It is easy to see that the signal v_b can be written

$$v_b = \underbrace{(B \cos \phi) \sin \omega t}_{\text{term 1}} + \underbrace{(B \sin \phi) \cos \omega t}_{\text{term 2}} \qquad (2.7c)$$

Term 1 represents a signal which is completely correlated with v_a (same time function, no phase difference), while term 2 corresponds to a signal which is uncorrelated with v_a (as it is described by a completely different function of time). It is obvious that the amplitudes of these signals vary according to the value of Φ, and sometimes the two terms may even cancel.

This example makes it possible to introduce a technique which is widely used in noise analysis: *when we are dealing with two partially correlated noise generators n_a and n_b, the first can be split into two generators n_{a1} and n_{a2}, such that n_{a1} be completely correlated with n_b and n_{a2} is uncorrelated with n_b.*

■ **Noise Signals.** Noise signals (or fluctuations) cannot be described by mathematical functions of time, only on a probabilistic basis. Hence, it is rather difficult to apply the previous time domain approach in order to look for correlation. From a practical standpoint, *checking whether two fluctuations are correlated is equivalent to check whether they have common physical origin.*

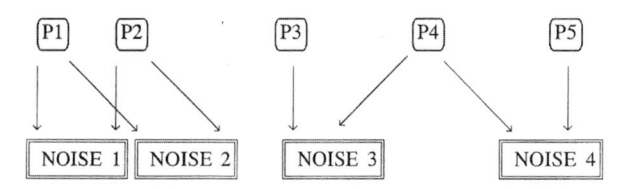

Fig. 2.2. Checking for correlation: noise signals 1 and 2 are completely correlated, 3 and 4 are partially correlated, but 1 and 3 (or 2 and 4, or 1 and 4, or 2 and 3) are independent

For instance, if a fluctuation arises mainly from a particular physical process which is also partially involved in the generation of another fluctuation, these fluctuations will to a certain extent obviously show a similar behavior. They are said to be *partially correlated*.

All possible situations are shown in Fig. 2.2, where the physical noise sources are denoted by P1, P2, P3, P4, and P5. Note that they represent either real physical processes, or equivalent noise generators introduced for computational convenience.

■ **Correlation Coefficient.** In order to appreciate to what extent two noise signals are correlated, we need a measure. This is provided by the correlation coefficient. Without any loss of generality, let us consider the harmonic signals v_a, v_b described by (2.7a) and (2.7b), which are applied in series to a 1-Ω resistor (Fig. 2.3). We are interested in evaluating the total power dissipated in this resistor.

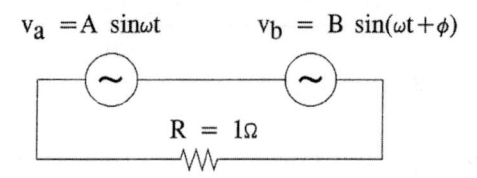

Fig. 2.3. Superposition of two signals

According to Sect. 2.2.2, since $R = 1\,\Omega$, this is the *normalized* power and we have to compute the mean square value of the total voltage v across the load

$$\overline{v^2} = \overline{(v_a + v_b)^2} = \overline{v_a^2} + 2\,\overline{v_a v_b} + \overline{v_b^2} \tag{2.8a}$$

The right hand expansion is possible because the mean value of a sum is always equal to the sum of the means (*Caution*: for a product, this is true if and only if the signals v_a and v_b are uncorrelated!). Upon substituting expressions (2.7a) and (2.7b) into (2.8a), we get

$$\overline{v^2} = \frac{1}{T}\int_0^T \left(A^2 \sin^2 \omega t + 2AB \sin \omega t \, \sin(\omega t + \phi) + B^2 \sin^2(\omega t + \phi)\right) dt \tag{2.8b}$$

After some algebra, the final result is

$$\boxed{\overline{v^2} = \frac{A^2}{2} + AB\,\cos\phi + \frac{B^2}{2}} \tag{2.8c}$$

Three special cases must be considered:

1) If $\phi = 0$ the waveforms are said to be coherent (or completely correlated). Equation (2.8c) becomes

$$\overline{v^2} = V_{\text{rms}}^2 = \left(\frac{A}{\sqrt{2}} + \frac{B}{\sqrt{2}}\right)^2 \tag{2.8d}$$

which proves that *for two completely correlated signals the rms value of the total voltage is the sum of the rms values.*

2) If $\phi = \pi/2$, the signals are uncorrelated. In this case

$$\overline{v^2} = \frac{A^2}{2} + \frac{B^2}{2} \tag{2.8e}$$

For uncorrelated signals, the total normalized power dissipated in the load is equal to the sum of the individual normalized powers.

3) If $0 < \phi < \pi/2$, the signals are partially correlated. The correlation coefficient is given by

$$c = \frac{\overline{v_a v_b}}{\sqrt{\overline{v_a^2}\, \overline{v_b^2}}} \tag{2.9}$$

Using definition (2.2) together with (2.4), we get

$$c = \frac{\overline{v_a v_b}}{(v_a)_{\text{rms}}(v_b)_{\text{rms}}} = \frac{(AB\cos\phi)/2}{(A/\sqrt{2})(B/\sqrt{2})} = \cos\phi \tag{2.10}$$

The final form of expression (2.10) proves that the magnitude of the correlation coefficient always lies between 0 (for uncorrelated signals) and 1 (for completely correlated signals). The possible value -1 can be interpreted as a special case of completely correlated signals in antiphase, which therefore subtract, instead of adding.

2.3 Fluctuations: Probabilistic Approach

2.3.1 One Random Variable

■ **Fluctuation.** Any physical quantity observed over some time interval exhibits inherent fluctuations about its mean value. These fluctuations may be interpreted as random processes. For many physical phenomena, the instantaneous value of the observed quantity follows a Gaussian (or normal) distribution law, and we have strong arguments (based on the central limit theorem) to justify this kind of distribution.

■ **Random Variable.** In electrical circuits, the voltage and current fluctuations are described by means of random variables. By definition, *a random variable is a rule that assigns numerical values to the outcomes of a chance experiment*. Note that from the standpoint of classical mathematics, the concept of a random variable is in fact a function.

There are two categories of random variables: *discrete* (if they can take only specific allowed values) and *continuous* (if they take any value within a given interval). For instance, flipping a coin is a chance experiment, for which normally there are only two possible results: "head" or "tail". By establishing a correspondence between these results and "0" and "1" respectively, we introduce a discrete random variable. Also, the fluctuating number of charge carriers inside the volume of a semiconductor is a statistical process, and the number of carriers may be viewed as a discrete random variable.

Consider now the experiment of throwing a dart at a target. Each time, we measure the distance between the impact point and the center of the target, and this value constitutes a random variable. Here the distance can take any value, and consequently we have a *continuous* random variable.

■ **Random Process.** If the outcome of an experiment is not a number, but a random function of time (referred as a *sample function*), we have a *random process*. The set of all sample functions resulting from many similar experiments is called an *ensemble*.

For instance, the thermal noise generated by a particular resistor constitutes a random process. The waveform of the noise voltage produced by this resistor constitutes a sample function. The set of all noise voltages produced by many similar resistors constitutes an ensemble (Fig. 2.4).

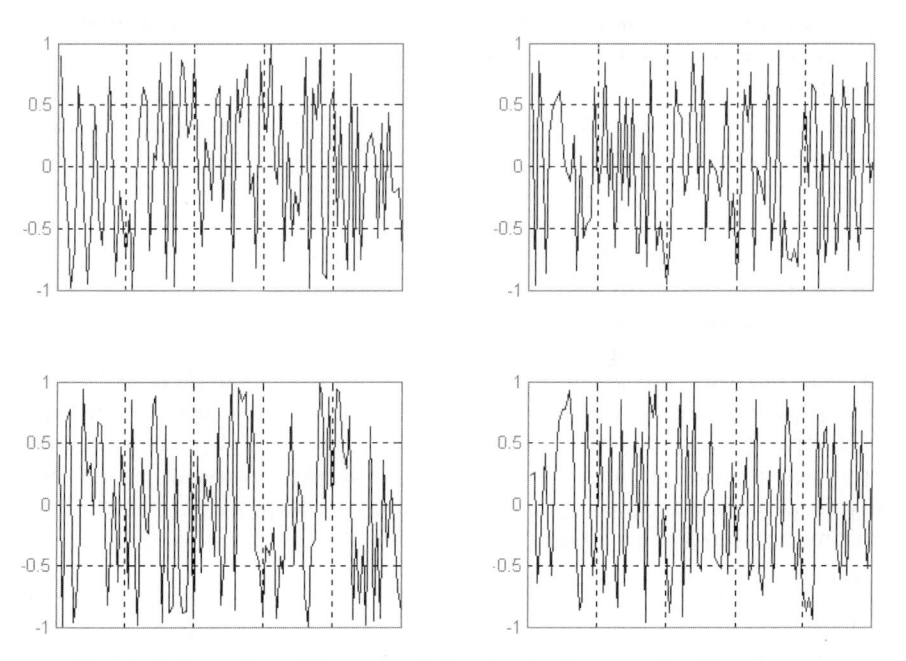

Fig. 2.4. Ensemble of 4 samples belonging to the same random process

2.3 Fluctuations: Probabilistic Approach

■ **Probability Density Function (PDF).** In the case of electrical noise we may adopt as the fluctuating quantity the instantaneous value (denoted by X) of the voltage or current. We are interested in evaluating the probability ΔP that the variable X describing the fluctuation takes a value in the range from x to x+Δx.

Assume an ideal situation, where we are able to measure the value of this quantity of interest at each instant of time. After measuring a large number of instantaneous values, every measured value x is plotted as a point on the x-axis.

Next suppose that we divide the x-axis into equal elementary intervals of length Δx and count the number of points in each interval. In this way the probability density function can be represented by [21]

$$f(X) = \lim_{\substack{\Delta x \to 0 \\ N \to \infty}} \frac{(\text{Number of points in the interval } \Delta x) / \Delta x}{\text{Total number N of measured points}}$$

It follows that the probability of finding the current (or voltage) amplitude in an infinitesimal range dx around x is $\Delta P = f(X)\, dx$.

Generally f(X) cannot be determined experimentally, but sometimes it can be deduced from a study of the fluctuating quantity X.

■ **Cumulative Distribution Function (CDF).** This is defined as *the probability that the instantaneous value of the fluctuation at instant t_1 is less than some specified value x_1*; that is, $F_1(x_1, t_1) = P(X(t_1) \leq x_1)$. By definition,

$$F_1(x_1, t_1) = \int_{-\infty}^{x_1} f(x)\, dx \tag{2.11}$$

Note that the PDF is the derivative of the CDF.

■ **How to Describe Fluctuations.** All fluctuations share the property that their instantaneous amplitude cannot be predicted. This may arise from a lack of understanding of the phenomena, or a system so complicated that calculation is impractical, or some basic randomness in the physical context.

The tools required to *completely* characterize any fluctuation are the PDF and CDF. However, in the case of noise, it is rarely possible to use them, since in most situations they are not available. Therefore, we are forced to adopt an *incomplete description* based on averages (moments).

To be specific, let us consider a noise current i(t). There are two ways to compute its average value:

1) *Averaging over time*, by monitoring the instantaneous value of i(t) for a sufficiently long time, and then calculating the average (mean) value. This average will be denoted by $\bar{i}$.

2) *Averaging over an ensemble* of N identical systems (N being as large as possible), each delivering a similar noise current. Supposing that the measurement conditions are identical, we read the instantaneous value of i(t) at a definite instant of time. We then compute the average over the N systems at the given instant of time. This average will be denoted by ⟨i⟩.

Note that averaging over time is common to noise measurement techniques, but averaging over an ensemble is mainly used in noise computation.

■ **Ergodic Process.** The question which arises now is whether the two averaging procedures yield the same result. In the case of noise, both methods yield the same value, because the physical processes implied in noise generation are ergodic. By definition, *an ergodic process is one in which the statistics over a long time interval for any one system coincide with the statistics over an ensemble of systems at any instant in time.*

■ **Stationary Process.** By definition, *a stationary process is one which yields the same statistical parameters no matter where we place the time observation interval.*

Note that every ergodic process is also stationary, but the converse is not true.

■ **Notation.** We adopt the following notation:

Random variable	X
Value of the variable	x
Time average	$\overline{X}$
Ensemble average	⟨X⟩

■ **Averages.** Any average offers the advantage of characterizing a random quantity by means of a constant value.

By definition, the *n-th order average* (also called the *n-th order moment*) of the continuous random variable X whose PDF is denoted by f(x) is

$$m_n(X) = \overline{X^n(t)} = \int_{-\infty}^{+\infty} x^n\, f(x)\, dx \qquad (2.12)$$

The most often encountered averages are m_1 (the arithmetic mean) and m_2 (the mean square).

For thermal noise, the mean value $m_1 = \overline{X}$ is always zero and is not representative; the first nonzero mean is $m_2 = \overline{X^2}$. Whenever m_1 differs from zero, it is common practice to use a *random* characteristic of the fluctuation, the *deviation* of the random variable from its mean value, i.e. $(X - \overline{X})$. The second-order average of this quantity (or the *mean square deviation*)

2.3 Fluctuations: Probabilistic Approach

$$m_2(X - \bar{x}) = \text{var}\{X\} = \overline{(X - \overline{X})^2} = \overline{X^2} - (\overline{X})^2 \qquad (2.13)$$

is called the *variance*. It is also possible to define the quantity

$$\sigma_x = \sqrt{\text{var}(x)} \qquad (2.14)$$

which is called *standard deviation*. Hence, the standard deviation is the root mean square deviation.

■ **Physical Meaning.** For *ergodic processes*, it is useful to stress the physical meaning of some of previous quantities. For instance:

- The mean value $\overline{X(t)} = \langle X^2(t) \rangle$ corresponds to the *DC component* of the fluctuation.
- The square of the first-order average: $[\overline{X(t)}]^2 = \langle X(t) \rangle^2$ gives the *normalized DC power* (the DC power developed in a unit resistance).
- The second-order average: $\overline{X^2(t)} = \langle X^2(t) \rangle$ can be identified with the *normalized total power* (the total power dissipated in a unit resistance).
- The variance: $\text{var}\{X(t)\} = \overline{X^2(t)} - [\overline{X(t)}]^2$ is the *normalized power in the AC component* (the AC power dissipated in a unit resistor).
- The standard deviation $\sigma_x = \sqrt{\text{var}(X)}$ can be interpreted as the *rms value of the AC component*. Note that expression (2.13) can also be written

$$\overline{X^2(t)} = \text{var}\{X(t)\} + [\overline{X(t)}]^2$$

which simply means that the total normalized power is the sum of the AC normalized power and the DC normalized power.

■ **Characterization of Discrete Random Variables.** For a discrete random variable the PDF is obtained by studying the statistics of the fluctuating quantity. There are three special distribution laws that are widely encountered in practice: the binomial distribution, the Poisson distribution, and the normal distribution.

□ **Binomial Distribution.** Consider an experiment that yields only two possible results: A (with probability p) and B with probability (1–p). Repeating this experiment m times, the probability of obtaining the result A in exactly n of the trials is

$$\boxed{P(n) = \frac{m!}{n!\,(m-n)!}\, p^n\, (1-p)^{m-n}} \qquad (2.15)$$

The definition of average yields

$$\bar{n} = mp; \quad \text{var}\{n\} = mp\,(1-p).$$

28   2 Fundamental Concepts

☐ **Poisson Distribution.** Next consider a sequence of independent events that occur randomly at a low average rate λ. The probability P(n) of having n occurrences during a given lapse of time T is

$$P(n) = \frac{(\lambda T)^n \, \exp(-\lambda T)}{n!} \tag{2.16}$$

In this case: var{n} = λ.

☐ **Normal Distribution.** Many natural phenomena have this particular distribution. Assume a sequence of independent events occurring randomly at an average rate λ, which this time is high. The probability P(n) of having n occurrences in a given unit time is

$$P(n) = \frac{1}{\sqrt{(2\pi\sigma^2)}} \exp\left\{-\frac{(n-\lambda)^2}{2\sigma^2}\right\} \tag{2.17}$$

where σ^2 is the variance of n.

It can be shown that both the binomial distribution and the Poisson distribution approach a normal distribution when n becomes large. For the special case $\sigma^2 = \lambda$, the normal distribution is called a *Gaussian distribution*. It has the interesting property that no matter how large the fluctuation amplitude considered there is a nonzero (but very small) probability of exceeding it during observation.

■ **Continuous Random Variables.** For a continuous random variable X with mean value zero, (2.17) becomes

$$dP(X) = \frac{1}{\sqrt{(2\pi\sigma^2)}} \exp\left(-\frac{X^2}{2\sigma^2}\right) dX \tag{2.18}$$

■ **Note**

1) A physical noise process described by (2.18) is called Gaussian.
2) Almost all fluctuating noise voltages and currents have a PDF with a Gaussian distribution.
3) The main reason why most fluctuating quantities present a normal distribution is that each fluctuation can be viewed as the macroscopic sum of a huge number of microscopic independent random variables. We can then apply the central limit theorem.

■ **Central Limit Theorem.** Denote by $X_1, X_2, \ldots, X_n$ a set of independent random variables with means $m_1, m_2, \ldots, m_n$ and variances $\sigma_1^2, \sigma_2^2, \ldots, \sigma_n^2$, respectively. As n increases, the PDF of the sum $Y = \sum_{i=1}^{n} X_i$ approaches a normal distribution, with

$$\text{mean} \quad m = \sum_{i=1}^{n} m_i \quad \text{and} \quad \text{variance} \quad \sigma^2 = \sum_{i=1}^{n} \sigma_i$$

(provided that no X_i dominates the sum).

Note that all means involved in the theorem are ensemble averages (but the processes are ergodic).

2.3.2 Two Random Variables

■ **Comment.** The two-dimensional approach represents more of a practical necessity than a mathematical complication. As a matter of fact, when describing a fluctuation using the one-dimensional approach it is impossible to obtain information about its power.

■ **Two Random Variables.** In order to access the power spectrum, we need to consider the statistics of pairs of amplitude values, taken at two different instants in time. In this way the two random variables introduced are *independent*, since the value of the second does not depend on the value of the first.

■ **Joint Probability Density Function (JPDF).** By extension, the *joint PDF is the function which by multiplication with the infinitesimal area (dx dy) gives the probability that the first variable lies between x and (x+dx), while the second simultaneously lies between y and (y+dy)*.

■ **Joint Cumulative Distribution Function (JCDF).** By extension, the *joint CDF is the probability that the instantaneous value of the first variable is less than or equal to some specified x, while the second is simultaneously less than or equal to some specified y*:

$$F(x,y) = \int_{-\infty}^{x} \int_{-\infty}^{y} f(x,y) \, dx \, dy \tag{2.19}$$

Here f(x,y) represents the joint probability density function (JPDF).

■ **Averages.** These are computed in the same way, except that we now sum over two variables (for discrete random variables) or by double integration (for continuous random variables).

For instance, the joint average $\overline{X^i Y^k}$ is given by

$$m_{ik}(x,y) = \overline{X^i Y^k} = \int_{-\infty}^{+\infty} \int_{-\infty}^{+\infty} x^i \, y^k \, f(x,y) \, dx \, dy \tag{2.20}$$

where f(x,y) is the JPDF. Note that

$$m_{00} = 1 \tag{2.21a}$$

$$m_{01} = \overline{Y} = \int_{-\infty}^{+\infty}\int_{-\infty}^{+\infty} y\, f(x,y)\, dx\, dy \tag{2.21b}$$

$$m_{10} = \overline{X} = \int_{-\infty}^{+\infty}\int_{-\infty}^{+\infty} x\, f(x,y)\, dx\, dy \tag{2.21c}$$

Since usually $\overline{X} = \overline{Y} = 0$ for electrical fluctuations, the first meaningful quantities are $\overline{X^2}$, $\overline{Y^2}$ and $\overline{XY}$.

As already shown, we may introduce the deviations $(X - \overline{X})$ and $(Y - \overline{Y})$, for which the joint average of order i, k respectively is

$$\mu_{ik} = \overline{(X - \overline{X})^i\,(Y - \overline{Y})^k} = \int_{-\infty}^{+\infty}\int_{-\infty}^{+\infty} (x - \overline{x})^i\,(y - \overline{y})^k\, f(x,y)\, dx\, dy \tag{2.22}$$

The most important averages are

$$\mu_{20} = \overline{(X - \overline{X})^2} \quad \text{and} \quad \mu_{02} = \overline{(Y - \overline{Y})^2}$$

known as the *variances* of X and Y, respectively.

■ **Covariance.** By definition, the *covariance* of the two quantities X and Y is

$$\mu_{11} = \overline{(X - \overline{X})\,(Y - \overline{Y})} = \overline{XY} - \overline{X}\,\overline{Y} \tag{2.23}$$

It is possible to prove that the covariance vanishes if X and Y are statistically independent; their average product is then equal to the product of the averages.

■ **Correlation.** *The correlation describes a relationship between two fluctuating quantities X and Y which have a natural tendency to influence one another, this tendency not just being a chance result.*

If two fluctuations are statistically independent (i.e., the first doesn't influence the second, and vice versa), then they are said to be *uncorrelated*.

An intuitive way of judging whether the quantities X and Y are correlated is to plot all pairs (x,y) as two-dimensional points [22]. Several outcomes are possible:

- If the points are uniformly scattered over the plane (as in Fig. 2.5a), we may infer that X and Y are uncorrelated.

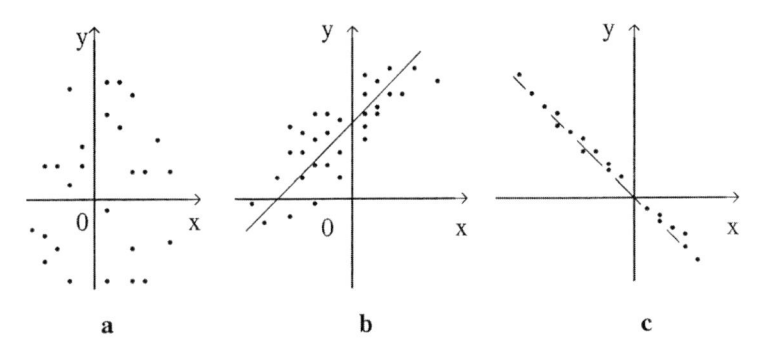

Fig. 2.5. Various degrees of correlation for two random variables: **a** uncorrelated; **b** weak correlation; **c** strong linear correlation

– If the points cluster, we must admit that a correlation exists. For instance, the points may be distributed more or less about a curve describing their functional dependence. Figure 2.5b illustrates weak linear correlation; Fig. 2.5c shows points located near a straight line, with strong linear correlation.

☐ **Correlation Coefficient.** By definition, *the correlation coefficient of two fluctuating quantities X and Y is the ratio of: (1) their covariance to (2) the product of their standard deviations.*

$$c = \mu_{11} / \sigma_x \sigma_y \tag{2.24}$$

This is often called the *normalized covariance* because

$$-1 \le c \le +1 \tag{2.25}$$

Note that both the covariance and the correlation coefficient characterize the functional dependence between quantities X and Y. However, in practice the latter is preferred, due to its limited range of possible values.

Several situations can occur:

– $c = 0$, X and Y uncorrelated (independent);
– $c = 1$, X and Y completely correlated;
– $|c| < 1$, X and Y partially correlated.

☐ **Partial Correlation.** If two fluctuations X and Y are partially correlated, one of them (Y, say) can be expressed as a sum of two terms, the first being completely correlated with X and the second (denoted by Z) being uncorrelated (see Sect. 2.2.5):

$$Y = aX + Z, \quad a = \text{const.} \tag{2.26a}$$

2 Fundamental Concepts

If:
$$\overline{X} = \overline{Y} = \overline{Z} = 0 \qquad (2.26b)$$

then the correlation coefficient becomes

$$c = (\text{sgn } a)\left(1 + \overline{x^2}/\overline{a^2 x^2}\right)^{1/2} \qquad (2.27)$$

and it has the same sign as the constant a.

☐ **Remark.** Decomposing a fluctuation into a sum of two terms has no particular physical significance, since the variable to be split (X or Y) is chosen arbitrarily; furthermore, the solution is not unique.

☐ **Autocorrelation.** In practice, we are often interested in assessing how fast a fluctuation evolve in time. A first attempt might be to use its deviation from the mean, but this is of course also random. A better choice is to use the autocorrelation function, a nonrandom characteristic of the fluctuation.

To be specific, suppose that measuring the signal delivered by a noise source at two different instants yields

$$x = V(t_1), \quad y = V(t_1 + \tau)$$

If the fluctuation V(t) is stationary, the statistics don't depend on t_1, just on τ. In this case, *the autocorrelation function is defined to be the mean product of V(t₁) and V(t₁ +τ)*:

$$R(\tau) = \overline{v(t_1)\,v(t_1 + \tau)} \qquad (2.28)$$

As it is obvious from Fig. 2.6, the width of the main peak of the autocorrelation function increases when the fluctuations evolve slowly in time; a narrow plot corresponds to faster evolution [26].

According to the physical interpretation given in Sect. 2.2.2, it follows that *the autocorrelation function is equal to the total power developed by the signal across unit resistance.*

In conclusion, the autocorrelation function and the power of the signal are not independent quantities.

☐ **Properties.** As stated in [28], the autocorrelation function has the following properties:

- it is always an even function;
- it has a maximum value at τ = 0;
- often the autocorrelation function is a monotonic function, tending to zero for large values of τ.

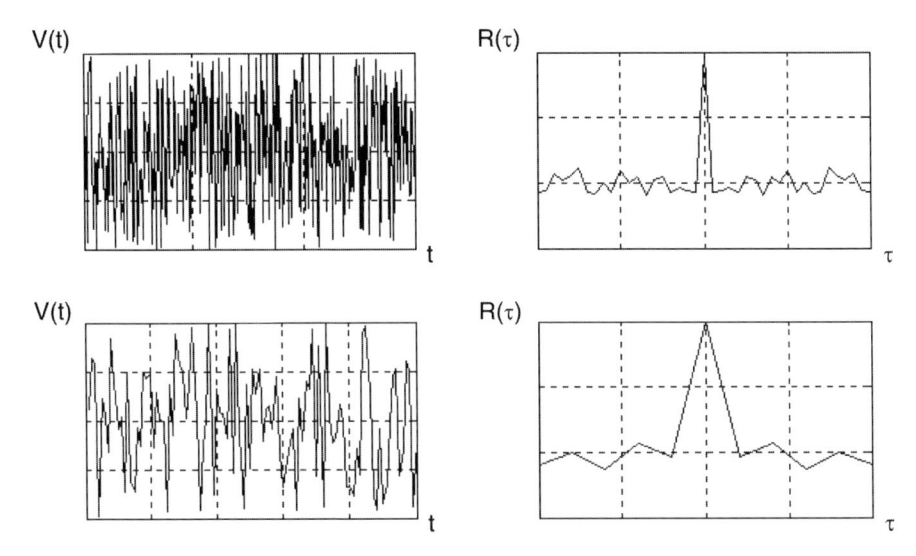

Fig. 2.6. Autocorrelation functions for two different samples

2.3.3 Energy and Power Spectra

■ **Fourier Transform.** Any signal f(t) can be expressed as a sum of sinusoidal components of specified frequencies (harmonics). In this way, we obtain the frequency-domain representation F(f) of the signal f(t) by means of the direct Fourier transform

$$F(f) = \int_{-\infty}^{+\infty} f(t) \exp(-j2\pi ft) \, dt = \mathscr{F}[f(t)] \tag{2.29a}$$

The inverse transform, whose corresponding operator is $\mathscr{F}^{-1}$, has the form

$$f(t) = \int_{-\infty}^{+\infty} F(f) \exp(j2\pi ft) \, df = \mathscr{F}^{-1}[F(f)] \tag{2.29b}$$

Note the perfect symmetry between the time and frequency variables.

☐ **Notation.** In the time domain, all quantities of interest are denoted by lower-case letters, while in the frequency domain they are represented by upper-case letters.

☐ **Property.** A remarkable property of the Fourier transform is that of *symmetry*. For a real f(t), we have

$$F^*(f) = F(-f) \tag{2.30a}$$

where ∗ denotes the conjugate of the complex quantity. It follows that

$$|F(f)| = \sqrt{F(f)\ F^*(f)} = |F(-f)| \qquad (2.30b)$$

■ **Spectrum of Aperiodic Signals.** For aperiodic signals, the amplitude and phase spectra, denoted by $|F(f)|$ and $\angle F(f)$ respectively, are both continuous (in contrast to periodic signals, whose spectra have discrete lines).

■ **Power Spectral Density.** *The power spectral density is a real, even, nonnegative function S_f which yields the total average power per ohm when integrated over the frequency domain.* Therefore, the power spectral density is the relative distribution of signal power throughout the spectrum.

If the signal v(t) has a constant rms value and is applied to unit resistance, we may reasonably write over a bandwidth Δf

$$\frac{V_{rms}^2}{\Delta f} = \left(\frac{V_{rms}}{\sqrt{\Delta f}}\right)^2 = S_f(V) \qquad (2.31)$$

The quantity $V_{rms}/\sqrt{\Delta f}$ is called *voltage spectral density*; it is expressed in $V/\sqrt{Hz}$. In the same way, it is equally possible to define a *current spectral density*, expressed in $A/\sqrt{Hz}$.

■ **Remark.** According to [19], the spectral density is defined to be the distribution of the average power with respect to *angular frequency* $\omega = 2\pi f$. To avoid confusion, this quantity will be denoted by S_ω. It is obvious that for the same average power

$$\boxed{S_\omega = \frac{1}{2\pi} S_f} \qquad (2.32)$$

■ **Normalized Energy.** For finite-duration signals, we may introduce the concept of normalized energy [21]. By definition, *the normalized energy of a signal v(t) is the total energy dissipated per ohm*:

$$W = \int_{-\infty}^{+\infty} v^2(t)\ dt \qquad (2.33)$$

(provided that this integral exists and has a finite value).

■ **Parseval's Theorem.** This theorem establishes the correspondence between the product of two signals (in the time domain) and the product of their Fourier transforms (in the frequency domain):

$$\int_{-\infty}^{+\infty} x_1(t)\ x_2(t)\ dt = \int_{-\infty}^{+\infty} X_1^*(f)\ X_2(f)\ df \qquad (2.34)$$

The special case $x_1(t) = x_2(t) = v(t)$ ($v(t)$ may, in general, be complex) refers to the energy of the signal

$$W = \int_{-\infty}^{+\infty} |v(t)|^2 \, dt = \frac{1}{2\pi} \int_{-\infty}^{+\infty} F(j\omega)F^*(j\omega) \, d\omega = \frac{1}{2\pi} \int_{-\infty}^{+\infty} |F(j\omega)|^2 \, d\omega \quad (2.35)$$

where use has been made of the fact that the product of a complex number and its conjugate is equal to the square of its magnitude.

Expression (2.35) is very important. Note that *the total energy of a signal can be computed in two different ways: 1) by integrating the quantity $|F(j\omega)|^2$ over the spectrum; 2) by integrating the squared value over the time domain.*

Therefore, (2.35) establishes an equivalence between the frequency spectrum and the squared value of the signal.

■ **Energy Spectral Density.** This represents the way in which the energy of an aperiodic signal is distributed over the spectrum. The right-hand side of expression (2.35) leads to the energy spectral density

$$S_\omega(W) = \frac{1}{2\pi}|F(j\omega)|^2 \quad (2.36)$$

By definition, *the energy spectral density is a real, even, nonnegative function which integrated over the domain of all ω, yields the total average energy per ohm.*

Note that the spectral energy is modified when the signal passes through a two-port.

■ **Concluding Remark.** It is important to note that both power and energy spectral densities are defined with three implicit conditions:

- *unit load* is assumed;
- the quantity of interest is the *average* power or energy;
- *unit bandwidth* is considered.

2.3.4 Fourier Analysis of Fluctuations

■ **Comment.** A fluctuating quantity $x(t)$ is a variety of aperiodic signal; hence, expressions (2.31) to (2.35) should be adopted.

■ **Bilateral and Unilateral Power Spectra.** Traditionally, in signal theory the power of deterministic signals is calculated in the frequency domain, which extends from $-\infty$ to $+\infty$. As this spectrum includes positive and negative frequencies, it is called *bilateral*. In noise theory, we are concerned only by frequencies extending from 0 to $+\infty$. Since this time only positive frequencies are considered, the spectrum is said to be *unilateral*.

36   2 Fundamental Concepts

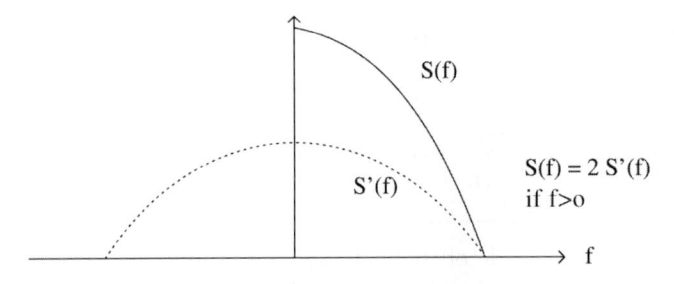

Fig. 2.7. Relationship between the bilateral power density S'(f) and the unilateral power density S(f)

The relationship between the bilateral power density (S') and the unilateral power density (S) is given in Fig. 2.7.

■ **Wiener-Khintchine Theorem.** Consider a fluctuation x(t), whose autocorrelation function R(τ) exists and is established by (2.28). Then, according to [25], *the power spectral density S_f of the fluctuation and the autocorrelation function are related by a Fourier transform*:

$$S_f(x) = \mathcal{F}[R(\tau)] = \int_{-\infty}^{+\infty} R(\tau) \, \exp(-j\omega\tau) \, d\tau \qquad (2.37)$$

$$R(\tau) = \mathcal{F}^{-1}[S_f(x)] = \int_{-\infty}^{+\infty} S_f(x) \, \exp(+j\omega\tau) \, df \qquad (2.38)$$

■ **Measurement.** Experimentally it is easier to obtain R(τ) (rather than S_f) by measuring the average power of the fluctuation and then using Parseval's theorem. Then $S_f(x)$ can be derived by Fourier transformation of the autocorrelation function. However, there are several conditions imposed during measurement, which must be clearly specified:

- The terminals at which the fluctuation appears must be loaded by a high impedance in order to simulate the open-circuit condition. The intrinsic noise of this impedance must either be low, or be known in advance.
- The load must be connected through a filter of narrow bandwidth Δf in order to maintain constant power density and constant input impedance. A possible solution consists in adopting a high-Q tuned circuit.
- The average power must be monitored over a time interval much greater than $1/\Delta f$.

■ **Superposition of Fluctuations.** In practice this situation is encountered often, since the observed noise results from the superposition of many elementary fluctuations. We wish to evaluate the power resulting from the superposition of two partially correlated signals x(t) and y(t), i.e.

2.3 Fluctuations: Probabilistic Approach

$$z(t) = x(t) + y(t).$$

We take the mean square value of each side:

$$\overline{z^2(t)} = \overline{\{x(t) + y(t)\}^2} = \overline{x^2(t)} + 2\,\overline{x(t)\,y(t)} + \overline{y^2(t)}$$

yielding:

$$\boxed{\overline{z^2(t)} = P_1 + 2P_{12} + P_2} \qquad (2.39)$$

where P_1 and P_2 are called the *self-power* of fluctuations $x(t)$, $y(t)$, respectively, and P_{12} represents the *cross-power*. When $x(t)$ and $y(t)$ are uncorrelated (they originate from distinct noise processes), $P_{12} = 0$ and the two powers simply add.

■ **Cross-Correlation Function.** According to [19], *the cross-correlation function $R_{xy}(\tau)$ of two random signals $x(t)$ and $y(t)$ is the time average of the product $x(t)\,y(t+\tau)$*. If $x(t)$ is the random input of a linear system and $y(t)$ is the response, then $R_{xy}(\tau)$ is the inverse Laplace transform of the transfer function of the system.

Two random processes are said to be *orthogonal* if $R_{xy}(\tau) = 0$ for all τ.

For instance, if two stationary processes are statistically independent and at least one of them has zero mean, then they are orthogonal (but the converse in not true!).

■ **Cross-Power Spectral Density.** By definition, *the cross-power spectral density of two stationary random processes $x(t)$ and $y(t)$ is the Fourier transform of their cross-correlation function*

$$S_f(xy) = \mathcal{F}[R_{xy}(\tau)] \qquad (2.40)$$

□ **Physical Meaning.** This definition is precise, but the physical meaning is not self-evident. To improve insight, consider the setup of Fig. 2.8 in which the band-pass filters are assumed to be ideal and centered on an adjustable frequency f_o. The outputs of the two filters are connected to a multiplier, and consequently the output signal (whose average is accumulated over a long interval) is proportional to the product of the input signals. Here $\mathcal{R}e$ denotes the real part and $\mathcal{I}m$ the imaginary part.

Now, for two signals $s_1(t)$ and $s_2(t)$ with Fourier transforms S_1 and S_2 respectively, the cross-power $P_{12} = \overline{S_1^* S_2}$ is a complex quantity whose real and imaginary parts may be measured as in Fig. 2.8.

If the same signal $s_1(t)$ is applied to both inputs, we get the real and imaginary parts of the self-power spectral density $\overline{S_1^* S_1}$. According to Sect. 2.2.5, if the signal is shifted by 90° the resulting signal is uncorrelated with the original one. We conclude that *the imaginary part of any self-power spectral density is always zero*.

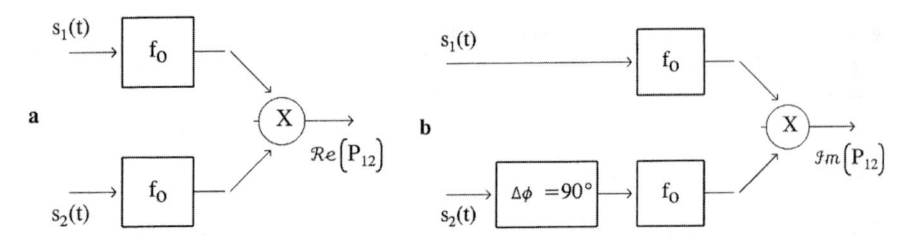

Fig. 2.8. Measuring: **a** the real part, and **b** the imaginary part of the cross-power at a specified frequency f_o

If we permute the signals $s_1(t)$ and $s_2(t)$ in Fig. 2.8, we get the cross-power spectral density P_{21}. Nevertheless, $P_{12} = P_{21}^*$.

■ **Linear System.** In the transmission of fluctuations through linear systems, we need information about the output. Assume a linear, lumped, time invariant system whose transfer function is $H(j\omega)$. At the input we apply the fluctuation $x(t)$, with known mean and spectral density $S_f(x)$. The spectral density of the fluctuation $y(t)$ obtained at the output is [24]

$$S_f(y) = H(j2\pi f)\ H^*(j2\pi f)\ S_f(x) = |H(j2\pi f)|^2\ S_f(x) \qquad (2.41)$$

Expression (2.41) is very important, since it states that *for a linear system the spectrum of the output noise is equal to the spectrum of input noise multiplied by the square modulus (magnitude) of the transfer function.*

The output autocorrelation function is just the inverse Fourier transform of $S_f(y)$

$$R_y(\tau) = \mathscr{F}^{-1}\{S_f(y)\} = \int_{-\infty}^{+\infty} |H(j2\pi f)|^2\ S_f(x) \exp(j2\pi/\tau)\ df \qquad (2.42)$$

■ **Note.** We emphasize several important points:

1) It is possible to prove that the cross-correlation function of input x with output y is equal to the convolution of input autocorrelation function with the impulse response:

$$R_{xy}(\tau) = h(\tau) \circ R_x(\tau) \qquad (2.43)$$

2) The cross-power spectral density is the complex quantity

$$S_f(xy) = H(j2\pi f)\ S_f(x) \qquad (2.44)$$

3) As we have seen

$$S_f(xy) = S_f^*(xy) \qquad (2.45)$$

2.3.5 Correlation Matrix

■ **Self- and Cross-Power Spectra.** Consider a fluctuation z(t) which is the sum of two partially correlated fluctuations x(t) and y(t):

$$z(t) = x(t) + y(t)$$

For computational convenience, the first step is to truncate all fluctuations at $t = \tau$; at the end of the procedure, we take the limit of the resulting expressions as $\tau \to \infty$.

In terms of Fourier transforms, the total power is

$$|Z|^2 = (X+Y)(X^* + Y^*) = |X|^2 + |Y|^2 + XY^* + X^*Y \qquad (2.46)$$

The bilateral spectrum of the resulting fluctuation is [25]

$$S'_f(Z) = \lim_{\tau \to \infty} \frac{|Z|^2}{\tau} = S'_f(XX) + S'_f(YY) + S'_f(XY) + S'_f(YX) \qquad (2.47)$$

where

$$\boxed{S'_f(XX) = \lim_{\tau \to \infty} \frac{\overline{X^*X}}{\tau}} \quad \text{and} \quad \boxed{S'_f(YY) = \lim_{\tau \to \infty} \frac{\overline{Y^*Y}}{\tau}} \qquad (2.48)$$

are the *self-power spectral densities* of fluctuations x(t) and y(t), respectively. In a similar manner,

$$\boxed{S'_f(XY) = \lim_{\tau \to \infty} \frac{\overline{X^*Y}}{\tau}} \quad \text{and} \quad \boxed{S'_f(YX) = \lim_{\tau \to \infty} \frac{\overline{Y^*X}}{\tau}} \qquad (2.49)$$

are, by definition, the *cross-power spectral densities* of fluctuations x(t) and y(t).

☐ **Properties**

- the self-power spectral densities are *real* quantities;
- the cross-power spectral densities are *complex conjugate* quantities.

☐ **Consequence.** Practical circuits always contain nonzero real-part impedances. Therefore, the real parts of the cross-power spectral densities are even functions, while their imaginary parts are odd functions. Equation (2.47) becomes

$$S'_f(Z) = S'_f(XX) + S'_f(YY) + 2\mathcal{R}e[S'_f(XY)] \qquad (2.50a)$$

Since $\mathscr{R}e[S'_f(XY)]$ is an even function, this relation also holds for the unilateral spectra, i.e.

$$S_f(Z) = S_f(XX) + S_f(YY) + 2\mathscr{R}e[S_f(XY)] \quad (2.50b)$$

☐ **Example.** To illustrate a rarely encountered situation when only the imaginary part of the cross-power spectral density is meaningful, Van der Ziel proposes [25] the circuit of Fig. 2.9. The noise current generator is in parallel with the capacitor C and the noise voltage generator is in series with both.

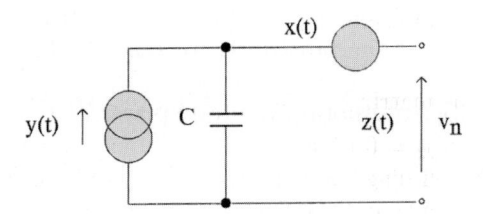

Fig. 2.9. Circuit with 2 partially correlated generators

For $0 \leq t \leq \tau$, we obtain the Fourier transforms of the truncated x(t), y(t), and z(t); then, considering only the n-th order terms, we may write

$$c_n = a_n + b_n/j\omega C$$

where a_n, b_n, and c_n are the coefficients of fluctuations x(t), y(t), and z(t), respectively. Hence:

$$\overline{c_n c_n^*} = 2\,\overline{(a_n + b_n/j\omega C)(a_n^* - b_n^*/j\omega C)}$$

from which

$$S_f(Z) = S_f(XX) + S_f(YY)/(\omega^2 C^2) - 2\mathscr{I}m[S_f(XY)]/(\omega C)$$

To conclude, if the circuit is purely reactive, only the imaginary part of the cross-power spectral density is meaningful.

■ **Wiener-Khintchine Theorem (Two Random Variables).** Extending the formula obtained for a single variable, we may write

$$S_f(XX) = 2 \int_{-\infty}^{+\infty} \overline{X(t)\,X(t+s)}\,\exp(j\omega s)\,ds \quad (2.51a)$$

$$S_f(YY) = 2 \int_{-\infty}^{+\infty} \overline{Y(t)\,Y(t+s)}\,\exp(j\omega s)\,ds \quad (2.51b)$$

$$S_f(XY) = 2 \int_{-\infty}^{+\infty} \overline{X(t)\,Y(t+s)} \, \exp(j\omega s) \, ds \qquad (2.51c)$$

■ **Correlation Matrix.** Two partially correlated fluctuations are described by their correlation matrix C, which, at a given frequency, is defined with respect to their *self-* and *cross-power spectral densities*:

$$C = \begin{bmatrix} C_{11} & C_{12} \\ C_{21} & C_{22} \end{bmatrix} = \begin{bmatrix} S_f(XX) & S_f(XY) \\ S_f(YX) & S_f(YY) \end{bmatrix} \qquad (2.52)$$

□ **Properties.** This matrix has the following properties:

1) $\Im m\, C_{11} = \Im m\, C_{22} = 0$ (diagonal elements are always real).
2) $C_{12} = C_{21}^*$ (the off-diagonal elements are complex conjugate).
   This means that matrix C is *Hermitian*. Other properties are
3) $C_{11} \geq 0$ and $C_{22} \geq 0$.
4) $det\,\{\,C\,\} = C_{11}C_{22} - |C_{12}|^2 \geq 0$.

Properties 1) and 2) have been proved in Sect. 2.3.5.

□ **Extension.** For n partially correlated fluctuations, the correlation matrix is the following n-order array:

$$C = \begin{bmatrix} S_f(X_1X_1) & S_f(X_1X_2) & \cdots & S_f(X_1X_n) \\ S_f(X_2X_1) & S_f(X_2X_2) & \cdots & S_f(X_2X_n) \\ & & \cdots & \\ S_f(X_nX_1) & S_f(X_nX_2) & \cdots & S_f(X_nX_n) \end{bmatrix} \qquad (2.53)$$

■ **Kleckner's Correlation Coefficient.** The terminal noise behavior of a linear multiport is generally investigated by means of a set of n equivalent generators situated at the external terminals. Since the various noise mechanisms are located inside the multiport (rather than at the ports), the noise generators should be correlated, and they are described by their correlation matrix (2.53). Kleckner proposed the following definition for the correlation coefficients, which usually are frequency-dependent complex quantities [27]:

$$\Gamma_\omega(i,j) = \frac{S_\omega(i,j)}{\sqrt{S_\omega(i,i)\,S_\omega(j,j)}} \qquad (2.54)$$

They have the following properties resulting from the definition:

- $\Gamma_{-\omega}(i,j) = \Gamma_\omega^*(i,j) = \Gamma_\omega(j,i)$;
- $|\Gamma_\omega(i,j)| \leq 1$ for all ω;

- the quantity $|\Gamma_\omega(i,j)|$ represents the "total" correlation between generators i and j;
- the quantity $\Re\{\Gamma_\omega(i,j)\}$ corresponds to the "effective" correlation (on a power density basis) between sources i and j. When $\Re\{\Gamma_\omega(i,j)\} = 0$, there is zero power interaction.

☐ **Discussion.** Expressed in the time domain, the correlation coefficient is always a real quantity. However, in the frequency domain, it often also has an imaginary part. This simply means that for two partially correlated fluctuations, a component of one may be deduced from the other by means of a shifting circuit (a differentiator or an integrator circuit).

Kleckner's correlation coefficient is useful for investigating possible noise reduction, since its phase relative to 180° indicates the phase shift required to maximize noise cancellation.

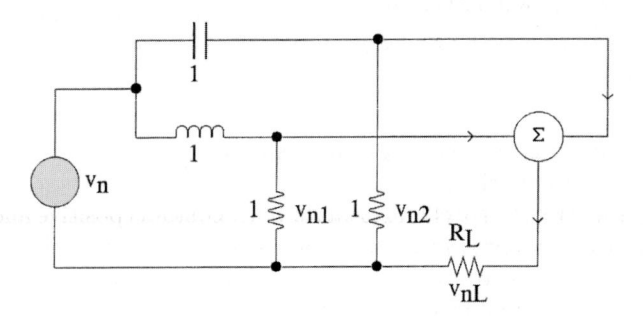

Fig. 2.10. Example of a noisy circuit

☐ **Example.** Consider the circuit in Fig. 2.10, where v_n is a white-noise generator with a constant power spectral density denoted by K. All elements values are normalized, and the load is R_L. Determine

1. The power spectral density of the noise across each resistor.
2. Kleckner's correlation coefficients.

Solution

1. We can write

$$v_{n1} = \frac{1}{1+j\omega} v_n, \quad \text{so} \quad |v_{n1}|^2 = \frac{1}{1+\omega^2} v_n^2$$

Also:

$$v_{n2} = \frac{1}{1+1/j\omega} v_n, \quad \text{so} \quad |v_{n2}|^2 = \frac{\omega^2}{1+\omega^2} v_n^2$$

2.3 Fluctuations: Probabilistic Approach

The noise power spectral densities are respectively

$$S_\omega(1,1) = \overline{v_{n1}v_{n1}^*} = \overline{|v_{n1}|^2} = \frac{\overline{v_n^2}}{1+\omega^2} = \frac{K}{1+\omega^2}$$

$$S_\omega(2,2) = \overline{v_{n2}v_{n2}^*} = \overline{|v_{n2}|^2} = \frac{\omega^2}{1+\omega^2}\overline{v_n^2} = \frac{\omega^2}{1+\omega^2}K$$

$$S_\omega(1,2) = \overline{v_{n1}v_{n2}^*} = \frac{1}{1+j\omega}\overline{v_n}\frac{1}{1-1/j\omega}\overline{v_n} = \frac{-j\omega}{1+\omega^2}K$$

Note that the power spectral density across the load is K (that is, the sum of $S_\omega(1,1)$ and $S_\omega(2,2)$).

2. $\Gamma_\omega(1,2) = -j$ ($\phi = -90°$).

This simply means that fluctuations v_{n1} and v_{n2} are totally correlated, since $|\Gamma_\omega(1,2)| = 1$ (which is not a surprise, as they originate from the same source v_n!). As the real part of Kleckner's coefficient is zero, this denotes zero power interaction.

Summary

- Two fluctuations are said to be *partially correlated* if they have at least a common physical origin.
- In signal theory, the spectrum is *bilateral* (it contains positive and negative frequencies). However, in noise theory widespread convention chooses to consider only positive frequencies; hence, the spectrum is *unilateral*.
- For the same amount of white noise, the height of the unilateral spectrum is twice that of the corresponding bilateral spectrum.
- The *Fourier transform* establishes the correspondence between two possible signal specifications: time domain and frequency domain.
- Partial time-domain signal specification requires the following quantities: average value, mean square value, form factor, and peak factor.
- To ensure a consistent description, in the time domain the *time average* is used, while for the frequency domain we use the *ensemble average*. Nevertheless, electrical noise is ergodic, so these averages yield identical results.
- The mean random signal squared per unit frequency band is called the *spectral density*.
- The *correlation matrix* C is defined with respect to the self- and cross-power spectral densities.
- The off-diagonal elements of C are, in general, complex quantities, while the diagonal elements are real numbers. The correlation matrix is Hermitian.
- For a linear system the spectrum of the output noise is equal to the spectrum of input noise multiplied by the squared absolute magnitude of the transfer function.

3

Physical Noise Sources

> "Be not afeard: the isle is full of noises, sounds
> and sweet airs, that give delight, and hurt not"
>
> (W. Shakespeare, Tempest)

3.1 Introduction

In order to understand why intrinsic noise is important, consider a communication system in which the transmitter and the channel are both assumed to be ideal (noiseless). The signal that reaches the receiver input is exactly the image of the transmitted signal, except that it is attenuated during propagation in the channel. After passing through the receiver this signal is amplified, but something happens: its content has been affected by the intrinsic noise of the receiver. This noise is the result of all fluctuations occurring in the receiver circuits. It is added to the useful signal and may modify its content. In a situation where the received signal is very weak and the noise strong, the transmitted information may be completely lost. This demonstrates why intrinsic noise is a key issue in all circuits that process weak signals.

The level of intrinsic noise is related to its mean square value. Unfortunately, the instantaneous value cannot be predicted, since the noise is the result of a random process. Additional valuable information may be found from its spectrum, but the spectral bandwidth largely depends on the physical processes that generated the fluctuations. For this reason, emphasis is given in this chapter to the study of the main noise sources encountered in any electronic circuit.

3.2 Noise Sources

3.2.1 Thermal Noise

■ **Brief History.** Random motion of particles of pollen in a fluid was discovered by R. Brown in 1827, and is now known as *Brownian motion*. In 1906, A. Einstein predicted that the Brownian motion of free electrons inside a piece of metal (that is in thermal equilibrium) would produce a fluctuating electromotive force at its ends (Fig. 3.1). This phenomenon was first observed by J.B. Johnson in 1928, and its power spectrum was computed by H.T. Nyquist the same year.

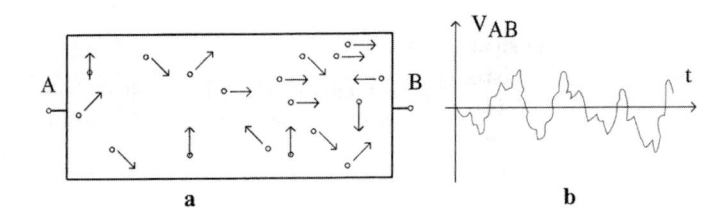

Fig. 3.1. a Spontaneous clustering of free electrons at one end, **b** Thermal noise voltage versus time

■ **Origin.** The physical origin of this noise is the thermal motion of free electrons inside a piece of conductive material, which is totally random.

■ **Explanation.** Figure 3.1a shows a piece of conductive material where some free electrons are indicated. Random thermal motion of electrons inside the material leads to temporary agglomeration of carriers at one end or the other. From a macroscopic standpoint, this means that the potential of end contact B will be more negative than the potential of end contact A. In other words, a potential difference V_{AB} appears, whose polarity and magnitude are fluctuating. This is a *thermal noise* voltage. A possible pattern of its variation versus time is given in Fig. 3.1b.

As there is no reason to have a permanent accumulation of charge at either terminal, the mean value of the fluctuating voltage V_{AB} must obviously be zero.

■ **Comment.** Thermal noise is a consequence of the discrete nature of charge and matter. At a microscopic level, thermal motion is a general property of matter, regardless of temperature.

Physical systems containing a huge number of identical particles have a large number of degrees of freedom, corresponding to the number of ways in which energy can be stored in the system. When we engage in a macroscopic description of the system (i.e., by means of a few physical quantities, like voltage v, current i, etc.), only the energy stored in degrees of freedom corresponding to macroscopic quantities is of interest. Widespread convention chooses to lump all remaining energy under the label "thermal energy" and the remaining degrees of freedom as "thermal degrees of freedom."

According to [29], at equilibrium we must recognize that an interaction exists between the macroscopic quantities and the thermal degrees of freedom. The flow of energy from macroscopic to thermal degrees of freedom corresponds to *dissipation*. The reverse flow, from thermal to macroscopic degrees of freedom, manifests itself as *fluctuations* in the observed physical quantities (i.e., voltage or current).

■ **Remarks**

- Thermal noise in a physical resistor is perceived as a fluctuation in the electrical current (if the resistor is in a closed loop) or in the electrical voltage across its terminals (if the resistor is open-circuited). In both situations the DC component of the fluctuation is zero.
- Thermal noise does not depend on the applied voltage, since for usual values of the electric field, the extra energy supplied to the free electrons by the field is negligible with respect to their thermal energy.

■ **Nyquist's Theorem.** This theorem states that for linear resistances in thermal equilibrium at temperature T, the current or voltage fluctuations are quite independent of the conduction mechanisms, type of material, and shape and geometry of the resistor. The generated noise depends exclusively upon the value of the resistance and its temperature T (given in kelvins). In the literature, four equivalent statements are encountered:

1) The *noise voltage spectral density*, under open-circuit condition, is a constant quantity, given by

$$\boxed{S(V_n) = \frac{\overline{v_n^2}}{\Delta f} = 4kTR} \quad [V^2/Hz] \qquad (3.1)$$

2) The *noise current spectral density*, under short-circuit condition, is a constant quantity, given by

$$\boxed{S(I_n) = \frac{\overline{i_n^2}}{\Delta f} = \frac{4kT}{R} = 4kTG} \quad [A^2/Hz] \qquad (3.2)$$

where G denotes the conductance (the reciprocal of R).

3) The *available noise power spectral density* (i.e., the noise power delivered to an identical resistor R, per unit bandwidth) is a constant quantity that depends solely on temperature

$$\boxed{S_p = kT} \quad [W/Hz] \qquad (3.3)$$

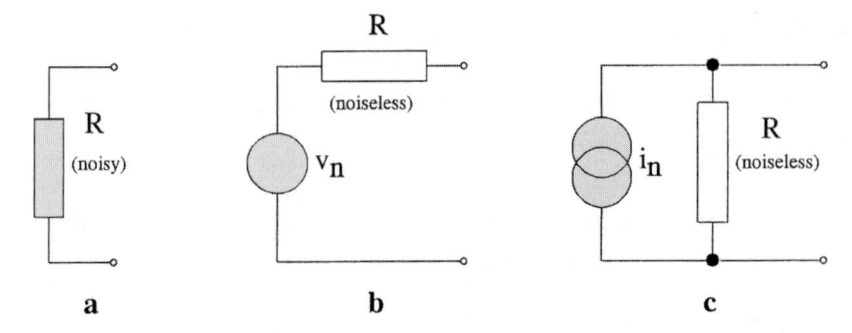

Fig. 3.2. **a** Noisy resistor, **b** Thévenin equivalent circuit, **c** Norton equivalent circuit

4) A noisy resistor is modeled by means of an identical noiseless resistor and a lumped external noise source. Two equivalent circuits are presented in Fig. 3.2, where

$$\boxed{\overline{v_n^2} = 4kTR\,\Delta f} \quad \text{(Thévenin model)} \qquad (3.4)$$

and:

$$\boxed{\overline{i_n^2} = 4kT\,\Delta f/R} \quad \text{(Norton model)} \qquad (3.5)$$

In all previous expressions, k is Boltzmann's constant (k = 1.38·10^{-23} J/K) and Δf is the bandwidth within which the noise is measured (usually imposed by the measurement equipment).

Note that both representations are valid only to model the noise of a resistor *as it is perceived at its terminals*. Under no circumstances (3.4) and (3.5) describe the noise *inside* the resistor itself!

■ **Note**

1) The previous four statements are equivalent, and any of them can be used to prove the remaining three.
2) Nyquist's theorem is valid only for *linear* systems, in *thermal equilibrium* with their environment.
3) Equations (3.1) and (3.2) show that the spectral density of the thermal noise is constant with respect to frequency (at least up to frequencies at which the quantum correction is imposed). Like white light, which contains all visible components in equal proportion, thermal noise is called *white noise*.
4) Equation (3.3) suggests that the total available noise power of a resistor is infinite if the bandwidth is also infinite. However, as Oliver proved, the quantum correction limits it to a finite value given by $(\pi kT)^2/6\,h$ [W], (h is Planck's constant, 6.62·10^{-34} Js).

3.2 Noise Sources

■ **Extension.** It is interesting to take into account several special cases:

☐ **Arbitrary Impedance.** Consider an arbitrary impedance $Z = R + jX$, in thermal equilibrium. Then (3.2) becomes

$$S(I_n) = 4kT\,\mathcal{R}e\{Z^{-1}\} = \frac{4kTR}{R^2 + X^2} \qquad (3.6)$$

where $\mathcal{R}e$ denotes the real part of its argument.

☐ **One-Port Network.** For a linear two-terminal network containing only noisy resistors, capacitors, and inductors, the noise voltage (or current) at its terminals can be calculated by substituting each resistor with its appropriate model (as given in Fig. 3.2). Next, each individual contribution at the output is calculated and the resulting mean square values are finally added. A much simpler approach consists in evaluating the real part of the complex impedance Z seen when looking into the terminals and then with expression (3.4)

$$\boxed{\overline{v_n^2} = 4kT \int_0^{+\infty} \mathcal{R}e\{Z\}\,df} \qquad (3.7)$$

Equation (3.7) is often called *Nyquist's formula*. For a one-port containing only resistors, this equation obviously reduces to

$$\overline{v_n^2} = 4kTR_{eq}\,\Delta f \qquad (3.8)$$

where R_{eq} denotes the equivalent resistance of the network.

☐ **High Frequency and Low Temperature.** As previously stated, Nyquist's theorem is not valid up to arbitrarily high frequency, since by integrating (3.3) over the entire frequency domain, we obtain infinite noise power. Therefore, a quantum correction is imposed. It has been shown [30] that in all situations with

$$hf/kT \gg 1 \qquad (3.9)$$

(with f in Hz and T in kelvins), (3.3) becomes

$$\boxed{S(p) = \frac{hf}{\exp(hf/kT) - 1}} \qquad (3.10)$$

We introduce the Planck factor

$$\mathcal{K} = \frac{hf/kT}{\exp(hf/kT) - 1} \qquad (3.11)$$

The mean square value of the open-circuit noise voltage delivered by a resistor can be written

$$\overline{v_n^2} = 4kT \int_{f_1}^{f_2} R(f)\ \mathcal{H}(f)\ df \quad [v^2] \qquad (3.12)$$

where f_1 and f_2 define the frequency range.

☐ **Non-reciprocal Networks.** In the case of linear multiports containing some non-reciprocal devices, Thevenin's and Nyquist's theorems must be formulated in a different manner. Twiss proved that the impedance appearing in the previous theorems must be replaced by a linear combination of some elements of the impedance matrix used to describe the non-reciprocal network [31].

☐ **Distributed Circuits.** Nyquist's theorem can also be applied to linear, dissipative, and distributed systems, such as a transmission line or waveguide.

We need only consider a finite, elementary section of the distributed structure [32]; the noise power spectral density per unit length (propagating in one direction) is $2\alpha kT$, α being the attenuation constant of the line or waveguide per unit length. Haus investigated the case of lines with a nonuniform temperature distribution; he also showed that fluctuations in three-dimensions distributed media can be simulated by means of a noise current generator added to Ampere's law (in Maxwell's equations) [33].

☐ **Nonlinear Circuits.** There are compelling reasons to believe that Nyquist's theorem cannot be directly applied to nonlinear circuits [29]. Nevertheless, if the nonlinear system is in thermal equilibrium and the resulting fluctuations are small, we can still apply this theorem to the small-signal input impedance of the circuit. Gupta has proposed a theorem yielding a noise model of a nonlinear resistor in the form

$$\overline{v_n^2} = 4kT\ \Delta f \left(\frac{dV}{dI} + \frac{1}{2} I \frac{d^2V}{dI^2} \right) \bigg|_{I=I_{CC}} \qquad (3.13)$$

Here V and I are the terminal voltage and current of the nonlinear resistor, and I_{CC} is the steady bias current.

■ **Properties**

- The frequency distribution of thermal noise power is uniform, at least up to $f_c = 0.15\,kT \cdot 10^{34}$ [Hz].
- The instantaneous amplitude of the thermal noise has a Gaussian distribution. The mean value of the fluctuation is zero, and the mean square value is established with (3.4) and (3.5).
- The peak factor is roughly 4 (if only peaks occurring at least 0.01% of the monitoring time are considered).

■ **Measurement.** Frequently, to detect the mean square value of the thermal noise, a wideband oscilloscope can be used. All we have to do is to evaluate the peak-to-peak value, ignoring one or two of the most important peaks on the screen; then, this value is divided by 8.

■ **Where Is This Noise Important?** Thermal noise is generated only in *dissipative* systems (for instance a purely reactive device does not generate thermal noise). Therefore, it is associated with all resistors and lightly doped semiconductor layers.

3.2.2 Diffusion Noise

■ **Origin.** The physical origin of diffusion noise is carrier velocity fluctuations caused by collisions.

■ **Explanation.** This noise is related to the diffusion process that results from nonuniform carrier distribution. If carrier density increases at one end of a semiconductor (for instance by illumination), it is a natural tendency for the carriers to move towards the opposite end, where the population is reduced. This is a *diffusion process*. Note that no applied voltage is necessary to sustain it.

However, during their trip the electrons are scattered via collisions with the lattice or with ionized impurity atoms. As a result, carrier velocities are modified, and this represents a perturbation of the regular diffusion tendency. As scattering occurs randomly, the instantaneous value of the diffusion current is also random. This is the mechanism of diffusion noise.

■ **Model.** Becking was the first to model diffusion noise, followed by Van der Ziel. They considered an n-type semiconductor with an electron density gradient; the proposed approach [34] consists in partitioning the volume of semiconductor into elementary rectangular boxes ($\Delta x \, \Delta y \, \Delta z$). Due to collisions, electrons are able to pass from one box to an adjacent box, this being considered a series of independent events.

Finally, the current spectral density of the diffusion noise is

$$S(I) = 4q^2 \, D_n \, n(x) \, \frac{\Delta y \, \Delta z}{\Delta x} \qquad (3.14)$$

where q is the elementary charge, n(x) the electron concentration on the x axis, D_n the diffusion constant, and Δx, Δy, Δz are the dimensions of the elementary box.

■ **Remark.** From a microscopic point of view, if Einstein's relation

$$D_n = \frac{kT}{q} \mu_n \qquad (3.15)$$

holds (μ_n being the electron mobility and D_n the electron diffusion constant), the current spectral density becomes

$$S(I) = \frac{4kT}{\Delta R} \qquad (3.16)$$

ΔR being the resistance of the elementary box ($\Delta x, \Delta y, \Delta z$), given by

$$\Delta R = \frac{\Delta x}{q\mu_n\, n(x)\, \Delta y\, \Delta y} \qquad (3.17)$$

Comparing (3.16) with (3.2) it is obvious that *for majority carriers only, diffusion noise reduces to thermal noise, if Einstein's relation holds*. From a macroscopic standpoint, we must inspect the device current-voltage relationship: if it follows Ohm's law, the noise is thermal; otherwise, it is diffusion noise.

■ **Properties**

- In contrast to thermal noise, which is due to random motion of carriers, diffusion noise is related to fluctuations induced by random collisions.
- The spectrum of this noise is relatively flat.

■ **Where Is This Noise Important?** Diffusion noise is dominant in all devices where the I-V characteristic does not follow Ohm's law. A typical example is the channel of a FET (or MESFET) transistor: if the bias is such that Ohm's law is valid (it operates in the linear region), the channel generates essentially thermal noise. But if the transistor operates in the saturation region, the channel is producing diffusion noise.

■ **Conclusion.** Diffusion noise represents a more general process than thermal noise. Nevertheless, if Einstein's relation (or its macroscopic equivalent, Ohm's law) holds, diffusion noise becomes thermal noise.

3.2.3 Shot Noise

■ **Origin.** The physical origin of shot noise is the discrete nature of charge carriers.

■ **Explanation.** In any electronic device containing a potential barrier, the current through the device is limited only to those electrons that possess enough energy to cross the potential barrier. The passage of charge carriers (electrons) across this barrier constitutes a series of independent, random events.

To be specific, consider a stream of electrons in vacuum, between two electrodes A and B (Fig. 3.3a). The electrons are emitted by electrode A and

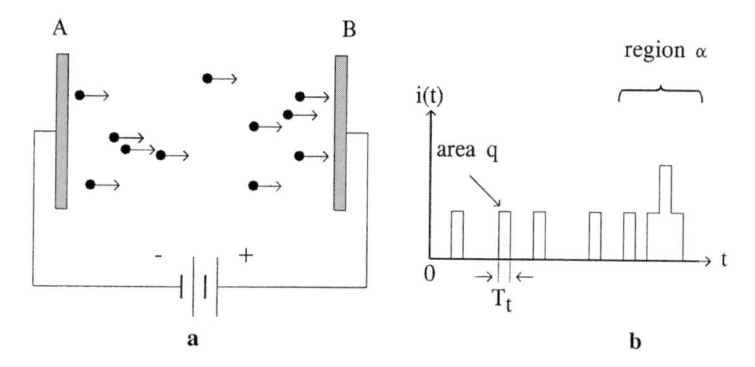

Fig. 3.3. a Electron clustering during flight between two plates; b elementary current pulses

collected by electrode B, which has a positive potential with respect to A (as in a photomultiplier tube).

The number of electrons crossing a specific plane (let us say, plane B) will fluctuate from one short time period to the next, due to the random emission rate of electrons at plane A, and also due to the random distribution of individual velocities. Every time an electron crosses plane B, an elementary current pulse appears in the external circuit. Its area is equal to the elementary charge q and T_t corresponds to the transit time (the average time needed to cross the barrier). Region α corresponds to a cluster of 5 electrons reaching plane B.

Although from a macroscopic perspective the "steady" current seems to be constant, from a microscopic perspective its instantaneous value fluctuates about the mean. This noise mechanism is referred to as *shot noise* and is essentially due to the discrete (granular) nature of electric charge.

■ **Remark.** It is important to note that shot noise appears in all devices collecting a flow of electrical particles. The implicit condition is that the particles must have *ballistic trajectories*, i.e., there are no interactions during flight. This assumes that the density of charge carriers is low and the external field high. Otherwise, the randomness of their position and velocity is reduced by interactions (repulsion between identical charged particles or collisions with lattice atoms), and the shot noise is "smoothed."

■ **Analysis.** W. Schottky first deduced the expression for shot noise in a vacuum diode [35]; this was the first attempt to evaluate noise in an electronic device.

The classic approach to estimating shot noise relies on Carson's theorem [34], where the elementary event is identified to be the arrival of an electron at the collecting terminal. Without taking transit time into account, the instantaneous current is the sum of elementary Dirac pulses of weight q

3 Physical Noise Sources

$$I(t) = q \sum_i \delta(t - t_i) \qquad (3.18)$$

whose mean value is

$$\overline{I(t)} = I_o = \lambda\, q \qquad (3.19)$$

where λ is the average number of electrons collected per second. The (unilateral) current spectral density of shot noise is

$$\boxed{S(I) = 2qI_o} \quad [\text{A}^2/\text{Hz}] \qquad (3.20)$$

The spectrum is obviously white (however, if the transit time of electrons is considered, this is no longer the case and a cut-off frequency $\omega_c \cong 3.5/T_t$ emerges). Remember that expression (3.20) was deduced for charge carriers on *ballistic trajectories*.

■ **Model.** Shot noise is modeled as a noise current generator whose mean square value is

$$\boxed{\overline{i_n^2(t)} = 2qI_o\, \Delta f} \quad [\text{A}^2] \qquad (3.21)$$

where I_o represents the average (DC) current flowing through the device. Again, Δf is the measurement system bandwidth and q is the elementary charge (q = $1.6 \cdot 10^{-19}$ C).

■ **Shot Noise in PN Junctions.** The current of a PN junction essentially results from injection of *minority* carriers into the bulk region, followed by their diffusion and recombination. To calculate the noise, we may diffusion noise sources and generation-recombination noise sources for the minority carriers take into account. This is the *collective approach*.

However, the noise can be related to emission across the junction potential barrier; the passage of each carrier constitutes an independent random event, and the Campbell's theorem is of great help in evaluating shot noise. This is the *corpuscular approach*. Van der Ziel proved that the two approaches are equivalent [34].

Another approach less desirable (according to [36], as it does not describe the noise mechanism, and only accidentally leads to correct results) starts with the observation that for independent noise sources, the mean square values of their currents (or voltages) are additive. Traditionally, the junction current is expressed as

$$I = I_S \left(\exp\left(\frac{qV}{kT}\right) - 1 \right) \qquad (3.22)$$

where V is the junction applied voltage and I_S is the saturation current. Common practice is to decompose current I into two currents, namely

$I_S \exp(qV/kT)$ and $-I_S$. Assuming that the two yield independent fluctuations,

$$\overline{i_{n,tot}^2} = \left(2qI_S \exp\left(\frac{qV}{kT}\right) + 2qI_S\right) \Delta f = 2q(I + 2I_S) \Delta f \qquad (3.23)$$

Using the small-signal low-frequency conductance of a forward biased junction ($I \gg I_S$)

$$g_m = \frac{dI}{dV} = \frac{qI}{kT}$$

expression (3.23) becomes

$$\overline{i_{n,tot}^2} \simeq 2qI \, \Delta f = 2kT \, g_m \, \Delta f \quad [A^2] \qquad (3.24)$$

Comparison with (3.5) suggests that the shot noise of a forward biased junction is half the thermal noise generated by a resistor equal to the junction small-signal resistance.

■ **Shot Noise in Metal-Semiconductor Junctions.** The current flow in metal-semiconductor junctions (Schottky diodes) is due to *majority* carriers. Two categories of carriers must be taken into account:

– Carriers going from metal into semiconductor, which encounter a potential barrier of height E_o. These produce the current $-I_S$, which is almost independent of the applied voltage (V).
– Carriers going from semiconductor into metal, which encounter a barrier of height $q(\Phi_c - V)$, where Φ_c is the contact potential and V the voltage applied to the metal. Finally, as the overall current is the sum of these components, the total current spectral density is found to be

$$S(I_{tot}) = 2q(I + 2I_S) \qquad (3.25)$$

or

$$S(I_{tot}) = 2kT \, g_m \, (I + 2I_S)/(I + I_S) \qquad (3.26)$$

■ **Note**

1) Unlike thermal noise, shot noise doesn't depend on temperature.
2) Since shot noise depends only on the DC current through the device, modifying the bias current represents an easy way to control the noise level. This property is particularly useful in the design of calibrated noise sources.
3) A junction diode always generates shot noise. In this case the potential barrier is associated with its depletion layer. Expression (3.21) is valid for both forward bias and reverse bias. However, *since fluctuations are increasingly smoothed out as the average current increases*, in forward-biased junctions the noise current is much lower than predicted by (3.20).

■ Properties

- The power distribution versus frequency is uniform (white noise), at least up to frequencies close to the reciprocal of the transit time.
- Shot noise exhibits a Gaussian distribution of its instantaneous amplitudes.
- The ratio of fluctuations to the average current can be reduced by increasing the number of electrons reaching the collector terminal.
- Shot noise could hypothetically be reduced if the magnitude of an individual charge could be diminished.

■ Where Is This Noise Important?
Shot noise is always generated in photomultipliers and all vacuum devices (where electrons move on ballistic trajectories), as well as in junction diodes. Shot noise effects are better perceived in reverse-biased diodes (or in forward-biased junctions operating at very low currents).

■ Conclusion.
Shot noise is always associated with charge carriers crossing a potential barrier. In many situations (particularly in a forward-biased diode), shot noise is "smoothed" by the space charge and the transit time of carriers. Shot noise effects are stronger when the average current through the device is very low.

3.2.4 Quantum Noise

■ Origin.
The quantified (discrete) nature of electromagnetic radiation represents the fundamental origin of quantum noise.

■ Explanation.
According to quantum mechanics, electromagnetic energy is radiated or absorbed in small discrete quantities, called *photons*. The energy of a photon is related to the frequency of its associated radiation by

$$E = hf \qquad (3.27)$$

where h is Planck's constant (h = $6.63 \cdot 10^{-34}$ Js).

Consider a system exchanging thermal or optical energy with its environment (e.g., a bolometer or an optical detector, as in Fig. 3.4).

The bilateral flow of energy between the detector and its environment has a fluctuation, according to the random number of photons released or absorbed per second. This is called *quantum noise*.

■ Heisenberg Principle.
Consider two conjugate canonical quantities, like the energy E of a given system and the time t during which the system is reckoned to possess this energy. The Heisenberg uncertainty principle states that these quantities cannot be simultaneously accurately measured. The minimum uncertainty is given by

$$\Delta E \, \Delta t = h/4\pi \qquad (3.28)$$

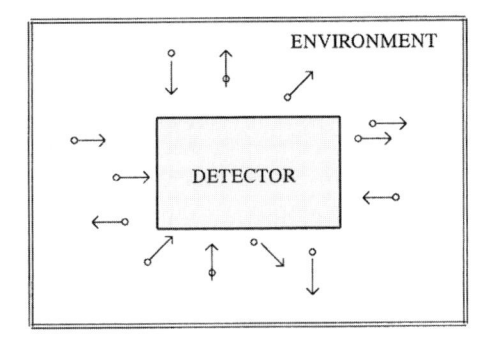

Fig. 3.4. Photon exchange between the detector and its environment

Hence, for a photon, the more accurately measured the energy, the less accurately the instant of time at which it has this energy is known, and vice versa.

■ **Remark.** Quantum noise is quite negligible at usual engineering frequencies, but it becomes important in and above the infrared region of the spectrum. It also shows up at very low temperatures (several kelvins), where thermal noise is less significant. Recall the quantum correction to the Nyquist power spectral density of a resistor

$$S(p) = \frac{hf}{\exp(hf/kT) - 1} \quad [W/Hz] \qquad (3.29)$$

This equation suggests that at very high frequencies, where $hf \gg kT$ (as in optical communications), the noise of the system might tend to zero. This does not occur, because the minimum noise limit is established not by (3.29), but by the Heisenberg principle.

■ **Paradox.** Assume an ideal amplifier, perfectly linear and noiseless, with gain G and propagation time τ. Its input is excited by some source, and at the output we connect an ideal detector (able to measure the output energy E_o and corresponding time t_o with optimal accuracy: $\Delta E_o \Delta t_o = h/4\pi$). The corresponding input energy E_i and its associated time t_i are related to the output values by:

$$E_i = E_o / G \quad \text{and} \quad t_i = t_o - \tau$$

Hence, the minimum uncertainty at the input is less then $h/4\pi$, because

$$\Delta E_i \, \Delta t_i = \Delta E_o \, \Delta t_o / G = h/4\pi G \qquad (3.30)$$

The final result of (3.30) is inconsistent with Heisenberg principle and the only way to resolve this paradox is to conclude that it is impossible to have a completely noiseless linear amplifier. The amplifier *must* add some noise, in order to satisfy the limit at the input imposed by the uncertainty principle.

3 Physical Noise Sources

■ **Solution.** Oliver [30] has shown that in the best case, the amplifier must add a Gaussian noise with power spectral density

$$S_a(p) = hf \quad [W/Hz] \qquad (3.31)$$

This is exactly the *quantum noise*, and it is possible to prove that the input noise temperature of the amplifier cannot be less than:

$$T_{min} = \left(\ln \frac{2 - 1/G}{1 - 1/G}\right)^{-1} \frac{hf}{k} \qquad (3.32)$$

■ **Note**

1) From this vantage point, quantum noise seems to be a consequence of the uncertainty principle. Applying the quantum correction to the generalized Nyquist theorem (3.10), one obtains

$$\boxed{S(p) = \left(\frac{1}{2} + \frac{1}{\exp(hf/kT) - 1}\right) hf} \qquad (3.33)$$

2) Masers and lasers can operate very close to the fundamental noise limit. Quantum noise generated in these devices is due mainly to incoherent radiation associated with random transitions between energy levels.

■ **Where Is This Noise Important?** This type of noise is negligible up to infrared frequencies. It becomes important only at extremely high frequencies ($f \geq 10^{15}$ Hz), where the thermal noise spectrum begins to roll off.

3.2.5 Generation-Recombination Noise (G-R Noise)

■ **Origin.** This noise is related to statistical fluctuations in the population of charge carriers due to random generation, random recombination, random trapping, and the release of carriers in a semiconductor.

■ **Explanation.** Every time a covalent bond is broken, an electron-hole pair is generated. To break a covalent bond, a small amount of extra energy is needed. For instance, this can be supplied either thermally or by illuminating the semiconductor surface. Since the flow of energy (phonons or photons) is quantized, and consequently nonuniform, the generation of charge carriers is a random process, both in space and in time.

Opposing charge generation is the simultaneous process of pair recombination. This takes place every time an electron meets a hole. The circumstances of the encounter depend on the Brownian motion of carriers, and thus recombination also has a random nature. However, on average, generation must balance recombination.

3.2 Noise Sources

Note that traps in the bulk or on the surface of a semiconductor play an important role. Since electrons and holes are captured, and then released after a variable lapse of time, additional fluctuation in the population of charge carriers is induced. All crystal lattice defects (including impurity atoms or molecules that contaminate the surface of the semiconductor during fabrication) act as traps.

■ **Calculation.** As a consequence of fluctuation in charge carrier density, the current through the device also fluctuates. When this process is described by means of Langevin's method [34], the following differential equation is obtained:

$$\frac{d(\Delta N)}{dt} = -\frac{\Delta N}{\tau_o} + H(t) \tag{3.34a}$$

where ΔN represents the fluctuation in the number of carriers, τ_o is the average lifetime of added carriers, and $H(t)$ is a random white noise source that describes the fluctuation. After some algebra we find the power spectrum of the G-R noise

$$\boxed{S(p) = 4 \, \overline{\Delta N^2} \, \frac{1}{1 + \omega^2 \tau_o^2}} \quad [\text{W/Hz}] \tag{3.34b}$$

■ **Remark.** To accurately compute the power spectral density it is necessary to consider the lifetime constants associated with all fluctuating quantities (i.e., number of electrons, number of holes, number of ionized impurities, etc.). Expression (3.34b) takes only *majority carriers* into account and is therefore inaccurate.

■ **Model.** G-R noise is modeled by means of a noise current generator, with power spectral density given by (3.34b), and the mean square value

$$\overline{i_n^2} = \frac{4 \, \overline{I^2} \, \tau}{\overline{N}(1 + \omega^2 \tau^2)} \tag{3.34c}$$

Here $\overline{N}$ is the average number of carriers, $\overline{I}$ is the average (DC) current through the device, and τ is time constant characteristic of the generation-recombination process. We can conclude that the level of G-R noise is proportional to the square of the average (DC) current through the device.

■ **Properties**

– Since τ may be as brief as 1 ns, it follows that the G-R noise spectrum is flat up to a frequency of about $1/\tau$. Beyond this point, it falls at about 20 dB per decade.

- G-R noise induces fluctuations in the conductivity of the material because the number of charge carriers is fluctuating; in the case of thermal noise, the number of carriers is roughly constant, but their spatial distribution fluctuates.

■ **Where Is This Noise Important?** As a general rule, G-R noise is not important in semiconductors with common levels of doping, since the conductance is mainly determined by carriers released from impurity atoms, which greatly dominate those that are thermally or optically generated.

G-R noise shows up in regions where carriers concentrations are relatively weak, for instance, in intrinsic semiconductors, lightly doped semiconductors, and in the space charge layer of every junction.

3.2.6 1/f Noise (Flicker Noise or Excess Noise)

■ **Origin.** The physical origin of this noise has yet to be understood; it also occurs in many non-electrical situations (membranes in ionic solutions, and even in cosmic systems). No generally accepted explanation is available, but one general property of 1/f noise is that its spectrum is frequency-dependent (increasing with decreasing frequency), sometimes extending to frequencies as low as 10^{-4} Hz.

■ **Explanation.** This type of noise represents a rather general process encountered in nonequilibrium systems (the flow of a DC current through a device represents an example of a nonequilibrium system).

In electronic devices, flicker noise is always conditioned by the existence of a DC current in a discontinuous medium. It is assumed to be due either to defects affecting the semiconductor lattice (including unwanted impurity atoms), or to interactions between charge carriers and the surface energy states of the semiconductor [37, 38].

Imperfect contacts (such as those between granules of carbon in carbon resistors) can also generate 1/f noise. In this case the 1/f noise is equivalently called "excess noise," since it adds to the thermal noise of the resistor. Note that a wirewound resistor has no excess noise.

■ **Model.** There have been many attempts to model flicker noise. For instance, McWhorter relates 1/f noise to surface states, and introduces a theoretical model based on the time constants associated with the recombination process. Hooge proposed an empirical approach suggesting that 1/f noise may be related to bulk effects in dissipative media. Handel developed a quantum 1/f noise theory, in which 1/f noise is conditioned by the emission of photons each time a charge carrier collides with the lattice.

Bliek proposed a new model for 1/f noise, starting with the assumption that this kind of noise is produced by the fluctuation of a parameter u, subject to the diffusion equation

$$\frac{\partial u}{\partial t} = D \, \nabla^2 u \qquad (3.35a)$$

Here ∇ is the Laplacian operator and D is a constant. This equation is the same as that used to describe thermal-conductivity or carrier diffusion. Despite the fact that this equation does not lead to an accurate representation of flicker noise, it has been proved [39] that for finite-dimension systems, the power spectral density has a 1/f region covering several decades. Below and above this region the frequency dependence is of the form f^{-k}, k being an integer or a real number.

More recently, Forbes treats flicker noise generated in semiconductor materials (like GaAs) as a bulk phenomenon associated with localized high-frequency variations and long-range low-frequency fluctuations, the lowest frequency being limited only by the bulk volume. In his theory [40], any distributed system of large extent, high resistance, and some capacity, will generate flicker noise when disturbed.

In any case, the 1/f noise is modeled by a current generator with mean square value

$$\overline{i_n^2} = \frac{K \, I^\alpha}{f^n} \, \Delta f \qquad (3.35b)$$

Here K is a device-dependent constant; α is a constant that range from 0.5 to 2; n = 1 for pink noise characterized by constant power per octave (although n may be 2 for fluctuations in the Earth's rotational speed, and 2.7 for galactic radiation); Δf represents the bandwidth of the measurement system and I is the DC current passing through the device.

According to Hooge, flicker noise in homogeneous semiconductors can be characterized by a parameter α such that

$$\frac{S_R}{R^2} = \frac{\alpha}{fN} \qquad (3.35c)$$

with S_R being the power spectral density of the noise in the resistance R, f the measurement frequency, and N the total number of free charge carriers. Consequently, α corresponds to the normalized contribution to the relative noise of a single electron, per unit bandwidth (assuming that the contributions of other electrons are independent). An average value of $2 \cdot 10^{-3}$ was first proposed for α. It subsequently turned out that α also depends on the quality of the crystal; in perfect material it can be taken 2 or 3 orders of magnitude lower.

■ Properties

- The PDF of flicker noise amplitudes is not Gaussian.
- Its spectrum is frequency-dependent (increasing with decreasing frequency, never reaching frequency zero).
- This noise is more prominent whenever electric currents are due to a very small number of charge carriers.

– Tests performed on a large number of samples show that 1/f noise varies from sample to sample, for devices fabricated in the same run. The level is related to the quality of the tested device; *high 1/f noise means poor quality and low reliability.*

■ **Flicker Noise in Bipolar Transistors.** In bipolar transistors 1/f noise is mainly produced by traps (lattice defects or unwanted impurities) situated in the emitter-base region. Carriers are trapped, then released, in a random manner. The time constants of these phenomena are relatively long and this explains why the noise spectral density is more important at low frequency.

According to Van der Ziel, the flicker noise current is

$$\overline{i^2} = K_1 \frac{I^a}{f^b} \delta f \qquad (3.36)$$

In this equation, for any sample, K_1 is a constant that depends weakly on temperature; I is the forward current through the device; δf is the elementary bandwidth centered on f; constant a is in the range 0.5 to 2, and constant b is close to unity.

For b = 1 the power spectral density of flicker noise is shown in Fig. 3.5. We may reasonably represent this noise as a current generator in parallel with the emitter junction.

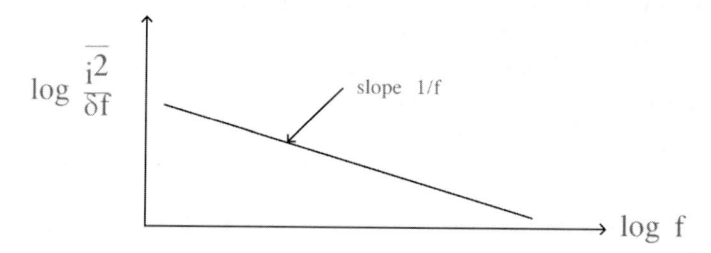

Fig. 3.5. Power spectral density of 1/f noise (b = 1)

It has been noted in practice that PNP transistors have a less significant 1/f noise than NPN transistors. Furthermore, the noise level of devices fabricated on silicon depends on the crystal orientation. For instance, a (100)-surface is more favorable than (111), since the surface-state density is lower.

□ **Test.** To check that this noise is generated at the base–emitter junction, Jaeger and Brodersen fabricated transistors with an additional terminal, corresponding to a metallic contact situated above the emitter)base junction [41]. The carrier density on the surface can be controlled locally by applying an appropriate voltage to this contact; the flicker noise level can be varied in this way.

■ **Flicker Noise in Carbon Resistors.** No 1/f noise is generated if no DC current is flowing through the resistor. More details on this topic are provided in Chap. 7.

■ **Flicker Noise in Integrated Resistors.** According to [44], the 1/f noise voltage developed in integrated resistors has the general form

$$\overline{v^2} = K_R \frac{R_\square^2}{A_R} V_{DC}^2 \frac{\delta f}{f} \qquad (3.37)$$

where V_{DC} is the DC voltage across the resistor, $R_\square$ is the sheet resistance, A_R is the area of the resistor, and K_R is a technological constant. For a diffused or ion-implanted resistor, $K_R \cong 5 \cdot 10^{-24} \, S^2 \, cm^2$, while for thick-film resistors, it is roughly 10 times greater.

■ **Flicker Noise in Integrated GaAs Resistors.** Recently, it has been proved that this type of resistor exhibits 1/f noise whose level is proportional to the square of the applied voltage.

■ **Flicker Noise in MOSFET.** Typically, the drain current exhibits fluctuations with a 1/f spectrum. This is due to fluctuations in the population of charge carriers in the channel, a consequence of carrier trapping in the surface states situated at the SiO_2–Si interface.

Van der Ziel has established that for a MOSFET device the surface-state density at the Fermi level is the only parameter that influences 1/f noise. Hence, the only way to reduce flicker noise significantly is to lower the surface-state density in the vicinity of the Fermi level.

He has shown that 1/f noise increases with decreasing temperature, since the density of surface states increases toward the conduction band.

For MOS transistors operating in strong inversion, flicker noise does not depend on the gate bias, because the surface potential varies very slowly with gate charge. In this case the only way to significantly lower the noise level is to modify the device geometry.

■ **Where Is This Noise Important?** The more heterogeneous the structure of the device, higher the expected 1/f noise. Any inhomogeneity (in the lattice structure or materials) as well as contamination during processing, poor ohmic contacts, etc. can increase the flicker noise.

Flicker noise is low in devices with homogeneous structure and a significant volume of material (for instance a 1-W resistor is less noisy than a 0.25-W resistor of the same value and type).

In monolithic circuits, down-scaling of devices is favorable to increasing the 1/f noise. In recent CMOS processes the corner frequency of 1/f noise tends to increase up to 1 MHz.

Contrary to common belief, the effects of 1/f noise are not limited to low frequency, since it can be up-converted by an existing nonlinearity, as in active mixers, frequency dividers, and voltage-controlled oscillators.

3.2.7 Popcorn (Burst) Noise

■ **Origin.** The physical origin of this noise is not well understood, but it seems to be associated with shallow, heavily doped emitter junctions. Historically, Montgomery (1952) was the first to note that some germanium NPN transistors (all fabricated in the same run) exhibit popcorn noise, especially when previously subjected to high bias voltages.

■ **Explanation.** This noise is observed in many planar-diffused devices: tunnel diodes, bipolar transistors, integrated circuits, film resistors, and so forth. Its onset is associated with contamination by heavy metallic atoms during processing, or to crystallographic damage of regions close to junctions [34]. It seems that the appearance and disappearance of bursts is associated with single-trap activity in a region with few free carriers (like the depletion region of the emitter junction). Such devices usually have small dimensions and are subject to high fields and current densities.

■ **Model.** Previous work by Chenette, then Brodersen and Jaeger, revealed that the popcorn noise source is located near the emitter junction, much closer to the base contact than the flicker noise source.

For a bipolar transistor, burst noise is modeled as a current generator with mean square value

$$\overline{I^2} = \frac{K_1 I_B \Delta f}{1 + \pi^2 f^2 / 4a^2} \quad (3.38)$$

where a represents the number of bursts observed per second, K_1 is a constant, and I_B is the base current of the transistor.

■ **Properties**

– Popcorn noise shows up as bursts in the collector current at a random rate, several hundred per second to one every few minutes. Burst width is variable, between several microseconds and a few minutes. For a particular sample, the amplitude remains the same (as in Fig. 3.6a), but sometimes the bursts can overlap (Fig. 3.6b).

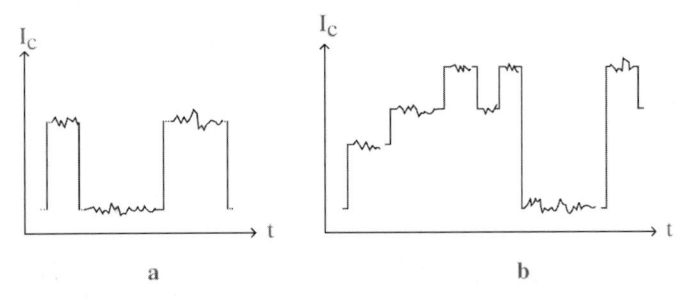

Fig. 3.6. Typical waveforms for popcorn (burst) noise

- The popcorn noise spectrum is proportional to $1/f^2$. Consequently, it affects mainly the low-frequency domain, like flicker noise.
- The instantaneous amplitude of popcorn noise is 2 to 100 times that of the thermal noise. For a given device, this amplitude is constant, since it is determined only by the specific defect of the junction.
- For a given sample, this noise can appear or disappear at random.
- Among all devices fabricated in a single run, this noise might affect only a few. Usually this implies poor quality.

■ **Comment.** Although in the 1970s the term "burst noise" was popular, in the late 1980s there was a tendency to call it "RTS noise" ("Random Telegraph Signal noise"). The term "RTS noise" now seems to be reserved for clean submicron devices, while "burst noise" is reserved for poorer-quality devices of larger size [43].

■ **Where Is This Noise Important?** Progress in manufacturing technology has almost completely eliminated burst noise (in large devices). It is still present in submicron bipolar or MOS transistors with lattice structure damage in sensitive areas.

3.2.8 Avalanche Noise

■ **Origin.** The physical origin of this noise is the process of carrier multiplication due to impact ionization in a reverse-biased PN junction.

■ **Explanation.** Under reverse bias, when the applied voltage is higher, the electric field inside the depletion layer is augmented. Consequently, the minority carriers (holes in the N region and electrons in the P region) are accelerated more and more, and their energy is increased. Colliding with the neutral atoms of the lattice (Fig. 3.7), they are able to generate by impact one or more electron-hole pairs.

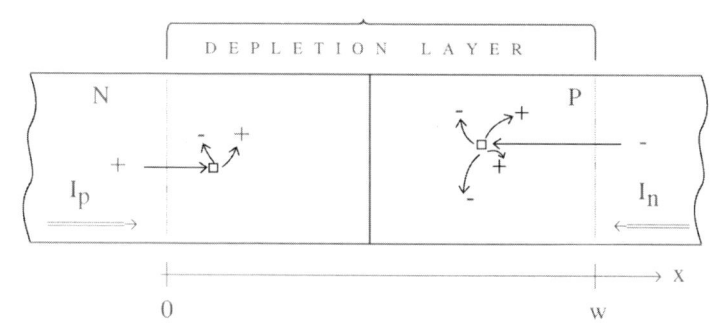

Fig. 3.7. Carrier avalanche multiplication process ($\square$ = neutral atom of the lattice; + = hole; − = free electron)

66   3 Physical Noise Sources

In turn, these pairs of newly generated carriers are accelerated and collisionally produce additional pairs of charge carriers. Moreover, when there are imperfections in the crystal structure, microplasma is formed in these low-volume regions and the current through the junction is dramatically increased [42]. If the avalanche process cannot be controlled, breakdown occurs.

■ **Features.** The regions in which microplasma appears have a random distribution, but they all exhibit the following properties:

- a diameter between 1 μm and 2 μm;
- a high associated series resistance (about $10\,\mathrm{k\Omega}$);
- high noise generation;
- low current intensity (of order tens of μA).

■ **Model.** In Sect. 3.2.3 it has been shown that every carrier crossing the junction induces an elementary current pulse. Superposing all elementary current pulses gives the total current, which fluctuates according to the number of carriers traversing the junction potential barrier per second.

The situation of avalanche-generated carriers crossing the junction is more complicated, since the carriers are multiplied by a factor M, which is a random variable of both space (denoted by x in Fig. 3.7) and time. Furthermore, the probability of generating new pairs by collision is not the same for electrons and holes. Another point is that breakdown does not occur simultaneously overall.

Assuming several simplifications (for instance, the M factor is independent of x), Ambrozy [42] has found the mean square value of the avalanche noise current

$$\overline{i^2} = 2qI_o\,\overline{M^2}\,\Delta f \qquad (3.39)$$

This makes it possible to express the current spectral density as

$$\boxed{S(I) = 2qI_o\,\overline{M^2}} \quad [A^2/Hz] \qquad (3.40)$$

where

$$I_o = I_p(0) + I_n(w) + qA\int_0^w g(x)\,dx \qquad (3.41)$$

Here $I_p(0)$ represents the hole current at x=0, $I_n(w)$ is the electron current at x=w, g(x) represents the number of charge pairs generated at coordinate x per unit volume per second, and A is the junction area.

With the adopted simplifications, the avalanche noise spectrum is white.

■ **Note**

1) This noise can also be found in structures consisting of two metallic conductors in contact at microscopic points.

2) Equation (3.40) applies to reverse voltages higher than 8 V, where the breakdown mechanism comes into play. For reverse voltages lower than 5 V, breakdown is the result of the Zener effect. If the reverse voltage lies between 5 V and 8 V, it has been found that

$$S(I) = 2qI_0\left\{1 + f(M^*)\right\} \quad (3.42)$$

$f(M^*)$ being a function of the averaged ionized rates. In this case, primary carriers are generated by the tunnel effect, and the secondary carriers by collision.

■ **Properties**

– Avalanche noise has a typical waveform with several levels separated by a few millivolts; at first there is fast but random switching among all levels. However, as the reverse current becomes more and more significant, the highest level becomes dominant. A possible explanation is the crossing of the barrier in the "packet mode."
– The spectrum of avalanche noise is white.

■ **Where Is This Noise Important?** Avalanche noise mainly affects reverse-biased diodes operating at more than 8 V.

3.3 Conclusion

The various noise mechanisms discussed here are present in semiconductor devices to a greater or lesser extent, depending on their structure, bias, temperature, operating frequency, and so forth. Table 3.1 gives a summary.

Summary

- Every resistor generates thermal noise. Purely reactive devices have no thermal noise.
- In noise calculation the temperature is always expressed in kelvins.
- Thermal noise depends on temperature, resistance, and bandwidth, but not on current.
- The thermal noise of a passive one-port is equal to the noise generated by a resistor with the same value as the real part of the one-port input impedance.
- Shot noise is associated with a current flowing through a potential barrier.
- Shot noise depends on the DC current, but not on temperature.
- The power spectra of thermal and shot noise are constant up to very high frequencies, where a roll-off is observed.

Table 3.1. Various types of noise in most encountered devices

	Zero bias	Low frequency (with bias)	High frequency (with bias)	$f > 1\,\text{THz}$
Resistor	thermal	1/f	thermal	thermal quantum
Diode	thermal	shot 1/f G-R avalanche	shot avalanche	
Bipolar transistor	thermal	shot 1/f G-R popcorn	thermal shot	
FET	thermal	thermal diffusion shot G-R	thermal diffusion	
MOS	thermal	1/f thermal diffusion		
Laser				quantum

- 1/f noise is generated by the flow of a DC current through an inhomogeneous structure. The more homogeneous the material, the lower the 1/f noise. Its spectrum is not white, but pink (power spectral density is inversely proportional to frequency).
- For many samples produced in the same run, flicker noise is different from one sample to another. 1/f noise is used as a tool to evaluate the quality of processing.
- Generally, 1/f noise is inversely proportional to the volume of the conductive medium. For planar devices, this noise is inversely proportional to the size of the device.
- 1/f noise shows up mainly at low frequency. However, up-conversion is responsible for translating it to high or very high frequency.
- Low-frequency noise is important in the phase noise of RF circuits, such as voltage-controlled oscillators.
- Popcorn noise is eliminated by improving the quality of processing.
- Avalanche noise appears in semiconductor junctions operating at high reverse bias.

4

Noise Parameters

"I hate definitions"

(Benjamin Disraeli)

4.1 Introduction

The noise generated by an electronic circuit has various origins. The user perceives it mainly at the output, and his problem is to appreciate how strong this noise is. Two possible approaches exist:

1) We may use the *correlation matrix*, as introduced by (2.53), provided that all internal noise sources are totally characterized. This is a frequency domain description, which has the advantage of being complete, but the resulting values of the spectral densities are not practical since they are of the order of 10^{-15} to 10^{-24} W/Hz.
2) Often it is preferable to represent the noise of a circuit referred to thermal noise. We recall that a noisy resistor can be represented either by its Thévenin model, with an equivalent noise voltage generator

$$\overline{v_n^2} = 4kTR\ \Delta f \qquad (3.4)$$

or by its Norton model, having an equivalent noise current generator

$$\overline{i_n^2} = 4kT\ \Delta f/R \qquad (3.5)$$

The equivalence between a thermal noise source and a non-thermal one requires that both sources deliver the same power. Mathematically, there are two possibilities to adjust the noise level in (3.4) or (3.5):

– We may keep T equal to the physical system temperature and modify the resistance (conductance) value. In this way we introduce the concept of *equivalent noise resistance (or conductance)*.

70   4 Noise Parameters

– Another possibility consists in keeping the resistance at its physical value, but modifying the temperature in order to reach the imposed noise level. This leads to the concept of *equivalent noise temperature*.

It is important to note that in both cases the implicit condition is that the noise delivered by the non-thermal source must have a *white spectrum*, exactly like the thermal noise source.

Globally, the second approach is more intuitive; besides, the quantities used to characterize the intrinsic noise (such as noise resistance, noise temperature, noise factor, etc.) are much more easily measured and have "nice" values.

Another important point to remember is that all noise parameters are frequency dependent. Hence, they are specified either at a certain frequency (*spot values*) or they are defined in a frequency band, by means of certain statistics.

The purpose of this chapter is to review the background of circuit theory needed to introduce various noise parameters, and then to quote the most accepted definitions for each.

4.2 Definitions Concerning Electrical Power and Bandwidth

4.2.1 Normalized Power

■ **Purpose.** This section is devoted to recalling some fundamental concepts from network theory in order to introduce the notions of power and bandwidth, which are pertinent to noise parameter definition.

■ **Convention.** Harmonic signals can be expressed in terms of either their amplitudes or their rms values. The latter is more suited to power calculation and will be preferred. To be specific, consider a harmonic signal (current or voltage) whose rms value is denoted by X; then

$$x(t) = X\sqrt{2}\cos(\omega t + \theta) \tag{4.1}$$

■ **Complex Representation.** The trigonometric function of (4.1) can be developed by means of Euler's formula, i.e.

$$\cos(\omega t + \theta) = \frac{1}{2}\Big(\exp(j\omega t)\exp(j\theta) + \exp(-j\omega t)\exp(-j\theta)\Big)$$

The *complex amplitude* is defined as

$$\mathscr{X} = X\sqrt{2}\exp(j\theta) \tag{4.2a}$$

4.2 Definitions Concerning Electrical Power and Bandwidth

and its complex conjugate

$$\mathcal{X}^* = X\sqrt{2}\,\exp(-j\theta) \tag{4.2b}$$

Hence, (4.1) can be written

$$\boxed{x(t) = \frac{1}{2}\Big(\mathcal{X}\exp(j\omega t) + \mathcal{X}^*\exp(-j\omega t)\Big)} \tag{4.3a}$$

which shows that *the original signal is equal to half the sum of the complex quantity $\mathcal{X}\exp(j\omega t)$ and its conjugate*.

Note also that

$$\boxed{x(t) = \mathfrak{Re}\Big(\mathcal{X}\exp(j\omega t)\Big)} \tag{4.3b}$$

which means that, alternatively, *the original signal can be obtained by taking the real part of the complex quantity $\mathcal{X}\exp(j\omega t)$*.

Finally, any of (4.1), (4.3a), and (4.3b) can represent the signal x(t).

■ **Remark.** In the following sections, complex quantities will no longer be denoted by script fonts, but rather by standard ones. However, the "real" or "complex" nature of the quantity will be evident from the context.

■ **Normalized Power.** Let us denote by V the effective (rms) value of the signal v(t) applied across a resistor R. In this case the power dissipated is

$$P = \frac{V^2}{R} = RI^2 \tag{4.4}$$

where I is the effective (rms) value of the current through the resistor. By definition, *the normalized power is the power dissipated by a one-ohm resistance*. According to (4.4), it is numerically equal to either V^2 or I^2. Since the mean square value is equal to the square of the rms value (see (2.5)), we deduce that *the normalized power is numerically equal to the mean square value of the signal*, i.e.

$$\boxed{P_{\text{normalized}} \doteq \overline{v^2} \doteq \overline{i^2}} \tag{4.5}$$

where "$\doteq$" denotes "numerically equal" (regardless of dimensions).

■ **Remark.** When adopting the complex representation of a signal, remember that for any complex quantity X (with conjugate X^*),

$$XX^* \equiv |X|^2 \tag{4.6a}$$

$$(X + X^*) \equiv 2\,\mathfrak{Re}\,(X) \tag{4.6b}$$

where $\mathfrak{Re}$ denotes the real part.

4.2.2 Various Definitions of Power

■ Time Domain Definitions

☐ **Instantaneous Power.** For a one-port, *this is the product of the instantaneous voltage across the port terminals and the instantaneous current through it*:

$$p(t) = v(t)\, i(t) \tag{4.7}$$

If the signal is sinusoidal, with rms values V and I

$$v(t) = V\sqrt{2}\, \cos(\omega t + \theta) \quad \text{and} \quad i(t) = I\sqrt{2}\, \cos(\omega t + \theta - \phi)$$

the instantaneous power becomes

$$\boxed{p(t) = VI\, \cos\phi + VI\, \cos(2\omega t + 2\theta - \phi)} \tag{4.8}$$

It is easy to see that this power is negative twice in each period, when the passive one-port returns power to the source. Only for a purely resistive one-port ($\phi = 0$) is the instantaneous power always positive.

☐ **Active Power.** Is defined as *the average of the instantaneous power*, i.e.

$$\boxed{P_{act} = \frac{1}{T}\int_0^T p(t)\, dt = VI\, \cos\phi} \tag{4.9}$$

☐ **Fluctuating Power.** The second term on the right-hand side of (4.8) corresponds to the fluctuating power:

$$\boxed{P_f(t) = VI\, \cos(2\omega t + 2\theta - \phi)} \tag{4.10}$$

☐ **Reactive Power.** The reactive power is

$$\boxed{Q = VI\, \sin\phi} \tag{4.11}$$

☐ **Apparent Power.** Is *the maximum value of the active power* (that is, the value obtained when $\cos\phi = 1$)

$$\boxed{\mathscr{S} = VI = \sqrt{P^2 + Q^2}} \tag{4.12}$$

4.2 Definitions Concerning Electrical Power and Bandwidth

■ **Frequency Domain Definitions**

☐ **Complex Power.** *Is the product of the complex voltage V and the complex conjugate of the current I, both at the terminal pair considered:*

$$\boxed{S = V\, I^*} \qquad (4.13)$$

It is easy to prove that its real part is just the active power, and its imaginary part corresponds to the reactive power. By means of identity (4.6a), (4.13) can be represented in a different but equivalent form

$$S = ZII^* = Z|I|^2 \qquad (4.14a)$$

or

$$S = VY^*V^* = Y^*|V|^2 \qquad (4.14b)$$

where use has been made of the fact that $V = ZI$, Z being the impedance seen at the terminal pair, and $Y = 1/Z$.

☐ **Complex Conjugate Power.** This power is

$$S^* = V^*I \qquad (4.15)$$

☐ **Average Power.** By definition, the average power is

$$\boxed{P_{av} = \mathcal{R}e(VI^*) = \frac{1}{2}(VI^* + V^*I)} \qquad (4.16)$$

where use has been made of the identity (4.6b).

☐ **Average Power Absorbed by a Multiport.** Consider a multiport, with impedance matrix Z and current vector I (corresponding to the terminal currents). The total average power absorbed by the multiport is

$$\boxed{P_{av} = \frac{1}{2} I^+ \left(Z + Z^+ \right) I} \qquad (4.17)$$

Here I^+, Z^+ are called the Hermitian conjugates of matrices I and Z. (The Hermitian conjugate of matrix A is obtained in two steps: 1) take the complex conjugate of A, 2) transpose the resulting matrix.)

☐ **Average Power of a Random Signal.** Consider the noise current I_n at one terminal pair of a multiport. Using the expression (4.14a), the average power dissipated in a 1-Ω resistance can be written as

$$P_{av} = \overline{I_n\, I_n^*} \qquad (4.18)$$

This represents the *average normalized power*. Suppose now that we expand the current I_n in a truncated Fourier series. Equation (4.18) applies to each term of the sum, and in this way we obtain the average normalized power spectral density. Since

$$I_n^*(j\omega) = I_n(-j\omega)$$

(4.18) yields

$$\boxed{S(P_{av}) = \overline{I_n(j\omega)\ I_n(-j\omega)} = \overline{I_n^2}} \qquad (4.19)$$

In the case of thermal noise produced by a conductance G,

$$S(P_{av}) = \overline{I_n^2} = 4kTG \qquad (4.20)$$

4.2.3 Available Power and Available Power Gain

■ **Available Power.** As stated in [45], *this is the maximum power that can be transferred from the source to a load at a specified frequency.*

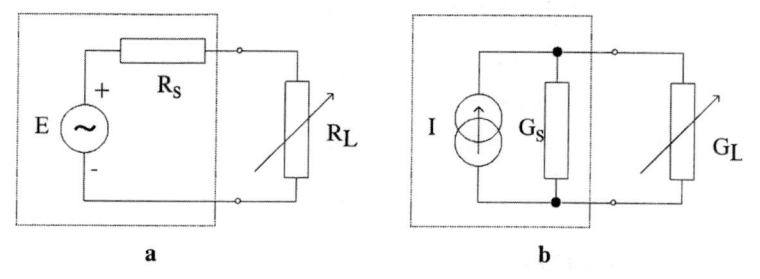

Fig. 4.1. Source and adjustable load: **a** Thévenin representation; **b** Norton representation

Consider a source with internal resistance R_s, and an adjustable load R_L (Fig. 4.1a). The power absorbed by the load is

$$P_L = \frac{R_L}{(R_L + R_s)^2} E^2 \qquad (4.21a)$$

It is easy to prove that the maximum of P_L is obtained when $R_L = R_s$. In this case the load is *matched*, and half the total power of the source is transferred to the load, while the other half is dissipated by its own internal resistance. The maximum of P_L represents the *available power*, denoted by P_a. Obviously

$$\boxed{P_a = Max.\,(P_L) = \frac{E^2}{4R_s} = \frac{I^2}{4G_s}} \qquad (4.21b)$$

4.2 Definitions Concerning Electrical Power and Bandwidth

In the general case, when impedances are considered instead of load and source resistances, maximum power transfer occurs when the load impedance is the conjugate of the source impedance, i.e., $\boxed{Z_L = Z_s^*}$. Note the restriction that the source impedance must always have a positive real part.

☐ **Arbitrary One-Terminal Pair Network.** Haus and Adler have proposed a slightly different definition of available power, which offers the benefit of being easily extended to negative resistances. According to [46], the available power is defined as *the greatest power that can be drawn from the source by arbitrary variation of its terminal current (or voltage)*. Let Z be the internal (or driving) impedance of the one-port, with $R = \Re e(Z) > 0$; the proposed definition yields

$$\boxed{P_a = \frac{1}{4}\frac{E\,E^*}{R} = \frac{1}{2}\frac{E\,E^*}{Z+Z^*} \quad \text{for } R > 0} \qquad (4.22)$$

provided that the load is matched and the source is non-random. However, if the source is random, averaging should be considered; hence

$$\boxed{<P_a> = \frac{1}{4}\frac{E\,E^*}{R} = \frac{1}{2}\frac{E\,E^*}{Z+Z^*}} \qquad (4.23)$$

☐ **Remark.** Equations (4.21b) and (4.22) prove that *the available power is a characteristic of the source*, since it doesn't depend on the load.

☐ **Particular Situations**

- Resistor – The thermal noise of a single resistor has an available power

$$P_a = \frac{4kTR\,\Delta f}{4R} = kT\Delta f \quad [W] \qquad (4.24)$$

   which is consistent with (3.3).
- Diode – If the load of the diode is purely resistive and is matched to its internal small-signal conductance, the available noise power is

$$P_a = kT\Delta f/2 \quad [W] \qquad (4.25)$$

■ **Available Power Gain.** Consider an active two-port, described by its Y-parameters, inserted between a signal source of internal admittance $Y_s = G_s + jB_s$ and the load Y_L (Fig. 4.2).

According to [45], the available power gain G_a is *the ratio of: (1) the available power from the output terminals of the two-port (P_{ao}), under specified input termination, to (2) the available power from the input generator (P_{ag})*:

4 Noise Parameters

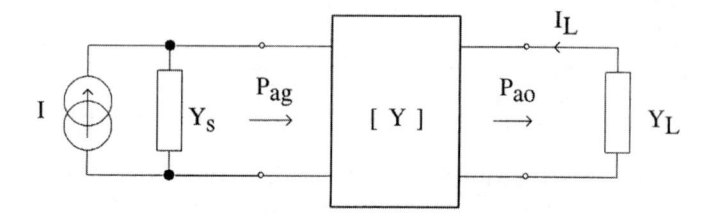

Fig. 4.2. Avaiable power gain of a two-port

$$G_a = \frac{P_{ao}}{P_{ag}} \qquad (4.26)$$

In terms of the Y-parameters of the two-port, the available power gain is [57]

$$G_a = \frac{|y_{21}|^2 G_s}{\mathcal{R}e(y_{22})|y_{11} + Y_s| - \mathcal{R}e(y_{12} y_{21}(y_{11} + Y_s)^*)} \qquad (4.27)$$

This equation shows that if the source admittance is variable, G_a is also variable. Of course the available power gain exhibits a maximum for a particular value of the source admittance.

☐ **Resistive Two-Port with Resistive Terminations.** Consider a sinusoidal source of internal resistance R_s and voltage E, connected to the input of a two-port, with output resistance R_o and output voltage E_o (Fig. 4.3).

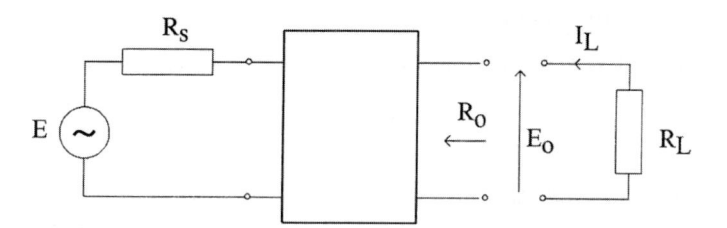

Fig. 4.3. Two-port with resistive terminations

According to the definition, the available power gain is

$$G_a = \frac{E_o^2 / 4R_o}{E^2 / 4R_s} = \frac{E_o^2 / R_s}{E^2 / R_o} = A_v^2 \frac{R_s}{R_o} \qquad (4.28)$$

where A_v is the voltage gain related to the generator.

Equation (4.28) shows that *the available power gain does not depend on the load.* It depends only on the generator internal resistance and its match to the input resistance of the two-port (which establishes the value of E_o).

In conclusion, the available power gain *is not* a two-port parameter, since it also depends on the signal source.

Suppose now that the input generator is identified with the noise source V_n, which models the thermal noise of the resistor R_s; in this case (4.28) still applies. The output available power is now the noise power N_o:

$$N_o = \frac{E_o^2}{4R_o} = \frac{(A_v v_n)^2}{4R_o} = \frac{A_v^2 \; 4kTR_s \; \Delta f}{4R_o} = G_a kT \; \Delta f \qquad (4.29)$$

□ **Note.** A useful property of the available gain is that *its value remains the same for signal and noise*. This makes it particularly attractive for noise calculation.

□ **Cascade of Several Amplifiers.** We have already seen that the available power gain depends on the two-port parameters, as well as on how the signal source is coupled to the two-port. Assume a chain of N amplifiers, where the available power gain of the k-th amplifier is determined by the output resistance of the (k − 1)-th stage, which actually plays the role of its signal source. Hence

$$[G_a]_t = G_{a1} \; G_{a2} \; \ldots \; G_{aN} = \frac{P_1}{P_a} \frac{P_1}{P_2} \; \ldots \; \frac{P_N}{P_{N-1}} = \frac{P_N}{P_a} \qquad (4.30)$$

Here P_k (k =1, 2 ...N) is the available power at the output of the k-th two-port, and P_a represents the available power of the generator.

4.2.4 Exchangeable Power and Exchangeable Power Gain

■ **Discussion.** Consider again the signal source and the adjustable load R_L shown in Fig. 4.1a. The power transferred to the load is

$$P_L = \frac{E^2 \; R_L}{(R_L + R_s)^2} \qquad (4.31)$$

Under the conditions $R_L > 0$ and $R_s > 0$, this power reaches a maximum when $R_L = R_s$. In this way has been introduced the concept of available power.

Now consider that no constraint is imposed on the load and source resistance. Then the maximum possible power is reached when $R_L = -R_s$, for which $P_L \to \infty$. However, this value is neither an extremum, nor a stationary value of the function described by (4.31). In order to preserve the stationary property with respect to the terminal current and consequently the spirit of (4.22), the concept of *exchangeable power* is required.

■ **Exchangeable Power.** Assume a sinusoidal signal source having internal admittance Y_s. Its Norton equivalent has an ideal current generator I in parallel with Y_s. According to (4.22), the available power is

4 Noise Parameters

$$P_a = \frac{1}{4} \frac{I\,I^*}{\Re e(Y_s)} \quad \text{when} \quad \Re e(Y_s) > 0 \qquad (4.32)$$

When $\Re e(Y_s) < 0$, the *exchangeable power* (denoted by P_e) is the stationary (extremum) value of the power flow per unit bandwidth, from or to a port under arbitrary variation of the terminal current or voltage. Therefore

$$P_e = \frac{1}{4} \frac{I\,I^*}{\Re e(Y_s)} \quad \text{when} \quad \Re e(Y_s) < 0 \qquad (4.33)$$

Note that whenever $\Re e(Y_s) > 0$, the exchangeable power becomes available power. The special case $\Re e(Y_s) < 0$ indicates that the power is *extracted* from a matched load Y_s^* (which acts as an energy source).

■ **Remark.** The same active one-port can provide power to the load (when $\Re e(Y_s) > 0$) or absorb power from the load (if $\Re e(Y_s) < 0$). In both situations, the maximum transferred power can be calculated using (4.32) or (4.33).

■ **Exchangeable Power Gain.** Since the right-hand side of (4.32) is the same as that of (4.33), the expression used to calculate the exchangeable power gain must be identical to that for the available power gain.

According to [45], *the exchangeable power gain of a linear two-port (denoted by G_e) is the ratio of: (1) the exchangeable signal power of the output port, to (2) the exchangeable signal power of the source connected to the input port.*

If the input signal source is modeled as an ideal current generator I_s in parallel with the source admittance Y_s, and the two-port output is represented as an ideal current generator I_o in parallel with the output admittance Y_o, we may write

$$G_e = \frac{\Re e(Y_s)\,I_o\,I_o^*}{\Re e(Y_o)\,I_s\,I_s^*} \qquad (4.34)$$

In this equation, either $\Re e(Y_s)$ or $\Re e(Y_o)$ is negative; if they have the same sign, (4.34) leads to the available power gain.

☐ **Random Signals.** Expression (4.33) has been introduced for a periodic signal. If instead we have a random signal, the only difference is that the average of the product must be considered

$$P_e = \frac{1}{4} \frac{\overline{I_n\,I_n^*}}{\Re e(Y_s)} = \frac{1}{4} \frac{\overline{|I_n|^2}}{\Re e(Y_s)} \qquad (4.35)$$

4.2 Definitions Concerning Electrical Power and Bandwidth

☐ **Thermal Noise.** Assuming that we wish to calculate the exchangeable power of the thermal noise produced by a conductance G, we have

$$P_e = \frac{1}{4} \frac{\overline{|I_n|^2}}{\mathcal{R}e(Y_s)} = \frac{1}{4} \frac{4kT\,|G|\,\Delta f}{G} = \begin{cases} +kT\Delta f & \text{if } G > 0 \\ -kT\Delta f & \text{if } G < 0 \end{cases} \quad (4.36)$$

■ **Conclusion.** The exchangeable power (or gain) represents an extension of the available power (or gain) concept that makes provision for negative resistances.

4.2.5 Various Gain Definitions

Many gain definitions can be found in the technical literature referring to two-ports. For convenience, they are recalled here:

- *Voltage gain*:

$$A_v = \frac{\text{output voltage}}{\text{input voltage}}$$

- *Current gain*:

$$A_i = \frac{\text{output current}}{\text{input current}}$$

- *Direct power gain*:

$$G_p = \frac{\text{power delivered to the load}}{\text{power delivered to the two-port input}}$$

- *Insertion power gain*:

$$G_i = \frac{\text{power delivered to the load with two-port inserted}}{\text{power directly delivered to the load without two-port}}$$

- *Transducer power gain*:

$$G_t = \frac{\text{power delivered to the load}}{\text{available power from the signal source}}$$

- *Available power gain*:

$$G_a = \frac{\text{available power at the two-port output}}{\text{available power of the signal source}}$$

- *Exchangeable power gain*:

$$G_a = \frac{\text{exchangeable power at the two-port output}}{\text{exchangeable power of the signal source}}$$

4.2.6 Noise Bandwidth

■ **Why Noise Bandwidth?** As noise results from the superposition of many random signals, it seems natural to define the bandwidth in a different manner from the usual −3 dB for harmonic signals.

■ **Signal Bandwidth.** According to [45], for a signal-transmitting system, *the bandwidth is the range of frequencies within which performance, with respect to some characteristic, falls within specific limits.*

For amplifiers or tuned circuits, the bandwidth (or passband), denoted B, is commonly defined at the points where the response is 3 dB less than the reference value (0.707 rms voltage ratio or half-power). If the frequency response has a symmetrical curve, the center frequency f_0 is related to the low-cutoff frequency f_l and high-cutoff frequency f_h by

$$f_0 = \sqrt{f_l \, f_h} \tag{4.37}$$

Often, sufficient accuracy is obtained by taking

$$f_0 \cong (f_l + f_h)/2 \tag{4.38}$$

■ **Equivalent Noise Bandwidth.** *Is the bandwidth (Δf) of an ideal circuit (with rectangular power transfer characteristic) such that the total transmitted noise power is equal to that transmitted by the actual circuit.*

The definition is illustrated in Fig. 4.4, where G(f) denotes the actual power gain response, having the same area under the curve as the area of the rectangle of height G_0 and width Δf.

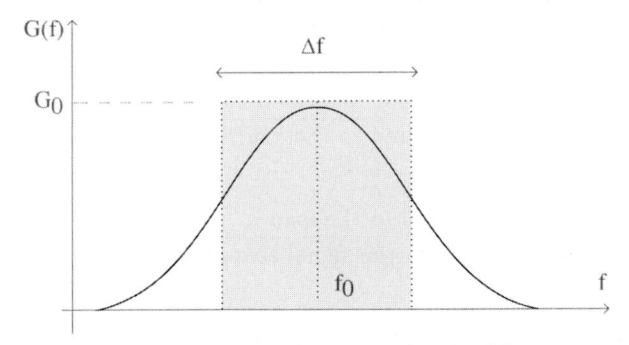

Fig. 4.4. Equivalent noise bandwidth

In this way, the equivalence between total noise power (defined for the actual response) and total noise power transmitted by the equivalent bandwidth is established. This equivalence can be expressed in mathematical terms as

$$P_{tot} = G_0 kT \, \Delta f = kT \int_0^{+\infty} G(f) \, df \tag{4.39}$$

from which:

$$\Delta f = \frac{1}{G_0} \int_0^{+\infty} G(f) \, df \quad (4.40)$$

■ **IEEE Definition.** According to [45], the equivalent noise bandwidth is *the frequency interval, determined by the response frequency characteristics of the system, that defines the noise power transmitted from a noise source of specified characteristics.*

■ **Remarks**

- The concept of noise bandwidth is rather an artifice that is useful in practice.
- The power gain quoted in the definition is the *available power gain* (it is the only one that does not depend on the load!).
- Equation (4.40) is consistent with the unilateral spectrum concept, since integration takes place over the domain $(0, +\infty)$.
- As the equivalence requires a rectangular characteristic, this implies uniform noise distribution.
- If the power gain has a single pole (low-pass characteristic), the equivalent noise bandwidth is $(\pi/2)f_c$, where f_c denotes the cutoff frequency.
- If the actual power gain plot has several peaks (as in Fig. 4.5), it is not easy to select the center frequency at which the height of the rectangle should be defined. The choice is not unique [48]; two possible equivalent bandwidths Δf_1 and Δf_2 are shown in Fig. 4.5, depending on the arbitrary selection of the "maximum" (G_{01} or G_{02}).

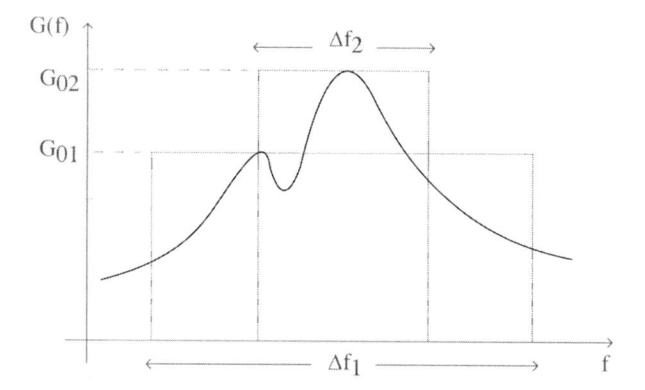

Fig. 4.5. Equivalent noise bandwith for an irregular response

82   4 Noise Parameters

■ **Experimental Method.** Conventionally, noise bandwidth is determined graphically. In practice it is easier to measure the voltage gain of the amplifier, denoted by A_V, versus frequency, than the power gain. Suppose that its maximum is A_{vo}. Since the power gain is proportional to the square of the voltage gain, (4.40) yields

$$\Delta f = \frac{1}{A_{vo}^2} \int_0^\infty |A_v(f)^2| \, df \qquad (4.41)$$

■ **Multistage Amplifiers.** Noise bandwidth is not the same as the classical −3 dB bandwidth. Usually, for multistage amplifiers, we know the −3 dB bandwidth (B) rather than the equivalent noise bandwidth Δf. For this reason, it is common to consider that $\Delta f \cong B$. In order to estimate the error made when doing this, we need an expression relating the two quantities. For a cascade of m identical stages, each having n distinct poles, this expression is

$$\Delta f = \frac{B}{\sqrt{2^{1/n} - 1}} \int_0^\infty \left(\frac{1}{1 + x^{2n}} \right)^m dx \qquad (4.42)$$

For various values of m (with n = 1) and for different values of n (with m = 1), the ratio $\Delta f/B$ is calculated; the results are presented in Table 4.1 and Table 4.2, respectively.

Table 4.1. m identical stages, each with one pole

m stages (n = 1)	$\Delta f/B$
1	1.571
2	1.222
3	1.155
4	1.13
5	1.11
6	1.10
∞	1.06

Table 4.2. One-stage amplifier with several poles

n poles (m = 1)	$\Delta f/B$
1	1.571
2	1.111
3	1.05
4	1.025
5	1.02
6	1.01
∞	1.00

It is easy to see that as soon as m (or n) is greater than 4, noise bandwidth can be identified with the −3 dB bandwidth, without much loss of accuracy. For a transmission system, where the number of stages is considerably greater than 4, Δf is taken equal to the −3 dB bandwidth.

4.3 Noise Parameters of Linear One-Ports (Spot Values)

■ **Comment.** By *spot noise parameters* we mean those parameters whose values are pertinent to an elementary noise bandwidth of 1 Hz, situated about a specific frequency. It follows that they are related conceptually to the power spectral densities. However, widespread convention chooses to employ the powers (instead of their spectral densities counterparts), but in order to compensate for the "narrow-band condition," powers are calculated within an elementary bandwidth δf. Ultimately, δf is set equal to unity, so it disappears from the final expressions.

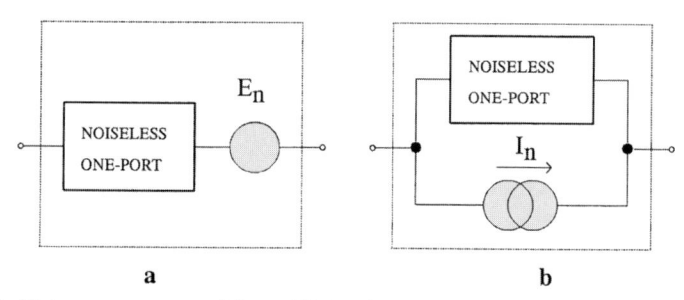

Fig. 4.6. Noisy one-port models: **a** Thévenin representation; **b** Norton representation

■ **Model.** A noisy one-port is modeled as a noisy resistor, with either its Thévenin or Norton equivalent circuit (Fig. 4.6).

Let us denote by $S(E_n)$, $S(I_n)$ the spectral densities of the noise generators E_n and I_n, respectively; thus

$$\boxed{S(E_n) = |Z|^2 \, S(I_n)} \tag{4.43}$$

with

$$E_n = Z \, I_n \tag{4.44}$$

Z being the impedance looking into the one-port terminals. Expression (4.43) is important since it establishes the relation between the spectral densities of the two representations.

4.3.1 Equivalent Noise Resistance

■ **Nielsen Definition.** If a one-port delivers a mean square noise voltage $\overline{v_n^2}$, then the equivalent noise resistance R_n is defined as *the value of a hypothetical resistor which, maintained at the reference temperature $T_o = 290\,K$, will produce the same amount of noise as the actual one-port* [47].

It follows that

$$\boxed{R_n = \overline{v_n^2} \,/\, 4kT_o \, \Delta f} \tag{4.45}$$

4 Noise Parameters

■ **Savelli Definition.** The equivalent noise resistance R_n is defined as *the value of a hypothetical resistor which, maintained at the same temperature T as the actual one-port, will produce the same amount of noise as the actual one-port* [49].

In this case

$$R_n = \overline{v_n^2} \;/\; 4kT \,\Delta f \tag{4.46}$$

■ **IEEE Definition.** The equivalent noise resistance R_n *is a quantitative representation in resistance units of the spectral density S_v of a noise voltage generator at a specified frequency* [45].

Note:

1) The relation between the equivalent noise resistance and the spectral density S_V of the noise generator is

$$R_n = (\pi S_v) \;/\; kT_o \tag{4.47}$$

with $T_o = 290\,\text{K}$.

2) The equivalent noise resistance in terms of the mean square value $\overline{v_n^2}$, within a small frequency interval δf, is

$$R_n = \overline{v_n^2} \;/\; 4kT_o \,\delta f \tag{4.48}$$

which is consistent with Nielsen definition.

■ **Remarks**

- The proposed definitions don't imply the existence of a physical resistor of value R_n. The equivalent noise resistor is instead a theoretical concept.
- If the noisy one-port is already at the reference temperature (290 K), then the first and second definitions are equivalent.
- Van der Ziel [50] adopted the expression (4.45), but he defined T_o as "a fixed reference temperature," which does not necessary mean 290 K.
- Very likely, in (4.47) S_V represents the bilateral spectral density with respect to the angular frequency ω.

■ **Comments**

- It is important to see that the concept of equivalent noise resistance can apply only to noise with a flat spectral density (white noise); otherwise, the equivalence is inconsistent.
- In order to ensure consistency between the definitions proposed by Nielsen and Savelli (which are more intuitive) and the IEEE definition, it is necessary to set $\Delta f = 1$ in (4.45) and (4.46).

4.3.2 Equivalent Noise Current

■ **Discussion.** According to Schottky theorem, the mean square value of the noise current of a saturated diode is

$$\overline{i_d^2} = 2qI_D \, \Delta f \tag{4.49}$$

where I_D denotes the average (direct) current. This expression can be used to establish an equivalence between the noise of a one-port and that of a diode.

■ **Van der Ziel Definition.** According to [50], *the equivalent noise current of the one-port is equal to the current of a hypothetical saturated diode which would generate noise with the same spectral density as the actual one-port.*

Suppose that the actual one-port delivers a mean square noise current $\overline{i^2}$ within an elementary frequency interval δf; then the equivalent noise current, denoted by I_n, is

$$\boxed{I_n = \overline{i^2} \, / \, 2q \, \delta f} \tag{4.50a}$$

The meaning of this parameter is explained by Van der Ziel in the following way: if the noise power of the actual one-port were measured and a saturated diode generating a current I_n were connected in parallel with the one-port, then the noise output power of the system would double.

This explanation suggests that the main reason that we have such a variety of noise parameters (many of them conceptually redundant) is the steady accumulation of noise measurement techniques over the last half century.

Again, letting $\delta f = 1$ in (4.50a) yields a consistent definition of the equivalent noise current.

■ **IEEE Definition.** According to [45], *the equivalent noise current is a quantitative representation in current units of the spectral density of a noise current generator, at a specified frequency.*

Note: the relation between the equivalent noise current I_n and the spectral density S_i of the noise current generator is

$$\boxed{I_n = (2\pi S_i) \, / \, q} \tag{4.50b}$$

As for the equivalent noise resistance, note that the spectral density is assumed to be bilateral and defined with respect to the angular frequency ω.

4.3.3 Noise Temperature

■ **Comment.** As we have seen, the available noise power of a resistor does not depend on its value, just its temperature (see (4.24)). This suggests that we can express the noise of any one-port (regardless of its physical origins) only by controlling the fictitious temperature T of a hypothetical resistor. Note that T is not necessarily equal to the actual temperature of the device.

■ **Benett Definition.** According to [51], *the noise temperature of a single-port device, at a specified frequency, is the temperature at which the thermal noise power is equal to the amount of available noise power from the output of the device.*

■ **Note.** For a resistor, the noise temperature is equal to its actual temperature, while for a diode the noise temperature may be quite different from the physical temperature of the diode. A uniform temperature throughout the passive system is assumed.

■ **Standard Reference Temperature.** In noise theory and noise measurements, the standard reference temperature is

$$\boxed{T_o = 290\,\text{K}}$$

For this value, $kT/q = 0.025$ V.

■ **Savelli Definition.** This definition establishes an equivalence between the actual noise delivered by a one-port (which may be of nonthermal nature) and the noise of a hypothetical one-port, which is only thermal.

According to [49], the noise temperature is *the temperature T_{eq} at which the hypothetical one-port generates the same amount of thermal noise power within the small frequency interval δf as the noise power of the actual one-port.*

■ **IEEE Definition.** According to [45], *the noise temperature at a pair of terminals and at a specific frequency is the temperature of a passive system with an available noise power per unit bandwidth equal to that of the actual terminals.*

If extended to any port, the definition becomes: *the noise temperature of a port and at a selected frequency is the ratio of the exchangeable power density divided by Boltzmann's constant*:

$$\boxed{T = \frac{P_e/\Delta f}{k} = 7.25 \cdot 10^{22}\frac{P_e}{\Delta f}} \tag{4.51}$$

■ **Comment.** The noise temperature has the sign of $\mathfrak{Re}(Z)$, Z being the impedance seen at the port in question. T is negative for a port whose internal impedance has a negative real part.

■ **Interconnected One-Ports.** Suppose that several one-ports having different noise temperatures are connected in series or in parallel (Fig. 4.7). In

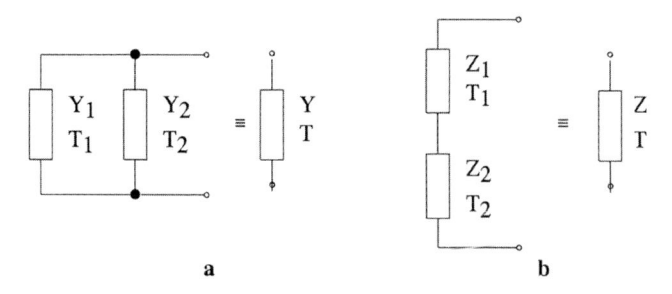

4.3 Noise Parameters of Linear One-Ports (Spot Values)   87

Fig. 4.7. Associating several one-ports in **a** parallel; **b** series

each case an equivalent one-port results. In order to find the noise temperature of the equivalent one-port, we assume that the noise power produced at its terminals (within an elementary 1–Hz bandwidth) is the same as the noise power generated by the actual network resulting from the interconnection of several one-ports.

For the parallel connection (Fig. 4.7a), we may write

$$4kT_1G_1 + 4kT_2G_2 = 4kTG \quad \text{with} \quad G_i = \Re e(Y_i), \quad i = 1, 2$$

Since $G = G_1 + G_2$, we obtain

$$T = \frac{G_1T_1 + G_2T_2}{G_1 + G_2} \tag{4.52}$$

For the series connection (Fig. 4.7b) we have

$$4kT_1R_1 + 4kT_2R_2 = 4kTR \quad \text{with} \quad R_i = \Re e(Z_i), \quad i = 1, 2$$

As $R = R_1 + R_2$, we obtain

$$T = \frac{R_1T_1 + R_2T_2}{R_1 + R_2} \tag{4.53}$$

In any case, for positive or negative T_1, T_2, or T, it can be shown [53] that

$$\frac{1}{T_1} < \frac{1}{T} < \frac{1}{T_2} \quad \text{if} \quad \frac{1}{T_1} < \frac{1}{T_2} \tag{4.54}$$

4.3.4 Noise Ratio

■ **Savelli Definition.** The noise ratio $\mathcal{N}$ of a one-port is *the ratio of: (1) the noise spectral density (or mean square value) generated by the actual one-port, to (2) the same quantity generated by a hypothetical one-port of identical impedance which produces only thermal noise* [49]. Therefore

88   4 Noise Parameters

$$\mathcal{N} = \frac{\overline{E^2}}{4kTR\ \Delta f} = \frac{\overline{I^2}}{4kTG\ \Delta f} = \frac{R_n}{R} = \frac{q}{2kT}\frac{T_n}{G} \qquad (4.55a)$$

where R, G are the real parts of the impedance Z and admittance $Y = Z^{-1}$, respectively.

■ **Benett Definition.** According to [51], *the noise ratio of a single-port device is the ratio of: (1) the equivalent noise temperature T, to (2) the actual temperature T_a of the device (in thermal equilibrium)*:

$$\mathcal{N} = \frac{T}{T_r} = \frac{S_p}{kT_a} \qquad (4.55b)$$

where S_p denotes the power spectral density expressed in units of kT.

■ **Remarks**

– If the one-port is purely resistive, then $\mathcal{N} = 1$ and the thermal noise has not been augmented by the device.
– The quantity $(\mathcal{N} - 1)$ is called the *excess noise ratio*, since it indicates the amount of noise generated in excess of the thermal noise.
– $\mathcal{N}$ is usually expressed in decibels (by taking twenty times the logarithm to the base 10).

4.3.5 Noise Equivalent Power (NEP)

■ **Outline.** This is a specific parameter of optical or thermal detectors, which is used in order to evaluate their quantum noise level.

■ **Definition.** According to [52], the noise equivalent power (NEP) is *the input signal power (light power) that produces an output electrical signal equal to the noise output present when no input is applied, per unit bandwidth*. It is computed with the expression

$$\text{NEP} = 4\sqrt{A\sigma_s kT^5\ \Delta f} \qquad (4.56)$$

where A is the detector's absorbing area, T is the detector temperature (supposed to be the same as that of its environment), Δf is the measuring system bandwidth, k is the Boltzmann's constant and σ_s is the Stefan-Boltzmann's constant ($\sigma_s = 5.67 \cdot 10^{-8}$ W / m^2 K^4).

4.3.6 Signal-to-Noise Ratio (S/N Ratio)

■ **Convention.** The acronym for the signal-to-noise ratio, throughout this book, is *S/N ratio*.

■ **Definition.** For a noiseless one-port Z' connected to a generator, *the S/N ratio, at a specific frequency, is the ratio of exchangeable power of the signal (P_{es}) to the exchangeable power of the noise (P_{en}), both produced by the generator*:

$$S/N = P_{es} / P_{en} \qquad (4.57)$$

4.4 Noise Parameters of Linear One-Ports (Average Values)

■ **Comment.** The previously introduced parameters refer to the noise of a single-port device, within some small bandwidth δf (ideally 1 Hz). It is assumed that within that interval all electrical and noise parameters of the device are constant.

■ **Wide Bandwidth.** Many practical devices, however, operate over a wide bandwidth, and their parameters are functions of frequency. How can we then estimate the noise parameters over a fairly wide range of frequencies?
As pointed out in [47], there are two possible approaches:

1) We may divide the spectrum into a number of elementary bands δf, such that the assumption of constant parameters might still be valid. Next, the noise parameters are determined within each elementary band and their values are plotted versus frequency. Despite its accuracy, this approach fails to provide a single value that allows easy global characterization of the system noise.
2) Another possible approach is to define average noise parameters (over the bandwidth). All average parameters are denoted by an overbar, and their units are the same as for the spot noise parameters.

■ **Note.** Average and spot noise values are identical provided that the gain is constant with respect to frequency and all internal noise sources have a flat spectrum.

4.5 Noise Parameters of Linear Two-Ports (Spot Values)

4.5.1 Equivalent Input Noise

■ **E_n–I_n Model.** The internal noise of an amplifier is often lumped into two noise generators E_n and I_n, located at the input of a hypothetical amplifier identical to the actual one, except that it is noiseless. Since both E_n and I_n must represent the effects of all internal fluctuations, their evolution obviously

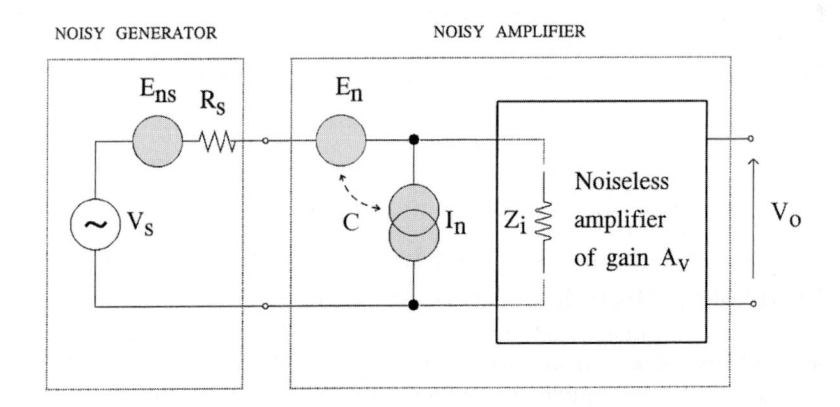

Fig. 4.8. Modeling a noisy amplifier and its signal source

cannot be completely independent. They are said to be correlated, and the correlation coefficient C is a complex quantity.

The input signal source is denoted by subscript "s" and its associated thermal noise is made explicit (Fig. 4.8). This representation makes it possible to evaluate the overall noise of the system made up of the amplifier and its signal source combined.

■ **Problem Formulation.** Given an amplifier and its signal source, find the equivalent noise generator (denoted E_{ni}) which, connected to the input of an identical but noise-free system, would produce the same effect as all noise generators E_{ns}, E_n, and I_n.

■ **Comment.** In this way, noise is referred to the input (across V_s) and two benefits are expected: 1) noise and signal are known at the same location, so that the signal-to-noise ratio (S/N ratio) can be conveniently computed; 2) as the remaining network is noiseless, it is easier to appreciate the effect of minor circuit modifications on the output noise.

Traditionally, E_{ni} is called *equivalent input noise voltage*.

■ **Solution.** It is paramount to understand that the input impedance Z_i has not thermal noise (since Z_i belongs to a noiseless module). It can be shown [52] that the equivalent input noise voltage is

$$\overline{E_{ni}^2} = \overline{E_{ns}^2} + \overline{E_n^2} + \overline{I_n^2}\, R_s^2 + 2C\, \overline{E_n\, I_n}\, R_s \qquad (4.58)$$

C being the correlation coefficient between E_n and I_n.

■ **Remark.** Since E_{ni} is independent of the amplifier gain and its input impedance, this parameter is very useful in practice to compare noise of various amplifiers and devices.

■ **Note.** Despite the straightforward procedure to measure E_n (imposing $R_s = 0$) or I_n (with high R_s), the equivalent input noise voltage is not easily calculated, since the correlation coefficient is not known. This explains why for many electron devices (field-effect transistors, operational amplifiers, etc.) noise in data books is specified only by the values E_n and I_n (assumed to be uncorrelated). Ultimately, the spread in E_n, I_n values from one sample to another may have more of an affect on the resulting E_{ni} than the error made by assuming $C = 0$.

■ **Equivalent Noise Resistance.** According to [45], this is *the value of a resistor which, when connected to the input of a hypothetical noiseless amplifier with the same gain and bandwidth, produces the same output noise.*

This concept is equivalent to the input noise voltage, provided the amplifier noise spectrum is flat from zero up to very high frequencies (which is rarely the case!). Remember that the equivalent noise resistance is not a physical resistor located anywhere in the circuit.

4.5.2 Signal-to-Noise Ratio

■ **Comment.** It is worthwhile for a receiver to have a S/N ratio as high as possible. However, a poor S/N ratio does not necessary mean a poor-quality receiver, because it may happen that the S/N ratio of the signal picked up by the antenna was already low.

■ **Benett Definition.** *The S/N ratio at any point, at a specified frequency, is defined as the ratio of: (1) signal power, to (2) noise power.* It is usually expressed in decibels:

$$S/N = 10\log\left(V_s^2 \,/\, \overline{V_n^2}\right) \quad [\text{dB}]$$

■ **Comment.** The S/N ratio may be defined either at a specified frequency (as stated previously) or over a wide bandwidth (as for mobile communication systems). In the former case, we use the power spectral densities, and consequently a normalized S/N ratio is obtained. In the latter case, the total noise power must be considered and the S/N ratio is a function of the bandwidth of the transmission system. However, we must bear in mind that if the bandwidth is greater than required for signal transmission, the S/N ratio can be seriously degraded (since outside the required bandwidth, the circuit continues to add noise, but the signal power remains constant).

■ **IEEE Definition.** According to [45], the S/N ratio is *the ratio of the value of the signal to that of the noise (the two being expressed in a consistent way, as for example peak signal to peak noise ratio, rms signal to rms noise ratio, peak-to-peak signal to peak-to-peak noise ratio, etc.).*

An exception occurs in television transmission, where the S/N ratio is defined as the ratio of: (1) the maximum peak-to- peak voltage of the video signal, including synchronizing pulse, to (2) the rms value of the noise (because of the difficulty of defining the rms value of the video signal or the peak-to-peak value of random noise).

☐ **Comment.** This ratio is expressed in terms of peak values in the case of impulse noise, and in terms of rms value in the case of random noise.

■ **Caution.** Attention must be paid to the following points:
- For a given physical system, the numerical values of the S/N ratio obtained using Benett's or the IEEE definition are not necessarily the same.
- Whenever possible, Benett definition is preferred, since it is consistent with the noise factor definition (which is introduced in terms of *powers*).

4.5.3 Input Noise Temperature

■ **Pettai Definition.** Consider a linear two-port, whose input is connected to a source having the frequency f_s and the internal resistance R_s.

According to Pettai [48], *the input noise temperature T_e (expressed in kelvins) is that signal source temperature which, when connected to a noise-free equivalent of the two-port (Fig. 4.9a), results in an available output noise power per unit bandwidth N_o equal to that of the actual noisy two-port, excited by a noise-free identical source (Fig. 4.9b).*

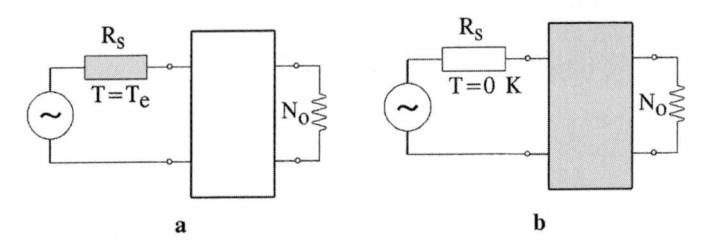

Fig. 4.9. Illustrating the definition of the input noise temperature T_e

■ **Computation.** Consider the two-port shown in Fig. 4.10; its input is driven by a source with available noise power $kT_s\delta f$ (T_s being the noise temperature of the signal source and δf an elementary bandwidth about the operating frequency f).

If the two-port were noiseless, the noise at the output would be the input noise power multiplied by the available power gain of the two-port, i.e., $(kT_s\delta f)G_a$. However, the actual two-port adds its own internal noise, and the available noise power at the output is augmented:

$$N_o = (kT_s\delta f)G_a + N_n \qquad (4.59a)$$

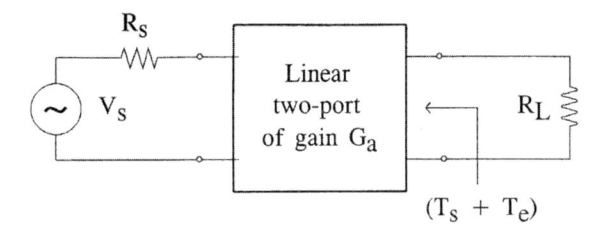

4.5 Noise Parameters of Linear Two-Ports (Spot Values)

Fig. 4.10. Noise temperature of a two-port

where N_n denotes the *excess noise power* (the noise power added by the two-port). Expression (4.59a) can be equally well written

$$N_o = (kT_s \delta f)G_a + kT_n \delta f \tag{4.59b}$$

T_n being the noise temperature corresponding to noise power N_n. Note that *the summation of noise powers is permissible because the noise of the source and two-port are uncorrelated.*

If $(kG_a \delta f)$ is taken as a common factor, we have

$$N_o = kG_a \delta f(T_s + T_n / G_a) = kG_a \delta f(T_s + T_e) \tag{4.59c}$$

On the right-hand side, by definition

$$\boxed{T_e = T_n / G_a} \tag{4.60}$$

is the *equivalent input noise temperature* or simply the *input noise temperature* (since dividing by G_a refers it to the input).

This concept allows us to represent the noise of any linear two-port and its associated signal source as in Fig. 4.11. Note that the output noise power expression suggests that the input noise temperatures T_s and T_e are additive.

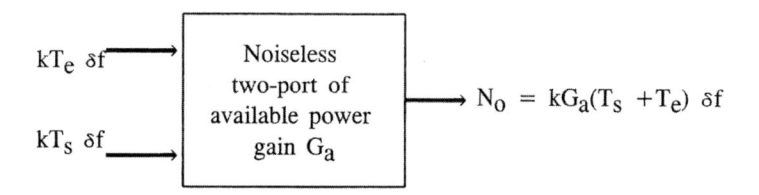

Fig. 4.11. Noise model of a linear two-port

■ Remarks

– The value of the equivalent input noise temperature is frequency dependent.
– When introducing the concept of input noise temperature, the load is assumed to be ideal (noiseless).

- The input noise temperature does not depend on T_s, but only on the source internal impedance (which determines the available gain G_a).
- In the case of non-linear two-ports, several input frequencies may correspond to a single output frequency, or vice versa. For any pair of frequencies (one at the input, the other at the output) an input noise temperature can be defined.
- This parameter is useful for comparing the noise of two different two-ports: the one with a lower input noise temperature adds less noise to the incoming signal.
- The benefit of using this parameter is that noise temperatures are additive. For instance, if the signal source has an equivalent noise temperature T_s and the two-port an input noise temperature T_e, then the overall equivalent input temperature is $T_{eq} = T_s + T_e$.

■ **Note.** By means of the input noise temperature, we can separate the noise from the electrical circuit of the two-port, exactly as we did for a noisy resistor (Fig. 3.1).

■ **Measurement of T_e.** Despite the fact that T_e is a fictitious quantity, it is easily measured. A simplified setup is proposed in Fig. 4.12, where the amplifier input is first connected to a "hot" noise source (of noise temperature T_h).

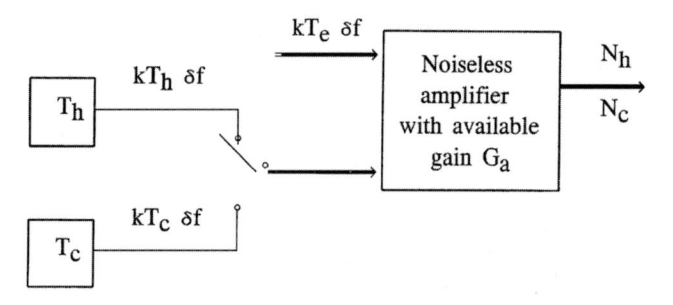

Fig. 4.12. Measurement of the input noise temperature

In this case, the output noise power is

$$N_h = kT_h G_a \, \delta f + kT_e G_a \, \delta f = kG_a \, \delta f (T_h + T_e) \tag{4.61}$$

Next, the "hot" noise source is replaced by a "cold" (low noise) source, with an equivalent noise temperature $T_c < T_h$. Equation (4.61) becomes

$$N_c = kG_a \, \delta f (T_c + T_e) \tag{4.62}$$

The quantity $Y = N_h/N_c > 1$ is defined (called the *Y-factor*) and by substituting (4.61) and (4.62) into the Y-factor expression one obtains

4.5 Noise Parameters of Linear Two-Ports (Spot Values)

$$\boxed{T_e = \frac{T_h - YT_c}{Y - 1}} \qquad (4.63)$$

Equation (4.63) can be used to find the input noise temperature of the two-port, provided that in Fig. 4.12 both noise sources *have exactly the same internal impedance*.

■ **Two-Ports in Cascade.** Consider a cascade of N two-ports (Fig. 4.13). Each two-port is characterized by its available power gain G_i and its input noise temperature T_{ei}. We wish to compute the overall input noise temperature T_e.

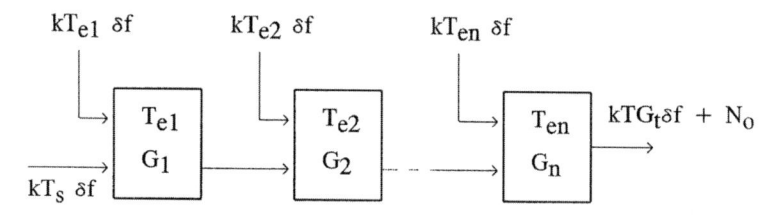

Fig. 4.13. Cascade of several two-ports

The excess noise individual contributions of the two-ports are summed at the cascade output. Hence

$$\left((kT_{e1}G_1G_2 \ldots G_n) + (kT_{e2}G_2 \ldots G_n) + \ldots + (kT_{en}G_n)\right)\delta f \qquad (4.64)$$

The overall available power gain is

$$G_t = G_1G_2 \ldots G_n \qquad (4.65)$$

(It should be stressed that the available power gain of stage i is measured with the (i − 1)-th stage as generator.)

Then, applying the previous approach, the overall input noise temperature T_e is obtained by dividing (4.64) by $kG_t\delta f$:

$$\boxed{T_e = T_{e1} + \frac{T_{e2}}{G_1} + \frac{T_{e3}}{G_1G_2} + \ldots + \frac{T_{en}}{G_1G_2\ldots G_{n-1}}} \qquad (4.66)$$

This result shows that if the gain of the first stage is very high, the noise contributions of the following stages are negligible, as long as all G_i are greater than unity. However, if one of the following stages attenuates the signal (instead of amplifying it), its individual contribution becomes relevant and may jeopardize the overall noise performance.

■ **Concluding Remark.** *The first stage of an amplifier should be chosen with the lowest possible noise and as much gain as possible.*

4.5.4 Operating Noise Temperature

■ **Comment.** Equation (4.59b) demonstrates that the output noise of any linear two-port depends on both T_e and T_s. Sometimes we are able to control T_e, but in many practical situations we have no influence on the noise arriving at the input of the receiver (for instance, the noise collected by an antenna). Therefore, it would be highly desirable to have an unique parameter to characterize the overall noise performance of the receiver and its antenna, instead of using the two different parameters T_e and T_s. This unique parameter is the operating noise temperature, denoted by T_{op} and expressed in kelvins.

■ **Definition.** The ultimate goal of computing noise performance is to arrive at a S/N ratio. Suppose that the signal output power spectral density is denoted by S_o, and S_i is the input available signal power spectral density. It is clear that $S_o = S_i G_t$, G_t being the transducer power gain (see Sect. 4.2.4).

In a similar way, the total output noise power spectral density N_o can be referred to the input by dividing it by G_t; this yields the operating noise temperature.

As stated in reference [45], *the operating noise temperature is the temperature in kelvins given by*

$$\boxed{T_{op} = \frac{N_o}{kG_t}} \qquad (4.67)$$

where N_o is the output noise power per unit bandwidth at a specified output frequency. Note that G_t *is the ratio of: (1) the signal power delivered to the output circuit at the specified output frequency, to (2) the signal power available at the corresponding input frequency or frequencies at its accessible input terminations.*

■ **Remarks**

– The S/N ratio at the output is

$$\frac{S_o}{N_o} = \frac{S_o / G_t}{N_o / G_t} = \frac{S_i}{kT_{op}}$$

– In any linear two-port with a single input and a single output frequency, matched at the output port, the operating noise temperature is

$$\boxed{T_{op} = T_e + T_s} \qquad (4.68)$$

– In a nonlinear system, T_{op} may depend on the signal level.
– The total output noise power has the following components:
  • the noise of the signal source (modeled by T_s) transmitted at the output;

- the noise produced by the two-port itself (modeled by T_e) transmitted at the output;
- the load noise flowing towards the two-port output that is reflected at the output port (due to the eventual mismatch).

For an RF or microwave amplifier, the last component is largely negligible, since the load is matched to the output.

■ Concluding Remarks

- The characterization of noise by the input noise temperature or operating noise temperature is particularly useful when the noise power level is low and the frequency is as high as 1000 GHz. Instead of expressing the noise power in infinitesimal fractions of watt (for instance, a resistor at $T = 300$ K has a noise power spectral density of order 10^{-21} W/Hz), it is more convenient to specify them in kelvins.
- The input noise temperature (T_e) is independent of the load noise, and therefore is consistent with the concept of noise factor. As we shall see later, these two quantities are equivalent.
- The operating noise temperature (T_{op}) is a concept which makes it possible to characterize a communication system globally, because the noise of the load *is* taken into account.

4.5.5 Effective Noise Temperature

■ **Comment.** This quantity, denoted by T_{eff}, is mostly employed in the field of (satellite) communication systems.

■ **Definition.** With respect to the output noise power spectral density, the effective noise temperature is defined at a given frequency as

$$\boxed{T_{eff} = \frac{S(N_o)}{k}} \qquad (4.69)$$

■ **Note.** The main difference between T_{eff} and T_{op} is that *T_{eff} is defined at the output, while T_{op} is referred to the input*. However, both parameters consider the noise generated into the load.

4.5.6 Noise Factor (Noise Figure)

■ **Discussion.** The noise factor (noise figure) is one of the most commonly used (perhaps overused) terms. It is a single number that characterizes the noise of a two-port, *provided that the source impedance (or admittance) is specified*. Without this information, this parameter is worthless, since the noise factor is conceptually based upon the comparison between the noise of the two-port and the noise of the signal source.

4 Noise Parameters

■ **Traditional Approach.** In order to introduce the concept of noise factor it is assumed that all noisy devices belonging to the actual two-port are at the same temperature, which is the standard reference temperature T_o. With this assumption, we compare the actual two-port with a fictitious two-port, that has an identical physical structure, except that the latter is noiseless. We suppose that both two-ports have the same amount of noise at their input (i.e., the same signal source).

Under these assumptions, we compare the available noise power at the output (P_{ano}), delivered by the actual two-port, with the same quantity (denoted P_{an}) obtained at the output of the fictitious two-port. (The former quantity can be measured, but the latter must be calculated with the rules of network theory and the basic concepts introduced in Chap. 2.)

The noise factor, denoted by F, is defined as

$$F = \frac{P_{ano}}{P_{an}} = \frac{P_{Rs} + P_{TP}}{P_{RS}} = 1 + \frac{P_{TP}}{P_{RS}} \qquad (4.70)$$

where P_{Rs} is the output noise power due to the noise of the source resistance R_s, which has been amplified by the noiseless two-port, and P_{TP} is the two-port contribution to the output noise power when its input is connected to a noiseless resistor R_s.

■ **Properties.** The noise factor introduced with expression (4.70) has several interesting properties:

– This parameter does not include the contribution of the load to the output noise.
– The noise factor closely depends on the internal resistance of the signal source.
– A noise-free two-port necessarily has unity noise factor.
– A noisy two-port always adds its own noise to that of the source. This contribution is estimated with the quantity (F–1). In other words, *the noise factor is always greater than unity.*

■ **Noise Factor or Noise Figure?** Although these terms seem to be synonymous, there are several points to be observed:

– The Institute of Radio Engineers (IRE) Standards Committee stated in 1952 that they leave the choice of name to the user's preference, because both terms had become well-established in current usage.
– In 1957, IRE reviewed the definition and recommended (without formal obligation) the term "noise factor" where it is desired to emphasize that the noise figure is a function of input frequency. This point of view is actually maintained by the Institute of Electrical and Electronics Engineers (IEEE). Another important modification has been the substitution in the definition of the term "transducer" for "two-port", because the latter is more general.

4.5 Noise Parameters of Linear Two-Ports (Spot Values)

– As no unanimity exists, Pettai [48] and Ott [54] proposed adopting a tradition which reserves the term *noise factor* (denoted by F) for numerical ratios, while the term *noise figure* (denoted by NF) is employed for the decibel version:

$$\boxed{\text{Noise Figure NF} = 10 \log (\text{Noise Factor F})}$$

In the present book, this point of view will be adopted.

■ **Note.** E.W. Herold and D.O. North of RCA, and H.T. Friis of Bell Telephone Laboratories were the early pioneers in working with the noise factor (in the 1940s) as a method of evaluating noise in vacuum tubes. They proposed two equivalent definitions: one structured on the ratio of the total noise, referred to the input, to the source thermal noise (North), and the other based on the degradation of the S/N ratio when the signal passes through the two-port (Friis).

Both are accepted by IEEE and will now be detailed.

■ **North Definition.** As stated in [45], *the noise factor of a two-port transducer at a specified input frequency is the ratio of: (a) the total noise power per unit bandwidth N_o at a corresponding output frequency available at the output port, when the noise temperature of its input termination is standard (290 K) at all frequencies, to (b) that portion of (a), denoted by N'_o, engendered at the input frequency by the input termination at the standard noise temperature (290 K).*

Part (a) is illustrated in Fig. 4.14a and part (b) in Fig. 4.14b. Note that for part (b) the two-port is assumed to be noise-free, since only the contribution of the input termination is considered.

$$F = \frac{N_o}{N'_o} = \frac{N_o}{kT_o G_a}$$

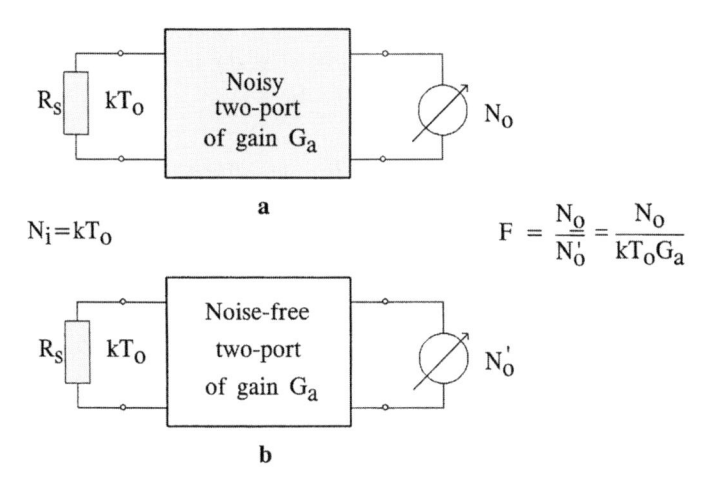

Fig. 4.14. Illustration of North definition

4 Noise Parameters

Therefore F may be expressed as

$$F = \frac{N_o}{N'_o} = \frac{N_o\,(T = 290\,\text{K})}{kT_o\,G_a} = 1 + \frac{N_n}{kT_o\,G_a} > 1 \qquad (4.71)$$

where N_n denotes the excess noise power (the noise added by the two-port) and G_a is the available power gain of the two-port.

■ **Remark.** As indicated in [45], the phrase "available at the output port" may be replaced by "delivered by the system into an output termination." As a consequence, the available power gain can be replaced by the *transducer power gain* (which can easily be measured), and the output matching condition can be relaxed.

■ **Friis Definition.** According to [55], *the noise factor F of a two-port, at a specified frequency, is the ratio of: (1) the available S/N ratio at the signal generator termination (when the temperature of the input termination is 290 K and the bandwidth is limited by the receiver), to (2) the available S/N ratio at its output terminals.*

$S_i \Longrightarrow$ | Noisy two-port of gain G_a | $\Longrightarrow S_o = G_a S_i$
$N_i = kT_o \rightarrow$ | | $\rightarrow N_o = G_a(kT_o) + N_n$

$$F = \frac{S_i/N_i}{S_o/N_o}$$

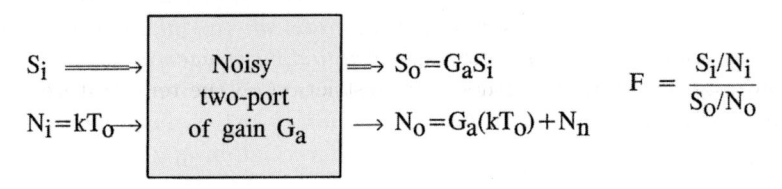

Fig. 4.15. Illustration of Friis definition

This definition is illustrated in Fig. 4.15, and accordingly F is a measure of the degradation of the input S/N ratio that occurs when the signal passes through the two-port.

Thus:

$$F = \frac{S_i/N_i}{S_o/N_o} = \frac{S_i/kT_o}{S_o/N_o} \qquad (4.72)$$

where S_i, S_o are the available power of the signal (per unit bandwidth) at the input and output. Similarly, $N_i = kT_o$ and N_o represent the available noise power, per unit bandwidth, at the input and output (the load is assumed to be noiseless).

Since the available power gain of the two-port is defined as

$$G_a = S_o\,/\,S_i,$$

it is possible to rewrite (4.72) in an equivalent form

$$\boxed{F = \frac{1}{G_a}\,\frac{N_o}{kT_o}} \qquad (4.73)$$

4.5 Noise Parameters of Linear Two-Ports (Spot Values)

This expression demonstrates the equivalence of the North and Friis definitions, provided the two-port is *linear* (i.e., *its power gain is the same for signal and noise*).

■ **Note.** For heterodyne systems there are several output frequencies corresponding to a single input frequency and vice versa. For each pair of corresponding frequencies a noise factor is defined. *Caution*: spurious contributions such as those from image frequency conversion must be avoided.

■ **Comments**

- The concept of noise factor is meaningless without additional information on the source internal impedance.
- This parameter may be expressed either as a ratio (*noise factor*) or in decibels (*noise figure*). In the latter case:

$$\boxed{F_{dB} = 10 \, \log \, F} \qquad (4.74)$$

- The noise factor is defined at the standard reference temperature (290 K), and *it is meaningful provided that everybody uses the same reference*. Therefore, it is not so general as the noise temperature, where only the noise power must be known, without any restriction on the temperature.

■ **Caution.** *The noise factor provides a comparison between the noise generated by the two-port and the noise of the signal source, and not an absolute estimate of the two-port noise.* For instance, if we artificially augment the resistance of the signal source (and consequently its thermal noise), (4.70) predicts that the noise factor decreases, and we may erroneously think that the two-port is suddenly less noisy, despite evidence that the output noise power has been augmented!

For this reason, never insert a resistor between the signal source and the input of the two-port, because the S/N ratio will certainly deteriorate, despite improvement in the noise factor value.

In practice, the only interest in working with the noise factor is to select the least noisy two-port from among several candidates, *provided the signal source is the same*.

■ **Limitations.** The noise factor definition involves several restrictions:

- If the internal impedance of the signal source is purely reactive, its noise vanishes and the resulting noise factor becomes infinite.
- When the noise added by the two-port is negligible with respect to that of the source, the noise factor is the ratio of two nearly equal quantities. This may lead to unacceptable errors.

4 Noise Parameters

- The noise factor value depends on the signal frequency, bias, temperature, and signal source impedance. Comparing two noise factors without having identical conditions in all such respects is quite useless.

■ **Relationship Between F and T_e.** In single response systems (the term *single response* means that to each output frequency there corresponds only one input frequency) this relationship is established starting with the North definition, which for an elementary bandwidth δf becomes

$$F = \frac{N_o}{kT_o G_a\ \delta f} = \frac{N_n + kT_o G_a\ \delta f}{kT_o G_a\ \delta f} = 1 + \frac{N_n\ /\ G_a}{kT_o\ \delta f} \qquad (4.75)$$

From (4.59a),

$$N_n = kT_n\ \delta f$$

and therefore

$$F = 1 + \frac{kT_n\ \delta f / G_a}{kT_o\ \delta f}$$

This equation can be represented in an equivalent form:

$$\boxed{F = 1 + \frac{T_e}{T_o}} \qquad (4.76)$$

where T_e is the equivalent input noise temperature. Alternatively, we can write

$$\boxed{T_e = T_o(F - 1)} \qquad (4.77)$$

■ **Remark.** For low-noise systems, where the noise factor normally lies between 1 and 1.6, a more convenient way to express the noise is in terms of the noise temperature, which provides an extended scale (for the previous values of F, T_e is in the range 0 to 175 K).

■ **Other Definitions.** In early technical publications there was no unanimity on the value of the standard reference temperature T_o (as it appears in (4.71)), and consequently slightly different values resulted for the noise factor. Sometimes, in particular fields, we still find such deviations. For instance, Robinson adopted $T_o = 293$ K while Montgomery had a preference for $T_o = 300$ K; Lawson and Uhlenbeck chose $T_o = 292$ K, and Norton adopted $T_o = 288.39$ K, etc.

Finally, the reader must pay attention when comparing the noise performance of several competing systems and ask which value of reference temperature is being used, especially when the standard reference temperature is not clearly specified.

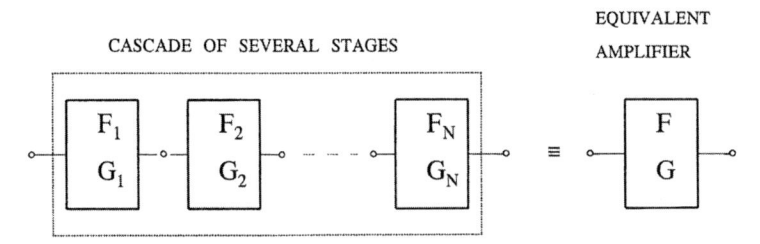

4.5 Noise Parameters of Linear Two-Ports (Spot Values)

Fig. 4.16. Computing the overall noise factor for a cascade

■ **Cascade of Several Two-Ports.** Consider N two-ports connected as in Fig. 4.16, whose individual gains and noise factors are known. We wish to calculate the overall noise factor, denoted by F.

This problem is very important, since the noise generated in the front-end of a receiver is amplified by the following stages (which may exhibit high gain) and consequently contributes most to the output noise.

In order to deduce the overall noise factor, we must make use of (4.77) and replace all noise temperatures in expression (4.66) with their noise factor counterparts; in this way we obtain the following expression, which is often called Friis formula:

$$F = F_1 + \frac{F_2 - 1}{G_1} + \frac{F_3 - 1}{G_1 G_2} + \ldots + \frac{F_N - 1}{G_1 G_2 \ldots G_{N-1}} \qquad (4.78)$$

■ **Conditions.** Equation (4.78) is valid under the following conditions:

- The noise factor of stage i (F_i) must be evaluated with the $(i-1)$-th stage as generator.
- All cascaded two-ports are assumed to be linear (i.e., the input and the output frequencies are the same).
- The real part of the output impedance of each two-port must be a positive quantity.
- Each two-port must add noise.

■ **Conclusion.** Equation (4.78) shows that we must avoid having an attenuator at the front-end of the cascade, lest the overall noise factor would be augmented (the contributions of the following stages, instead of being divided by a gain $G_1 > 1$, would be multiplied by the loss $L_1 = 1/G_1 > 1$.)

4.5.7 Operating Noise Factor

■ **Note.** The concept of operating noise factor was introduced by North in 1942. Actually, it is seldom used and the IEEE Standard Dictionary of Electrical and Electronics Terms does not include it. However, for the sake of completeness, it will be explained here.

■ **Definition of Mumford and Scheibe.** According to [55], *the operating noise factor F_{op} is the ratio of: (a) the available output noise power, in an elementary bandwidth δf, at the actual temperature of the two-port, to (b) that part of (a) due solely to the input signal source at the standard reference temperature.*

Thus

$$F_{op} = \frac{N_o}{G_t(kT_o\,\delta f)} = \frac{T_{op}}{T_o} \qquad (4.79)$$

where T_{op} is the operating noise temperature of the two-port, $T_o = 290\,\mathrm{K}$, and G_t is the transducer power gain.

■ **Note.** With respect to North definition of the noise factor, the temperature of the two-port is not assumed to be $T_o = 290\,\mathrm{K}$ (hence, it can be the actual temperature of the two- port); another difference is that in (4.79) the *transducer* power gain is used, instead of the *available* power gain required by expression (4.71).

4.5.8 Extended Noise Factor

■ **Necessity.** North definition of the noise factor is quite satisfactory, as long as both the resistance of the signal source and the output resistance of the two-port are positive quantities. However, in many practical situations it may happen that negative output resistance occurs somewhere in a chain of amplifiers.

Concerning the signal source, a priori its resistance is always positive, except in the case of several cascaded stages. Here, the output resistance of the i-th stage plays the role of the equivalent source resistance for the (i+1)-th stage. Depending on the particular configurations of adjacent stages and the operating frequency, the output resistance may also be a negative quantity. Hence, the following stage is driven from a negative resistance source.

Finally, it seems wise to consider the possibility of extending the noise factor to situations where at least one of the two resistances involved in the definition may be negative.

■ **Discussion.** If negative values appear for the source and/or the output resistance, we can no longer use the concept of available power in the noise factor definition. According to Sect. 4.2.3, the "available power" concept must be replaced by "exchangeable power." The benefit to expect here is that for positive resistances the exchangeable power turns into available power (so nothing changes!), but for negative resistances the exchangeable power is also negative and its extremum has the same positive, finite value.

■ **Definition.** The extended noise factor F_e is

$$F_e = 1 + \frac{N_{en}}{G_e(kT_o\,\delta f)} \qquad (4.80)$$

4.5 Noise Parameters of Linear Two-Ports (Spot Values)

which is quite similar to the classical definition (4.71), except that now N_{en} represents the exchangeable output noise power obtained when the input signal source is maintained at the standard reference temperature $T_o = 290\,K$; G_e denotes the exchangeable power gain.

■ **Comments**

– This approach can be applied to expression (4.72) (Friis definition), and we obtain:
$$F_e = \frac{S_{ei}/N_{ei}}{S_{eo}/N_{eo}}, \quad R_s > 0 \quad (4.81)$$
where subscript "e" stands for "exchangeable" and "i", "o" denote "input" and "output", respectively.
Note that if the source resistance is negative, (4.81) is no longer valid. This shows that Friis definition is not so versatile as the North definition (and explains the preference of the IEEE to adopt the latter).

– The sign affecting the extended noise factor, considering all possible situations, is given in Table 4.3, where R_o denotes the output resistance.
A negative power can be interpreted as a power flowing into the source, rather than out of it. The quantity $(F_e - 1)$ is obviously always negative when $R_s < 0$.

Table 4.3. Sign relationships for $(F_e - 1)$

R_s	R_o	G_e	N_{en}	$F_e - 1$
+	+	+	+	+
+	−	−	−	+
−	+	−	+	−
−	−	+	−	−

■ **Cascaded Stages.** The extended noise factor is a useful concept, since it can be substituted directly into (4.78), without ambiguity, *provided that the source resistance is positive*. Therefore

$$F_e = F_{e1} + \frac{F_{e2} - 1}{G_1} + \frac{F_{e3} - 1}{G_1 G_2} + \ldots + \frac{F_{en} - 1}{G_1 G_2 \ldots G_{n-1}}, \quad R_s > 0 \quad (4.82)$$

Note that all gains involved in (4.82) are of the "exchangeable gain" type. Also, in expression (4.78) we have imposed the constraint that each two-port of the cascade must *add* noise. It can be shown that this condition is automatically satisfied by (4.82).

4.5.9 Noise Measure

■ **Comment.** This quantity has an obscure physical significance but is useful whenever the noise factor F and/or the available gain G of a particular amplifier have values close to unity. It is often employed in practice, but is not included in the IEEE Standard Dictionary of Electrical and Electronics Terms.

■ **Definition of Haus and Adler.** For a two-port with available power gain G and noise factor F, the noise measure (M) is [46]

$$\boxed{M = \frac{F-1}{1-1/G}} \qquad (4.83)$$

In order to circumvent difficulties when the input and output admittances of two-ports have negative real parts, Fukui [57] has shown a preference to use the exchangeable gain and extended noise factor instead of the available gain and noise factor in (4.83).

■ **Approach.** Van der Ziel proposed the following approach [50] in order to legitimize the choice of expression (4.83). Suppose we have several identical amplifiers with available gain G_1 slightly greater than unity and noise factor F_1. In order to amplify the signal, more than one stage is required, but having more stages also implies more noise.

The question which arises is "How much more noise?". In order to answer, consider (4.78), which applied to this particular case gives

$$F = 1 + (F_1 - 1) + \frac{F_1 - 1}{G_1} + \frac{F_1 - 1}{G_1^2} + \frac{F_1 - 1}{G_1^3} + \dots \qquad (4.84)$$

whose limit is

$$F = 1 + \frac{F_1 - 1}{1 - 1/G_1} = 1 + M \qquad (4.85)$$

as soon as the number of amplifiers in the cascade is sufficiently large. Hence, M obviously represents a good measure for the added noise.

■ **King Definition.** As stated in [58], *the noise measure is the excess noise factor of a cascade containing an infinity of identical two-ports.*

■ **Comments**

- The excess noise ratio is equal to the quantity (F–1).
- If the available power gain is high, there is no need to introduce the noise measure.

■ **Property.** A noteworthy feature of the noise measure has been pointed out by Haus and Adler. They have shown that when cascading two stages with individual noise measures $M_1 > M_2$, the best choice is to place amplifier 2 before amplifier 1.

They also proved that when cascading several stages, *the least noisy resulting amplifier is the one with the lowest possible value not for F, but for M.*

4.6 Average Values of Two-Port Noise Parameters

■ **Comment.** Our main interest in average values is that in practice we always measure *average* values for noise parameters, even if the setup operates in a narrow bandwidth.

Another motivation is that in order to characterize a communication system, the noise performance *over a band of frequencies* (and not a spot value) is often required.

Nevertheless, the approaches proposed in Sect. 4.4 remain valid.

■ **Average Input Noise Temperature.** This quantity is defined as *the ratio of: (a) the added noise output power* N_n, *to (b) the product of the average available power gain* $\overline{G_a}$, *Boltzmann constant, and the bandwidth* B_o. Thus

$$\overline{T_e} = N_n/(k\overline{G_a}B_o) \qquad (4.86a)$$

Note that the noise power N_n is measured over bandwidth B_o, and $\overline{G_a}$ is evaluated over the same bandwidth by means of the equality

$$\overline{G_a}B_o = \int_{f_1}^{f_2} G_a(f)\, df \qquad (4.86b)$$

In practical situations, as long as we are dealing with a cascade of several amplifiers, the noise equivalent bandwidth Δf can be identified with the signal bandwidth B_o.

■ **Average Noise Factor.** Concerning the average noise factor $\overline{F}$, Mumford and Scheibe [55] have proposed the following expression which relates it to the spot noise figure $F(f)$:

$$\overline{F} = \left\{\int_0^{\infty} F(f)G(f)df\right\} \Big/ \left\{\int_0^{\infty} G(f)df\right\} \qquad (4.87)$$

where f is the input frequency and $G(f)$ is the spot value of the available power gain. Note that we now need additional information about the frequency distribution of both $F(f)$ and $G(f)$.

■ **Conclusion.** Ultimately, the average values of the noise parameters are more important than their spot-values counterparts, since they correspond better to what is really measured (or required) in practice.

4.7 Noise of a Linear Multiport

4.7.1 Output Noise Power

■ **Problem Formulation.** Assume a linear multiport (Fig. 4.17) with n input ports and a single output port to which the load Z_L is connected. Let us denote by E_i (i = 1, 2,..n) the open-circuit noise voltages generated by the impedances Z_i, while E_L represents the open-circuit output noise voltage. We wish to calculate the noise power spectral density at the output at a specified frequency.

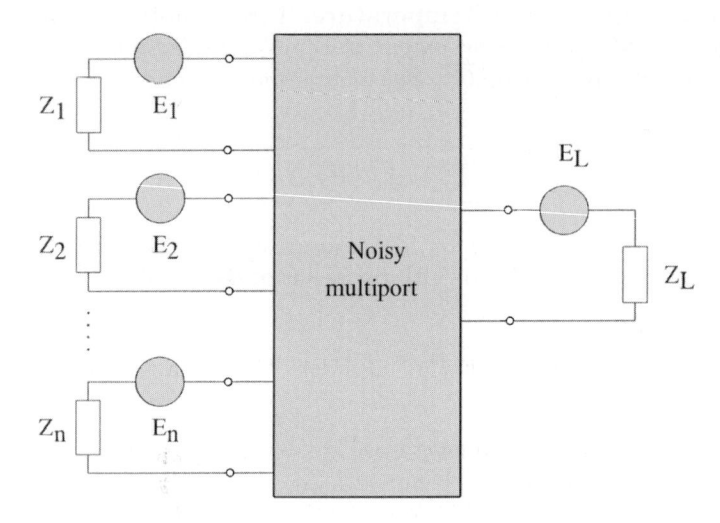

Fig. 4.17. Linear multiport with one output

■ **Solution.** The output noise power has two components:
1) the available noise powers of impedances Z_i connected at the inputs, which are transmitted through the multiport to the output port;
2) other noise sources (such as the noise generated inside the multiport, the noise resulting from eventual frequency conversion, and the noise emerging from the load, back-reflected by the output port).

Note that these components are uncorrelated.

Let us denote by $S(N_{eo})$ the power spectral density of component (1) and by $S(N_N)$ that of component (2). Therefore, the power spectral density of the total output power N_{oL} is

$$S(N_{oL}) = S(N_{eo}) + S(N_N) \tag{4.88}$$

Generally, p responses may exist, each being described by its equivalent input noise temperature T_{ei}; assuming that the p responses are uncorrelated, we can write

4.7 Noise of a Linear Multiport

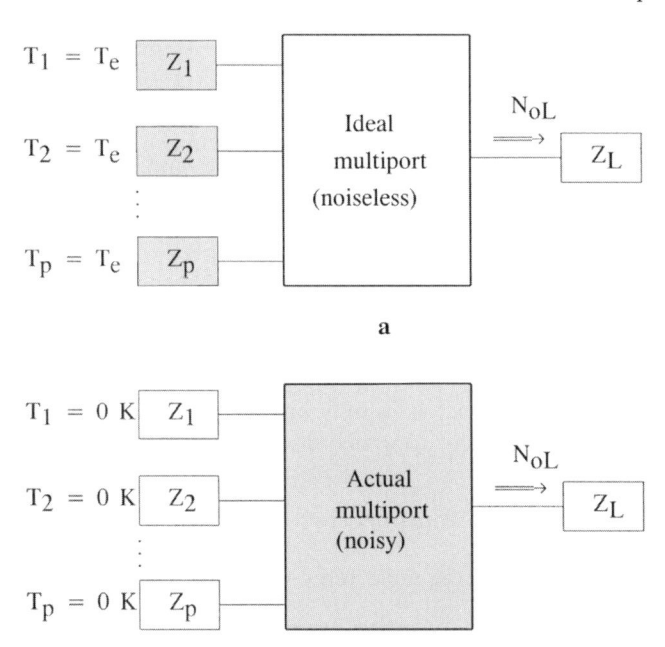

Fig. 4.18. Illustrating the definition of T_e (for a multiport)

$$S(N_{eo}) = k(T_{e1}G_1 + T_{e2}G_2 + \ldots T_{ep}G_p) \tag{4.89}$$

G_i being the transducer power gain associated with the response i. In a similar way, we may express the noise contribution of the multiport with respect to the equivalent input temperature T_e (which is common to all responses):

$$S(N_N) = kT_e(G_1 + G_2 + \ldots + G_p) + S(N_L) \tag{4.90}$$

Here N_L is a term pertaining to the load contribution (whose noise is transmitted towards the output port and then reflected back). Generally N_L is quite negligible, except when the multiport has low gain; hence, neglecting N_L and upon substituting (4.90) and (4.89) into (4.88), we obtain

$$\boxed{S(N_{oL}) = k\Big(G_1(T_{e1} + T_e) + G_2(T_{e2} + T_e) + \ldots + G_n(T_{en} + T_e)\Big)} \tag{4.91}$$

This represents the *output noise power spectral density*.

4.7.2 Input Noise Temperature

■ **Definition.** As stated in [59], *the input noise temperature T_e (of a multiport transducer with one port designated as output) is the noise temperature* *in kelvins which, assigned simultaneously to the specified impedance terminations of all ports (except the output) of a noise-free equivalent of the multiport (Fig. 4.18a), would yield the same available power per unit bandwidth at a specified output frequency at the output as that of the actual multiport connected to noise-free equivalent terminations at all ports except the output (Fig. 4.18b).*

■ **Note**

1) Despite the fact that the definition requires the concept of noise-free devices, this does not affect one's ability to measure the input noise temperature.
2) This definition is restricted to only one output frequency. When the average $\overline{T}_e$ is required, the approach described in Sect. 4.6 can be applied.

4.7.3 Operating Noise Temperature

■ **Definition.** The operating noise temperature of a multiport, expressed in kelvins, is given by

$$T_{op} = \frac{S(N_{oL})}{k\, G_t} \tag{4.92}$$

where $S(N_{oL})$ is the noise power spectral density at the output (as stated by (4.91)) and G_t is the transducer power gain of the multiport, defined between the port where the input signal is applied and the output.

■ **Average Value.** In practice, when the frequency response of G_t is almost rectangular, we prefer to employ the average operating noise temperature, which expressed in kelvins is

$$\overline{T_{op}} = \frac{N_{oL}}{k\, B_o\, G_t} \tag{4.93}$$

Here B_o is the bandwidth of the output signal (for a heterodyne system, it is the bandwidth of the intermediate frequency stage).

■ **Concluding Remark.** If degradation of information due to noise is the main criterion, the communication system with the smallest T_{op} will provide the best performance.

4.8 Conclusion

■ **Measurement.** In practice, the only measurable quantity is the noise power. It is common practice to measure the output noise power, instead of input noise power, for two reasons:

- the noise level at the output is more important, hence better accuracy is expected;
- connecting the measurement equipment at the output avoids degrading the eventual input symmetry and/or shielding.

If required, the input noise level can be deduced by dividing the output level by the gain of the system under test.

■ **Conditions.** Whenever we wish to measure the noise power and a wattmeter is available, several conditions are imposed on the meter:

- The wattmeter must be sensitive to the noise power level.
- Its bandwidth must be at least 10 times the equivalent noise bandwidth.
- Its peak factor must by greater than 4.

■ **Parameters.** Equations (4.59), (4.67), and (4.73) relate the output noise power to the input temperature, operating temperature, and noise factor, respectively. Hence, as soon as the output noise power is measured, any of the previous parameters can be deduced.

In practice, one of the most common mistakes is to use a single parameter to characterize transducer noise. Although this may seem reasonable (as the parameter in question is directly related to the output noise power), as a matter of fact, we lose any potential information on how this power is modified by variations of the source admittance, bias, or frequency. For this reason, characterization with a single parameter is *incomplete*.

An overview of all noise parameters is presented in Tables 4.4 (one-port networks) and 4.5 (two-ports). In most cases, the associated ideal value is the value obtained for a noiseless network. The asymptotic value is the value obtained from a hypothetical network that contributes infinite noise power.

Table 4.4. One-port noise parameters

Parameter	Notation	Ideal value	Asymptotic value
Equivalent noise resistance	R_n	0	∞
Equivalent noise current	I_n	0	∞
Noise temperature	T	0	∞
Noise ratio	N	1	∞

Summary

- Whenever a signal passes through a two-port, the output power is the input power multiplied by the gain; whenever noise passes through a two-port, the output noise power is the input power multiplied by the gain, to which *the internal noise of the two-port is added*.

Table 4.5. One-port noise parameters

Parameter	Notation	Ideal value	Asympt. value	Source noise accounted for	Load noise accounted for
Equivalent noise resistance	R_n	0	∞	yes	no
S/N ratio	S/N	∞	0	yes	no
Equivalent input noise temperature	T_e	0	∞	yes	no
Operating noise temperature	T_{op}	0	∞	yes	yes
Noise factor	F	1	∞	yes	no
Noise mesure	M	0	∞	yes	no

- The input S/N ratio is *always* degraded when a signal is transmitted through a two-port.
- If several stages are cascaded, the front-end stage has a paramount influence on global noise performance. It must not be a lossy stage.
- Low-noise systems are better characterized by the equivalent input noise temperature.
- The equivalent input noise temperature is not an intrinsic property of the two-port alone, but also depends on the internal impedance of the signal generator.
- The input noise temperature is defined in terms of *available* output noise power.
- The operating noise temperature is based on *total* output noise power.
- The standard reference temperature is always 290 K.
- The noise factor merely compares the thermal noise of the signal source with the internal noise of the two-port. It is helpful only when choosing the least noisy transducer for a given signal source.
- With Friis definition, the noise factor is a measure of degradation of the S/N ratio.
- The noise factor depends on the source admittance, frequency of operation, and bias.

5
Noise Analysis of Linear Circuits

> "Rules and models destroy genius and art"
>
> (William Hazlitt)

5.1 Introduction

■ **Noise Calculation.** We have already explained that inside an electronic circuit the noise sources are scattered throughout. As a matter of fact, the noise of any resistor is described by a thermal noise source and an excess noise source; similarly, any transistor has at least three noise sources. Even for small circuits, the classical approach of substituting a noise model for each device followed by calculation of the output noise power with the network theory techniques, leads to serious difficulties. For medium- or large-scale circuits, this approach obviously becomes quite impractical.

This is why macro-modeling is performed, i.e., the overall circuit is regarded as a one-port, two-port or multiport, with physical structure identical to the actual circuit, except that it is noiseless. Hence, in order to reproduce the noise behavior of the actual circuit, we must connect either noise voltage generators or noise current generators to its ports. Convenient values must then be assigned to the physical parameters describing their noise, such as the equivalent noise resistance, noise conductance, and noise temperature.

■ **Conditions.** This way of simulating the internal noise of a multiport by means of lumped generators connected to the ports is possible under the following conditions:

– The circuit must be linear. This means that inside the circuit the signal and noise add without interacting, and the gain is the same for signal and noise.

– Any input signal must produce an output signal. If this were not the case, there would be no coupling between the considered input and output; consequently, the noise generated inside the multiport could not reach the output terminals.

■ **General Remark.** During noise analysis, it is almost impossible to avoid mixing quantities specific to the frequency domain (like impedances and admittances) with functions of time (describing the evolution of signal sources). This amalgamation is undesirable and may suggest some lack of consistency.

A possible solution consists in substituting for all impedances their inverse Laplace transforms; in this way calculation can be carried out in the time domain. However, this requires convolution integrals, leading to tedious computation, and for that reason the approach is not practical.

The other possibility is to transform all quantities to the frequency domain; all sources are then described by their power spectral densities, but we then obtain cumbersome and difficult to interpret expressions, especially for correlated noise sources.

Finally, it is obvious that for calculation by hand, hybrid expressions containing both frequency- and time-domain quantities are unavoidable.

5.2 Noise Models of One-Port Circuits

■ **Equivalent Circuit or Model?** Nowadays, when computer simulation has become an essential tool in almost every field, the term "model" is overused and often misused. In electrical and electronic engineering we must carefully distinguish between what is traditionally considered an equivalent circuit and what is called a model.

As stated in the IEEE Dictionary of Electrical and Electronics Terms, an equivalent circuit is *an arrangement of circuit elements that has electrical characteristics, over a specified frequency band, equivalent to those of a different circuit or device.*

In many applications, for convenience of analysis the equivalent circuit replaces the actual circuit or device, which is much more complicated. Note that in this case we write down equations and deduce *expressions* for the desired performance.

In computer simulation our primary concern is to obtain *numerical values* (not expressions) for performances. Therefore, an equivalent circuit alone is not of much help unless we assign appropriate numerical values to all elements. *An equivalent circuit, together with a set of particular numerical values assigned to all elements, is called a model.*

In conclusion, when modeling is required, not only must a suitable equivalent circuit be selected, but so also must a method to deduce appropriate numerical values for its elements.

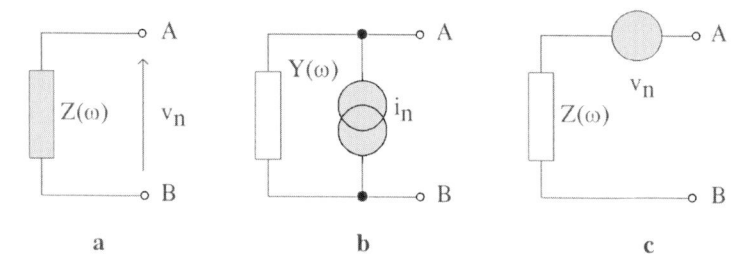

Fig. 5.1. **a** Noisy one-port; **b** Schottky equivalent circuit; **c** Thévenin equivalent circuit

5.2.1 One-Port at Uniform Temperature

■ **Passive, Linear, and Reciprocal Network.** Consider the one-port network in Fig. 5.1a, which has a terminal impedance

$$Z(\omega) = R(\omega) + jX(\omega) \tag{5.1}$$

The equivalent admittance is

$$Y(\omega) = 1/Z(\omega) = G(\omega) + jB(\omega) \tag{5.2}$$

Assume that fluctuations across the terminals A and B are due only to several thermal noise sources within the one-port, which is in thermal equilibrium.

The representation shown in Fig. 5.1b was proposed in 1918 by Walter Schottky in order to represent the noise of a vacuum diode; it consists of adding a noise current generator in parallel with an ideal (noiseless) identical one-port.

Of course, by means of Thévenin theorem, its dual circuit (Fig. 5.1c) can be deduced.

Assume that the noise sources are all thermal, with the same temperature. The mean square terminal noise voltage over a bandwidth Δf about the operating frequency f can be calculated with Nyquist formula

$$\boxed{\overline{v_n^2} = 4kTR\,\Delta f \quad \text{with} \quad R = \mathcal{R}e[Z]} \tag{5.3a}$$

The equivalent noise current is given by

$$\boxed{\overline{i_n^2} = 4kTG\,\Delta f \quad \text{with} \quad G = \mathcal{R}e[Y]} \tag{5.3b}$$

Equations (5.3) are not valid if the circuit is nonlinear or it contains additional nonthermal noise sources.

116   5 Noise Analysis of Linear Circuits

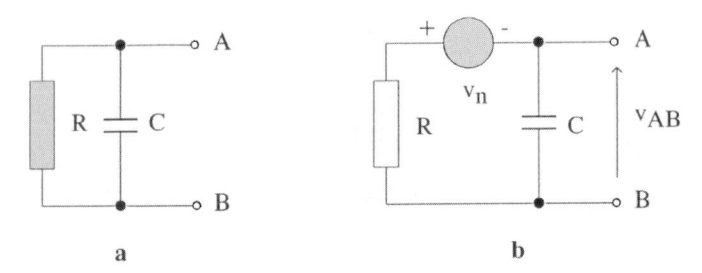

Fig. 5.2. a Single-pole RC network; b Its equivalent circuit

■ **Example.** Consider a single-pole RC circuit (Fig. 5.2a), where we wish to determine the noise at terminals A-B. The only noise source is thermal, and according to (5.3a), $\overline{v_n^2} = 4kTR\ \Delta f$.

Applying the voltage divider expression to the equivalent circuit of Fig. 5.2b, we obtain the transfer function

$$H = \frac{v_{AB}}{v_n} = \frac{1}{1 + j\omega RC} \qquad (5.4)$$

According to (2.41), the output noise normalized power equals the input noise normalized power times the square of the magnitude of the transfer function H, i.e.,

$$\overline{v_{AB}^2} = \overline{v_n^2}|H|^2 = \frac{\overline{v_n^2}}{|(1 + j\omega RC)|^2} = \frac{4kTR\ \Delta f}{1 + \omega^2 C^2 R^2} \qquad (5.5)$$

The final form of (5.5) shows that the low-frequency components of the noise voltage reach the output almost without attenuation, but the high-frequency components are practically short-circuited by the capacitor C. If the value of R is greater, the noise voltage is also greater but the cutoff frequency is reduced.

Now suppose that we want to calculate the total noise power; we must integrate the previous result over the entire frequency domain, i.e.,

$$\int_0^\infty \overline{v_{AB}^2}(f)\ df = \frac{kT}{C} \qquad (5.6)$$

This is a surprising result, since it suggests that the total output noise power does not depend on the noise source itself (R); moreover, it becomes unbounded as soon as C = 0 (the capacitor is removed).

■ **Contribution of Pyati.** In order to explain this paradox, Pyati [60] suggests that due to quantum effects, the resistor does not act as a white noise source at $f > 10^9$ Hz or so. Using (3.10), a quantum correction for the spectral density is necessary and adopting G = 1/R, he finds

5.2 Noise Models of One-Port Circuits

$$\overline{v_{AB}^2} = 4Gh \int_0^\infty \frac{f \, df}{(\exp(hf/kT) - 1)(G^2 + 4\pi^2 f^2 C^2)} \quad (5.7a)$$

With the help of integral tables, the following results are obtained:

$$\overline{v_{AB}^2} = \frac{2\pi^2 k^2 T^2 R}{3h} \left(12x^2 \left(\ln x - \psi(x) - \frac{1}{2x} \right) \right) \quad \text{for} \quad 0 \leq C < \infty \quad (5.7b)$$

$$\overline{v_{AB}^2} = \frac{kT}{C} \left(2x \left(\ln x - \psi(x) - \frac{1}{2x} \right) \right) \quad \text{for} \quad 0 \leq R < \infty \quad (5.7c)$$

where $x = h/(4\pi^2 kTRC)$ and $\psi(x) = -\gamma - \frac{1}{x} + \sum_{n=1}^{\infty} \frac{x}{n(x+n)}$, γ being Euler's constant ($\gamma = 0.57221...$). Equations (5.7) are more appropriate to describe the noise behavior of an RC circuit, because when R and C take the extreme values of 0 and ∞, consistent results are obtained.

■ **Model of Van Nie.** The main limitation of the classical model of a noisy resistor (Fig. 3.1) consists in its inability to determine the noise current through the resistor itself. For instance, the Norton or Thévenin representations are suitable only for evaluating the *external* noise performance of the resistor. Consequently, they cannot be used to describe the noise current through resistive elements such as linearized detector circuits.

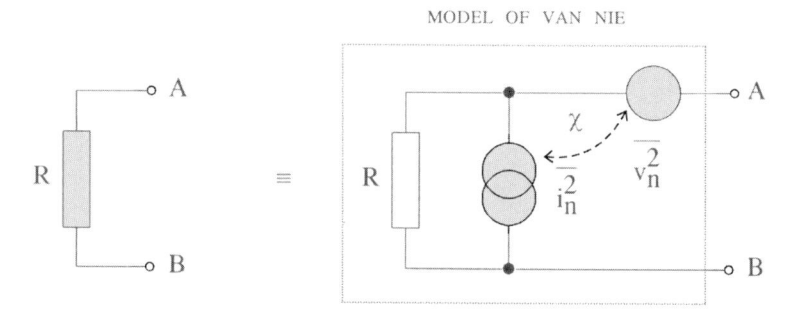

Fig. 5.3. Noisy resistor representation, according to Van Nie

To remedy this situation, Van Nie [61] proposed a resistor model containing two correlated noise generators (Fig. 5.3). These generators must satisfy the following conditions:

1) The generators must be independent of the external circuit.
2) The open-circuit noise voltage must be consistent with the classical expression $4kTR\Delta f$.

3) The short-circuit noise current must be consistent with the traditional expression 4kTGΔf, where $G = 1/R$.
4) The noise current in R must be independent of the external circuit, provided that they are in thermal equilibrium.

Note that requirements 2 and 3 make it possible to recover the Thévenin and Norton representations of a noisy resistor, when a model with a single noise generator is employed.

The correlated noise generators of the resistor R are defined by

$$\overline{v_n^2} = 4kTR_{eq} \; \Delta f \qquad (5.8a)$$

$$\overline{i_n^2} = 4kTG_{eq} \; \Delta f \qquad (5.8b)$$

Let $\chi = \xi + j\zeta$ be their correlation coefficient, such that

$$\langle i_n v_n^* + i_n^* v_n \rangle = 4kT\xi \; \Delta f \quad \text{and} \quad \langle i_n v_n^* - i_n^* v_n \rangle = 4kT\zeta \; \Delta f \qquad (5.8c)$$

Applying (5.8) to a simple resistor, after some algebra it can be proven from the previous requirements that

$$\boxed{R_{eq} = R, \quad G_{eq} = G \quad \text{and} \quad \xi = 1, \; \zeta = 0} \qquad (5.9)$$

☐ **Generalization.** For an arbitrary noisy one-port (Fig. 5.4) with admittance $Y = G + j\omega C$ (both G and C are functions of frequency, as for a diode), the Van Nie approach yields

$$R_{eq} = \mathcal{R}e\{1/Y\}, \quad G_{eq} = \mathcal{R}e\{Y\}, \quad \text{and} \quad \chi = Y/Y^* \qquad (5.10)$$

In this case, the noise current through G is independent of the external circuit.

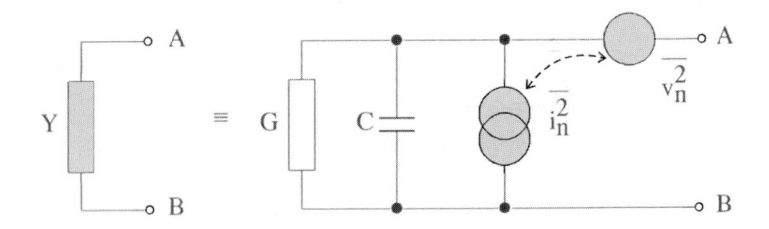

Fig. 5.4. Noisy one-port representation with two correlated noise generators

5.2.2 One-Port at Different Temperatures

■ **Explanation.** Although this case seems to be rather theoretical, in practice it can be encountered whenever the temperature is nonuniformly distributed inside a physical system. For instance, a satellite may have a sensor located in outer space (in sun or in the dark), while the circuits processing the signal from the sensor are inside. They are connected by a transmission line, which is not in thermal equilibrium, since it experiences a significant temperature gradient. The system sensor and its associated transmission line are identified to a one-port whose elements are at different temperatures.

A direct application of the Thévenin representation for the transmission line would be useless unless we are able to define an effective temperature for the equivalent resistance of the system that takes into account the weighted contributions of the individual resistors at various temperatures. The following rule can be used to find it.

■ **Pierce's Rule.** This statement, formulated by J.R. Pierce, applies to linear, passive, and reciprocal one-ports, and is often used to determine the noise temperature of an antenna [62]. It states that *for a transmitting antenna, where a_1 denotes the amount of power absorbed (after all reflections are taken into account) by a body at temperature T_1, a_2 is the amount of power absorbed by a body at temperature T_2, a_3 is the amount of power absorbed by a body at temperature T_3, etc., the noise temperature T_a of the radiation resistance (as a source of thermal noise) is*

$$T_a = a_1 T_1 + a_2 T_2 + a_3 T_3 + \ldots \tag{5.11a}$$

■ **Pettai's Reformulation.** Pettai [63] modified Pierce's rule in the following way: *if a unit of power is delivered to a linear, passive one-port, and if a fraction a_1 is absorbed by the resistance R_1 at temperature T_1, a fraction a_2 is absorbed by R_2 at temperature T_2, a fraction a_3 is absorbed by R_3 at temperature T_3, etc., then the effective input temperature of the one-port is*

$$\boxed{T_{\text{eff}} = a_1 T_1 + a_2 T_2 + a_3 T_3 + \ldots} \tag{5.11b}$$

with

$$a_1 + a_2 + a_3 + \ldots = 1 \tag{5.11c}$$

■ **Application.** Consider an attenuator with loss L, matched to resistance R_o at a uniform temperature T_2. Assume that its input is fed by a resistor R_o at a different temperature T_1 (Fig. 5.5).

1. Determine the effective output temperature.
2. Find the input noise temperature of the attenuator.

120   5 Noise Analysis of Linear Circuits

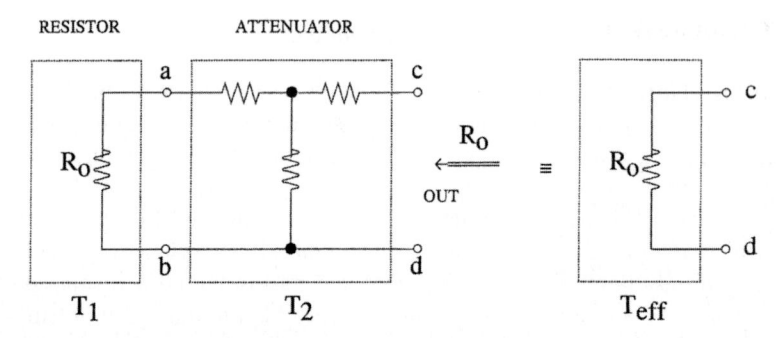

Fig. 5.5. Attenuator matched to resistance R_o

☐ **Solution**

1. The noise power generated inside the external resistor R_o adds to the noise power generated by the attenuator itself (since the two are uncorrelated). We must find the hypothetical temperature T_{eff} which, attributed to the output resistance R_o, would produce the same amount of noise as the actual system.

   Applying the Pierce's rule (reformulated by Pettai), we suppose that *unit* noise power is applied to terminals c-d of the system. Let us denote by a_1 the fraction of power absorbed by the input resistor R_o and a_2 the fraction of power absorbed by the attenuator. Thus

   $$T_{eff} = a_1 T_1 + a_2 T_2 \quad \text{with} \quad a_1 + a_2 = 1$$

   Because the loss of the attenuator is L, it follows that the fraction of the original power reaching the resistor R_o must be

   $$a_1 = \frac{1}{L}$$

   As a result, the remaining power is absorbed by the attenuator:

   $$a_2 = 1 - a_1 = \left(1 - \frac{1}{L}\right)$$

   Substituting these expressions into Pierce's formula, we obtain

   $$T_{eff} = \frac{1}{L} T_1 + \left(1 - \frac{1}{L}\right) T_2 \qquad (5.12)$$

2. If both sides of (5.12) are multiplied by Boltzmann's constant, then

   $$kT_{eff} = \frac{k}{L}\left(T_1 + (L-1) T_2\right)$$

   Identifying this with (4.59c), the input noise temperature of the matched attenuator is

   $$T_e = (L-1) T_2 \qquad (5.13)$$

5.2.3 Conclusion

In representing the noise of a one-port circuit, two situations can arise:

- If the one-port is in thermal equilibrium (i.e., all its elements have the same temperature), the Thévenin representation ($\overline{v_n^2} = 4kTR_{eq}\Delta f$) or the Norton representation ($\overline{i_n^2} = 4kTG_{eq}\Delta f$) can be successfully used to model its noise. In this case $T_{eff} = T$.
- Whenever the one-port is not in thermal equilibrium (its resistors R_j have different temperatures T_j), the Thévenin or Norton model still holds, provided that we replace the temperature T with the effective temperature T_{eff} (which must be previously calculated with Pierce's rule).

5.3 Time Domain Analysis of Noisy Two-Ports

5.3.1 Background

This section relates exclusively to linear, passive two-ports. We recall that a *port* is defined as *a pair of terminals by means of which current may entry or exit a circuit, and across which a voltage exists.*

■ **Convention.** For any linear, passive two-port the voltages and currents at the ports are defined according to Fig. 5.6.

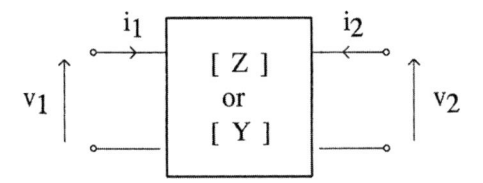

Fig. 5.6. Two-port network

■ **Description.** The electrical behavior of the two-port is described by means of matrices Z, Y or A:

$$\begin{bmatrix} v_1 \\ v_2 \end{bmatrix} = \begin{bmatrix} z_{11} & z_{12} \\ z_{21} & z_{22} \end{bmatrix} \begin{bmatrix} i_1 \\ i_2 \end{bmatrix} \tag{5.14}$$

$$\begin{bmatrix} i_1 \\ i_2 \end{bmatrix} = \begin{bmatrix} y_{11} & y_{12} \\ y_{21} & y_{22} \end{bmatrix} \begin{bmatrix} v_1 \\ v_2 \end{bmatrix} \tag{5.15}$$

$$\begin{bmatrix} v_1 \\ i_1 \end{bmatrix} = \begin{bmatrix} A & B \\ C & D \end{bmatrix} \begin{bmatrix} v_2 \\ -i_2 \end{bmatrix} \tag{5.16}$$

122   5 Noise Analysis of Linear Circuits

where the open-circuit impedance parameters z_{jk}, the short-circuit admittance parameters y_{jk} and the chain parameters ABCD are usually frequency dependent.

If two two-ports (designated by indices 1 and 2, respectively) are interconnected, the resulting two-port has the following properties:

- If the two-ports are connected in series, then the overall open-circuit impedance matrix is $Z = Z_1 + Z_2$.
- If the two-ports are connected in parallel, then the overall short-circuit admittance matrix is $Y = Y_1 + Y_2$.
- If the two-ports are connected in cascade, then the overall chain matrix is $A = A_1 \, A_2$.

Table 5.1 summarizes the transformations that change the representation (note that $\Delta = AD - BC$).

Table 5.1. Transformation of two-port parameters

	Z	Y	ABCD
Z	$\begin{bmatrix} z_{11} & z_{12} \\ z_{21} & z_{22} \end{bmatrix}$	$\begin{bmatrix} \dfrac{y_{22}}{\|y\|} & \dfrac{-y_{12}}{\|y\|} \\ \dfrac{-y_{21}}{\|y\|} & \dfrac{y_{11}}{\|y\|} \end{bmatrix}$	$\begin{bmatrix} \dfrac{A}{C} & \dfrac{\Delta}{C} \\ \dfrac{1}{C} & \dfrac{D}{C} \end{bmatrix}$
Y	$\begin{bmatrix} \dfrac{z_{22}}{\|z\|} & \dfrac{-z_{12}}{\|z\|} \\ \dfrac{-z_{21}}{\|z\|} & \dfrac{z_{11}}{\|z\|} \end{bmatrix}$	$\begin{bmatrix} y_{11} & y_{12} \\ y_{21} & y_{22} \end{bmatrix}$	$\begin{bmatrix} \dfrac{D}{B} & \dfrac{-\Delta}{B} \\ \dfrac{-1}{B} & \dfrac{A}{B} \end{bmatrix}$
ABCD	$\begin{bmatrix} \dfrac{z_{11}}{z_{21}} & \dfrac{\|z\|}{z_{21}} \\ \dfrac{1}{z_{21}} & \dfrac{z_{22}}{z_{21}} \end{bmatrix}$	$\begin{bmatrix} \dfrac{-y_{22}}{y_{21}} & \dfrac{-1}{y_{21}} \\ \dfrac{-\|y\|}{y_{21}} & \dfrac{-y_{11}}{y_{21}} \end{bmatrix}$	$\begin{bmatrix} A & B \\ C & D \end{bmatrix}$

■ **Two-Port with Independent Sources.** If in addition to passive elements (R, L, C etc.) the two-port also contains independent voltage or current sources, it cannot be modeled solely by means of the matrices Z, Y, or ABCD. These matrices are still used to represent the passive network, but two (fictitious) equivalent generators must be added at the terminals to account for the power delivered by the internal sources at the terminals. Consequently, (5.14), (5.15), and (5.16) become

$$\begin{bmatrix} v_1 \\ v_2 \end{bmatrix} = \begin{bmatrix} Z_{11} & Z_{12} \\ Z_{21} & Z_{22} \end{bmatrix} \begin{bmatrix} i_1 \\ i_2 \end{bmatrix} + \begin{bmatrix} E_1 \\ E_2 \end{bmatrix} \quad (5.17)$$

$$\begin{bmatrix} i_1 \\ i_2 \end{bmatrix} = \begin{bmatrix} y_{11} & y_{12} \\ y_{21} & y_{22} \end{bmatrix} \begin{bmatrix} v_1 \\ v_2 \end{bmatrix} + \begin{bmatrix} I_1 \\ I_2 \end{bmatrix} \quad (5.18)$$

$$\begin{bmatrix} v_1 \\ i_1 \end{bmatrix} = \begin{bmatrix} A & B \\ C & D \end{bmatrix} \begin{bmatrix} v_2 \\ -i_2 \end{bmatrix} + \begin{bmatrix} E \\ I \end{bmatrix} \quad (5.19)$$

Equations (5.17)–(5.19) suggest the equivalent circuits of an active two-port presented in Fig. 5.7. The choice among them is decided by ease of calculation in any particular application.

The values of the equivalent generators connected to the ports are deduced as suggested by (5.17) and (5.18): E_1 and E_2 are obtained by measuring the port voltages under open- circuit conditions and I_1, I_2 by measuring the short-circuit port currents. However, measuring the values of E and I (Fig. 5.7c) is not so straightforward.

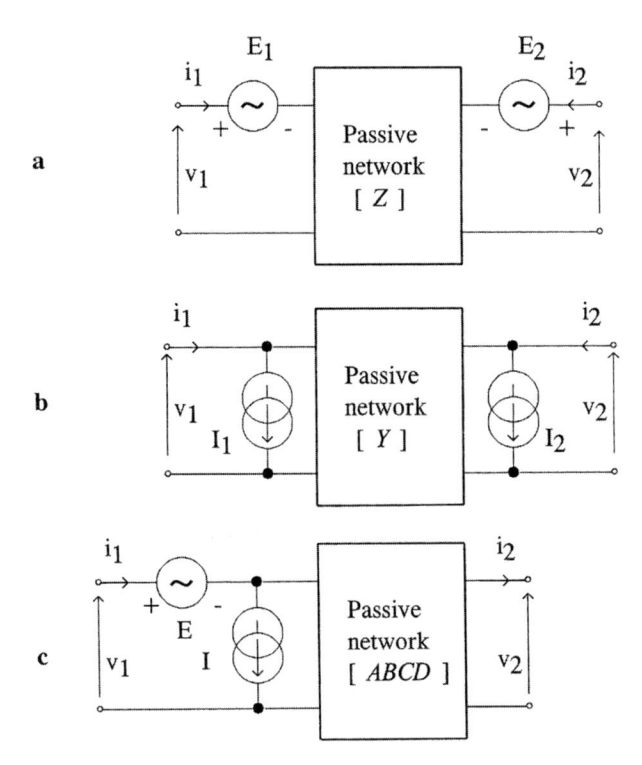

Fig. 5.7. Various representations of a two-port with independent internal sources

124   5 Noise Analysis of Linear Circuits

■ **Note.** Generally, the equivalent generators connected to the ports are not independent, since their expressions (derived via Thévenin's or Norton's theorem) rely upon the same variables (the internal independent sources). Therefore their time evolution must follow, to a certain degree, a similar pattern. The equivalent signal generators are said to be *partially coherent* or *partially synchronized*.

5.3.2 Noisy Two-Port

■ **Approach.** A two-port containing several noise sources can be regarded as a two-port with independent sources, except that the internal sources are no longer harmonic, but random.

The representation of any noisy two-port requires four complex parameters (the Z-, Y-, or A-matrix) to describe the passive but hypothetical noise-free two-port, and two real parameters corresponding to the equivalent noise generators. Since these generators are always *partially correlated*, the correlation coefficient is a complex quantity; hence, the noise of a two-port can be represented by four real parameters.

It is important to remember that the ability of an equivalent network to represent the noise behavior of a two-port is no guarantee that it resembles the actual configuration responsible for noise.

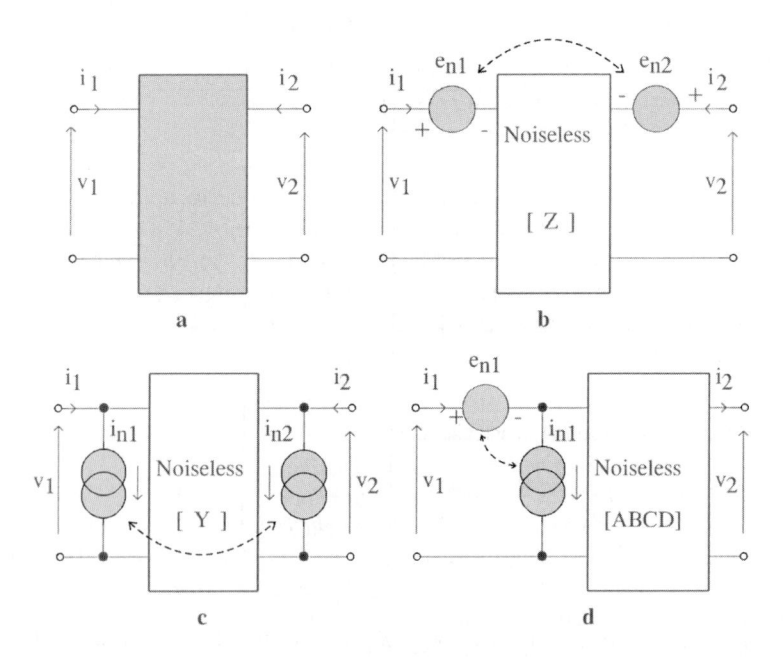

Fig. 5.8. a Noisy two-port; **b** open circuit independance representation; **c** short circuit admittance representation; **d** chain parameters representation (dotted arrows suggest correlation)

■ **Remark.** The reader may wonder why the noise generators in Fig. 5.8 are oriented, because, after all, they represent fluctuations, whose polarity varies randomly. It should be stressed that the sign convention is applied to the noise generators only to accommodate Kirchhoff's laws. As the deduced quantities (currents or voltages) are ultimately squared, the imputed signs are not important.

■ **Discussion.** Regardless of the adopted representation, four real quantities are required to completely describe the noise of a two-port. These quantities are the values assigned to the two noise generators, as well as the real and imaginary parts of the correlation coefficient relating them.

Thus, it is important to find the correlation coefficient. This may be done either in the time domain (by means of Montgomery's theorems) or in the frequency domain (the noise generators being described by their self- or cross-power spectral densities).

■ **Montgomery's First Theorem.** According to [64], *if two noise currents (or voltages) partly originate from a common noise source and partly from independent sources, and if α is the fraction of power in the first due to the common source and β the fraction of power in the second due to the common source, then the correlation between the two quantities is the geometric mean of α and β.*

■ **Montgomery's Second Theorem.** *The correlation between two noise currents (or voltages) is not changed by passing one or both currents through linear networks that have real transfer functions.*

■ **Comment.** The second theorem explains why the model of Fig. 5.8d has traditionally been preferred. This model separates the noise generators (ahead of the two-port) from the ideal (noiseless) two-port itself. As soon as the correlation coefficient has been determined at the input, the same value is preserved at the output (provided the two-port is linear and has a real transfer function).

5.3.3 Model of Rothe and Dahlke

■ **Model Description.** This model [65] adopts the chain matrix representation of the noisy linear two-port (Fig. 5.8d); the effect of all internal noise sources is lumped at the input port into an equivalent noise voltage generator e and an equivalent noise current generator i. As stated previously, a correlation should exist between the two generators.

According to the technique discussed in Sect. 2.2.5, the noise voltage generator can be split in two components:

$$e = e_n + e'$$

The first (e_n) is assumed to be independent of i, but the second (e′) is completely correlated with the noise current generator. Thus, this second term must be proportional to i; as the proportionality factor, Rothe and Dahlke introduced the *complex correlation impedance*

$$Z_{cor} = R_{cor} + j\, X_{cor} \qquad (5.20)$$

such that

$$e = e_n + i\, Z_{cor} \qquad (5.21)$$

Applying the duality principle, we can decompose the noise current generator into two components (one being totally correlated with the noise voltage generator), and thereby define a *complex correlation admittance*:

$$Y_{cor} = G_{cor} + j\, B_{cor} \qquad (5.22)$$

Hence, we can write

$$i = i_n + e\, Y_{cor} \qquad (5.23)$$

In the circuit of Fig. 5.9a the voltages are related by Kirchhoff's voltage law:

$$v_1 = e + v_1'$$

Substituting (5.21), we have

$$v_1 = e_n + i\, Z_{cor} + v_1' = e_n + (i_1 - i_1')Z_{cor} + v_1' \qquad (5.24a)$$

Similarly, for the dual case

$$i_1 = i + i_1' = i_n + e\, Y_{cor} + i_1' = i_n + (v_1 - v_1')Y_{cor} + i_1' \qquad (5.24b)$$

Equations (5.24) can be written in a different but equivalent form:

$$v_1 = v_1' + e_n + i_1\, Z_{cor} - i_1'\, Z_{cor} \qquad (5.25a)$$

$$i_1 = i_1' + i_n + v_1\, Y_{cor} - v_1'\, Y_{cor} \qquad (5.25b)$$

Equations (5.25) represent the mathematical model of Rothe and Dahlke; the corresponding circuits are shown in Fig. 5.9b and 5.9c. Traditionally, they are called the T- and Π-models of Rothe and Dahlke.

■ Comments

- The immittances Z_{cor} and Y_{cor} are both assumed to be noiseless. Sometimes this condition is suggested by writing $T = 0\,K$.
- Immittances $-Z_{cor}$ and $-Y_{cor}$ result from negative terms in (5.25). Often, their inclusion is viewed as a mean to compensate their positive counterparts, in order to avoid signal attenuation and preserve the transfer properties of the two-port. Note that from a signal standpoint the equivalent noise generators are assumed to be ideal i.e., their internal resistance is either zero (for the voltage generator) or infinite (for the current generator); hence, no attenuation is expected from them.

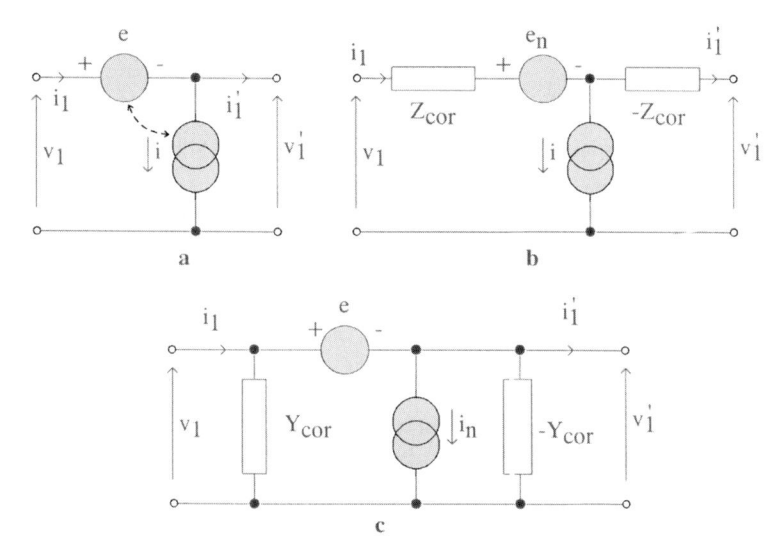

Fig. 5.9. Noisy two-port representations: **a** with correlated sources; **b** and **c** models proposed by Rothe and Dahlke

■ **Parameters.** We can relate the noise voltage and current generators of the Π-model (Fig. 5.9c) to their equivalent noise resistance R_n and equivalent noise conductance G_n:

$$\overline{e^2} = 4kT_o \, \Delta f \, R_n, \quad \overline{i_n^2} = 4kT_o \, \Delta f \, G_n \tag{5.26a}$$

Similarly, for the dual T-model (Fig. 5.9b) we have

$$\overline{e_n^2} = 4kT_o \, \Delta f \, r_n, \quad \overline{i^2} = 4kT_o \, \Delta f \, g_n \tag{5.26b}$$

Finally *the two-port noise behavior can be described with a set of three noise parameters R_n, G_n, and Y_{cor} (or r_n, g_n, Z_{cor}).* (Note that there are actually four real noise parameters, since Y_{cor} and Z_{cor} are complex quantities, each defined by a real and imaginary part.) The transformation rules between these sets of parameters are presented in Table 5.2.

■ **Noisy Two-Port Description.** A general property of any linear two-port is the dependence of its noise factor F on the source admittance $Y_s = G_s + jB_s$; this function exhibits a minimum, denoted F_o, called its *optimum* (or *minimum*) *noise factor*.

The particular value of the source admittance for which this minimum is attained is called *optimum source admittance* (denoted by $Y_o = G_o + jB_o$, where G_o is the *optimum source conductance* and B_o the *optimum source susceptance*).

It follows that the noise of any linear two-port is *completely* described by the four parameters F_o, G_o, B_o, and R_n, sometimes called *the standard noise parameters*.

Table 5.2. Transformation rules between dual forms

From Π to T	From T to Π
$g_n = G_n + R_n\|Y_{cor}\|^2$	$R_n = r_n + g_n\|Z_{cor}\|^2$
$r_n = \dfrac{G_n}{\|Y_{cor}\|^2 + (G_n/R_n)}$	$G_n = \dfrac{r_n}{\|Z_{cor}\|^2 + (r_n/g_n)}$
$Z_{cor} = \dfrac{Y^*_{cor}}{\|Y_{cor}\|^2 + (G_n/R_n)}$	$Y_{cor} = \dfrac{Z^*_{cor}}{\|Z_{cor}\|^2 + (r_n/g_n)}$

Once again, we must resist the temptation of describing the two-port noise with a single parameter (its noise factor), which only provides a *partial* description.

■ **Calculation.** In order to deduce relationships among the classical noise parameters (F_o, G_o, B_o, R_n) and those of the Rothe and Dahlke model, the first step is to write the equation for the noise factor. Next, the derivative with respect to the signal source admittance is taken and expressions for the minimum noise factor F_o and optimum admittance Y_o are found.

Hence, for the Π-model and recalling the North's definition (the noise factor F is the ratio of the actual noise power at the output to the output noise power that would be found if the signal source were the only noisy element), we obtain

$$F = 1 + \frac{G_n + R_n\left((G_s + G_{cor})^2 + (B_s + B_{cor})^2\right)}{G_s} \quad (5.27a)$$

In the same way, the T-model yields

$$F = 1 + \frac{r_n + g_n\left((R_s + R_{cor})^2 + (X_s + X_{cor})^2\right)}{R_s} \quad (5.27b)$$

Taking derivatives and doing the algebra, we obtain the results in Table 5.3.

It can be shown that

$$G_{cor} = \frac{F_o - 1}{2R_n} - G_o, \quad B_{cor} = -B_o, \quad G_n = R_n(G_o^2 - G_{cor}^2) \quad (5.28)$$

■ **Conclusion.** The model of Rothe and Dahlke is important because it suppresses the correlation which exists between equivalent noise generators. This correlation is replaced by a complex immittance, which provides more facility when analysis is carried out with the classical concepts of network theory.

Table 5.3. Relationships among parameters

Π-model	T-model
$B_o = -B_{cor}$	$X_o = -X_{cor}$
$G_o = \sqrt{(G_n/R_n) + G_{cor}^2}$	$R_o = \sqrt{(r_n/g_n) + R_{cor}^2}$
$F_o = 1 + 2R_n(G_o + G_{cor})$	$F_o = 1 + 2g_n(R_o + R_{cor})$
$F = F_o + \dfrac{R_n}{G_s}|Y_s - Y_o|^2$	$F = F_o + \dfrac{g_n}{R_s}|Z_s - Z_o|^2$

Another benefit of this model is that the expressions derived for its noise parameters do not include the electrical parameters of the two-port.

5.3.4 Basic Relationships Among Noise Parameters

■ **Fundamental Equation.** Upon substituting (5.28) into (5.27a), one obtains

$$F = F_o + \frac{R_n}{G_s}\left((G_s - G_o)^2 + (B_s - B_o)^2\right) \qquad (5.29)$$

Equation (5.29), relating the noise factor to the source admittance, is very important, and will be called the *fundamental equation* of the noisy two-port.

■ **Significance**

- F_o is the minimum noise factor expected from a given two-port, which is obtained when the signal source is perfectly matched ($Y_s = Y_o$).
- R_n is a positive quantity, expressed in resistance units, which determines how fast the noise factor deteriorates when the two-port is supplied from a suboptimal source admittance. It follows that the noise factor of a two-port with small R_n is relatively insensitive to variations in the signal source admittance. In practice, low R_n is essential for broadband operation, where high tolerance in the input match is worthwhile and highly desirable.
- G_o and B_o are the optimal values of the real and imaginary part of the source admittance. Note carefully that these always differ from the values that yield maximum power gain. As a general rule in any two-port *maximum power gain and minimum noise does not occur simultaneously*.

■ **Note.** If it were possible to separately tune the real and imaginary parts of the source admittance Y_s, it would be easy to adjust B_s to B_o and G_s to G_o, and consequently reach the optimum noise factor. Unfortunately, this is not the case, and the only possibility we can (sometimes) use is to insert a matching transformer (or network) between the source and two-port. In this way, we achieve a matching condition for minimum noise (which does not automatically lead to matching for maximum power gain).

■ **Lange's Contribution.** It is common practice to consider R_n exclusively as an equivalent noise resistance. Lange [66] was the first to point out that R_n is instead a constant that indicates how fast the noise factor deteriorates when the two-port is excited by a mismatched signal source. To facilitate the characterization of devices above 1 GHz, he proposed instead the parameter N, defined by

$$\boxed{N = R_n \, G_o} \qquad (5.30a)$$

The fundamental equation, expressed in terms of this parameter N is

$$F = F_o + N\frac{|Z_s - Z_o|^2}{R_s R_o} \quad \text{or} \quad F = F_o + N\frac{|Y_s - Y_o|^2}{G_s G_o} \qquad (5.30b)$$

(where $Z_s = R_s + jX_s$ is the source impedance and Z_o its optimal value).

The attractive features of parameter N are:

- The dual forms (5.30b) are perfectly symmetrical.
- N and F_o are functions only of the intrinsic device (chip), and are not influenced by the package (provided the package parasitics are weak).
- In a transmission line system, N is independent of the location of the reference plane (hence N can be deduced from VSWR data only).
- If several identical transistors operate in parallel, the whole ensemble has the same N and F_o as the individual devices. In integrated circuits, a category of devices that differ only in their active areas would have the same N and F_o.
- Inserting a lossless input network will modify neither N nor F_o.

■ **Wiatr's Inequality.** Wiatr has proven the following equivalent inequalities for a physical two-port:

$$\boxed{R_n \geq \frac{F_o - 1}{4G_o}} \qquad (5.31a)$$

$$\boxed{F_o - 1 \leq 4N} \qquad (5.31b)$$

$$\boxed{T_{e\ min} \leq 4NT_o} \qquad (5.31c)$$

The constraints (5.31) are important whenever we need a coherent set of noise parameters for a hypothetical transistor used for simulation purposes.

■ **Relationship Between S/N Ratio and F.** The signal-to-noise (S/N) ratio can be expressed in terms of the noise figure, provided that the latter is measured with the same signal source:

$$\text{S/N} = 10\log\left(\frac{V_s^2}{4kTR_s}\right) - F_{dB}\bigg|_{\text{same } R_s} \quad (5.32)$$

where V_s denotes the rms value of the signal source, and R_s is its internal resistance.

■ **Noise Surface.** If the investigated two-port is a simple transistor, it is interesting to plot the surface predicted by the fundamental equation (5.29).

Widespread convention chooses to represent it as a symmetric or asymmetric paraboloid (Fig. 5.10a and 5.10b).

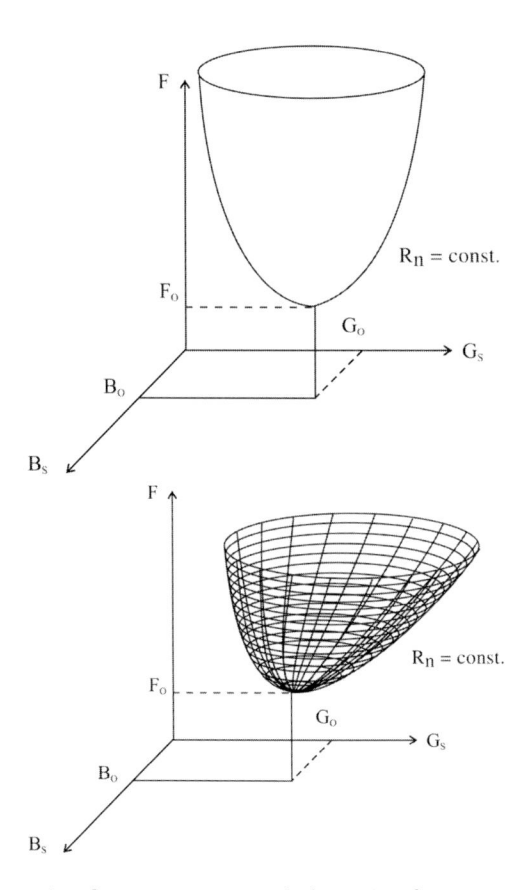

Fig. 5.10. Conventional representation of the noise factor versus the real and imaginary part of the source admittance

132   5 Noise Analysis of Linear Circuits

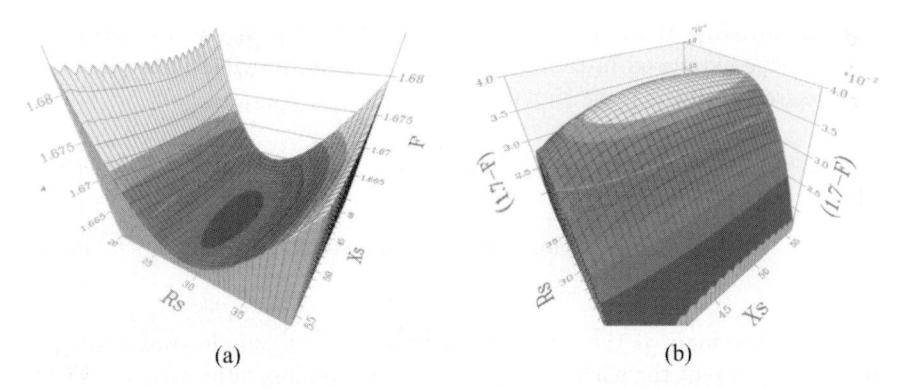

Fig. 5.11. a True shape of the noise surface; b inverted surface

If we take a particular transistor (for instance, the NE045 at f = 8 GHz) and choose several pairs of values for G_s and B_s, we obtain the surface plotted in Fig. 5.11a. Its appearance is quite different from the conventional representations, and it looks merely like a river bed, which somewhere has a shallow minimum. In order to locate the minimum, we invert the surface; this operation transforms the minimum into a maximum (Fig. 5.11b). This investigation has been repeated for several transistor types and at several frequencies; the shape of the surface remains the same, despite a wide range of values adopted for the noise parameters F_o, R_n, G_o, and B_o.

It is important to note the features of the actual surface:

- It is not laterally closed, like a paraboloid, but open.
- Its minimum is not at all so prominent as in the conventional representation, but rather shallow (note the 10^{-2} scale factor affecting the $(1.7 - F)$-axis in Fig. 5.11b). The gradients of the noise surface defined with respect to R_n, $R_s = 1/G_s$ and $X_s = 1/B_s$ have been computed at several points [75]. For instance, at point M where $R_s = 27\,\Omega$, $X_s = -50\,\Omega$ and $R_n = 4.5\,\Omega$ the gradients have been found to be

$$\left|\frac{\partial F}{\partial X_s}\right| = 2.11 \cdot 10^{-3} \quad \left|\frac{\partial F}{\partial R_s}\right| = 2.91 \cdot 10^{-3} \quad \left|\frac{\partial F}{\partial R_s}\right| = 22.23 \cdot 10^{-3}$$

These extremely low values explain why searching for the minimum with traditional optimization techniques leads to unacceptable errors.

■ **Noise Parameters Extraction.** Usually we extract the device noise parameters from a set of data obtained by repetitively measuring F for different values assigned to the source admittance. Applying (5.29) to each point, a nonlinear algebraic system of equations results, whose exact solution is presented in [76]. In practice, several extraction methods are employed, based on low or high redundancy (i.e., more than 4 measurement points collected) and data-smoothing techniques [77–82]. All methods have pros and cons in terms of simplicity, accuracy, and the requested number of data points. A good comparative evaluation of most methods is proposed in [83].

■ **Useful Expressions.** Equations (5.30b) can be formulated more generally:

$$F = F_o + \frac{N\,|Z_s - Z_o|^2}{\mathcal{R}e\{Z_s\}\,\mathcal{R}e\{Z_o\}} \tag{5.33a}$$

and

$$F = F_o + \frac{N\,|Y_s - Y_o|^2}{\mathcal{R}e\{Y_s\}\,\mathcal{R}e\{Y_o\}} \tag{5.33b}$$

Another possibility is to replace F by the equivalent input temperature in the fundamental equation; this yields

$$T_e = T_{e\,min} + T_o \frac{R_n}{G_s}\,|Y_s - Y_o|^2 \tag{5.33c}$$

■ **Microwave Representations.** For microwave applications, (5.33) must be related instead to reflection coefficients (rather than immittances), since the former are directly measurable quantities. If Γ_s denotes the reflection coefficient measured at the source end, then

$$F = F_o + 4\,\frac{R_n}{Z_R}\,\frac{|\Gamma_s - \Gamma_o|^2}{|1 + \Gamma_o|^2(1 - |\Gamma_s|^2)} \tag{5.34a}$$

$$T_e = T_{e\,min} + 4T_o\,\frac{R_n}{Z_R}\,\frac{|\Gamma_s - \Gamma_o|^2}{|1 + \Gamma_o|^2(1 - |\Gamma_s|^2)} \tag{5.34b}$$

where the optimal reflection coefficient Γ_o corresponds to the optimal source admittance Y_o. Recall that

$$\Gamma_o = \frac{Y_r - Y_o}{Y_r + Y_o} \quad \text{and} \quad N = \frac{|\Gamma_s - \Gamma_o|^2}{1 - |\Gamma_s|^2} \tag{5.34c}$$

The reference (or characteristic) impedance of the system is denoted by $Z_R = 1/Y_R$ (usually 50 Ω).

5.4 Correlation Matrices

5.4.1 Linear Two-Port Circuits

■ **Noise Matrix.** In the frequency domain, any fluctuation can be described by its average power in a bandwidth Δf centered on the frequency f. By definition, *the noise matrix N of two noise sources S_1 and S_2 is a two-dimensional array whose elements are their self- and cross-average powers.* The noise matrix is obtained by taking the ensemble average of the product of the source vector with its Hermitian conjugate. Recall that obtaining the Hermitian conjugate S^+ of any vector S comprises two steps:

1. transpose S
2. take its complex conjugate.

Therefore

$$N = \langle SS^+ \rangle = \left\langle \begin{bmatrix} S_1 \\ S_2 \end{bmatrix} [S_1^* \ S_2^*] \right\rangle = \begin{bmatrix} \langle S_1 S_1^* \rangle & \langle S_1 S_2^* \rangle \\ \langle S_2 S_1^* \rangle & \langle S_2 S_2^* \rangle \end{bmatrix} \quad (5.35)$$

Note that the ensemble average is used, since in the frequency domain the fluctuation can be decomposed into an infinite sum of sinusoidal components; averaging is carried out over the ensemble of components.

As stated in [67], the mean of the complex cross-product of two fluctuations v and i is related to their cross-power spectral density according to

$$\langle VI^* \rangle = 4\pi \ \Delta f \ S_\omega(iv) \quad (5.36)$$

Recalling (2.32), (5.36) becomes

$$\langle VI^* \rangle = 2\Delta f \ S_f(iv) \quad (5.37)$$

We conclude that for fluctuations S_i and S_j originating from two ergodic processes,

$$\boxed{\langle S_i S_j^* \rangle = \overline{S_i S_j^*} = 2\Delta f \ S_f(ji)} \quad (5.38)$$

■ **Correlation Matrix.** Consider two fluctuations denoted by S_1 and S_2; by means of expression (5.38) we can now express the correlation matrix in terms of the noise matrix:

$$C = \frac{1}{2\Delta f} N = \frac{1}{2\Delta f} \begin{bmatrix} \langle S_1 S_1^* \rangle & \langle S_1 S_2^* \rangle \\ \langle S_2 S_1^* \rangle & \langle S_2 S_2^* \rangle \end{bmatrix} \quad (5.39)$$

Here S_1 and S_2 can equally be the equivalent noise generators in the Y-representation of the noisy two-port.

Note that correlation matrices are obtained either through noise measurements or theoretical considerations.

5.4 Correlation Matrices

□ **Examples.** For linear passive two-ports containing only thermal noise sources, Hillbrand and Russer [68] stated that the correlation matrices in the impedance and admittance representations are respectively

$$\mathscr{C}_z = 2\text{kT} \, \mathscr{Re}\{Z\} \tag{5.40a}$$

$$\mathscr{C}_y = 2\text{kT} \, \mathscr{Re}\{Y\} \tag{5.40b}$$

where Z and Y are the open-circuit impedance matrix and short-circuit admittance matrix of the two-port. In the special case of a single series or parallel resistor, (5.40) reduce to Nyquist's formula.

■ **Chain Representation.** For the chain representation of a two-port (Fig. 5.8d), the noise matrix is

$$N = \begin{bmatrix} \langle e_n e_n^* \rangle & \langle e_n i_n^* \rangle \\ \langle i_n e_n^* \rangle & \langle i_n i_n^* \rangle \end{bmatrix} = 2 \, \Delta f \, \mathscr{C}_A \tag{5.41a}$$

where $\mathscr{C}_A$ is the correlation matrix. In situations where the correlation matrix cannot be deduced from theoretical considerations, measurement of noise performance remains the only possibility. Extracting the four noise parameters F_o, R_n, G_o, and B_o from an ensemble of measured data, matrix $\mathscr{C}_A$ can be written

$$\mathscr{C}_A = 2\text{kT} \, C_A \tag{5.41b}$$

where C_A is the *normalized chain correlation matrix*, given by

$$C_A = \begin{bmatrix} C_{A11} & C_{A12} \\ C_{A21} & C_{A22} \end{bmatrix} = \begin{bmatrix} R_n & \dfrac{F_o - 1}{2} - R_n Y_o^* \\ \dfrac{F_o - 1}{2} - R_n Y_o & R_n |Y_o|^2 \end{bmatrix} \tag{5.41c}$$

Conversely, if the normalized correlation matrix is known, the four noise parameters can be deduced from its elements [68]:

$$R_n = C_{A11} \tag{5.42a}$$

$$Y_o = G_o + jB_o = \sqrt{\dfrac{C_{A22}}{C_{A11}} - \left\{\mathscr{Im}\dfrac{C_{A12}}{C_{A11}}\right\}^2} + j \, \mathscr{Im}\dfrac{C_{A12}}{C_{A11}} \tag{5.42b}$$

$$F_o = 1 + 2\Big(\mathscr{Re}(C_{A12}) + C_{A11} \, G_o\Big) \tag{5.42c}$$

The noise factor expressed in terms of the normalized chain correlation matrix elements is

$$F = 1 + \mathscr{Re}\left(\dfrac{C_{A11}}{R_s} + C_{A12} + C_{A21} + C_{A22} \, R_s\right) \tag{5.42d}$$

5 Noise Analysis of Linear Circuits

■ **Admittance Representation.** This representation is important because it is the favorite tool when analysis is based upon a nodal approach. By definition, the noise matrix is

$$N = \begin{bmatrix} \langle i_{n1} i_{n1}^* \rangle & \langle i_{n1} i_{n2}^* \rangle \\ \langle i_{n2} i_{n1}^* \rangle & \langle i_{n2} i_{n2}^* \rangle \end{bmatrix} = 2\,\Delta f\,\mathscr{C}_I \qquad (5.43a)$$

where the notation of Fig. 5.8c is employed. The correlation matrix is

$$\mathscr{C}_I = 2kT\,C_I \qquad (5.43b)$$

with C_I being the *normalized admittance correlation matrix*. Attention should be paid to the fact that, for this representation, the matrix C_I depends not solely on the noise parameters, but equally on the Y-parameters of the two-port:

$$C_I = \begin{bmatrix} C_{I11} & C_{I12} \\ C_{I21} & C_{I22} \end{bmatrix}$$

$$= \begin{bmatrix} G_n + |y_{11} - Y_{cor}|^2 R_n & Y_{21}^*(y_{11} - Y_{cor})R_n \\ y_{21}(y_{11} - Y_{cor})^* R_n & |y_{21}|^2 R_n \end{bmatrix} \qquad (5.43c)$$

Assuming that the noise and correlation matrices have been previously evaluated, the standard noise parameters of the two-port are given by [71]

$$R_n = \frac{1}{|y_{21}|^2} \frac{\langle i_{n2} i_{n2}^* \rangle}{4kT_o\,\Delta f} = \frac{1}{|y_{21}|^2} C_{I22} \qquad (5.44a)$$

$$Y_{cor} = G_{cor} + jB_{cor} = y_{11} - y_{21} \frac{\langle I_1 I_2^* \rangle}{\langle I_2 I_2^* \rangle} = y_{11} - y_{21} \frac{C_{I12}}{C_{I22}} \qquad (5.44b)$$

$$G_n = \frac{\langle I_1 I_1^* \rangle}{4kT_o\,\Delta f} - \frac{\langle I_1 I_2^* \rangle}{\langle I_2 I_2^* \rangle} \frac{\langle I_1^* I_2 \rangle}{4kT_o\,\Delta f} = C_{I11} - \frac{C_{I12}}{C_{I22}} C_{I21} \qquad (5.44c)$$

Equations (5.44), combined with the relations presented in Table 5.3, make it possible to deduce the standard four noise parameters F_o, R_n, G_o, and B_o. The relationship between the noise factor and the source admittance is

$$4kT_o\,\Delta f(F-1) = \langle i_{n1} i_{n1}^* \rangle + \left| \frac{y_{11} + Y_s}{y_{21}} \right|^2 \langle i_{n2} i_{n2}^* \rangle$$

$$-2\mathfrak{Re}\left(\frac{y_{11} + Y_s}{y_{21}} \langle i_{n1}^* i_{n2} \rangle \right) \qquad (5.44d)$$

■ **Noise Wave Representation.** At microwave frequencies, the wave representation of noise is preferred [72, 73], because it is compatible with the

5.4 Correlation Matrices

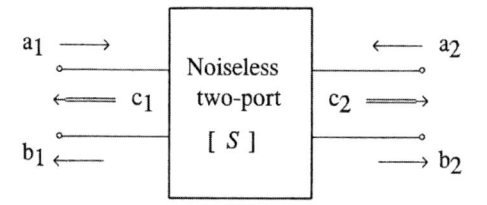

Fig. 5.12. S-parameter and noise-waves representation of a two-port

scattering-parameter description of microwave circuits. The noise produced by a two-port is represented using waves c_1 and c_2 that emanate from its ports (Fig. 5.12). The signal waves to the ports are denoted by
a_1 – input incident wave b_1 – input reflected wave
a_2 – output incident wave b_2 – output reflected wave
The S-matrix description is

$$\begin{bmatrix} b_1 \\ b_2 \end{bmatrix} = \begin{bmatrix} s_{11} & s_{12} \\ s_{21} & s_{22} \end{bmatrix} \begin{bmatrix} a_1 \\ a_2 \end{bmatrix} + \begin{bmatrix} c_1 \\ c_2 \end{bmatrix} \quad (5.45a)$$

The noise waves are time-varying complex random quantities characterized by the (non-normalized) correlation matrix

$$C_S = \left\langle \begin{bmatrix} c_1 \\ c_2 \end{bmatrix} \begin{bmatrix} c_1 \\ c_2 \end{bmatrix}^+ \right\rangle = \begin{bmatrix} \langle c_1 c_1^* \rangle & \langle c_1 c_2^* \rangle \\ \langle c_2 c_1^* \rangle & \langle c_2 c_2^* \rangle \end{bmatrix} = \begin{bmatrix} C_{S11} & C_{S12} \\ C_{S21} & C_{S22} \end{bmatrix} \quad (5.45b)$$

In terms of $T_{e\,min}$, Γ_o and R_n, the noise wave correlation matrix elements are [74]

$$C_{S11} = kT_{e\,min}(|s_{11}|^2 - 1) + \frac{4kT_o R_n}{Z_R} \frac{|1 - s_{11}\Gamma_o|^2}{|1 + \Gamma_o|^2} \quad (5.45c)$$

$$C_{S22} = |s_{21}|^2 \left(kT_{e\,min} + \frac{4kT_o R_n}{Z_R} \frac{|\Gamma_o|^2}{|1 + \Gamma_o|^2} \right) \quad (5.45d)$$

$$C_{S12} = \frac{4kT_o R_n}{Z_R} \frac{(-s_{21}^* \Gamma_o^*)}{|1 + \Gamma_o|^2} + \frac{s_{11}}{s_{21}} C_{S22} \quad (5.45e)$$

Z_R being the characteristic (or reference) impedance (usually, 50 Ω). The inverse relations (the four noise parameters derived from the correlation matrix) are

$$R_n = \frac{Z_R}{4kT_o} \left[C_{S11} - 2\Re e \left(C_{S12} \frac{1 + s_{11}}{s_{21}} \right) + C_{S22} \left| \frac{1 + s_{11}}{s_{21}} \right|^2 \right] \quad (5.45f)$$

$$kT_{e\,min} = \frac{C_{S22} - \zeta |\Gamma_o|^2}{|s_{21}|^2 |1 + \Gamma_o|^2} \quad (5.45g)$$

138   5 Noise Analysis of Linear Circuits

$$\Gamma_o = \frac{\eta}{2}\left(1 - \sqrt{1 - \frac{4}{|\eta|^2}}\right) \tag{5.45h}$$

$$\eta = \frac{C_{S22} + \zeta}{C_{S22}s_{11} - C_{S12}s_{21}} \tag{5.45i}$$

$$\zeta = C_{S11}|s_{21}|^2 - 2\Re e\left(C_{S12}s_{21}s_{12}^*\right) + C_{S22}|s_{11}|^* \tag{5.45j}$$

The noise factor, expressed in terms of the noise wave matrix elements, is [74]

$$F = 1 + \frac{C_{S11}\left|\dfrac{s_{21}\Gamma_G}{1 - s_{11}\Gamma_G}\right|^2 + C_{S22} + 2\Re e\left(C_{S12}\left|\dfrac{s_{21}\Gamma_G}{1 - s_{11}\Gamma_G}\right|\right)}{(1 - |\Gamma_G|^2)\left|\dfrac{s_{21}}{1 - s_{11}\Gamma_G}\right|^2} \tag{5.45k}$$

where Γ_G is the reflection coefficient at the generator end.

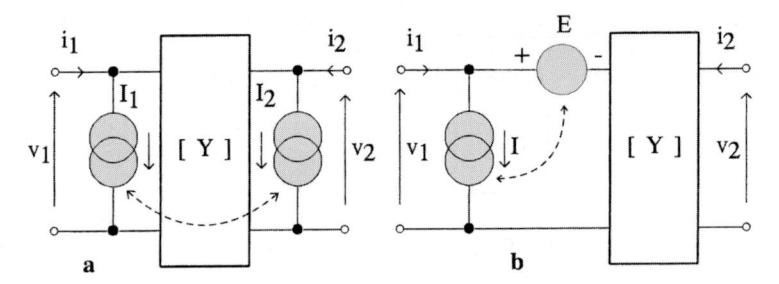

Fig. 5.13. Possible representations of a noisy two-port (*dotted arrows* suggest correlation)

■ **Application.** Consider a noisy two-port in terms of its admittance representation (Fig. 5.13a).

For computational convenience, it is preferred to adopt instead the chain representation of Fig. 5.13b, while maintaining the electrical characterization of the two-port with its Y-parameters. We wish to find the relationship between the elements of the noise matrix of the chain representation (Fig. 5.13b) and the elements of the noise matrix of the admittance representation (Fig. 5.13a).

□ **Solution**

Taking into account the definition (5.41a), the objective is to derive expressions for

$$\langle EE^*\rangle, \ \langle II^*\rangle \quad \text{and} \quad \langle EI^*\rangle = \langle IE^*\rangle$$

5.4 Correlation Matrices

in terms of the (known) noise powers

$$\langle I_1 I_1^* \rangle, \langle I_2 I_2^* \rangle \quad \text{and} \quad \langle I_1 I_2^* \rangle = \langle I_2 I_1^* \rangle$$

This problem is a typical case in which the noise generators must be oriented.

The starting point is the classical system describing the representation of Fig. 5.13a (which we call representation **a**):

$$\begin{cases} i_1 = y_{11} v_1 + y_{12} v_2 + I_1 \\ i_2 = y_{21} v_1 + y_{22} v_2 + I_2 \end{cases} \quad (5.46a)$$

As representation **b** (Fig. 5.13b) must be equivalent to representation **a**, its mathematical model must be derived from (5.46a), under appropriate conditions. These conditions can be found by inspection:

1) Since representation **b** has no output noise generator, the corresponding condition is to let $I_2 = 0$.
2) Comparing the two circuits, we must substitute $(v_1 + E)$ for v_1 to derive representation **b** from **a**.

Implementing these conditions in (5.46a), we have

$$\begin{cases} i_1 = y_{11}(v_1 + E) + y_{12} v_2 + I \\ i_2 = y_{21}(v_1 + E) + y_{22} v_2 \end{cases} \quad (5.46b)$$

Identifying each equation of (5.46a) with the corresponding one in (5.46b), we find after some algebra

$$\begin{cases} E = I_2 / y_{21} \\ I = I_1 - I_2 \, y_{11} / y_{21} \end{cases} \quad \text{so} \quad \begin{cases} E^* = I_2^* / y_{21}^* \\ I^* = I_1^* - I_2^* \, y_{11}^* / y_{21}^* \end{cases} \quad (5.46c)$$

We then calculate the required noise powers, recalling that

- The product of a complex quantity and its conjugate is its squared absolute magnitude.
- The conjugate of a sum is the sum of the conjugates.

Therefore:

$$\langle E E^* \rangle = \langle I_2 I_2^* \rangle \frac{1}{y_{21} y_{21}^*} = \langle I_2 I_2^* \rangle \frac{1}{|y_{21}|^2} \quad (5.46d)$$

$$\langle I E^* \rangle = \langle I_2^* (1/y_{21}^*) (I_1 - I_2 y_{11}/y_{21}) \rangle$$
$$= \langle I_1 I_2^* \rangle (1/y_{21}^*) - \langle I_2 I_2^* \rangle y_{11}/|y_{21}|^2 \quad (5.46e)$$

$$I I^* = I_1 I_1^* - I_1^* I_2 (y_{11}/y_{21}) - I_1 I_2^* (y_{11}^*/y_{21}^*) + I_2 I_2^* |y_{11}/y_{21}|^2 \quad (5.46f)$$

As the sum of a complex quantity and its conjugate is equal to twice its real part, expression (5.46f) can be written in the equivalent form

$$\langle II^* \rangle = \langle I_1 I_1^* \rangle - 2\mathfrak{Re}\left(\langle I_1^* I_2 \rangle (y_{11}/y_{21})\right) + \langle I_2 I_2^* \rangle |y_{11}/y_{21}|^2 \quad (5.46g)$$

Thus, (5.46d), (5.46e), and (5.46g) represent the solution.

5.4.2 Linear Multiport Circuits

■ **Multiport Representation.** At any frequency, a linear active multiport is completely specified with respect to its ports by its Z-matrix representation (Fig. 5.14) and the open-circuit port voltages E_k. Note that the equivalent generators E_k are not independent, since they lump the effects of the same internal sources. Finally, a complete description requires the open-circuit impedance matrix Z and the vector

$$E = [E_1 \ E_2 \ \ldots \ E_p]^t$$

As for two-ports, we may equally well consider the Y-matrix representation with a set of short-circuit terminal current generators.

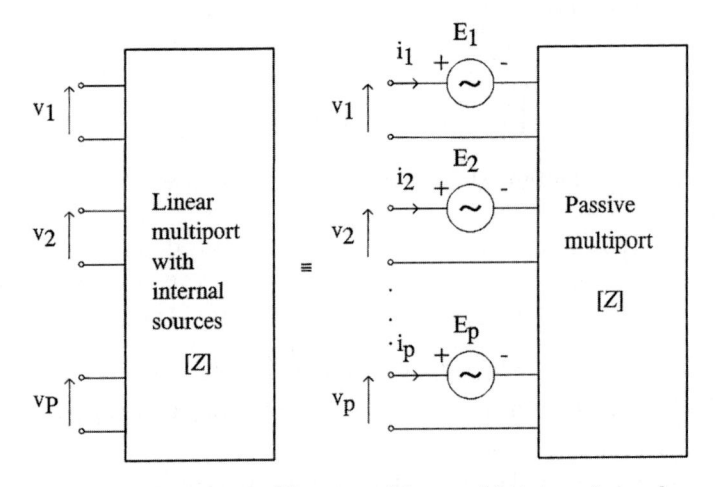

Fig. 5.14. Representation of linear multiport with internal signal sources

■ **Representation of Noisy Multiports.** The only difference with respect to the previous case is that now the internal sources are no longer harmonic, but random.

With respect to the terminal pairs, the noisy multiport is equivalent with an identical, but noiseless, multiport (described by means of its Z-matrix), which has equivalent noise generators in series at its ports, all partially correlated (Fig. 5.15).

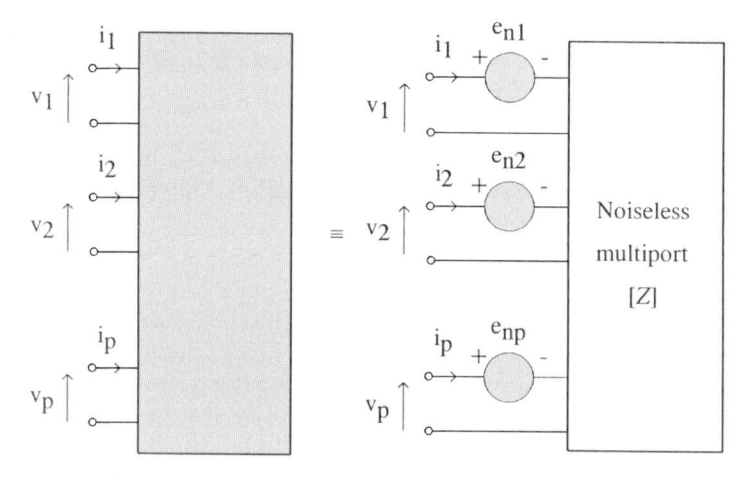

Fig. 5.15. Representation of a noisy linear multiport

The equivalent noise generators at the terminal pairs are specified by their self- and cross-average powers $\langle e_{ni}e_{nk}^*\rangle$, which are the constituent elements of the noise matrix:

$$\langle EE^+\rangle = \begin{bmatrix} \langle e_{n1}e_{n1}^*\rangle & \langle e_{n1}e_{n2}^*\rangle & \cdots & \langle e_{n1}e_{np}^*\rangle \\ \langle e_{n2}e_{n1}^*\rangle & \langle e_{n2}e_{n2}^*\rangle & \cdots & \langle e_{n2}e_{np}^*\rangle \\ \cdots & \cdots & \cdots & \cdots \\ \langle e_{np}e_{n1}^*\rangle & \langle e_{np}e_{n2}^*\rangle & \cdots & \langle e_{np}e_{np}^*\rangle \end{bmatrix} \tag{5.47}$$

(E^+ being the Hermitian conjugate of E, obtained by taking the complex conjugate of E and transposing it).

Recalling that only *positive frequencies* are considered in noise theory, Haus and Adler [69] proved that for a passive dissipative multiport in thermal equilibrium

$$\boxed{\langle EE^+\rangle = 2kT\,\Delta f\,(Z+Z^+)} \tag{5.48}$$

Twiss [70] proved that in the dual case (Y-matrix representation),

$$\boxed{\langle II^+\rangle = 2kT\,\Delta f\,(Y+Y^+)} \tag{5.49}$$

Note that (5.48) and (5.49) are consistent with (5.40a) and (5.40b), respectively.

■ **Characteristic-Noise Matrix.** This noise parameter, denoted by N_z, was first introduced by Haus and Adler [69]. If a Z-matrix representation of the multiport is adopted, the characteristic-noise matrix is given by

$$N_Z = -\frac{1}{2}\,(Z+Z^+)^{-1}\langle EE^+\rangle \tag{5.50}$$

Haus and Adler also proved that for a passive dissipative multiport at equilibrium temperature T, this matrix is diagonal, with all eigenvalues equal to $-kT\,\Delta f$.

■ **Note.** The minus sign of the eigenvalues is explained by the fact that in Fig. 5.15 the currents are oriented in the direction opposite the generators; the same notation applies to noise generators.

5.5 Conclusion

Frequency domain analysis of noisy circuits is a more powerful tool than time domain analysis, since in the latter only two-ports can be considered. Only frequency domain analysis is easily extensible to multiports.

From a different standpoint, time domain analysis is suitable for hand calculation (especially due to the faculties of the Rothe and Dahlke model), while frequency domain noise analysis is almost always performed with dedicated software.

Summary

- The noise of any two-port circuit can be represented by means of two equivalent noise generators, which are partially correlated, situated at its terminal pairs.
- The model of Rothe and Dahlke is suited to time domain analysis. Its chief advantage is that it transforms the pair of correlated generators into an uncorrelated pair and a correlation immittance.
- A complete noise characterization of any linear two-port requires a set of four noise parameters.
- Frequency domain analysis is performed by means of the correlation and noise matrices.
- The constituent elements of the correlation matrix are the self- and cross-power noise spectral densities of the equivalent generators.
- The constituent elements of the noise matrix are the self- and cross-average noise powers of the equivalent generators.

6

Frequency Domain Noise Analysis of Linear Multiports

> "Multiplication is vexation,
> Division is as bad;
> The Rule of three doth puzzle me,
> And Practice drives me mad"
>
> (Elizabethan MS, dated 1570)

6.1 Introduction

■ **Purpose.** In this chapter we present two approaches to computing the noise performance of a given circuit, regarded as a two-port with respect to its input and output terminals. The first is suitable for a large class of passive circuits, while the second is devoted to both passive and active circuits of arbitrary topology. Both will be discussed.

■ **Computer-Aided Noise Analysis.** Although both approaches can be implemented in a computer program, due to its generality attention will be paid to the latter. The following comments refer exclusively to it.

□ **Constraints.** One of the most important point when building a computer program is to specify how the circuit submitted for analysis is to be described. This information is paramount, because not only are all the following steps in software design structured upon it, but the choice of computing algorithm often depends equally on the adopted description.

Bearing in mind that SPICE is widely employed, and that it is likely that all potential users of noise analysis programs are already familiar with it, it seems desirable to observe the following general requirements:

– The program must accept circuits with any internal topology.
– A branch-node description of the circuit is imposed.

– The noise of each active device (transistor) is described by means of its four noise parameters {F_o, R_n, G_o, and B_o} or any other equivalent set. The electrical performance of the active device must instead be specified by means of easily measured quantities (such as the conventional Y- or S-parameters), rather than by equivalent circuits.

□ **Hypothesis.** It is assumed that the circuit operates in the linear region and all its passive devices generate only thermal noise.

□ **Results.** The expected results of noise analysis are the noise parameters of the overall circuit regarded as a two-port. It is desirable to have an equivalent description of the overall two-port in terms of its correlation matrix.

6.2 Method of Hillbrand and Russer

■ **History.** This method, published in 1976, is devoted to passive circuits [84]. The starting point is the prior work of Haus and Adler [85] who stated that any noisy two-port is equivalent to an identical (but noiseless) two-port with two additional noise generators at its ports.

This representation is described in the frequency domain by the correlation matrix of the noise generators rather than by voltages and currents.

6.2.1 Description

■ **Procedure.** The method of Hillbrand and Russer applies to circuits composed of elementary two-port configurations with known electrical and noise performances.

The analysis is based on the correlation matrix concept; the two-port to be analyzed is decomposed into basic (elementary) two-ports, whose electrical performance and noise correlation matrices are determined first. Analysis then proceeds by interconnecting these two-ports until the correlation matrix of the original circuit is finally obtained.

Depending on how the elementary two-ports are interconnected (in series, in parallel, or cascaded), the Z-matrix, Y-matrix, or chain matrix description is selected. As an example, consider two elementary two-ports, denoted by subscripts 1 and 2; the global correlation matrix is related to the correlation matrices of the original two-ports by

$$\mathcal{C}_Z = \mathcal{C}_{Z1} + \mathcal{C}_{Z2} \quad \text{(series connection)} \tag{6.1}$$

$$\mathcal{C}_Y = \mathcal{C}_{Y1} + \mathcal{C}_{Y2} \quad \text{(parallel connection)} \tag{6.2}$$

$$\mathcal{C}_A = A_1 \, \mathcal{C}_{A2} A_1^+ + \mathcal{C}_{A1} \quad \text{(cascade connection)} \tag{6.3}$$

A is the chain matrix of the two-port and A^+ its Hermitian conjugate.

6.2 Method of Hillbrand and Russer

Depending on the representation, the correlation matrix of each elementary two-port is evaluated with one of the following expressions:

$$\mathscr{C}_Z = 2kT\,\Re e(Z) \tag{6.4}$$

$$\mathscr{C}_Y = 2kT\,\Re e(Y) \tag{6.5}$$

$$\mathscr{C}_A = 2kT \begin{bmatrix} R_n & \dfrac{F_o - 1}{2} - R_n Y_o^* \\ \dfrac{F_o - 1}{2} - R_n Y_o & R_n |Y_o|^2 \end{bmatrix} \tag{6.6}$$

In the course of the analysis, it may happen that two or more different representations exist; for a consistent calculation, some must be transformed into the preferred form. The new correlation matrix (denoted by $\mathscr{C}'$) can be found in terms of the original correlation matrix (denoted by $\mathscr{C}$) with the formula

$$\mathscr{C}' = \mathscr{T}\,\mathscr{C}\,\mathscr{T}^+ \tag{6.7}$$

where $\mathscr{T}$ is called the *transformation matrix*, selected according to Table 6.1.

Table 6.1. Transformation matrices

$\mathscr{T}$ to	from admittance	from impedance	from chain
admittance	$\begin{bmatrix} 1 & 0 \\ 0 & 1 \end{bmatrix}$	$\begin{bmatrix} y_{11} & y_{12} \\ y_{21} & y_{22} \end{bmatrix}$	$\begin{bmatrix} -y_{11} & 1 \\ -y_{21} & 0 \end{bmatrix}$
impedance	$\begin{bmatrix} z_{11} & z_{12} \\ z_{21} & z_{22} \end{bmatrix}$	$\begin{bmatrix} 1 & 0 \\ 0 & 1 \end{bmatrix}$	$\begin{bmatrix} 1 & -z_{12} \\ 0 & -z_{22} \end{bmatrix}$
chain	$\begin{bmatrix} 0 & a_{12} \\ 1 & a_{22} \end{bmatrix}$	$\begin{bmatrix} 1 & -a_{11} \\ 0 & -a_{21} \end{bmatrix}$	$\begin{bmatrix} 1 & 0 \\ 0 & 1 \end{bmatrix}$

Note that the transformation matrices depend only upon the electrical configuration, not the noise of the two-port. A simple technique for deducing them is suggested in [84].

Therefore, by decomposing the circuit submitted to noise analysis into a cascade of two-ports, the resulting chain correlation matrix can be determined. Finally, the noise parameters of the overall two-port can be computed with the resulting $\mathscr{C}_A$ matrix by means of expressions (5.42).

■ **Limitation.** The method of Hillbrand and Russer can be applied only to a certain class of circuits, namely those that can be decomposed into interconnected elementary two-ports. For medium- or large-scale circuits this method becomes impractical.

146   6 Frequency Domain Noise Analysis of Linear Multiports

Note also that it is implicitly assumed that there is no correlation between the noise sources of various two-ports. As long as the basic two-ports correspond to different physical devices, this is always true. But when they correspond to elements of the same device model, problems can arise, since the noise of two (or more) elements might have a common physical origin.

6.2.2 Noise Parameters of an Attenuating Pad

■ **Configuration.** The electrical circuit of the attenuator is presented in Fig. 6.1.

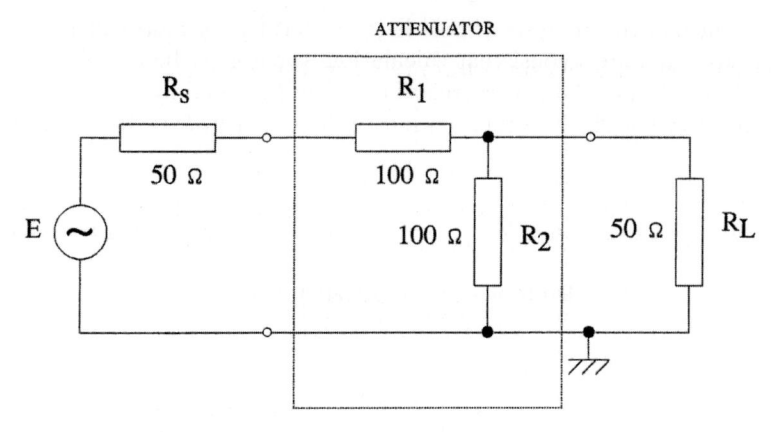

Fig. 6.1. Attenuator loaded by 50-Ω resistors

■ **Computation.** In order to evaluate the noise parameters, we must decompose the circuit into two elementary two-ports connected in cascade, as depicted in Fig. 6.2.

Since the two basic two-ports are connected in cascade, the objective is to calculate the global chain correlation matrix with (6.3). Consequently, the chain matrices of both two-ports are required, as well as their individual correlation matrices. The calculation is carried out in four steps.

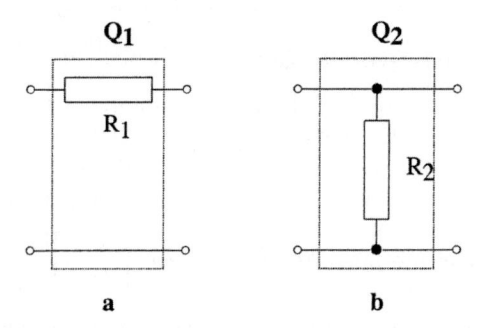

Fig. 6.2. Basic two ports

6.2 Method of Hillbrand and Russer

☐ **Step 1: Chain Description of Q1.** For the reader's convenience, two-port Q1 (Fig. 6.2) is electrically characterized by its Y-matrix

$$Y_1 = \begin{bmatrix} 1/R_1 & -1/R_1 \\ -1/R_1 & 1/R_1 \end{bmatrix} \tag{6.8}$$

The correlation matrix is deduced with the aid of expression (6.5)

$$\mathscr{C}_{Y1} = 2kT \, \mathfrak{Re}\{Y_1\} = 2kT \begin{bmatrix} 1/R_1 & -1/R_1 \\ -1/R_1 & 1/R_1 \end{bmatrix} \tag{6.9a}$$

The chain matrix is obtained from (6.8), by using Table 5.1

$$A_1 = \begin{bmatrix} 1 & R_1 \\ 0 & 1 \end{bmatrix} \tag{6.9b}$$

The chain correlation matrix of Q1 is determined with (6.7), by properly selecting the transformation matrix from Table 6.1, namely

$$T_1 = \begin{bmatrix} 0 & a_{12} \\ 1 & a_{22} \end{bmatrix} \tag{6.9c}$$

Hence

$$\mathscr{C}_{A1} = 2kT \begin{bmatrix} 0 & R_1 \\ 1 & 1 \end{bmatrix} \begin{bmatrix} 1/R_1 & -1/R_1 \\ -1/R_1 & 1/R_1 \end{bmatrix} \begin{bmatrix} 0 & 1 \\ R_1 & 1 \end{bmatrix}$$

$$= 2kT \begin{bmatrix} R_1 & 0 \\ 0 & 0 \end{bmatrix} \tag{6.10}$$

☐ **Step 2: Chain Description of Q2.** For Q2 (Fig. 6.2b), it is easier to use the Z-matrix:

$$Z_2 = \begin{bmatrix} R_2 & R_2 \\ R_2 & R_2 \end{bmatrix} \tag{6.11a}$$

from which we deduce (with Table 5.1)

$$A_2 = \begin{bmatrix} 1 & 0 \\ 1/R_2 & 1 \end{bmatrix} \tag{6.11b}$$

and also

$$\mathscr{C}_{Z2} = 2kT \, \mathfrak{Re}\{Z_2\} = 2kT \begin{bmatrix} R_2 & R_2 \\ R_2 & R_2 \end{bmatrix} \tag{6.11c}$$

In order to find the chain correlation matrix of Q2, we select from Table 6.1 the appropriate transformation matrix, which is

148   6 Frequency Domain Noise Analysis of Linear Multiports

$$T_2 = \begin{bmatrix} 1 & -a_{11} \\ 0 & -a_{21} \end{bmatrix} = \begin{bmatrix} 1 & -1 \\ 0 & -1/R_2 \end{bmatrix} \tag{6.11d}$$

Finally, applying (6.7),

$$\mathcal{C}_{A2} = 2kT \begin{bmatrix} 1 & -1 \\ 0 & -1/R_2 \end{bmatrix} \begin{bmatrix} R_2 & R_2 \\ R_2 & R_2 \end{bmatrix} \begin{bmatrix} 1 & 0 \\ -1 & -1/R_2 \end{bmatrix} \tag{6.12}$$

$$= 2kT \begin{bmatrix} 0 & 0 \\ 0 & 1/R_2 \end{bmatrix}$$

☐ **Step 3: Overall Chain Matrix.** The noise of the cascade Q1 and Q2 is described by (6.3); substituting expressions (6.9b), (6.12), and (6.10) into (6.3), one obtains

$$\mathcal{C}_{A2} = 2kT \left\{ \begin{bmatrix} 1 & R_1 \\ 0 & 1 \end{bmatrix} \begin{bmatrix} 0 & 0 \\ 0 & 1/R_2 \end{bmatrix} \begin{bmatrix} 1 & 0 \\ R_1 & 1 \end{bmatrix} + \begin{bmatrix} R_1 & 0 \\ 0 & 0 \end{bmatrix} \right\}$$

$$= 2kT \begin{bmatrix} (R_1^2/R_2) + R_1 & R_1/R_2 \\ R_1/R_2 & 1/R_2 \end{bmatrix}$$

The normalized correlation matrix of the global two-port is

$$C_A = \begin{bmatrix} (R_1^2/R_2) + R_1 & R_1/R_2 \\ R_1/R_2 & 1/R_2 \end{bmatrix} \tag{6.13}$$

☐ **Step 4: Calculation of Noise Parameters.** With expressions (5.42), we find

$$R_n = (R_1^2/R_2) + R_1 \tag{6.14a}$$

$$Y_o = \sqrt{1/(R_1^2 + R_1 R_2)} \tag{6.14b}$$

$$F_o = 1 + 2 \left(\frac{R_1}{R_2} + \left(\frac{R_1^2}{R_2} + R_1 \right) \sqrt{\frac{1}{R_1^2 + R_1 R_2}} \right) \tag{6.14c}$$

and the noise factor is

$$F = 1 + \left(\frac{1}{R_s} \left(\frac{R_1^2}{R_2} + R_1 \right) + \frac{R_1}{R_2} + \frac{R_1}{R_2} + \frac{R_s}{R_2} \right) \tag{6.14d}$$

In the special case: $R_1 = R_2 = R$, one obtains

$$R_n = 2R \tag{6.15a}$$

$$Y_o = 1/R\sqrt{2} \tag{6.15b}$$

$$F_o = 1 + 2(1 + \sqrt{2}) \tag{6.15c}$$

$$F = 1 + \left(\frac{2R}{R_s} + \frac{2R + R_s}{R}\right) \quad (6.16)$$

Considering the proposed numerical values (given in Fig. 6.1), the following noise parameters are obtained:
$R_n = 200\,\Omega$
$Y_o = (7.071 + j\,0)\,\text{mS}$
$F_o = 5.8284$ (or $7.6555\,\text{dB}$)
$F = 7.5$ (or $8.7506\,\text{dB}$)

6.2.3 Collection of Elementary Passive Circuits

In this section we present the results obtained by applying the method of Hillbrand and Russer to several elementary configurations of passive two-ports.

■ **Series Impedance**

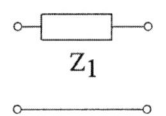

Z_1

$$C_A = \begin{bmatrix} \mathcal{R}e(Z_1) & 0 \\ 0 & 0 \end{bmatrix}$$

R_n	$\mathcal{R}e(Z_1)$
Y_o	0
N	0
F_o	1

■ **Parallel Impedance**

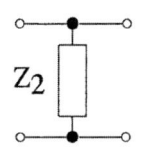

Z_2

$$C_A = \begin{bmatrix} 0 & 0 \\ 0 & \dfrac{\mathcal{R}e(Z_2)}{Z_2 Z_2^*} \end{bmatrix}$$

R_n	0
Y_o	undefined
N	0
F_o	1

■ **Voltage-Controlled Current Source**

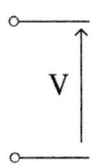

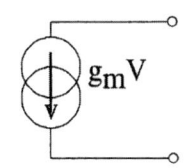

V, $g_m V$

R_n	0
Y_o	0
N	0
F_o	1

■ **Gamma Section**

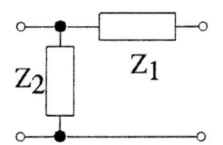

Z_2, Z_1

$$C_A = \begin{bmatrix} \mathcal{R}e(Z_1) & \mathcal{R}e(Z_1)/Z_2^* \\ \mathcal{R}e(Z_1)/Z_2 & \dfrac{1}{Z_2 Z_2^*}\bigl(\mathcal{R}e(Z_1) + \mathcal{R}e(Z_2)\bigr) \end{bmatrix}$$

■ Reverse Gamma Section

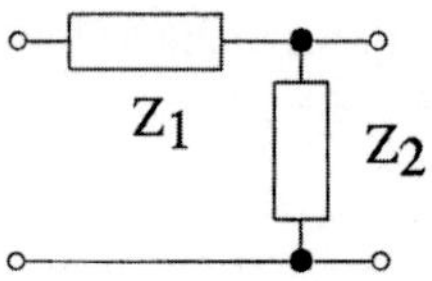

$$C_A = \begin{bmatrix} \dfrac{|Z_1|^2}{|Z_2|^2}\mathfrak{Re}(Z_2) + \mathfrak{Re}(Z_1) & \dfrac{Z_1\,\mathfrak{Re}(Z_2)}{Z_2 Z_2^*} \\[2ex] \dfrac{Z_1^*\,\mathfrak{Re}(Z_2)}{Z_2 Z_2^*} & \dfrac{\mathfrak{Re}(Z_2)}{Z_2 Z_2^*} \end{bmatrix}$$

■ Unbalanced T Section

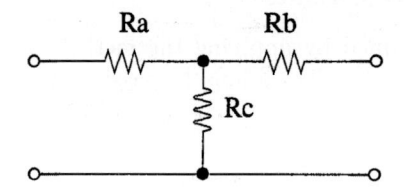

R_n	$\dfrac{Ra}{Rc}\left(2Rb + \dfrac{Ra\,Rb}{Rc} + Ra\right) + Rb + Ra$
Y_o	$\sqrt{\dfrac{\dfrac{Rb}{Rc^2} + \dfrac{1}{Rc}}{\dfrac{Ra}{Rc}\left(2Rb + Rb\dfrac{Ra}{Rc} + Ra\right) + Rb + Ra}}$
N	$\sqrt{\left(\dfrac{Rb}{Rc^2} + \dfrac{1}{Rc}\right)\left(\dfrac{Ra}{Rc}\left(2Rb + Rb\dfrac{Ra}{Rc} + Ra\right) + Rb + Ra\right)}$
F_o	$1 + 2\left(\dfrac{Rb}{Rc} + \dfrac{RaRb}{Rc^2} + \dfrac{Ra}{Rc} + \sqrt{\left(\dfrac{Rb}{Rc^2} + \dfrac{1}{Rc}\right)\left(\dfrac{Ra}{Rc}\left(2Rb + \dfrac{RaRb}{Rc} + Ra\right) + Rb + Ra\right)}\right)$

■ Unbalanced Π Section

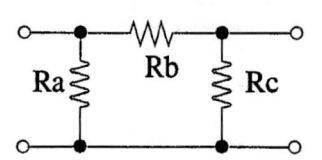

R_n	$\dfrac{Rb^2}{Rc} + Rb$
Y_o	$\sqrt{\dfrac{2 + \dfrac{Rc+Rb}{Ra} + \dfrac{Rc+Ra}{Rb}}{Ra(Rb+Rc)}}$
N	$\sqrt{\left(\dfrac{Rb^2}{RcRa^2} + \dfrac{2Rb}{RaRc} + \dfrac{1}{Rc} + \dfrac{Rb}{Ra^2} + \dfrac{1}{Ra}\right)\left(Rb + \dfrac{Rb^2}{Rc}\right)}$
F_o	$1 + 2\left(\dfrac{Rb}{Ra} + \dfrac{Rb^2}{RaRc} + \dfrac{Rb}{Rc} + \sqrt{\left(\dfrac{Rb^2}{RcRa^2} + \dfrac{2Rb}{RaRc} + \dfrac{1}{Rc} + \dfrac{Rb}{Ra^2} + \dfrac{1}{Ra}\right)\left(Rb + \dfrac{Rb^2}{Rc}\right)}\right)$

■ Balanced X Section

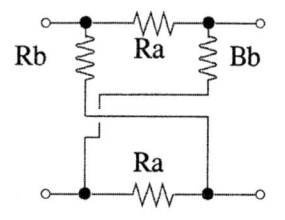

R_n	$\dfrac{(Ra+Rb)}{2}\left(\dfrac{(Ra+Rb)^2}{(Ra-Rb)^2} - 1\right)$
Y_o	$2\sqrt{\dfrac{1}{(Ra+Rb)^2 - (Ra-Rb)^2}}$
N	$(Ra+Rb)\left(\dfrac{(Ra+Rb)^2}{(Ra-Rb)^2} - 1\right)\sqrt{\dfrac{1}{(Ra+Rb)^2 - (Ra-Rb)^2}}$
F_o	$1 + 2\left(\dfrac{(Ra+Rb)^2}{(Ra-Rb)^2} - 1\right)\left(1 + \sqrt{\dfrac{(Ra+Rb)^2}{(Ra+Rb)^2 - (Ra-Rb)^2}}\right)$

152   6 Frequency Domain Noise Analysis of Linear Multiports

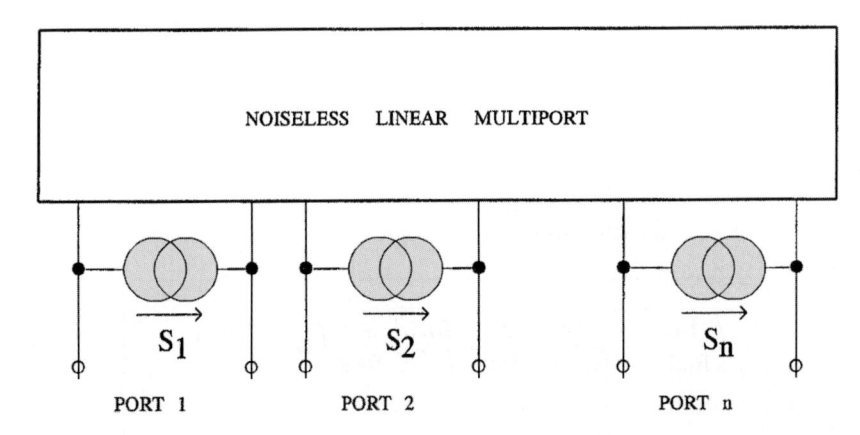

Fig. 6.3. Noisy multiport, Norton representation

6.3 Noise Analysis of Linear Multiport Circuits

■ **Model.** Generally, the noise sources of a linear multiport are associated with its resistors (which generate thermal noise) or active devices (transistors), producing thermal, flicker, diffusion, shot, or GR noise. As previously suggested, it is more appropriate to lump the effects of all individual noise sources into equivalent current noise generators connected across the terminal pairs of an identical, but noiseless, multiport. The reader may well wonder why current noise generators are preferred over voltage noise generators. As a matter of fact, noise analysis cannot be dissociated from electrical simulation of the circuit, and almost all available simulators are structured on a nodal or modified nodal approach. These methods employ admittance matrices to describe the circuit, and the current generators are consistent with this environment.

For this reason, the equivalent circuit of Fig. 6.3 (which is simply the Norton equivalent for one-ports, extended to multiports) is preferred [86–88].

■ **Problem Formulation.** Consider a linear multiport with n external ports that incorporates m active devices. Each active device is viewed as a two-port, electrically described by its Y-admittance matrix; its noise is characterized by the four spot noise parameters given at each operating frequency.

The objective is to find the correlation matrix of the equivalent noise generators at the terminal pairs, denoted by S_i (i=1, 2,..., n).

■ **Approach.** The starting point is the de-embedding of all active devices from the multiport; in this way additional ports are created, two for each active device (Fig. 6.4). Since the remaining network is purely passive and lossy, it generates only thermal noise. This noise results from statistically independent random processes with respect to those of the active devices; as

6.3 Noise Analysis of Linear Multiport Circuits

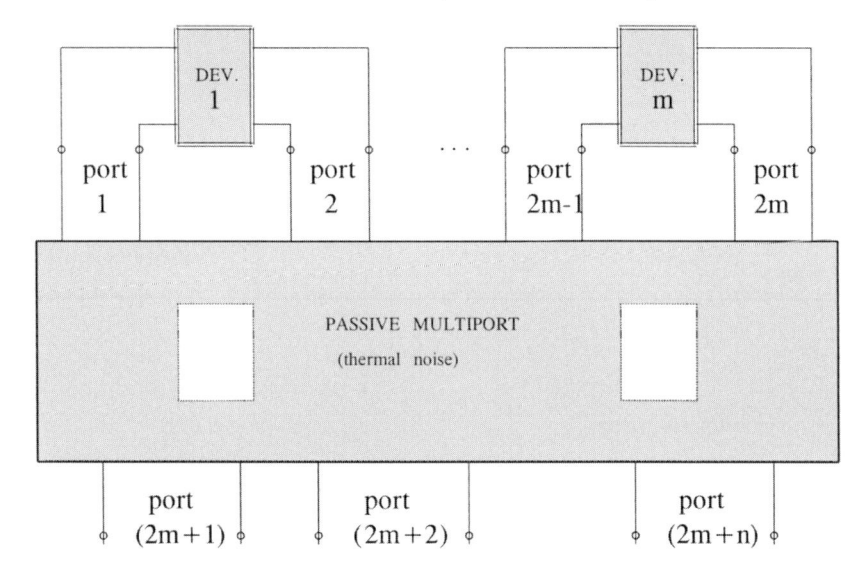

Fig. 6.4. De-embedding the active devices

a consequence, the noise contribution of the passive multiport is *uncorrelated* with the noise contribution of the active devices. Thus, the total noise power results by summing the two contributions.

The next step [86] consists in representing each noisy device by its Norton equivalent counterpart (for a two-port, according to Fig. 5.8c, and for the multiport, as in Fig. 6.3). Thus we obtain the representation shown in Fig. 6.5, where two types of noise equivalent sources are apparent:

- The generators N_k (k = 1, 2,..., 2m+n), which model the thermal noise of the lossy passive multiport. Since each lumps the effects of the same internal sources, they are correlated, and their associated noise matrix is

$$N_N = \left[\langle N_p N_q^* \rangle\right] = 2kT\,\Delta f\,(Y + Y^+) \quad (6.17)$$

- The generators denoted by J_i (i = 1, 2,..., 2m), which model the noise of the active devices (not necessary thermal). Note that each pair J_i and J_{i+1} belonging to the noise model of the same device is correlated (but each pair is uncorrelated with the others!).

■ **Method.** To compute the total noise power at the output port [86], the Y-matrix of the multiport is partitioned as follows

$$Y = \left[\begin{array}{c|c} Y_{dd} & Y_{de} \\ \hline Y_{ed} & Y_{ee} \end{array}\right] \quad (6.18)$$

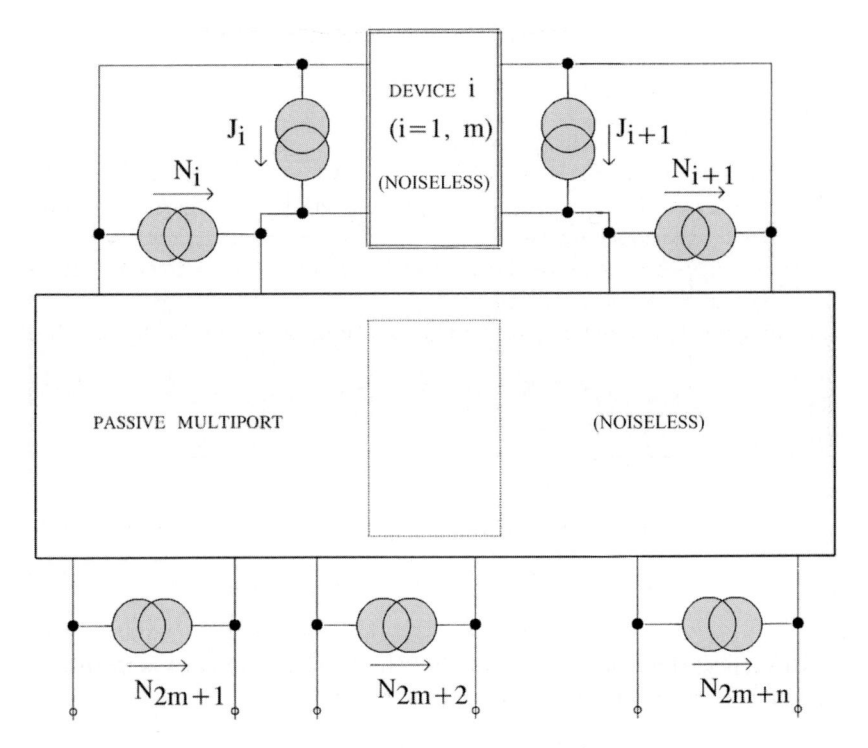

154   6 Frequency Domain Noise Analysis of Linear Multiports

Fig. 6.5. Norton equivalent of the noisy multiport (after de-embedding active devices)

where subscript "d" refers to the 2 m device ports and subscript "e" to the external ports (input and output).

Next, the network equations are written in the form

$$\begin{cases} I_d = Y_{dd}\,V_d + Y_{de}\,V_e + N_d & \text{(6.19a)} \\ I_e = Y_{ed}\,V_d + Y_{ee}\,V_e + N_e & \text{(6.19b)} \\ I_d = -y\,V_d - J & \text{(6.19c)} \end{cases}$$

where

$V = [V_d \mid V_e]^t$ is the vector of port voltages;
$I = [I_d \mid I_e]^t$ is the vector of port currents;
$N = [N_d \mid N_e]^t$ is the vector of thermal noise sources;
J is the vector of all noise generators associated with active devices.

Matrix y (6.19c) is the *diagonal sum of the individual device admittance matrices*, i.e.

6.3 Noise Analysis of Linear Multiport Circuits

$$y = \begin{bmatrix} y^{(1)} & 0 & \cdot & 0 \\ 0 & y^{(2)} & \cdot & 0 \\ \cdot & & \cdot & \cdot \\ 0 & 0 & \cdot & y^{(m)} \end{bmatrix} \quad (6.20)$$

The classical procedure to estimate the value of the Norton generator applies a shortcircuit to the port in question and "measures" the resulting current through the shortcircuit.

Consequently, to estimate the port current vector, we let $V_e = 0$ in system (6.19); let S be the resulting vector I_e.

Eliminating I_d from the transformed equations of system (6.19), we obtain

$$\begin{cases} -y\ V_d - J = Y_{dd}\ V_d + N_d & (6.21a) \\ \\ S = Y_{ed}\ V_d + N_e & (6.21b) \end{cases}$$

Next, V_d is eliminated by combining the equations of (6.21); we obtain

$$S = H_N\ N + H_J\ J \quad (6.22)$$

where the first term of the sum represents the noise contribution of the lossy (passivated) multiport, and the second represents the noise contribution of all active devices. From (6.21) we conclude that

$$H_J = -Y_{ed}\ (Y_{dd} + y)^{-1} \quad (6.23)$$

and

$$H_N = \begin{bmatrix} H_J \mid I_n \end{bmatrix} \quad (6.24)$$

I_n being the identity matrix of order n (as a matter of fact, the array denoted by H_N is obtained by augmenting H_J with I_n).

Since the equivalent generators N and J are statistically independent, the overall noise matrix is obtained by summing their contributions:

$$\langle SS^+ \rangle = H_N \langle NN^+ \rangle H_N^+ + H_J \langle JJ^+ \rangle H_J^+ \quad (6.25)$$

(superscript "+" denotes the transpose of the complex conjugate quantity).

Using expression (6.17), (6.25) may be represented in a different but equivalent form

$$\langle SS^+ \rangle = 2kT\ \Delta f\ \left(H_N(Y + Y^+)H_N^+ + 2H_J C_J H_J^+ \right) \quad (6.26)$$

where C_J is the diagonal sum of the individual device noise correlation matrices $C_J^{(i)}$ (constructed exactly in the same way as the matrix y).

Finally, the normalized noise correlation matrix is

$$\boxed{C_S = H_N(Y + Y^+)H_N^+ + 2H_J C_J H_J^+} \quad (6.27)$$

156   6 Frequency Domain Noise Analysis of Linear Multiports

■ **Note**

1) Equation (6.27) provides separate explicit expressions for the noise contribution of the active devices and the thermal noise of the lossy passive multiport.
2) Whenever the circuit has only one input and one output (i.e., n = 2), the four spot noise parameters of the global circuit can be calculated in terms of matrix C_S (see expressions (5.44)).
3) If the circuit has more than one input (or output), the only way to completely describe the multiport noise is by means of the normalized correlation matrix C_S.
4) The global admittance matrix (denoted by Y_L) of the multiport is a side-product of the calculation and is given by

$$Y_L = Y_{ee} + H_J Y_{de} \qquad (6.28)$$

Provided that the impedances connected to the external ports are known (this being the case for RF and microwave circuits, where they are equal to 50 Ω), it is possible to calculate the voltage and power gain. Furthermore, if Y_L is available, the S-parameters of the overall circuit can be determined, using the conventional transformation expressions.

6.4 Algorithm

■ **General Considerations.** The investigated circuit may belong to one of the following categories:

a) passive or active RF circuits;
b) passive or active microwave circuits (assuming that the microwave circuit can be modeled as a lumped equivalent, with a branch-node description).

In both cases, the active devices are characterized either by their measured Y (case a) or S parameters (case b).

■ **Algorithm.** The steps of the proposed algorithm [87, 88] are as follows:
Step 1: The active devices are de-embedded in order to separate the active part from the passive part of the circuit. The additional ports created during this process are designated as voltage ports and the subscript d is appended to them. The external ports (input and output) are labeled by subscript e and they are also treated as voltage ports.
Step 2: With the formulation of n-port constraint matrices developed by Lin [89], the passive multiport is analyzed to find its Y-matrix.
Step 3: The Y-matrix is partitioned according to subscripts d and e (see (6.18)).
Step 4: For each active device, the admittance matrix $y^{(i)}$ and the noise correlation matrix $C_J^{(i)}$ are calculated, starting with their given or measured data and using expression (5.43c).

Step 5: The diagonal sum y of all individual device admittance matrices, as well as the diagonal sum C_J of all individual noise correlation matrices $C_J^{(i)}$ are constructed.

Step 6: The matrix T defined by

$$T = Y_{\text{ed}} \left(Y_{\text{dd}} + \mathrm{y} \right)^{-1} \tag{6.29}$$

is calculated.

Step 7: The admittance matrix Y_{L0} of the global multiport, with the active devices embedded and the external port open-circuited, is computed as

$$Y_{L0} = Y_{\text{ee}} - T\, Y_{\text{de}} \tag{6.30}$$

Next, the admittance matrix Y_L of the global multiport, with external ports loaded by the characteristic resistance $R_R = 50\,\Omega$, is obtained:

$$Y_L = Y_{L0} + \frac{1}{R_R}\, I_2 \tag{6.31}$$

Here I_2 denotes the identity matrix of order 2.

Step 8: The impedance matrix Z_L of the multiport is evaluated (the external ports being loaded by $R_R = 50\,\Omega$):

$$Z_L = (Y_L)^{-1} \tag{6.32}$$

Step 9: The row matrices defined by

$$UJ = [0\ 1]\, Z_L\, T \tag{6.33a}$$

$$UN = [UJ \mid (0-1)\cdot Z_L] \tag{6.33b}$$

are calculated. Note that UJ is a triple product, while UN is obtained by augmenting UJ with two columns.

Step 10: Compute the individual contributions to the total output noise power of the passive lossy multiport (denoted by PP) and of the active devices (referred as PA):

$$\text{PP} = UN\ Y\ UN^+ \tag{6.34a}$$

$$\text{PA} = UJ\ C_J\ UJ^+ \tag{6.34b}$$

where UN^+, UJ^+ are the transposed complex conjugates of UN and UJ, respectively.

Step 11: Compute the noise factor F (referred to the selected input/output ports)

$$F = 1 + R_R\, \frac{\mathfrak{Re}(\text{PP}) + \text{PA}}{|Z_{L21}|^2} \tag{6.35}$$

Step 12: Compute the normalized correlation matrix with (6.27).

158   6 Frequency Domain Noise Analysis of Linear Multiports

Step 13: Compute the electrical performance of the circuit: the voltage gain, the power gain, its S matrix. Also, find the four noise parameters of the global circuit, expressed as (R_n, G_n, Y_{cor}), (F_o, R_n, G_o, B_o) and/or (F_o, R_n, Γ_o), with (5.44), (5.34c) and Table 5.3.

■ **Implementation.** This algorithm has been implemented in program NOF, which was developed to analyze the noise performance of linear microwave circuits. This program can be run as a DOS application on any personal computer having a 486, Pentium 1, Pentium 2, or Pentium 3 processor.

The maximum size of the circuit which can be analyzed is 40 nodes, 80 branches, and 5 active devices.

The repertoire of elements implemented in NOF includes resistors (with or without thermal noise), inductors, capacitors, transmission lines, active two-ports, and all types of controlled sources.

NOF is equipped with its own input/output editor (menu driven) to create and modify the input file, as well as to plot the results.

6.5 Example

6.5.1 FET Microwave Amplifier

■ **Configuration.** The FET amplifier in Fig. 6.6 is intended for 2 – 10 GHz operation. Adjacent to the circuit are the electrical and noise parameters of the FET.

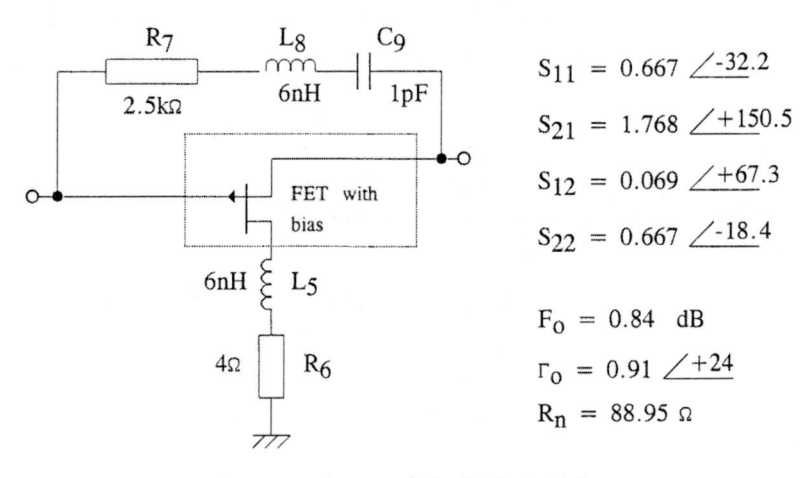

$S_{11} = 0.667 \angle -32.2$
$S_{21} = 1.768 \angle +150.5$
$S_{12} = 0.069 \angle +67.3$
$S_{22} = 0.667 \angle -18.4$

$F_o = 0.84$ dB
$\Gamma_o = 0.91 \angle +24$
$R_n = 88.95\ \Omega$

Fig. 6.6. A 2–10 GHz FET amplifier

■ **Circuit Model.** The active device is replaced by a two-port described by its S parameters, and the nodes are numbered.

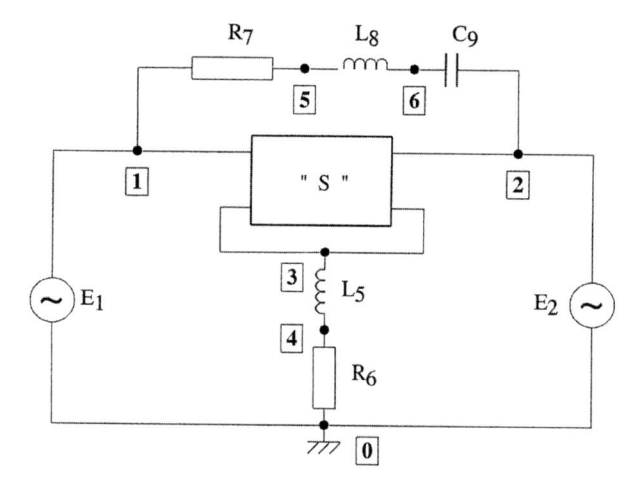

Fig. 6.7. Modeling the FET amplifier

As usual, ground number is designated 0; successive integers in increasing order are assigned to all nodes, the first two integers (1 and 2) being reserved for the input and output, respectively.

The input and output ports must be specified as voltage ports (Fig. 6.7).

The branches are numbered according to their subscripts (for instance, R_7 denotes branch number 7); each port is described by a single branch.

■ **Checking the Model.** Suppose now that the FET is de-embedded and the resulting ports are also specified as voltage ports. We must check whether there are any loops containing only independent voltage sources. If YES, the model is inconsistent and needs to be slightly modified.

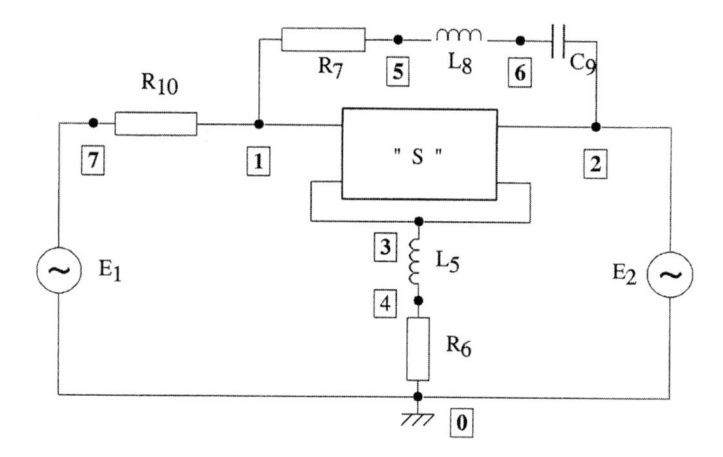

Fig. 6.8. Model to be analyzed

160   6 Frequency Domain Noise Analysis of Linear Multiports

In the case of Fig. 6.7, the loop passing through the nodes **0-1-3-2-0** contains only voltage ports, and the model is inconsistent, since its Y matrix does not exist. We must break this loop by introducing a resistor (R_{10}) in series (Fig. 6.8). Its value must be low enough to simulate a short circuit (for instance, of order $10^{-3}\,\Omega$) and thus not perturb the electrical behavior of the circuit. This new circuit model can then be analyzed by NOF.

■ **Input Data File.** The input data file is presented here; note the low value of resistor R_{10} ($10^{-4}\,\Omega$).

Example
```
4
0
1   0   7   ' E'   (0.      ,      0.)   0   (0.   ,      0.)
2   0   2   ' E'   (0.      ,      0.)   0   (0.   ,      0.)
3   1   3   ' QS'  (0.667   ,   -32.20)  4   (0.069,   67.30)
4   2   3   ' QS'  (1.768   ,   150.50)  3   (0.667,  -18.40)
5   4   3   ' L'   (0.      ,    6.E-9)  0   (0.   ,      0.)
6   4   0   ' R'   (4.      ,      0.)   0   (0.   ,      0.)
7   1   5   ' L'   (0.      ,    1.E-9)  0   (0.   ,      0.)
8   5   6   ' C'   (0.      ,   1.E-12)  0   (0.   ,      0.)
9   2   6   ' R'   (2500.   ,      0.)   0   (0.   ,      0.)
10  7   1   ' R'   (1.E-4.  ,      0.)   0   (0.   ,      0.)
0   0   0   '  '   (0.      ,      0.)   0   (0.   ,      0.)
0   0   0   '  '   (0.      ,      0.)   0   (0.   ,      0.)
Noise Parameters
0.84      88.95      0.91     24.03
STOP
```

■ **Output Data File.** NOF yields the following results:

FREQUENCY OF OPERATION = 4.000 GHz

R E S U L T S
****

NOISE CONTRIB. OF PASSIVE NETWORK
$$= .1158563E+01 + j\,-.1007071E+02$$

NOISE INJECTED BY ACTIVE DEVICES
$$= .1886914E+01 + j\,.9302455E-16$$

SPOT NOISE FIGURE OF THE CIRCUIT (lin) F = .1733980E+01
SPOT NOISE FIGURE OF THE CIRCUIT (dB) F = .2390440E+01 dB

VOLTAGE GAIN (complex) = .33511E–01+j .31469E+00
VOLTAGE GAIN (modulus) = .3164659E+00
VOLTAGE GAIN (in dB) = –.9993460E+01 dB

TRANSDUCER POWER GAIN = .3319413E+00
TRANSDUCER POWER GAIN = –.4789387E+01 dB

CORRELATION MATRIX OF THE EQUIVALENT NOISE SOURCES
****
(the circuit is regarded as a noiseless two-port, having an equivalent noise current source connected to each port)

.2439696E–02 + j .1668494E–12   .6030939E–03 + j .3291525E–03
.6030939E–03 + j –.3291525E–03   .2743691E–02 + j .9905174E–14

THE EQUIVALENT NOISE PARAMETERS OF THE CIRCUIT
****
Noise resistance Rn = 30.9061321 Ohm
Correlation conduct. GCOR = .7889518 mS
Correlation suscept. BCOR = 2.4712220 mS
Noise conductance Gn = 1.1338211 mS
Noise figure (ratio) F = 1.7339795
Noise figure (in dB) F = 2.3904397 dB
Optimum source conduct. G0 = 6.1080606 mS
Optimum source suscept. B0 = –2.4712218 mS
Min. noise fig. (ratio) F0 = 1.4263200
Min. noise fig. (in dB) F0 = 1.5421696 dB
Opt.source reflection coef.= .5380424 (mag.) 15.493911 (deg.)

Equivalent S Parameters of the circuit
****
	Magnitude	Arg. (degrees)
S11	.821668	–4.026257
S12	.146026	62.208545
S21	.576144	82.105468
S22	.861735	–3.813636

Stop – Program terminated.

6.5.2 Concluding Remark

One of the most important problems for any algorithm for computer-aided analysis of electronic circuits is to avoid matrix inversions. From long time it has been known that this is a very risky operation, which affects the accuracy of computation in quite unexpected ways [90]. Attention has been paid in developing the algorithm to implement as few matrix inversions as possible.

In contrast to other similar algorithms, NOF contains only two matrix inversions. After each inversion, an accuracy test is automatically performed, and if unacceptable loss of accuracy results, a message is displayed on the screen to alert the user.

Summary

- To completely characterize the noise of any circuit regarded as a two-port, we must find the following four noise parameters: F_o, R_n, G_o, and B_o (or any other equivalent set).
- For simple, linear passive circuits, the method of Hillbrand and Russer can be successfully employed. It offers the additional benefit of understanding how various circuit elements affect the noise performance.
- For linear active circuits, the only way to compute noise performance is to employ an appropriate software package.

7

Noise Models of Electronic Devices

> "Hereafter, when they come to model Heaven
> And calculate the stars, how they will wield
> The mighty frame, how build, unbuild, contrive
> To save appearances .."
>
> (Milton, Paradise Lost)

7.1 Resistor Noise

■ **Background.** In this section the noise mechanisms of fixed resistors are discussed. *Resistors are passive devices, discrete or integrated, fabricated by depositing a resistive material on an insulating substrate.*

The value of the resistance is given by the classical relation

$$R = \rho \frac{l}{s} \tag{7.1}$$

where ρ is the resistivity of the material, l the length of the resistive layer, and s its cross-sectional area.

■ **Fabrication.** There are several types of fixed value resistors:

- *Wirewound resistors.* These are obtained by wounding a wire around an insulator (glass or ceramic). The wire ends are soldered to the leads and the entire structure is coated with a protective glaze. They are relatively expensive and limited to lower values of resistance, but they offer accurate values and are very stable with time.
- *Metal foil resistors.* In this case the resistive conductor is a bulk metal foil bonded to a dielectric substrate; the desired value is obtained by etching a serpentine pattern. Fine adjustment is performed with a laser beam. Finally, the structure is glazed for protection.

- *Thin film resistors.* These are obtained by vacuum deposition of specific resistive alloys (nickel-chromium, carbon-boron, tantalum, or various oxides) on a ceramic substrate. Film thickness is around 5000 Å, and is often spiraled to increase the value of the resistance. To produce cheaper resistors, carbon layers can be used instead. *Deposited carbon resistors* are obtained in this way.
- *Carbon composition resistors,* produced by mixing carbon particles with a special binder, molded around lead wires, and put in a furnace to sinter and solidify.
- *Thick film resistors (cermet),* made by screen deposition of a resistive ink composed of ceramic and metal powders (such as silver, chromium, palladium, and glass), followed by heating in a furnace at about 800°C. At this temperature, the solids melt to form the resistive film, the organic binder evaporates, and the film is bonded to the substrate. The film thickness is of order 10^{-2} mm.

Note that *integrated resistors* are either monolithic (a buried layer of semiconductor in an opposite substrate type) thin-film or thick-film types.

■ **Noise Mechanisms.** Any resistor exhibits two kinds of noise: thermal noise and excess noise.

Thermal noise is associated with any dissipative structure; according to (3.4), its mean square value is given by

$$\overline{v_{nth}^2} = 4kTR\,\Delta f \qquad (7.2)$$

Excess noise (or 1/f noise) is related to the phenomena appearing when a DC current flows in a discontinuous conductor (like a carbon compound resistor, for instance). In this case the lines of current flow are nonuniformly distributed inside the structure, mainly due to the differing resistivity of the carbon particles and the binder. Between adjacent carbon particles micro-arcs may appear, randomly distributed in space and time. These micro-discharges are responsible for the excess noise.

Another region where excess noise can appear is in the ohmic contacts connecting the bulk of the resistor itself to its leads.

Excess noise has a 1/f spectrum; this implies that there is constant power per frequency decade. Experimentally, the excess noise voltage density of a resistor R at frequency f is found to be

$$S(v_{nex}) = \frac{C\,I_{dc}^2\,R^2}{f} = \frac{C\,V_{dc}^2}{f} \qquad (7.3)$$

C is a constant that depends on the manufacturing process, I_{dc} the DC current in the resistor, and V_{dc} the DC voltage across it. Note that the *excess noise voltage is proportional to the DC voltage drop across the resistor.*

7.1 Resistor Noise

As thermal noise and excess noise are uncorrelated (because they arise from different physical processes), the total mean square noise voltage over a bandwidth $\Delta f = (f_2 - f_1)$ is the sum of the two contributions, i.e.

$$\overline{v_n^2} = \int_{f_1}^{f_2} \frac{C\,V_{dc}^2}{f}\,df + 4kTR(f_2 - f_1) \tag{7.4a}$$

$$\overline{v_n^2} = C\,V_{dc}^2\,\ln\frac{f_2}{f_1} + 4kTR(f_2 - f_1) \tag{7.4b}$$

Expression (7.4b) is interesting, since it shows that the contribution of excess noise (first term) does not depend on the resistance value, but only on the DC voltage drop across it, in contrast with the contribution of thermal noise (second term), which depends on the resistance. This means that *resistances of different values, technologically alike, exhibit the same excess noise in the same bandwidth, provided that the DC voltage across their terminals is the same.*

According to (7.4b), the normalized power of the excess noise depends only on the ratio of upper and lower cutoff frequencies, and is therefore constant in each frequency octave or frequency decade. It follows that the excess noise voltage over one frequency decade is

$$\overline{v_{nex}^2} = C\,V_{dc}^2\,\ln 10 \tag{7.5}$$

A plot of the total noise power versus frequency is given in Fig. 7.1.

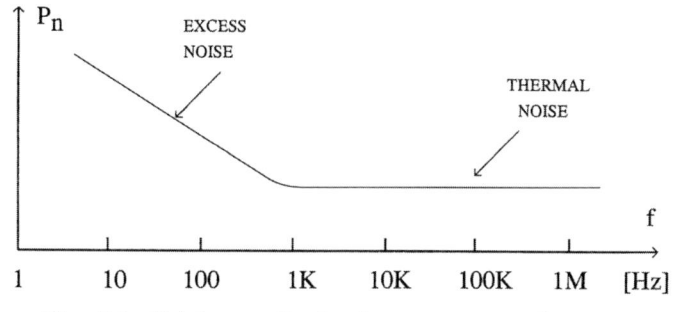

Fig. 7.1. Total normalized noise power versus frequency

☐ **Noise Index.** The noise index (NI) is a parameter used to express the amount of excess noise in a resistor.

According to [91], the *noise index is defined in one decade of frequency* $(f_2 = 10f_1)$ as the ratio of: (a) the rms value (in μV) of the v_{nex}, to (b) the DC drop across the resistor (in V).

Thus

$$\text{NI} = \frac{V_{nex}}{V_{dc}}\left[\frac{\mu V}{V}\right] \tag{7.6a}$$

Often the noise index is expressed in decibels (since in this form, it tends to have a normal distribution law, when samples of "identical" resistors are measured):

$$[\text{NI}]_{\text{dB}} = 10 \log \frac{\overline{v_{nex}^2}}{V_{dc}^2} = 20 \log(\text{NI}) \tag{7.6b}$$

where v_{nex} is expressed in microvolts per frequency decade.

The ratio $(v_{nex})/(V_{dc})$ is usually less than unity and consequently the noise index expressed in decibels has negative values.

☐ **Evaluating Constant C.** To find the expression relating the constant C to the noise index, we must first square (7.6a):

$$\overline{v_{nex}^2} = (\text{NI})^2 \, V_{dc}^2 \tag{7.6c}$$

Then, combining it with (7.5), we obtain

$$C = \frac{(\text{NI})^2}{\ln 10} = \frac{10^{([\text{NI}]_{\text{dB}}/10)}}{\ln 10} \tag{7.6d}$$

As soon as the expression for constant C is available, we can substitute it into (7.3) and get the excess noise, at a specified frequency f, relative to the noise index of the resistor and the DC voltage drop at its terminals

$$\overline{v_{nex}^2} = \frac{V_{dc}^2 \, 10^{([\text{NI}]_{\text{dB}}/10)}}{f \, \ln 10} \; [\mu V^2] \tag{7.6e}$$

■ **Application.** Let us consider a 33-kΩ carbon resistor at T = 300 K, with a NI of –1 dB, inserted into a circuit where the DC drop across the resistor terminals is 10 V. Find the total noise produced by this resistor, in the frequency range 10 Hz–10 kHz.

Solution

A NI = –1 dB corresponds to a ratio $\frac{V_{nex}}{RI_{dc}} = 0.89$. This means that the excess noise is 0.89 µV for each volt of DC per frequency decade. Subject to a DC of 10 V, the resistor will produce an excess noise of 8.9 µV / decade.

The imposed frequency bandwidth comprises three decades, and there is constant noise power per each decade of frequency. Since the noise of each decade is uncorrelated with the noise of the remaining ones, according to (2.8e) the total noise in three decades is

$$v_{nex3}^2 = 8.9^2 + 8.9^2 + 8.9^2 \; [\mu V^2]$$

$$v_{nex3} \cong 15.41 \; \mu V$$

The thermal noise delivered by the resistor in the imposed bandwidth is

$$v_{nth} = \sqrt{4kTR \, \Delta f} \cong 2.34 \; \mu V$$

7.1 Resistor Noise

Applying (7.5), the total noise power is the sum of the individual contributions, and finally

$$v_n = \sqrt{v_{nex3}^2 + v_{nth}^2} \cong 15.58 \ \mu V$$

Comparing v_n to v_{nex3}, it is obvious that in this case the thermal noise contribution is quite negligible relative to the excess noise, which is dominant.

■ **Remark.** Assume that a resistor generating excess noise is subject to an AC signal of frequency f_o, instead of DC. In this case, noise similar to 1/f noise can be observed in both sidebands centered on f_o. This kind of noise is often called *1/Δf noise*, because its power spectral density is proportional to the ratio $1/|f_o - f|$. Furthermore, the power spectral density is proportional to the mean square value of the AC current flowing in the resistor.

■ **Selecting a Low-Noise Resistor.** In all applications where the signal is weak and a DC bias is applied to resistors, it is important to *select low noise index resistors*. Whenever this information is present in a data sheet, remember that there is always a 20-dB spread in the noise index, due to the impossibility of controlling the processing quality of each sample (high noise is an indication of some manufacturing defect).

The designer must select the type of resistor, keeping in mind that the amount of excess noise mainly depends on the type of processing and the quality of the manufacturing process. As a general rule, the more homogeneous the structure of the resistive conductor, the less excess noise will be expected.

It follows that the carbon deposited or carbon composition resistors have significant excess noise, due to their inhomogeneity (carbon granules and binder); metal thin-film resistors have a considerably more homogeneous structure and consequently are less noisy. The best choice is the bulk metal foil or wirewound resistors, provided that for both categories the quality of the ohmic contact is high. Special attention must be paid to the inductive effect associated with wirewounded resistors; if it is not compensated (or shielded), pick-up of parasitic signals can occur and their level can greatly surpass the usual excess noise!

As for integrated resistors, a priori the thin-film category is less noisy than the thick-film one. The former results by depositing resistive material in vacuum almost atom by atom (resulting in a higher regular pattern), while the latter uses a process similar to printing with resistive ink.

■ **Additional Criteria.** There are additional factors to be considered when selecting a low-noise resistor, such as:

- A high-power resistor (hence, one with greater volume) is preferred to a low-power equivalent one. The smaller size of the latter makes it more sensitive to processing defects.

168   7 Noise Models of Electronic Devices

Table 7.1. Average noise indices of commercial resistors

IB[dB]	−40 −30 −20 −10 0 +10
Discrete resistors	
Carbon composition	(−20 to +10)
Deposited carbon	(−25 to 0)
Metal foil	(−40 to −30)
Wirewound	(−40)
Integrated resistors	
Thin-film	(−25 to −5)
Thick-film	(−15 to +10)
Value	
less than 100 Ω	(−40 to −15)
less than 100 KΩ	(−20 to +10)
greater than 100 KΩ	(−10 to +10)

– As a general rule, a high-value resistor has a higher noise index than a low-value one, assuming that both are rated for the same power and manufactured in the same way (see Table 7.1). For instance, in carbon resistors, a higher value is associated with a low concentration of carbon granules and more binder (relative to a lower-value one), and consequently the structure is more inhomogeneous.
– Helical resistors show less excess noise than similar nonhelical ones (in value and size). This is due to the different resistivity of the conductive layer, which must be higher for a nonhelical resistor in order to ensure the same resistance value as for the helical type. However, a helical resistor is prone to parasitics, due to its associated inductive effect.
– It has been noted that a resistor with significant excess noise is less reliable and tends to be less stable with aging.

■ **Integrated Polysilicon Resistors.** Polysilicon resistors are the preferred choice of analog integrated circuit designers. The excess noise of medium and large polysilicon resistors is [136]

$$\frac{S(I)}{I^2} = \frac{1}{WL} \frac{\alpha}{f} \qquad (7.7)$$

where W and L are the width and length of the resistor, and α is a constant which includes all technology-dependent parameters and the temperature. According to data reported in [136], p-type polysilicon resistors are the preferred choice for low-noise circuit design. The following remarks also apply:

– minimizing the DC offset current will reduce 1/f noise;
– changing the amplitude of signals applied to resistors has no impact on the S/N ratio.

■ **Concluding Remark.** Especially at low frequency (less than 100 kHz), excess noise is dominant relative to thermal noise. As a general rule, the total measured noise of a good quality resistor can exceed the computed value of its thermal noise by 10% to 100% .

Recall that *if no DC bias is applied to a resistor, there is no excess noise.*

Table 7.1 compares the average noise indices of various categories of resistors. Due to the spread in noise indices from one sample to another, the proposed intervals are only approximate.

7.2 Capacitor Noise

■ **Background.** *Capacitors are passive discrete devices made up of two metal plates separated by a dielectric layer.* For a parallel plate capacitor, if A is the area of the metal plate and d the thickness of the dielectric, the resulting capacitance is

$$C = \varepsilon_o \, \varepsilon_r \, \frac{A}{d} \tag{7.8}$$

where $\varepsilon_o = 8.85 \cdot 10^{-12} \, \text{F m}^{-1}$ is the vacuum permittivity and ε_r is the relative permittivity of the insulator.

■ **Classification.** According to their quality, there are several types of capacitors:

– *Low loss capacitors*, where the dielectric is glass, ceramic, mica, or high quality organic compounds. They are very stable with aging or temperature and are high-voltage rated, but their capacitance is rather limited to low values (up to 10 nF).
– *Medium loss capacitors* consist of alternating layers of metal foil and one or more layers of paper saturated with oil or flexible plastic films. In order to reach high values while preserving reasonable volume, they are rolled and encapsulated. Due to this, an associated parasitic inductive effect is observed, and they are also less stable.
– *Electrolytic capacitors*, which offer the largest possible capacitance per volume unit. They employ tantalum or aluminum, both metals having the remarkable property of producing very thin oxide layers with high relative permittivity at the anode. To contact the oxide over its entire surface, a solid or liquid substance (electrolyte) is used, which is mainly responsible for the electrical properties of the capacitor. Their capacitance can be as high as $10^4 \, \mu\text{F}$, for a reasonable volume; their breakdown voltage is relatively low (tens of volts). They suffer from high leakage currents; low temperatures reduce performance and high temperatures dry them out; and they have moderate stability with aging.

170   7 Noise Models of Electronic Devices

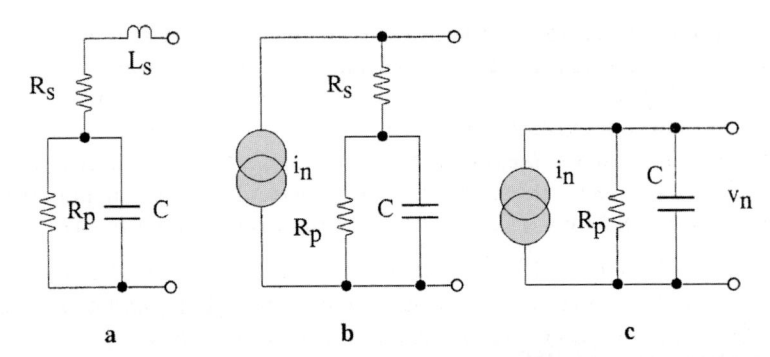

Fig. 7.2. Various representations of a capacitor: **a** electrical equivalent circuit; **b** noise equivalent circuit; **c** simplified noise equivalent circuit

■ **Equivalent Circuit.** Each capacitor has a series resistance R_s (associated with the electrolyte, ohmic contacts, and leads) and a shunt leakage resistance R_p (Fig. 7.2a). In the case of an electrolytic capacitor operating at the ambient temperature, R_s may be of order $1\,\Omega$ (but at $-40°C$, it may rise to about $4\,\Omega$).

The value of R_p depends mainly on the type of dielectric and the processing quality; for instance, a styrene capacitor may have R_p around several $M\Omega$, but for a paper equivalent capacitor this value may be reduced by a factor of ten. Especially for large-value film capacitors, parasitic inductance must be included in their equivalent circuit.

■ **Noise.** An ideal (lossless) capacitor has no thermal noise, since a pure reactance is not dissipative. A real capacitor has some losses, but the associated thermal noise is negligible. The reason is twofold:

– The series resistance is rather low.
– The leakage resistance R_p is in parallel with the capacitance C (Fig. 7.2a), and according to (5.5) the thermal noise is dominant only at very low frequency (since the practical values of the time constant $R_p C$ are very high).

The noise model of the capacitor includes a noise current generator in parallel with its electrical counterpart (Fig. 7.2b). For the simplified model (Fig. 7.2c), the noise generator i_n can be viewed as the thermal current noise source associated with the resistor R_p, to produce the output fluctuation v_n.

Under the assumption that the loss tangent in the dielectric is frequency independent, Van der Ziel proved [92] that the power noise spectrum has a $1/f$ distribution.

■ **Practical Considerations**

– To improve the low-frequency response of electronic circuits, high-value capacitors (bypass or coupling) are frequently required. Often, electrolytic

capacitors are adopted in practice. Tantalum capacitors are mainly recommended because they are less noisy than aluminum types (where losses are more significant).
– Excess noise occasionally manifests itself in the coupling electrolytic capacitors. Sometimes, transients cause a reverse bias at the terminals of a coupling capacitor, which generates bursts of noise for a time between a few minutes and several hours [91].

This problem is typical of multistage amplifiers; when powered on, an intermediate stage may turn on faster than its neighbors, and consequently, for a few moments, its coupling capacitor is reverse-biased.

■ **Solutions.** In order to avoid these problems, Motchenbacher and Connelly [91] suggest several solutions:

- The circuit must be designed in such a way that for any possible transient, there is no reverse bias on any coupling capacitor.
- Otherwise, put a high-quality silicon diode (with low saturation current) across each coupling capacitor in such a way that a reverse bias forward biases the diode.
- To obtain a nonpolar electrolytic capacitor two identical electrolytic capacitors are connected in series, back to back. Their values must be twice the desired value; for any applied bias, at least one of the capacitors takes the proper voltage drop.

7.3 Inductor Noise

■ **Intrinsic Noise.** If noise is defined as *any random signal (fluctuation) superimposed on the processed information that tends to mask its contents*, then an ideal (lossless) inductor has no intrinsic noise.

However, a real inductor always has a series resistance (associated with the windings), and this generates thermal noise. Actually, the level of thermal noise is quite negligible, as the winding wire resistance is seldom higher than $100\,\Omega$.

■ **External Noise.** From the point of view of external noise, any inductor is subject to two categories of interference:

– So called *inductive noise*. This noise is traditionally associated with voltage transients across any inductor, when the current in it varies abruptly. Due to inherent parasitic coupling, this voltage can perturb other circuits and consequently acts as a disturbing signal (interfering signals are detailed in part II). Note that this kind of noise is not random, since its waveform can be predicted.

172   7 Noise Models of Electronic Devices

– Another problem with inductors results from their inherent ability to pick up parasitic magnetic fields and transform them into interfering signals. Although these signals do not originate in the inductor itself, they appear at the terminals as a noise generated inside.

To conclude, attention must be paid to inductors: whenever we cannot avoid them, they must be properly shielded; furthermore, their inductive noise must be carefully controlled.

7.4 Noise in Junction Diodes

7.4.1 Ideal PN Junctions

■ **Definition.** *A junction diode consists of two bulk (P and N) highly doped regions in close contact and fitted at their extremities by two ohmic contacts.*

A semiconductor junction is said to be *ideal* if we may neglect:

– Generation and recombination of charge carriers in the depletion layer.
– All surface effects.

■ **Background.** Any semiconductor device has one or more PN junctions. For an ideal PN junction, the voltage-current relationship is given by the Shockley equation

$$I = I_s \left(\exp\left(\frac{qV}{kT}\right) - 1 \right) \tag{7.9}$$

where I_s denotes the saturation current and q is the elementary charge (q = $1.6 \cdot 10^{-19}$ C). Formally, this expression can be represented in a different but equivalent form:

$$I = \underbrace{(I + I_s)}_{\text{majority carriers}} - \underbrace{I_s}_{\text{minority carriers}} \tag{7.10}$$

$$= I_s \exp\left(\frac{qV}{kT}\right) - I_s$$

The junction conductance G is defined by

$$G = \frac{\partial I}{\partial V} = \frac{q}{kT}(I + I_s) \tag{7.11}$$

■ **Theory of Buckingham and Faulkner.** The starting point is the Shockley theory, according to which the currents crossing the junction (Fig. 7.3) are I_F (majority carrier flow) and I_R (minority carrier flow). The former is a diffusion current, while the latter is a drift current. They are considerably

7.4 Noise in Junction Diodes

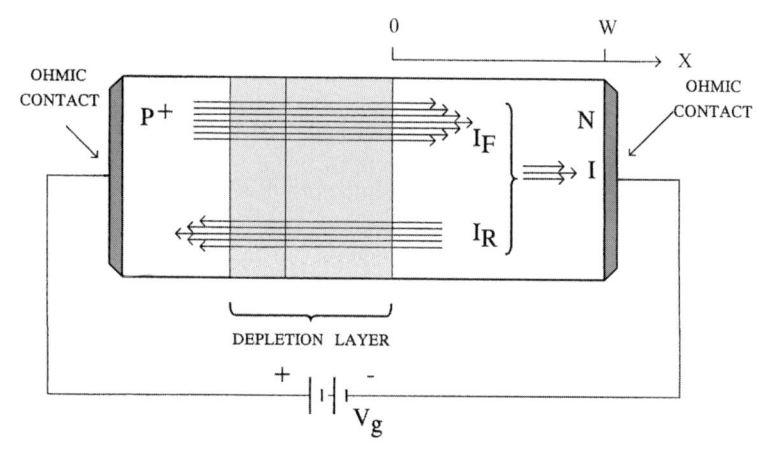

Fig. 7.3. Current I_F and I_R in a P^+N junction

larger than the terminal current I (which is equal to I_F–I_R), and they are given by the following expressions [93]:

$$I_F = \frac{q\,D\,A\,p_n}{\overline{l_f}} \exp \frac{qV}{kT} \tag{7.12a}$$

with:
- p_n – equilibrium concentration of holes on the N side
- D – diffusion constant for holes
- A – cross-sectional area of the junction
- $\overline{l_f}$ – mean free path of holes
- V – effective voltage at the junction level (which is the applied voltage less the voltage drop in the bulk regions and ohmic contacts).

The minority carrier flow is

$$I_R = qDA\,p_o/\overline{l_f} \tag{7.12b}$$

where p_o is the hole density at $x = 0$.

The role of these currents is to maintain equilibrium between the majority carrier density on one side of the separating plane at $x = 0$ and the minority carrier density on the other side.

Obviously I_F cannot be identified with $(I_s \exp \frac{qV}{kT})$, nor can I_R with I_s.

Following a detailed study of the physical phenomena (based on the fact that the bulk diffusion process is the bottleneck in the flow of charge carriers), Buckingham and Faulkner described the main noise mechanisms of an ideal junction:

– thermal fluctuations in the minority carrier flow
– bulk region minority carrier generation-recombination.

174   7 Noise Models of Electronic Devices

Both mechanisms came into play in the bulk regions close to the edges of the depletion layer, and are statistically independent (hence their power spectra add).

Although fluctuations in the terminal current I are still related to random variation of the number of carriers crossing the depletion layer, these fluctuations are not a direct result of shot noise associated with the currents I_F and I_R. In fact, the fluctuations of these huge currents are the result of carrier diffusion in the P and N bulk regions (far from the transition region). Moreover, currents I_F and I_R are responsible for large charge accumulation in the area close to the depletion layer, and this alters the electric field of the depletion layer so as to almost completely hide shot noise effects (see also the discussion of Sect. 3.2.3). This process is very efficient and consequently, as Robinson pointed out [94], the shot noise associated with currents I_F and I_R make almost no contribution to the total noise of the junction.

Finally, Buckingham and Faulkner established the noise spectral density of the ideal semiconductor junction to be

$$S(I_n) = 2q(I + 2I_s) + 4kT(G_2 - G) \qquad (7.12c)$$

Here the dynamic conductance G_2 is related to the high-frequency behavior of the carrier diffusion process, and G is the low-frequency value of G_2 (given by (7.11)).

This equation globally represents the noise of an ideal junction, and yields the simplified model of Fig. 7.4a.

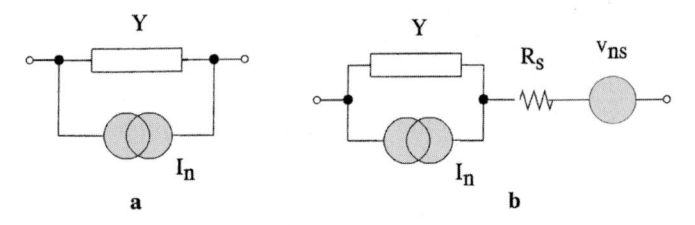

Fig. 7.4. Noise models of an ideal PN junction: **a** simplified; **b** series resistance included

7.4.2 Forward-Biased Diodes

■ **Comment.** Equation (7.9) suggests that all ideal diodes have the same current-voltage characteristic; furthermore, it predicts that for any ideal diode, with saturation current 1 nA at T = 300 K and forward-biased with V = 0.6 V, the resulting current should be over 10 A! Anyone familiar with semiconductor devices will know that the current in a silicon forward-biased diode is typically several miliamps (at V = 0.6 V) and that it depends on the type of diode and the manufacturing process.

7.4 Noise in Junction Diodes

Equation (7.9) obviously cannot correctly describe the behavior of real semiconductor diodes, because something has been neglected, namely generation-recombination events in the depleted layer and nearby bulk regions.

■ **Note.** In contrast to an ideal PN junction, a real diode model contains a series resistance R_s, which corresponds both to the bulk regions P and N and to the ohmic contacts.

■ **Noise.** It is traditionally assumed that the noise of a forward-biased real diode has three components:

1) *Shot noise*, which is directly related to the DC current through the diode, as stated by (3.21). However, according to the theory of Buckingham and Faulkner, the noise level predicted by (3.21) is so small, as to be negligible compared to the total noise of the diode. Consequently, shot noise is not the dominant mechanism of noise in a forward-biased PN junction.

2) *1/f noise*, due mainly to carriers that fall into traps (contamination centers located at the surface of the device or dislocations existing in the bulk), remain for a time, and then relax. As trapping and relaxing are both randomly distributed events (in space and time), fluctuations appear in the terminal current.

   This kind of noise is closely related to technological defects; for many samples manufactured in the same run, it differs from one sample to another. A high level of 1/f noise indicates a low quality and a less reliable device.

   If the surface is passivated during processing with a thin oxide layer and the fabricated device hermetically sealed in its package, excess noise is considerably reduced.

3) *Thermal noise*, produced by the series resistance R_s. Due to the bipolar nature of charge carriers (electrons and holes), this resistance is nonlinear in the current I through the device. A truncated Taylor series expansion yields

$$R_s = R_{so} + I(\partial R_s/\partial I) \qquad (7.13)$$

where $\partial R_s/\partial I$ is negative. At high frequency, this derivative becomes a complex quantity, and we may write instead

$$Z_s = R'_{so} + j\omega L_s \qquad (7.14)$$

Therefore, the thermal noise is given by

$$\overline{v_s^2} = 4kT\,\mathcal{R}e(Z_s)\,\Delta f \qquad (7.15)$$

■ **Comments**

– Semiconductor diodes operating at high injection levels can be modeled by means of transmission lines [92] to take into account the effect of stored charge in the device.

– Schottky diodes exhibit additional 1/f noise, due both to fluctuation of carrier mobility and the current injected by the metal. A two-phase tunnel effect can be observed, related to generation-recombination centers located in the depleted region.

7.4.3 Reverse-Biased Diodes

■ **Noise.** There are several noise sources in a reverse-biased diode:

– *Shot noise* of the reverse current. As the reverse current becomes very low, shot noise becomes more evident.
– *Avalanche noise*, when the diode operates near the breakdown region. Depending on the diode type, two breakdown mechanisms exist, the Zener effect and the avalanche multiplication of carriers.

☐ **Zener Effect.** This effect is mainly observed in highly-doped diodes, with a very thin depleted layer. When the reverse voltage reaches several volts, the electric field is high enough to stimulate tunneling (i.e., penetration of the potential barrier by accelerated minority carriers). This phenomenon, responsible for shot noise and very little excess noise, is pertinent to Zener diodes operating at low voltages (less than about 5 V). Zener diodes are used as reference diodes in low-noise circuits and low-noise regulators.

☐ **Avalanche Breakdown.** This mechanism appears in diodes with a larger depleted layer width, and generates multistate noise, as described in Sect. 3.2.8. Avalanche breakdown is typical of diodes whose breakdown voltage is in the range 5 V to 50 V.

The noise level is much more significant than the noise produced in a Zener diode, and varies from sample to sample for devices manufactured in the same production run. If necessary, low-noise units should be selected.

■ **Remark.** Avalanche noise can be distinguished from shot noise by observing the (negative) sign of the slope of the noise spectral density versus the applied DC bias.

■ **Advice.** Whenever a circuit requires a reference voltage of less than 5 V, the most acceptable solution it to choose a Zener diode. However, if the required reference is between 5 V and 50 V, we may either use two Zener diodes (maximum) in series, or select (at the proper operating current level) a low-noise avalanche unit.

Regardless of the adopted solution, it is highly recommended that be added across the diode a high-value, good-quality capacitor to reduce the noise level.

7.4.4 PSPICE Model

■ **Electrical Model.** The diode model of PSPICE includes a series resistance (RS/*area*) with the intrinsic junction. The parameter called *area* is under the user's control, and represents a scale factor for the saturation current (IS), the series resistance (RS), zero-bias junction capacitance (CJO), and the breakdown voltage (BV). Its default value is 1.

■ **Noise Model.** Noise power is computed in a 1-Hz bandwidth (hence, this is the power spectral density), by taking into account the following contributions:

– *Thermal noise* of the series resistance RS:

$$S(I_s) = 4kT \, / \, (RS/area) \tag{7.16}$$

– *Shot noise* and *1/f noise* produced by the intrinsic junction:

$$S(I_s) = 2q \, I_d + KF \, (I_d)^{AF}/FREQ \tag{7.17}$$

In (7.17) KF is the coefficient of the 1/f noise (default value is 0), AF is another user-controlled coefficient (default value is 1), and FREQ denotes the operating frequency.

7.5 Battery Noise

■ **Comment.** Sometimes batteries are used to supply electronic equipment (for instance, portable communication systems, devices for spaceflight, satellites, smart weapons, remote control, etc.) or low-frequency noise measurement systems (where commercial solid-state regulated sources introduce an unacceptable level of noise).

Another possible application involves low-noise amplifiers, where using batteries avoids ground loops and pickup of parasitic signals by an AC power supply. A typical example is an outdoor receiving antenna, where both the preamplifier and its associated batteries are placed in the same shielded enclosure, right near the antenna, but relatively far from the indoor receiving equipment.

In practice, when coupling between equipment via the internal resistance of a single power supply is to be avoided, using batteries instead (which have negligible internal resistance) may be a good solution.

■ **Noise.** As a general rule, a battery is not noisy, for two main reasons:

1) Its internal series resistance is negligible.
2) A battery acts like a negligible capacitive reactance, in parallel with its internal resistance. Therefore, the inherent weak internal noise is even further reduced.

However, the noise level of a battery is seriously increased if the supplied DC current exceeds a certain value I_{MAX}, which depends on the type and capacity of the battery. Manufacturers indicate that I_{MAX} is approximately equal to the current at which 100 hours of full operation is sufficient to completely discharge the battery.

If the risk of exhausting a battery during operation exists, the solution is to bypass the battery with a high-value, low-loss capacitor, in order to reduce its noise.

7.6 Noise in Bipolar Transistors

■ **Brief History.** In 1900, Sir John A. Fleming invented the thermionic valve, which was the basis for all vacuum electron devices. The vacuum tube is an active device in which a heated cathode produces a flow of electrons, collected by a distant positive anode, all placed in a low-pressure sealed envelope. The flight of electrons between the electrodes can be modified or modulated by interposing other electrodes (grids) to control the anode current. These devices were used to amplify weak signals or to generate oscillations, and they enabled the development of the earliest radio communications and broadcast services. However, soon after their implementation, it was discovered that their ability to amplify weak signals was limited by their own shot noise.

On June 26 1918, after remarkable research accomplished under the terribly difficult conditions that prevailed in defeated Germany after the First World War, Walter Schottky published the first paper dedicated to noise in active electron devices, entitled "Über spontane Stromschwankungen in verschiedenen Elektrizitätsleitern."

The invention of point contact transistors in 1948 by Walter Brattain, John Bardeen, and William Shockley at Bell Laboratories marked the advent of solid-state devices. The point contact transistor consisted of two closely spaced metallic wires, implanted in a germanium substrate (the base); this was the first transistor structure. One year later, Shockley published his pioneering paper entitled "The theory of p-n junctions in semiconductors and p-n junction transistors," which marked the beginning of the second age of electronics.

Although semiconductor junctions and transistors were clearly very promising devices, their main problems were soon discovered. Perhaps the most undesirable feature of point contact transistors was their high noise; for instance, in 1950 noise factors in the range 50 − 70 dB were reported at 1 kHz for PNP germanium transistors!

In 1952, Montgomery found that NPN point contact transistors had much lower noise (18 to 25 dB at 1 kHz); in the same period, measurements made in the 20 Hz − 20 kHz band revealed a strong 1/f noise component, which was bias dependent.

7.6 Noise in Bipolar Transistors

Progress in technology (namely the advent of indium-alloy transistors, which can be considered the second generation of bipolar transistors) resulted in a reduction of the 1/f noise level. For instance, in 1953 Montgomery and Clark reported that at frequencies higher than 1 kHz, germanium alloy transistors had negligible 1/f noise. But at lower frequencies the noise was still so high that ubiquitous hisses, hums, and pops systematically poisoned the quality of the first generation of transistor radio receivers.

Notwithstanding the 1/f noise, good agreement with calculation was observed for the shot noise and thermal noise associated with the bulk regions of second generation transistors. These noise mechanisms were recognized to be the limiting factors, and consequently attention was paid to the study of shot noise. The pioneering work in the field was published in 1955 by Van der Ziel in a paper entitled "Theory of shot noise in junction diodes and junction transistors." One year later Giacoletto developed and published the first common-emitter noise-equivalent circuit of the transistor.

Note that in a short lapse of time after the invention of junction transistor, the investigation of its intrinsic noise mechanisms had been considerably advanced.

7.6.1 Preliminary Remarks

■ **Approach.** To investigate the noise of bipolar transistors, two approaches exist:

- *The physical approach*, based on the study at a microscopic level of transport mechanisms of charge carriers. The expected result is the evaluation of the fluctuation of the current through the device.
- *The equivalent circuit approach.* If the fluctuations in currents and voltages are regarded as internally generated signals with very low amplitude, then an equivalent electrical circuit is needed (like the Ebers-Moll or Giacoletto circuit), to which noise sources are added, one for each noise mechanism. Note that we must not automatically attribute a thermal noise source to every resistor in the equivalent circuit, just to those that are undoubtedly dissipative! For instance, consider the element r_{ce} of Giacoletto equivalent circuit: this resistor simulates the transport of carriers between emitter and collector, but it is not dissipative; hence, it has no thermal noise source. In any event, a physical understanding of device operation is obviously required; consequently, we shall in the following recall some background theory of transistor operation.

Comparing the two approaches, the former has the advantage of being general and accurate; its main drawback is its laboriousness. The latter is straightforward, but has the usual shortcomings associated with any equivalent circuit (second-order effects are neglected and frequency-dependent validity).

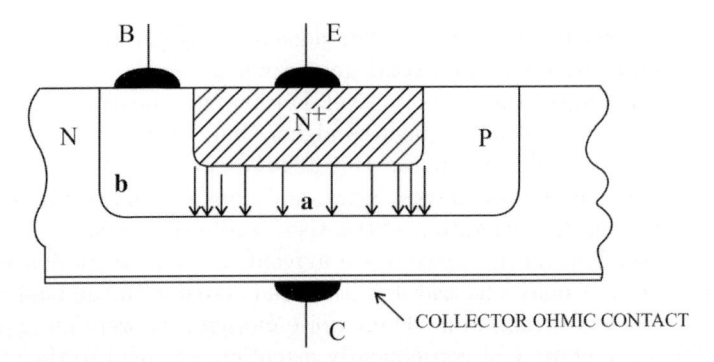

Fig. 7.5. Cross section of a discrete diffused NPN transistor

■ **Device Structure.** Figure 7.5 presents a cross section of a discrete diffused NPN transistor (the ohmic contacts are represented symbolically). An integrated NPN transistor has an identical structure, except the ohmic contacts, which are now true interconnecting metal paths, all lie on the upper side.

Two regions of the base are of interest:

- the active base (called also intrinsic base) denoted by **a**, which is traversed (vertically) by the emitter-to-collector flow of electrons;
- the inactive (or extrinsic) base denoted by **b**, which corresponds to the bulk P region. This region is technologically unavoidable and serves to transmit the bias from the base contact to the active base.

A base current exists through the thin and lightly doped active base to supply recombination current to the central region of the active base; this current must flow through both regions **a** and **b** laterally to the base contact. Therefore, the base-current flow causes a lateral voltage drop in the bulk regions (right and left), which leads to two consequences:

– The polarity of this drop is such that the central portions of the emitter junction have less forward bias than the outer portions; hence, a crowding of the emitter current lines is observed on the outer portions of the emitter.
– Another consequence is the reduction of the effective bias of the active region (**a**) with respect to the applied external voltage by an amount equal to the lateral voltage drop. To take this effect into account, a base resistance R_b is introduced into the equivalent circuit in series with the base lead of the intrinsic transistor.

7.6.2 Physical Aspects of Noise

■ **Robinson's Theory.** Robinson [94] developed a noise model of bipolar transistors based on the evaluation of noise spectral densities.

7.6 Noise in Bipolar Transistors

Since the collector current is a drift current arising from charge carriers which randomly cross the base-collector junction, shot noise exists, and its power spectral density is

$$S_C = 2q\, I_C \qquad (7.18)$$

The emitter current also displays shot noise, but the emitter current also contains a fluctuation produced by electrons that fail to cross the base. It seems reasonable to think that some electrons, after crossing the emitter-base junction are unable to travel further and re-enter the emitter, due to hazards (collisions with the lattice or other charge carriers). Therefore, for the emitter current we may write

$$S_E = 2q\, I_E + 4kT\, \Delta G_E \qquad (7.19)$$

where ΔG_E is a conductance simulating the re-entering.

In order to estimate the correlation between the collector and emitter currents, it is useful to note that each electron contributing to the collector current was previously implicated in the emitter current. Let us simplify, assuming that the transit time τ through the base is the same for all electrons. Hence, a Fourier component of the collector current at frequency f will be totally correlated with the corresponding component of the emitter current at an earlier time $(t-\tau)$. This suggests that

$$\langle i_e i_c^* \rangle = \langle i_c i_c^* \rangle \exp(j\omega\tau) \qquad (7.20)$$

and

$$\langle i_e^* i_c \rangle = \langle i_c^* i_c \rangle \exp(-j\omega\tau) \qquad (7.21)$$

In actual operation, the transit time through the base is not identical for all electrons, and we must adopt a mean value for the cross-power spectral density:

$$S_{EC} = \langle i_e^* i_c \rangle = S_C\, \overline{\exp(-j\omega\tau)} \qquad (7.22)$$

From a traditional point of view, the current i_e is related to the collector current i_c according to the following expressions:

$$i_c = \alpha_o\, i_e \qquad \text{(low-frequency response)} \qquad (7.23)$$

and

$$i_c = \alpha_o\, \overline{\exp(-j\omega\tau)}\, i_e \qquad \text{(high-frequency response)} \qquad (7.24)$$

The high-frequency current gain can be expressed as

$$\alpha = \alpha_o\, \overline{\exp(-j\omega\tau)} \qquad (7.25)$$

Using (7.25) and (7.22), we have

$$S_{EC} = 2q\, I_C\, \frac{\alpha}{\alpha_o} \qquad (7.26)$$

7 Noise Models of Electronic Devices

The base current fluctuations are heavily dependent on the fluctuations of i_e and i_c, because $i_b = i_e - i_c$; applying (2.39) to the difference of two fluctuations, we obtain

$$S_B = S_E + S_C - S_{EC} - S_{EC}^*$$
$$= 2q\,I_E + 2q\,I_C\left(1 - \frac{\alpha + \alpha^*}{\alpha_o}\right) + 4kT\,\Delta G_E \quad (7.27)$$

Equation (7.27) can be represented in a different but equivalent form

$$\boxed{S_B = 2q\,I_B + 2q\,I_C\left(2 - \frac{\alpha + \alpha^*}{\alpha_o}\right) + 4kT\,\Delta G_E} \quad (7.28)$$

The cross correlation spectrum associated with i_b and i_c is $S_{BC} = S_{EC} - S_C$:

$$\boxed{S_{BC} = 2q\,I_C\left(\frac{\alpha}{\alpha_o} - 1\right)} \quad (7.29)$$

At low frequencies (where $\alpha \cong \alpha_o$), the expressions in Table 7.2 are obtained. Note that i_b and i_c are uncorrelated, but as soon as the frequency increases, the base and collector noise currents become partially correlated, according to (7.29).

Table 7.2. Low-frequency noise of a bipolar transistor

Power spectral densities	Cross-power spectral densities
$S_E \cong 2q\,I_E$	$S_{EC} \cong 2q\,I_C$
$S_C = 2q\,I_C$	
$S_B \cong 2q\,I_B$	$S_{BC} \cong 0$

Since i_c and i_b determine i_e, it is no longer necessary to investigate the noise behavior of i_e, and the equivalent representation of Fig. 7.6 can be adopted. Together with the previous expressions, this represents the *Robinson noise model* of bipolar transistors. Note that instead of the ideal (noiseless) transistor shown in Fig. 7.6, any traditional equivalent circuit can be employed.

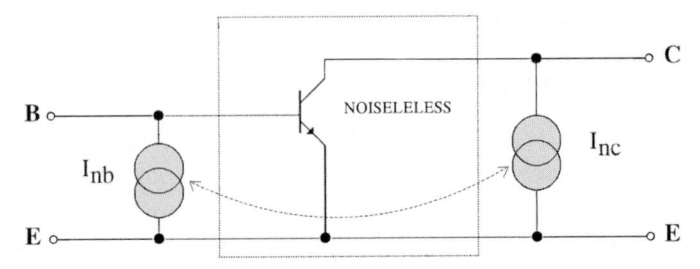

Fig. 7.6. Robinson's noise model of bipolar transistors

7.6 Noise in Bipolar Transistors

■ **Buckingham's Contribution.** Buckingham corrected the previous results by taking charge carrier recombination into account in the depleted layer of the emitter-base junction [93]. He obtained the following results:

$$S_E = 4q\, I_{ES} \left(\frac{G_E}{G_{Eo}} - \frac{1}{2} \right) + 2q\, I_{BR} \tag{7.30}$$

$$S_C = 2q\, I_C \tag{7.31}$$

$$S_B = 2q\, I_C \left(\frac{1}{\beta_o} + \frac{2G_E - (\alpha Y_E + \alpha^* Y_E^*)}{\alpha_o\, G_{Eo}} - \frac{2(1-\alpha_o)}{\alpha_o} \right) \tag{7.32}$$

where β_o is the low-frequency current gain in common-emitter configuration, which takes the recombination process into account. In the following, the term "ideal" is used in connection with the absence of recombination in the depleted layer.

The remaining notation is:

G_E – ideal conductance of the emitter-base junction
G_{Eo} – low-frequency value of conductance G_E
$I_{ES} = I_E - I_{BR}$ (I_E being the steady emitter current)
I_{BR} – depletion layer recombination component of I_B
Y_E – ideal admittance of the emitter-base junction

The cross-power spectral densities are

$$S_{EC} = -2q\, I_C\, \frac{\alpha}{\alpha_o}\, \frac{Y_E}{G_{Eo}} \tag{7.33}$$

$$S_{BC} = -2q\, I_C \left(1 - \frac{\alpha}{\alpha_o}\, \frac{Y_E}{G_{Eo}} \right) \tag{7.34}$$

$$S_{BE} = -2q\, I_C \left\{ \frac{2G_E}{\alpha_o G_{Eo}} - \frac{2}{\alpha_o} + \frac{1}{\beta_o} + 1 - \frac{\alpha^* Y_E^*}{\alpha_o G_{Eo}} \right\} \tag{7.35}$$

Note that for low-frequency operation ($Y_E \cong G_{Eo}$ and $\alpha \cong \alpha_o$) the expressions in Table 7.2 are recovered.

■ **Contribution of Niu.** In reference [137], it is emphasized that *since the transition of carriers across the collector-base region is a drift process, no shot noise can be associated with this junction alone.* Hence, the shot noise of the collector current is merely the consequence of the emitter current, which already has a shot noise component.

■ **Thermal Noise.** Thermal noise is associated with all regions of the device where real dissipative resistances exist (namely the intrinsic base resistance R_b), as well as with the ohmic contacts. The main problem is to evaluate these resistances, since in most cases they are nonlinear (i.e., the resistance depends on the current flowing through it).

184   7 Noise Models of Electronic Devices

■ **1/f Noise.** As pointed out for junction diodes, the contribution of surface trapping centers to 1/f noise can be reduced by careful surface processing during manufacturing. However, the contribution of random recombination of minority carriers in the depleted layers remains. This effect is observed in the base noise current at low frequency, and it adds to the base shot noise

$$S_B = 2q\, I_B\, (1 + f_o/f) \tag{7.36}$$

The value of the constant f_o depends on the particular sample of transistor, but as a general rule it is around 1 kHz.

■ **Concluding Remark.** In this section the internal noise sources of a bipolar transistor have been presented, together with the corresponding expressions for the power spectral densities. This approach is valuable because it enables one to calculate the noise factor of the transistor.

The difficulty is to determine certain quantities which appear in the power spectral density expressions, which are not easily obtained by measurement. This explains the need to introduce new models, based on the equivalent circuit approach.

7.6.3 Nielsen's Model

■ **Introduction.** In 1957, Nielsen proposed a low-frequency model [96] suited to manual calculation of the noise figure. The starting point was Van der Ziel's model, which has been improved; however, it still neglects the frequency dependence of the emitter-base impedance.

■ **Noise Sources.** Apart from 1/f noise, transistor noise is assumed to arise from thermal noise of the base resistance, diffusion, and recombination processes in the base region.

■ **Configuration.** Figure 7.7 presents the noise equivalent circuit for the common emitter configuration; it is made up by connecting the following noise generators to the T-equivalent circuit: thermal noise generators e_{ne} and e_{nb} (associated with the emitter and base resistance, respectively) and a collector noise generator (i_{nc}).

■ **Equations.** Since R_b and R_s are physical resistors, their thermal noise is given by:

$$\overline{e_{nb}^2} = 4kTR_b\, \Delta f \quad \text{and} \quad \overline{e_{ns}^2} = 4kTR_s\, \Delta f \tag{7.37}$$

Attention must be paid to r_e, which is not a dissipative resistance, because it represents the dynamic resistance (slope) of the base-emitter impedance:

$$r_e = kT\, /\, qI_E$$

7.6 Noise in Bipolar Transistors

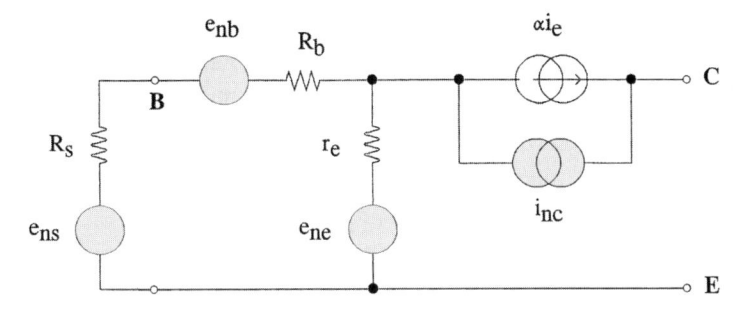

Fig. 7.7. Transistor noise equivalent circuit, according to Nielsen

Its noise can be expressed as

$$\overline{i_{ne}^2} = 2q\, I_E\, \Delta f \tag{7.38}$$

or as a noise voltage generator

$$\overline{e_{ne}^2} = 2kTr_e\, \Delta f \tag{7.39}$$

According to [96], the collector noise generator is

$$\boxed{\overline{i_{nc}^2} = 2q\, I_C \left(1 - \frac{|\alpha|^2}{\alpha_o}\right) \Delta f} \tag{7.40}$$

where α is the short-circuit current gain in common base configuration, and α_o represents its low-frequency value.

■ **Discussion.** Expression (7.40) is the key to Nielsen's model, since it simulates the frequency dependence of the transistor noise.

As a result of experimental work, it has been established that the noise factor of a bipolar transistor has a frequency characteristic with three distinct regions:

a) a low-frequency region where the noise factor increases when frequency decreases (as a result of the 1/f noise contribution);
b) a mid-frequency flat region where the noise factor is essentially constant;
c) a high-frequency region where the noise factor increases.

Now, in (7.40) we note that when the frequency increases, α decreases and the collector noise becomes more significant. At high frequencies, α approaches zero and the expression (7.40) approaches the shot noise (3.21).

At low frequencies, $\alpha \cong \alpha_o$ and the collector noise is reduced by a factor of $(1 - \alpha_o)$.

186   7 Noise Models of Electronic Devices

■ **Results.** Using Nielsen's model, the noise factor expression is

$$F = 1 + \frac{R_b}{R_s} + \frac{r_e}{2R_s} + \left\{1 - \alpha_o + \frac{f^2}{f_\alpha^2}\right\} \frac{(R_s + R_b + r_e)^2}{2\alpha_o R_s r_e} \qquad (7.41)$$

where f_α denotes the $-3\,\mathrm{dB}$ cutoff frequency of the current gain α

$$\alpha = \frac{\alpha}{1 + jf/f_\alpha} \qquad (7.42)$$

Nielsen computed the noise factor of the common-emitter (CE), common-base (CB) and common-collector (CC) circuits. According to his results, the low-frequency behavior is the same, but the high-frequency behavior is different (above f_α, the noise factor of the CE and CB stages still increases, while for the CC stage it remains essentially constant).

■ **Note**

1) By differentiating (7.41) with respect to R_s, the minimum noise factor can be found, as well as the optimal source resistance.
2) Nielsen's model predicts a frequency dependence of F due only to the transit time τ (which does not depend on the collector current). This may be a good explanation for the discrepancies observed between measured and computed values of F.
3) Expression (7.41) was obtained under the hypothesis that the signal source is purely resistive ($X_s = 0$).

7.6.4 Hawkins's Model [97]

■ **Introduction.** In 1976, Hawkins proposed some modifications to improve Nielsen's model (especially the high- frequency behavior). In 1993, Pucel and Rohde added to the previous development an exact expression to compute parameter R_n [98].

■ **Configuration.** The starting point is the statement that in microwave bipolar transistors the high-frequency behavior is mainly determined by the current-dependent time constant of the emitter-base junction. Nielsen's model is based on the assumption that the high-frequency performance is controlled by the base transit time, which does not depend on the injection level. To correct this, Hawkins added a capacitor C_e to Nielsen's model to simulate the emitter-base junction capacitance (Fig. 7.8). Furthermore, the signal source is assumed to have a complex internal impedance Z_s.

■ **Equations.** Expressions (7.37) to (7.40) remain the same.

■ **Results.** Applying North's definition of the noise factor, Hawkins found the following result:

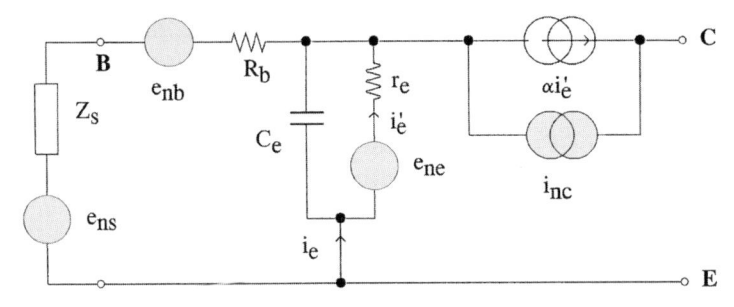

Fig. 7.8. Hawkins' noise model of a bipolar transistor

$$F = 1 + \frac{R_b}{R_s} + \frac{r_e}{2R_s} + \left\{\frac{\alpha_o}{|\alpha|^2} - 1\right\} \frac{(R_s + R_b + r_e)^2 + X_s^2}{2r_e R_s} + \\ + \frac{\alpha_o}{|\alpha|^2} \frac{r_e}{2R_s} \left\{\omega^2 C_e^2 X_s^2 - 2\omega C_e X_s + \omega^2 C_e^2 (R_s + R_b)^2\right\}$$
(7.43)

Differentiating this equation with respect to X_s and putting $\partial F/\partial X_s = 0$, the optimum source reactance is found to be

$$X_o = \frac{\alpha_o}{|\alpha|^2} \frac{\omega C_e r_e^2}{a}$$
(7.44)

Next, (7.43) is differentiated with respect to R_s, and setting $\partial F/\partial R_s = 0$, the optimum source resistance is found, as well as the optimum noise factor F_o corresponding to $R_s = R_o$ and $X_s = X_o$:

$$R_o^2 = R_b^2 - X_o^2 + \frac{\alpha_o}{|\alpha|^2} \frac{r_e(2R_b + r_e)}{a}$$
(7.45)

$$F_o = a \frac{R_b + R_s}{r_e} + \frac{\alpha_o}{|\alpha|^2}$$
(7.46)

In (7.44), (7.45), and (7.46), a denotes the coefficient

$$a = \left\{1 - \frac{|\alpha|^2}{\alpha_o} + \omega^2 C_e^2 r_e^2\right\} \frac{\alpha_o}{|\alpha|^2}$$
(7.47)

■ Note

1) Hawkins' model is valid in the low- and mid-frequency range, because all parasitic capacitances (except C_e) are neglected, as well as all parasitic resistances (except R_b). Furthermore, the inherent delay associated with the current gain α at high frequencies is also neglected.
2) In the special case $X_s = 0$, the first four terms of (7.43) are identical to Nielsen's model prediction.

188   7 Noise Models of Electronic Devices

3) The only noise parameter missing in Hawkins' model is R_n; in fact, its expression cannot be found with the proposed approach.

■ **Contribution of Pucel and Rohde.** With the method of Hillbrand and Russer discussed in Sect. 6.2, Pucel and Rohde found an expression for the parameter R_n [98]. The first step was to represent Hawkins' model in the form of a Rothe and Dahlke equivalent two-port; then, the noise global correlation matrix is delivered and the element located on its first row and first-column is identified to R_n. If f_b is the α cutoff frequency for the CB configuration, then:

$$R_n = R_b \left(D - \frac{1}{\beta_o} \right) + \frac{r_e}{2} \left(D + \left(\frac{R_b}{r_e} \right)^2 (1 - \alpha_o + k_b + k_e + k_f) \right) \quad (7.48)$$

where

$$k_b = (f/f_b)^2, \qquad k_e = \left[f(2\pi C_e r_e) \right]^2 \quad (7.49)$$

$$D = \left(1 + (k_b)^2 \right) / \alpha_o^2 \quad \text{and} \quad k_f = \left(\frac{1}{\beta_o} + k_b k_e \right)^2 \quad (7.50)$$

■ **Conclusions.** Hawkins' model yields good agreement with measurements of minimum noise figure over a wide range of emitter current at mid-frequencies. With the contribution of Pucel and Rohde, its main benefit is to completely describe the noise of bipolar transistors by computing the four noise parameters (F_o, R_n, R_o, and X_o). However, its main limitations are

– at low frequencies, 1/f noise is neglected;
– at high frequencies, all parasitics (except C_e and R_b) are neglected.

7.6.5 Model of Motchenbacher and Fitchen [99]

■ **Introduction.** In 1973, the first version of this model was proposed; it is useful for computing the transistor equivalent input noise.

■ **Configuration.** This model is based on Giacoletto's equivalent circuit (Fig. 7.9), by neglecting elements $C_{b'c}$ and $r_{b'c}$. We recall that $r_{bb'}$ is the bulk resistance of the inactive base (point B' denotes the active base).

■ **Noise Sources.** The noise mechanisms taken into account are

- shot noise, for the base and collector currents
- thermal noise associated with R_s, R_L and $r_{bb'}$
- 1/f noise current, which has been observed experimentally to flow through r_{bb}.

7.6 Noise in Bipolar Transistors

Fig. 7.9. Giacoletto's equivalent circuit

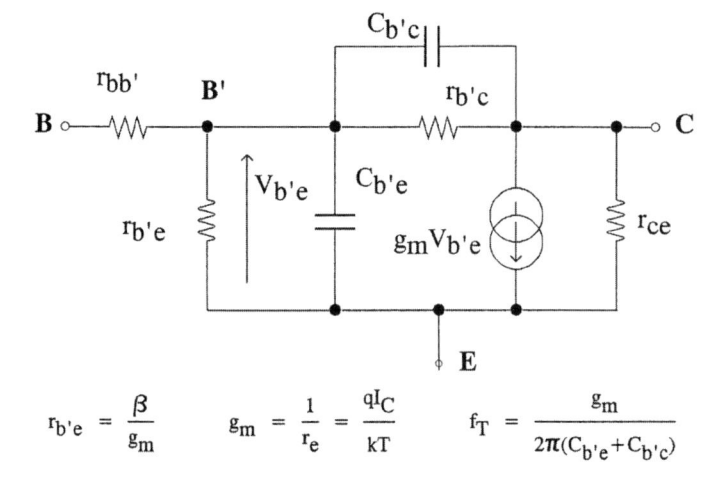

$$r_{b'e} = \frac{\beta}{g_m} \qquad g_m = \frac{1}{r_e} = \frac{qI_C}{kT} \qquad f_T = \frac{g_m}{2\pi(C_{b'e}+C_{b'c})}$$

$$C_{b'c} = A\,(C_1/V_{CE})^{1/3}$$

C_1 - constant related to the impurity profile of the junction

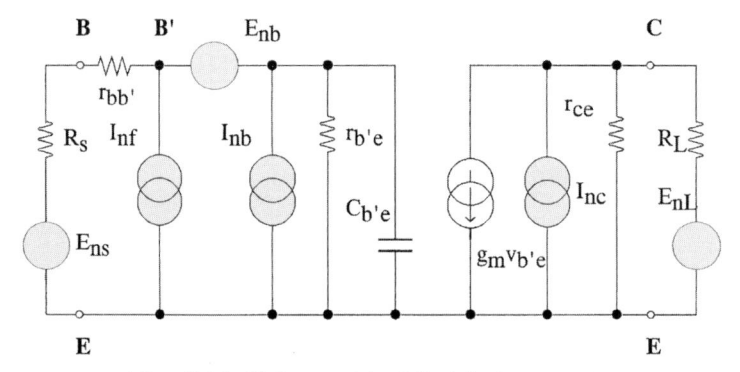

Fig. 7.10. Noise model of the bipolar transistor

The noise model is obtained by adding the corresponding noise generators to Giacoletto's equivalent circuit (Fig. 7.10).

■ **Equations.** The expressions for the noise generators are

$$\overline{E_{nb}^2} = 4kTr_{bb'}\,\Delta f \qquad \overline{E_{nL}^2} = 4kTR_L\,\Delta f \qquad \overline{E_{ns}^2} = 4kTR_s\,\Delta f \qquad (7.51)$$

$$\overline{I_{nb}^2} = 2qI_B\,\Delta f \qquad \overline{I_{nc}^2} = 2qI_C\,\Delta f \qquad \overline{I_{nf}^2} = 2qI_B^\gamma\,\frac{f_L}{f}\,\Delta f \qquad (7.52)$$

In the last equation of (7.52), γ is a coefficient ranging between 1 and 2 (often taken to be 1) and the corner frequency f_L is a constant for a particular sample (for the most common transistor types, its value is between 1.7 kHz and 7 MHz).

■ **Remark.** To obtain the noise voltage corresponding to the 1/f generator, we must multiply I_{nf} by the resistance through which the current is flowing ($r_{bb'}$). However, it has been observed experimentally that for common transistors we obtain excessive values. Therefore, to match measurements, a reduced value of $r_{bb'}$ is proposed for computing the resulting fluctuation. For instance, we may use $r_b = r_{bb'}/2$ instead (r_b is often called the *effective base resistance*).

■ **Results.** The ultimate purpose is to evaluate the transistor equivalent input noise (introduced in Sect. 4.5.1), in order to derive the signal-to-noise ratio.

The equivalent input noise generator E_{ni} is found with the following procedure:

1. Compute the total output noise.
2. Evaluate the voltage gain referred to the signal source.
3. Divide the total output noise by the voltage gain previously obtained.

Letting $a = (r_{bb'} + R_s)$, the power spectral density becomes

$$\frac{\overline{E_{ni}^2}}{\Delta f} = 4kTa + 2qI_B a^2 + 2q \frac{I_C}{\beta_o^2}(a + r_{b'e})^2 + \\ 2qI_B^\gamma \frac{f_L}{f}(R_s + r_b)^2 + 2qI_C \left(\frac{a_f}{f_T}\right)^2 \quad (7.53)$$

■ **Note**

1) Expression (7.53) takes the noise produced by the transistor into account, as well as the noise of the signal source.
2) Expression (7.53) is not exact, because both $C_{b'c}$ and r_{bc} have been neglected. It is reasonable to think that the high-frequency response will be less accurate (especially near the cutoff frequency).

■ **E_n–I_n Model.** As soon as the equivalent input noise has been found, the next step is to build the E_n–I_n model. This is merely the two-port noise representation given in Fig. 5.8d. We recall that to obtain any model, we need not only choose the circuit topology, but also assign numerical values to all elements (or at least to indicate computational expressions or measuring techniques to derive them).

7.6 Noise in Bipolar Transistors

According to Sect. 4.5.1 (see expression (4.58)), imposing the condition $R_s = 0$, the expression for the power spectral density of the noise generator E_n can be obtained. Inserting it into (7.53) yields

$$S(E_n) \cong 4kTr_{bb'} + 2q\frac{I_C}{\beta_o^2}(r_{bb'} + r_{b'e})^2 + 2qI_B^\gamma \frac{f_L}{f}r_b^2 + 2qI_C\left(\frac{r_{bb'}f}{f_T}\right)^2 \quad (7.54)$$

Assuming then that R_s is very large, we divide by R_s and take the limit of expression (7.53), yielding the power spectral density of the noise generator I_n

$$S(I_n) = 2qI_B + 2qI_B^\gamma \frac{f_L}{f} + 2qI_C\left(\frac{f}{f_T}\right)^2 \quad (7.55)$$

The minimum noise factor is

$$F_o = 1 + \sqrt{\frac{2r_{bb'}}{r_{b'e}} + \frac{1}{\beta_o}} \quad (7.56)$$

and the optimum source resistance (evaluated outside the area where 1/f noise is dominant) becomes

$$R_o = \sqrt{\frac{0.05\,\beta\,r_{bb'}}{I_C} + \frac{(0.025)^2 \beta}{I_C^2}} \quad (7.57)$$

■ **Concluding Remarks.** For typical transistors, it has been observed that:

– The values of E_n and I_n vary with the operating point and frequency. This is quite normal as the expressions for E_n and I_n are directly related by the collector current, and indirectly by means of the equivalent circuit elements (whose values depend on bias).
– As a general rule, E_n is reduced by increasing I_C, but I_n increases with I_C.
– For low-noise design and typical low-power, low-frequency transistors, the best performance is obtained when
$$10\,k\Omega < R_s < 100\,k\Omega \text{ and } 1\,\mu A < I_C < 100\,\mu A.$$
– The main limitation of the E_n–I_n model is that the noise generators E_n and I_n are assumed to be uncorrelated (although this is not true!).

7.6.6 Fukui Model [100]

■ **Introduction.** In 1966, Fukui proposed a noise model for microwave bipolar transistors, together with expressions for the noise parameters. For the first time the effect of header parasitics were considered, in order to account for the discrepancy observed between theory and experiment at frequencies higher than 1 GHz.

192   7 Noise Models of Electronic Devices

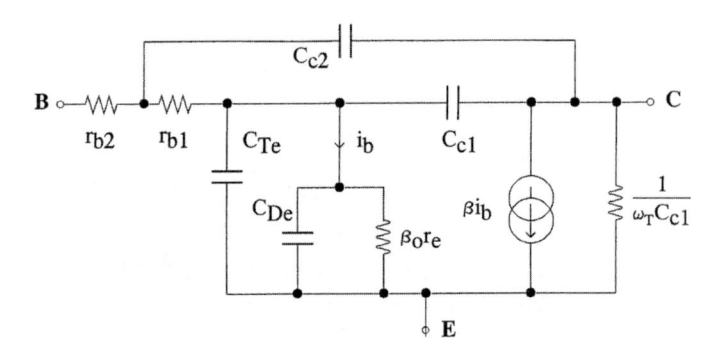

Fig. 7.11. Equaivalent circuit for the microwave transistor wafer in CE configuration

■ **Equivalent Circuit.** For a microwave transistor wafer in the common-emitter configuration, a new equivalent circuit was proposed (Fig. 7.11).
In this circuit the following notation is employed:

C_{De} – emitter diffusion capacitance
C_{Dc} – collector diffusion capacitance
r_{b1} (r_{b2}) – inner (outer) base resistance
C_{Te} – E–B transition region capacitance
C_{Tc1} (C_{Tc2}) – inner (outer) C–B transition region capacitance
C_{c1} – inner C–B capacitance ($= C_{Dc} + C_{Tc1}$)
C_{c2} – outer C–B capacitance ($= C_{Tc2}$)
ω_T – total unity gain angular frequency

■ **Noise.** The noise model is obtained by adding a shot noise source in the collector, another shot noise source in the base, and a thermal noise source associated with the internal base resistance (denoted here $r_b = r_{b1} + r_{b2}$). 1/f noise is ignored.
Of course, these noise sources are uncorrelated with one another, and they are described by the classical equations

$$\overline{i_{nb}^2} = 2qI_B \, \Delta f \qquad \overline{i_{nc}^2} = 2qI_C \, \Delta f \qquad \overline{e_{nb}^2} = 4kTr_b \, \Delta f \qquad (7.58)$$

The noise model (Fig. 7.12) includes these noise sources, as well as the thermal noise of the signal source (represented in the Norton representation). For computational facility, C_{c1} and C_{c2} are neglected; Fukui explains that up to L-band, their omission does not affect accuracy (since they provide negative feedback, which is assumed to be largely favorable to noise reduction).
Header parasitics C_{BE}, C_{CE}, L_{B1}, L_{B2}, L_E and L_C are determined from the transistor Y-parameters characterization.

7.6 Noise in Bipolar Transistors

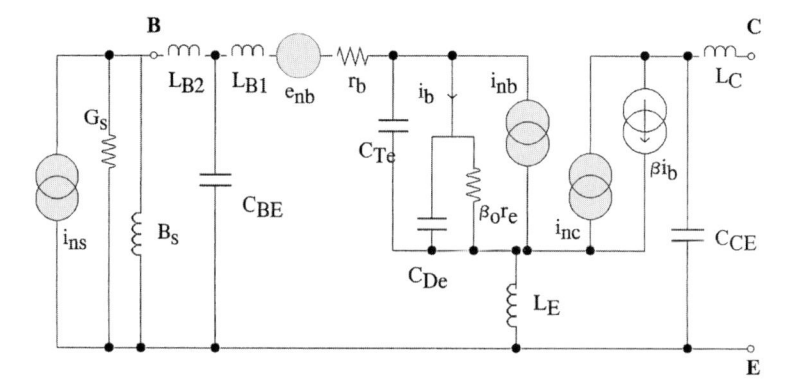

Fig. 7.12. Fukui's microwave transistor noise model

■ **Results.** Applying North's definition of the noise factor, Fukui obtained

$$F = \mathcal{A} + \mathcal{B}G_s + \frac{\mathcal{C} + \mathcal{B}B_s + \mathcal{D}B_s}{G_s} \quad (7.59)$$

Coefficients $\mathcal{A}, \mathcal{B}, \mathcal{C}, \mathcal{D}$ are computed with respect to the following quantities:

$$a = r_b\left(1 + \frac{1}{\beta_o}\right) + \frac{kT}{2qI_C} +$$

$$\left\{1 + \left[\frac{qI_C}{kT}r_b\right]^2 \left(\frac{1}{\beta_{cc}} + \frac{1}{\beta_o^2} + \frac{\omega^2}{\omega_T^2}\left(1 + \omega^2(L_{b1} + L_{B2})^2/r_b^2\right)\right)\right\}$$

$$b = \omega(L_{B1} + L_{B2} + L_E); \quad c = 1 - \omega^2 L_{B1} C_{BE}; \quad d = \omega^2 L_{B1}(L_{B2} + L_E)$$

$$e = \frac{1}{\beta_{cc}} + \frac{1}{\beta_o^2} + (\omega/\omega_T)^2 \qquad g = 1 - \omega^2 C_{BE}(L_{B2} + L_E)$$

Then

$$\mathcal{A} = 1 + \frac{1}{\beta_o} + \frac{qr_bI_C}{kT}e \quad (7.60a)$$

$$\mathcal{B} = ac^2 + \frac{qI_B}{2kT}[b - \omega C_{BE}d]^2 + \frac{qI_C}{2kT}\left\{\left(\frac{1}{\beta_o} + \frac{\omega^2}{\omega_T^2}\right)(\omega L_{B1})^2 + \right.$$

$$\left. + 2\left(\frac{1}{\beta_o^2} + \frac{\omega^2}{\omega_T^2}\right)\left(1 + \frac{C_{BE}}{C_e}\right)cd - \frac{2kT}{qI_C}\frac{\omega}{\omega_T}bc\right\} \quad (7.60b)$$

$$\mathscr{C} = \frac{qI_C}{2kT}\left(\frac{g}{\beta_{cc}} + \frac{1}{\beta_o^2} + \frac{\omega^2}{\omega_T^2}\left(1 + \frac{2C_{BE}}{C_E}\right)\right)g + (\omega C_{BE})^2\left\{r_b\left(1 + \frac{1}{\beta_o}\right) + \right.$$

$$\left. + \frac{kT}{2qI_C}\left(1 + \left(\frac{qr_bI_C}{kT}\right)^2\right)e\right\} \tag{7.60c}$$

$$\mathscr{D} = \frac{\omega}{\omega_T} + 2\omega C_{BE}\left\{r_b\left(1 + \frac{1}{\beta_o}\right) + \frac{kT}{2qI_C}\left(1 + \left(\frac{qr_bI_C}{kT}\right)^2 e\right)\right\}c - $$

$$-\frac{qI_C}{kT}\left(eg + \frac{\omega^2}{\omega_T^2}\frac{2C_{BE}}{C_e}\right)(b - d\omega C_{BE}) \tag{7.60d}$$

Finally, the transistor noise parameters are

$$F_o = \mathscr{A} + \sqrt{4\mathscr{B}\mathscr{C} - \mathscr{D}^2} \tag{7.61a}$$

$$G_o = \frac{\sqrt{4\mathscr{B}\mathscr{C} - \mathscr{D}^2}}{2\mathscr{B}}, \quad B_o = -\frac{\mathscr{D}}{2\mathscr{B}} \tag{7.61b}$$

$$R_n = \mathscr{B} \tag{7.61c}$$

■ **Note**

1) Expression (7.59) contains only one term ($\mathscr{A}$) not related to the header parasitics.
2) With increasing frequency, the contribution of terms $1/\beta_{cc}$, $1/\beta_o$, and $1/\beta_o^2$ in (7.61a) vanishes; neglecting header parasitics, a quick estimate of high-frequency minimum noise figure yields

$$F_o|_{HF} \cong 1 + x\left(1 + \sqrt{1 + \frac{2}{x}}\right) \tag{7.62}$$

with

$$x = \frac{qr_bI_C}{kT}\left(\frac{\omega}{\omega_T}\right)^2 \tag{7.63}$$

This demonstrates that the high-frequency value of F_o increases monotonically with x, which in turn depends on I_C.

Consequently, lowering I_C may reduce noise, but this also reduces the transition frequency f_T and increases the base resistance. Therefore, it is likely that F_o will reach a minimum for a certain value of I_C, which is obtained by differentiating (7.62) with respect to I_C. Assuming that

$$r_b = r_{b\infty} + s/I_C$$

where $r_{b\infty}$ is the hypothetical value of r_b at $1/I_C = 0$ and s is the slope of the r_b versus $1/I_C$ line, the minimum F_o is obtained for

$$(I_C)_{opt} = 0.08\, f_1\, C_E \left(1 + \sqrt{1 + (50s\, /\, f_1\, C_E\, r_{b\infty})}\right) \text{ [mA]} \quad (7.64a)$$

(f_1 expressed in GHz, C_E in pF, and $r_{b\infty}$ in Ω). The following variables have been introduced in (7.64a):

$$\frac{1}{\omega_1} = \frac{1}{\omega_T} - \frac{kT}{qI_C} C_E \quad \text{and} \quad C_E = C_{Te} + C_{BE} \quad (7.64b)$$

■ **Conclusion.** The expressions for noise parameters deduced by Fukui reveal the following effects due to header parasitics:

- the parasitics of high quality headers have practically no effect on the minimum noise factor;
- in contrast, R_n, G_o, and B_o are strongly dependent on both the header parasitics and the parasitic inductances of the transistor leads. At frequencies close to f_T, these parasitics in conjunction with the wafer parameters can give rise to a series resonance in the frequency characteristics of R_n, G_o, and B_o.

7.6.7 Model of Voinigescu et al.

■ **Introduction.** This model, proposed in 1997 [135], belongs to the class of analytic models. It is fully scalable and has been tested on high-speed Si and SiGe circuits. The main idea is to simplify noise matching in silicon RF circuit design by focusing the optimization on the size of transistors, rather than on the matching circuit around a transistor.

■ **Noise Equivalent Circuit.** This is structured upon the model of Robinson (Fig. 7.6), to which the thermal noise sources of R_E, R_B, and R_C (the series emitter, base, and collector resistances) are added. The shot noise generators in the base and collector are assumed to be uncorrelated (this is equivalent to neglecting the imaginary part of the transconductance, which, according to the authors, is a reasonable simplification up to $f_T/2$).

■ **Noise Model.** The approach is based on the chain correlation matrix and its companion set of (5.42). The device size is typically tailored to the emitter length (denoted by l_E) to achieve optimal low-noise matching. Parameter R_n scales as $1/l_E$, while the optimal noise resistance scales as l_E and decreases with increasing frequency. For the bias currents and frequency range used in RF design, the four noise parameters are

$$R_n \cong \frac{n^2 V_T}{2I_C} + (R_E + R_B) \quad (7.65a)$$

$$Y_o \cong \frac{f}{f_T R_n}\left(\sqrt{A + \frac{n^2 f_T^2}{4\beta_o f^2}} - j\frac{n}{2}\right) \qquad (7.65b)$$

$$R_o \cong \frac{R_n f_T}{f}\frac{\sqrt{A + \frac{n^2 f_T^2}{4\beta_o f^2}}}{A + \frac{n^2}{4}\left(1 + \frac{f_T^2}{\beta_o f^2}\right)} \qquad (7.65c)$$

$$F_o \cong 1 + \frac{n}{\beta_o} + \frac{f}{f_T}\sqrt{A + \frac{n^2 f_T^2}{\beta_o f^2}} \qquad (7.65d)$$

with

$$A = \frac{I_C}{2V_T}(R_E + R_B)(1 + f_T^2/\beta_o f^2) \qquad (7.65e)$$

The following notation has been used: $V_T = kT/q$ (thermal voltage), β_o is the DC current gain, n is the collector current ideality factor (approximately equal to 1), and f_T is the transition frequency.

■ **Concluding Remarks**

- As long as the reverse form factor of the emitter stripe (l_E/w_E) exceeds 10, it is stated in [135] that F_o remains invariant to changes in emitter length, and increases almost linearly with frequency.
- For low-noise design, R_o should be set equal to $Z_o = 50\,\Omega$ at the desired frequency of operation.

7.6.8 PSPICE Model

■ **Electrical Model.** For simulation purposes, the intrinsic bipolar transistor in PSPICE is an extended Gummel-Poon model. There are 40 parameters describing this model, to which default values are provided.

To improve simulation accuracy, ohmic resistances in series with the collector (RC/*area*) and emitter (RE/*area*) are added; the base resistance is evaluated as a function of base current. Note that *area* represents the relative device area (it plays the role of a scaling factor, whose default value is 1).

For NPN and PNP transistors, an isolation junction capacitance is provided between the intrinsic collector and the substrate node (pertinent to integrated circuits).

■ **Noise Model.** The noise model includes the following noise sources, described by their power spectral densities:

- Thermal noise, associated with all parasitic series resistances:

$$S(I_c) = 4kT / (RC/area) \qquad (7.66a)$$

7.6 Noise in Bipolar Transistors

$$S(I_b) = 4kT / (R_b) \tag{7.66b}$$

$$S(I_e) = 4kT / (RE/area) \tag{7.66c}$$

– Two noise sources (accounting for shot and 1/f noise) in the collector:

$$S(I_C) = 2qI_C + KF(I_C)^{AF} / FREQ \tag{7.67a}$$

and a similar pair in the base:

$$S(I_B) = 2qI_B + KF(I_B)^{AF} / FREQ \tag{7.67b}$$

Model parameters:

KF – flicker noise coefficient (default is 0)
AF – flicker noise exponent (default is 1)
FREQ – frequency

It is assumed that the collector and the base noise are uncorrelated (this is true only at low frequency).

■ **Contribution of Zillmann and Herzel [103].** The high-frequency noise of the model was improved by replacing the thermal noise of R_b with a quasi-thermal noise generator in the input circuit

$$S(i_{gin}) = 4kT\, \mathfrak{Re}\{y_{in}\} \tag{7.68a}$$

where y_{in} is the complex input admittance of the transistor. The noise current source corresponds to (7.68a), connected to the external base and the internal emitter node. Neglecting the emitter series resistance, the input admittance is approximately given by

$$y_{in} \cong \left[(1/r_{b'e} + j\omega C_{b'e})^{-1} + R_b\right]^{-1} \tag{7.68b}$$

Similarly, the power spectral density of the thermal noise generated in the collector is

$$S(i_{cout}) = 4kT_C\, \mathfrak{Re}\{y_{out}\} \tag{7.68c}$$

where y_{out} is the noisy output admittance and T_C represents the effective collector temperature (which may exceed the lattice temperature due to carrier heating in the base–collector space charge region).

Finally, the noise factor can be evaluated with the equation

$$F = 1 + \frac{1}{4kTg_G}\left(2q|I_B| + 4kT\,\mathfrak{Re}\{y_{in}\} + 2q|I_C| + \right.$$

$$\left. 4kT_C\,\mathfrak{Re}\{y_{out}\}\frac{1}{|G|^2}\right) \tag{7.68d}$$

where G is the current gain of the terminated two-port and g_G is the real part of the signal source admittance.

■ **Results.** PSPICE calculates the noise contributions from each device and does an rms sum at the node designated as output. The equivalent input noise is also available.

■ **Conclusion.** The bipolar transistor model adopted by PSPICE accurately simulates electrical performance at both low and high frequencies.

Noise analysis accuracy is satisfactory only at low frequencies, for several reasons:

- the correlation between collector and base noise is neglected;
- except for the isolation substrate junction, no other parasitics are taken into account;
- the four spot noise parameters (F_o, R_n, R_o, and X_o) of the transistor are not available.

The contribution of Zillmann and Herzel considerably improves the high-frequency accuracy of the traditional SPICE noise model.

7.6.9 Conclusion

The various noise models are compared in Table 7.3.

Table 7.3. Noise models for bipolar transistors

Model	Computed noise parameters	Frequency domain
Nielsen	F	~1K
Hawkins	F_o, R_n, G_o, B_o	~1K–1M
Motchenbacher	E_{ni}	0–1M
Fukui	F_o, R_n, G_o, B_o	0–1M
Voinigescu	F_o, R_n, G_o, B_o	1M–1G
PSPICE	E_{ni}, E_{no}	0–1K

Note:
1) All models are restricted to small-signal operation.
2) If the frequency domain of validity includes 0 Hz, this means that 1/f noise is considered. Otherwise, the indicated limits of validity are approximate.

7.6 Noise in Bipolar Transistors   199

3) All models in Table 7.3 have the following features:
   - Each model is valid over a well defined frequency range, which is not easy to explain, due to the various simplifications adopted during computation.
   - Each model contains a lot of "odd" parameters not currently found in transistor data files; they must either be deduced from measurements, or calculated from the available specifications. The former technique is preferred, since it allows for the inherent spread in values from one sample to another.
   - Since the models of Hawkins, Fukui, and Voinigescu yield the four noise parameters F_o, R_n, G_o, and B_o (or an equivalent set), they are recommended when a matching condition is imposed (as in microwave amplifiers). In contrast, the models leading to the equivalent input noise (Motchenbacher or PSPICE model) are appropriate in low-frequency applications, where the load is not necessarily matched (for instance, a sensor followed by its amplifier).

7.6.10 Noise in Heterojunction Bipolar Transistors (HBTs)

■ **Introduction.** In 1951 Shockley and Kroemer proposed using heterostructures to build a better bipolar transistor. Several decades later, improvements in materials and technology provided the basis for fabricating these devices.

The HBT is a device that makes use of the improved transport properties of compound semiconductors. It is well-established that in classical bipolar junction transistors, one of the most important limiting factor is the emitter injection efficiency (in an NPN structure, the maximum current gain is limited by the hole component injected by the base into the emitter). Another limiting factor is the carrier transit time through the base.

A practical solution to all these problems consists in making use of an abrupt heterojunction, such that the emitter region has a wider gap than the base. One of the first candidates was the AlGaAs/GaAs heterostructure, followed by GaN and SiGe heterostructures. Recently fabricated InGaP/InGaAs HBTs show improved reliability and higher breakdown voltage than their AlGaAs/GaAs counterparts.

■ **Structure.** The cross section of an InGaP/InGaAs HBT fabricated on a 635-μm thick semi-insulating GaAs wafer [131] is shown in Fig. 7.13. Due to bandgap engineering of the emitter-base junction, hot electrons are injected into the base and consequently cross the base layer at high speed (almost ballistically). The transit time is dramatically reduced. At the same time, the injection of holes from the base region into the emitter is reduced.

■ **Small-Signal Equivalent Circuit.** Figure 7.14 details the small-signal equivalent circuit of an HBT. According to [131], typical numerical values for an InGaP/InGaAs based HBT at $I_C = 5\,\text{mA}$ are also indicated in the

200   7 Noise Models of Electronic Devices

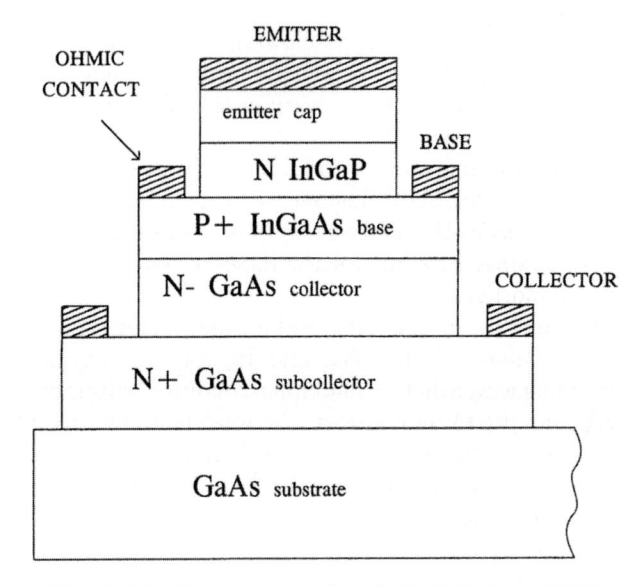

Fig. 7.13. Cross section of an InGaP/InGaAs HBT

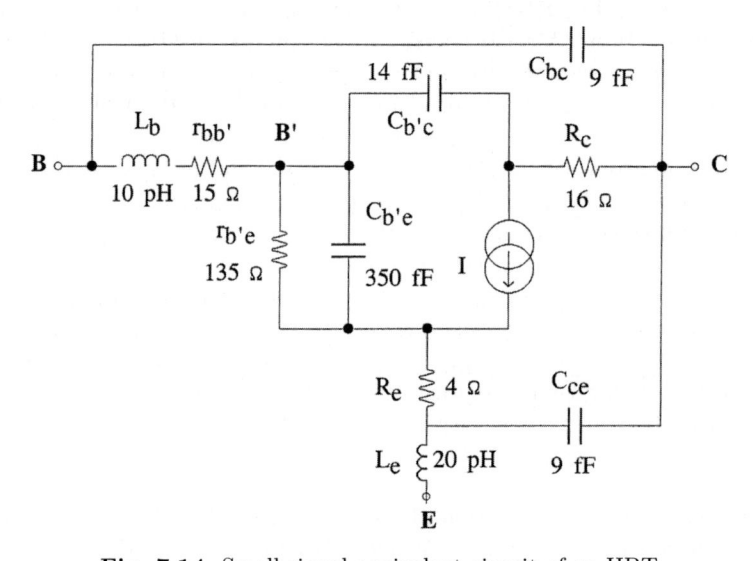

Fig. 7.14. Small-signal equivalent circuit of an HBT

figure. Note the value of the base spreading resistance, which is considerably lower than in a classical bipolar transistor. This has a favorable effect on the high-frequency behavior, as well as yielding very low values of junction capacitance. Despite ultrathin layers employed in the transistor structure, the ohmic resistances of the emitter and base are kept as low as usual.

7.6 Noise in Bipolar Transistors   201

■ **Noise.** One of the most important benefits of the HBT is its reduced intrinsic flicker noise. This is explained by its exceptional heterointerface quality, with very few dislocations and low density of interface states.

On the other hand, if the base resistance $r_{bb'}$ is reduced, the associated thermal noise must consequently be reduced. Also considering that the base width is typically around 70 nm and that the injected electrons are hot electrons traveling near ballistically, we expect that diffusion noise is no longer dominant. Finally, at high frequencies, the main noise mechanisms are shot noise, G–R noise, and thermal noise. Nevertheless, the total noise level must be considerably reduced with respect to classical bipolar transistors.

■ **Noise Model.** To the best of the author's knowledge, a general noise model for HBTs has yet to be published. Several tentative models exist, but they cover either particular aspects (like low-frequency noise sources [132]) or are dedicated to devices fabricated in a particular material (such as SiGe, which is widely used in RF applications [133]).

The noise sources considered are the collector current noise, the base current noise, and ultimately the thermal noise associated with the base spreading resistance.

The model configuration is that of Robinson (Fig. 7.6). According to [132], the problem specific to HBT which also complicates empirical extraction of the noise coefficients is the temperature dependence of the base current amplification factor ($\beta = I_C/I_B$). It is assumed that this relation is of the form

$$\beta = \beta_o \exp(\Delta E_g / kT) \tag{7.69}$$

where β_o denotes the the current amplification factor for a homojunction transistor and ΔE_g is the band gap between the two materials in the base-emitter junction.

A Y-parameter noise model is proposed in [133]; the Y-parameters of the transistors are measured or obtained by simulation. For the chain correlation matrix representation proposed in Fig. 5.8d, the noise spectral power densities are

$$S(i_n) = 2qI_B + \frac{2qI_C}{|h_{21}|^2} \tag{7.70a}$$

$$S(e_n) = 4kTr_{bb'} + \frac{2qI_C}{|y_{21}|^2} \tag{7.70b}$$

$$S(e_n i_n^*) = \frac{2qI_C\, y_{11}^*}{|h_{21}|^2} \tag{7.70c}$$

where h_{21} is the AC current gain and y_{11}, y_{21} are the input admittance and transfer admittance, respectively. In this way, the correlation matrix is known; next, the four noise parameters are deduced with expressions (5.42). Note that in this model, the G–R noise is not taken into account.

■ **Conclusion.** The Y-parameter based noise model shows that the input current noise generator can be reduced by increasing β and f_T. This translates into optimization of the SiGe profiles for higher β and f_T under the fundamental constraint of SiGe film stability. For the best low-noise profile, $F_o = 0.2$ dB at 2 GHz has been reported [133], with an associated gain of 13 dB.

7.7 Noise of Junction Field Effect Transistors (JFETs)

7.7.1 Background

■ **Definition.** The term field effect transistor (FET) refers to a class of transistors in which current flows from a *source* to a *drain* via a *channel* whose resistance is controlled by applying a voltage to a *gate*.

The FET is a unipolar device (the term *unipolar* denotes a device in which current results from the flow of charge carriers of only one polarity: electrons or holes). The name *field effect* suggests that the current flow is controlled by an electric field set up in the device by an externally applied voltage to the gate.

■ **Classification.** There are three types of FET:

- JFET (**J**unction **F**ield **E**ffect **T**ransistor), in which the junction gate (reverse biased) controls the resistance of the channel by means of the width of the depletion layer. Currently these are fabricated in silicon as discrete devices or as monolithic integrated circuits.
- MOS or MOSFET (**M**etal **O**xide **S**emiconductor), where the gate is isolated from the channel by a thin silicon dioxide layer. These are also fabricated in silicon (as discrete devices or in monolithic integrated circuits).
- MESFET (**ME**tal **S**emiconductor **F**ield **E**ffect **T**ransistor), where the gate electrode forms a Schottky barrier contact with a doped semiconductor region. These transistors are usually fabricated in GaAs. Belonging to this family is the modulation-doped AlGaAs/GaAs transistor, also called HFET (**H**eterostructure **FET**) or TEGFET (**T**wo-**D**imensional **E**lectron **G**as **F**ield **E**ffect **T**ransistor), or MODFET (**MO**dulation **D**oped **F**ield **E**ffect **T**ransistor), or HEMT (**H**igh **E**lectron **M**obility **T**ransistor).

In the following, a separate section will be devoted to the noise of each category.

■ **Device Structure.** Figure 7.15 shows a simplified diagram of a generic FET: a bar of N-type material has regions of P material embedded in each side. The two P regions are joined electrically and the common connection is the gate. Ohmic contacts are made to the two ends of the bar, known as

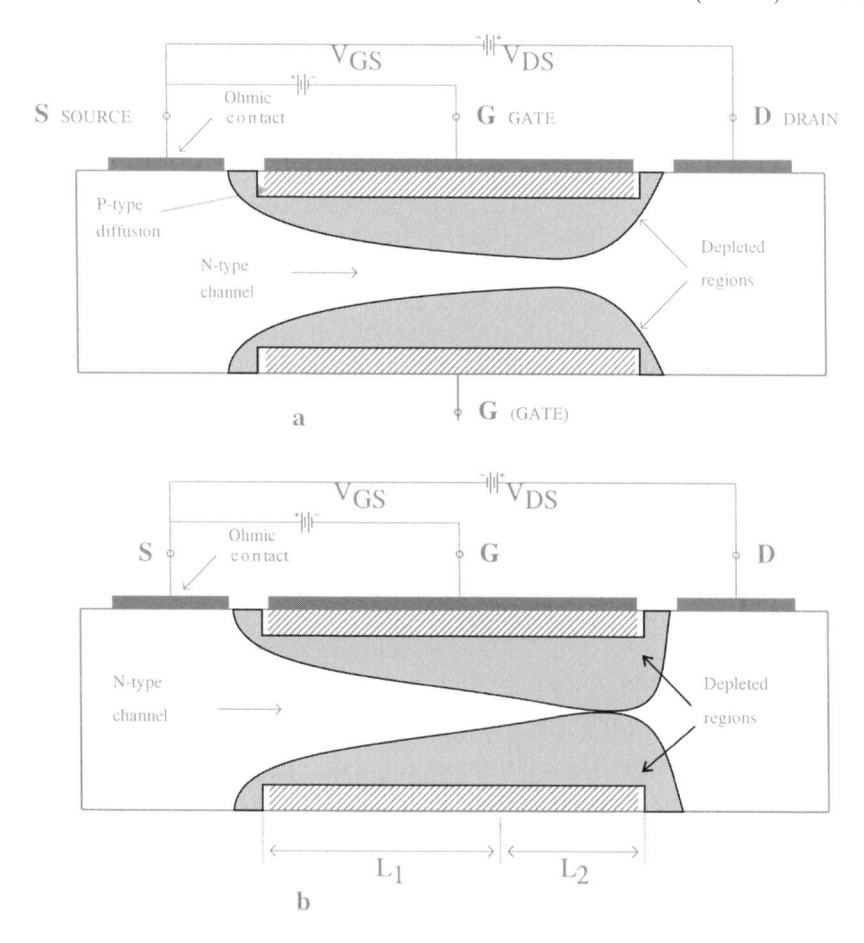

Fig. 7.15. Schematic cross section of a generic JFET: **a** linear operation mode; **b** saturated operation mode

source (the terminal through which charge carriers enter the bar) and the *drain* (the terminal through which charge carriers leave the bar).

In the case of MOS, the gate is isolated from the channel, by a thin oxide layer; the substrate appears as a fourth terminal (often connected to ground). A capacitor is produced under the metal of the gate, the second plate being the semiconductor itself. Most electrical properties of MOS transistors are determined by the quality of this capacitor.

The structure of a MESFET is similar to that of a JFET, except for the semiconductor of the gate, which is replaced by a thin metallic layer.

■ **JFET Operation.** Assuming an N-type semiconductor bar, current flow results from the transport of electrons between source and drain, across the channel. As a general rule, the gate-channel junction is reverse-biased; it follows that the input current is negligible, and the JFET will exhibit high power gain.

Recall that an ideal depleted region has no charge carriers, and its conductivity is therefore zero. Also recall that as reverse bias applied to a junction increases, so does the thickness of its depleted layer.

For the JFET shown in Fig. 7.15a, the reverse bias causes a pair of depleted regions to form, which mainly expand into the channel (because this is more lightly doped than the gate). Depending on the applied gate voltage, these depleted regions encroach upon the channel to a greater or lesser extent. Hence, the effective width of the channel will progressively decrease with increasing reverse bias: the channel acts as a variable resistor (whose cross-sectional area is voltage-controlled). As the cross-sectional area shrinks, the electric field produced by V_{DS} becomes stronger, and consequently the electron average velocity increases. This corresponds to *linear mode* operation. Note that the depletion regions are broader near the drain, due to the voltage drop produced by current flowing along the channel, which is superimposed on the gate voltage, so the bias is enhanced at the drain edge.

If the gate voltage is increased enough, the depleted regions can meet within some section (L_2) inside the channel (Fig. 7.15b). This condition is called *pinch-off*. Inside L_2 the velocities of electrons are no longer proportional to the electric field produced by V_{DS}, and the drain current *saturates*. This corresponds to operation in the *saturated mode*.

■ **Material Properties.** JFET's are usually fabricated in silicon, but MESFET's and HEMT's are gallium arsenide (GaAs). A comparison between the physical properties of these two semiconductors is presented in Table 7.4. Due to high electron mobility in GaAs, the bulk resistances are reduced relative to Si (for devices with the same geometry); hence, the thermal noise of MESFET's and HEMT's is expected to be considerably reduced, provided that heat dissipation is not a problem. Since the thermal conductivity of GaAs is not as good as that of Si, it is possible to reduce thermal resistance and avoid overheating by thinning the GaAs substrate.

Table 7.4. Comparison of GaAs and Si

Feature	GaAs	Si	Units
Electron mobility ($N_D=10^{17}$ cm^{-1})	4000	800	cm^{2}V^{-1}s^{-1}
Electron saturation velocity	$1.4 \cdot 10^7$	$6.5 \cdot 10^6$	cm s^{-1}
Hole mobility ($N_A=10^{17}$ cm^{-1})	250	350	cm^{2}V^{-1}s^{-1}
Dielectric constant	12.6	12	
Intrinsic resistivity	10^9	10^6	ohm cm
Band gap	1.43	1.12	eV
Thermal conductivity	0.9	1.5	W cm^{-1}K^{-1}

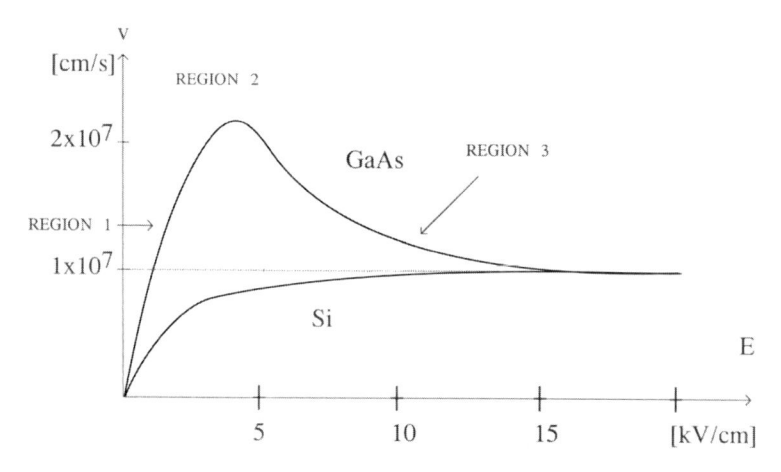

7.7 Noise of Junction Field Effect Transistors (JFETs)

Fig. 7.16. Electron drift velocity versus electric field

Another important feature is the electron drift velocity-electric field relationship. The plots in Fig. 7.16 show that GaAs exhibits a peak at about 3 KV/cm, while for Si the variation is close to monotonic (both materials exhibit velocity saturation at high field). This explains the difficulty of modeling the transport of electrons along the MESFET's channel, and consequently noise evaluation.

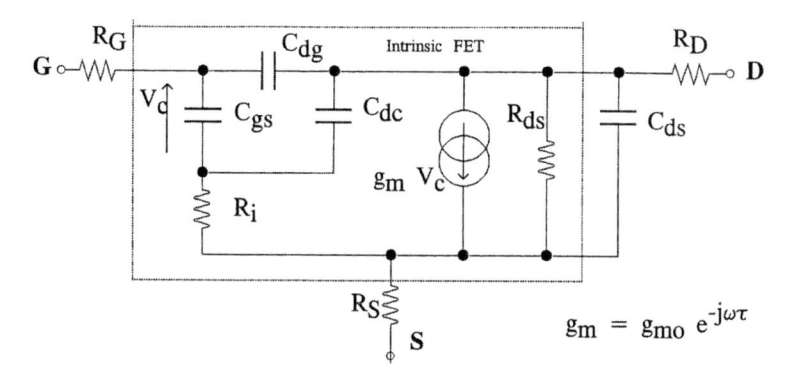

Fig. 7.17. Small-signal equivalent circuit of a FET

■ **Small-Signal Equivalent Circuit.** The FET channel should be accurately modeled as a distributed RC network; nevertheless, at frequencies lower than 12 GHz, the lumped circuit shown in Fig. 7.17 can still be successfully employed [101]. The intrinsic FET (in dashed box) contains the following elements (typical values for a MESFET with a 1 μm×500 μm gate are shown in parenthesis):

C_{dg} – drain-gate (fringing) capacitance *(0.014 pF)*
C_{dc} – drain-channel capacitance *(0.02 pF)*
C_{gs} – gate-source (depletion layer) capacitance (the total gate-channel capacitance is $C_{gc} = C_{dg} + C_{gs}$) *(0.62 pF)*
R_i – that part of the channel resistance across which C_{gs} charges (approximately $1/g_{mo}$) *(2.6 Ω)*
R_{ds} – drain (output) resistance *(400 Ω)*
g_{mo} – low-frequency transconductance *(53 mS)*
τ – delay *(5.0 ps)*

To completely model the device, the following extrinsic elements must be considered:

C_{ds} – drain-source (fringing) capacitance *(0.12 pF)*
R_G – gate metal resistance *(2.9 Ω)*
R_S – source-channel resistance including contact resistance *(2.0 Ω)*
R_D – drain-channel resistance including contact resistance *(3.0 Ω)*
L_G – inductance of gate contact *(0.05 nH)*
L_D – inductance of drain contact *(0.05 nH)*
L_S – inductance of source contact *(0.04 nH)*

7.7.2 Noise Mechanisms

■ **Sources.** In JFETs noise arises from two categories of sources: 1) noise sources associated with the intrinsic FET; 2) noise sources associated with extrinsic elements.

As the second category generates thermal noise evaluated with Nyquist's equation, attention will be paid to the specific aspects of the first.

□ **Channel Noise.** Depending on the bias applied to the gate, two situations can arise:

- The channel operates in linear mode (REGION 1 in Fig. 7.16): in this case the charge carriers are in thermal equilibrium with the lattice (Einstein relation holds), and from a macroscopic point of view Ohm's law is satisfied. The channel generates *thermal noise*, evaluated according to Nyquist's equation.
- If pinch-off occurs, the channel operates in saturated mode (REGION 2): the carriers are no longer in thermal equilibrium with the lattice (due to their increased velocity and energy, they are "hot" electrons). Their behavior is instead determined by collisions with lattice atoms, and consequently Einstein relation no longer holds, nor does Ohm's law. In this case the channel generates *diffusion noise*.

7.7 Noise of Junction Field Effect Transistors (JFETs)

■ **Gate Noise.** This noise has two components:

1) *Noise induced in the gate by the channel*; indeed, due to the capacitive coupling between gate and channel, any fluctuation in the drain current causes a fluctuation in the gate charge. It follows that the induced gate noise must be correlated with the channel noise, at least partially. The inherent capacitive coupling has two consequences: 1) the correlation coefficient is an imaginary quantity; 2) the power spectral density is proportional to frequency.

2) *Shot noise of the residual gate current*; the JFET has a gate current [95]

$$I_G = I_o + I_{Gc} + I_{Gt}$$

where I_o is due to minority carriers thermally generated in the channel; I_{Gc} corresponds to minority carriers generated by impact ionization in the channel, and I_{Gt} is due to electron-hole pairs generated in the depleted regions of the channel. Note that in a silicon JFET at low temperature I_{Gt} predominates.

All of these currents generate shot noise; their fluctuations are uncorrelated (since they arise from different physical processes).

■ **1/f Noise.** The main mechanism responsible for 1/f noise is the generation and recombination of carriers in the pair of depleted regions that extend into the channel. The generation centers (impurity atoms or lattice defects) randomly generate charge carriers, and thus randomly switch from a neutral electric state to a charged state, and vice versa. This fluctuation in the charge of various centers induces local modulation of the channel width. Consequently, the drain current fluctuates, and this is the main source of excess noise. Note that as the generation centers are equally involved in producing leakage gate current, *a good practical rule in selecting a low 1/f noise JFET is to choose a sample with low gate current.*

Finally, this kind of noise is not dominant in JFETs, where conduction occurs in the bulk, not at the surface; therefore, in a first-order model 1/f noise can be neglected.

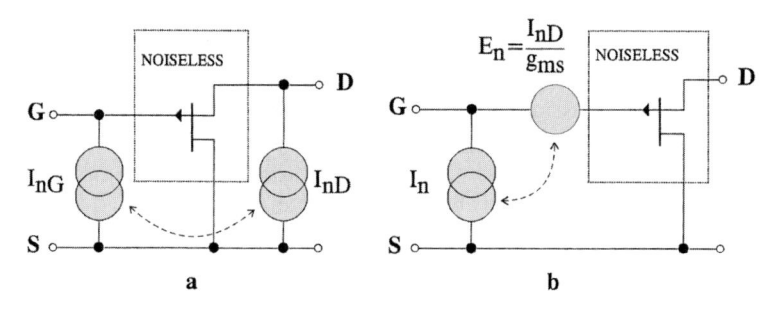

Fig. 7.18. a Physical representation of JFET noise; b Noise referred to the input

■ **Noise Equivalent Circuit.** It is convenient for noise calculation to represent all internal noise sources by means of two partially correlated current generators (Fig. 7.18a), one connected across the input and the other across the output of a noiseless JFET.

The equivalent representation given in Fig. 7.18b is particularly useful for manual calculation of the noise factor.

7.7.3 Van der Ziel Model

■ **Introduction.** In 1962, the first version of this model was proposed [102]. The JFET is assumed to be a planar device made of P-type material, with an one-dimensional channel operating in the linear mode. Van der Ziel adopted Shockley's basic assumption that the field in the channel can be treated as a superposition of two one-dimensional fields: a longitudinal field corresponding to carriers flowing between source and drain, and a transverse field corresponding to depleted regions in the channel. This field decomposition is possible provided that the channel cross section varies "slowly" along the channel (the implication being that the channel length must be at least three times the channel thickness). This is what is traditionally called the *gradual channel approximation*.

■ **Noise Model.** For the representation in Fig. 7.18a, the short-circuit drain noise current is given by

$$\overline{I_{nD}^2} = 4kT\, g_{do}\, P\, \Delta f \qquad (7.71a)$$

where P is a noise coefficient describing how the channel noise is related to the charge fluctuation. It depends on device geometry and bias. T is the lattice temperature and g_{do} is the channel conductance at $V_{DS} = 0$. In practice, g_{do} is difficult to measure, since for $V_{DS} = 0$ the drain-to-source current is also zero. (According to [127], the source conductance in saturation is $-g_{do}$, and hence g_{do} can be determined indirectly by measuring the source conductance.)

The fluctuation induced in the gate is

$$\overline{I_{nG}^2} = 4kT\, \frac{(\omega C_{gs})^2}{g_{max}}\, R\, \Delta f \qquad (7.71b)$$

with g_{max} denoting the maximum conductance of the channel. The noise coefficient R is related to the noise induced by the channel in the gate; its value depends on transistor geometry and bias (at 300 K it is around 0.5).

The drain and gate noise currents are partially correlated (note that they would be completely correlated if the channel had constant cross-sectional area along its length, a situation that arises if, for instance, $V_{DS} = 0$). The correlation coefficient, denoted by C, is imaginary (see Sect. 7.7.2)

$$jC = \frac{\overline{I_{nG}^* I_{nD}}}{\sqrt{\overline{I_{nG}^2}\, \overline{I_{nD}^2}}} \qquad (7.71c)$$

Its numerical value depends only on the applied bias.

7.7 Noise of Junction Field Effect Transistors (JFETs)

■ **Conclusion.** This model contains three noise coefficients (P, R, and C), which must be fit to the measured data. The validity of this model is restricted to JFETs with a long channel operating in the linear mode. Nevertheless, this model can be extended to MOS transistors operating in weak, medium, or strong inversion.

7.7.4 Robinson Model

■ **Introduction.** In 1973, Robinson established a physical noise model of the JFET [94]. He followed Van der Ziel's approach, taking care to relate the channel noise to the transconductance instead of the conductance. He suggested accounting for noise generated in the pinch-off region by means of a different channel temperature (depending on position x).

■ **Simplification.** As in Van der Ziel model, he proposed neglecting the noise contribution of the pinch-off region and considering only the thermal noise of the channel.

■ **Noise Model.** For the representation in Fig. 7.18a and standard bias conditions, Robinson deduced the following expressions:

$$S(I_{nD}) \cong \frac{2}{3} 4kT \, g_m \qquad (7.72)$$

$$S(I_{nG}) = \frac{4}{15} \frac{\omega^2 C^2}{g_m} 4kT \qquad (7.73)$$

$$S(I_{nG} I_{nD}^*) = \frac{j\omega C \, 4kT}{6} \qquad (7.74)$$

where C denotes the input capacitance in common source configuration. Note that the contribution of shot noise is neglected.

A crude estimate of the squared correlation coefficient yields 5/32 (which corresponds to a value of about 0.4).

If the configuration proposed in Fig. 7.18b is adopted, then the following equations apply:

$$S(E_n) = \frac{2}{3} \frac{4kT}{g_m} \qquad (7.75)$$

$$S(I_n) = \frac{4}{15} \frac{\omega^2 C^2}{g_m} 4kT + 2qI_G \qquad (7.76)$$

$$S(I_n E_n) = -\frac{1}{6} \frac{j\omega C \, 4kT}{g_m} \qquad (7.77)$$

■ **Buckingham's Contribution.** Buckingham subsequently evaluated the noise factor expression [93] for the noise model described by (7.75) to (7.77)

$$F = 1 + \frac{S(I_n)\,R_s}{4kT} + \frac{S(E_n)}{4kT\,R_s} \tag{7.78}$$

Its minimum with respect to R_s is

$$F_o = 1 + \left(\frac{2}{3}\right)^{1/2} \frac{\omega C}{g_m} \tag{7.79}$$

7.7.5 Ambrozy Model

■ **Introduction.** This model was published in 1982 [95] and represents a physical model based on Van der Ziel's approach. One noteworthy feature of his contribution is the unified treatment of JFET and MOS devices.

■ **Noise Model.** Adopting the representation of Fig. 7.18a, Ambrozy established the following equations:

$$\overline{I_{nD}^2} = 4kT\,g_{ms}\,K_d(\eta)\,\Delta f \tag{7.80}$$

$$\overline{I_{nG}^2} = 4kT\,\frac{(\omega C_{gc})^2}{g_{ms}}\,K_g(\eta)\,\Delta f \tag{7.81}$$

$$\overline{I_{nG}^* I_{nD}} = 4kT\,j\omega C_{gc}\,K_{dg}(\eta)\,\Delta f \tag{7.82}$$

For a JFET (in the linear mode of operation), $K_d(\eta)$ depends on the dopant profile and geometry of the device. In MOS, $\eta = V_{DS}/(V_{GS} - V_T)$ and $g_{ms} = \gamma(V_{GS} - V_T)$.

The noise coefficients K_d, K_g, and K_{dg} are independent of the operating point (for JFETs) and belong to the following domains:
$1 > K_d > 2/3 \quad 1/12 < K_g < 16/135 \quad 0 < K_{dg} < 1/9 \quad \text{for} \quad 0 < \eta < 1$.

Ambrozy also estimated the power spectral density of the shot noise associated with the gate current

$$S(I_{nG}) = 2q\left(I_o + \frac{2}{3}I_{Gc} + I_{Gt}\right) \tag{7.83}$$

The notation of Sect. 7.7.2 is employed.

7.7.6 Bruncke Model

■ **Introduction.** This is a simple, empirical model that contains only one noise coefficient to fit. It was published in 1966 [104].

■ **Approach.** Assuming the small-signal equivalent circuit of a JFET (Fig. 7.17), the input equivalent circuit is deduced. If $C_{dg} \ll C_{gs}$, the input circuit reduces to that of Fig. 7.19a. The circuit is then transformed, letting Q_1 be the quality factor of the equivalent configuration.

7.7 Noise of Junction Field Effect Transistors (JFETs)

Fig. 7.19. Input equivalent circuits of a JFET

To model the noise at the output, Bruncke considered the equivalent noise resistance $R_n = Q/g_m$ in series with the gate; Q is a noise coefficient to fit to the measured data (for practical silicon JFETs manufactured in the 1960s it was essentially constant and close to 0.7). Nevertheless, it was found that for frequencies between 10 MHz and 4 GHz, modern JFETs and MESFETs agree well with this approach.

■ **Noise Model.** The noise model proposed by Bruncke and Van der Ziel is given in Fig. 7.20; note that the thermal noise of the signal source is also considered.

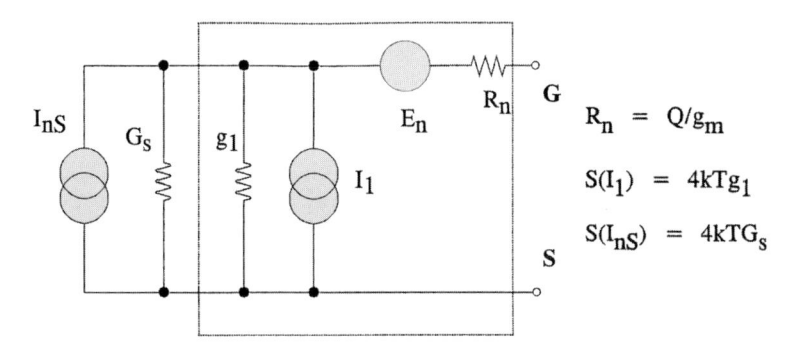

Fig. 7.20. JFET noise model proposed by Bruncke

Here R_n is just one of the four standard noise parameters involved in the classical equation (5.29). Letting $A = g_1 R_n$, for JFETs in silicon, we obtain

$$F = 1 + \frac{g_1}{G_s} + \frac{R_n}{G_s}(G_s + g_1)^2 \qquad (7.84)$$

$$F_o = 1 + 2A + 2\sqrt{A + A^2} \qquad (7.85)$$

$$G_o = g_1\sqrt{1 + \frac{1}{A}} \quad \text{and} \quad B_o = -\omega C_1' = -\frac{Q_1^2}{1 + Q_1^2}\omega C_1 \qquad (7.86)$$

212   7 Noise Models of Electronic Devices

■ **Conclusion.** This model requires only one noise coefficient adjustment to completely model the noise ($Q \cong 0.7$). This is an attractive feature, since it reduces the measurement effort.

7.7.7 PSPICE Model

■ **Electrical Model.** The JFET is modeled as an intrinsic JFET, described by the equations relating I_{DS} to bias and device geometry, and the nonlinear junction capacitances C_{gd} and C_{gs}. Ohmic resistances R_G, $R_S/area$, and $R_D/area$ are respectively added in series with the gate, the source, and the drain. Parameter *area* is a scaling factor.

■ **Noise Model.** The following noise contributions are taken into account, all considered uncorrelated and expressed as power spectral densities (1-Hz bandwidth):

– Thermal noise associated with ohmic resistances:

$$S(I_{nS}) = 4kT / (RS/area) \quad \text{and} \quad S(I_{nD}) = 4kT / (RD/area) \quad (7.87)$$

– Noise of the intrinsic FET (channel thermal noise and 1/f noise):

$$S(I_{nD}) = 4kT\, g_m \frac{2}{3} + KF\, (I_{DS})^{AF} / FREQ \quad (7.88)$$

where KF is a coefficient (default 0), AF an exponent (default 1), and FREQ is the operating frequency.

■ **Concluding Remark.** This model closely follows Robinson model, neglecting the induced gate noise. It is expected that the simulation results will be more optimistic than measured data.

7.7.8 Conclusion

The main noise mechanisms of a generic JFET are:

1) *Thermal noise*, associated with all parasitic resistances.
2) Noise of the intrinsic FET, namely:
   - *Thermal noise* of the channel when it operates in the linear mode (or when the pinch-off region can be neglected relative to the channel length, as in long-channel devices).
   - *Diffusion noise* when it operates in the saturated mode. This noise is significant in short-channel devices.
   - *Induced gate noise* correlated with the channel noise, since the former is induced across the gate junction capacitance.

The most widely adopted global noise description of the intrinsic FET is by means of the normalized correlation matrix

$$\begin{bmatrix} \overline{I_{nG}I_{nG}^*} & \overline{I_{nG}I_{nD}^*} \\ \overline{I_{nG}^*I_{nD}} & \overline{I_{nD}I_{nD}^*} \end{bmatrix} = \begin{bmatrix} \dfrac{\omega^2 C_{gs}^2 R}{g_m} & \omega C_{gs}\sqrt{RP}C \\ \omega C_{gs}\sqrt{RP}C^* & g_m P \end{bmatrix} \quad (7.89)$$

where P, R, C are noise coefficients fit to the measured data. In practice, since it is difficult to evaluate the correlation coefficient, the usual approach is to neglect it.

In conclusion, JFETs are used primarily for low-power, low-frequency applications where high input resistance is desirable. For high-frequency applications they are not appropriate, because they do not use high-mobility materials and the charge stored in the gate is important.

7.8 Noise in MOS Transistors

7.8.1 Introduction

■ **Noise Mechanisms.** Noise in MOS transistors results mainly from channel noise, 1/f noise, and G–R noise in the depleted region.

■ **Physical Operation.** Consider a generic NMOS enhancement transistor (P-type substrate, N-type drain and source), as depicted in Fig. 7.21. The diffusion regions are heavily doped, and the oxide layer, grown on the surface of the substrate, is very thin (0.02 to 0.1 µm).

When no bias is applied to the gate, no current flows between S and D, because the transistor acts like two back-to-back diodes; globally the MOS acts like an *open switch*.

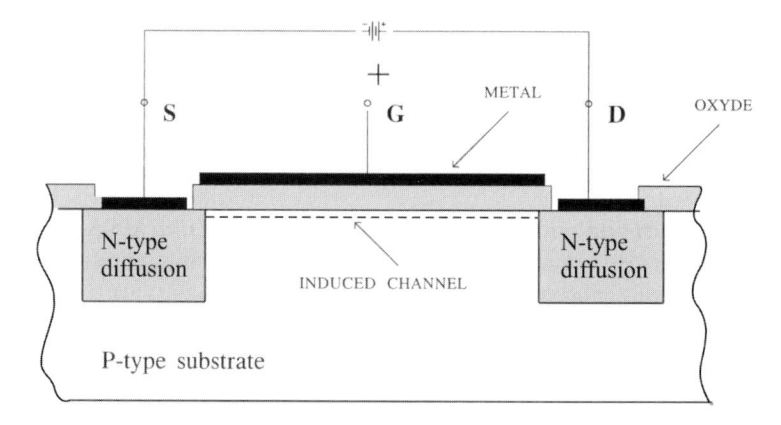

Fig. 7.21. Cross section of an enhancement N-channel MOS

214   7 Noise Models of Electronic Devices

Consider next that a positive potential is applied to the gate (V_G), which is progressively increased. The electric field across the oxide layer will repel free holes from the region under the gate and attract electrons (from the N-diffusion regions, where they are abundant). When V_G reaches a threshold value (V_T), a sufficient number of free electrons accumulate near the surface between the substrate and the gate, and an N-channel is effectively created. Current can flow between S and D, and the MOS acts now like a *closed switch*.

■ **Note**

1) The conducting channel appears almost on the surface separating the substrate and the oxide, not in the bulk. Therefore, 1/f noise is expected to be stronger than that of a JFET, where conduction occurs in the semiconductor bulk. We expect that this noise can be considerably reduced if a high-quality oxide-semiconductor interface is obtained during processing.
2) In contrast with the JFET, where the gate voltage modifies the cross-sectional area of the channel, in MOS transistors the gate voltage instead controls the charge carrier concentration inside the channel. Since the channel is enhanced in electrons, the MOS is called *enhancement MOS*.

■ **Noise.** Concerning MOS noise, note that

– 1/f noise is more important than in a JFET. At the boundary between the Si and SiO_2, additional electron energy states exist, which can randomly trap and release free electrons from the channel, producing flicker (1/f) noise. The fluctuation of the number of carriers inside the channel causes a fluctuation of the surface potential, which in turn produces a variation in charge carrier mobility. *Both fluctuation in the number of carriers and fluctuation in carrier mobility are responsible for increasing 1/f noise.*
– Van der Ziel found that the noise of the channel is stronger at the source end.
– A strong correlation exists between the 1/f noise of the gate and that of the drain. As stated in [92], the former can be compensated (at least theoretically!) by adding an appropriate capacitor between gate and ground.

7.8.2 PSPICE Model

■ **Electrical Model.** The device is modeled as an intrinsic MOS with ohmic resistances R_D, R_S, R_G, and R_B in series with the drain, source, gate, and bulk, respectively, and a shunt resistance R_{DS} in parallel with the channel. The set of equations describing the electrical model of the intrinsic MOS is selected according to the value assigned to the variable LEVEL:

LEVEL = 1 is the Shichman-Hodges model
LEVEL = 2 is a geometry-based, analytic model
LEVEL = 3 is a semi-empirical, short-channel model.

■ **Noise Model.** Noise is computed in a 1-Hz bandwidth with the following spectral densities:

- *Thermal noise* of ohmic resistances R_D, R_S, R_G, and R_B:

$$S(I_{nj}) = 4kT/R_j \quad \text{with} \quad j = D, G, S, B \tag{7.90a}$$

- *Intrinsic noise*, mainly due to channel thermal noise and 1/f noise. This is represented by an equivalent current generator located between the drain and source of the device:

$$\overline{I_{nD}^2} = \left(4kT\, g_m\, \frac{2}{3} + KF\, (I_D)^{AF} / FREQ\, Kchan\right) \Delta f \tag{7.90b}$$

where

$$g_m = \partial I_D / \partial V_{GS} \tag{7.90c}$$

$$Kchan = (L - 2LD)^2 (\text{permittivity of SiO}_2)/TOX \tag{7.90d}$$

and

FREQ – operating frequency
TOX – oxide thickness (default is infinite)
L – channel length
LD – lateral diffusion length (default is 0)
KF – flicker noise coefficient (default is 0)
AF – flicker noise exponent (default is 1)

■ **Limitations.** The noise model has several limitations:

- it is too simplistic;
- it cannot be used to simulate noise in the linear mode, since for $V_{DS} = 0$, the channel thermal noise predicted by (7.90b) vanishes;
- it is not versatile (it cannot be applied to all models selected by LEVEL).

■ **Contribution of Knoblinger, Klein, and Tiebout.** The thermal channel noise (the first term on the right-hand side of (7.90b)) is implemented in other SPICE models with the following equation:

$$\overline{I_{nD}^2} = \frac{8}{3} kT\, (g_m + g_{ds} + g_{mb}) \Delta f \tag{7.90e}$$

g_m is the gate transconductance, g_{ds} is the channel conductance, and g_{mb} is the bulk transconductance.

A much more general formula is

$$\overline{I_{nD}^2} = \left(\frac{4kT}{L^2} \mu Q_{inv}\right) \Delta f \tag{7.90f}$$

where Q_{inv} is the inversion layer charge, L is the channel length, and µ is the mobility.

216   7 Noise Models of Electronic Devices

According to [134], an effective mobility μ_{eff} is recommended for use in (7.90f) to account for the influence of vertical and lateral electric fields; for instance,

$$\mu_{eff} = \mu_s / \left(1 + \left(\frac{E_x}{E_{crit}}\right)^\beta\right)^{1/\beta} \quad (7.90g)$$

with μ_s the surface mobility in the BSIM3v3 model, $E_x = V_{DS_{eff}}/L_{eff}$ the lateral electric field, $E_{crit} = v_{sat}/\mu_s$, v_{sat} the saturation velocity, $\beta = 2$ for electrons, and

$$V_{DS_{eff}} = \begin{cases} V_{DS} & \text{in the linear region} \\ V_{DS_{sat}} & \text{in saturation} \end{cases} \quad (7.90h)$$

A method to extract thermal channel noise out of RF measurement data is proposed in [134].

7.8.3 Model of Nicollini, Pancini, and Pernici

■ **Introduction.** Published in 1987, this is an empirical model, which improves upon (7.90b) in order to guarantee noise simulation accuracy in both saturation and linear operation mode [105].

■ **Noise Model.** New expressions are proposed for the channel noise

$$S(I_{nD}) = \begin{cases} \frac{2}{3} 4kT(g_m + g_{mb} + g_{do}) \left(\frac{3}{2} - \frac{V_{DS}/2}{V_{Dsat}}\right) & \text{if } V_{DS} < V_{Dsat} \\ \frac{2}{3} 4kT(g_m + g_{mb} + g_{do}) & \text{if } V_{DS} \geq V_{Dsat} \end{cases} \quad (7.91a)$$

and for the flicker noise

$$\overline{I_{nD}^2} = K(V_{DS}) \frac{K_F I_D^{AF} (T/T_o)^{BF} \left(1 + CF \left(\frac{g_{mb}}{g_m}\right)^2\right)}{C_{ox} L^{DF}} \sqrt{M} \, \Delta f \quad (7.91b)$$

with

$$M = \frac{1 + K_1 f}{(1 + K_2 f)(1 + K_3 f)(1 + K_4 f)} \quad (7.91c)$$

$$K(V_{DS}) = \begin{cases} 0.5(1 + V_{DS}/V_{Dsat}) & \text{if } V_{DS} < V_{Dsat} \\ 1 & \text{if } V_{DS} > V_{Dsat} \end{cases} \quad (7.91d)$$

In these equations, C_{ox} is the gate capacitance per unit area, $T_o = 300\,\text{K}$, f is the frequency, $g_{mb} = \partial I_D/\partial V_{BS}$ is the bulk transconductance, $g_{do} = \partial I_D/\partial V_{DS}$ is the channel conductance, and V_{Dsat} corresponds to the V_{DS} value separating the linear and saturated regions of the I–V curve.

For a given manufacturing process, the nine noise coefficients to adjust are AF, BF, CF, DF, K$_1$, K$_2$, K$_3$, K$_4$, and K$_F$. The authors proposed the following defaults:

AF = 1, BF = 0, CF = 0, DF = 2
K$_1$ = 0, K$_2$ = 10, K$_3$ = 10, K$_4$ = 0, K$_F$ = 10^{-30}

■ **Contribution of Fox [106].** The most serious deficiency in the model of Nicollini et al. is that it yields incorrect results when used with SPICE level 1 or level 3 models. Therefore, Fox proposed the following equation

$$\overline{I_{nD}^2}/\Delta f = \alpha 4kT\, g_{do} \tag{7.91e}$$

where

$$\alpha = \begin{cases} \dfrac{1 - v + (v^2/3)}{1 - v/2} & \text{if } V_{DS} < V_{Dsat} \\ 2/3 & \text{if } V_{DS} \geq V_{Dsat} \end{cases} \tag{7.91f}$$

Here $v = V_{DS}/V_{Dsat}$, g_{do} represents the drain conductance with $V_{DS} = 0$ (in all SPICE models, neglecting short-channel effects, $g_{do} = \beta(V_{GS} - V_T)$, β constant).

Fox's contribution yields good agreement with measured data for long-channel devices.

■ **Integrated MOS.** For an integrated device, before computing the thermal noise of R$_g$ it is necessary to estimate the gate ohmic resistance with the expression

$$R_g = R_\square W/3n^2 L \tag{7.91g}$$

R$_\square$ denotes the sheet resistance in units of Ω/square, W is the width of the gate, L is the length of the gate, and n is the number of fingers connected together at one end in the interdigitated structure (CMOS technology).

7.8.4 Model of Wang, Hellums, and Sodini

■ **Introduction.** The model proposed by the authors (1994) is able to predict the thermal noise of both long- and short-channel MOS, in linear and saturated mode.

■ **Noise Model.** As stated in [107], the drain thermal noise current in the linear mode (triode region) has a power spectral density

$$S(I_{nD}) = 4kT\mu_{ef}^2 C_{ox}^2 \frac{W^2}{L^2 I_D}\left((V_{GS}-V_T)^2 V_{DS} - \alpha_x(V_{GS}-V_T)V_{DS}^2 + \frac{\alpha_x^2}{3}V_{DS}^3\right)$$

$$-4kT\frac{W\,C_{ox}\,\mu_{ef}}{L^2\,E_c}\left((V_{GS}-V_T)V_{DS} - \frac{\alpha_x}{2}V_{DS}^2\right) \tag{7.92}$$

218   7 Noise Models of Electronic Devices

In the saturated mode of operation, (7.93) still applies, provided that V_{DS} is replaced by V_{Dsat}, and L by L_{ef}.

The following notation has been used:

L, W – length, width of the channel
C_{ox} – gate-oxide capacitance
V_T – threshold voltage
α_x – represents bulk effect on threshold voltage
E_c – critical field at which carriers are velocity-saturated
μ_{ef} – effective surface mobility
L_{ef} – effective electrical channel length (channel length between drain and source minus extension of drain depletion region into the channel).

7.8.5 Lee's Scalable Model

■ **Introduction.** This model emerged in the late 1990s, with the advent of full custom CMOS integrated circuits, where the favorite approach used to achieve noise matching is to appropriately tailor the device geometries.

■ **Noise Model.** The starting point is Van der Ziel model (7.71), where for long-channel devices it is known that $2/3 \leq P \leq 1$, and its value depends on bias and channel length. For short channel devices P may be as high as 2 or 3, $R \cong 4/3$, and $C \cong 0.395$. Assuming that the distributed and Miller effects may be neglected, the expressions for the current noise power spectral densities have been derived [139] and the expressions of the four noise parameters have been found by Lee [140] as follows

$$F_o \cong 1 + \frac{\omega}{\omega_T}\sqrt{P\delta\zeta(1-|C|^2)} \tag{7.93a}$$

$$R_n \cong \frac{P\, g_{d0}}{g_m^2} \tag{7.93b}$$

$$G_o \cong \frac{g_m\, \omega C_{gs}}{g_{d0}}\sqrt{\frac{\delta\zeta\,(1-|C|^2)}{P}} \tag{7.93c}$$

$$B_o \cong -\omega C_{gs}\left(1 - C\,\frac{g_m}{g_{d0}}\sqrt{\frac{\delta\zeta}{P}}\right) \tag{7.93d}$$

where g_{d0} is the zero-bias drain conductance and g_m the transconductance.

■ **Discussion.** In the previous expressions g_{d0}, g_m and C_{gs} scale linearly with the device width, while noise coefficients P, δ, ζ, and C are width independent.

Equations (7.93) suggest that G_o and B_o are proportional to the device width W, but R_n is proportional to W^{-1}, and F_o has no width dependence.

This means that larger devices are more recommended to reduce the noise figure, but a compromise must be found with the power budget. Note also that ω_T is proportional to L_{eff}^{-1}, so shorter devices with large widths offer less noise and enough gain over a large bandwidth.

7.8.6 Conclusion

At medium- and high-frequencies, the MOS noise is about the same as JFET noise. However, MOS high-frequency gain is considerably lower than that of the JFET.

For low-noise, low-frequency applications, MOS transistors are not recommended, because their 1/f noise is more significant than that of JFET devices. Flicker noise is the limiting performance factor in CMOS circuits, where downscaling the devices degrades the signal-to-noise ratio.

In the microwave range, the electrical and noise performance of silicon MOS devices are not yet satisfactory.

7.9 Noise of MESFET Transistors

7.9.1 Introduction

■ **Overview.** The MESFET is the key device in microwave monolithic integrated circuits (MMIC). It combines relatively simple geometry with excellent electrical and noise performance.

High-speed circuits, microwave low-noise amplifiers, microwave power amplifiers, broad-band receivers, and data communication systems all require MESFETs whose S parameters and noise parameters are available over a wide bandwidth. Usually, this information results from a significant effort to collect and process measured data. However, in the millimeter range, no full noise characterization is possible, due to the lack of suitable test equipment. On the other hand, the accuracy of measurements in actual very low-noise devices is limited. For both reasons, it is of paramount importance to develop accurate models for MESFETs.

A good model must not only be able to fit the actual measured noise data, but also to accurately predict the values of noise parameters at higher frequencies, where measurements are not accessible. Ideally, the model should make it possible to extrapolate noise parameters measured in X-, Ku-, or K-band into the millimeter wave frequency range.

■ **Classification.** Any MESFET noise model belongs to one of the following categories:

- *Physical models*, based on numerical or analytic solution of the transport equations for charge carriers in the device, from which voltage and current fluctuations can be deduced (as with the model of Pucel-Haus-Statz,

Brooks, Cappy, etc.). Despite their complexity, these offer a deep understanding of device operation and noise mechanisms. Generally, they require knowledge of a large number of theoretical parameters (related to processing, physical properties of the material, and device geometry), which are not readily available for purchased devices. Hence, the resulting noise coefficients must be fit to measured data.
– *Empirical models*, based on the small-signal equivalent circuit, to which several noise generators are added. According to the available knowledge on noise mechanisms these generators are conveniently placed. They are described by means of several noise coefficients, to be adjusted in order to fit the measured data. This is the case of Fukui's model, Podell's model, Pospieszalski's model and others.
– *Hybrid models*, which are mainly used in electrical simulators (like SPICE). The best-understood part of the model is described by physical equations, while some noise generators are still needed to account for the less well understood mechanisms. Their power spectral densities are established with semi-empirical expressions containing noise coefficients and their companion default values (a typical example is the 1/f noise description).

It is important to note that the common point to all categories is *the necessity to adjust some coefficients according to the measured noise data*. Also note that due to the impossibility of perfectly controlling all parameters during device manufacture, an inherent spreading of values in the noise coefficients is expected (their actual values may vary from device to device).

■ **Another Classification.** According to the operating frequency band, the models are grouped into two classes:

- *Lumped models*, which are employed at frequencies where the corresponding wavelength is much greater than the transistor dimensions. In this case, propagation phenomena inside the device structure can be neglected.
- *Distributed models*, which are appropriate when the device dimensions become of the same order of magnitude as the wavelength. Generally, the device is "sliced" into several sections, each one fairly thin relative to the wavelength, so that propagation effects can be neglected. A lumped equivalent circuit is found for each section and all equivalent circuits are finally interconnected with appropriate transmission lines. This is the philosophy of noise models in the millimeter range, such as Heinrich's model, the Escotte-Mollier model or the Hickson-Gardner-Paul model.

7.9.2 Physical Aspects

■ **State of the Art.** MESFET amplifiers, mixers, oscillators, switches, and modulators for microwave applications are widely used due to their high gain and low-noise properties. Continuing progress has improved the noise figure of the MESFETs, which is actually close to 2 dB at 35 GHz.

7.9 Noise of MESFET Transistors 221

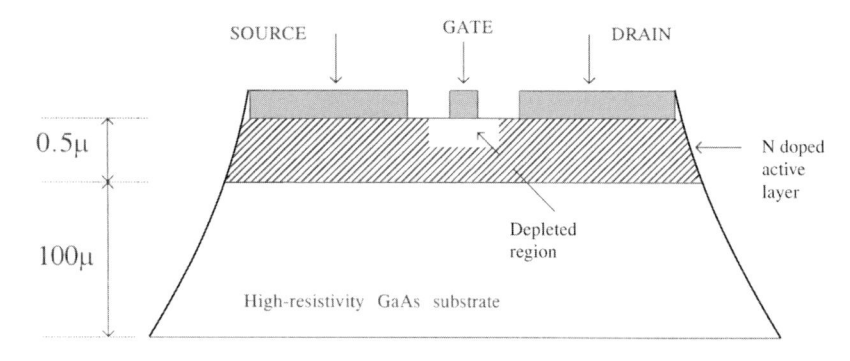

Fig. 7.22. Simplified view of the cross section of a GaAs MESFET

■ **Device Description.** The simplified structure of a GaAs MESFET is shown in Fig. 7.22. The order of magnitude of the doping density in the N active layer is $10^{23}\,\mathrm{m}^{-3}$, while in the substrate it is several orders of magnitude lower. Due to its high resistivity, the substrate is sometimes called "semi-insulating substrate"; as a consequence, most of the current within the device flows in the N layer (although some substrate current may exist). The N layer is deposited by epitaxy (for discrete manufactured devices) or by ion implantation (in MMIC devices). The Schottky barrier gate electrode is formed by deposition of a metal (such as gold) on the epitaxial layer. The gate length is usually between 0.1 μm and 1 μm. Note also that under the metal of the drain (source), a heavily doped N layer can be deposited to reduce the drain (source) ohmic resistance.

■ **MESFET Operation.** The active N layer is divided into two regions: the depleted region under the gate and the active channel, where the current flows. The applied gate voltage controls the penetration depth of the depletion region into the channel, and therefore the channel effective cross section, its resistance, and finally the drain-source current. For $V_G = 0$ and low values of V_{DS}, the drift velocity of electrons is proportional to the electric field (device operation is confined to REGION 1 of Fig. 7.16).

When the applied V_{DS} increases, the velocity of electrons begins to saturate, reaching a peak, and then the velocity decreases with increasing field to the hard saturated region. In REGION 2 all happens as the equivalent temperature of the accelerated electrons would be much more higher as the lattice temperature (they are called "hot" electrons). Since in any collision with lattice atoms the hot electrons lose energy, we expect the noise mechanisms to be different from those of REGION 1. Additional phenomena must be considered in REGION 3:

– Due to the multi-valley structure of the energy bands of GaAs, when an electron acquires enough energy it is transferred from the central valley to a satellite valley, where its effective mass increases and the mobility drops

largely below the central valley value. Consequently, the electron velocity is drastically reduced and its contribution to the total current vanishes. This is the mechanism of *intervalley scattering noise*.
- The electron mobility decreases even further, due to high density of ionized donors in the active layer.

■ **Small-Signal Equivalent Circuit.** This is the same as for a JFET (Fig. 7.17), except that in the microwave frequency range, parasitic inductances must be added in series with R_S, R_G, and R_D.

■ **Noise.** Since in all microwave circuits MESFETs are biased above pinch-off, conduction is non-ohmic and mobility is a function of the electric field. Therefore, the channel generates diffusion noise rather than thermal noise.

Of course, the thermal noise of ohmic resistances R_G, R_D, and R_S must be taken into account.

As for 1/f noise, since its spectrum affects only low frequencies, we might expect that it can be neglected in MESFETs operating in microwave circuits. Nevertheless, if the MESFET is employed in a mixer or oscillator, 1/f noise can be up-converted and $1/\Delta f$ noise can result.

Finally, the noise performance of the MESFET depends mainly on the drift velocity of the charge carriers in the channel in saturation (Fig. 7.16). Also, the gate length is the most significant factor determining device performance, at a given operating frequency. For instance, for a gate length less than 1 μm, the nonuniform distribution of the electric field inside the channel plays an important role. The channel being very short, the carriers actually spend less time in the channel and do not reach the drift velocity imposed by equilibrium conditions; consequently, they cluster near the drain edge.

As stated in [101], device optimization for low-noise performance requires

- a reduction in gate length
- an increase in transconductance g_m
- a reduction in gate-to-source capacitance C_{gs}
- minimization of R_S and R_G.

■ **Remark.** The noise models which will be next introduced are valid at frequencies where the transit time of charge carriers in the channel may be neglected.

7.9.3 Bächtold Model

■ **Introduction.** This model represents an extension of the JFET model proposed by Van der Ziel. It takes intervalley scattering noise into account and hence corrects Van der Ziel model for nonconstant mobility of carriers. The noise coefficients P, R, and C are expressed in terms of the gate voltage [108, 109].

■ **Configuration.** The intrinsic MESFET small-signal equivalent circuit (Fig. 7.17) is simplified by neglecting all elements except C_{gs}, R_i, R_{ds}, and the voltage-controlled current source.

■ **Noise Model.** The noise sources of the intrinsic transistor are thermal channel noise and induced gate noise. Using this model the following expression results for the minimum noise figure of the *intrinsic* MESFET

$$F_o = 1 + 2\sqrt{PR(1-C^2)}\,\frac{f}{f_T} + 2g_m R_i P\left(1 - C\sqrt{\frac{P}{R}}\right)\left(\frac{f}{f_T}\right)^2 \quad (7.94a)$$

where the transition frequency is

$$f_T = 10^3\, g_m\, /\, 2\pi C_{gs} \quad [\text{GHz}] \quad (7.94b)$$

C_{gs} being expressed in pF and g_m in S. P and R coefficients depend on the bias conditions.

7.9.4 Model of Pucel, Haus, and Statz

■ **Introduction.** This model, published in 1972, represents a cornerstone in the development of various FET noise models, due to its comprehensive treatment of signal and noise properties [110]. The authors adopt Shockley's analysis based on the fact that the channel of a microwave GaAs MESFET contains two distinct sections (Fig. 7.15a): the linear (or ohmic) region between the source and pinch-off region, and the saturated section (between the pinch-off area and the drain, where due to the high electric field, velocity saturation is reached). Since Van der Ziel's analysis does not hold in the saturated section, the original contribution of Pucel, Haus, and Statz was to develop a noise model for this section (which roughly corresponds to REGION 2 in Fig. 7.16).

■ **Approach.** All microwave MESFETs operate above the pinch-off point. Consequently, the channel of the transistor can be separated into two longitudinal sections:

– Section I, (of length L_1 in Fig. 7.15b) corresponding to ohmic conductivity;
– Section II, (of length L_2 in Fig. 7.15b) where the velocity saturation applies.

The distribution of potential along these sections is established, and by integrating the equations the expression for the current is obtained. Then, the small-signal equivalent parameters are deduced from their definitions.

■ **Small-Signal Equivalent Circuit.** Initially, analysis is simplified by neglecting C_{dg}, C_{ds}, R_{ds}, the transit time τ of the carriers within the channel, and all parasitics.

224   7 Noise Models of Electronic Devices

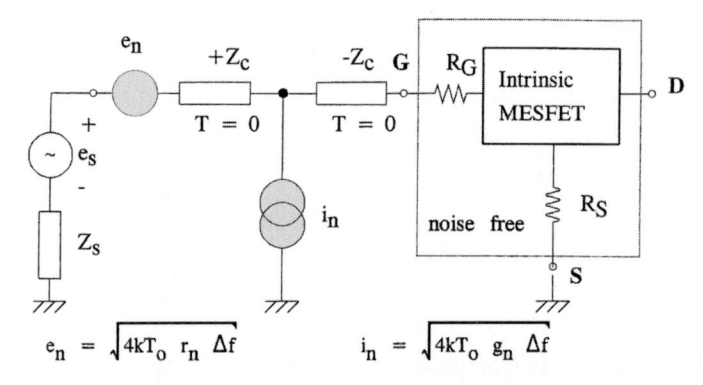

Fig. 7.23. The MESFET noise model proposed by Pucel, Haus, and Statz

■ **Noise.** The noise equivalent circuit (Fig. 7.23) is based upon the Rothe and Dahlke representation. The Van der Ziel approach (improved by Bächtold) of thermal noise generated in section I is used; in section II, the noise is attributed to the generation of dipole layers, rather than Johnson noise. The open-circuit drain voltage fluctuation produced by sources in section I and section II is computed. The short-circuit gate current fluctuation produced by sources in section I and section II are separately calculated, as well as the correlation coefficient. Applying North's definition, the noise factor is

$$F = 1 + (1/r_s)\left(r_n + g_n|Z_s + Z_c|^2\right) \quad (7.95a)$$

with r_s being the real part of the internal impedance Z_s of the signal source.

The noise parameters r_n, g_n, and Z_c (pertinent to the model of Rothe and Dahlke) are

$$r_n = (R_s + R_G) + K_r \frac{1 + (\omega C_{gs} R_i)^2}{g_m} \quad (7.95b)$$

$$g_n = K_g \frac{(\omega C_{gs})^2}{g_m} \quad (7.95c)$$

$$Z_c = (R_s + R_G) + \frac{K_c}{Y_{11}} \quad (7.95d)$$

with $Y_{11} = (R_i + 1/j\omega C_{gs})^{-1}$. K_g, K_r, and K_c are functions of noise coefficients P, R, and C, as in the following expressions:

$$K_g = P\left(\left(1 - C\sqrt{R/P}\right)^2 + (1 - C^2)R/P\right) \quad (7.95e)$$

$$K_c = \frac{1 - C\sqrt{R/P}}{\left(1 - C\sqrt{R/P}\right)^2 + (1 - C^2)R/P} \quad (7.95f)$$

$$K_r = \frac{R(1-C^2)}{\left(1 - C\sqrt{R/P}\right)^2 + (1-C^2)R/P} \quad (7.95\text{g})$$

Recall that *coefficient P describes the channel noise due to charge fluctuations, and R describes the gate noise induced by the channel noise; the two noise sources are correlated with correlation coefficient jC.*

The dimensionless coefficients $P = P_1 + P_2$, $R = R_1 + R_2$, and $C = C(P_1, P_2, R_1, R_2)$ are related only to transistor geometry and indirectly to bias. Subscript 1 refers to section I and subscript 2 to section II of the channel.

The minimum noise factor is

$$F_o = 1 + 2g_n\left(R_c + \sqrt{R_c^2 + r_n/g_n}\right) \quad (7.95\text{h})$$

R_c being the real part of the correlation impedance Z_c.

■ **Note**

1) Pucel, Haus, and Statz proposed to determine the gate charging resistance R_i by assuming the product $R_i C_{gs}$ to be approximately proportional to the transit time τ through the channel.
2) Expressions (7.95) are very useful in designing low-noise MESFETs; they can also be used to select the optimum bias for low-noise operation in microwave circuit design.
3) Expressions (7.95), together with the equations given in [110] relating K_g, K_r, and K_c to the applied bias, have the following properties:
   - F_o decreases when the transition frequency f_T increases (i.e., a short gate MESFET has a reduced noise factor);
   - above 18 GHz, the minimum noise factor increases with frequency, with a slope of 0.5 dB/oct.
   - the optimum bias for the low-noise operation of a GaAs MESFET is $I_{DS} \cong I_{DSS}/8$ and $V_{DS} \cong 4\,\text{V}$;
   - F_o is diminished drastically if the channel temperature T decreases (a practical solution is to cool the MESFET).
4) For a real MESFET device (parasitics included) Pucel, Haus, and Statz found the following expression for the minimum noise factor:

$$F_o = 1 + 2(\omega C_{gs}/g_m)\sqrt{K_g\left(K_r + g_m(R_G + R_S)\right)}$$
$$+ 2(\omega C_{gs}/g_m)^2 \left(K_g\, g_m(R_G + R_S + K_c R_i)\right) + \ldots \quad (7.95\text{i})$$

Since both g_m and the sum $(R_G + R_S)$ appear in the coefficients of the terms containing ω and ω^2, their influence in the microwave frequency range must be significant.

■ **Conclusion.** Pucel, Haus, and Statz proposed the most detailed and elegant theory of noise mechanisms in FETs. They accounted for saturation

226   7 Noise Models of Electronic Devices

effects by considering a two-section model, which offers a deep understanding of MESFET operation and noise mechanisms. The only limitation of this model is the fact that the resulting expressions are cumbersome and contain coefficients which cannot be extracted reliably from measurement data.

7.9.5 Fukui Model

■ **Introduction.** The model proposed by Fukui in 1979 is a simple empirical model, verified experimentally and widely used in the design of MESFETs [111, 112].

■ **Noise Model.** The minimum noise factor of a GaAs MESFET was found to be

$$F_o = 1 + K_f \frac{f}{f_T} \sqrt{g_m(R_G + R_S)} \qquad (7.96a)$$

or

$$F_o = 1 + K_l L f \sqrt{g_m(R_G + R_S)} \qquad (7.96b)$$

where K_f is a fitting coefficient of approximately 2.5 which depends on the channel material; K_l is another fitting coefficient whose value is roughly 0.27, when the gate length L is given in micrometers.

Comparing (7.96a) with its analog for a bipolar transistor (7.62), it is obvious that for a MESFET the frequency dependence of the parameter F_o is *linear*, while for a bipolar junction transistor it is *quadratic*. This means that the noise performance of a MESFET is less affected by increasing frequency, and explains why MESFETs are largely preferred in the microwave industry.

■ **Noise Equivalent Circuit.** The noise equivalent circuit is presented in Fig. 7.24.

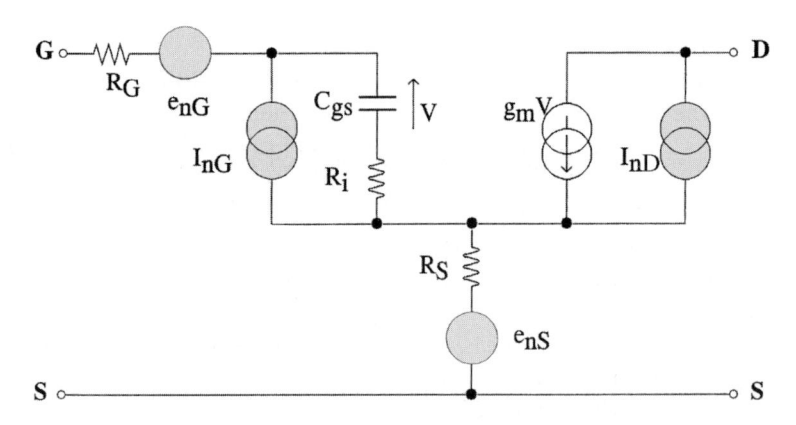

Fig. 7.24. MESFET noise equivalent circuit proposed by Fukui

7.9 Noise of MESFET Transistors

■ **Equations.** The Fukui modeling procedure contains several steps:

1. Derivation of expressions for noise parameters in terms of equivalent circuit elements:

$$F_o = 1 + k_1 f C_{gs} \sqrt{\frac{R_G + R_S}{g_m}} \qquad (7.97a)$$

$$R_n = k_2 / g_m^2 \qquad (7.97b)$$

$$R_o = k_3 \left(R_G + R_S + 1/4g_m \right) \qquad (7.97c)$$

$$X_o = k_4 / f C_{gs} \qquad (7.97d)$$

where k_1, k_2, k_3, and k_4 are fitting coefficients and f is the frequency (for instance, for GaAs MESFETs with 2 µm-long gates, it was found that $k_1 = 0.016$, $k_2 = 0.03$, $k_3 = 2.2$, $k_4 = 160$).

2. Determination of semi-empirical expressions for transconductance, gate-source capacitance and cutoff frequency as a function of device geometry and processing parameters:

$$g_m \cong k_5 \, Z \left(\frac{N}{a_L} \right)^{1/3} \qquad [\Omega^{-1}] \qquad (7.97e)$$

$$C_{gs} \cong k_6 \, Z \left(\frac{NL^2}{a} \right)^{1/3} \qquad [pF] \qquad (7.97f)$$

$$f_T \cong \frac{10^3 \, g_m}{2\pi C_{gs}} = \frac{9.4}{L} \qquad [GHz] \qquad (7.97g)$$

The fitting coefficients k_5 and k_6 were found to be 0.02 and 0.34, respectively, for MESFETs under zero-bias gate condition.

3. The simplified expressions for parasitic resistances are

$$R_G = \frac{17 \, z^2}{hLZ} \qquad [\Omega] \qquad (7.97h)$$

$$R_S = \frac{1}{Z} \left(\frac{2.1}{a^{0.5} \, N^{0.66}} + \frac{1.1 \, L_{SG}}{(a - a_s) N^{0.82}} \right) \qquad [\Omega] \qquad (7.97i)$$

with h being the gate metallization height, L_{SG} the distance between the source and gate electrodes, a_s the depletion layer thickness at the surface, and a the effective channel thickness under the source electrode, all expressed in micrometers.

The geometrical and material parameters are as follows: N is the free carrier concentration in the active channel in units of 10^{16} cm^{-3}, Z is the total device width in mm, z is the unit gate width in mm, and L is the gate length in µm.

4. Derivation of expressions for device noise parameters

$$F_o = 1 + 0.038f \left(\frac{NL^5}{a}\right)^{1/6} (SUM)^{1/2} \quad (7.97j)$$

$$R_n = 75\, Z^{-2} \left(\frac{aL}{N}\right)^{2/3} \quad [\Omega] \quad (7.97k)$$

$$R_o = 2.2\, Z^{-1} \left(12.5 \left(\frac{aL}{N}\right)^{1/3} + SUM\right) \quad [\Omega] \quad (7.97l)$$

$$X_o = \frac{450}{f\,Z}\left(\frac{a}{NL^2}\right)^{1/3} \quad [\Omega] \quad (7.97m)$$

where

$$SUM = \frac{17z^2}{hL} + \frac{2.1}{a^{0.5}\,N^{0.66}} + \frac{1.1\,L_{SG}}{(a-a_s)N^{0.82}} \quad (7.97n)$$

and f is the frequency in GHz.

■ **Note**

1) Expression (7.97k) suggests that a small R_n value is obtained with a short gate device and a heavily doped thin active channel. These conditions are inconsistent with obtaining low values of F_o; consequently, a compromise in choosing a and N must be found. Remember that for broadband applications, a low F_o is necessary, but this does not necessarily guarantee low F (due to the unavoidable mismatch at the band edges), *unless a small value R_n is assured.*
2) When designing a MESFET, it is important to have a low value of R_G (relative to R_S), this condition being desirable for minimum F_o and high power gain.

■ **Conclusion.** This model, which is in widespread use, contains practical expressions for the four noise parameters in terms of device geometry and material parameters, parasitics also being taken into account. Its only limitation is to neglect the transit time τ and therefore to restrict the validity with respect to the frequency range.

7.9.6 Podell Model

■ **Introduction.** The model proposed by Podell in 1981 [113], represents a development of the Bruncke's JFET model to simulate the noise of GaAs MESFETs. The main difficulty results from the lack of accuracy when using the single adjustment coefficient Q (for instance, when I_{DS} is about 5% of I_{DSS}, an additional output noise can be observed, which is greater than the value predicted by Bruncke's equation $R_n = Q/g_m$).

7.9 Noise of MESFET Transistors

■ **Approach.** The noise is described by two uncorrelated generators:

– At the input of the device, we have a thermal generator lumping the effect of various resistances in the gate-source loop. This noise is frequency-dependent, and can be estimated from the transistor equivalent circuit.
– At the output, we have a noise generator, which is frequency-independent. This noise is a function of the drain current and voltage, and was modeled by means of coefficient R_n (according to Bruncke).

■ **Noise Model.** Instead of parameter Q, Podell proposed to use a function F_1 such as

$$R_n = \frac{1}{g_m} F_1(I_D, V_{DS}) \tag{7.98a}$$

Experimentally, he found that $F_1(I_D, V_{DS})$ is an exponential function of $I = I_{DS}/I_{DSS}$ and a weak function of V_{DS}. It was established empirically that

$$R_n = \frac{K_0}{|g_m|} \exp(K_2 I) \tag{7.98b}$$

where K_0 and K_2 are empirical constants obtained by fitting the measured data. It is assumed that $|g_m|$ is frequency-independent. The following remarks apply:

– K_2 is frequency-independent.
– K_0 tends to be higher for shorter gate MESFETs.
– Above 4 GHz, K_0 seems also to be frequency-independent.

Then, he computed the four MESFET noise parameters. Since no losses are assumed in his noise model, he concluded that *a single measurement of F_o at any frequency, along with the small-signal equivalent circuit of the MESFET, can be used to predict F_o at any other frequency.*

■ **Contribution of Gupta [114].** Gupta proposed a noise equivalent circuit for on-wafer submicron-gate-length MESFETs, whose elements are frequency independent. By including the signal source impedance and the input matching network, he computed the optimum (minimum) noise figure. He proposed to relax the adopted hypothesis of uncorrelated I_{nD} and I_{nG} noise generators (which is required in the model of Bruncke or Podell) by considering a high correlation between the sources.

The correlation coefficient is assumed to have a *real part* negligible small, but a significant magnitude. This actually corresponds better to reality, since capacitive coupling exists between drain and gate, and consequently the correlation coefficient must be imaginary. By carefully selecting the real part of the input matching network, he proved that F_o can be found with a single on-wafer measurement point, with no individual tuning of devices.

230   7 Noise Models of Electronic Devices

Gupta and Greiling [115] took subsequently into account the effect of parasitics of interconnections for a discrete packaged MESFET.

■ **Conclusion.** Podell improved the Bruncke model by introducing (7.98b); then, assuming that the magnitude of the correlation coefficient between the noise current generators (Fig. 7.18a) is zero, the MESFET can be characterized with a single measurement point.

7.9.7 Heinrich Model

■ **Introduction.** This is a physical distributed model, published in 1989 [116]. Heinrich was motivated by the gap between existing small-signal models (covering only the range up to 20 GHz) and MESFET capabilities (up to about 70 GHz at that time).

■ **Approach.** In order to remediate the situation, Heinrich suggested

– accounting for wave propagation along the S, G, and D electrodes (by modeling them as distributed structures);
– considering the transit time influence on the transconductance, as well as the intrinsic capacitance C_{dsi} in the small-signal equivalent circuit.

For the distributed noise analysis, Heinrich proposed to "slice" the device structure along the S, G, D electrodes in several sections, each one (Fig. 7.25) being characterized by the correlation matrix of the noise equivalent generators ΔE_n and ΔI_n. The noise generated by different sections is uncorrelated.

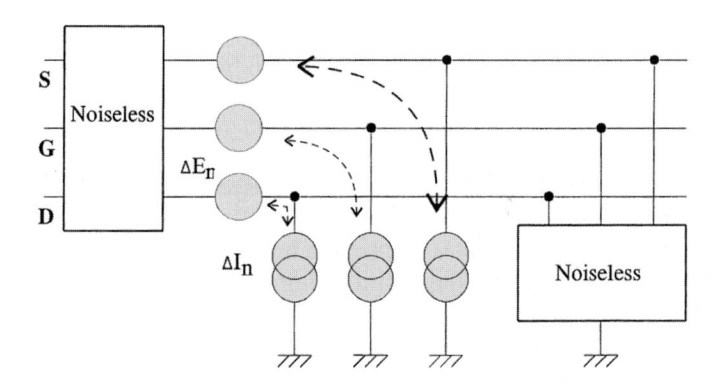

Fig. 7.25. Noise model of an elementary gate with section

■ **Noise Model.** Noise analysis is performed by applying to each elementary gate width section the method of Hillbrand and Russer (Sect. 6.2). Since the noise produced by different sections is uncorrelated, the noise correlation matrix of the whole MESFET can be evaluated by superposing the contributions

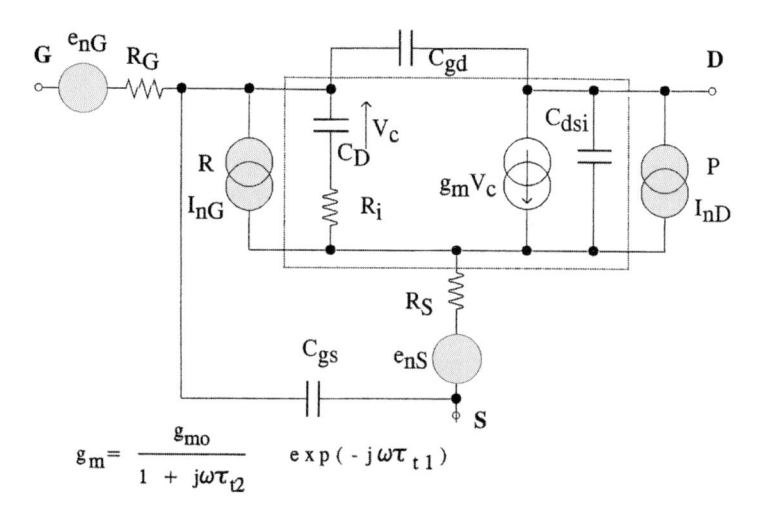

Fig. 7.26. The improved MESFET lumped noise model proposed by Heinrich

of all sections. The resulting integral expression was solved numerically using an iterative trapezoidal rule. According to the results obtained, the extended lumped noise model of Fig. 7.26 is proposed, where the noise generators are defined to be

$$\overline{I_{nD}^2} = 4kT\, g_{mo}\, P\, \Delta f \tag{7.99a}$$

$$\overline{I_{nG}^2} = 4kT\, \frac{(\omega C_D)^2}{g_{mo}}\, R\, \Delta f \tag{7.99b}$$

$$\overline{I_{nG}^* I_{nD}} = 4kT\, \omega C_D\, \Delta f\, jC_{cor} \sqrt{RP} \tag{7.99c}$$

with $P = 1.1$, $R = 0.5$, and $C_{cor} = 0.7$.

■ **Results.** The four noise parameters of the MESFET are determined with the model shown in Fig. 7.26; to simplify computation, the exponential function was replaced by a power series expansion in ω (second-order approximation). After some algebraic manipulations, the optimum noise figure is obtained

$$F_o = 1 + 2R_N \left(G_K + \sqrt{G_K^2 + \frac{G_N}{R_N}} \right) \tag{7.100}$$

with

$$R_N = \frac{1}{|c|^2}\, N_{22} \tag{7.101}$$

$$G_N = \frac{(\omega C_D)^2}{|b|^2}\left(N_{11} - \frac{1}{N_{22}}|N_{12}|^2\right) \tag{7.102}$$

7 Noise Models of Electronic Devices

$$G_K = \omega C_D \, \mathcal{I}m \left(\frac{1}{b} \left(c \frac{N_{12}}{N_{22}} - a \right) \right) \tag{7.103}$$

$$N_{11} = R_G |a|^2 + \frac{1}{R_S} |g|^2 + \frac{R}{g_{mo}} |h|^2 + g_{mo} |g|^{2P} - 2\sqrt{RP} \mathcal{R}e(Cgh^*) \tag{7.104}$$

$$N_{22} = R_G |c|^2 + \frac{1}{R_S} |d|^2 + (\omega C_D)^2 \frac{R}{g_{mo}} |e|^2 + g_{mo} |f|^{2P} + 2\omega C_D \sqrt{RP} \mathcal{I}m(e^* fC) \tag{7.105}$$

$$N_{12} = R_G ac^* + \frac{1}{R_S} gd^* + C^* \sqrt{RP} hf^* - g_{mo} P gf^* + j\omega C_D \left(ge^* C \sqrt{RP} - \frac{R}{g_{mo}} he^* \right) \tag{7.106}$$

$$a = Y_1 \frac{1}{1 + j\omega\tau_i} + \frac{C_{gs} + C_{gd}}{C_D} (Y_1 + Y_2) \tag{7.107}$$

$$b = R_G \Big((Y_1 + Y_2) Y_3 + Y_1 Y_e \Big) \tag{7.108}$$

$$c = g_m Y_1 - j\omega \Big(C_{dsi} Y_2 + C_{gd} (Y_1 + Y_2) \Big) \tag{7.109}$$

$$d = R_G \Big(g_m Y_3 + j\omega (C_{dsi} (Y_e + Y_3) + C_{gd} Y_e) \Big) \tag{7.110}$$

$$e = (-R_G) \Big(g_m (Y_1 + Y_3) + j\omega (C_{dsi} Y_3 - C_{gd} (Y_1 + g_m)) \Big) \tag{7.111}$$

$$f = R_G \left(\frac{1}{R_S} (Y_e + Y_3) + Y_e \left(\frac{1}{R_G} + j\omega C_{gs} \right) \right) \tag{7.112}$$

$$g = 1/(1 + j\omega\tau_1) \tag{7.113}$$

$$h = Y_1 + g_m \tag{7.114}$$

$$Y_1 = j\omega C_{dsi} + 1/R_S \tag{7.115}$$

$$Y_2 = Y_e + g_m \tag{7.116}$$

$$Y_3 = 1/R_G = j\omega(C_{gs} + C_{gd}) \tag{7.117}$$

$$Y_e = j\omega C_D \, g \tag{7.118}$$

$$g_m = g_{mo} \frac{1}{1 + j\omega\tau_{t2}} \exp(-j\omega\tau_{t1}) \tag{7.119a}$$

$$\tau_1 = C_D R_i \tag{7.119b}$$

Here R_N, G_N, G_K, N_{11}, and N_{22} are real numbers; the remaining quantities are complex and the asterisk denotes complex conjugation.

■ **Conclusion.** The model proposed by Heinrich predicts accurately the minimum noise figure up to 60 GHz, when second-order approximation in the power series expansion is considered. Note that a first-order approximation would seriously restrict the frequency range of validity to a maximum of 20 GHz.

7.9.8 Model of Escotte and Mollier

■ **Introduction.** This model was proposed in 1990, in an attempt to improve with a semi-distributed structure the accuracy of simulation of MESFETs operating in the millimeter-wave range [117].

■ **Approach.** The transistor model is composed of N identical sections sliced along the actual MESFET gate width (w); N is chosen so that the ratio $w/N\lambda_g$ (λ_g being the propagation wavelength on the transistor electrodes) is of the order of a few percent. Under this condition, each section of the actual transistor can be represented by a lumped two-port model.

The N sections are interconnected by means of transmission lines, simulating the metallic electrodes of gate and drain (Fig. 7.27). The noise generated by each section is described with the noise coefficients P, R, and C (Van der Ziel JFET model).

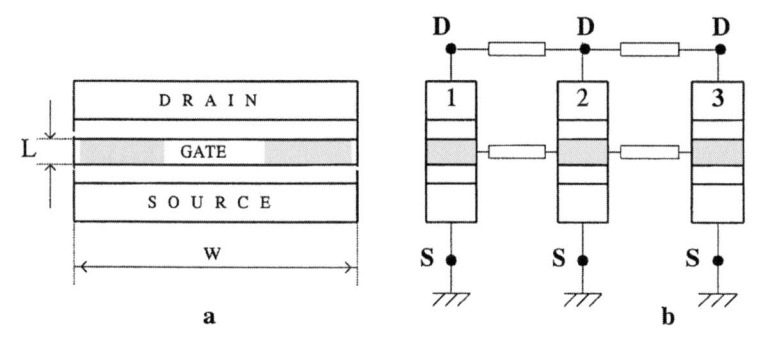

Fig. 7.27. a Top view of the MESFET structure; **b** sliced model having N = 3 sections

■ **Noise Model.** The key point is to determine the noise coefficients P, R, and C. As they are weakly frequency-dependent, the model is expected to accurately predict the device behavior at higher frequencies than those at which data has been collected.

Step 1: Experimentally determine the MESFET noise figure for several input impedances. The four noise parameters F_o, R_n, R_o, and X_o are extracted with the least-squares fit proposed by Mitama and Katoh [138].

Step 2: Find an analytic expression for the noise factor of the device, composed of several "sliced" sections. The noise of each section is described by means of the intrinsic noise generators I_{nG} and I_{nD} and the thermal noise of the extrinsic resistances. As the noise generated by one section is not correlated with the noise of the others, their contributions can be added. By applying North's definition, the expression of F is deduced. Then, by differentiating it with respect to the real and imaginary parts of the source impedance $Z_g = R_g + jX_g$, two expressions are obtained. The cancellation of the two expressions yields R_o and X_o. Substitution of R_o and X_o for R_g and X_g gives the minimum noise factor F_o.

Step 3: Determine the coefficients P, R, and C by fitting the expressions for F_o, R_n, R_o, and X_o with the experimental data obtained during *Step 1*.

■ **Conclusion.** This model is able to accurately predict the transistor S and noise parameters up to 40 GHz.

7.9.9 HSPICE Model

■ **Introduction.** Some versions of SPICE include a model based on the same equivalent circuit, to simulate MESFETs and JFETs (the only difference being the material's parameters, which correspond either to silicon for JFETs, or to GaAs for MESFETs).

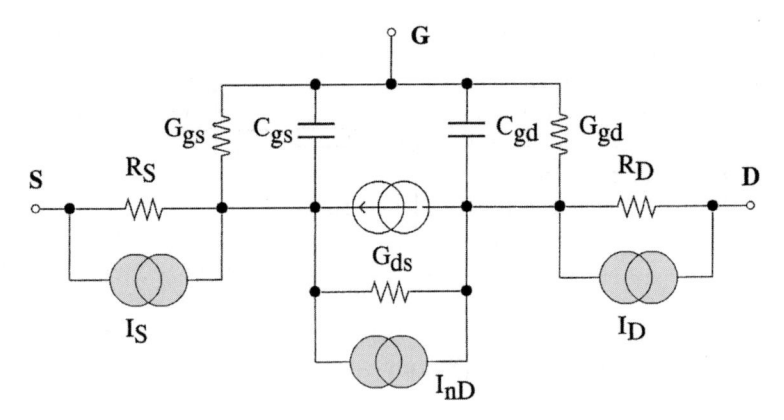

Fig. 7.28. MESFET noise model, employed in HSPICE

■ **Noise Model.** Since noise analysis is always preceded by AC analysis, the noise model is derived from the small-signal equivalent circuit (Fig. 7.28), to which the following noise sources are added:

– Thermal noise of the source and drain resistances:

$$S(I_S) = 4kT / R_S \quad \text{and} \quad S(I_D) = 4kT / R_D \qquad (7.120)$$

– The channel noise denoted by I_{nD}, which originates from the thermal noise of the channel (I_{th}) and the flicker noise (I_f):

$$I_{th} = \begin{cases} \left(\dfrac{8KT\, g_m}{3}\right)^{1/2} & \text{if NLEV} < 3 \\[2ex] \left(\dfrac{8KT}{3}\text{BETA}_{\text{eff}}(V_{GS}-\text{VTO})\dfrac{1+\alpha+\alpha^2}{1+\alpha}\text{GDSNOI}\right)^{1/2} & \text{if NLEV} = 3 \end{cases}$$
$$(7.121)$$

$$I_f = \left(\frac{KF\ I_{DS}^{AF}}{f}\right)^{1/2} \qquad (7.122)$$

with

$$\alpha = \begin{cases} 1 - \dfrac{V_{DS}}{V_{GS} - VTO} & \text{in the ohmic region} \\ 0 & \text{in saturation} \end{cases}$$

The following notation has been introduced:

AF	– Flicker noise exponent	*(default to 1.0)*
KF	– Flicker noise coefficient	*(default to 0.0)*
	(recommended values between 10^{-19} and 10^{-25})	
NLEV	– Parameter to select noise equation	*(default to 2)*
GDSNOI	– Shot noise coefficient	*(default to 1.0)*
	(to be used only with NLEV=3)	

■ **Conclusion.** The noise model proposed in HSPICE is not appropriate. Adopting the same topology for the JFET model (dedicated to low-frequency applications) and the MESFET model (mainly used in the microwave range) is a very risky choice. In reality, it is rather difficult to simulate both types of transistors with the same equivalent circuit.

Moreover, the diffusion noise which is paramount to MESFET operation above pinch-off is completely neglected, as well as the thermal noise of the gate resistance.

Although there are a lot of available valuable electrical models of MESFET for circuit simulation (for a comprehensive account, see [118]), efforts are still needed to improve the SPICE family noise models.

7.9.10 Conclusion

The noise in any MESFET has several components:

a) *Thermal noise*, generated by the parasitic resistances of source, drain, and gate.
b) *Channel noise*, which is essentially thermal if the device is operating in the linear mode, or diffusion noise in the saturation mode.
c) *Gate noise*, which arises from the channel noise induced in the gate (across the channel-to-gate capacitance), and which is then amplified by the device. Due to the capacitive coupling, this noise has a power spectral density proportional to frequency.
d) *Flicker (1/f) noise*, originating from random carrier generation-recombination in the lattice imperfections or contaminating impurities. The noise power spectrum has $1/f^n$ behavior, n being close to unity.

236   7 Noise Models of Electronic Devices

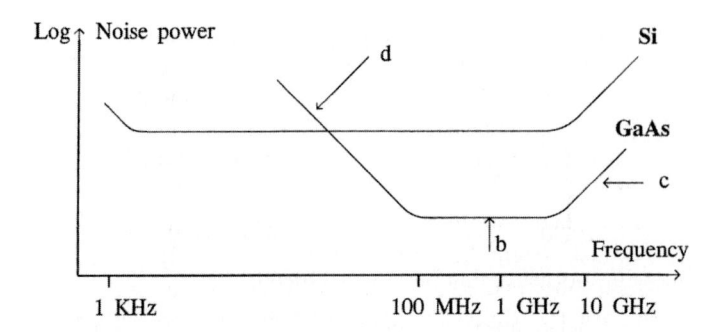

Fig. 7.29. Comparison of Si and GaAs MEFSETs

Superimposing all these contributions, plots of noise power versus frequency for silicon and GaAs MESFETs are given in Fig. 7.29 (it is assumed that component (a) is negligible).

It is obvious that compared to silicon, GaAs offers better noise behavior (due to improved carrier mobility), with one notable exception: the corner frequency of the 1/f noise is around 100 MHz, instead of 1 kHz for silicon devices. This explains why in all broadband, low-noise systems, both types of MESFETs are commonly encountered.

7.10 Noise of HEMT Transistors

7.10.1 Introduction

■ **Overview.** The high electron mobility transistor (HEMT) emerged in the early 1980s, thanks to laboratory progress in growing epitaxial monocrystal layers. The HEMT takes advantage of band-gap engineering in III–V semiconductors by utilizing a heterostructure in the gate junction, which is grown free of imperfections.

■ **Discussion.** We have already seen that the flow of charge carriers between source and drain in a MESFET structure occurs in a doped channel. High doping is essential to increase the concentration of charge carriers, but an unwanted side-effect is the degradation of the mobility, due to strong ionized impurity scattering of carriers (note also that the lattice defects resulting from doping the semiconductor, by diffusion or ion implantation techniques, are responsible for trapping). This fact becomes dominant in submicrometer structures, where the gate length diminishes and the doping level must be increased. In order to obtain transistors with high electron mobility, the adopted solution consists in introducing a heterojunction between the low-doped channel and a doped semiconductor layer which supplies the channel with free electrons. This is achieved by doping solely the material with the largest band gap.

In this way, the thin low-doped channel offers a high electron density in a region free of material defects, where the rate of ionized impurity scattering and trapping of carriers is drastically reduced. Consequently, for deep submicron devices, cutoff frequencies in excess of 300 GHz have been achieved, and the highest reported oscillation frequency of a HEMT oscillator is 213 GHz.

■ **Device Structure.** As shown in Fig. 7.30, a thin AlGaAs layer (4) is physically separated from the semi-insulating layer (2) by means of a very thin AlGaAs alloy undoped layer (3). This structure constitutes a heterojunction. Therefore, the two-dimensional electron gas channel (2DEG) is spatially separated from the carrier supply layer (4).

■ **Principle of Operation.** According to Anderson's explanation, due to the relative position of the heterojunction energy bands, a large number of electrons are stimulated to migrate from layer 4 to layer 2, across the very thin layer 3 [119]. Since the transfer of electrons in the reverse direction is very unlikely (due to the barrier of the heterojunction), a significant number of free electrons accumulate against the interface between layers 2 and 3. These constitute the so-called "2DEG" (two-dimensional electron gas, since the region where they accumulate has practically no thickness). In this way, a spatial separation is created between the heavily doped layer (4) supplying free electrons and the channel (2DEG) where the flow of electrons takes place. Increasing the thickness of layer 3 is expected to improve the separation and consequently the electron mobility; however, too thick a layer restrains the migration of free electrons from layer 4 into the channel. Therefore, an optimum value of the layer 3 thickness must be found; a good trade-off between these constraints is indicated in Fig. 7.30.

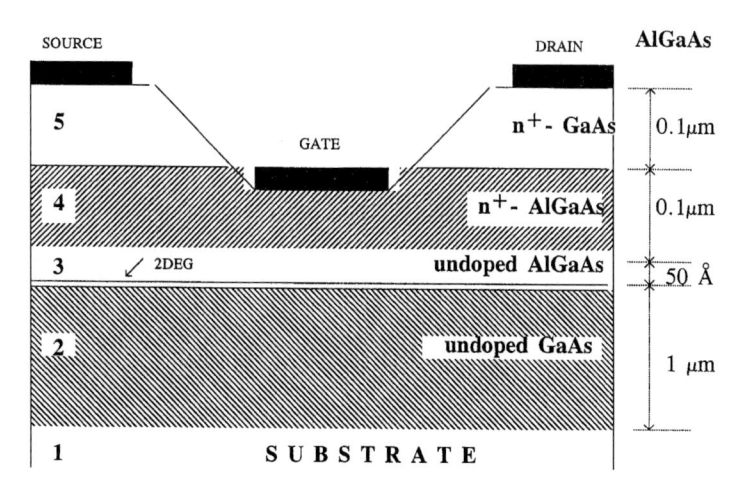

Fig. 7.30. Cross section in a HEMT (MODFET or HFET) structure

To conclude, the channel is located at the interface separating layers 2 and 3; the depleted region, controlled by the gate voltage, appears under the gate, and partially extends into the layers 4 and 3.

■ **Comparison.** With respect to MESFETs, HEMTs have two main advantages:

- Since the channel is created in an undoped region, the carrier density and mobility are both improved. This translates into a high transconductance value and a very high cutoff frequency.
- Due to highly doped layers 4 and 5, the parasitic resistances of drain and source are reduced.

Both effects suggest that the overall noise generated by a HEMT device must be lower than that of a MESFET.

■ **Small-Signal Equivalent Circuit.** Figure 7.31 presents the small-signal equivalent circuit of a HEMT device, which is similar to that of a MESFET. The only difference is the obligation to take into account the parasitic inductances of source, gate, and drain, because the device operates at higher frequencies.

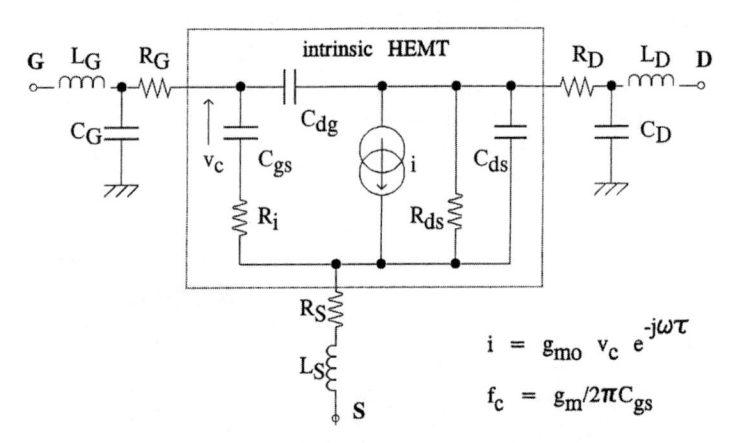

$$i = g_{mo} v_c e^{-j\omega\tau}$$

$$f_c = g_m/2\pi C_{gs}$$

Fig. 7.31. Small-signal equivalent circuit of a HEMT

■ **Physical Aspects of Noise.** As in any semiconductor device, there are five dominant noise mechanisms: thermal noise, diffusion noise, shot noise, G–R noise, and flicker noise.

It is common practice to consider for HEMT devices only the thermal noise of R_D, R_S, and the diffusion noise of the channel. The flicker noise, which shows up primarily at frequencies lower than 1 MHz, is often disregarded, since most HEMT devices are dedicated to microwave applications.

Nevertheless, there are at least two situations in which 1/f noise still affects microwave circuit operation by means of side-effects like up-conversion occurring in mixers and the phase noise of HEMT oscillators.

7.10.2 Cappy Model

■ **Introduction.** This is a one-dimensional physical model which was developed in the late 1980s [120, 121].

■ **Approach.** In order to find expressions for the four noise parameters of the transistor, the following steps are required:

Step 1. Find the DC characteristics and the small-signal equivalent circuit for each bias point.

Step 2. Determine the noise of the intrinsic transistor: the equivalent drain and gate noise generators (as well as the correlation between them).

Step 3. Add the extrinsic noise sources.

Step 4. Calculate the four noise parameters.

■ **Comment.** During *Step 1*, the source-to-drain space is divided into incremental sections; the fundamental transport equations are established for each section, with appropriate boundary conditions. By integrating the electric field and the carrier density over the whole channel length, the drain-to-source voltage and the stored charge are computed. As soon as the DC electrical characteristics are available, the small-signal intrinsic parameters are computed by a perturbation method.

■ **Noise Analysis.** The numerical modeling performed during *Step 1* makes it possible to obtain both the intrinsic noise sources and the correlation coefficient. With the approach of Pucel, Haus, and Statz (Sect. 7.9.4) the following expressions are deduced

$$\overline{I_{nD}^2} = 4kT\,\Delta f\,\frac{g_d^2 + \omega^2 C_{gd}^2}{g_d^2}\,\frac{g_m}{C_{gs}}\,L\,(\alpha w + \beta I_{DS}) \qquad (7.123)$$

$$\overline{I_{nD}^2} = \frac{2kT\,C_{gs}^2}{g_m}\,\omega^2\,\Delta f \qquad (7.124)$$

In these equations, L and w are the gate length and width, and α and β are two fitting coefficients, which are almost independent of the active layer geometry and material properties. Their recommended values are
$$\alpha = 2\cdot 10^5\,\text{pF/cm}^2 \text{ and } \beta = 1.25\cdot 10^2\,\text{pF/mA/cm}.$$
The correlation coefficient lies between j0.7 and j0.8.

240   7 Noise Models of Electronic Devices

■ **Remark.** For frequencies satisfying $\omega C_{gd} \ll g_d$, the results are in good agreement with those predicted by the Pucel, Haus, and Statz model.

■ **Noise Model.** Neglecting the feedback capacitance C_{dg}, the standard four noise parameters of the HEMT (minimum noise figure F_o, noise conductance G_n and optimum source impedance Z_o) are

$$F_o = 1 + 2\sqrt{P}\,\frac{f}{f_c}\,\sqrt{g_m(R_S + R_G)} \qquad (7.125a)$$

or

$$F_o = 1 + 2\sqrt{\frac{I_{DS}}{E_c\,L}}\,\frac{f}{f_c}\,\sqrt{R_S + R_G} \qquad (7.125b)$$

$$G_n = P g_m (f/f_c)^2 \qquad (7.125c)$$

$$Z_o = \sqrt{\frac{g_m(R_S + R_G)}{P}}\,\frac{1}{\omega C_{gs}} + \frac{1}{j\omega C_{gs}} \qquad (7.125d)$$

with

$$P = \frac{I_{DS}}{E_c\,L\,g_m} \qquad (7.125e)$$

In these expressions, f_c is the cutoff frequency and E_c is the critical field of an idealized V–E relationship.

■ **Comment.** Equation (7.125a) demonstrates that the minimum noise factor can be lowered by reducing g_m (hence, by using a large epilayer 2). However, expression (7.125b) no longer contains g_m, and this constraint disappears. As stated in [121], from an experimental point of view, it is rather difficult to separate these two approaches, and a compromise must be found.

■ **Conclusion.** The strong points of this model are:

- deep understanding of how the electrical and technological parameters influence noise quantities;
- all four noise parameters of the device are available.

7.10.3 Pospieszalski Model

■ **Introduction.** This model was developed in the late 1980s [122]. Basically it can be considered a simple but accurate empirical model that contains only two fitting coefficients. It is widely used for HEMTs in microwave circuits operating at room and cryogenic temperatures.

■ **Approach.** The model proposed by Pospieszalski for the noise of an intrinsic device needs only two frequency-independent fitting coefficients, in addition to the small-signal equivalent circuit, to predict the standard four noise parameters at any frequency. These two fitting coefficients are

7.10 Noise of HEMT Transistors

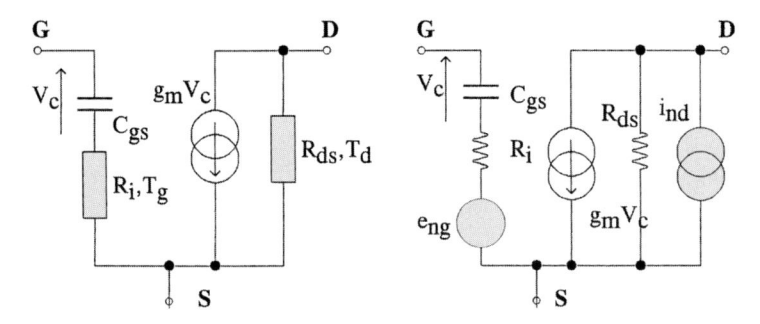

Fig. 7.32. Noise equivalent circuits of a HEMT chip

1) The equivalent temperature T_g of the intrinsic gate resistance R_i (briefly called *equivalent gate temperature*). The corresponding noise generator is denoted e_{ng} (Fig. 7.32) and has a spectral density $S(e_{ng}) = 4kT_g R_i$.
2) The equivalent temperature T_d of the drain conductance R_{ds} (briefly called *equivalent drain temperature*).

The noise equivalent generators are assumed to be uncorrelated; the noise due to parasitic resistances of source, gate, and drain is essentially thermal, and depends on the ambient temperature.

■ **Noise Analysis.** The noise parameters of the device chip under various representations (Figs. 5.8c and 5.8d, and a third representation consisting of minimum noise temperature T_{min}, optimal source impedance Z_o, and noise conductance g_n) are obtained. The main results are

$$X_o = 1/\omega C_{gs} \tag{7.126}$$

$$R_o = \sqrt{\left(\frac{f_T}{f}\right)^2 \frac{R_i}{g_{ds}} \frac{T_g}{T_d} + R_i^2} \tag{7.127}$$

$$T_{min} = 2\frac{f}{f_T}\sqrt{g_{ds}R_i T_g T_d + \left(\frac{f}{f_T}R_i g_{ds} T_d\right)^2} + 2\left(\frac{f}{f_T}\right)^2 R_i g_{ds} T_d \tag{7.128}$$

$$g_n = \left(\frac{f}{f_T}\right)^2 \frac{g_{ds} T_d}{T_o} \tag{7.129}$$

$$\frac{4NT_o}{T_{min}} = \frac{2}{1+(R_i/R_o)} \tag{7.130}$$

$$R_n = \frac{T_g}{T_o}R_i + \frac{T_d}{T_o}\frac{g_{ds}}{g_m^2}\left(1+(\omega C_{gs}R_i)^2\right) \tag{7.131}$$

$$cor = \rho\sqrt{R_n g_n} = \frac{T_d}{T_o}\frac{g_{ds}}{g_m^2}\left(\omega^2 C_{gs}^2 R_i + j\omega C_{gs}\right) \tag{7.132}$$

where $g_{ds} = 1/R_{ds}$ and $f_T = g_m/(2\pi C_{gs})$.

242   7 Noise Models of Electronic Devices

■ **Discussion.** The previous expressions simplify if certain conditions are imposed. For instance, assuming that $\dfrac{f}{f_T} \ll \sqrt{\dfrac{1}{R_i\, g_{ds}}\dfrac{T_g}{T_d}}$, (7.127) and (7.128) become

$$R_o \cong \frac{f_T}{f}\sqrt{\frac{R_i}{g_{ds}}\frac{T_g}{T_d}} \tag{7.133}$$

$$T_{min} \cong 2\frac{f}{f_T}\sqrt{g_{ds}R_i T_g T_d} \tag{7.134}$$

Hence,

$$\frac{4NT_o}{T_{min}} \cong 2 \tag{7.135}$$

Another interesting situation occurs when $T_g \to 0$. Then

$$T_{min} \cong 4\left(\frac{f}{f_T}\right)^2 R_i\, g_{ds}\, T_d, \quad R_o = R_i, \quad \text{and} \quad \frac{4NT_o}{T_{min}} \cong 1 \tag{7.136}$$

It follows that the Pospieszalski model predicts the measured ratio of $4NT_o/T_{min}$ at any frequency to be

$$\boxed{1 \le \frac{4NT_o}{T_{min}} < 2} \tag{7.137}$$

Expression (7.137) provides a fast check of model validity.

■ **Note**

1) The only fitting coefficients of this model are T_g and T_d, which need to be determined from measurement. If these noise coefficients are known (in addition to the small-signal equivalent circuit of the transistor), the noise parameters are available with (7.126) to (7.132).
2) If the transit time τ associated with the transconductance g_m is taken into account, the final expressions for the noise parameters T_{min}, Z_o and g_n are not modified.
3) The minimum value of the noise measure is not changed by adding the feedback capacitance C_{gd} and the transit time τ to the small-signal equivalent circuit.

■ **Interpretation.** Despite the fact that T_d and T_g have been introduced as fitting coefficients, one may wonder what physical significance might be attributed to them.

☐ **Drain Temperature T_d** *is an equivalent temperature of the output impedance of the device if the gate at the chip terminal is open-circuited.*

Practical values of the order of several thousands kelvins are reported at room temperature. This noise coefficient may be viewed also as the average temperature of "hot" electrons, since T_d directly depends on the electric field applied to the channel.

☐ **Gate Temperature T_g** interpretation poses greater difficulties due to the impossibility of providing an accurate value for R_i. In actuality, we can accurately fit the S parameter data of a HEMT with quite different values of R_i, the only caution being to properly adjust each time the values of remaining small-signal parameters. This demonstrates that T_g remains an adjustment coefficient, whose physical interpretation largely depends on the possibility of accurately determining R_i. Nevertheless, Pospieszalski proposes to consider it as the electron temperature averaged over the length of the channel, in a direction perpendicular to the channel. In this view, T_g is not so field-dependent, and its value is close to the ambient temperature; consequently, *it is no longer necessary to measure it*.

■ **Comparison.** In the low-frequency approximation, the equivalence between the noise coefficients, denoted by K, of the model of Pucel, Haus, and Statz and T_d, T_g is established:

$$K_g = \frac{T_d}{T_o} \frac{g_{ds}}{g_m}, \quad K_r = g_m R_i \frac{T_g}{T_o}, \quad K_c = 1 \quad (7.138)$$

A complete comparison of the Pucel, Haus, and Statz model with respect to Pospieszalski model is proposed in [123].

■ **Conclusion.** From a practical point of view, the Pospieszalski model is simple, consistent with the small-signal equivalent circuit, and easy to implement in any CAD package. The extraction of T_g and T_d from measured data is very robust and does not suffer from large scatter [123]. The model noise coefficients are frequency-independent above 1 GHz and they can be successfully used to predict the noise parameters over a wide frequency range. Below 1 GHz, only a single noise coefficient (T_d) is needed, and this greatly reduces measurement effort. Note that this last feature constitutes the starting point of the measurement method proposed by Tasker et al. [124].

7.10.4 Model of Hickson, Gardner, and Paul

■ **Introduction.** This model, developed in the early 1990s [125], takes the inherent distributed nature of the intrinsic millimeter-wave HEMT into account, by dividing the active region into a number of slices (similar to the approach of Heinrich or Escotte and Mollier). Each slice is characterized by applying the noise model of Pospieszalski. The novelty of this model is to make the noise parameters of the first slice different from those of the remaining slices (to account for the nonuniform field distribution). The slices

244   7 Noise Models of Electronic Devices

are connected by lossy transmission lines. However, it is assumed that every slice has the same transconductance g_m.

■ **Noise Model.** The device is divided into N slices (Fig. 7.33); each slice is considered an elementary intrinsic HEMT, whose small-signal equivalent circuit is detailed in Fig. 7.34.

Integer N must be chosen such that each slice is short relative to the guided wavelength at the gate and drain electrodes at the maximum operating frequency. L_1 and R_1 are the parasitic inductance and resistance of the gate electrode, while L_2 and R_2 correspond to the drain electrode.

Values of various elements of the noise model are optimized to fit the transistor S parameters from 2 to 40 GHz, and the noise parameters from 14 to 26 GHz.

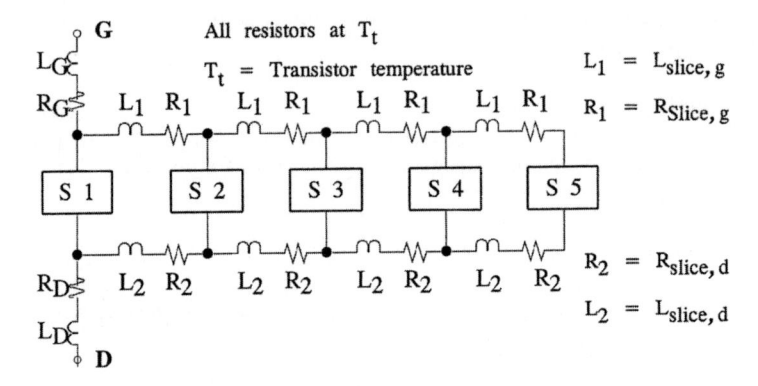

Fig. 7.33. Overall slice model (for N = 5)

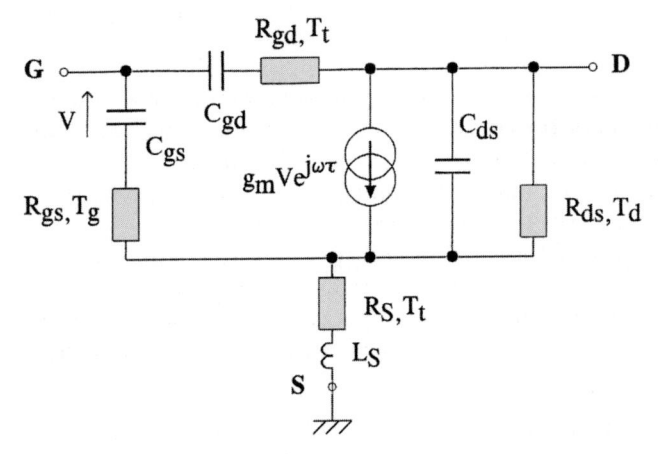

Fig. 7.34. Individual slice noise model (T_1 denotes transistor temperature)

Table 7.5. Noise model of the HEMT chip JS8900-AS ($T_t = 25.5°C$)

Parameter	Slice 1	Slices 2, 3, 4, 5
T_g	1.0°C	69.1°C
T_d	188.3°C	1589.1°C

■ **Results.** A sample of the results obtained for the Toshiba type JS8900-AS HEMT chip, is presented in Table 7.5.

■ **Concluding Remarks.** Note that the gate temperature is close to T_t (as stated in the approach of Pospieszalski), but the drain temperature is considerably higher. Note also the difference between the numerical values obtained for the noise coefficients of the first slice in comparison to those of the remaining ones. This suggests that there are significant differences in the field structure in the gate feed region, and substantiates the proposed approach.

7.10.5 Model of Klepser, Bergamaschi, Schefer, Diskus, Patrick, and Bächtold

■ **Introduction.** In the early 1990s, exceptionally high cutoff frequencies and low noise properties were reported for InP based HEMTs. Inasmuch as low-noise integrated microwave circuit design starts by simulating the circuit to be built, it became necessary to develop good models for both electrical and noise analysis. This model was published in 1995, and is essentially an analytic, bias-dependent noise model for InP HEMTs [126].

■ **Approach.** Noise analysis is based on the Pucel, Haus, and Statz approach and requires the following steps:

1. S- and noise-parameter measurement at various bias points.
2. Extraction of the equivalent circuit elements at all bias points.
3. Evaluation of the noise coefficients P, R, and C.
4. Mathematical fitting of the large-signal model to the measured bias dependence of the equivalent circuit elements.
5. Calculation of the bias dependence of P, R, and C by combining their analytic expressions with the large-signal model.
6. Fitting the calculated bias dependence with elementary mathematical functions to obtain a semi-empirical, bias-dependent model.

■ **Noise Analysis.** The proposed noise equivalent circuit is detailed in Fig. 7.35. Thermal noise is associated with all parasitic resistances R_G, R_S, and R_D; noise coefficients P, R, and C are defined according to the model of

246   7 Noise Models of Electronic Devices

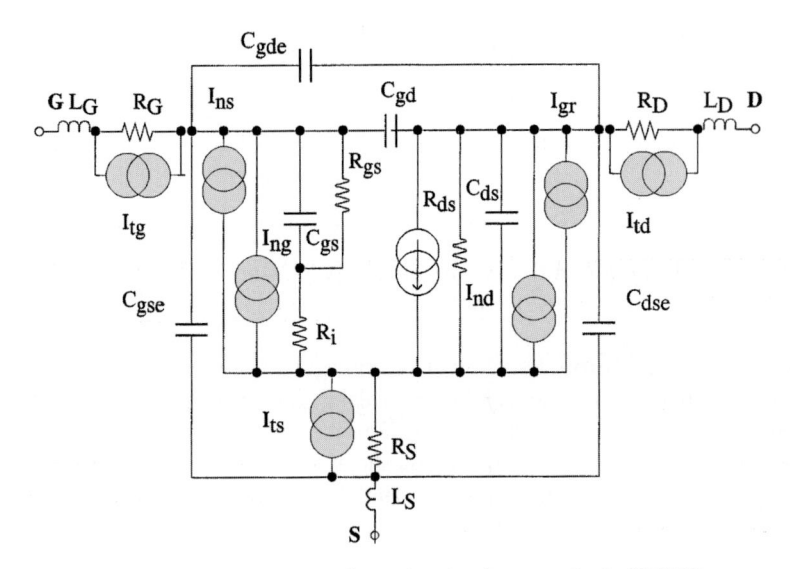

Fig. 7.35. Noise equivalent circuit of an extrinsic HEMT

Bächtold (or expressions (7.71)). The fundamental two-section channel modeling introduced by Pucel, Haus, and Statz requires one to estimate P, R, and C in the ohmic region I (subscript 1) and region II (subscript 2), where electrons travel at their saturation velocity. Various calculations show that for submicrometer HEMTs, the correlation coefficient is roughly constant, and close to unity; the authors set its value at 0.95.

In addition to the previous intrinsic noise sources, the authors introduced the following noise sources to account for the nonideal HEMT:

– Gate leakage current, responsible for shot noise

$$\overline{I_{ns}^2} = 2qI_g\,\Delta f \qquad (7.139)$$

– The G–R processes, responsible for frequency-dependent noise in the channel

$$\overline{I_{gr}^2} = \frac{|I_{gro}|^2}{1+(f/f_{gro})^2} \qquad (7.140)$$

where I_{gro} denotes the low-frequency value and f_{gro} is the cutoff frequency associated with the G-R processes.

■ **Bias Dependence.** Bias dependence of the noise sources is described in terms of the bias voltages V_{GS} and V_{DS}. The following expressions have been proposed:

$$P_1 = \frac{P_1'}{(V_{GS}-V_{off})^{0.75}\,V_{DS}} \qquad (7.141)$$

$$P_2 = P'_2 (V_{GS} - V_{off})\sqrt{V_{DS}} + P''_2 (V_{GS} - V_{off})^5 V_{DS}^4 \quad (7.142)$$

$$R = R' (V_{GS} - V_{off})^2 V_{DS}^2 \quad (7.143)$$

V_{off} denotes the pinch-off voltage, and P'_1, P'_2, P''_2 and R' are fitting coefficients.

The leakage gate current bias dependence is

$$|I_g| = \left| I_{gds} \left(\exp\left(\frac{V_{DS} - V_{GS}}{nkT}\right) - 1 \right) \right| + \left| I_{gss} \left(\exp\left(\frac{V_{DS}}{n'kT}\right) - 1 \right) \right| \quad (7.144)$$

and for the G-R noise current

$$I_{gro} = \sqrt{V_{GS} - V_{off}} \left(I_{gr1} + I_{gr2} V_{DS}^\gamma \right) \quad (7.145)$$

where I_{gds}, n and I_{gss}, n' are the saturation currents and ideality factors of the forward gate-drain heterojunction and gate-source Schottky diodes, respectively. Note that I_{gr1}, I_{gr2}, and γ are fitting parameters.

■ **Concluding Remarks.** Since the G-R noise and shot noise of the gate leakage current are not neglected, this model is well suited for broadband noise description of an InP-based HEMT. Expressions (7.141)–(7.145) establish the bias dependence of the noise sources, and therefore enable one to choose the optimum bias point. According to the authors experience, for frequencies above 10 GHz the minimum noise figure is obtained when $V_{GS} = -0.3$ V and $V_{DS} = 0.6$ V. Another strong point of this model is that it is possible to predict noise source behavior as a function of device scaling. As a general conclusion, all noise sources (Fig. 7.35) show a linear increase with the gate width of the transistor (it is preferable to use multi-finger gates for gate widths above 70 µm).

7.10.6 Conclusion

An overview of all models discussed is detailed in Fig. 7.36, which must be viewed as a navigation chart. Note that

- Classification in two categories (physical and empirical models) is based on the dominant feature. Realistically, every physical model also relies on certain fitting coefficients, and every so-called empirical model is based on a theoretical approach. Perhaps the only real difference between these two classes is that the fitting coefficients of the empirical models have no physical meaning.
- Arrows indicate the logical order of models; for instance, the MESFET model of Podell requires the JFET model proposed by Bruncke. This must not be viewed as an attempt to establish a historical hierarchy, but rather an indication to the reader of the background required to understand a particular model.

248   7 Noise Models of Electronic Devices

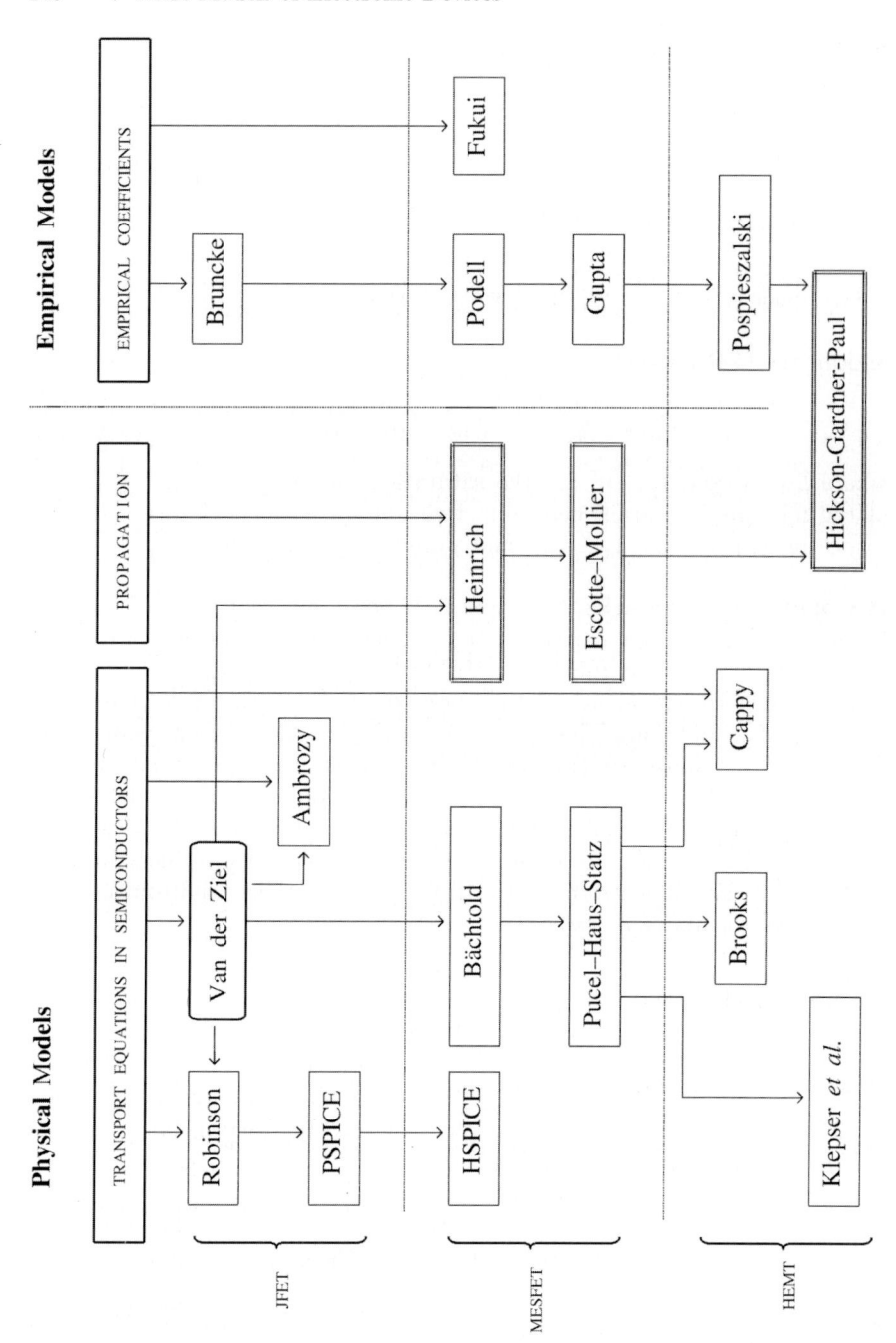

Fig. 7.36. Overview of field effect transistor models (double box indicates semi-distributed models)

- There in no "universal" model, that supplants all the others. Each model has its strong points, as well as its weaknesses, and often working to improve the weakness of a particular model yields a new model. Consequently, it has been considered worthwhile to present a collection of the most important models, hoping that the reader will properly select the most appropriate one for his particular application.

7.11 Noise in Operational Amplifiers

■ **Introduction.** A detailed analysis of noise in operational amplifiers is not an easy task, since noise sources are associated with all internal resistors and transistors. Due to the increasing complexity of electrical circuits in actual operational amplifiers, a description based on the lumped equivalent input voltage and current noise remains the only possibility. Consequently, attention will be paid to elaboration of *noise macromodels* consistent with the noise description proposed by manufacturers in the data sheet.

■ **Physical Aspects.** As in any amplifier, noise in the input stage is critical. In operational amplifiers, the first stage is usually a differential amplifier whose transistors operate at low injection levels and consequently with low values for parameter β. This worsens the noise problem. On the other hand, the fact that the bandwidth of the operational amplifier is limited by its internal poles has a favorable effect on the overall noise power.

■ **Noise Models.** Consider the typical configuration of an inverting amplifier (Fig. 7.37a) and its companion noise equivalent circuit (Fig. 7.37b).

V_{n1} and V_{n2} denote the thermal noise of resistors $R_{s1} = R_1 R_f/(R_1 + R_f)$ and R_{s2}, respectively. The internal noise of the operational amplifier is lumped into the equivalent generators E_{n1}, E_{n2}, I_{n1}, I_{n2}.

Noise analysis is based on the approach introduced in Sect. 4.5.1. Hence, the equivalent input noise voltage is found to be

$$E_{ni}^2 = 4kT\Delta f(R_{s1} + R_{s2}) + E_{n1}^2 + E_{n2}^2 + (I_{n1}R_{s1})^2 + (I_{n2}R_{s2})^2 \quad (7.146a)$$

This equation corresponds to the noise model depicted in Fig. 7.38a.

As stated in [128], we may introduce a new noise generator denoted by E_n', such as $(E_n')^2 = E_{n1}^2 + E_{n2}^2$. Equation (7.146a) can be rewritten as

$$E_{ni}^2 = 4kT\Delta f(R_{s1} + R_{s2}) + (E_n')^2 + (I_{n1}R_{s1})^2 + (I_{n2}R_{s2})^2 \quad (7.146b)$$

Now, only three noise equivalent generators are needed to model the noise of the operational amplifier, namely E_n', I_{n1}, and I_{n2} (Fig. 7.38b).

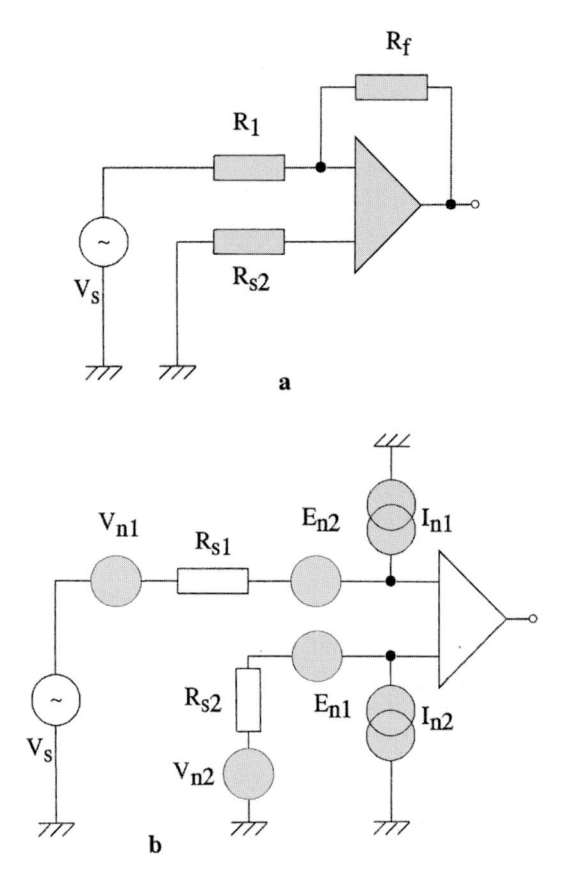

Fig. 7.37. a Inverting amplifier circuit; b noise model

Assuming that $R_{s1} = R_{s2} = R_s$ (which is a practical condition to minimize the offset), we let $(I'_n)^2 = I_{n1}^2 + I_{n2}^2$, and (7.146b) becomes

$$E_{ni}^2 = 8kT\Delta f R_s + (E'_n)^2 + (I'_{n1} R_s)^2 \qquad (7.146c)$$

This means that only two noise equivalent generators E'_n and I'_n are required to model the noise of the operational amplifier (Fig. 7.38c).

■ **Noise Specification.** Operational amplifier manufacturers use various methods to specify the noise of their products:

- Often the values for E_n and I_n at each input (Fig. 7.37a) are proposed. However, if perfect symmetry is assumed at the input, $E_{n1} = E_{n2} = E_n$ and $I_{n1} = I_{n2} = I_n$.
- Sometimes the noise voltage E'_n is proposed, together with that value of the noise current that applies to each input separately: I_{n1} and I_{n2} (model of Fig. 7.38b).

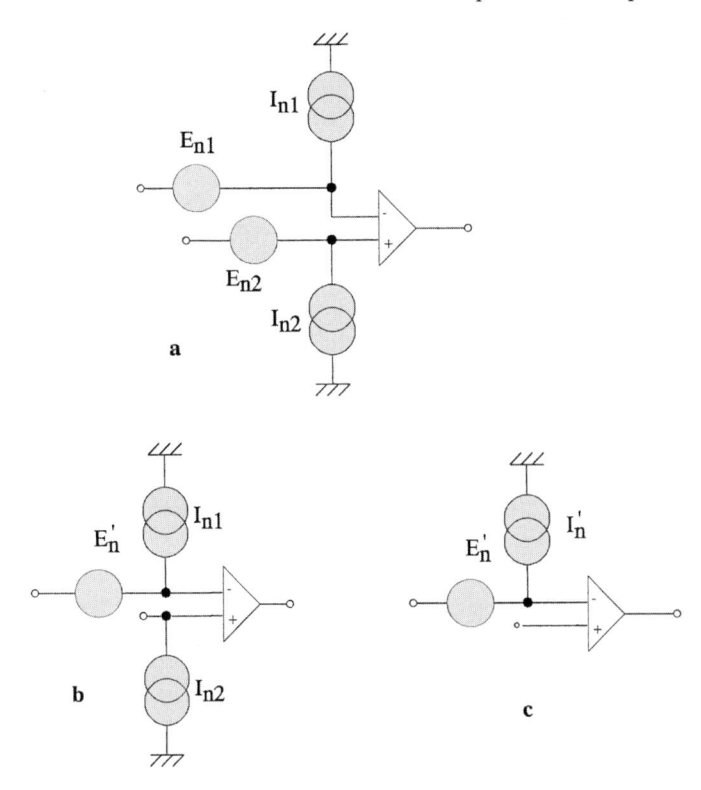

Fig. 7.38. Various noise models of an operational amplifier

– The values of E'_n and I'_n applied only to one input (Fig. 7.38c) are indicated. The magnitude of the noise generators shown in Fig. 7.38c relates to the individual generators of the model detailed in Fig. 7.38a:

$$E'_n = \sqrt{2}\,E_n \quad \text{and} \quad I'_n = \sqrt{2}\,I_n \qquad (7.147)$$

Since no universal convention exists, *the task of the user is to understand the noise model underlying the values proposed in the data sheet.*

■ **Noise Analysis.** Assuming a 1/f component, the power spectral densities of the equivalent noise generators are given by the following equations [129]:

$$S(I_n) = S_i(1 + f_i/f) \qquad (7.148)$$

$$S(E_n) = S_e(1 + f_e/f) \qquad (7.149)$$

where S_i, S_e, f_i, and f_e are constants.

For the inverting or non-inverting configuration, assuming that the noise of the signal source emerges either from R_{s1} or from R_{s2} (but not simultaneously from both) the noise factor is established with the expression [128]

252   7 Noise Models of Electronic Devices

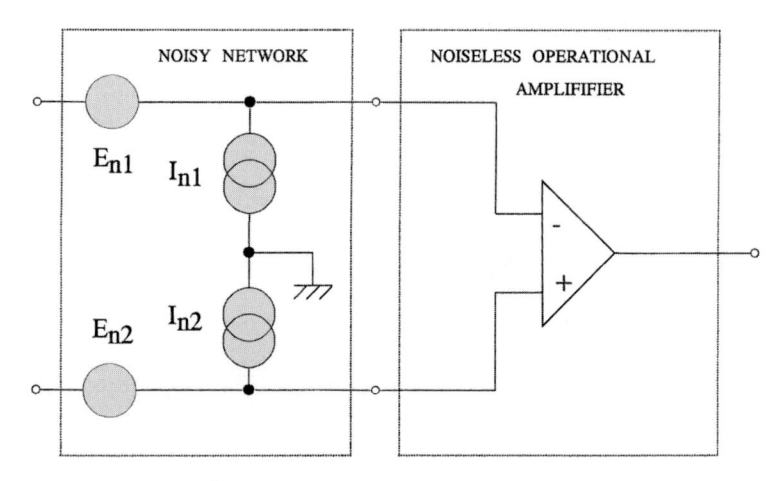

Fig. 7.39. The proposed SPICE noise macromodel

$$F = 2 + \frac{(E'_n)^2 + (I'_n R_s)^2}{4kTR_s \, \Delta f} \qquad (7.150)$$

In a differential amplifier (both inputs driven), the thermal noise of the signal source arises from both R_{s1} and R_{s2} (therefore $2R_s$) and the noise factor is

$$F = 1 + \frac{(E'_n)^2 + (I'_n R_s)^2}{8kTR_s \, \Delta f} \qquad (7.151)$$

■ **SPICE Model.** In [130], a noise macromodel for an operational amplifier is proposed. The configuration is shown in Fig. 7.39, where E_{n1}, E_{n2} are correlated noise generators. They are of type VCVS (voltage-controlled voltage sources), both controlled by the same noise voltage. Each accounts for half the equivalent input noise voltage specified in the data sheet. The noise current generators I_{n1}, I_{n2} are uncorrelated, and they are of type VCCS (voltage-controlled current sources).

As calibrated noise sources, the best choice is to adopt diodes, because their 1/f noise component is useful to simulate the excess noise of the operational amplifier. Their detailed modeling procedure is similar to that discussed in Case Study 17.2 and 17.3.

■ **Conclusion.** Often the noise of an operational amplifier is modeled by means of two lumped generators, located at one or both inputs. Their frequency-dependent values can be obtained either by measurement or from the data sheet of the manufacturer.

Part II

Interfering Signals

8
External Noise

"Though this be madness, yet there is method in't"

(W. Shakespeare)

■ **Objective.** The purpose of this chapter is to present external noise sources and to detail the interference coupling mechanisms.

■ **Basic Terms.** In this section, the term *system of interest* denotes the ensemble of electronic circuits whose noise performance is under investigation. For example, the sensor together with its preamplifier or the transmitter-receiver ensemble of a communication system may both be regarded as systems of interest.

Considering the location of the noise source with respect to the system of interest, the noise may be *internal* or *intrinsic* (source located inside the system), or *external* (noise source outside the system). Often external noise is called *interfering signals*.

External noise sources are either natural (such as solar noise, galactic noise, atmospheric noise) or man-made (which include industrial noise, electric motors, arc welders, switches, broadcast communication systems, mobile phones, etc.).

■ **Comparison.** Compared to intrinsic noise, whose main feature is its randomness, many external noise sources are rather deterministic in nature.

A typical example is crosstalk: signals which are deterministic (and useful) in one communication channel can couple into another channel, where they are unwanted, and mask the content of the signals there, so they are treated as external noise.

Another example is high-voltage power lines, which can induce 50-Hz (60-Hz) signals and their harmonics into the systems exposed to their electromagnetic fields. These signals are also deterministic in nature, the only unknowns being their amplitude and phase, which largely depend on the coupling between the noise source and the victim.

In communication systems operating at frequencies up to 50 MHz, external noise picked up by the receiving antenna is generally greater than the internally generated noise, and it therefore represents the main limiting factor. Above this limit, the situation is reversed, and internal noise becomes the limiting factor.

8.1 External Noise Sources

■ **Classification.** External noise sources can be grouped into two categories:

1. Natural noise sources,
2. Man-made noise sources.

Each category includes several items, which are detailed in Table 8.1, where "ISM equipment" stands for "Industrial, Scientific, and Medical equipment".

8.1.1 Natural Noise Sources

■ **Comment.** For convenience, atmospheric noise, precipitation static, solar noise, Galactic noise and hot-Earth noise are all grouped together under "Sky noise." In practice, all quoted noise sources combine with the celestial background in the radiation pattern of the receiving antenna.

Magnetic storms induced by solar flares have the ability to induce damaging surges in power-line voltage, destroying electrical equipment over a huge area. They also dramatically affect signal propagation and sky noise.

■ **Atmospheric Noise.** This is defined as *noise having its source in natural atmospheric phenomena, mainly lightning discharges in thunderstorms.* Their location is time-variable, depending on time of the day, season of the year, weather, altitude, and geographical latitude. As a general rule, they are more frequently encountered in the equatorial region than at temperate latitudes and above. However, the electromagnetic waves produced by thunderstorms propagate at thousands of kilometers via ionospheric skywave.

In the time domain, this noise is characterized by large spikes against a background of short random pulses.

Its frequency spectrum extends up to 20 MHz and the spectral density is proportional to $1/f$. Consequently, it mainly affects long-range navigation systems (maritime radio), terrestrial radio broadcasting stations (LW, MW, and SW) and to a considerably lesser extent, FM and TV reception.

■ **Precipitation Static.** *This kind of noise is encountered in rain, snow, hail, and dust storms in the vicinity of the receiving antenna.* Its frequency

Table 8.1. Miscellaneous external noise sources

Natural noise sources	Sky noise	atmospheric noise solar noise Galactic noise hot-Earth noise precipitation static
	Magnetic storms	
	Telluric currents	
Man-made noise sources	Industrial (electromagnetic) noise sources	spark plugs in ignition systems arc welders electric motors and switches high-voltage transmission lines 220 V / 50 Hz supply lines neon signs radio and TV broadcast services ISM equipment cellular telephones mobile radios household appliances
	Electrostatic noise sources	triboelectric effect piezoelectric effect
	Non-electric noise sources	galvanic action electrochemical effect Seebeck effect contact noise

spectrum peaks below 10 MHz. It can be substantially reduced by eliminating sharp metallic points from the antenna and its surroundings, and by providing paths to drain static charges that build up on an antenna and in its vicinity during storms.

■ **Galactic Noise.** According to [141], this is defined as *noise at radio frequencies caused by disturbances that originate outside the Earth or its atmosphere.*

Galactic noise sources can be grouped into two classes: discrete sources and distributed sources.

In the former category the chief source is the Sun, together with thousands of known discrete sources, such as supernova remnant Cassiopeia A, one of the most intense sources of cosmic radio emission as viewed from Earth. The Sun is the most powerful noise source, with its temperature of about 6000°C and its proximity to Earth. Its energy is radiated in a continuous mode, and the frequency spectrum mainly covers the range from several MHz up to several GHz. During quiescent periods, the Sun's noise temperature is about

258   8 External Noise

700,000 K at 200 MHz, and about 6000 K at 30 GHz. However, during sunspot and solar-flare activity these values are considerably higher.

Also belonging to this category are the quasars, which emit copious quantities of energy (usually as powerful radio waves), often in a continuous mode, although the signals arriving at Earth are perceptible only by sensitive radiotelescopes.

The *distributed noise sources* are the ionized interstellar gas clouds in our Galaxy and a considerable number of extragalactic sources known as radio galaxies. Depending on emission mechanisms, the distributed noise sources are thermal or nonthermal [142]. So-called thermal noise sources are associated with random encounters of electrons and ions in gas clouds, mostly ionized hydrogen. Nonthermal noise sources (also called synchrotron radiation) involve electrons moving in magnetic fields. This is a general galactic phenomenon, encountered even in interstellar space.

As a general rule, cosmic radio noise covers the frequency range from 15 MHz up to 100 GHz, with a predominance between 40 MHz and 250 MHz. It is observed that it reaches a maximum when the receiving antenna points toward the center of the Milky Way.

■ **Sky Noise.** This term lumps the effects of all previous noise sources. Experimentally, it has been found that for a receiving antenna pointing toward the zenith (elevation angle of 90°), the typical variation of the sky noise temperature versus frequency looks like the plot of Fig. 8.1. Two peaks appear: the first is situated near 22 GHz (where the atmospheric absorption of radio waves reaches a maximum, as a consequence of water vapor), and the second is located at about 60 GHz, due to oxygen absorption. (It is well known that both oxygen and water vapor can absorb energy from a radio wave due to the permanent electric dipole moment of their respective molecules.)

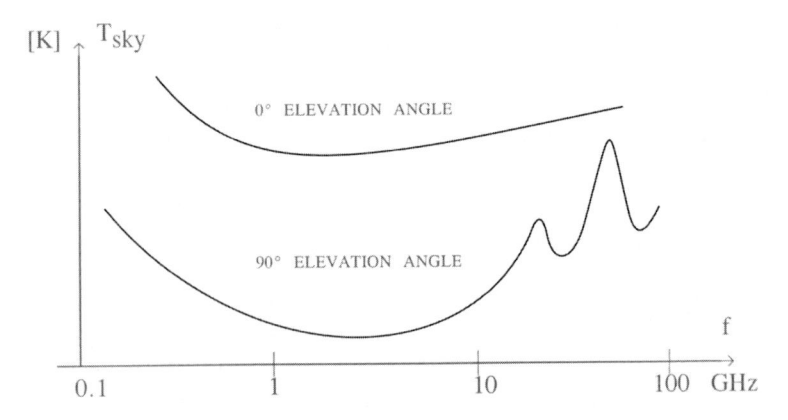

Fig. 8.1. Sky noise temperature versus frequency

8.1 External Noise Sources 259

Table 8.2. Approximate limits of sky noise temperature

	Maximum	Average	Minimum
Sky noise temperature	$1450\lambda^2$	$100\lambda^{2.4}$	$58\lambda^2$

Globally, in the frequency range $1 - 10\,\text{GHz}$ the received noise power is at a minimum (this is also called the "space communication window").

Of course, the sky temperature depends strongly on antenna orientation: if the elevation angle is low (approaching zero), the sky temperature increases because the antenna looks through much more atmosphere.

According to various publications, Table 8.2 groups the approximate limits of sky noise temperature for the VHF range (30 MHz to 300 MHz), where λ denotes the corresponding wavelength, expressed in meters.

■ **Remark.** In microwave communication systems, the actual noise power received by the antenna is more important than the value predicted by Table 8.2. To understand why, consider a steerable dish antenna pointing toward a satellite.

Ideally (Fig. 8.2a), the beamwidth of the antenna should exactly cover the angular diameter of the satellite, to avoid noise picked up from regions outside the target.

However, in practice, the beamwidth of any single practical ground based antenna is far larger then the minuscule angular extent of the target satellite (Fig. 8.2b). In this case the antenna "sees" more sky noise than in the ideal situation and this explains why the values predicted by Table 8.2 are lower limits.

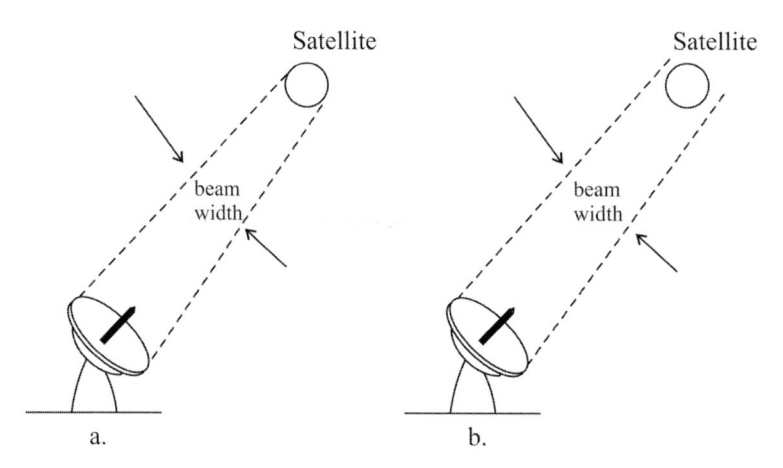

Fig. 8.2. Beamwidth of an antenna relative to the angular diameter of a satellite: **a** ideal case; **b** real case

8.1.2 Man-Made Noise Sources

■ **Classification.** Depending on origin, these noise sources are grouped in three categories:

1. Electromagnetic sources,
2. Electrostatic sources,
3. Non-electric sources.

■ **Electromagnetic Noise Sources**

□ **Automotive Ignition Systems.** There are two major sources: spark plugs and the current flowing through the ignition system. Both are responsible for radiated electromagnetic energy, which comes in bursts of short duration pulses (nanoseconds), the burst width ranging from microseconds to milliseconds. The frequency of the bursts depends on the number of cylinders in the motor and the angular motor speed (RPM). It can be reduced by using spark plugs with a built-in interference suppressor and shielding the entire ignition system (when possible).

□ **Arc Welders.** Typically, arc welders use an RF arc whose fundamental frequency is around 2.8 MHz [143]. Their spectrum covers the 3 kHz to 250 MHz frequency range, but the fundamental remains at a significant level even at a distance of several hundred meters. In radio receivers, this noise is perceived as a "frying" noise. The emission is considerably reduced by improving welder grounding, using short welding leads, shielding wiring, and avoiding proximity to power lines.

□ **Electric Motors.** All high-power motors involved in electric transport systems (underground, trains, conveyor belts, elevators, etc.) generate noise when switched, but also do so in steady-state operation. Switching produces transients which can reach several hundred volts as a result of current interruption in an inductive load. In the steady state, motor brushes are responsible for arc production, which increases with aging. Besides radiation, these sparks generate spurious signals that are conducted and distributed to nearby systems by the power supply lines. The same problem (although less aggressive) appears in all household appliances that use electric motors (washing machines, vacuum cleaners, ventilators, etc.). This noise can be reduced by inspecting the motor brushes and changing them when necessary, as well as by adding a capacitor of about 1 µF in parallel, to suppress sparks. Note that massive introduction of microprocessors to control the operation of equipment has also a favorable effect on noise generation.

□ **High-Voltage Transmission Lines.** Noise from transmission lines peaks at 50 (60) Hz, and it can cause interference at distances of several hundred meters. It is especially perceptible in AM receivers. Transients associated with switching of loads occur in bursts, and have rise times of a few nanoseconds; amplitude spikes larger than 2 KV are seldom observed. Another major source of noise is the Corona effect, which consists in a large number of discharges around the conductors of a power line. This occurs when the electric field around the conductor exceeds the value required to ionize the ambient gas (air), but is insufficient to cause a spark. Discharges are initiated by the presence of small irregularities in the conductor surface (like dust, pollen, snow, ice crystals, etc.) and the resulting noise mainly affects AM communication systems.

☐ **AC Supply Lines.** The 220 (or 110) V supply lines connect all the rooms of a building in a power distribution network, as well as all nearby buildings. Besides its proper 50 (or 60) Hz fields and the transients caused by switching various loads, the mains wiring constitutes an excellent antenna, which picks up noise radiated in one room from perturbing equipment and delivers it to all other rooms sharing the same line. Hence, it propagates perturbations from one site to another. In order to protect sensitive equipment, filters and surge suppressors must be provided so that the bulk of energy is absorbed before it claims victims. Various types of surge suppressors exist, such as gas-discharge devices (which can handle high power, but are slow) and semiconductor devices (using Zener diodes).

☐ **Fluorescent Lamps and Neon Signs.** Noise is generated in two distinct areas:

1) In the ionized gas column, which presents a small but fluctuating resistance when the light is on.
2) In the associated circuitry, which includes a starter. Usually, the starter is made of a bimetallic strip, which bends when the temperature changes and abruptly breaks the current flowing through an inductor. A voltage spike occurs, which is used to trigger the discharge; however, this spike is also a source of interference for nearby systems.

☐ **ISM Equipment.** This category includes industrial equipment (such as relay-controlled devices, electrical switching gear, laser cutters, microwave ovens, etc.), scientific equipment (for instance, all sorts of computer facilities), and medical equipment for intensive care units, physical therapy facilities, electrosurgical units, diathermy, CAT scanners, etc. The frequency spectrum of these noise sources can extend up to several megahertz or even gigahertz.

☐ **Radio, Television, and Radar Transmitters.** These are intentional emitters of electromagnetic waves that can interfere with systems not intended for any form of reception. All such transmitters have considerable power, since they must cover a large area. Less powerful (but no less harmful) are electromagnetic waves emitted by CB transmitters, cellular phones, mobile radios, portable computers, and so forth.

☐ **Power Supplies.** The major noise sources belonging to this category are DC/DC converters and switching-mode power supplies. Both employ switching transistors operating at frequencies up to 100 kHz (but with actual V-MOS devices, the tendency is to increase the switching frequency, for higher efficiency). The frequency spectrum is dominated by the fundamental and its harmonics, but it extends well above the switching frequency fundamental.

■ **Electrostatic Noise Sources**

☐ **Triboelectric Effect.** This entails *generating electrostatic charges of opposite sign when two materials are rubbed together and separated*, leaving one positively charged and the other negatively charged [144]. The term *triboelectricity* refers to electricity produced by friction of two dissimilar solids (as by sliding). In practice, this phenomenon affects mainly the dielectric material within a coaxial cable. When the cable bends, the metallic conductors slide along the dielectric used to separate them (if the dielectric does not maintain permanent contact with the metallic parts). A charge accumulation appears in the equivalent capacitor formed between the metallic shield and the inner conductor, separated by the insulator. This charge fluctuates according to the rhythm of mechanical flexing of the cable and hence acts as a noise source. This phenomenon is especially pertinent when the coaxial cable is used to connect a generator with high internal impedance to a high-value load (like an electrometer). In this situation, the discharge of the equivalent capacitor through the terminal impedances is slow, and additional flexing of the cable causes additional charge to accumulate. For instance, a coaxial cable terminated by 10-MΩ resistances generates noise voltages by intermittent flexing which fluctuate during 50% of the monitoring time around a few millivolts; however, if the terminal resistances are lowered to 1 MΩ, the noise voltage level decreases to several hundred microvolts. If the cable is terminated by low impedances, triboelectricity is no longer a factor.

This kind of noise is critical in cables employed in vehicles, satellite or airborne instruments, rockets, and military applications, where vibration is unavoidable. The best solution is to reduce vibration whenever possible; otherwise, special low-noise cable can be used, where friction is reduced by an additional layer of graphite (Fig. 8.3).

☐ **Piezoelectric Effect.** This is defined as *the generation of a potential difference in a crystal when a strain is introduced*. In piezoelectric materials the converse effect is also observed, namely that a strain results from the application of an electric field [145]. In practice, some circuit board materials exhibit this effect. Consequently, they are vibration-sensitive, and noise voltages can

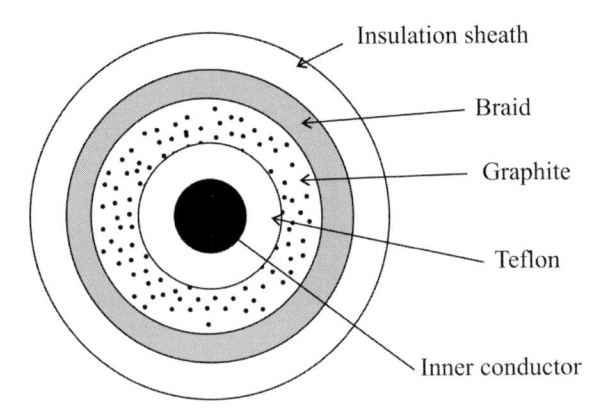

Fig. 8.3. Low-noise coaxial cable (cross-section)

appear between conductors connected to opposite sides, or between tracks situated on opposite sides. To avoid this kind of noise, the only solution is to carefully select circuit boards employing insulators that do not exhibit the piezoelectric effect.

■ **Noise of Non-electrical Origin.** Under this label are grouped non-electric phenomena that generate voltage or current fluctuations. Among them, the most important ones are:

☐ **Galvanic Action.** Whenever two dissimilar metals are used in the same connecting path, *an e.m.f. at the contact area appears due to the electrochemical action of the two metals.* This e.m.f. will normally keep a constant value determined by the position of the metals in the galvanic series. However, in the presence of water vapor or moisture (that penetrates the contact region), the metal surface (especially at the anode) is degraded by corrosion. The rate of corrosion depends on the relative positions of the metals in the galvanic series (the closer they are, the slower the corrosion), the moisture level and temperature. As a consequence, the e.m.f. fluctuates according to external conditions, and this noise can corrupt the useful signal.

☐ **Electrochemical Contamination.** This phenomenon is chiefly encountered on printed circuit boards, where *fluctuating voltages can appear across two metallic tracks separated by a contaminated insulating material* (which as an electrolyte). Often the contamination is the result of poorly cleaning the flux from a circuit board after soldering, although small particles of oxides and abrasives can also contaminate the surface during fabrication. The remedy is to carefully clean the circuit board after soldering to remove all residues. Sealing and coating the assembly with a protective film represents the final step in circuit board fabrication.

□ **Seebeck Effect.** According to [146], *this is the production of an e.m.f. in a circuit comprising two dissimilar metals when the two sides of the junction are at different temperatures.* Assume a junction of two metals, denoted by 1 and 2, such that metal 2 is at higher temperature than metal 1. The junction voltage difference appearing is proportional to (T_2-T_1). Fluctuation in (T_2-T_1) can result from turbulent convection of air or from local variations in temperature. As a result, fluctuation in the voltage difference is also observed, which becomes a noise source.

Two types of metal junctions are commonly encountered in electronic circuits:

- gold plating on connectors and copper traces on a circuit board;
- a copper trace with tin solder applied. The latter might be locally overheated, for instance by a lead from a power device connected (soldered) to the point in question.

Since it is impossible to avoid junctions between dissimilar metals, it is recommended that they be selected so that the Seebeck effect is as small as possible. For instance, a junction of copper and a tin-lead solder alloy has a Seebeck voltage in the range $1-3\,\mu V/^\circ C$. However, if a tin-cadmium alloy is used instead, this value drops to only $0.3\,\mu V/^\circ C$. Finally, good ventilation of the equipment and better heat evacuation has a favorable effect on this noise source.

□ **Poor Mechanical Contacts.** Typically, such contacts are parts of a relay, switch, or connector that is engaged or disengaged to open or close some electrical circuit [146]. Heavy-duty operation of such devices leads to oxidation and mechanical deterioration of the contact surfaces. The contact pressure becomes lower and moisture and dust accumulate, forming thin films on the surface. Finally the contact resistance increases, and even worse, fluctuates. Consequently, the current flowing through the contact in question also fluctuates. The only remedy is to periodically clean the contact surface, or to completely replace the contacts whenever necessary.

□ **Poor Solder Joints.** Most insidious are solder joints, which despite their normal appearance, can either contain flux/paste residues or be affected by defects (formation of voids which contain flux, micro-cracks, etc.). As a consequence, the distributed electrical resistance inside the joint is nonuniform, the current density varies locally from one micro-region to another, and the area which supports the maximum current density risks overheating. Hence, the current becomes even more non-uniformly distributed, etc. Finally, fluctuating resistance is associated with the poor solder joint, at a level that depends on the current, temperature, aging, vibrations, etc.

Detecting eventual poor solder joints on a circuit board is a difficult task; once they are found, the only remedy is to resolder them manually.

8.2 Glossary of Terms

■ **Absorber.** According to [147], *a material that causes the irreversible conversion of the energy of an electromagnetic wave into another form of energy (normally heat) as a result of interaction with the absorbed material.*

■ **Absorption.** *The loss of electromagnetic energy due to conversion into heat or other forms of energy, as a result of interaction with matter.*

■ **Attenuation.** *Decrease in signal energy, usually expressed in decibels.*

■ **Balun.** *A device for transforming an unbalanced voltage to a balanced voltage or vice versa* [147]. The term "balun" is an acronym for "BALanced-to-UNbalanced".

■ **Bond.** *A reliable connection to assure the required electrical conductivity between conductive parts, to maintain a common electrical potential* [145].

■ **Bonding.** *A process of connecting metal structures together in order to achieve a low resistance contact* [143].

■ **Bus.** *A conductor, or group of conductors, that serve as a common connection for two or more circuits* [145].

■ **Crosstalk.** *Undesired energy (disturbance) appearing in a transmission path by mutual coupling with other transmission paths.*

■ **Electromagnetic Compatibility (EMC).** *The capability of electronic systems, equipment or devices to operate in their intended electromagnetic environment without suffering or causing unacceptable degradation of performance as a result of electromagnetic interference.* Note that EMC has two aspects: the *emission* of perturbations and the *susceptibility* to perturbations.

■ **Electromagnetic Environment.** *The totality of electromagnetic phenomena existing at a given location.* This term is pertinent to the electromagnetic energy which is *unintentionally* conducted/radiated away.

■ **Electromagnetic Gasket.** *A conductive insert usually between flanges, intended to provide electrical continuity across a joint and to prevent intrusion of electromagnetic waves.*

■ **Electromagnetic Immunity.** *A relative measure of a device or system's ability to withstand EMI exposure while maintaining a predefined performance level.*

- **Electromagnetic Interference (EMI).** *EMI is the degradation of the performance of a device, circuit, or system caused by an electromagnetic perturbation.*

- **Electromagnetic Perturbation.** *Any electromagnetic phenomenon able to alter the performance of an electronic device, circuit, or system.*

- **Electromagnetic Susceptibility.** *A relative measure of a device or system's inability to perform without degradation in the presence of an electromagnetic perturbation.* Lack of electromagnetic immunity.

- **Emission.** *The propagation of electromagnetic energy liberated by a source, by radiation, or conduction.*

- **Far-Field Region.** *The region of the field of a transmitting antenna where the angular field distribution is essentially independent of the distance from the antenna* [147].

- **Ground (Earth).** *A conductive system whose potential is taken as reference for all voltages in the circuit.* This may or may not be an actual connection to earth, but it could be connected to earth without disturbing the operation of the circuit in any way. Note that an electronic system may have several grounds, and some of them can even float (for instance, a TV receiver has 3 different grounds, so-called "signal grounds" or "local grounds"). The signal grounds may or may not be at earth potential. In contrast, the ground connected to earth is called "safety ground."

- **Ground Plane.** According to [145, 147], *a conducting flat surface or plate used (1) as a common reference point for circuit returns, as well as for electric or signal potentials; and (2) to reflect emitted electromagnetic waves.* Note that a ground plane can bend and follow a curved surface.

- **Interfering Signal.** *A signal that impairs the reception of a useful signal.*

- **Isolation.** *Separation of one section of a system from undesired influences of other sections.* This may be achieved either by galvanic separation, or by inserting a high impedance between the sections in question.

- **Near-Field Region.** *The region of the field of a transmitting antenna wherein the angular field distribution is dependent on the distance from the antenna* [147].

- **Plane Wave.** *A wave whose equiphase surfaces form a family of parallel planes.*

■ **Radiation.** *The propagation of energy through space or through a material in the form of electromagnetic waves.* Usually classified according to frequency, e.g., Hertzian, microwaves, infrared, visible, ultraviolet, X rays, gamma rays, etc.

■ **Reflection.** *The phenomenon in which a wave that strikes a medium of different characteristics is returned to the original medium with the angles of incidence and reflection equal and lying in the same plane* (for instance, at the air-metal interface of a shield, or more generally, at any abrupt discontinuity of refractive index).

■ **Relative Conductivity (σ_r).** *A material constant which indicates (relative to copper) the electrical conductivity of the material.*

■ **Relative Permeability (μ_r).** *A material constant which indicates (relative to copper) the ability of the material to concentrate magnetic field lines.*

■ **Shield.** *A conducting barrier separating sources from receptors to reduce the effects of source electromagnetic fields upon receptors* [147].

■ **Shielding Effectiveness.** *An insertion loss measure of the ability of a shield to exclude or confine electromagnetic waves, usually expressed as a ratio (in the frequency domain) of the incident to penetrating signal amplitudes, in decibels* [147].

■ **Shielding Enclosure.** As stated in [147], *a metallic mesh, sheet housing, or continuous conductive layer designed expressly for the purpose of separating the internal and external electromagnetic environments.*

8.3 Interference Problem

■ **Physical Approach.** Electromagnetic perturbations are present everywhere, and they are a constitutive part of our industrial civilization. Interference occurs when the situation depicted in Fig. 8.4 is encountered.

Fig. 8.4. Overview of the electromagnetic interference problem

Three conditions are simultaneously required: the presence of a source of perturbations (perturbing or aggressor system), the presence of a perturbed system (victim), and at least one coupling path to transmit the noise from source to victim.

It is important to know how the aggressor is coupled to the victim, since in many practical situations, reducing the coupling coefficient is the only way to fight interference.

■ **Classification.** Two main situations are encountered:

1) Both aggressor and victim belong to the same system or facility. In this case, the interference is *inside* the system of interest.
2) The source of perturbation belongs to one system and the victim to another (for example, the aggressor is a mainframe computer and the victim is a radio receiver). The interference then appears between two distinct systems.

Note that in some situations the same equipment is both aggressor and its own victim.

■ **Solution.** To reduce interference, there are three possible approaches:

- reduce the emission of perturbations at the source,
- reduce the electromagnetic susceptibility of the victim,
- reduce the coupling between aggressor and victim.

Each will be detailed in a distinct chapter.

8.4 Coupling Paths

■ **Background.** Accurate modeling of coupling paths requires solving Maxwell's equations, which are differential equations containing the derivatives of the electric and magnetic fields with respect to time and with respect to three orthogonal spatial variables. The main difficulty arises from complicated boundary conditions affecting these equations, since at the interface between two media, the E and H fields can be discontinuous. To overcome this difficulty, the space dependency of E and H is often neglected, analysis being carried out only in the time domain (which is accurate for systems of small electrical length). In this way, approximate solutions (as functions of time) are delivered. From a conceptual point of view, this approach is equivalent to adopting the following assumptions [148]:

- all electric fields are confined inside capacitors,
- all magnetic fields are confined inside inductors,
- dimensions of the circuit are small compared to the signal wavelength(s).

Hence, the coupling paths (which are actually distributed elements) are represented by means of their lumped equivalents. Besides a considerable simplification of the problem, the expected benefit is clarify how interference depends on the system parameters, layout, or package. This information is not available from the solution of Maxwell's equations, even if accurately solving them were possible.

Of course, the drawback of this approach is the limited accuracy of the solution (but it should be remembered that the ultimate goal of analysis is to check that the perturbation does not exceed an imposed threshold, rather than finding an exact solution!).

8.4.1 Methods of Noise Coupling

■ **Coupling Paths.** Coupling paths can be grouped under two categories: conduction paths and radiation paths. The electrical power lines represent the traditional example of conductive coupling, especially when both the aggressor and victim share the same power line. An illustration of both categories is shown in Fig. 8.5, where the perturbations generated by the motor are transported by the distribution network to the victim (radio receiver). The radiation path transports the perturbation produced by the ignition system of a nearby car engine to the antenna of the receiver through free-space.

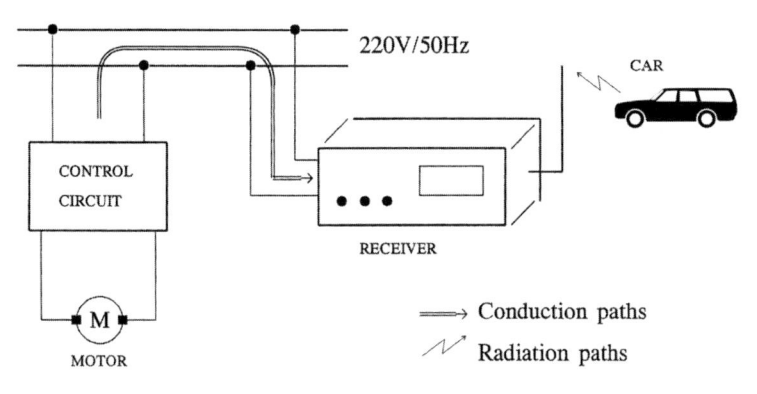

Fig. 8.5. Coupling paths in industrial environment

■ **Conducted Noise.** This kind of noise (also called conducted interference) is due to interfering signals that can propagate from source to victim via a conductive path. Depending upon the kind of conductive path involved, the commonest practical situations include:

☐ **AC Power Lines.** Consider the typical situation in which two different pieces of equipment connected to adjacent outlets share the same AC line

270   8 External Noise

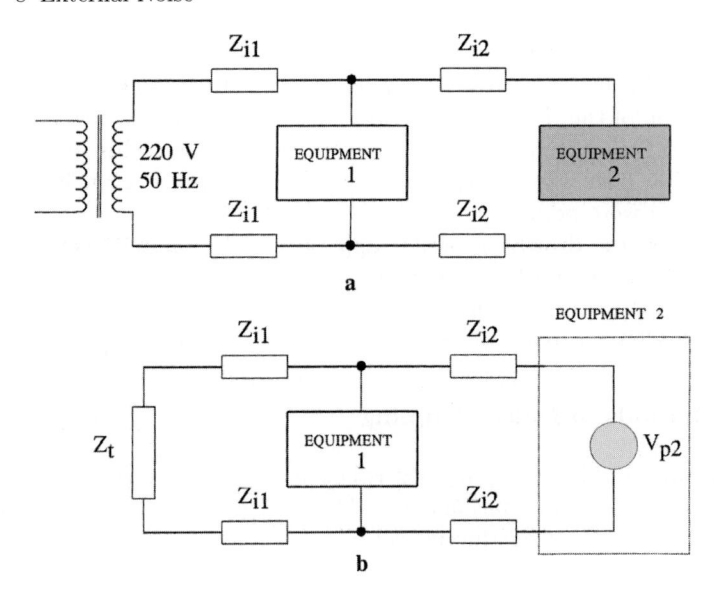

Fig. 8.6. a Conductive coupling through AC power supply; b equivalent circuit

(Fig. 8.6a). Let EQUIPMENT 2 be the source of perturbations and EQUIPMENT 1 the victim.

The equivalent circuit is shown in Fig. 8.6b, where Z_{i1}, Z_{i2} represent the impedances of the mains lines, which can on the average be approximated by 50 Ω resistances in parallel with 50 µH inductances [149]. Z_t is the impedance seen in the secondary of the transformer and V_{p2} represents the perturbations produced by the source. Applying the voltage divider formula, the amount of perturbing voltage reaching EQUIPMENT 1 is

$$V_{p1} = V_{p2} \frac{2Z_{i1} + Z_t}{2(Z_{i1} + Z_{i2}) + Z_t} \quad (8.1)$$

Note that V_{p1} increases when Z_{i2} is reduced (i.e., when the separating distance between the two units decreases). When $Z_{i2} \ll Z_{i1}$, $V_{p1} \cong V_{p2}$ and there is no attenuation of perturbations reaching the victim. The only remedy is to insert filters at the AC terminals of EQUIPMENT 1.

☐ **Common Ground Impedance.** Another type of conductive coupling is by means of a common ground impedance. Consider the circuit detailed in Fig. 8.7 involving two amplifiers, where M1, M2 are respectively a signal ground and the ground plane of the equipment. The connection between them is achieved by a short wire (or strip), which, at low frequency, acts like a short circuit. However, at higher frequencies the parasitic self inductance (L) of this wire and its associated resistance (R) give rise to an impedance which is no longer negligible.

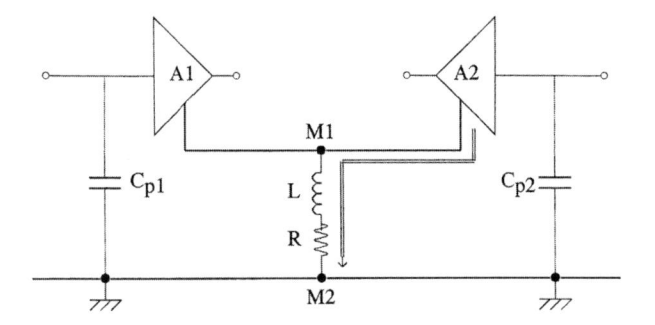

Fig. 8.7. Undesired coupling due to ground return impedance

Suppose that C_{p1} and C_{p2} are the stray capacitances between each input and ground; due to the current injected by the amplifier A2 into ground, a voltage drop appears on the series combination of L and R:

$$v = Ri + L\frac{di}{dt}$$

Despite the low value of L, the second term of the sum becomes dominant, especially when fast switching currents are flowing. This voltage transient is transmitted through C_{p1} and C_{p2} to the inputs and is then amplified, just like any useful signal.

■ **Radiated Noise.** Radiated noise is usually in the form of electromagnetic fields that escape from the source and reach the victim via propagation. This noise is transmitted by:

- *Capacitive coupling* (through the stray capacitances between wires or tracks on the circuit board), when electric field induction is dominant.
- *Inductive coupling* (mainly between cables or circuit board tracks), when magnetic field induction is dominant.
- *Electromagnetic coupling*, through the wires, cables, and circuit board traces of the victim system, which all act like small receiving antennas. Note that at the usual power levels, all involved fields drop off rapidly with distance, and are significant only close to the source.

☐ **Capacitive Coupling.** This mechanism is illustrated in Fig. 8.8, where the load Z_L is connected to the amplifier output by a wire (or a trace on the circuit board), which passes close to the wire (or trace) connected to the input of amplifier A2. The coupling capacitance between wires is distributed (Fig 8.8a); however, for simplicity it is represented in Fig. 8.8b as a lumped capacitor, labeled C_p.

The flow of current I_L through Z_L charges the left "plate" of C_p and as a consequence, the same amount of electric charge (but of opposite sign) will

272   8 External Noise

be induced on the right "plate". This is equivalent to a parasitic voltage V_p appearing at the input of amplifier A2, such that

$$V_p = C_p \frac{dV_L}{dt} (Z_{in} \parallel R_g) \qquad (8.2)$$

where Z_{in} represents the input impedance of A2 and R_g is the resistance of the signal generator V_g. Of course, reducing C_p (by increasing the separation between wires or traces) can help to reduce V_p, but the unexpected conclusion from (8.2) is that *a low input impedance of the victim circuit increases immunity to perturbations transmitted via capacitive coupling.*

If either of the grounds M1 or M2 is floating, the only difference is that the stray capacitance between the point in question and ground will be seen in series with C_p. Finally, the best solutions to reduce this type of interference are:

- decrease the coupling capacitance C_p by increasing the separation of traces or by shielding (for instance, introduce a ground trace between the traces in question);
- reduce the input impedance of the victim circuit (when possible).

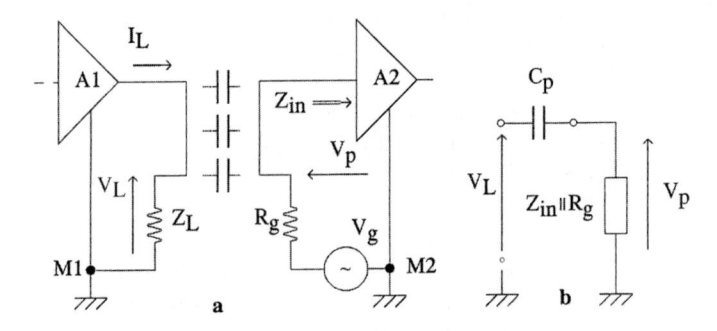

Fig. 8.8. a Capacitive coupling between wires; b Equivalent circuit

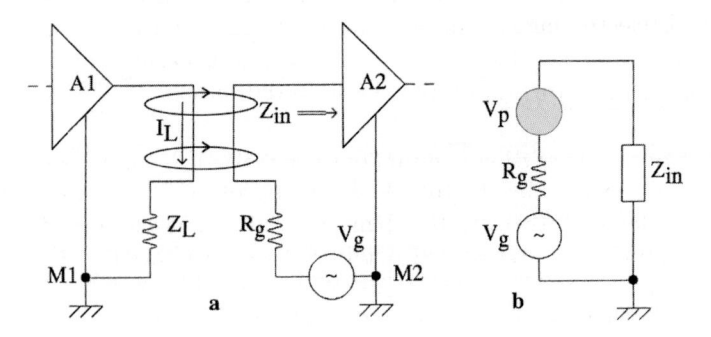

Fig. 8.9. a Magnetic coupling; b equivalent circuit

☐ **Magnetic Coupling.** This kind of coupling is illustrated in Fig. 8.9, where the current flowing through the output loop of amplifier A1 produces a magnetic field whose lines intersect the input loop of amplifier A2 [149].

Consequently, an induced parasitic voltage V_p appears in the input loop, which can be evaluated as

$$V_p = -M \frac{dI_L}{dt} \qquad (8.3)$$

M being the mutual inductance between the loops (which depends on both loop areas, their orientation, and their separation). In the equivalent circuit, the perturbing voltage V_p is in series with the signal generator V_g, hence the noise is added to the useful signal. Note that V_p is unaffected by whether or not M1 and M2 are floating or connected to ground.

To decrease the noise induced by magnetic coupling, three solutions may be considered:

- limit the area of the victim loop by properly designing the layout;
- reduce the magnetic field by shielding or by decreasing the output current in the aggressor circuit;
- when possible, modify the loop orientations so that their planes become perpendicular.

☐ **Electromagnetic Coupling.** Traditionally, this term refers to coupling between an electromagnetic plane wave and a transmission line (recall that a plane wave has both fields perpendicular to the direction of propagation and perpendicular to each other). In the present case, the electromagnetic coupling appears between the intentionally emitted waves (like TV and broadcast services, radars, etc.) and each wire of the circuit, which acts like a receiving antenna.

■ **Concluding Remark.** In practice, whenever the emitting source is close to the victim circuit (near-field condition), the E and H fields can be dissociated and treated separately. However, if the aggressor and the victim are far from each other (far-field condition), the two fields must be simultaneously considered (electromagnetic coupling).

8.4.2 Coupling Modes

■ **Comment.** Whenever a non-random perturbation contaminates the useful signal, it seems natural to ask what the phase difference is between them, in order to appreciate the damage. Thus, differential-mode coupling and common-mode coupling will not have the same effect on the resulting interference.

Since both modes have their origin in the theory of differential amplifiers, it is important to recall some basic concepts.

274   8 External Noise

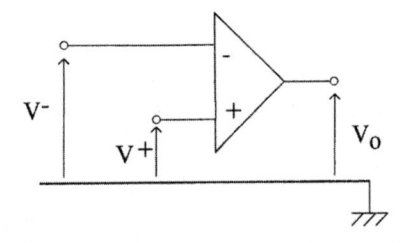

Fig. 8.10. Differential amplifier

■ **Ideal Differential Amplifier.** *An ideal differential amplifier is defined as a symmetrical circuit with two inputs and one output* (Fig. 8.10) *that amplifies the difference between two input signals.*

Let us denote by V^+ and V^- the input signals with respect to ground; according to the given definition, the output voltage is

$$V_o = A_d(V^+ - V^-) \qquad (8.4)$$

where A_d is called the *differential gain*. Typical values for A_d lie between 10^4 and 10^6.

Note that if the same voltage is simultaneously applied to V^+ and V^-, the output of the ideal differential amplifier is unaffected. Hence, the output voltage V_o is the same, regardless of the value of input signals, as long as their difference is the same. For instance, $V^+ = 10\,\mu V$, $V^- = -10\,\mu V$ would yield the same output as $V^+ = 1010\,\mu V$, $V^- = 990\,\mu V$.

■ **Actual Amplifier.** In practical amplifiers, the output depends not only on the difference between the input voltages, *but also on the average value of the input signals*. Briefly, this is due to the impossibility of achieving a perfect symmetry.

Of course, the average value of the input signals (called *common-mode voltage*) does not exert a significant influence, but nevertheless it can modify the expression (8.4):

$$V_o = A_d(V^+ - V^-) + A_c\left(\frac{V^+ - V^-}{2}\right) \qquad (8.5)$$

where A_c is called the *common-mode gain*. Typical values of A_c are between some fraction of unity and some small multiple; the greater the value of A_c, the more important the asymmetry of the amplifier.

As a figure of merit indicating amplifier symmetry, the *common-mode rejection ratio* (CMRR) is defined to be

$$\text{CMRR} = 20\log\frac{A_d}{A_c} \quad [\text{dB}] \qquad (8.6)$$

8.4 Coupling Paths 275

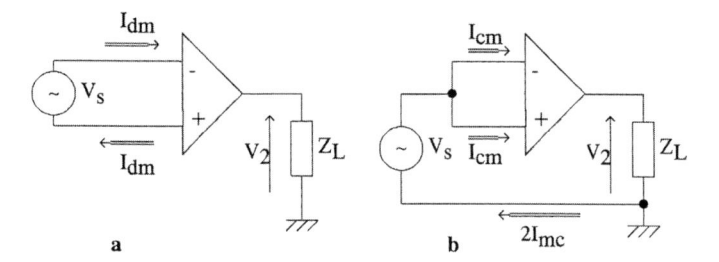

Fig. 8.11. a Differential-mode operation; b Common-mode operation

■ **Modes.** An operational amplifier operates in the differential mode (Fig. 8.11a) when the signal is applied between its inputs ($V_s = V^+ - V^-$). It is important to note that differential-mode input currents (denoted by I_{dm}) have *opposite directions*.

Common-mode operation occurs (Fig. 8.11b) when the voltage signal is *common* to both inputs. In this case the input currents have the *same direction*. Note that the return current (carried out by the ground) is twice as great and flows in the opposite direction.

■ **Coupling Modes.** In this section, the concept of coupling mode refers instead to the coupling between a radiated field and a victim circuit (although the same concepts may apply to conducted perturbations also). Each mode will be next detailed.

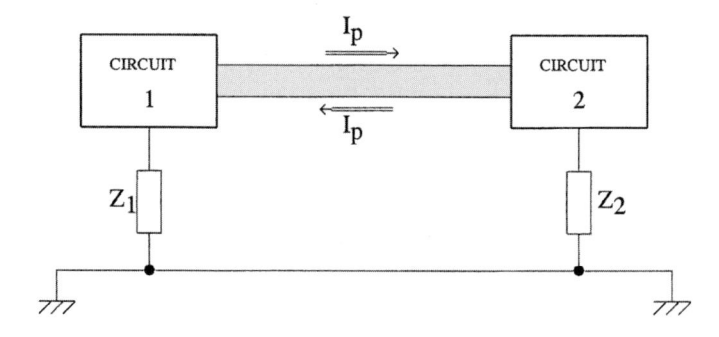

☐ **Differential Mode Coupling.** Similarly to the case illustrated in Fig. 8.11a, *an electromagnetic field is said to be coupled in differential mode if the induced currents flow in opposite directions*. Consider the situation depicted in Fig. 8.12, where the perturbing currents (I_p) are induced in the wires connecting two circuits belonging to the system of interest. The region of vulnerability is apparent.

Z_1 and Z_2 are the impedances to ground of circuits 1 and 2, respectively. They are either real impedances, or the equivalent impedances of the stray

Fig. 8.12. Differential-mode coupling

276   8 External Noise

capacitances to ground (when both circuits are floating), or the equivalent impedances of the bonding wires to ground. In any case, they have no influence on I_p.

☐ **Common-Mode Coupling.** Similarly to the general case shown in Fig. 8.11b, *a perturbing field is coupled in common mode if the resulting currents (apart from the return) all flow in the same direction.* This time the vulnerability area includes the ground plane; consequently, this time impedances Z_1, Z_2 affect both the amplitude and the spectrum of the induced currents (I_p).

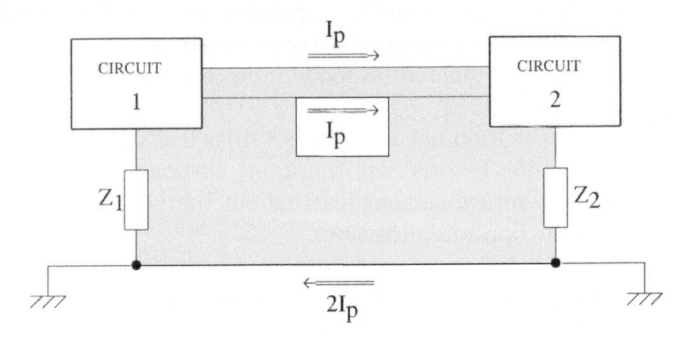

Fig. 8.13. Common-mode coupling

☐ **Antenna-Mode Coupling.** In this case, circuits 1 and 2, as well as the connecting wires (including the ground plane), all act as receiving antennas with respect to the perturbing field (Fig. 8.14). *The currents carried by the interconnecting wires and the ground plane all flow in the same direction.*

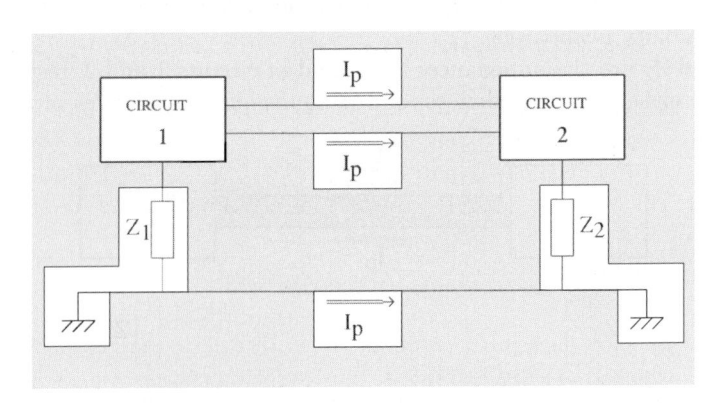

Fig. 8.14. Antenna-mode coupling

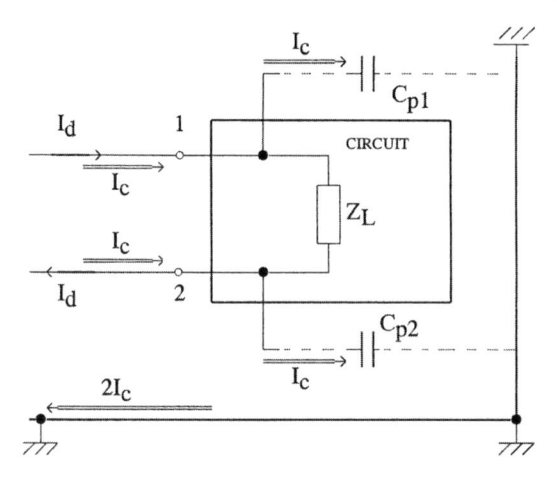

Fig. 8.15. Conversion between common- and differential-mode

■ **Conversion.** Interference may appear between the common-mode induced signals and the differential-mode signals (Fig. 8.15).

The useful signal is the voltage drop produced by the differential mode current I_d across the load impedance Z_L. However, due to the inherent stray capacitances to ground (C_{p1}, C_{p2}) which support the common mode currents (I_c), an additional differential-mode voltage may appear:

$$V_{ad} = I_c(Z_1 - Z_2) \tag{8.7}$$

where Z_1, Z_2 are the equivalent impedances of C_{p1}, C_{p2} respectively. Note that V_{ad} cancels if and only if the configuration is *perfectly balanced* (i.e., $Z_1 = Z_2$). The perfect balancing condition is not easily achieved in practice. Several solutions are proposed to ensure a well-balanced configuration:

- design a symmetrical layout (with identical trace configurations in the areas of C_{p1} and C_{p2});
- avoid the vicinity of any metallic body (shield, transformer etc.) that might introduce an asymmetry;
- whenever possible, add in parallel with C_{p1}, C_{p2} identical discrete capacitors (of higher value than the expected stray capacitances) to swamp the imbalance;
- use a common-mode choke.

■ **Remark.** The analysis of coupling between source and victim does not depend on the radiated or conducted nature of interfering signals.

Summary

- Noise sources can be grouped into three categories:
  1) intrinsic sources;
  2) man-made noise sources;
  3) environment noise sources.
- Sky noise includes cosmic noise, solar noise, and atmospheric noise.
- Electromagnetic perturbations are transmitted from source to victim either by radiation or conduction.
- There are three basic coupling modes:
  1) differential-mode coupling;
  2) common-mode coupling;
  3) antenna-mode coupling.
- Radiated noise can reach the victim in three ways:
  1) by capacitive coupling;
  2) by inductive coupling;
  3) by electromagnetic coupling.

9

Interference Reduction Methods

> "Physicians of the utmost fame,
> Were called at once, but when they came,
> They answered, as they took their fees,
> 'There is no cure for this disease'"
>
> (Hillaire Belloc)

The objective of this chapter is to present the most frequently encountered techniques employed to reduce interference at the circuit level. These techniques are versatile, since they are effective to reduce either the emission from offending circuits, or to improve immunity of victims, or to protect the interconnection between them.

9.1 Electromagnetic Shielding

Shielding is one of the chief methods to limit electromagnetic field penetration into the environment or into the system of interest. It works well at the device or circuit level, but can also be employed for protecting an entire system (or even rooms and buildings). Here attention will be paid to the protection of devices and circuits, since the latter aspect is traditionally related to the electromagnetic compatibility field.

This section presents the basic principles of shielding and solutions for handling the discontinuities in the shielding enclosures.

9.1.1 Background

■ **Near- and Far-Field.** The features of any electromagnetic field depend on the following factors:

- the properties of the source producing the field;
- the quality of the transmission medium;
- the distance between the source and the observation site.

The *near-field* is developed in the space region situated at less than $\lambda/2\pi$ from the source. The *far-field* is related to the space region extending from $\lambda/2\pi$ to infinity. In the vicinity of $\lambda/2\pi$, we have the so-called *transition-field region* [150].

For example, at a frequency of 100 Hz the boundary between near- and far-field is situated at 3000 km. This means that for man-made radiated noise (whose spectrum is primarily low-frequency), the victims are always in the near-field.

■ **Characteristic Impedance.** For a homogeneous material, the characteristic impedance is

$$Z_O = \sqrt{\frac{j\omega\mu}{\sigma + j\omega\varepsilon}} \quad (9.1)$$

where:
- ω is the angular frequency of the electromagnetic wave
- $\mu = \mu_o \mu_r$ is the permeability of the material
- σ is the conductivity of the material
- $\varepsilon = \varepsilon_o \varepsilon_r$ denotes the dielectric constant of the material
- $\mu_o = 4\pi \cdot 10^{-7}$ H/m (absolute permeability of free space)
- $\varepsilon_o = 10^{-9}/36\pi$ F/m (absolute permittivity of free space)

Two special situations can occur:

– For an insulator, $\sigma \ll j\omega\varepsilon$ and the characteristic impedance does not depend on frequency, since

$$Z_O \cong \sqrt{\frac{\mu}{\varepsilon}} \quad (9.1a)$$

For free space and a plane wave, upon substituting the numerical values one obtains $Z_O = 120\pi \cong 377\,\Omega$.

– For a conductor, $\sigma \gg j\omega\varepsilon$ and the characteristic impedance is also called *shield impedance* (Z_S). Hence:

$$Z_S = \sqrt{\frac{j\omega\mu}{\sigma}} = \sqrt{\frac{\pi f \mu}{\sigma}}\,(1+j) \quad (9.1b)$$

The important point to note here is that *the shield impedance has a resistive and a reactive (inductive) part.*

■ **Wave Impedance.** According to [150, 151], *the impedance of a wave impinging normally to a surface is the ratio of: (1) the tangential electric field E, to (2) the tangential magnetic field H* (where E and H are mutually perpendicular):

9.1 Electromagnetic Shielding

$$Z_W = \frac{E}{H} \quad (9.2)$$

■ **Wave Propagation.** When an electromagnetic wave propagates through a medium, its amplitude decays exponentially and finally Z_W approaches the characteristic impedance of the medium. For instance, the wave impedance in free space is

$$Z_W = k\sqrt{\frac{\mu}{\varepsilon}} = kZ_O \quad (9.3)$$

where ε and μ are the permittivity and permeability of free space. As stated in [150], coefficient k is equal either to $(\lambda/2\pi r)$ for a very high-impedance source in the near-field, or to $(2\pi r/\lambda)$ for a very low-impedance source in the near-field, r being the distance from the source of radiation. In the far-field, for any impedance of the source, $k \cong 1$.

■ **Field Distribution in Free Space.** This problem is solved by means of Maxwell's equations. The results obtained for two different types of sources are shown in Fig. 9.1 and Fig. 9.2.

Consider the situation depicted in Fig. 9.1, where the source antenna is an elementary short dipole. The current in the antenna is low (it is only the displacement current of the equivalent capacity between antenna and ground).

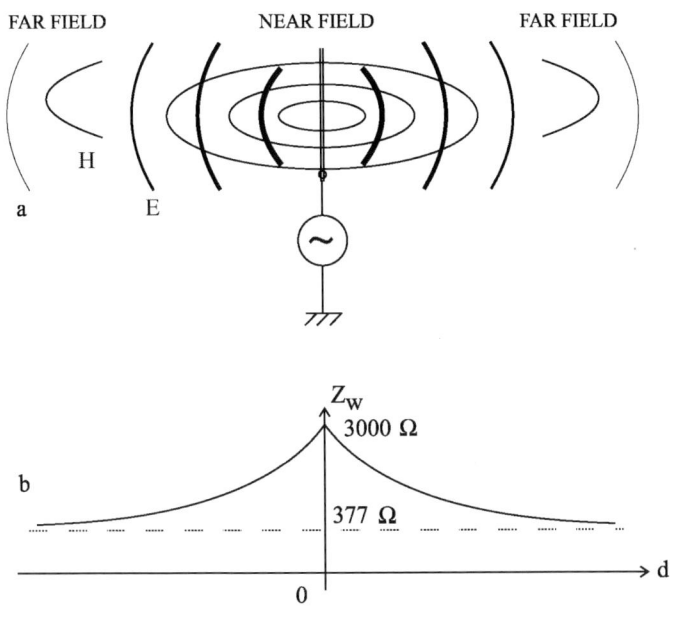

Fig. 9.1. High-impedance electric field source: **a** field distribution; **b** wave impedance versus distance

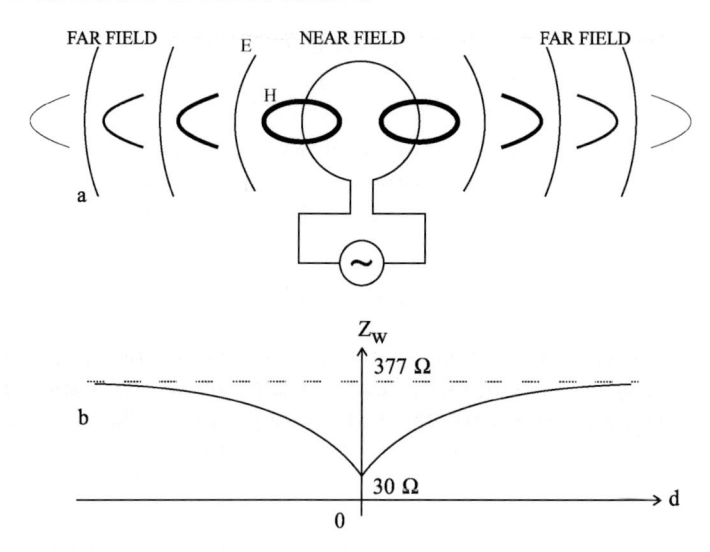

Fig. 9.2. Low-impedance magnetic field source: **a** field distribution; **b** wave impedance versus distance

Consequently, the source impedance $Z = V/I$ is high, and the wave impedance close to the source is also high (compared to 377 Ω). The electric field is very strong near the source, and decreases rapidly ($1/d^3$) with increasing in the distance d from the source. The associated magnetic field is relatively weak near the source and decreases as $1/d^2$.

The dual situation is illustrated in Fig. 9.2, where the radiating source feeds a loop antenna. The current flowing in the loop is significant, the source impedance $Z = V/I$ is low, and the wave impedance close to source is also low.

The magnetic field is strong close to the source, the electric field is weak, and so on.

Note that in both situations, far from the source the electromagnetic wave is a plane wave, of impedance 377 Ω. In the near-field region, the E and H fields can be treated separately.

■ **Remark.** The spectrum of industrial generated noise occupies the low-frequency region, with a dominant peak at 50 (60) Hz and its harmonics. The resulting wavelengths are very large, and consequently interference is always in the near-field. Therefore, *the effects of the electric and magnetic fields can be dissociated.*

■ **Skin Effect.** According to [152], this is *the phenomenon of nonuniform current distribution over the cross section of a conductor, caused by the temporal variation of the current in the conductor itself.* At high frequency, the current lines have the tendency to agglomerate just beneath the surface of the conductor.

9.1 Electromagnetic Shielding

■ **Skin Depth.** In a conductor carrying current, at a given frequency, the skin depth is *the thickness below the surface in which (1-1/e) or 63.2% of the total current is flowing* [150]. The skin depth, denoted by δ, is

$$\delta = \sqrt{\frac{2}{\omega\mu\sigma}} = 1/\sqrt{\pi f \mu \sigma} \quad (9.4a)$$

For practical purposes, we may take

$$\delta = 0.067/\sqrt{\mu_r \sigma_r f} \quad (9.4b)$$

where f is the frequency expressed in MHz, δ is expressed in mm, and μ_r, σ_r are the relative permeability and relative conductivity of the material with respect to copper (see Table 9.1). Note the strong frequency dependence of the relative permeability (Table 9.2). Note also that the permeability also depends on flux density (it drops when the flux increases).

Table 9.1. Conductivity and permeability of various metals relative to copper

Metal	σ_r	μ_r	Metal	σ_r	μ_r
Platinum	0.17	1	Iron (standard)	0.17	60
Gold	0.7	1	Iron (commercial)	0.17	200
Silver	1.05	1	Iron (purified)	0.17	5000
Lead	0.08	1	Steel	0.1	1000
Tin	0.15	1	Stainless steel	0.02	200
Copper	1	1	Silicon steel	0.038	1500
Chromium	0.66	1	Nickel	0.23	100
Aluminum	0.6	1	Nickel iron 50%	0.038	1000
Zinc	0.32	1	Permalloy	0.051	20000
Cadmium	0.23	1	Supermalloy	0.023	100000
Phosphor bronze	0.18	1	Mu metal (1 kHz)	0.029	20000

Table 9.2. Frequency dependence of the relative permeability

Frequency	μ_r of steel
≤ 100 kHz	1000
1 MHz	700
10 MHz	500
100 MHz	100
1 GHz	50

Whenever the current flowing inside a conductive material is induced by an incident electromagnetic wave (Fig. 9.3), this phenomenon also relates to the *absorption* of the wave energy by the conductor material.

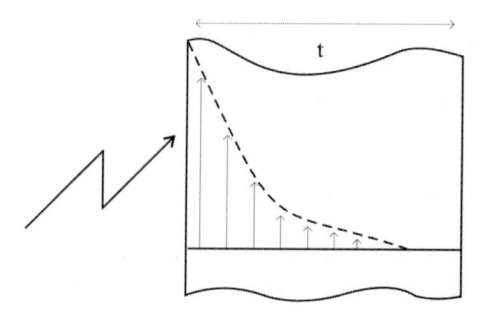

Fig. 9.3. Distribution of induced current inside the metal

■ **Absorption Loss.** *Absorption is a process by which an electromagnetic wave is reduced in intensity by interaction with the medium traversed.* In the situation illustrated in Fig. 9.3, the wave energy is partially converted into heat due to ohmic losses of the induced currents. This reduction in wave energy corresponds to absorption loss, which is evaluated with the formula

$$A = 20\,(t/\delta)\log(e) = 8.69\,(t/\delta) \quad [\text{dB}] \qquad (9.5)$$

where t represents the thickness of the metal sheet (shield) and δ is the skin depth at the wave frequency.

Equation (9.5) shows that *for a shield whose thickness is equal to one skin depth, the absorption loss is 8.69 dB.*

Generally, the absorption loss in a shield depends on the quality of the material and its thickness; for instance, the absorption loss of a 1-mm aluminum sheet is 3.2 dB at 1 kHz and 105 dB at 1 MHz.

■ **Wave Reflection.** *Reflection appears at the interface between two media due to a mismatch of characteristic impedances.* Consider the situation in which a plane wave is incident upon an ideal metallic wall. The wave electric field induces a current in the metal, whose associated magnetic field is such

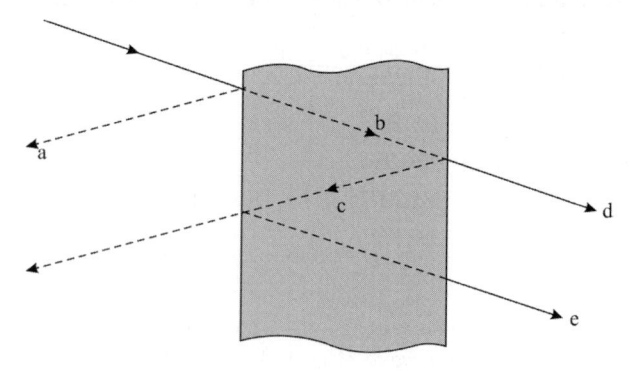

Fig. 9.4. Electromagnetic (plane) wave impinging on a metallic shield

that it tends to cancel the original magnetic field of the impinging wave in the region behind the shield. At the same time, this additional magnetic field doubles the original magnetic field in front of the shield. In this case the wave reflection is *total*.

In practice, part (a) of the incident wave is reflected by the first air-metal barrier and part (b) is transmitted through the metal sheet (Fig. 9.4). As a consequence, the reflection is only *partial*. When travelling through the metal, the wave undergoes absorption, before being internally reflected (part c) by the metal-air barrier upon egress. Finally, part (d) corresponds to the fraction of wave energy transmitted through the shield, while part (e) is the fraction of the wave undergoing re-reflection.

■ **Reflection Loss.** For either E- or H-field, the reflection loss is given by the expression

$$R = 20 \log \frac{|Z_W|}{4 |Z_S|} \quad [\text{dB}] \tag{9.6a}$$

where Z_W is the incident wave impedance and Z_S is the shield impedance. A good shield must reflect as much energy as possible from the incoming wave, hence it must have high reflection losses. From (9.6a), we deduce that *high reflection loss requires a significant mismatch between Z_W and Z_S*.

According to [153], when the metal thickness is much greater than the skin depth, reflection losses in the near-field can be evaluated with the following expressions:

$$R_e = 353.6 + 10 \log \frac{\sigma_r}{2.54 \, \mu_r \, r^2 \, f^3} \quad [\text{dB}] \tag{9.6b}$$

$$R_m = 20 \log \left(\frac{0.181}{r} \sqrt{\frac{\mu_r}{f \sigma_r}} + 0.053 r \sqrt{\frac{f \sigma_r}{\mu_r}} + 0.354 \right) \quad [\text{dB}] \tag{9.6c}$$

Equation (9.6a) is pertinent to reflection loss in an electric field (generated by a high-impedance source), while (9.6b) refers to reflection loss in a magnetic field (produced by a low-impedance source). The following notation has been adopted:

f – frequency [Hz]
r – distance from the emitting source
σ_r – relative conductivity (see Table 9.1)
μ_r – relative permeability (see Table 9.1)

For a plane wave (far-field condition), the reflection loss is

$$R_p = 168.2 + 10 \log \frac{\sigma_r}{\mu_r \, f} \quad [\text{dB}] \tag{9.6d}$$

When the radiating source is neither high-impedance nor low-impedance, but a source with characteristic impedance Z_O (such as a transmission line), the reflection loss is

$$R_o = 20 \log \frac{Z_O}{1.48 \cdot 10^{-3} \sqrt{\frac{\mu_r f}{\sigma_r}}} \quad [\text{dB}] \qquad (9.6e)$$

where the frequency is expressed in MHz.

■ **Remark.** If any of the expressions (9.6) yields a negative value for the reflection loss, we must instead adopt the value zero.

9.1.2 Circuit Shielding

■ **Explanation.** To reduce the effect of electric or magnetic fields, the most sensitive parts of circuits are protected by placing a conductive enclosure (shield) around them. For low-frequency applications, the shield is typically a metallic screen of appreciable thickness (1 mm or more). For high-frequency applications (over 30 MHz), as the skin depth decreases considerably, a thin metallic sheet or a thin conductive coating deposited on a plastic film is sufficient.

■ **Ideal Shield.** As stated in [151], an ideal shield is the conductive material that surrounds a circuit to attenuate any external field that might couple to the victim circuit. Ideally, *it is a seamless, six-sided cube with well-filtered penetrations occurring on only one area of the surface.*

■ **Practical Considerations.** In many applications, shielding can be extremely effective. However, the use of a shield can sometimes be counterproductive, due to undesirable changes in the shielded circuit parameters. A detailed discussion of such questions can be found in [154]. The decision to shield or not to shield must be made early in the project, paying attention to the following points:

- shielding is an expensive operation that requires additional tooling; consequently, it increases the cost of the final product.
- to be efficient, a shield must be complete (hermetically sealed). However, this is impossible to achieve in practice, since for any electronic equipment meter windows, apertures, and slots must be provided to pass cables or to allow ventilation.
- shielding increases the weight of the circuits. This can be important in aircraft and spacecraft applications.
- a shielding enclosure seriously limits accessibility to the protected circuits for periodic inspection and maintenance.

Restricting the discussion exclusively to circuit shielding, it is important to note that

- When a perturbing electromagnetic field is coupled to the victim in the differential mode, a shielding enclosure *is* the only solution (at any level, from integrated circuit to cable).

– Whenever the layout contains several distinct interfaces, shielding *may be* a solution. Note that if the interfaces are not too widely spread over the layout area, a ground plane may be as efficient as a shield.

■ General Rules

☐ **Rule 1.** *Only critical circuits have to be shielded.*

Otherwise, the overall cost of the equipment will dramatically increase without significant performance improvement.

☐ **Rule 2.** *Shielding is particularly effective when applied to radiating sources, rather than to victims.*

Supposing an ideal shield, two possibilities exist, as shown in Fig. 9.5.

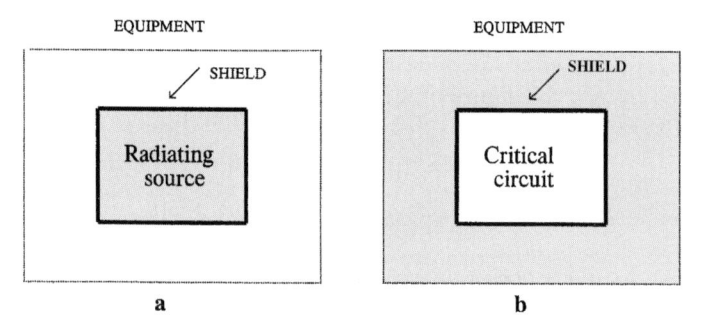

Fig. 9.5. a Shielding the source (no radiated noise outside); **b** Shielding the victim (no noise penetrating inside)

It is obvious that when shielding the victim is like curing the symptoms of the disease, while shielding the source is like curing the disease itself. In practice, since no ideal shield exists, both source and victim should be shielded.

☐ **Rule 3.** *Shielding effectiveness depends on the quality of mechanical implementation.*

Usually the shield is made from several panels joined together. Unfortunately, various imperfections can appear at seams (like mechanical distortions, paint residues, corrosion, etc.) which favor high contact resistance and electromagnetic leakage. Often mechanical implementation is the chief limiting factor in shielding efficiency, since apertures and ventilation slots are unavoidable.

☐ **Rule 4.** *The shield must be electrically connected to the ground of the circuit which is located inside it.*

This means that the signal ground of the protected circuit should be connected to the shield. When the physical dimensions of the shield are smaller than $\lambda/20$, a single connection point is adequate; otherwise, several connecting points, distributed on the most important dimension, should be considered.

■ **Guard Shield.** As stated in [158], *a guard shield is an internal floating shield surrounding the input section of an amplifier.* Note that effective shielding only results when the absolute potential of the guard is stabilized with respect to the input signal. Introducing a guard shield represents an efficient way to reduce input capacitance and leakage currents when the signal is delivered from a very high impedance source. Applications involving guard shields are proposed in Chap. 18.

■ **Shielding Effectiveness.** According to [152], *for a given external source, the shielding effectiveness is the ratio of electric or magnetic field strength at a point, before and after the placement of the shield in question.*

Denoted by SE, this figure of merit is calculated with the expression [150]

$$\text{SE} = 10 \, \log \, \frac{P_i}{P_t} \quad [\text{dB}] \tag{9.7}$$

where

P_i is the incident power density of the electromagnetic wave (measured at the point of observation, before shielding);

P_t is the transmitted power density of the electromagnetic wave (measured at the same point, after shield is in place).

Expression (9.7) holds for near- or far-field conditions, and can be written in a different but equivalent form if the power density expressed in W/m^2 is decomposed into the product of E [V/m] and H [A/m], i.e.

$$\text{SE} = 10 \, \log \, \frac{E_i H_i}{E_t H_t} \quad [\text{dB}] \tag{9.8a}$$

where

E_i, H_i – values of E and H fields before shielding;
E_t, H_t – values of E and H fields after shielding.

As long as the two fields are measured in the same medium, we may write for the electric fields

$$\text{SE} = 20 \, \log \, \frac{E_i}{E_t} \quad [\text{dB}] \tag{9.8b}$$

or for the magnetic fields

$$\text{SE} = 20 \, \log \, \frac{H_i}{H_t} \quad [\text{dB}] \tag{9.8c}$$

9.1 Electromagnetic Shielding

As a general rule, the shielding effectiveness depends upon the properties of the material employed, its geometry and thickness, and the inevitable holes, slots, covers, and windows. Electromagnetic sealing (by means of gaskets) plays an important role in reducing leakage due to imperfections when assembling various panels. All these factors are difficult to model and quantify for calculation purposes. Consequently, it is much easier to measure SE (according to (9.8)) than to predict it.

Nevertheless, equations are useful as a source of insight; hence, for an ideal shield, the total shielding effectiveness depends upon the absorption losses A and the reflection losses R in the following way [153–155]:

$$SE = A + R + B \quad [dB] \quad (9.9)$$

B being a term related to re-reflection (it is always negative). B can be neglected when 1) A > 9 dB; 2) equation (9.9) applies to an electric field; 3) for a plane wave.

The main conclusion emerging from (9.9) is that *SE improves when absorption and reflection losses increases.*

9.1.3 Estimating Shielding Effectiveness

■ **Ideal Shield.** Assuming a homogeneous shield of large dimensions without any electromagnetic leakage, the shielding effectiveness can be estimated by applying (9.9) to some special cases:

1a. Near-field, magnetic loop antenna (low-impedance source)
1b. Magnetic field, low frequency down to DC
2a. Near-field, dipole antenna (high-impedance source)
2b. Electric field, low frequency down to DC
3. Far-field, low- or high-impedance source

The corresponding expressions [150, 153, 155] are grouped in Table 9.3, where the resulting SE is expressed in decibels.

The following notation has been adopted:

μ_r – permeability of the shield metal relative to copper
σ_r – conductivity of the shield metal relative to copper
δ – skin depth, in mm
t – shield metal thickness, in mm
f – frequency in MHz
r – distance separating the source from the shield, in m.

■ **Practical Considerations**

– At any frequency, H-fields are low-impedance fields, while E-fields are high-impedance fields. Most shielding metals present a low impedance at low frequency, and a high impedance at high frequency. Since industrial noise

Table 9.3. Basic equations to compute SE of an ideal shield

Case	Condition	Expression
1a 1b	Strong H field	$SE = 131.43\, t\sqrt{f\mu_r\sigma_r} + 74.6 - 10\log \dfrac{\mu_r}{f\sigma_r r^2}$
2a 2b	Strong E field	$SE = 131.43\, t\sqrt{f\mu_r\sigma_r} + 141.7 - 10\log \dfrac{\mu_r f^3 r^2}{\sigma_r}$
3	Plane wave	$SE = 131.43\, t\sqrt{f\mu_r\sigma_r} + 108.1 - 10\log \dfrac{f\mu_r}{\sigma_r}$
1a	Extremely thin shield	$SE = 20\log(1 + 500\mu_r t/r)$ [dB] $(t/\delta \ll 1)$

sources are of low-frequency type, it follows that reflection losses are weak for magnetic fields, because the characteristic impedance of the shield is almost matched to the wave impedance. As a consequence, shielding effectiveness is mostly determined by absorption losses, and is therefore not so high.

In contrast, for the electric fields originating from industrial perturbations, the shield is much more efficient, since reflection losses are significant.

To conclude, *improving low-frequency protection against magnetic fields requires shields made of magnetic metals.*

- The critical features of any shield are its apertures and access covers. It is advisable to use conductive gaskets to reduce contact resistance between various elements. Welded assemblies are always preferred to those joined by screws, rivets, or any other fastener.

9.1.4 Performance Degradation

■ **Discussion.** Equation (9.9) presents the mechanisms of field penetration through a perfectly closed shield, of finite thickness but of infinite dimensions. Apart from the case of low-frequency magnetic fields or extremely thin shields, the equations in Table 9.3 lead to an over estimate of shielding effectiveness. In practice, the measured values of SE are lower, due to inherent discontinuities in shield structure. Globally, shielding effectiveness is reduced for two reasons:

- the presence of electromagnetic leakage;
- standing waves appearing at the resonant frequency of the enclosure.

Consequently, (9.9) is corrected in the following way:

$$SE = A + R + B - LK - SW \quad [dB] \tag{9.10}$$

LK is a term pertinent to leakage and SW to the effect of standing waves at cavity resonance, both expressed in decibels.

☐ **Electromagnetic Leakage.** Leakage results from the following factors:
- Poor mechanical assembly (recall that uniform soldering or welding of all seams is the best solution)
- Access covers and doors; these represent the critical point of any shielded enclosure
- Ventilation slots; there are unavoidable to evacuate the heat dissipated by power devices
- Windows for display devices, connecting plugs, LEDs, etc.
- Inhomogeneous regions inside the shield metal, due to folding, twisting, casting, etc.

In order to intuitively appreciate the degradation of SE, it is worth considering how the distribution of the induced current lines is modified by the presence of apertures. Figure 9.6 shows some distributions of current lines in the shield metal induced by a perturbing magnetic field [155, 157, 159].

The "best" configuration is that which least distorts the original flow (Fig. 9.6a) of current. For example, the apertures in Fig. 9.6c or d are preferred to that of Fig. 9.6b.

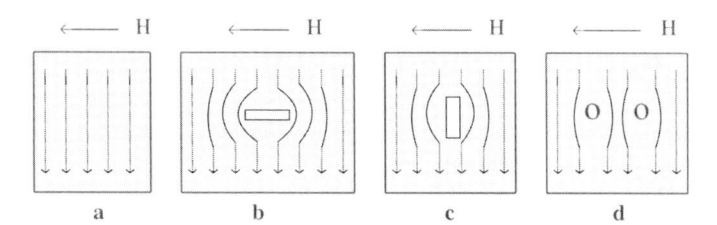

Fig. 9.6. Effect of various types of apertures on induced current-line distributions

Considering the impact of a rectangular aperture (Fig. 9.7) on shielding effectiveness, it must be remembered that reflection losses (R) and the absorption losses (A), expressed in decibels, both decrease with increasing frequency and aperture dimensions [150].

Note that what is called "aperture thickness" corresponds to panel thickness (for a homogeneous shield) or to material thickness around the aperture otherwise. In the latter case, increasing the thickness is especially beneficial for absorption losses, which become more significant. A practical solution is proposed in Fig. 9.8, where the aperture is shaped to form a waveguide (whose cutoff frequency is denoted by f_c).

9 Interference Reduction Methods

$$R = 97 - 20\log(Lf) + 20\log[1 + \ln(L/H)]$$

$$A = 30\, d/L \quad d = \text{aperture thickness}$$
$$f = \text{frequency in MHz}$$

Fig. 9.7. Panel with a rectangular aperture. For a circular aperture, with L = H = diameter, replace the term 97 with 99 in the expression for R

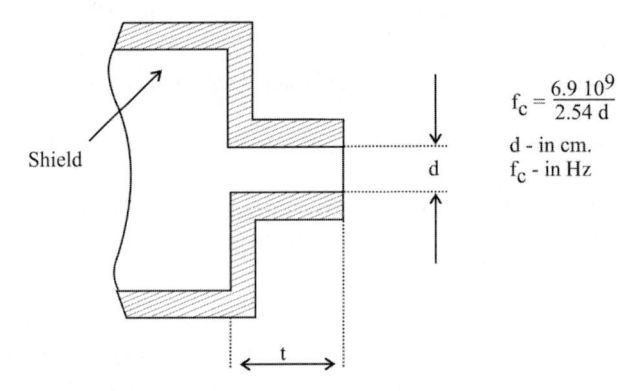

$$f_c = \frac{6.9 \cdot 10^9}{2.54\, d}$$

d - in cm.
f_c - in Hz

Fig. 9.8. Cross section of a cylindrical aperture

Operating this waveguide at $f \ll f_c$ results in additional wave attenuation. Under this condition and according to [156], the shielding effectiveness of the circular waveguide, with respect to a magnetic field, is

$$SE = 32\, \frac{t}{d} \quad [\text{dB}] \tag{9.11}$$

This formula shows that for a circular waveguide with $t \cong 3d$, the shielding effectiveness is close to 100 dB.

Generally, when the shield has several holes, its effectiveness is calculated with expression (9.9) modified to be

$$SE = R + A + FO \tag{9.12}$$

where FO represents a negative term comprising the "shadow effect." For typical configurations $FO \cong -5\,\text{dB}$.

To conclude, note that *apertures degrade the shielding of magnetic fields, rather than of electric fields.*

☐ **Standing Waves.** Due to imperfect shielding, an enclosed metal box is still penetrated by outside electromagnetic radiation. When the distance separating the opposite panels is $\lambda/2$, the box becomes a resonant cavity. The more conductive the metal, the higher the Q-factor of the resonant cavity.

According to [150], empty metal boxes with smooth internal walls yield quality factors between 3 and 10. The resonant frequencies of a metal rectangular box are established by the formula [153, 157]

$$f = 150\sqrt{(k/L)^2 + (m/H)^2 + (n/W)^2} \quad [\text{MHz}] \quad (9.13a)$$

where L is the length, W is the width, and H is the height of the box, all expressed in meters. Coefficients k, m, n are positive integers (no more than one can be zero at the same time).

Many resonant modes exist; below a specific frequency, for a cavity of volume V, the maximum number of modes is established by the equation:

$$N = 155 f^3 V \quad (9.13b)$$

f being expressed in GHz and V in m^3.

When the box has nearly equal dimensions on each axis, the frequencies associated with the TE$_{011}$, TE$_{101}$, and TE$_{110}$ modes are

$$f = 212/L = 212/H = 212/D \quad [\text{MHz}] \quad (9.14)$$

Around these frequencies, the distribution of E- and H-fields inside the box becomes strongly non-uniform: in the middle, the electric field is amplified and near the walls the magnetic field is augmented. Practical considerations can counteract this effect, for instance, if the box is not empty, but filled with various insulating and conducting items (corresponding to circuits, devices, boards, and cables) which reduce the resulting quality factor. Nevertheless, it is advisable to avoid operation near the resonant frequencies of the cavity, because any internal field reinforcement is equivalent to a considerable reduction in shield effectiveness.

■ **Cable Entries.** Cables and wires entering or leaving the shield can spoil shielding effectiveness. If a wire simply enters the shield enclosure (Fig. 9.9a), even when the entry hole is adjusted to the wire diameter, the wire acts as

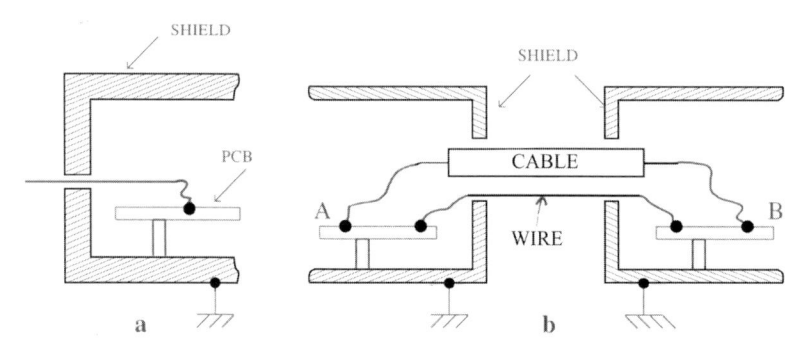

Fig. 9.9. Wrong way to pass cables and wires

a small antenna that collects the internal radiated energy and transmits it outside the box, or vice versa. This seriously affects shielding performance.

The situation depicted in Fig. 9.9b is also incorrect, since only one coaxial cable is employed to transmit signals between circuit boards A and B, the return path being a simple wire with no interface filtering (a common mistake is to think that filtering is unnecessary for a ground wire).

Actually, any perturbation induced in the ground wire will affect the signal using that wire as a return path. The solution consists in using two coaxial cables, one for the signal, the other for the return, both having their shields bonded to the enclosure.

Two conclusions can be drawn:

1) *All shielded cables entering a shielded enclosure must have their shields bounded to the enclosure.*
2) *All wires entering a shielded enclosure must be filtered.*

□ **Feedthrough Capacitor.** Any wire penetrating a shielded enclosure must be filtered for spurious signals by adopting a feedthrough capacitor (Fig. 9.10).

Essentially, this is a cylindrical capacitor whose dielectric around the central lead has high permittivity (Fig. 9.10a), and which is mounted by screwing or soldering its outer body directly to the chassis (Fig. 9.10b). Since the current to ground can spread through 360° around the body, the inductance associated with this terminal is very low. Note that to satisfy this condition, *the full circumference of outer body of the feedthrough capacitor must be bonded to the shield wall.*

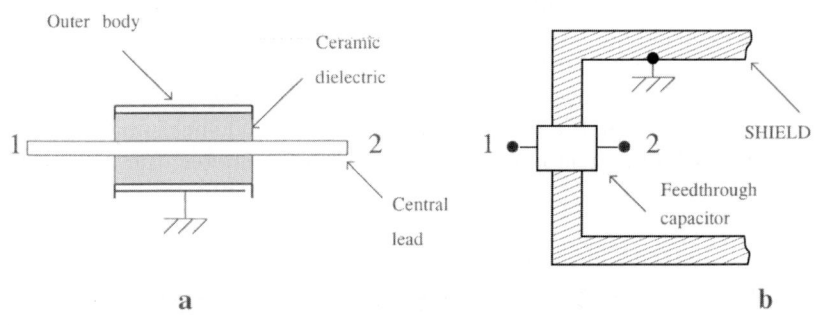

Fig. 9.10. Feedthrough capacitor: **a** cross section; **b** method of mounting in the chassis

The most commonly encountered symbols of the feedthrough capacitor are given in Fig. 9.11a, and its equivalent circuit is shown in Fig. 9.11b. The parasitic inductances (L) shown in the equivalent circuit contribute to low-pass filtering action. For reasonable dimensions, a value of 1 nF is obtained for the capacitance with an insulator having a dielectric constant higher than 3000; hence, a feedthrough capacitor is a relative expensive component.

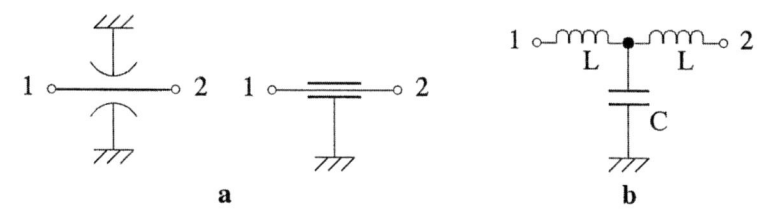

Fig. 9.11. Feedthrough capacitor: **a** most encountered symbols; **b** equivalent circuit

When correctly bonded to the metallic wall, HF filtering is efficient up to several GHz.

9.1.5 Conclusion

Shielding against electric fields is relatively easy and not too expensive. Problems appear when shielding is intended to protect against magnetic fields, especially at low frequency.

As a general rule, shielding effectiveness depends chiefly upon the quality of the mechanical implementation (the failure of seams to make good contact resistance); as soon as the shield is breached, performance drops significantly.

At frequencies above 1 MHz, shield effectiveness depends more upon leakage rather than shield thickness. As a consequence, all cables and wires entering the metallic box must be carefully filtered; otherwise, they can transmit the noise in either direction, in and out of the enclosure.

For large windows in the shield required by meters or display devices, it is advisable to use a second internal screen just behind the meter.

For ventilation slots it is better to select circular holes of small diameter (less than several mm); otherwise, provide grid covers for larger holes.

9.2 Filtering and Balancing

■ **Postulate.** *Sooner or later, any radiated interfering signal is transformed into a conducted interfering signal.*

To illustrate this statement, consider the case of a cellular phone used near a personal computer (PC). The electromagnetic radiation emitted by the phone can enter the mains line from which the PC is supplied. The induced spurious signals in the AC line enter the computer and can eventually interfere with the clock signals driving logic circuits in the PC, perturbing its normal operation.

Generally, conducted perturbations can reach victims either along the power lines or along a common ground plane. Protecting victims implies filtering the lines and cables connecting several systems, balancing circuits, and/or grounding circuits properly.

9.2.1 Filtering Interfering Signals

■ **Definition.** *A filter is a combination of lumped or distributed circuit elements arranged to selectively attenuate signals on the basis of their frequency.* They are used to control the spectral content of signal paths.

Traditionally, at the load terminals, a filter is regarded as a voltage divider whose ratio is frequency dependent. The concept of filter selectivity is based upon matching or mismatching of the load, this condition depending chiefly upon the operating frequency.

■ **Insertion Loss.** Insertion loss as a function of frequency is one of the most important parameters of a filter. The insertion loss is *the ratio of: (1) the power delivered to that part of the system following the filter before insertion of the filter, to (2) the power delivered to that part of the system following the filter after insertion of the filter.* Usually expressed in decibels, insertion loss (or attenuation) is computed with the expression

$$A_i = 10 \; \log \; \frac{|P_1|}{|P_2|} \tag{9.15a}$$

thus

$$A_i = 20 \; \log \; \frac{|V_1|}{|V_2|} + 10 \; \log \; \frac{|Z_2|}{|Z_1|} \tag{9.15b}$$

Subscript 1 denotes the filter input, subscript 2 refers to the filter output; when $Z_1 = Z_2$, the insertion loss can be expressed in terms of terminal voltages alone.

■ **Load Influence.** Filters are designed to operate into particular termination impedances. One of the most harmful effects of unknown filter termination is peaking in the filter response. Assume a low-pass filter intended to attenuate perturbations situated just above its cutoff frequency f_o, connected to a higher than expected impedance load (Fig. 9.12a). A peak appears in the response of the filter (Fig. 9.12b), whose amplitude depends on the damping factor of the resonant circuit L, C.

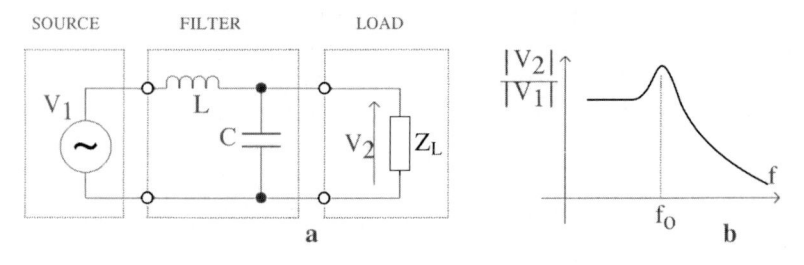

Fig. 9.12. a Low-pass filter; b Attenuation versus frequency

As a consequence, the filter is not effective in suppressing the unwanted signals of frequencies situated around f_o, worse, they are amplified instead. A possible solution is to place a ferrite bead around the wire connecting L, whose effect is to increase the total inductance. In this way, the peak will be shifted towards low frequencies (assuming that this is not harmful to normal operation). Note the following points:

- Any filter attenuating the incoming or outcoming noise also attenuates the useful signal. Ideally, the signal must be affected as little as possible.
- Peaking in the filter response can be avoided when the load has low impedance. This guarantees a high damping factor of the resonant circuit and reduced selectivity.

■ **Practical Filters.** As a general rule, the user does not control the design and fabrication of filters, but buys them instead. A wide selection is possible among commercial filters, which can be used either to reduce the conducted emission of perturbations from an offending circuit or to improve susceptibility of victims. The former case will be detailed in Chap. 10, while the latter is discussed in Chap. 12. The following gives an overview of the most common devices used for filtering.

□ **Line-to-Line Capacitors (X-Caps) and Line-to-Ground Capacitors (Y-Caps).** They are intended to filter the noise transmitted by power supply lines: the former for differential- mode coupling and the latter for common-mode coupling. For safety considerations, the insulating properties of Y-caps are particularly severe (the leakage current should not exceed 0.5 mA and must be certified), in order to prevent an accidental connection between the phase line and ground. Figure 9.13 presents a typical application, where they are used to filter the perturbations emitted by an offending circuit (such as a solid-state switching-mode power supply). Shielding is also provided, to protect the environment against radiated noise.

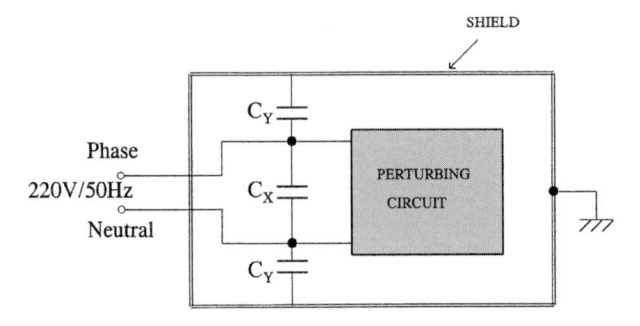

Fig. 9.13. Filtering the emission of conducted perturbations

☐ **Feedthrough Capacitors.** See Sect. 9.1.4.

☐ **Chokes and Ferrite Suppressors (Ferrite Beads).** These devices don't require a ground to function, since their action consists in introducing a series inductance which behaves like an elementary low-pass filter. They are attractive devices because they require no circuit redesign, and often no mechanical complication. Ferrite sleeves are easy to slip around a wire or cable; furthermore, it is straightforward to install a ferrite clamp on a ribbon cable. In any case, it is expected to increase the series inductance of the wire by at least a factor of several hundred. At DC and audio frequencies, ferrite suppressors offer a negligible additional series inductance, and therefore low insertion loss; however, they considerably attenuate the higher-frequency noise signals. Ferrite beads are lossy at high frequency, so that the perturbation energy is mostly absorbed in the suppressor rather than being passed along to another part of the circuit.

Common-mode chokes are used to block common-mode currents without affecting differential-mode currents. Figure 9.14a shows the equivalent circuit of a common-mode choke with two coils; some applications are suggested, such as blocking the common-mode noise of a DC supply (Fig. 9.14b) or of a signal line (Fig. 9.14c).

The operating principle is simple: since the currents flowing in the two coils are in phase, when a noise signal is induced in one conductor of the line, due to the common-mode choke, an equal in-phase noise signal is induced into the other conductor. With equal capacitances between the signal wires and surroundings, both undesired signals cancel reciprocally on any differential load.

When used as power-line filters, common-mode chokes are placed in a metal shield, to reduce the coupling of external fields into the inductors. A more elaborate filter (for suppressing both common- and differential-mode noise) is presented in Fig. 9.14d.

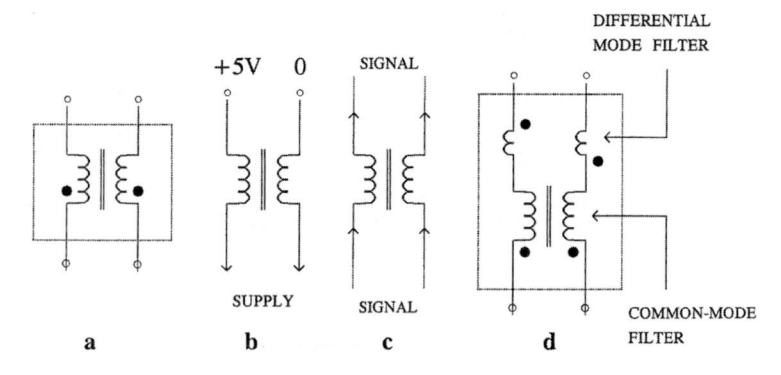

Fig. 9.14. Common-mode choke: **a** configuration; **b, c** various applications; **d** typical line filter

□ **Other Filters.** All filters employed in interference reduction are of low-pass type. Combining various generic low-pass configurations, more elaborate filters are obtained. Note that their performance depends strongly on the terminal impedances (see Case Study 18.7).

■ **Filter Selection.** When selecting a filter for a particular application, attention should be paid to the following specifications:

- maximum voltage and current ratings
- frequency range
- insertion loss expected
- size and weight of the filter
- miscellaneous electrical parameters, such as maximum leakage current, isolating resistance, maximum allowed voltage drop, etc.
- environmental operating parameters, such as temperature, vibrations, shocks, etc.

■ **Guidelines for Filter Mounting**

- when intended to suppress conducted emission, place the filter as close as possible to the source of noise.
- when intended to improve susceptibility, locate the filter as close as possible to the victim
- minimize the lengths of connecting wires, avoiding loops
- mount the filter inside the metal shielding enclosure of the victim circuit, directly on one wall (to maintain a good ohmic contact with ground)
- use shielded cables to transport the signal in and out of the filter. When the traces of a PCB are used for this purpose, provide good separation between the traces carrying signals into and out of the filter.

9.2.2 Balanced Circuits

■ **Definition.** As stated in [152], *a balanced circuit is a circuit having two sides that are electrically alike and symmetric with respect to a common reference, usually ground.* As a consequence, for an applied signal difference at the input, the signal relative to ground at equivalent points in the two sides must be opposite in polarity and equal in amplitude.

■ **Principle.** This technique contributes to eliminating stray pickup from outside the circuit. In order to be effective, balancing is applied at all levels: source, transmission path, and load (Fig. 9.15a). In this way, common-mode coupled noise gives rise to equal signals in both circuit (longitudinal) halves, which cancel in the load. Globally, the circuit can be considered a common-mode filter.

Whenever asymmetry of the load (or source) is unavoidable, the solution is to insert a balun (Fig. 9.15b) to correct the unbalance [159]. The balun is usually a transformer designed to accept 75-ohm unbalanced input (coaxial cable) and deliver the signal to a 300-ohm balanced twin lead.

■ **Limitations.** Several factors can upset the balance of circuits:

- nearby metallic obstacles (ground planes, transformers, metallic enclosures, cable connectors, filters, etc.)
- location and orientation of cards on the mothercard
- connection of load (or source) by means of coaxial cable (solution suggested in Fig. 9.15b).

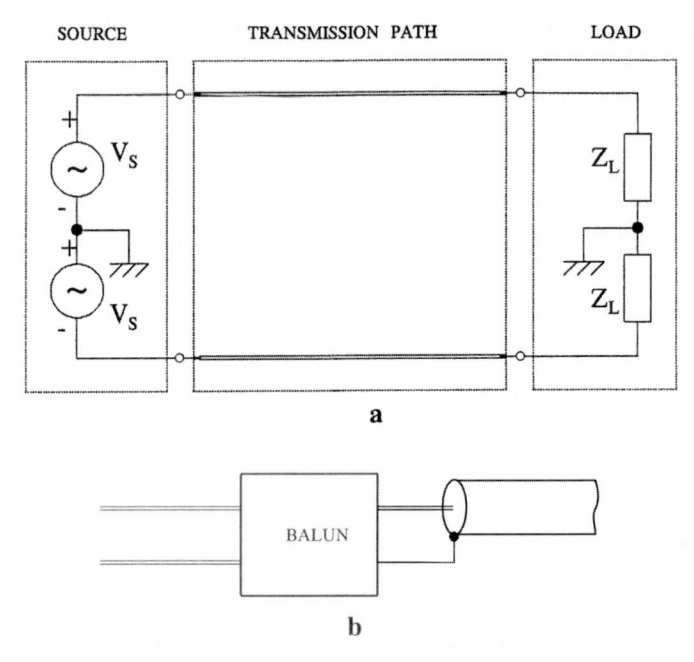

Fig. 9.15. a Overview of a balanced configuration configuration; b using a balun to drive an asymmetric load

For all these reasons, balancing remains a technique that is particularly powerful in low-frequency applications, where the effect of stray capacitance is not dominant. As soon as the frequency is increased, it is considerably more difficult to maintain the symmetry of a balanced system. Since this technique is mainly used to protect sensitive circuits against interference (hence, to increase immunity), more details can be found in Sect. 12.1.

9.3 Grounding and Bonding

9.3.1 Equipment Grounding

■ **Classification.** There are two categories of grounds: signal and earth grounds. *Signal grounds* serve as references for various voltages and are not necessarily at zero potential; *earth grounds*, also called *safety grounds*, are really connected to earth. The reason is the following: when the chassis, racks, or cabinets are floating, they can reach a relatively high potential, determined by the unknown values of stray capacitances. Hence, to prevent a shock hazard, they should be connected to a ground at zero (earth) potential.

■ **Definitions.** It is often stated that a ground is *a conductive system whose potential is taken as reference for all voltages in the circuit*. However, this definition does not take into account another important role of ground, namely to provide paths for the return currents.

Ott proposed another definition [155], namely *the signal ground is a low-impedance path for currents to return to the source*.

Consequently, the complete definition we propose is the following: *a ground is a conductive system whose potential is taken as reference for all voltages in the circuit and which provides a low-impedance path for return currents*.

Note that proper grounding represents one of the most important ways to provide protection against interference. It requires no additional devices, only additional skill.

■ **Myth.** One common myth is that electronic equipment always requires a safe earth ground for reliable operation. Despite the fact that there is no evidence for this condition, one wonders how electronic equipment installed in rubber-tire insulated cars, or on satellites and spacecraft (where an earth ground connection is out of question) can still operate reliably?

The answer is given in [152], where it is indicated that *a ground may be a conductive connection to a structure that serves a function similar to that of an earth ground (that is, a structure such as the framework of an air, space, or land vehicle that is not conductively connected to ground)*.

■ **Ideal Ground.** By definition, this is *a conductive system at zero potential which can source or sink an infinite amount of current, if necessary*. Note

– Since it maintains a constant potential, it is also sometimes known as an "equipotential conductor".
– The definition implies that the associated impedance of an ideal ground must be zero.

■ **Ground Impedance.** In practice, the ground is far from being an equipotential conductor, since its associated impedance is never zero. *Any* conductor has associated resistance and inductance; both can be estimated on the basis of the geometry and physical parameters of the conductor. For low-frequency applications, the corresponding impedance can turn out to be sufficiently low as to be negligible, but in high-frequency circuits (or fast logic circuits) where the skin effect dominates, the impedance of the ground system can be unexpectedly high (typically, hundreds of ohms).

Consider, for instance, a bonding wire to chassis whose associated impedance is

$$Z = R + j\omega L \qquad (9.16)$$

where R is the DC or AC resistance of the wire (including the skin effect, if necessary). The most harmful term in (9.16) is the self-inductance L, because its contribution to overall impedance increases with frequency.

To mitigate this effect, the first step is to reduce the self-inductance by increasing the wire diameter, since for a circular conductor

$$L = 0.002l \left(2.303 \log \frac{2l}{d} - 0.75\right) \quad [\mu H] \qquad (9.17a)$$

and for a rectangular strap

$$L = 0.002l \left(2.303 \log \frac{2l}{b+c} + 0.5 + 0.2255 \frac{b+c}{l}\right) \quad [\mu H] \qquad (9.17b)$$

In both equations, l is the conductor length, d is the diameter, and b and c the width and thickness respectively, all expressed in cm.

Due to the logarithmic dependence, increasing the diameter (or connecting several wires in parallel) is not practical. Hence, the only efficient solution is to reduce wire length.

☐ **Resonance.** Actually, the problem of parasitics associated with a ground wire is worsened by the resonance due to L and the stray capacitance (C_p) existing between the points where the wire is connected. A simplified equivalent circuit of the connecting wire is proposed in Fig. 9.16a, where the resonant frequency is evaluated with the traditional formula

$$f_o = \frac{1}{2\pi\sqrt{LC_P}} \qquad (9.18)$$

It is obvious that $|Z|$ has a maximum at f_o; up to f_o the behavior is inductive, and above f_o it is capacitive. The region of normal operation is very likely situated near the origin, where $|Z|$ is low. Actually, when a ground wire runs alongside a ground plane or chassis, a multi-peak frequency characteristic is observed due to the distributed LC structure (relative to ground). Some will be current-peaked (due to series resonance) and some will be voltage-peaked (due to parallel resonance).

Since many commercial applications (such as personal computers, cellular telephones, CB communication systems, etc.) employ signals whose frequency

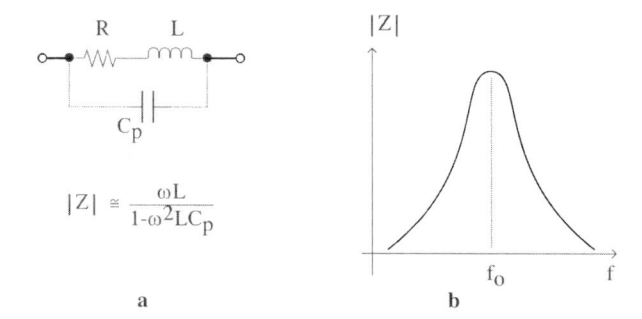

Fig. 9.16. a Simplified equivalent circuit of a short ground connection; **b** frequency dependence of $|Z|$

spectrum extends well above 100 MHz, resonances due to ground parasitics very likely fall inside the useful spectrum, and alter their operation in a quite unpredictable way.

■ **Grounding Techniques.** The various signal grounds existing in electronic equipment are ultimately connected to the same common point, i.e., the common ground (earth). Several grounding configurations are often encountered:

– single-point ground connection;
– multipoint ground connection;
– hybrid connection.

□ **Single-Point Ground Connection.** Because of its simplicity, one of the most popular grounding schemes is the *series connection* of all individual signal grounds (Fig. 9.17). The inductances associated with various conductors are displayed; note that there is also a resistance associated with each conductor (for simplicity, not shown on the figure).

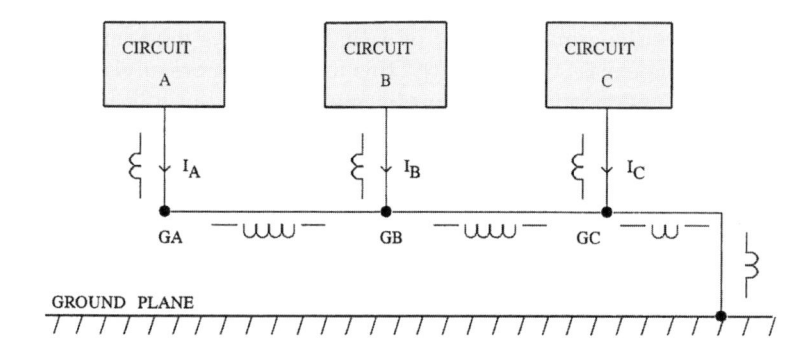

Fig. 9.17. Single-point, series ground connection

304   9 Interference Reduction Methods

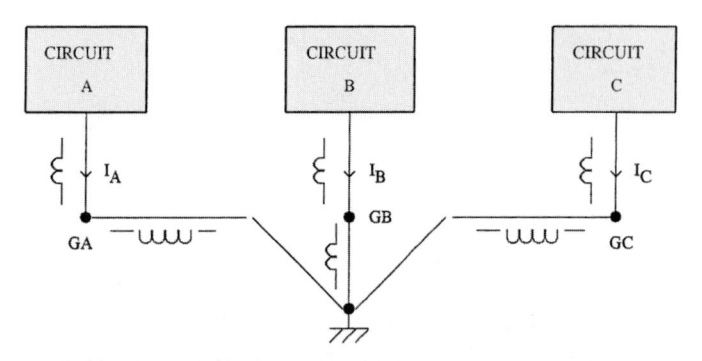

Fig. 9.18. Single-point, parallel ground connection

Due to the impedance of each path and the resulting voltage drop, the signal ground potentials will not have the same value, and it is expected that

$$V_{GA} > V_{GB} > V_{GC}$$

are all above the potential of the ground plane. This inequality shows that *the circuit with the highest electromagnetic susceptibility must be placed at position C.*

This grounding method is unsuitable from a noise standpoint, and is ill-advised, except for low-frequency applications (in the range from DC to several kHz). Its main drawback is the cross-coupling between ground currents from different circuits, because they all flow through a common impedance (the path between MC and the ground plane).

However, it can be still used, provided that

- circuits A, B, and C operate at the same power level (otherwise the ground current of the high-level stage will adversely affect the remaining stages)
- when A, B, and C are units in the same equipment, they are mechanically joined at all corners. This eliminates interunit ground cables (wires), but does not completely solve the problem of cross-coupling.

Another single-point grounding method is to connect the units in *parallel* (Fig. 9.18). Here the ground current of each circuit flows independently, and no cross-coupling exists. Consequently, this method is more advisable than a series ground connection for low-frequency applications, but it is mechanically cumbersome, due to the extra conductors (or extra layout space) required. Note also that since the lengths of the ground paths increase, additional parasitics are expected.

☐ **Multipoint Ground Connection.** This grounding method (Fig. 9.19) is especially employed in analog circuits operating at frequencies above 10 MHz or in high-speed digital circuits.

Here each of the return currents has its own path and a significant reduction in ground impedance is obtained. The reason is twofold:

9.3 Grounding and Bonding

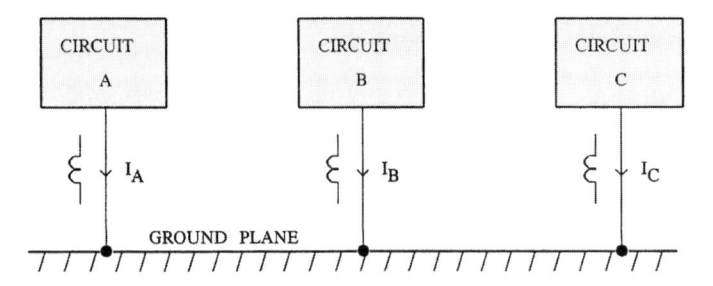

Fig. 9.19. Multipoint ground system

– each circuit is connected to the nearest point of the ground plane
– the ground plane offers a low-impedance path.

In practice, two techniques are widely employed:

a) the ground plane is gold- or silver-plated to reduce its skin resistance
b) when a ground plane is not available, a grid plane can be used instead.

☐ **Hybrid Connection.** This term refers to a grounding method whose configuration acts differently according to the range of the operating frequency. The technique is shown in Fig. 9.20, where the values of the grounding capacitors C are selected so that they act like short circuits at RF. For DC or LF signals, the system performs as a series single-point connection (because the capacitors are equivalent to open circuits).

Hence, for RF operation, the points GA, GB, and GC are directly connected to the nearest ground point through the low impedance of the capacitors. Note that capacitors C must be carefully selected for *low series parasitic inductance* (for instance, by using ceramic capacitors).

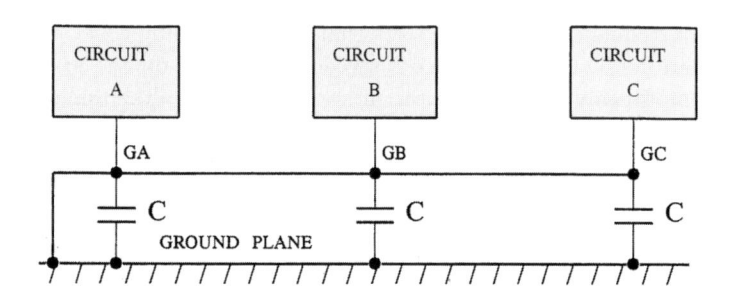

Fig. 9.20. Hybrid ground connection: acts as a series single-point system at LF and as a multipoint system at RF

9.3.2 Noise Related to Grounding

■ **Classification.** The grounds found in a variety of electronic equipment can be grouped into the following categories:

- signal grounds or quiet grounds (analog low-level circuits)
- noisy grounds (digital circuits, motors, relays, etc.)
- hardware ground (chassis, equipment racks, cabinets).

When applying the grounding methods discussed in Sect. 9.3.1, only grounds belonging to the *same category* can be tied together in a selected configuration. Finally, the three ground points corresponding to each category are connected at the mains level (the green–yellow conductor of the AC supply).

It is highly advisable to *draw an overall grounding map of the equipment*, and to update and verify it periodically.

■ **Noise.** Very seldom and only by sheer luck will two different ground points have the same potential. The rule is that *the potential difference between two grounds is never zero*.

To understand why this is not an accident but rather a general phenomenon, consider two points on the best available conductor, a ground plane. It is common practice to assume that their potential difference must be so small as to be negligible. However, according to [160], the impedance between two points on a plane is

$$Z_{GP} = (R_{DC} + j\, Z_{RF})\left(1 + |\tan(2\pi d/\lambda)|\right) \tag{9.19}$$

where R_{DC} is the DC resistance of the ground plane (using the Ω/sq. method), Z_{RF} is the impedance of the ground plane (using the Ω/sq. method), d is the distance separating the two points, and λ is the wavelength corresponding to the highest frequency of concern. Note that due to standing waves (the second term in brackets), the impedance between two points on the ground plane can be considerably higher than the almost negligible value predicted by the Ω/sq. method (the sum in parentheses). Equation (9.19) clarifies why the voltage between two different grounds is never zero (it usually lies between 1 and 100 mV).

Consider the case of two different equipment units, denoted by A and B, supplied from two different phases of the same AC mains. An offset voltage V_O exists between the two grounds (Fig. 9.21). This voltage contains residues of 50 (60) Hz and its harmonics, residues of radio frequencies, spurious signals, and all sorts of other trash. Globally, V_O represents a signature of the electrical activity in the area, and acts as an equivalent noise generator that contaminates the useful signal E_g.

To avoid interference, it is advisable [155] to eliminate (if possible!) one of the two grounds (for instance, by employing a differential input stage in

9.3 Grounding and Bonding

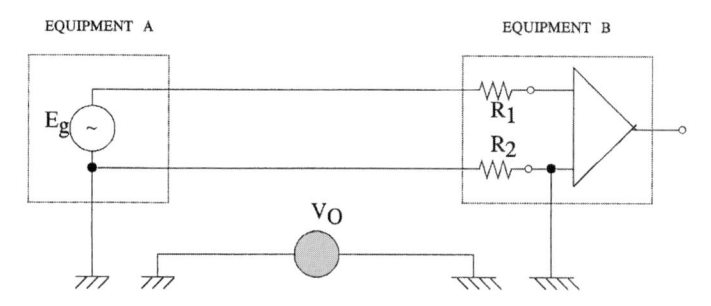

Fig. 9.21. Offset voltage between two different grounds

equipment B, or delivering the signal from equipment A by means of an output transformer).

■ **Ground Loops.** Ground loops appear when two or more points that are designated as local grounds (not necessarily having the same electric potential) must be connected to one another. A ground loop appears, and it can become a problem especially when its dimensions are significant, or when low-level analog circuits are involved. Furthermore, a ground loop is, like any loop, sensitive to radiated electromagnetic fields.

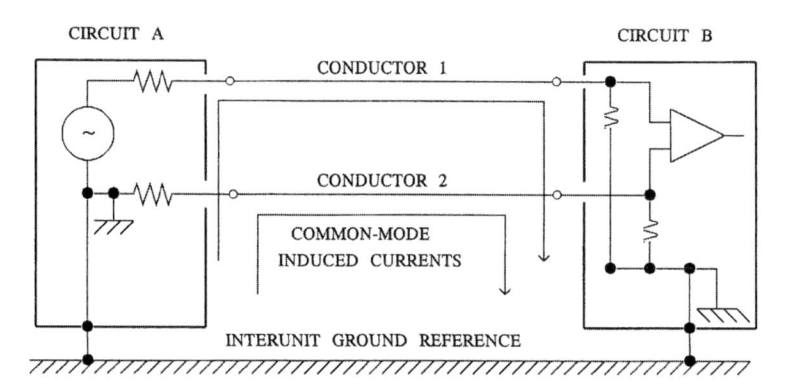

Fig. 9.22. Ground loop path

Consider the case illustrated in Fig. 9.22, where the ground planes of the two circuits are connected to their shields, which are in turn joined by bonding straps to the chassis, which plays the role of interunit ground reference. A conductive ground loop appears between each conductor transporting the signal and the ground reference. These loops can intercept the magnetic field radiated by a perturbing source, and consequently induced noise currents will appear in each conductor transporting the useful signal.

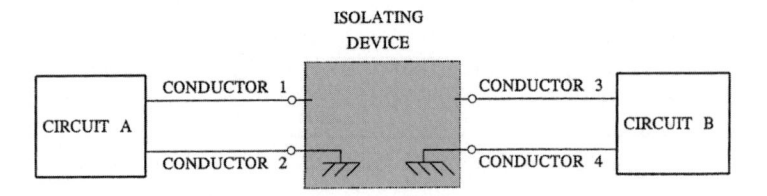

Fig. 9.23. Breaking the ground loop. Note that circuits A and B are galvanically isolated

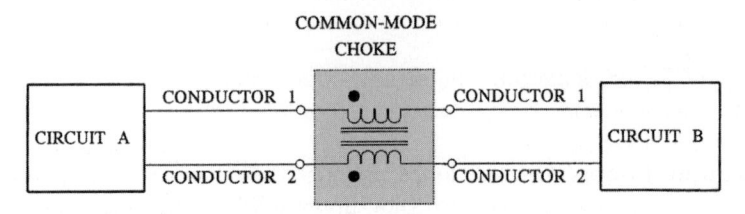

Fig. 9.24. Inserting a common-mode choke. Note that circuits A and B are *not* galvanically isolated

To prevent contamination of the useful signal, several solutions exist:

– Break the ground loop by introducing an isolator in the signal path (Fig. 9.23), which can either be a transformer, an optocoupler, or an isolating amplifier. Details concerning each option are given in Sect. 12.3.
– When breaking the ground loop is not possible or not practical, common-mode induced currents can be attenuated either by inserting common-mode chokes (Fig. 9.24) or by balancing the circuit.

9.3.3 Miscellaneous

- Any metallic box (shield, screen, chassis, rack, panel) must be firmly connected to ground. Otherwise, when left floating, a connection to ground is still made through stray capacitances, which are not under the user's control.
- To prevent radiation, any bonding strap or wire connected to ground must be kept shorter than $\lambda/20$ (λ being the wavelength of the highest expected frequency). No loops of grounding conductors are allowed.
- The conductors dedicated to ground return should pass as close as possible to a grounding plane, or even better, to the junction of two perpendicular metal walls (in which case better protection is expected against radiation).
- Sharp or 90° angles are forbidden in the ground traces of a printed circuit board (or in the connecting conductors), in order to avoid excessive radiation.

- Whenever a rectangular metallic strap is employed for bonding, the most advisable aspect ratio is 3 : 1; under special conditions, an aspect ratio of 5 : 1 may still be acceptable (it should be recalled that *the aspect ratio of a rectangular strap is defined as the ratio of the length to the width of its cross-section*).
- Due to their high clock frequency and fast rise times, digital circuits must be treated as RF circuits (including from a grounding standpoint).
- At frequencies below 1 MHz, a single-point ground suffices for connecting shields and metallic screens. In contrast, above 1 MHz, it is necessary to apply a multi-point grounding connection. Recall [157] that *at high frequencies there is no such thing as a single-point ground.*

9.4 Proper Use of Cables

■ **Discussion.** Cable optimization plays an important role in interference control, since cables are the longest components of any electronic equipment. When improperly employed, they can pick up or radiate interfering signals, degrading system noise immunity, noise emission, and crosstalk. Consequently, selecting the right type of cable for a given application and grounding it correctly is of paramount importance.

■ **Types of Cables.** The most commonly encountered types of cable are presented in Fig. 9.25. The first category includes cables for high-power, low-frequency applications (such as AC power supply or DC power supply, chassis ground connections, etc.). They are simply adjacent, non-twisted, non-screened, (hence, non-protected) wires in the cable bundle, of sufficient cross section to carry high currents (Fig. 9.25a).

Also belonging to this category are cables for low-power, low-frequency signals, which are similar except for the wire cross section, which is considerably smaller. They are more flexible than the former and are used to transmit audio signals, DC signals, low-frequency interfaces, ground connections, etc.

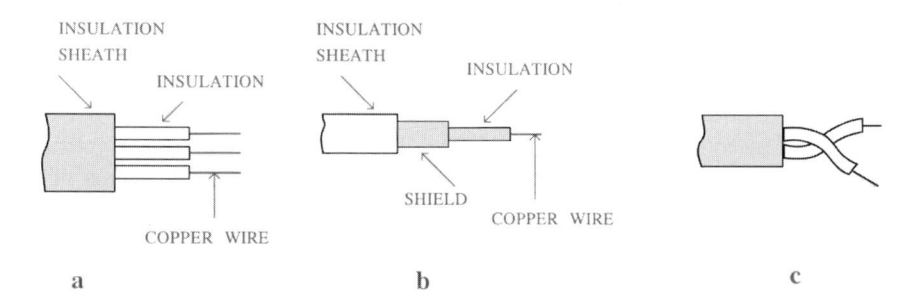

Fig. 9.25. Various types of cables: **a** AC supply cable; **b** coaxial cable; **c** shielded twisted pair

The second category includes coaxial cables (Fig. 9.25b), usually with a single braided shield (but coaxial cables with two braided shields are also commercially available). Depending on their geometry and size, their characteristic impedance is either $50\,\Omega$ or $75\,\Omega$. They are dedicated to RF and microwave applications (connecting the antenna to a TV set, microwave communication systems, etc.).

The last category includes shielded twisted pairs (Fig. 9.25c) intended for digital applications (data transmission). They provide good protection, but are more expensive and less flexible.

■ Remarks

- According to [157], to minimize crosstalk within a cable, it is recommended that the difference between the signal levels carried by different pairs not exceed 10 dB (voltage or current).
- Since a shielded twisted pair has more parasitic capacitance than a coaxial cable, the former can be useful up to 1 MHz, and the latter up to 100 MHz.

■ Types of Shielded Cable.
There are several types of commercially available screened cables [161]:

- Single-braid shields consist of wires woven into a braid, providing about 80% electromagnetic coverage, combined with reduced weight and high flexibility.
- Double-braided shields, with improved coverage, but the cables are stiffer, heavier, and more expensive.
- Laminated tape or foil shields, with drain wire. They provide full coverage, but the metallic foil or tape adds to cable weight and makes it stiffer. Shielding against electric fields is total, but protection against magnetic fields is less satisfactory.
- Composite tape and braid shields, which offer full protection and high-frequency performance, with moderate flexibility.

□ Grounding the Cable Shield.
If the shielding of the cable is improperly grounded, the advantages of using a shielded cable are lost. Ideally, the cable shield must be extended up to the chassis or ground plane and the connection should leave no gaps (360° around), so as to avoid coupling of ground currents to sensitive circuits through a high common impedance.

Often, because it is easier and faster, the outer conductor of a coaxial cable is connected to a shielding enclosure at a single point by means of a "pigtail connection" (Fig. 9.26a). This should be carefully avoided, since it impairs RF performance. The reason is twofold: the pigtail acts as a small antenna (radiating or receiving interfering signals), and secondly the parasitic inductance of the wire connection causes common-mode coupling and resonance with stray capacitances. The solution is to use an annular shielding (iris screen) connection or a BNC connector.

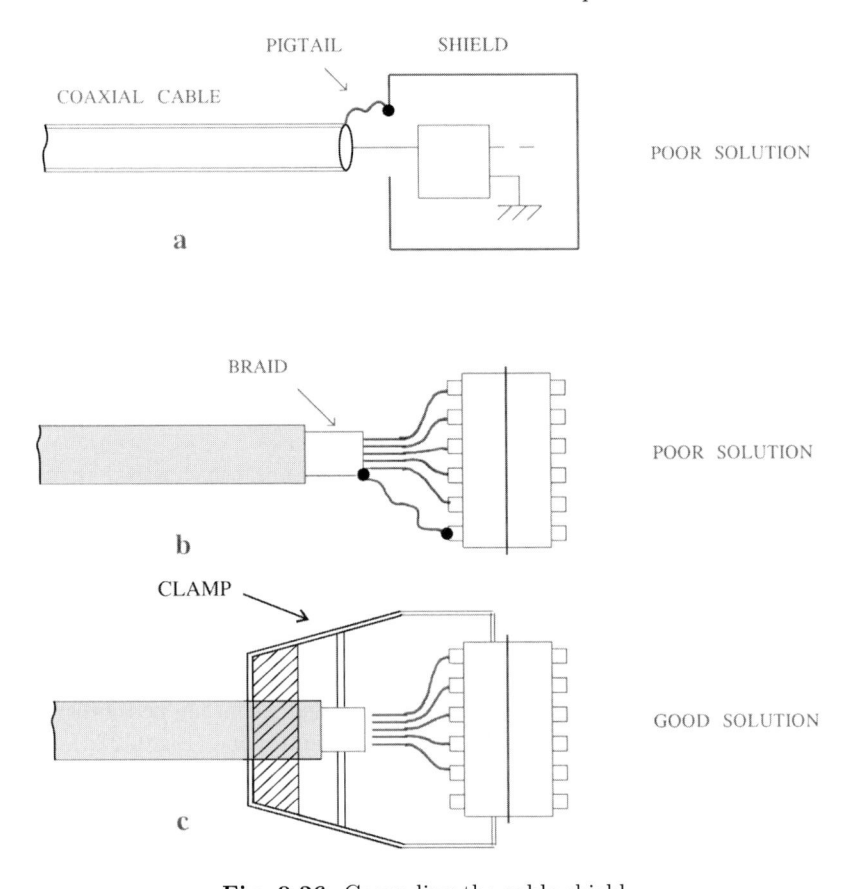

Fig. 9.26. Grounding the cable shield

This problem is also encountered when the cable is terminated with a connector; in Fig. 9.26b, the plane ground is accessed through a pigtail in series with the connector pin. Since the total series inductance is relatively high, this method should also be avoided.

The recommended solution is depicted in Fig. 9.26c, where the connector has a conductive clamp over the braided shield, this clamp being internally connected to the metallic enclosure of the connector, which by insertion is tied to the chassis ground.

■ **Ribbon Cables.** A ribbon cable is *a flat cable made up of several conductors laid parallel to one another*. Such cables represent an economical and widely encountered solution for parallel transmission of data inside metal boxes (such as over a computer bus).

The high-frequency performance of ribbon cables depends heavily on the positioning of ground conductors; ideally, each signal wire should have an adjacent ground return (Fig. 9.27b). However, this drastically reduces the num-

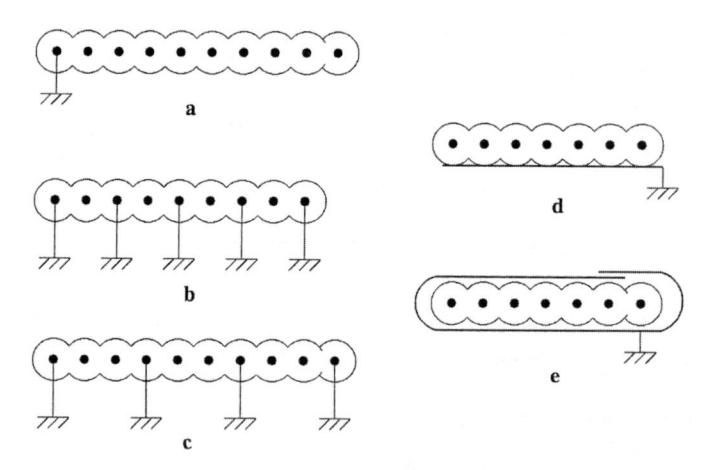

Fig. 9.27. Grounding positioning in ribbon cables

ber of available signal lines on a given ribbon cable. The opposite situation is shown in Fig. 9.27a, where (n−1) signal lines share the same ground return. This leads to high utilization, but crosstalk and common impedance coupling may become unacceptable. A compromise is attained with the configuration in Fig. 9.27c, where every third wire is grounded. Another possibility is to use ribbon cable bonded to a flexible metal ground plane (Fig. 9.27d), which offers very small loop areas and at the same time yields nearly constant line impedance. When radiation of interfering signals is critical, the best solution is the fully enclosed shield (Fig. 9.27e).

■ **Conclusion.** Selecting a particular type of cable for a given application represents an important way to control emission or to reduce the pickup of interfering signals. Equally important is the grounding of the cable shield, as well as the positioning of cables inside a metal enclosure.

Summary

- Many existing methods for preventing interference are employed simultaneously. For instance, shield effectiveness can be considerably impaired unless the cables and wires entering the metallic box are filtered. Moreover, the shield must be properly connected to ground to maintain shielding performance.
- Reflection losses are important for plane waves and electric fields.
- Reflection losses are negligible for low-frequency magnetic fields.
- Protection against low-frequency magnetic fields requires shields of high-permeability metals (iron, nickel-iron alloys, or mumetal).

- Protection against electric fields, plane waves, and high-frequency magnetic fields requires shields of high-conductivity metals (copper, gold-plated copper, etc.).
- An electronic system has at least three different grounds: 1) a "quiet" ground for analog circuits; 2) a "noisy" ground for digital circuits, electric motors, relays, switching-mode power supplies, etc.; 3) a safety ground for chassis, panels, racks, cabinets, etc.
- When two circuits (each with its own local ground) are connected together, either a ground loop appears (and is prone to common-mode noise current induction) or an offset voltage appears, which acts like a noise voltage (when the two grounds are not tied together). In both situations, noise is added to the useful signal.
- At a given frequency, the key factors affecting the performance of a cable shield are:
  - the shield type (foil, braid, tube)
  - its termination method (clamp, solder, pigtail, shielded connector)
  - the installation geometry (cable length, orientation).
- Avoid pigtails when terminating cable shields.

10
Methods of Reducing Emission of Interfering Signals

"I have done the deed. Didst thou not hear a noise?"

(W. Shakespeare, Macbeth)

■ **Objective.** This chapter is dedicated to the analysis of some of the most important sources of perturbation arising from operation of electrical and electronic equipment.

Any system prone to unintentional emission of electromagnetic energy is viewed as a *perturbing system*. Methods of limiting emission by perturbing systems will be discussed.

■ **Classification.** Based on its origin, the noise unintentionally emitted by an electronic system can be divided into the following categories:

- Disturbances associated with mains distribution (220 V / 50 Hz or 120 V / 60 Hz)
- Disturbances arising from DC power supplies
- Noise generated by mechanical contact switching
- Noise emitted by digital circuits
- Transformer noise
- Pulse noise due to electrostatic discharge.

Each category will be separately treated.

■ **Rules.** In order to reduce the emission of interfering signals during electronic circuit operation, the following basic rules must be considered:

□ **Rule 1.** *Minimize the frequency spectrum of the system of interest by controlling rise- and fall-times, as well as the pulse repetition rate.*

This rule applies mainly to logic circuits or digital communication systems. Improving performance by increasing clock speed has as its direct consequence a significant increase in frequency spectrum. Nevertheless, if a particular application can operate at a lower clock rate (without affecting overall performance), the level of emitted noise will decrease accordingly. Remember that there is a close relationship between the clock rate and the risk of interference.

☐ **Rule 2.** *In any electronic system, power conversion and control circuits are the principal sources of interfering signals.*

For instance, DC/DC converters and switching-mode power supplies are recognized as powerful sources of disturbances. Consequently, whenever possible, selecting a traditional DC power supply instead of a switching-mode supply will considerably reduce the associated noise. Also, whenever possible, it is advisable to replace the thyristors and triacs (whose exceptional dv/dt or di/dt performance exacerbates the noise problem) with power MOS devices.

10.1 Disturbances Associated with Mains Distribution

■ **Explanation.** Except for on-board systems, the public power network is the primary energy source to all electronic equipment. Since this network involves a large array of wires connecting various outlets to the primary system, it is important not to transform it into a noise transport vehicle. The generic configuration for supplying electronic equipment is shown in Fig. 10.1.

The conductor labeled EARTH (green or yellow-green wire) provides a current path for activating a circuit breaker, in order to limit damage due to a ground fault. Under normal operating conditions, this conductor carries no current.

The conductor labeled NEUTRAL is connected to supply ground and serves as a return path for current flowing through PHASE.

All conductors distributing mains power act as receiving antennas, and are sensitive to all sorts of surges (lightning discharges, etc.) and electromagnetic waves emitted by broadcast services (especially LW, MW, and SW stations, since at higher frequencies the supply lines act as a low-pass filter). At the same time, these conductors are contaminated by low-frequency disturbing signals (f < 1 MHz) resulting from transients, and switching-on and -off of electrical equipment, which are eventually supplied by the same line.

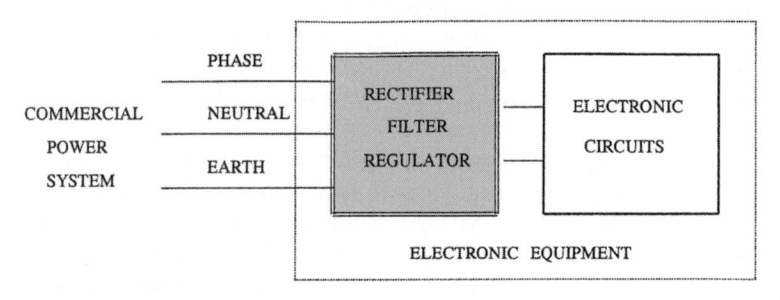

Fig. 10.1. Power source for electronic equipment

10.1 Disturbances Associated with Mains Distribution

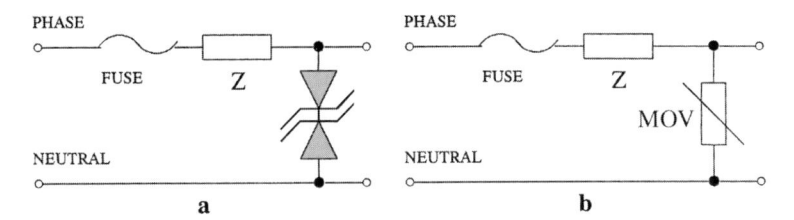

Fig. 10.2. Supressor circuits: **a** with bipolar avalanche diode, **b** with metal oxide varistor

■ **Transient and Surge Suppressors.** Voltage transients and current surges (whose rise times can be as short as a few nanoseconds) can be eliminated by employing a suppression device, such as a gas-discharge tube or a semiconductor suppressor, connected between the line and the ground [162]. The former can handle large transient currents (more than 10 kA); the latter is a threshold device (such as two back-to-back avalanche diodes or an MOV, a Metal Oxide Varistor).

The firing voltage of the gas tube (or the threshold voltage of the semiconductor device) is selected so to be a little higher than the peak voltage between the lines in normal operation. Practical circuits are sketched in Fig. 10.2.

In the absence of any transient, the protecting device presents an open circuit, since its internal impedance is much greater than the line impedance Z.

When a voltage peak appears on the line, the reverse-biased avalanche diode (or the MOV) begins to conduct. Since its internal dynamic resistance is much lower than the line impedance Z, the voltage divider relation makes it obvious that a drastic attenuation of any surge above the line voltage is obtained.

Note that for short transients the suppressor device must be high current rated for; for longer transients, fuse protection is needed.

Semiconductor devices exhibit progressive deterioration with repeated surges [162], and need to be replaced periodically. For better efficiency, they must be placed as close as possible to the protected equipment.

■ **Power Line Filters.** These are used to protect equipment against radio-frequency interference or to prevent line contamination by an offending circuit. Interference in the mains line has two components: common-mode and differential-mode coupled currents. Consequently, a typical power line filter must be effective in suppressing both.

Power line filters are installed at the power input of the equipment to be protected. Generally, they are low-pass configurations; a generic filter is illustrated in Fig. 10.3a. This filter is a combination of devices to block both differential- and common-mode noise. Common-mode interference is present in the two pairs (PHASE-EARTH) and (NEUTRAL-EARTH), while differential-mode interference is established on the pair (PHASE-NEUTRAL).

The common-mode inductances L are achieved by bifilar winding on the same ferrite toroidal core, so that the magnetic fluxes of the two sections

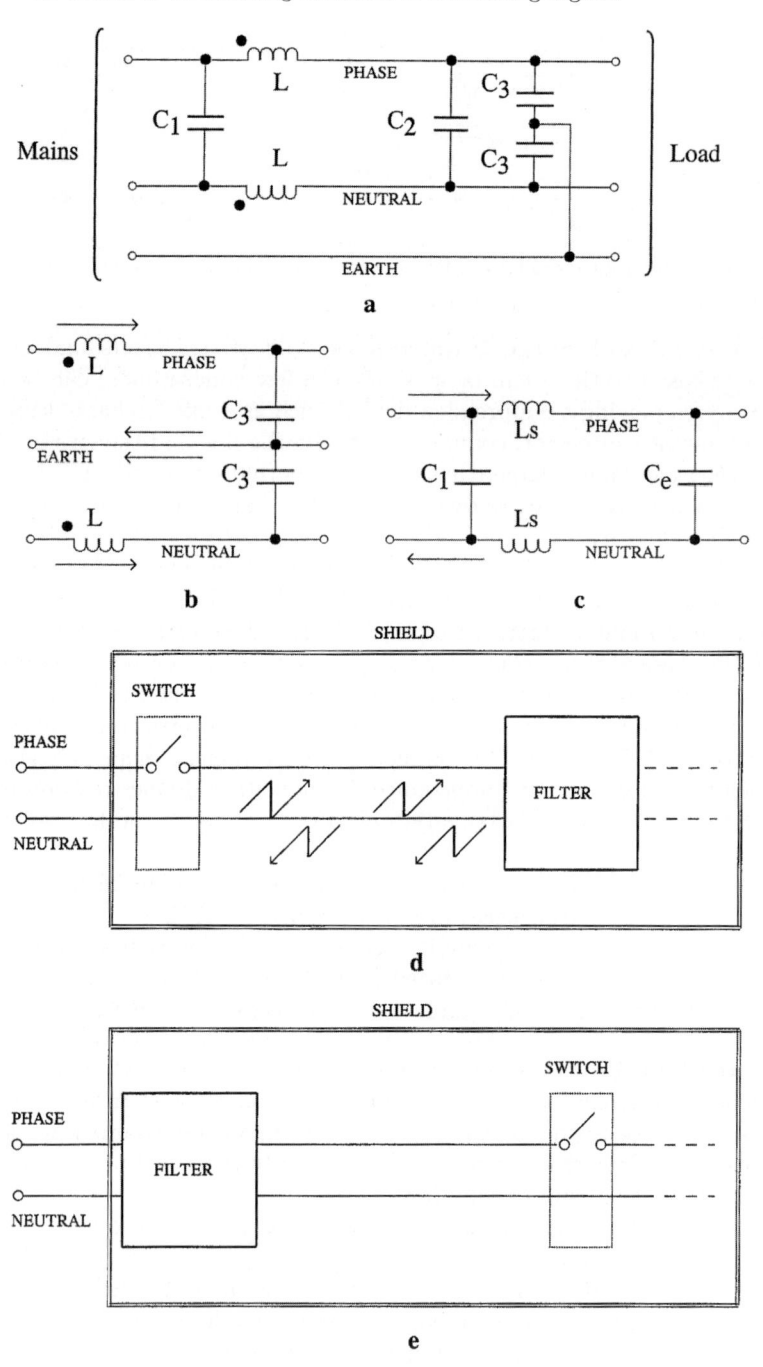

318 10 Methods of Reducing Emission of Interfering Signals

Fig. 10.3. a Power line filter; b common-mode equivalent cicrcuit; c differential-mode equivalent circuit; d unrecommended layout (noisy neutral); e good layout solution

cancel each other. In this way no saturation occurs in the ferrite core at the nominal operating current. Capacitors C_1, C_2 are X-caps, while C_3, C_4 are Y-caps (see Sect. 9.2.1).

The equivalent circuit for common-mode interference is shown in Fig. 10.3b. It should be remembered that the value of capacitors C_3 is limited by the leakage current to EARTH, which for safety reasons must be kept lower than a fraction of a milliamp. Consequently, to avoid filtering degradation in differential mode, a high value of C_2 is required.

The differential-mode equivalent filter is presented in Fig. 10.3c, where Ls denotes the stray inductances, which are effective in filtering the differential mode. This suggests that Ls should be not negligible, but their value depends chiefly on the method of constructing the choke. Note that $C_e = C_2 + 0.5 C_3$.

Most commercial power-line filters are shielded to prevent pick up of spurious magnetic flux.

☐ **Filter Installation.** Note that *a power line filter is effective only when the voltage is ON*. One of the most frequent mistakes is to place the filter after the mains ON/OFF switch (Fig. 10.3d), eventually on the rear panel, the switch being on the front panel. This requires long connecting wires from one panel to the other, which in the OFF state act like antennas; they radiate inside the shielded equipment the noise which has been induced in the external mains line. To avoid this, the filter has to be installed *before* the mains switch (Fig. 10.3e), directly on the front panel of the cabinet protecting the equipment. Mount the filter so that its input leads are as short as possible and separate them from the output leads (or better, insert a screen between them).

☐ **Transformer Shielding.** Many systems employ transformers to provide appropriate voltage levels for their power supplies. The problem is that a transformer can easily transmit transients and disturbances arriving on the mains line to the circuits connected to its secondary, due to its inherent capacitive coupling between the primary and secondary windings.

The solution consists in interposing a shield (a thin metallic foil connected to ground) between the two coils. It is better to use two metallic foils, the first (connected to AC EARTH) enclosing the primary coil and the second (connected to the signal ground of the system) covering the secondary coil (Fig. 10.4).

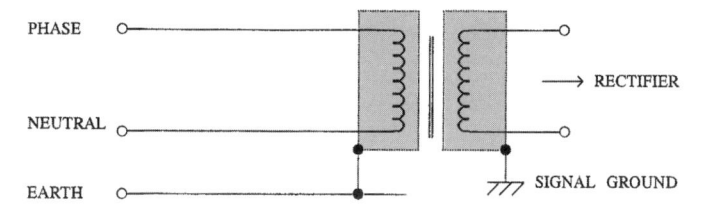

Fig. 10.4. Shielding the coils of a power transformer

10.2 Noise Arising from DC Power Supplies

■ **Definition.** The term *DC power supply* refers to a source of DC power that is operated from an AC power source (such as a 220 V/ 50 Hz line). Generally, *an ideal DC power supply is a voltage/current source of zero/infinite internal resistance.* In the following, we shall refer only to voltage sources.

In practice, the connecting wires, supply lines, switches and fuses are always responsible for inserting small series resistances; even the power supply itself is not a perfect circuit with zero internal resistance. Consequently, any DC power supply actually has some low internal resistance.

■ **Configuration.** A generic DC power supply consists of one or more of the following fundamental components [163]:

- A rectifier, to convert AC voltage into pulsating DC voltage.
- A low-pass filter, which extracts the average value of the waveform applied at its input and suppresses pulsations.
- A voltage regulator, to maintain almost constant output voltage, despite variations in input voltage or load current.

■ **Conditions.** Any DC voltage source must satisfy the following conditions:

- It must be able to maintain a constant voltage at its load terminals despite variations in the load current.
- The AC produced by some loads (such as oscillators) must not interfere with the DC of the supply.

Some of the commonest configurations will be presented next.

10.2.1 DC Power Supplies with Full-Wave Rectifiers

■ **Schematic.** A traditional topology is presented in Fig. 10.5, where a simple capacitive filter is employed.

□ **Operation.** In the absence of a filter, during each half-period of the sinusoidal input signal one diode is forward-biased, while the other is reverse-biased; consequently, the voltage on the load is a continuous series of positive half-cycles. If a capacitive filter is inserted, the transformer supplies a current only when the voltage on each secondary section exceeds the sum of the voltage across C and the 0.7 V-drop on the forward-biased diode. For practical values of components, the current provided by the transformer flows only during somewhat less than 20% of the period [164]. In the remaining interval, the current through the load is maintained by discharging the capacitor C. This explains the region of the waveform where the voltage drops below the average value V_o.

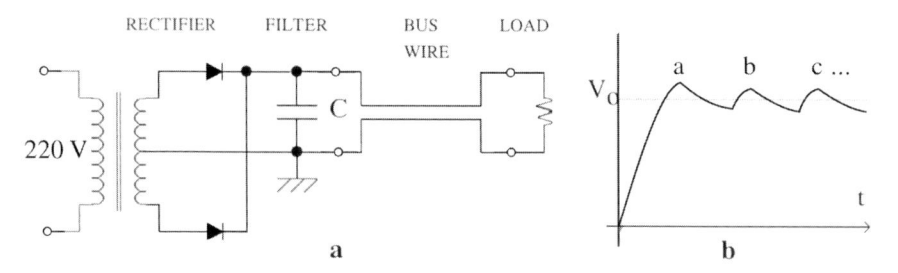

Fig. 10.5. DC supply with a full-wave rectifier: **a** circuit with a center-tapped transformer and two diodes; **b** voltage waveform

☐ **Noise Sources.** Several potential noise sources can be identified in the circuit shown in Fig. 10.5a:

1) When the power is turned on for the first time, the initial charging current of C may induce a potentially damaging peak (up to 10 A, depending on the capacitor value). This surge current produces excessive voltage drops on the BUS WIRE, as well as on all connections between rectifier and filter. These transients are perceived as noise. In order to limit them, several solutions can be considered:
   - The ground and load must be connected to the capacitor terminal instead of the center tap of the secondary.
   - The BUS WIRE must have large traces on the PCB (or wires with a large diameter).
   - When the rectifier is powered directly from the mains (without a transformer), two chokes are inserted in the AC supply lines to reduce the initial current peak (but the voltage amplitude across C will be reduced also!).
2) The current pulses associated with filter operation generate magnetic fluxes that are proportional to the loop area. To limit damage to nearby circuits, this area must be kept as small as possible.
3) The leakage inductance of the transformer, if excessive, can lead to parasitic magnetic fluxes that affect nearby circuits (they may perturb the voltage regulator). To reduce them, the two secondary windings must be made as symmetric as possible and almost identically coupled to the primary.

■ **Full-Wave Bridge Rectifier.** The generic topology of a full-wave bridge rectifier is presented in Fig. 10.6. Note that the transformer is simpler (and cheaper), since no center tap is required.

However, another potential problem appears here [164]: the internal parasitic capacities of the transformer coils are no longer connected to the load through the transformer center tap, but through the diodes. When the diodes

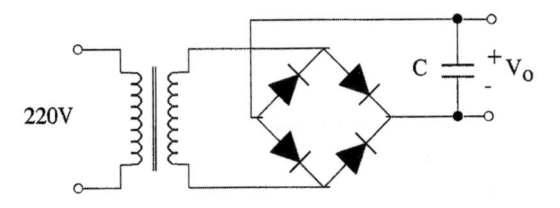

322 10 Methods of Reducing Emission of Interfering Signals

Fig. 10.6. Bridge rectifier

are in conduction, the connection is made through the small internal resistance of the forward-biased diodes and there is no problem. During a large fraction of the period, the diodes are reverse-biased, due to the voltage V_o across the filter capacitor. As a consequence, the connection is now made through the small junction capacitances of the reverse-biased diodes. In some applications involving sensitive loads this may be particularly troublesome.

To conclude, it is important to note that *when no transformer is used, the ground of the supply line must be isolated from the ground of the load.*

10.2.2 Voltage Regulators

■ **Configuration.** Generally, the voltage regulator is inserted between the filter and the load. Figure 10.7 shows the generic configuration of a series voltage regulator.

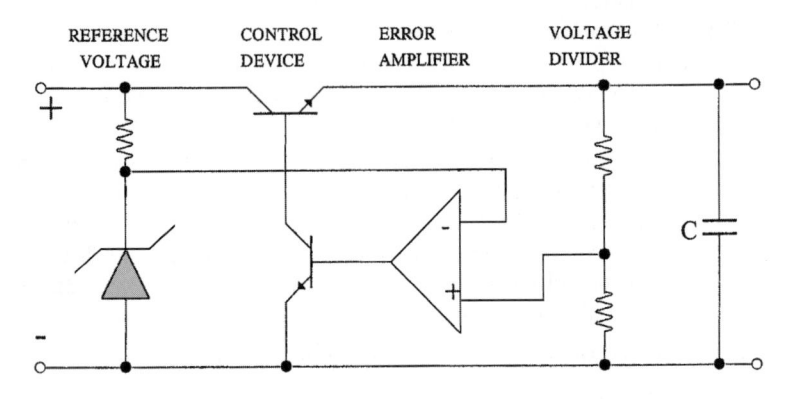

Fig. 10.7. Block diagram of a series voltage regulator

■ **Operation.** The error-amplifier compares the reference voltage (delivered by the Zener diode) with a fraction of the output voltage (sensed with the voltage divider) and amplifies the difference (also called "error difference"). At the output of the amplifier a control signal is generated, which modifies the bias of the power transistor (employed as a control device), in such a way that the voltage drop on it cancels the variation of output voltage. In this way, the output voltage is kept constant.

For instance, any tendency in the output voltage to increase is compensated by a larger voltage drop on the control device, so that the output voltage is ultimately kept constant. To be effective, a good regulator must satisfy two conditions:

- Good stability (no oscillations, overshoots, etc.).
- Fast response to any variation of the output voltage (within microseconds). To improve the response time, an output capacitor C is introduced, able to supply extra current to the load until the regulator circuit adjusts its response.

A high-performance voltage regulator has a negligible output internal resistance (at DC, as well as at high frequencies).

■ **Noise Sources.** An improperly designed (or operated) voltage regulator exhibits noise. Several reasons are listed below:

- The reference voltage is noisy. If this is the case, fluctuations in the reference voltage are transmitted after amplification to the output, where they act as noise. Increasing the value of C is not a solution.
- Noise (ripple) from the supply (or from outside) can contaminate the error amplifier or the reference voltage. This creates a big problem: the resulting output fluctuations cannot be filtered.
- When the regulator supplies a dynamic load (such as a logic circuit) it may happen that the regulator exhibits overshoot (i.e., a transient rise beyond regulated output limits, occurring when the AC power input is turned on or off, or for step changes in the load [166]). In the case of a logic circuit, the overshoot is synchronized with a logic command (or with the clock) and can contaminate nearby circuits, where it is perceived as noise.
- The load excursions exceed the regulator capabilities. For example, some regulators cannot accept an infinite load resistance; other regulators cannot operate with active loads (i.e., loads which are able not only to absorb current, but also to deliver a current during some intervals of time). In all these situations, noise appears at the regulator output.

■ **Conclusion.** Practical criteria to decide when a particular regulator is correctly operated are:

- The noise detected at the regulator output must not exceed the noise of the reference voltage (or the input noise).
- Noise associated with 50 Hz or its harmonics (like hum, ripple, etc.) must not be perceived at the output.

10.2.3 Ungrounded (Floating) Power Supplies

■ **Definition.** *A floating power supply is one that has no terminal at ground potential.*

■ **Ungrounded AC Power Supplies.** There are practical situations in which an ungrounded AC power supply is required. For instance, a shipboard power installation must have no ground, in order to avoid electrolysis.

In other applications, a floating AC supply offers the benefit of grounding the pipe containing the conductors carrying the AC current, to avoid shock hazards.

Generally, a floating power supply is more fault-tolerant, and this may be an attractive feature for some industrial applications, where AC current must be permanently supplied to ovens, furnaces, conveyor belts, etc.

However, ungrounded power supplies do exhibit much more noise than their grounded counterparts (especially common-mode coupled signals).

■ **Ungrounded DC Power Supplies.** These are required in some practical applications, such as providing bias to sensors with no allowed ground connection. A generic schematic diagram is shown in Fig. 10.8.

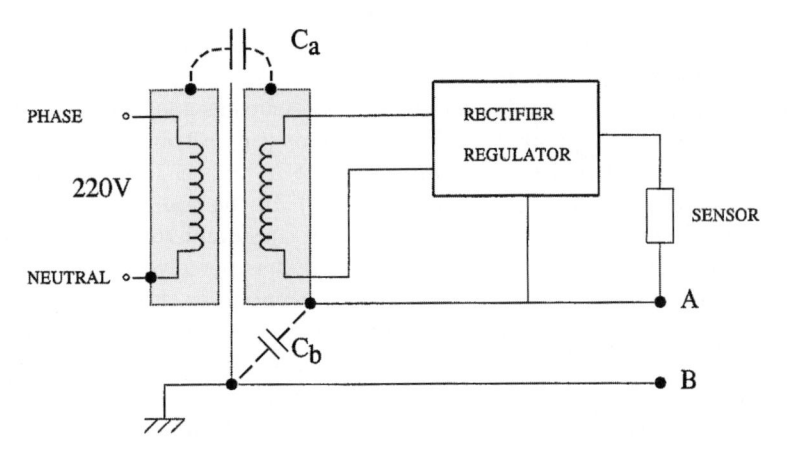

Fig. 10.8. Ungrounded DC power supply

To reduce capacitive coupling between the primary and secondary windings of the transformer, three shields are needed. Nevertheless, terminal A of the sensor can still weakly couple to ground via the stray capacitance C_b, as well as through C_a.

To check shielding efficiency, a resistor can be introduced between points A and B, and the leakage current through it can be measured (according to [164], this current must be less than 10 nA).

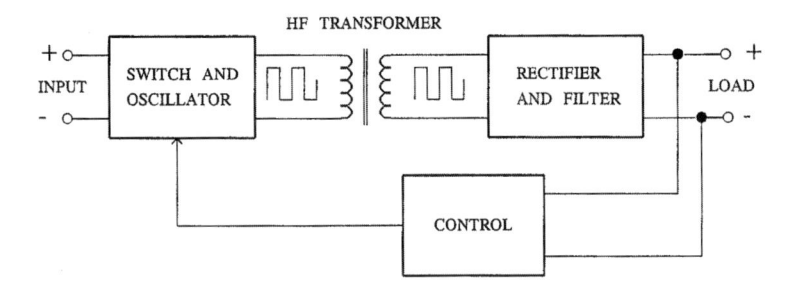

Fig. 10.9. Simplified block diagram of a switching supply

10.2.4 Switching-Mode Power Supplies

■ **Definition.** *A power supply in which the incoming mains voltage is passed through a switch that is turned on and off (at a relatively high frequency) to give rise to a signal that is easily transformed to a lower voltage, rectified, and smoothed.*

■ **Configuration.** There are many switching-supply and switching-regulator configurations. A simplified block diagram is presented in Fig. 10.9.

■ **Operation.** The oscillator controls a MOS switch which in the on state enables the DC current supplied to the input to flow through the primary coil. The CONTROL block contains a voltage reference, a voltage divider connected across the output, and an error amplifier (as in Sect. 10.2.2). The error voltage delivered by the amplifier controls the frequency (or duty cycle) of the signal generated by the oscillator in such a way that the voltage across the load is maintained constant.

The DC voltage applied to the INPUT is either supplied from a battery, or taken directly from the AC mains (after rectification and filtering). Since the oscillator frequency is at least several tens of kHz, the advantage is that the transformer can be made much smaller (less copper, less core) and consequently it is less expensive (recall that transformer cost represents a significant fraction of the complete power supply cost). Another advantage is that the LOAD voltage can be made higher than the INPUT voltage.

■ **Noise Sources.** There are several categories of disturbances associated with the switching process [166]:

– Ripple in the output voltage at the switching frequency (typically, in the range $10\,\text{mV} - 100\,\text{mV}$).
– Ripple (of the switching frequency) contaminating the input supply.
– Radiated noise at the switching frequency and its harmonics. This is merely radiated inductive noise, appearing when currents are rapidly switched through the parasitic inductances of conductors or traces on the PCB.

However, there are other contamination mechanisms specific to any particular realization (such as switching currents through the mica used to isolate switching power transistors from the chassis, or inductive noise associated with the output LC filter).

Possible solutions to reduce the noise are shielding of the switching supply (to control radiated interference) and filtering its supply line (to prevent contamination of nearby equipment powered from the same mains line). As general advice, whenever sensitive circuits are involved, it is better to use linear regulators instead.

10.2.5 Ripple Filtering

■ **Comment.** In any power supply, ripple is unavoidable. Ripple becomes troublesome when it couples to the input of low-signal amplifiers, where it is amplified just like any useful signal. Selecting high-value filter capacitors may ameliorate the problem, but it also worsens the noise associated with the initial surge of charging current. A good compromise is not easy to obtain, and often the solution consists in providing additional filtering to significantly lower the ripple amplitude.

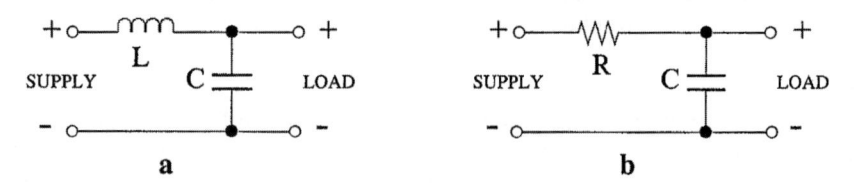

Fig. 10.10. Passive low-pass filters: **a** LC-type; **b** RC-type

■ **Passive Filters.** Generally, the passive filters employed to reduce ripple are of low-pass type. The commonest configurations are presented in Fig. 10.10. Note that these filters must be inserted on the DC bus just ahead of the circuit to be supplied (and protected). Here is a non-exhaustive list of potential problems arising when using them:

– Any configuration presents an equivalent series impedance, which adds to the internal impedance of the supply, impairing performance.
– The cutoff frequency of the filter must be lower than the ripple frequency (usually 100 Hz). This requires high-value capacitors that at the same time have low parasitic series resistance. An alternative solution is to increase the value of L or R, but neither is practical. High-value inductors are expensive, cumbersome, and prone to pick up spurious magnetic fields; high-value resistors introduce unacceptable DC voltage drops and must be rated for high-power.
– An LC filter resonates at $f_o = 1/2\pi\sqrt{LC}$, and at this frequency the filter can exhibit insertion gain instead of insertion loss.

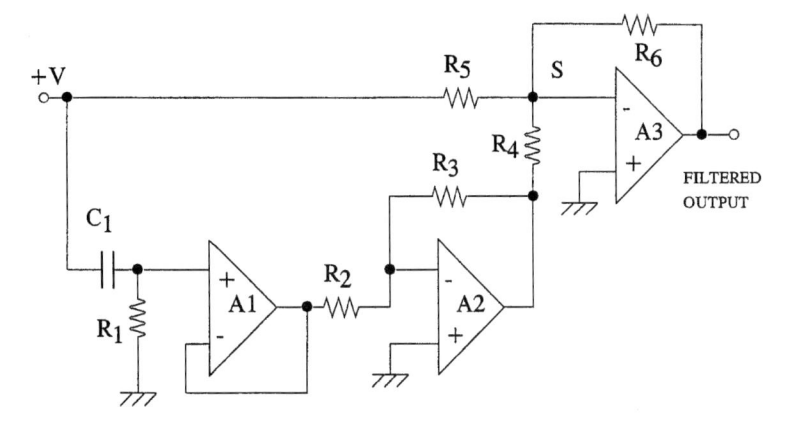

Fig. 10.11. Ripple active filter

■ **Active Filter.** Active filters offer reasonable size and weight when low frequencies are to be filtered out. A noise clipper circuit belonging to this category is proposed in Fig. 10.11 [167].

Its operation is as follows: the ripple superimposed on the voltage V is high-pass filtered (R_1 and C_1) and then applied to the input of the voltage-follower A1. After being amplified and phase-shifted by 180° by A2, the signal is applied to the summing point S, together with the original signal. If the gain of A2 exactly compensates the loss of the filter (R_1, C_1), then the incoming signals to point S must have equal amplitudes, but they are 180° out of phase. Consequently, at the output of A3 the ripple cancels. Problems can appear at relatively high-frequencies, where the behavior of amplifiers A1, A2, A3 is not identical.

10.3 Noise Generated by Mechanical Contact Switching

■ **Fundamentals.** Any closing or opening of an energized mechanical contact can initiate electrical breakdown. This breakdown has significant spectral content and can cause two kinds of problems:

- It represents a powerful source of interference. Noise can either be radiated by the arc established during contact operation, or conducted along the wiring paths carrying current to the mechanical switch.
- The discharge can damage the mechanical contact, by damaging its surface and hence shortening its lifetime (more accurately, by reducing the total number of operations, which, for a relay contact, may be rated at $10^4 - 10^5$).

Fortunately, cures that limit physical damage to contacts are also effective in reducing emitted noise.

When a contact that carries current opens, breakdown begins while the contacts are very close, but not touching each other, and will continue until the distance increases to the point that the available current is no longer able to sustain the voltage across the gap.

In the case of a closing contact subjected to a voltage in the off state, breakdown occurs when the contacts are at a critical distance, and continues until the contacts are closed.

In both cases, *contact bounce* is particularly damaging, because contact is not established at first, but only after a sequence of make-and-break rebounds. During this lapse of time, the distance separating the surfaces is critical, stimulating breakdown.

■ **Classification.** There are two different types of breakdown: 1) gas discharge and 2) arc discharge. Since the physical phenomena are distinct in the two cases, they will be treated separately.

10.3.1 Gas Discharge

■ **Explanation.** Gas discharges are also called "glow discharges" or "Townsend discharges." They appear in gas-filled tubes subjected to a sufficiently high electric field, and are due to emission of light from excited gas atoms. The firing voltage (which initiates the breakdown) depends on the gas pressure in the tube, the gas, its temperature, and above all the distance separating the electrodes. Primary ions in the gas (produced by thermal effects, natural radioactivity, light, etc.) are accelerated by the electric field and reach high kinetic energies; consequently, secondary ions are generated by collisions of accelerated ions with gas molecules. An avalanche breakdown process occurs and sustains itself even if the applied voltage is slightly reduced.

■ **Cure.** To prevent gas discharges between mechanical contacts, the voltage across the contacts should be limited to less than 300 V [168]. Since most practical applications in electronics involve much lower voltages, this kind of breakdown is not very likely to occur.

10.3.2 Arc Discharge

■ **Definition.** As stated in [165], this is *an electric discharge characterized by high cathode current densities and a low voltage drop at the cathode.*

■ **Condition.** This kind of breakdown results from field-stimulated emission of electrons from a cathode. This requires a gradient of 0.5 MV/cm, which might appear to be a safe limit never exceeded in practice. However, this is not true: consider, for instance, the situation in which the spacing between contacts is very small (a few tens of micrometers). Even at low voltages, the above limit can obviously be reached, and arcing occurs.

10.3 Noise Generated by Mechanical Contact Switching

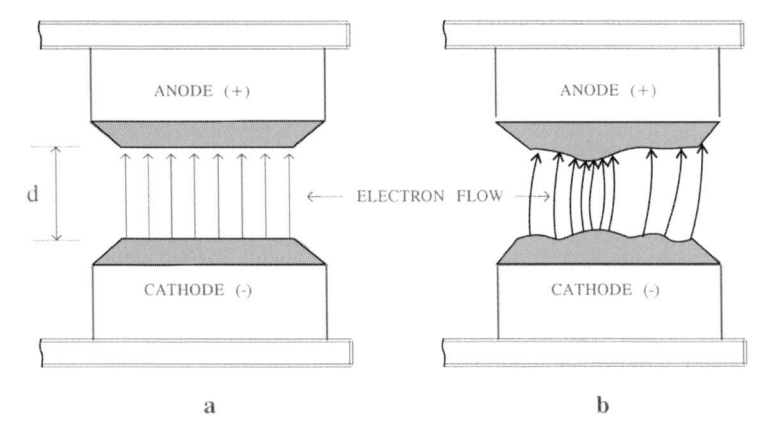

Fig. 10.12. a Ideal contact surfaces separated by distance d; b real contacts imply nonuniform electron flow, due to surface irregularities

■ **Explanation.** Generally, electrical breakdown of the gas in a gap between electrodes requires an avalanche process. As a general rule, any avalanche produces the material for at least another avalanche to occur, until a self-sustained discharge is established. The typical mechanism leading to the first avalanche, at atmospheric pressure, is the following: free electrons in any metal, near the surface, can occasionally be liberated by cosmic radiation or light into the surrounding space. If no electric field exists, the metal becomes positively charged and the liberated electrons are incited to return. However, an external electric field applied between the electrodes (Fig. 10.12a) can direct electrons from cathode to anode. Statistically, some will be accelerated enough to produce, in collisions with gas molecules, pairs of positive ions and free electrons. In turn, these secondary electrons can be accelerated and collide with other gas molecules causing new ionizations, and so forth.

Note that in the case of perfectly smooth, plane electrodes, the electric field is uniformly distributed, and hence for normal operating voltage ranges, the voltage gradient is rather low, so avalanches leading to arcing are less likely to occur. However, on a microscopic scale, no perfectly smooth surface exists (Fig. 10.12b).

The sharpest point on the cathode surface is prone to accumulate the most electric field lines bridging the air gap. In these micro-regions, free electrons are highly accelerated and arcing occurs. Local heating of the material is observed, perhaps to several thousand degrees. The metal melts locally and vaporizes, and a bridge of metal-vapor (or droplets of liquid metal) appears, which can transfer material from anode to cathode (if the arcs are of short duration) or vice versa (if they last long enough) [171]. The volume transferred per operation is very small, but numerous repetitions can augment existing irregularities. This damages the contact surfaces, which become more and more prone to arcing.

330   10 Methods of Reducing Emission of Interfering Signals

In addition to the previous explanation, there are many parameters that can influence arc intensity and duration (which typically lies between 1 and 10 µs), such as the velocity of contact closure or breaking, contact bounce, contact wear, external reactances, relative humidity, electrode cleanness, etc.

■ **Arc Suppression.** For some commonly materials, the minimum voltage (V_m) and the minimum current (I_m) required to sustain the shortest arc are given in Table 10.1, assuming operation in a clean, dry atmosphere (40% relative humidity). I_m depends on the relative humidity of the air and properties of electrodes; it is observed that it reaches a minimum at 40% relative humidity and increases at very low or great humidity [171].

Note that the contact hardness of the material is very important in electrode fabrication, since it determines the degree of surface degradation when a film of dust (containing hard particles) is deposited on the anode and cathode surfaces. Also, contact hardness decides what type of duty cycle the contact can sustain (low, medium or high). For instance, widia is specially recommended for high duty-cycle contacts.

It is interesting to note that these values are considerably reduced when the contact surfaces are rough or covered with films of dust, oxide, or impurities. For arcs between different metals, I_m is determined instead by the anode metal, while V_m is determined by the cathode metal.

There are two ways to prevent arcing:

1) Control the switch voltage to keep it below V_m.
2) Make arrangements such that the current is below I_m.

Table 10.1. Minimum arcing voltage and current

Material	V_m [V]	I_m [mA]	Contact hardness 10^8 N/m^2
Silver (Ag)	12	400	3 to 7
Gold (Au)	15	380	2 to 7
Palladium (Pd)	16	800	4 to 10
Platinum (Pt)	17.5	900	4 to 8
Nickel (Ni)	14	500	7 to 20
Copper (Cu)	13	430	4 to 7
Tungsten (W)	15	1100	12 to 40
Carbon (C)	20	30	5
Widia (WC)	14.5 – 16	650 – 1000	20
Pt + 10% Ir	20	740	6 to 20
Ag + 10% Pd	11	300	6
Bronze (Cu 8, Sn)	13.5	310	7

10.3 Noise Generated by Mechanical Contact Switching

■ **Contact Protection.** The expected lifetime of a switching contact is closely tied to its ability to rapidly inhibit an arc discharge whenever it occurs. As a general rule, contacts subject to AC are less damaged than those operating with DC, for several reasons:

- during every period, the AC voltage decreases below the threshold value that initiates an arc discharge;
- the polarity reversal favors equal erosion of anode and cathode;
- whenever the AC waveform crosses the axis, the arc is automatically extinguished.

Concerning the protection of contacts intended for DC circuits, the most critical situations involve either high inrush currents when the contacts close (e.g., due to large capacitive loads, motors, lamps whose hot filament resistance is up to eight times the cold resistance, etc.) or inductive loads that dump energy when the contacts open. Both must employ suppressors (quenching circuits), to reduce spark discharges and the associated radiated noise. Several possibilities are presented in Fig. 10.13; the basic idea is to connect a properly selected network N across the contact, that is able to absorb the energy of transients (Fig. 10.13a).

The simplest configuration is a high-value capacitor across the contacts (Fig. 10.13b); if it is initially discharged, when the contact open it will absorb significant current. Consequently, less current will be left for the switch itself, and arcing will be less likely. The problem here is that during the lapse of time when the switch is open, C is charged to voltage E through the load, but as soon as the switch is closed, C discharges violently through the contact. The peak current can be very high, since it is limited only by the ohmic resistance of the wiring; the spark is unavoidable.

In order to reduce the discharge current, the circuit of Fig. 10.13c is proposed, where the series resistance R must have a high value to avoid arcing when the contacts close, but sufficiently low to be effective when the

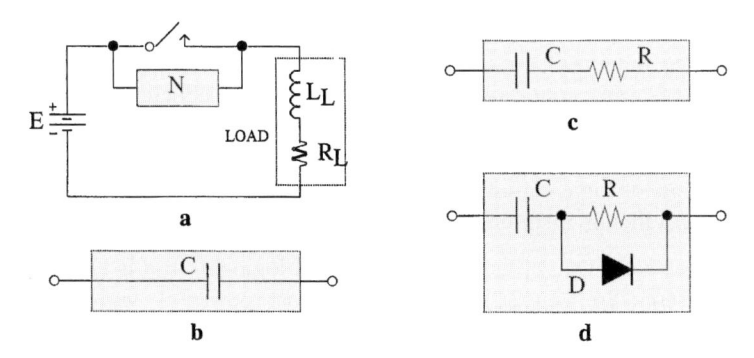

Fig. 10.13. a Contact protection with a reactive network N; b, c, d various configuration for network N

switch opens. This compromise is not always easy to attain. For this reason, the configuration shown in Fig. 10.13d is proposed instead, where a high-value R can be used. The diode acts as a short-circuit for R when the switch opens (so C has a short charging time constant), but when the switch closes the diode is reverse biased (by the voltage across C) and R is in series with the capacitor (the discharge time constant is long).

■ **Concluding Remarks.** Effective protection requires the following conditions [168]:

- Keep the contact voltage below 300 V to avoid glow discharge.
- The rate of voltage rise across the switch must be kept below 1 V/µs to prevent arc discharge. This condition is satisfied if the value of C is at least of 1 µF per ampere of load current.
- The diode in Fig. 10.13d should be properly rated (its breakdown voltage must be greater than E, and its peak current greater than the maximum load current).
- For the protection circuit shown in Fig. 10.13d, the current upon closure must be one-tenth of the arcing current. Consequently, R is chosen to be

$$R \geq \frac{10\,E}{I} \tag{10.1}$$

10.4 Noise Emitted by Digital Circuits

10.4.1 Introduction

■ **Comparison.** Noise originating in digital circuits is a real source of concern. Analog circuits are recognized to be much quieter for two reasons:

1) Signals generated and processed in logic circuits are close to rectangular, with fast transitions. At the usual high switching frequencies, the spectrum is much richer in harmonics than that of analog signals (with the notable exception of video signals in television circuits). It follows that parasitic elements associated with PCB traces or connecting cables (inductances and capacitances) have more harmful effects, due to the steep di/dt or dv/dt.
2) Except power amplifiers, the level of analog signals is considerably lower than that of logic signals; typical amplitudes of logic signals give rise to more significant interference.

For these reasons, the noise of high-speed logic circuits is significant and digital systems are generally among the most powerful aggressors in any electronic environment. Nevertheless, they can become their own first victim; for instance, in a microprocessor, the memory address lines and data lines can seriously perturb the operation of the processor itself if no countermeasures are taken.

■ **Remark.** It is well-known that logic circuit designers prefer to work in the time domain, where truth tables, transition maps, and waveform diagrams are their favorite tools. However, electromagnetic compatibility legal standards are all expressed in terms of frequency, and consequently the noise associated with digital circuits must be quantified in the frequency domain.

■ **Sources.** The most significant sources of interference in logic circuits are listed below, in decreasing order of importance:

- Processor clock(s), which deliver very high-frequency rectangular signals (actually reaching the GHz range), whose rise/fall times are often substantially faster than required by many applications. The clock is distributed to all integrated circuits located on the same PCB, and the tracks carrying it represent a powerful source of radiation and crosstalk;
- Fast transitions of signals produced by gates, flip-flops, registers, etc., especially when they belong to high-speed logic families;
- Reflections due to mismatch on the data lines;
- Abrupt changes in DC power supply currents when current demand changes suddenly.

It is important to note that all cited noise sources also need an environment favoring interference. For instance, a high-frequency clock is quite harmless provided that the traces carrying it are ideal (no associated inductance, no associated capacitance). Likewise, a parasitic capacitance introduces no delay when charged through a zero-resistance connection by a zero-resistance source. Another example is crosstalk, which vanishes if the mutual inductance and mutual capacitance of adjacent traces are both zero. We conclude that *whenever we are not able to significantly reduce noise emission by various sources, we must try as a last resort to improve the "environment," by reducing parasitics and coupling.*

■ **Can We Really Control Parasitics During Design?** Let us take a look at the typical design process for a logic system. Design work begins by considering the truth table and constraints, transition maps, and associated waveform diagrams. As soon as synthesis is complete, the system is simulated in order to optimize electrical performance. This is a time-consuming process, but it ultimately produces a final configuration and layout. In order to proceed with testing on the prototype, the PCB must then be fabricated. During the previous steps of synthesis and electrical simulation, it is generally assumed that the traces (or ground plane) have no parasitics, since it is impossible to calculate them *before* elaborating the layout and knowing the geometry of traces. At best, based on experience, the designer adopts during simulation provisional values for the capacitance of some traces where propagation delay is critical. However, these values must ultimately be reviewed, based on the actual layout configuration.

334   10 Methods of Reducing Emission of Interfering Signals

To the best of the author's knowledge, no commercial simulation package offers the ability to compute all parasitic values associated with the PCB traces and reapply them to the electrical simulator, to check the validity of the design. The only alternative is to test the fabricated prototype in the laboratory.

Thus, it is only at this last step that one discovers the harmful effects resulting from interconnections (e.g., crosstalk and malfunction due to self- and mutual capacitances and inductances, as well as the resistance of various traces).

To conclude, although highly desirable, it is not possible to control the parasitics of interconnections during design.

10.4.2 Inductive Noise

■ **Explanation.** The mechanism responsible for this kind of noise is the flow of rapidly switched current through the parasitic inductance of some trace.

It is well known that the voltage induced in an inductor L by a varying current I flowing through it is

$$V = -L \frac{dI}{dt} \quad (10.2)$$

Apart from the minus sign (Lenz's law), this relationship shows that the induced voltage is proportional to the rate at which the current varies and to the inductance itself.

In logic circuits, the rise and fall times of the switching currents generate perturbing voltages on the PCB traces (or interconnection cables), called *inductive noise*. To limit this noise, two approaches are possible:

– Reduce the slope of the current variation
– Reduce the inductance

■ **Example 1.** Consider a logic circuit employing only TTL gates. Typical values for the current drawn from the DC supply by a single gate are around 5 mA in the on state and 1 mA in the off state. The switching time is typically 2 ns. Assuming that the trace carrying this current has a self-inductance of 400 nH, the noise voltage given by (10.2) is 0.8 V, which is dangerously close to the noise margins of the TTL family. The risk of altering the state of the gate not by an intentional command, but due to the resulting inductive noise, is unacceptably high.

■ **Example 2.** Let us now consider the circuit in Fig. 10.14a [168], where the trace connecting the output of gate 1 to the input of gate 2 is represented by means of a 4-section lumped equivalent circuit. L_p, C_p, R_p denote the parasitic inductance, capacitance, and resistance of each section. Figure 10.14b

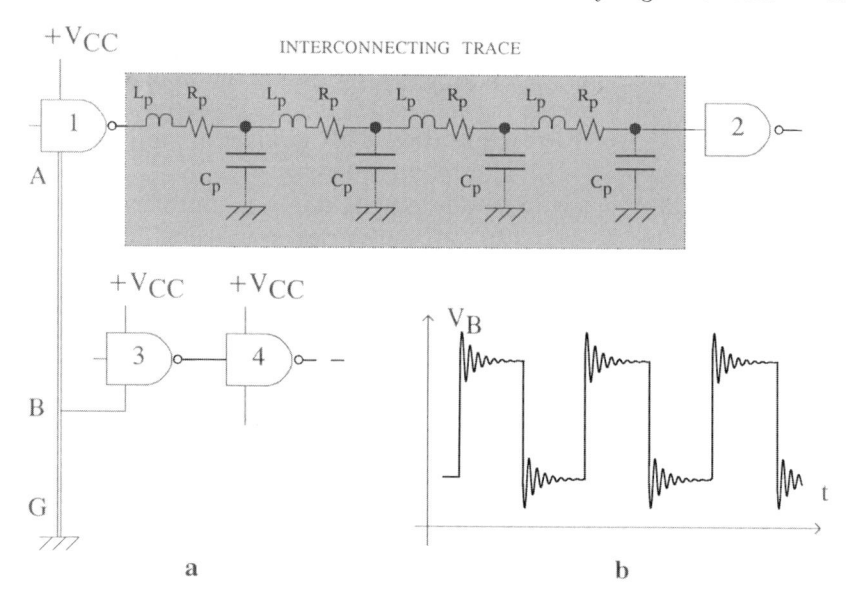

Fig. 10.14. Inductive noise resulting from the switching of gate 1: **a** circuit; **b** output waveform of gate 1

shows the waveform obtained by simulation with PSPICE for the proposed equivalent circuit of the trace, when driven by a near-rectangular input signal. Note that this waveform is obtained at the ouput, but similar waveforms appear between adjacent sections.

When the output of gate 1 switches from 0 to 1, all stray capacitances C_p charge to V_{CC}. Next, when the output of gate 1 switches from 1 to 0, all C_p must find a discharge path.

This path is: output of gate 1, point A, point B, and ground G; besides undesired radiation resulting from this loop, the discharge current generates in the trace connecting point A to ground (whose equivalent circuit is exactly the same as that shown for the interconnecting trace) an inductive noise voltage at point B, which is transmitted to the ground terminal of gate 3. Assuming that the output of gate 3 is 0 (connected to ground), the high amplitudes of the damped oscillations affecting every transition are also transmitted to the input of gate 4, possible triggering it accidentally.

■ **Solutions.** To limit inductive noise, several possibilities exist:

- The most effective approach is to reduce the parasitic inductance L_p of the connection. For a PCB trace of width w, the inductance per unit length is

$$l_p \cong 0.197 \ln \frac{2\pi h}{w} \quad [\text{nH/cm}] \qquad (10.3)$$

where h is the height above the path of the return current.

The total trace inductance is proportional to the length and has a logarithmic dependence on the width. Thus, increasing the width of the trace is not as effective as decreasing the length, but the designer often has little choice with regard to supply trace length, which is imposed by the component layout.

One possible approach is to parallel several traces to transport the supply (*Caution*: this means electrical, not geometrical paralleling!), because the total inductance of a system of two parallel traces is half the inductance of each trace, if mutual inductance is neglected. Consequently, the equivalent inductance of N identical inductors connected in parallel (each one of value L) is equal to L/N (if no mutual inductance is considered).

This explains why a ground plane (which can be viewed as the parallel combination of N $\rightarrow \infty$ inductances) is so effective in comparison with any ground trace, of any width.

- It has been shown in connection with the illustrative circuit of Fig. 10.14a, that another potential deleterious effect is due to radiation emanating from the transient current loop. In practice the loop area must be minimized by properly placing the components. A general rule can be deduced from expression (10.3): position the return current path as close as possible to the supply path (low h). It is best if the two paths can be symmetric. Even if a ground plane is used, problems can still emerge because *the return current will not necessarily take the path expected by the designer.*
- To limit oscillations in the switching waveform (Fig. 10.14b), the best solution is to insert a small (47 Ω) resistor in series with the gate output, which provides sufficient damping, without too much degradation in signal level.
- Last but not least, select the slowest logic family satisfying the specifications (see also Sect. 11.2.6). A general tendency is to adopt the same logic family throughout an entire design. A better idea is to use high-speed logic families (such as AC, AS, or F) where transitions must be fast, and lower-speed families (such as HC, LS, or CMOS) where fast transitions are not required.

To stress the importance of making the right choice, consider that to obtain the same level of differential-mode radiated noise, the loop area must be 400 cm^2 for CMOS logic circuits, 18 cm^2 for 74 HC logic circuits, and only 2.2 cm^2 for integrated circuits belonging to the 74 F family (assuming that all are driven by a 10 MHz clock).

Finally, the designer should resist the temptation to indiscriminately use the fastest, highest-performance family of logic circuits available.

10.4.3 Noise Related to Clock Radiation

■ **Explanation.** As a general rule, the clock delivers signals with very fast transitions and adequate power (the latter feature is required to guarantee the integrity of the clock signal, which must be delivered throughout the layout).

Traces transporting this rapidly-switched, powerful signal amount to radiating antennas. It has been noted in practice that clock noise can be 10 to 20 dB above the level of other disturbances radiated by the same logic system. In the spectrum, the most significant and harmful part extends up to the eighth harmonic of the clock frequency. Therefore, controlling emission from clock traces is certainly of paramount importance.

■ **Remedies**

1) *Whenever possible, deliberately reducing the rise/fall times of the clock signals* represents the best choice. Three methods can be employed, each with its drawbacks.
   – When the clock signal is carried by a circuit board trace (Fig. 10.15a), inserting a small-value capacitor C will certainly reduce the speed of transitions; however, the price to be paid is that the capacitor will augment the total capacitance between trace and ground. As a consequence, the charging/discharging current in the combined capacitance $(C + C_p)$ is augmented and the unfavorable effects described in Sect. 10.2.4 are reinforced.
   – Another possibility is to increase the series impedance of the interconnect, either by inserting a small series resistor at the output of the clock generator (trace on a PCB), or by adding a ferrite bead if a cable is employed (Fig. 10.15b). The series resistance will increase losses, while the added inductance may produce ringing.
   – The third possibility is to pass the clock signal, prior to distribution, through a low-quality buffer stage, which will slow down the transitions.
2) *Diminish the loop area of the interconnect carrying the clock signal.* All clock interconnects must have adjacent return paths. To prevent excessive noise emission, it is recommended to keep the clock loop area below a few square centimeters.
3) *Rigorously avoid using a single ground point connection; instead use a ground plane.* Since the goal is to minimize the total length of intercon-

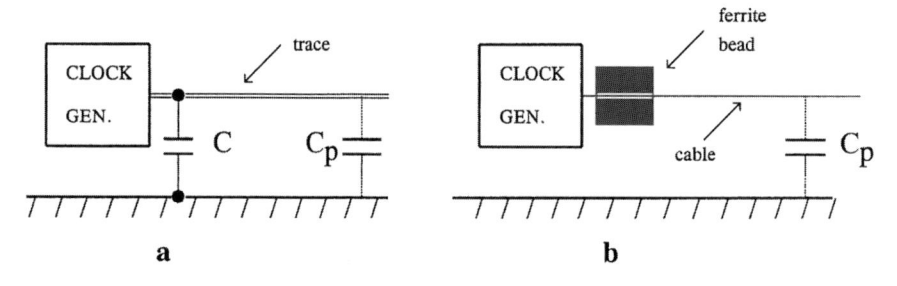

Fig. 10.15. Slowing clock transitions; **a** by inserting a parallel capacitor; **b** by mounting a ferrite bead on the cable

nects distributing the clock, a good policy is to partition the whole system into a few modules (cards) connected to a mother-board, each with its own clock generator.

10.4.4 Reflections Due to Mismatch on the Lines

■ **Explanation.** The current trend of constantly increasing clock frequency has as a consequence that all interconnects distributing it must be treated as transmission lines. Microwave circuit engineers have long since accepted the idea that any transmission line must be carefully matched at its input and output terminals. Unfortunately, this is not the case for logic circuit designers, because the mismatch was not at all a critical issue at lower clock frequencies.

Nevertheless, at current clock rates (up to a few GHz), the deleterious effects of the mismatch cannot be ignored. Figure 10.16 shows the degradation of quasi-rectangular waveforms when the traces carrying them are mismatched.

The damped oscillations result from the superposition of the signal propagating in one direction with the reflected signal traveling on the line in the opposite direction. Due to mismatch at both ends, multiple reflections appear. At the line resonance frequency dampened oscillations appear, whose amplitude is proportional to the mismatch (but their frequency depends on the electrical parameters of the line).

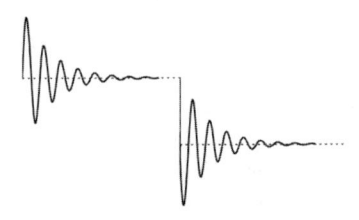

Fig. 10.16. Damped oscillations

■ **Remedy.** Of course, the best solution is to *terminate the line (trace) at both ends in its characteristic impedance*. However, in logic systems, a frequently encountered situation is that of a line connecting the output of one gate to the input of another. In this case the line actually sees complex terminal impedances at both ends, which change with input or output voltage level. It follows that matching the characteristic impedance at both ends is impossible, and this cannot be adopted as a major goal of the design process. The only remaining possibility is to respect the manufacturer's indications concerning maximum allowable trace length for each logic family. For instance, with the TTL 74 F family (whose circuits have typical rise/fall times of about 3 ns), the maximum allowed trace length is 25 cm (on an epoxy-glass substrate with relative permittivity of 4.5). In contrast, for the 74 AS series (where typical rise/fall times are of order 1.4 ns), this length is limited to 12.5 cm.

10.4.5 Abrupt Demand for DC Supply Current

■ **Explanation.** When several logic circuits switch at the same time, the current drawn from the supply exhibits abrupt changes. Like any step variation in the current flowing through an inductive load, this will induce inductive noise. Since the designer has no control over current transients, the only possibility is to limit the stray inductance of the connections.

Ideally, the DC supply traces must have a symmetric and parallel configuration with respect to the ground returns. However, in practice this is not always possible, and attention must be paid to two issues:

- Filter the ripple of the DC supply, especially if a switching-mode power source is employed. This has been explained in Sect. 10.2.5.
- Decouple the DC supplies, in order to minimize the loop area. Consider the situation illustrated in Fig. 10.17a, where a ground plane has been adopted, in order to reduce the ground stray inductance (L_{PM}) as much as possible. Nevertheless, when the gate output switches, the supplied current is abruptly altered from I_{OFF} to I_{ON} (or vice versa). This will cause inductive noise, mainly due to this fast transient (ΔI) flowing through $L_P \gg L_{PM}$. In this way a voltage transient appears on the trace (L_P), hence at point V_{CC} of the gate. If the gate output is 1, this noise voltage reaches the output, and eventually the input of the following gate, possibly switching it.

■ **Remedy.** Assuming that the length of the trace carrying the DC supply has already been optimized, the only solution is to add a decoupling capacitor (C_d), which essentially stores the charge (or delivers it) close to the gate, letting a less significant transient flow through L_P''. In Fig. 10.17a, the step current demand ΔI is carried by the large loop indicated there, while in Fig. 10.17b the most of ΔI flows through the much smaller loop $A - V_{CC} - \text{GND} - B - C_d$. Note that placement of C_d is critical; it must be as close as possible to the IC in question. The benefit is twofold: radiation from the loop carrying ΔI diminishes and the stray inductance of that portion of the track that supports much of ΔI flow (L_P') is considerably reduced. Overall, the noise associated with the pulse current ΔI is reduced.

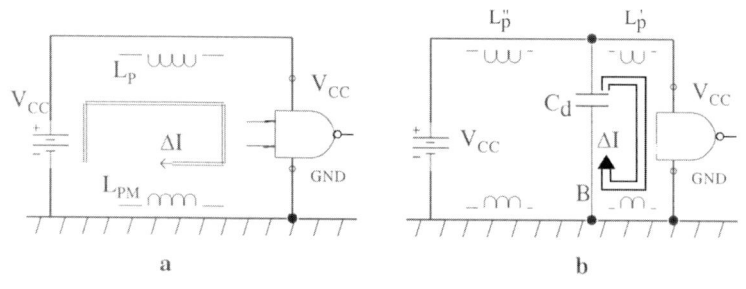

Fig. 10.17. Reducing the loop area of step transients in the DC current supply ($L_P'' \gg L_P'$)

340   10 Methods of Reducing Emission of Interfering Signals

■ **Selecting the Decoupling Capacitor.** The problem is that any capacitor also has a parasitic series inductance, which limits its effectiveness at high frequencies. Chiefly, this internal inductance originates in the lead inductance and also derives from the fabrication technology: an aluminized Mylar film rolled up into a capacitor inherently presents a high internal inductance, much more so than a metallized mica or ceramic disk capacitor. It follows that the designer must carefully select the type of capacitor to be used in decoupling logic circuits: small disk or multilayer ceramics are preferred. A good criterion is to check the manufacturer's data sheet as to whether normal operation is guaranteed up to 150 MHz [168].

The value of the decoupling capacitor C_d is estimated with the classic equation

$$C_d = \frac{dQ}{dV} = \frac{dI\, dt}{dV} \qquad (10.4)$$

For example, for a current step $dI = 40$ mA in a time $dt = 3$ ns, and an allowable DC voltage drop of up to 0.03 V, C_d must be at least 4 nF. One of the most common mistakes is to adopt a much greater value for C_d than necessary; it must be borne in mind that the greater the value of a capacitor, the greater its equivalent series inductance. The risk is that with a too large a value, the self-resonant frequency of the capacitor will fall well inside the operating bandwidth of the logic system, causing damage.

According to Ott [168], it has been found experimentally that optimum values for decoupling capacitors are between 470 pF and 1000 pF; the ultimate criterion is to keep the measured noise at integrated circuit terminals less than 500 mV.

10.5 Transformer Noise

10.5.1 Noise Sources

■ **Explanation.** There are two main noise mechanisms associated with any transformer:

1) Its own magnetic flux, which can escape through the shielding and can intersect nearby circuits.
2) Its principle of operation, which is based upon tight magnetic coupling between primary and secondary coils. Due to this, the noise carried by the mains line connected to its primary is transmitted to the secondary and consequently to the supplied circuits. Since the transformer acts as a low-pass filter, one might assume that only low-frequency noise, coupled in differential mode, could contaminate the secondary load. However, an actual transformer exhibits both magnetic and capacitive coupling between primary and secondary, and the latter is responsible for noise transmission *beyond the cut-off frequency of the transformer*. Consequently, the

main objective of protection must be to reduce capacitive coupling, which is much more damaging because it is effective in transmitting both differential and common-mode noise.

■ **Remark.** The most frequently encountered transformers in electronic equipment are either isolation transformers or power-supply transformers. Neither category is at all subject to the user's control, since once a transformer has been purchased it is not accessible to experimental modification or improvement. All we can do is to measure its performance.

■ **Differential-Mode Coupling Mechanism.** Consider the schematic diagram of Fig. 10.18, where C_1, C_2 lump the effects of mutual capacitance between the primary and secondary.

Let V_{p1} represents the differential-mode noise carried by the mains line and V_{p2} the noise voltage existing between NEUTRAL and EARTH. R_1 and R_2 denote the resistance of the connecting wires to NEUTRAL and to SIGNAL GROUND, respectively.

V_{p1} drives the following loop: PHASE, (C_1, C_2), (RECTIFIER & REGULATOR), amplifier, R_2, R_1 and NEUTRAL. Hence, a fraction of V_{p1} reaches the OUTPUT; in some special fields (such as medical applications) this can be dangerous.

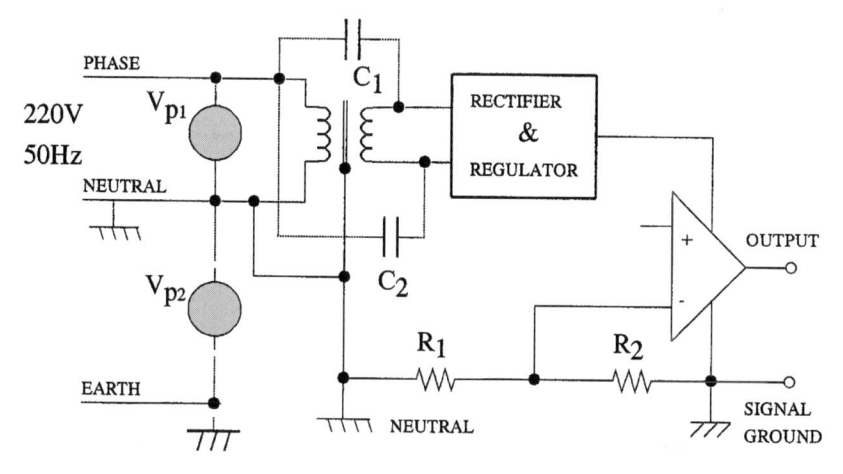

Fig. 10.18. Noise transmission by capacitive coupling through the power transformer

10.5.2 Solutions to Reduce Noise Coupling

■ **Discussion.** To reduce mutual capacitive coupling, two solutions are available:

1) Eliminate the transformer from the design (by adopting a battery supply or a rectifier connected directly to the mains).

342   10 Methods of Reducing Emission of Interfering Signals

2) Introduce one or more shields between the primary and secondary windings of the transformer. Several techniques exist [169]:
   – Faraday shields. The primary and secondary coils are wound on top of each other; a copper or aluminum foil laid onto a mylar foil is inserted between them.
   – Box shields. The shielded coils are constructed separately and then nested one inside the other or side by side.

■ **Single Shield Transformer** [164]. A Faraday shield connected to the chassis ground (EARTH) is inserted between the two windings (Fig. 10.19). It consists merely of a wrap of aluminum or copper over a coil. Attention must be paid not to form a closed loop with the copper foil of the Faraday shield (its edges must not touch each other), to avoid eddy currents in the shield volume. Since the shield cannot be perfectly closed, the residual capacitance between the windings (C_{DE}) corresponds to the gap between edges and it is responsible for differential-mode noise coupling.

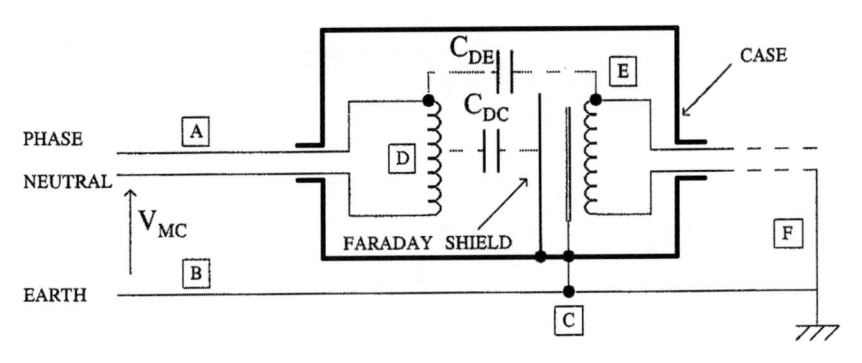

Fig. 10.19. Transformer with a single Faraday shield

The capacitance C_{DC} allows for a fraction of the common-mode noise voltage V_{MC} contaminating the line to reach the output, via the loops: (PHASE, primary coil, distributed C_{DC}, shield, C, B) and (NEUTRAL, primary coil, distributed C_{DC}, shield, C, B). Due to inherent asymmetries with respect to shield, these currents are not perfectly balanced, and their net difference induces a current in the secondary winding (this being an example of how common-mode noise can be converted into differential-mode noise).

■ **Transformer with Two Shields.** To further reduce common-mode noise transmission (particularly when a noisy load is connected to the secondary), it is necessary to add a second shield. This is superimposed on the first shield, but electrically isolated from it (Fig. 10.20).

The shield enclosing the primary is called the primary shield (S1), while S2 is the secondary shield. S1 is connected to the chassis ground (EARTH),

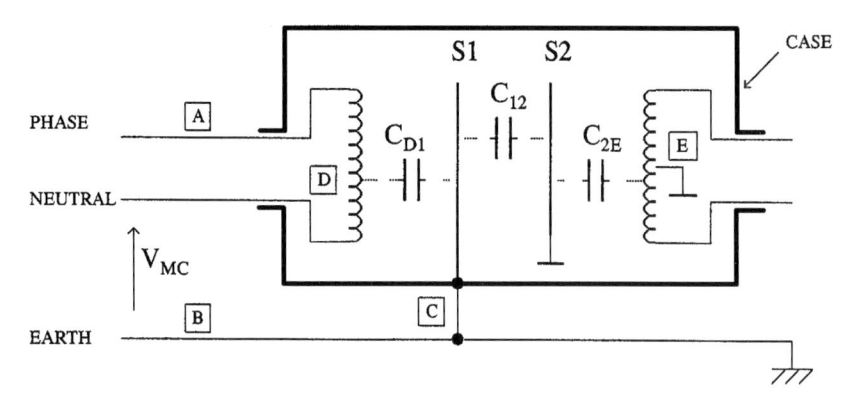

Fig. 10.20. Transformer with two shields

but S2 is tied to the nearest signal ground (for instance, the center tap of the secondary). For simplicity, the magnetic core of the transformer has not been shown, but it is connected to the safety EARTH (as in Fig. 10.19).

Since the distributed stray capacitances C_{D1}, C_{12}, C_{2E} are in series, the equivalent capacitance of this combination is less than any one of them. Therefore, the coupling between primary and secondary coils is considerably reduced and the common-mode current flowing through the loop: A, D, C, B is minimized. Further improvement can be obtained by adding a third shield.

■ **Remarks**

– Manufacturers often connect the centered tap of the secondary winding to the secondary shield.
– The benefit of inserting two shields is to reduce capacitive coupling between transformer windings; the side effect is a simultaneous reduction in the magnetic coupling between coils and consequently lower transformer efficiency.

■ **Conclusion.** Inserting one or two shields between the transformer primary and secondary windings has a favorable effect on common-mode noise transported by the mains. As a general principle, remember that *the common-mode currents flowing through a line significantly increase the radiated noise from that line, despite the relatively low values of these currents with respect to the differential-mode currents.*

10.6 Noise Due to Electrostatic Discharge

■ **Definition.** An electrostatic discharge *is a natural phenomenon in which a burst of electric charge is transferred between bodies at differing electrostatic potential, in proximity or through direct contact.*

■ **Comments.** The spectrum of this abrupt discharge is very broad and extends well into the very high frequency (VHF) region.

Viewed as an event, an electrostatic discharge (ESD) has two phases:

1) *The charging phase*, during which an electric charge is accumulated on objects.
2) *The discharge phase*, in which charge is abruptly transferred between objects or between an object and ground.

■ **Effects.** The burst of discharge current is responsible for one or more of the following consequences:

- irreversible damage to semiconductor devices through dielectric breakdown or heating effects;
- failures in logic circuits due to system reset, bit reversal, latch-up, erroneous toggles, timing faults, etc.;
- program and memory bit corruption;
- it generates fields which can couple to nearby sensitive circuits;
- can cause latent damage to sensitive devices, which may be apparent only after a long time or repeated ESDs;
- it produces electrical shock hazards to people or equipment located in the area.

Since many ESDs go undetected, there is a serious problem identifying them as a source of failure or damage. Clearly, ESD is to be avoided, or at least its effects must be limited.

10.6.1 Accumulation of Electrostatic Charge

■ **Explanation.** Charge is accumulated on objects in the following ways:

- by the triboelectric effect (insulating materials);
- by electrical induction (metals);
- by other means, such as evaporation, thermionic emission, photoelectric emission, piezo-charging, etc.

■ **Triboelectric Effect.** This effect, briefly discussed in Sect. 8.1.2, entails charge accumulation *on the surface* of insulators placed in contact through friction [170]. Generally, when one material is placed in contact with another, electrons tend to move to the material with the lower Fermi level. When they are separated, one material is negatively charged, the other acquires a positive charge. Rubbing them is just a way to increase the surfaces which are in close contact and consequently increase the charge transfer.

If the materials involved are metals (whose electrons have high mobility), during separation the charges migrate back to the metal from which they

10.6 Noise Due to Electrostatic Discharge

Table 10.2. Triboelectric series

1. Air	11. Cotton
2. Human skin	12. Wood
3. Asbestos	33. Amber
4. Glass	14. Hard rubber
5. Mica	15. Mylar
6. Human hair	16. Polyester
7. Nylon	17. Celluloid
8. Wool	58. Polyurethane
9. Silk	19. PVC
10. Paper	20. Teflon

were extracted. For this reason, the triboelectric effect is not observed in metals.

However, if the materials involved are insulators, the mobility of their electrons is exceedingly low, and after separation they can maintain their charge for hours.

Table 10.2 presents several dielectric materials arranged in a triboelectric series: the upper materials are more prone to become positively charged when in contact with materials lower down (which became negatively charged). The charge accumulated is proportional to the separation in the triboelectric series: for instance, rubbing nylon on wool produces less charge than rubbing nylon on cotton. However, other parameters can influence the amount of charge, such as the speed of separation, the pressure during friction, surface cleanliness, ambient humidity, etc.

□ **Examples.** In the real world, the triboelectric effect shows up when a person with insulating shoes walks on a carpet, or when sitting in a car (the friction of clothes with the surface of the car seat gives rise to charge separation), or even when removing tape (or labels) in dry atmosphere. Other situations where charge is accumulated by friction involve rolling furniture, wheelchairs, cooling fans, conveyor belts, paper movement in copiers, rockets in flight, etc.

■ **Induction Charging.** It is well known that any charged object generates an electric field around it. If a neutral conductor (B) is placed in its proximity, charge redistribution occurs on its surface (Fig. 10.21a). Note that an equilibrium exists, since the positive charges of B exactly balance the negative ones. Removing the object A, the conductor B is restored to its initial, non-polarized state.

Now consider the situation depicted in Fig. 10.21b, where the conductor is temporarily grounded by closing the switch K. A transient current i will flow to or from ground, which neutralizes one kind of charge. Removing the field produced by A does not restore the charge neutrality of object B, which

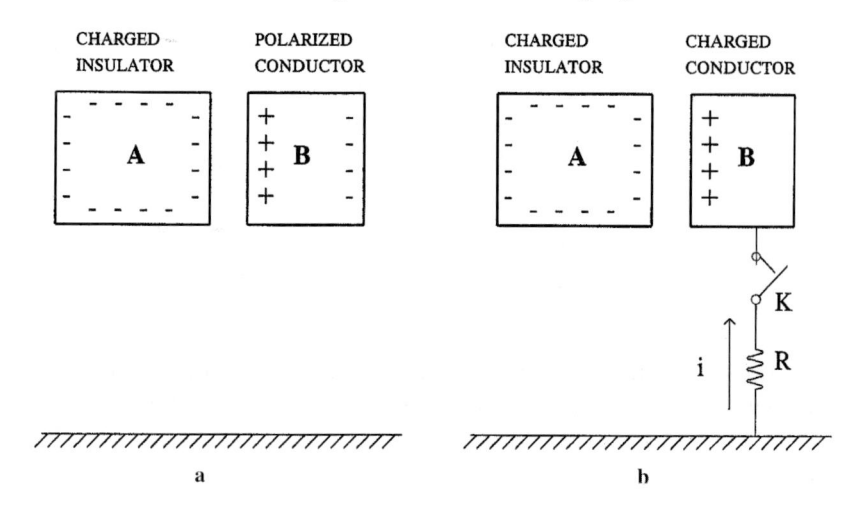

Fig. 10.21. Charging a conductor by electrical induction

retains a net charge, without being previously in contact with a charged item. This is the mechanism of induction charging.

Depending on the particular circumstances, note that resistance R can be as high as a few megohms, or as low as a few hundred ohms.

■ **Model of Human Body.** Even though humans are one of the most important sources of ESD, it is very difficult to construct a simple simulation model for ESD in general, and for the human body in particular. The problem is that any ESD event depends on a large number of electrical, geometrical, mechanical, and environmental parameters. One consequence is that repeatability of all implied parameters is almost never acquired, and each event is unique. Consequently, reliable models for simulation are difficult to build, and the numerical values of various elements involved are rather approximate.

The human body is modeled as a conductor covered by a thin insulator (the skin, provided that it is not humid!); with respect to ground, a typical range of values for the human body capacitance is 50 pF to 300 pF (depending on actual body shape, the surrounding geometry, and height above the floor) [170]. U.S. industry has adopted a simplified series RC ESD generator model for the human body, with a 200 pF capacitance and 1.5 KΩ resistance. Other significant electrical parameters of the human body are listed in Table 10.3.

■ **Human Body Charging.** Although body capacitance is rather low, the resulting voltages can be high. For instance, when a person walks on a carpet in a room with relative humidity 20%, the charge accumulation is estimated to be about 1 μC. For a body capacitance of 75 pF, the resulting voltage is

$$V = Q / C = 10^{-6} / 75 \cdot 10^{-12} \cong 13.3 \text{ kV}$$

10.6 Noise Due to Electrostatic Discharge 347

Table 10.3. Human body electrical characteristics

Parameter	Approximate value
Blood resistivity	$1.85\,\Omega\mathrm{m}$
Resistivity of internal tissues	$0.8\,\Omega\mathrm{m}$
Muscle resistivity	$15\,\Omega\mathrm{m}$
Skin resistivity	$(2\cdot 10^4 - 10^5)\,\Omega\mathrm{cm}^{-2}$
Body capacitance	$50 - 300\,\mathrm{pF}$

Remember that the maximum voltage a human can stand is 35 kV. Fortunately, the ESD spark is produced before reaching this threshold value, and the body is discharged!

10.6.2 Discharge Phase

■ **Explanation.** According to [168], any ESD event involves three steps:

1) Charging an insulator by friction.
2) By contact or induction, this charge is transferred to a moving conductor (human body, for instance).
3) When the moving conductor approaches a grounded conductor, discharge occurs.

Basically, discharge is stimulated by irregularities in the conductor surface, such as nonuniform shape, surface cleanliness, drops of water, etc. They are all responsible for a nonuniform electric field distribution; at sharp points, the electric field can concentrate and if the distance to the grounded object is small, breakdown (arcing) through the air is likely to occur (Fig. 10.22). The discharge may be repetitive, affecting the same place or nearby areas.

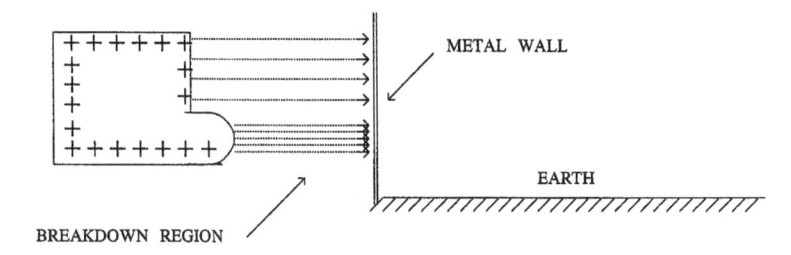

Fig. 10.22. Electrostatic discharge to a grounded metal wall

■ **Waveform.** A generic waveform of the discharge current between a person's finger and a grounded metal object is presented in Fig. 10.23.

It can be seen to contain an initial fast spike (which corresponds to the charge accumulated in the arm) followed by the main discharge which is much

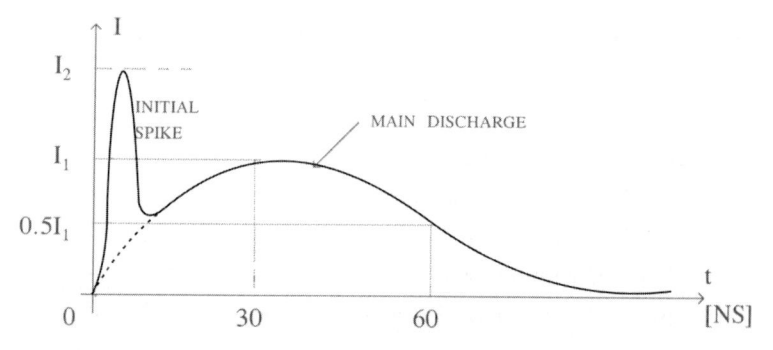

348   10 Methods of Reducing Emission of Interfering Signals

Fig. 10.23. Typical current waveform for a discharge involving a human body

more slower, corresponding to dumping of the body charge. The rise time of the initial spike may be less than 1 ns; it mainly depends upon the speed of the human body approaching the grounded metal object. This current spike is a powerful radiation source and has particularly damaging effects on nearby circuits. Its frequency spectrum is very rich and extends well above 100 MHz.

■ **Analytic Model.** The typical current waveform shown in Fig. 10.23 can be described with the following mathematical expression [162]:

$$I(t) = A\left(\exp(-t/2.2) - \exp(-t/2)\right) + B\left(\exp(-t/22) - \exp(-t/20)\right) \quad (10.5)$$

where t is expressed in ns. Constants A and B are chosen according to the assumed values of I_1 and I_2. For instance, taking $I_1 = 30$ A, $I_2 = 68$ A, and a rise time of the initial spike of 1.2 ns, $A = 1943$ and $B = 857$.

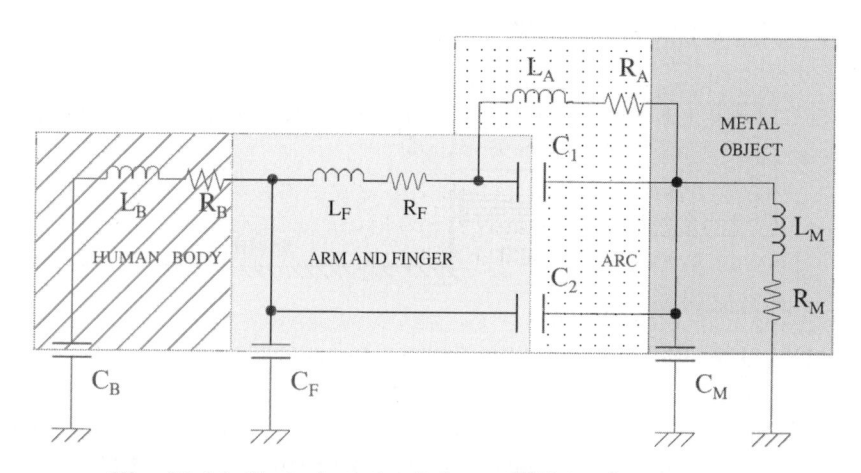

Fig. 10.24. Equivalent circuit for an ESD involving a person

10.6 Noise Due to Electrostatic Discharge

Table 10.4. Typical values of the human body model

Element	Meaning	Value range
R_B [kΩ]	Human body resistance	0.33 – 10
L_B [µH]	Human body inductance	0.4 – 2
C_B [pF]	Body-to-ground capacitance	60 – 300
R_F [Ω]	Arm-finger resistance	20 – 200
L_F [µH]	Arm-finger inductance	0.05 – 0.2
C_F [pF]	Arm-finger capacitance to ground	3 – 10

■ **Circuit Model for Electrostatic Discharge.** Assuming an ESD event between the finger of a person and a grounded metal object, the circuit model shown in Fig. 10.24 is proposed [162, 170].

C_B, C_F, and C_M are the capacitances with respect to ground of the body, arm-finger, and the metal object (the metal object being a doorknob, keyboard, chassis, etc.). C_1 and C_2 are inserted because the discharge always occurs *before* the finger touches the metal object; L_A and R_A model the arc discharge, which short-circuits the capacitance C_1.

The values of all elements depend on the particular circumstances of the ESD event; however, Table 10.4 presents the most widely accepted values of some elements of the equivalent circuit [170].

According to [162], the rise time of the initial spike depends on the ratio $(L_F+L_B)/(R_F+R_B)$, and it influences the far field; in contrast, the pulse width depends on the time constant $(R_F+R_B)(C_F+C_B)$, and it influences the near field.

10.6.3 Prevention and Control

"Prevention" means all arrangements and measures to be taken in order to avoid initiating ESD events at a particular site. Among them, the following aim at the prevention of charge accumulation:

- Humidity control; actually, any closed space that has to be protected against ESD must have a relative humidity of at least 50%.
- Employ air ionizers; this will contribute to elimination of accumulated charges.
- Use only antistatic carpets (or no carpets at all).
- Select conductive wheels for furniture and armchairs.
- Ensure personnel grounding by means of wrist straps and grounded work surfaces.

"Control" refers to measures to be adopted in order to mitigate ESD effects. Among them, note the following:

- Many components are manufactured with included built-in protection to avoid damage (typically, a shunt device to limit the voltage, without affecting normal operation of the component).
- Software protection is based upon redundancy, refreshing, checking, and restoring; also, long-term wait states are to be avoided.

Summary

- Conducted noise of the mains line can be eliminated by filtering or by employing surge suppressors.
- The noisiest power supplies are switching-mode supplies and DC/DC converters.
- Noise associated with switching of mechanical contacts has two sources: glow discharges and arc discharges. The latter is most likely to occur in electronic equipment.
- To avoid glow discharges, the voltage across contacts must be kept less than 300 V.
- To avoid arc discharges, the rising slope of the voltage across the contact must be kept slower than $1 \text{ V}/\mu\text{s}$.
- To control arc discharges, RC circuits can be employed, but the most effective is the RCD circuit.
- Logic circuits, with their fast-rising rectangular signals and high amplitudes, represent one of the most powerful sources of radiated and conducted noise.
- To limit the frequency spectrum of noise emitted by a logic circuit, the clock frequency, as well as the rise- and fall-times, must be carefully controlled; the design of the layout must minimize the stray inductance of critical paths and the position of data lines must be carefully selected; whenever possible, adopt less noisy power supplies.
- The capacitive coupling between transformer primary and secondary coils can be reduced by inserting one, two, or three Faraday shields between them.
- Electrostatic discharge represents one of the most powerful source of noise, and can cause irreversible damage to components and circuits. To avoid electrostatic charge buildup, control the ambient humidity, select an antistatic carpet, and use air ionizers.

11

Interconnection Modeling and Crosstalk

"The cure for this is not to sit still,
Or frost with a book by the fire;
But to take a large hoe and a shovel also,
And dig till you gently perspire"

(R. Kipling)

11.1 Introduction

■ **Definitions**

– **Interconnection** denotes *connections between and external to any functional item that forms a circuit or system of circuits.* Functional items include component parts, devices, subassemblies, and assemblies [172]. *Note*: In some textbooks [173, 174] the term "interconnect" is preferred (same meaning).
– **Track** refers to *an interconnecting metal path on an integrated circuit or printed circuit board* [175]. Note that the term **trace** is also employed for metal paths on printed circuit board (PCB) [176].
– **Crosstalk** is the undesired energy appearing in one signal path as a result of coupling from other (adjacent) signal paths [172].

■ **Examples of Interconnects.** Traces on a PCB, tracks on integrated circuits, cables, conductors, bonding wires, data buses, optical fibers, and waveguides are all examples of interconnects.

■ **Why Are Interconnects Important?** Assume a near-ideal situation, where all circuits and systems are perfectly protected against interference and their own noise is well contained by the shield enclosing each item. One might think that interference is no longer a problem, but this is not true, since the interfering signals can still contaminate the protected circuits by coupling to

352   11 Interconnection Modeling and Crosstalk

their interconnections. Considering that interconnects also have considerable length, often much greater than the circuit dimensions, it follows that they have significant susceptibility, and if system protection does not include them, the system is not immune.

Since the interfering signals usually couple to interconnections either through stray capacitances or stray inductances (or both), it is important to characterize the parasitics of tracks, traces, cables, and wires.

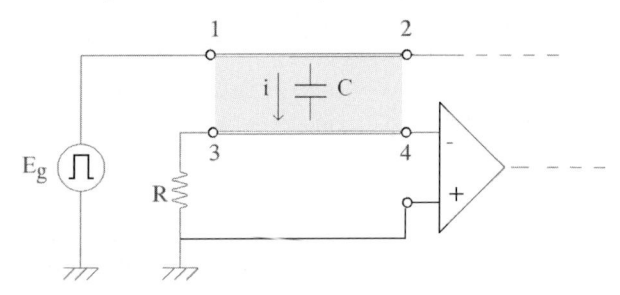

Fig. 11.1. Capacitively coupled interconnections

■ **Example.** Consider the circuit of Fig. 11.1, where the unintentional coupling between lines 1–2 and 3–4 is of concern.

The aggressor is line 1–2 and the victim is the amplifier input (fed by line 3–4), where a fraction of E_g arrives. Suppose that $C = 1\,\text{pF}$, $R = 10\,\text{k}\Omega$, and that E_g is the output of a logic gate, delivering a rectangular signal that switches between 0 and 5 V in 2 µs. In this case, the current flowing through the capacitor C is

$$i = C\,\frac{\mathrm{d}v}{\mathrm{d}t} = 10^{-12}\,\frac{5}{2 \cdot 10^{-6}} = 2.5\,\text{µA} \tag{11.1}$$

It is reasonable to assume that the input resistance of the operational amplifier is high (relative to $10\,\text{k}\Omega$); thus, almost all current i will flow through resistor R, and the crosstalk noise voltage reaching the amplifier input is

$$v_{\text{cr}} = R\,i = 25\,\text{mV} \tag{11.2}$$

Assuming that R represents the internal resistance of a sensor, it is likely that v_{cr} exceeds the useful signal delivered by the sensor!

This example shows how almost negligible capacitive coupling (of no more than 1 pF) can be completely overwhelming, when the aggressor is a fast-switching signal.

11.2 Interconnect Modeling

■ **Comment.** Any interconnecting conductor has resistance, inductance and capacitance to ground. Depending on the particular configuration, mutual capacitance and mutual inductance might be defined with respect to a similar conductor. For modeling purposes, all element values must be estimated.

11.2.1 Interconnect Resistance

■ **Calculation.** The resistance of a uniform track (trace) of width W, thickness T, and length L is evaluated with the classical expression

$$R = \rho \frac{L}{WT} = \left(\frac{\rho}{T}\right)\left(\frac{L}{W}\right) \quad [\Omega] \tag{11.3}$$

where ρ is the resistivity of the material.

■ **Sheet Resistance.** As the thickness of traces on a PCB (or of tracks on the integrated circuits) are uniform, we can introduce a new parameter

$$R_s = \rho / T \tag{11.4}$$

called sheet resistance and expressed in ohms per square ($\Omega/\square$). By definition, *the sheet resistance is the resistance per unit square of a thin-film material such as a metal or thin layer of semiconductor* [175]. It is a constant for a given material and a given process.

■ **Remarks**

– All squares, regardless of size, have the same resistance, which is just R_s.
– A rectangular film of length L and width W has a resistance equal to $(L/W)R_s$. Sometimes L/W is called the "aspect ratio," or simply the "ratio."
– A practical approach to calculating the resistance of a track is to decompose the rectangle into elementary squares; *the total resistance is obtained by multiplying R_s by the number of squares in series and dividing the result by the number of squares in parallel.*

■ **Examples.** Some of commonest configurations are given in Fig. 11.2.

Consider the rectangular film of Fig. 11.2a. The resistance measured between edges A and B is $(6/4) R_s$, while that between edges C and D is $(4/6) R_s$.

The resistance of non-rectangular shapes requires elaborate calculation (by solving Laplace's equation or by breaking the shape into several known configurations).

For a parallelogram, the ratio is given in Fig. 11.2b.

For right angle bends (Fig. 11.2c), the resistance of the corner square is approximately $0.5 R_s$; hence, between the bold edges the measured resistance is around $2.5 R_s$.

According to [173], the corner rectangle resistance (Fig. 11.2d) can be calculated with Bain's formula:

$$R \cong (0.46 + 0.1x) R_s \tag{11.5}$$

where $x = W_2/W_1$ ($W_2 > W_1$).

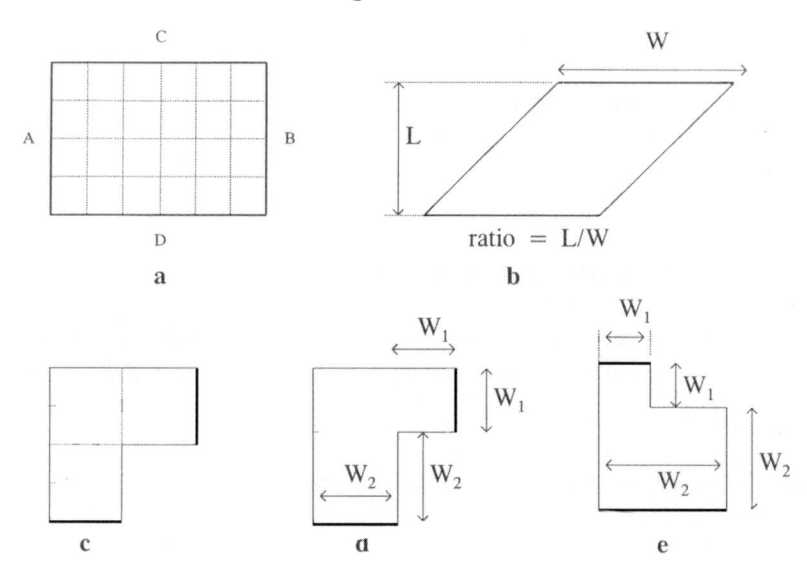

354   11 Interconnection Modeling and Crosstalk

Fig. 11.2. Various configurations often encountered in practice

For the shape in Fig. 11.2e, the resistance between the bold edges depends on the parameter x = W_2/W_1 ($W_2 > W_1$):

$$\begin{array}{c|c|c|c} x & 2 & 3 & 4 \\ \hline R & 2.25\ R_s & 2.5\ R_s & 2.65\ R_s \end{array} \quad (11.6)$$

■ **Comment.** In practice, one of the most important difficulties arises from the fact that a given trace (track) might contact several components instead of just two (as previously assumed).

11.2.2 Mutual Capacitance and Mutual Inductance

■ **Physical Considerations.** When signals are transmitted between two disjoint circuits A and B, two mechanisms are possibly involved in the electromagnetic interaction:

1) *Electric field coupling.* The signal in circuit A gives rise to voltage drops across various circuit components; as a consequence, the nodes of circuit A have different potentials with respect to ground. Lines of electric flux (normal to the conductor surface) are created by the potential difference. They originate in conductors of circuit A, and some of them end on conductors (Fig. 11.3a) of circuit B. The field-line distribution is nonuniform; in practice, they are more densely packed in regions where the difference in potentials is greater. By analogy with a typical capacitor, which confines most of the lines of electric flux to the interior of the component

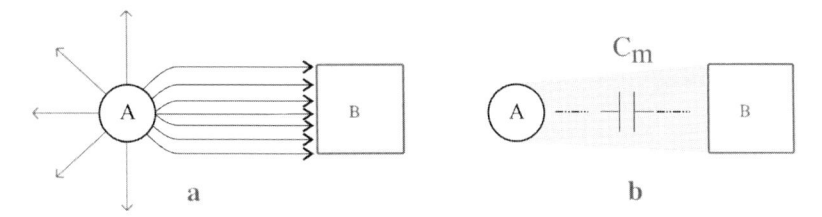

Fig. 11.3. a Electric flux lines distributed between 2 objects; b modeling the concentration of lines by a mutual capacitance

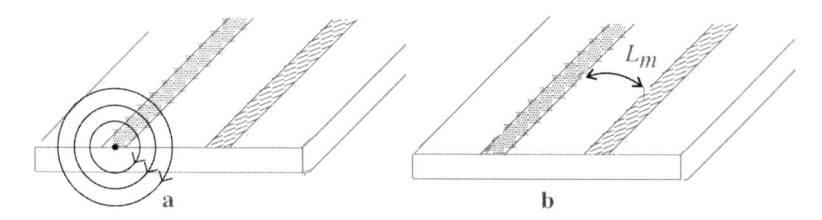

Fig. 11.4. a Distribution of magnetic flux lines; b modeling the concentration of magnetic lines by a mutual inductance

itself, corresponding to the space region where a significant concentration of electric flux lines is noted, a *mutual capacitance* C_m (see Fig. 11.3b) can be defined. This capacitance is also called "stray capacitance," and provides a convenient way to represent unintentional electric field coupling.

2) *Magnetic field coupling.* The signal in circuit A produces currents that flow through its components and interconnections. Each current is responsible for a magnetic field, represented by its lines of flux. These lines are closed, and encircle the conductor in which the current producing them flows; they can intercept a loop of circuit B (Fig. 11.4a). If the magnetic flux is time-varying, a voltage is induced in that loop.

In practice, inductors are components intentionally designed to concentrate most of their magnetic flux lines inside them; higher inductance corresponds to a higher density of magnetic flux. By analogy, in the space region separating circuits A and B where a strong concentration of magnetic flux lines is noted, a *mutual inductance* L_m can be defined (Fig. 11.4b). The mutual inductance between circuit A and B is responsible for magnetic coupling of signals from circuit A to circuit B.

■ **System of Two Conductors.** Consider the case of two conductors in an insulating medium, the first at potential V_1 and the second V_2, such that $V_1 > V_2 > 0$.

Conductor 1 is positively charged; conductor 2 is charged by induction, with negative charges on the side facing conductor 1 and positive charges on

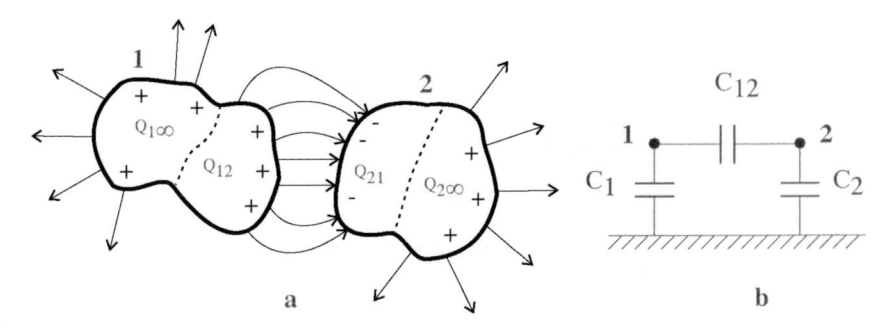

Fig. 11.5. a Distribution of field lines and charges in a two-conductor system; **b** equivalent capacitances

the opposite side (Fig. 11.5a). Some field lines originating in conductor 1 are directed to infinity, but most of them terminate on the negative charges of conductor 2. Due to the assumed charge neutrality condition of conductor 2, $|Q_{12}| = |Q_{21}|$.

The following capacitances can be defined [177]:

$$C_1 = \frac{Q_{1\infty}}{V_1} \tag{11.7a}$$

$$C_2 = \frac{Q_{2\infty}}{V_2} \tag{11.7b}$$

$$C_{12} = \frac{Q_{12}}{V_1 - V_2} \tag{11.7c}$$

$$C_{21} = \frac{-Q_{21}}{V_2 - V_1} = C_{12} \tag{11.7d}$$

The charges stored on each conductor are

$$\begin{cases} Q_1 = Q_{1\infty} + Q_{12} \\ Q_2 = Q_{2\infty} + Q_{21} \end{cases} \tag{11.8a}$$

Upon deducing charges from (11.7) and substituting them into (11.8), one obtains

$$\begin{cases} Q_1 = C_1 V_1 + C_{12}(V_1 - V_2) \\ Q_2 = C_2 V_2 + C_{21}(V_2 - V_1) \end{cases} \text{ or } \begin{cases} Q_1 = (C_1 + C_{12})V_1 - C_{12}V_2 \\ Q_2 = -C_{21}V_1 + (C_2 + C_{21})V_2 \end{cases} \tag{11.8b}$$

Note that (11.8b) completely describe the two-conductor system: if the charges are known, the potentials can be deduced, and vice versa.

■ **Capacitance Matrix.** Consider now a multiconductor system in an insulating medium (Fig. 11.6). By extension of (11.8b), this system can be described by the following equations:

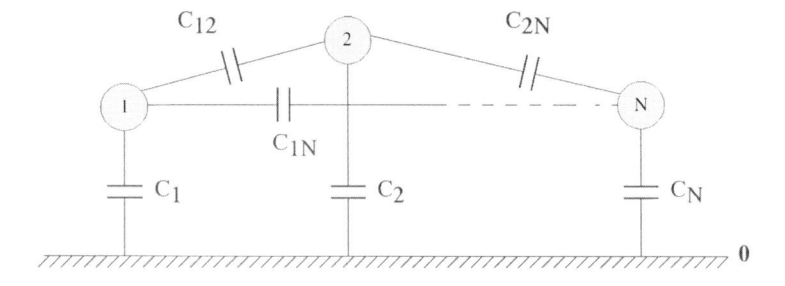

Fig. 11.6. An (N+1)-conductor system in an insulating medium

$$\begin{cases} Q_1 = (C_1 + C_{12} + \ldots + C_{1N})V_1 - C_{12}V_2 - \ldots - C_{1N}V_N \\ Q_2 = -C_{21}V_1 + (C_2 + C_{21} + \ldots + C_{2N})V_2 - \ldots - C_{2N}V_N \\ \ldots \\ Q_N = -C_{N1}V_1 - C_{N2}V_2 - \ldots + (C_{N1} + C_{N2} + \ldots + C_N)V_N \end{cases} \quad (11.9)$$

We use the notation $C_{ii} = (C_i + C_{i1} + \ldots + C_{iN})$, i= 1, ..., N. Thus, C_{ii} represents *the sum of all capacitances* associated with *conductor i*, and this is, by definition, the *self-capacitance* of conductor i. The remaining off-diagonal elements (C_{ij}) are said to be *mutual capacitances*. In more compact form, system (11.9) can be written

$$Q = CV \quad (11.10)$$

where Q is the stored charge vector, V is the vector of the potentials of various conductors (with respect to ground) and C is called the *capacitance matrix*.

■ **Approach.** For simple configurations, the characterization of mutual capacitances and mutual inductances is straightforward (based on the classical laws of electromagnetism). As soon as they are available, a lumped equivalent circuit can be found, which can be submitted to a general-purpose electrical simulator like SPICE. In this way the crosstalk noise can be computed.

■ **Remark.** Modeling the interconnect (which exhibits distributed stray capacitances and inductances) by means of lumped circuits considerably facilitates analysis. However, it should be remembered that *the equivalent circuit approach is valid if and only if the signal in the real structure has a single propagation mode (or a dominant mode).*

At low-frequencies there are no problems, but in high-frequency operation several modes can exist simultaneously. In all such situations, the equivalent circuit approach is of no help. Attention must be paid to how "high-frequency operation" is defined. Generally, whenever the signal wavelength has the same order of magnitude as the largest physical dimension of the circuit, it is customary to say that we are in the high-frequency domain. Nevertheless, low-rate switching signals, *which exhibit fast rise- and fall-times*, must also be treated as high-frequency signals.

11.2.3 Capacitance Estimation

■ **Comment.** It is beyond the scope of this book to present detailed calculations of self- and mutual capacitances by applying Maxwell's equations. Instead, several elementary interconnect configurations often encountered in practice are presented, with their associated evaluation formula.

■ **Self-Capacitance.** It must be recalled that for a single conductor, the concept of self-capacitance is defined as *the inherent distributed capacitance of the conductor in question with respect to ground (earth)*.

■ **Parallel-Plate Capacitor.** Capacitance between two plates of area A [cm^2], air-separated at distance d [cm] is

$$C = 0.0885 \, A/d \quad [pF] \tag{11.11}$$

If an insulator of relative permittivity ε is used instead of air, the resulting capacitance is ε times greater.

■ **Sphere.** For a sphere of radius r [cm], in free space, the self-capacitance is

$$C = 1.1 \, r \quad [pF] \tag{11.12}$$

Note that when the conductor has a rather complicated shape (like the human body, for example), the self-capacitance is estimated by taking an equivalent sphere with the same area.

■ **Cylinder.** The capacitance per unit length of a system containing two concentric circular cylinders of inner radius r_1, outer radius r_2, separated by air, is

$$C = 2\pi \, 0.0885 \, / \, \ln(r_2/r_1) \quad [pF/cm] \tag{11.13}$$

■ **Two-Wire System.** Two parallel conductors in air, each of diameter d, spaced D apart, have a capacitance per unit length

$$C = \pi \, 0.0885 \, / \, \cosh^{-1}(D/d) \quad [pF/cm] \tag{11.14}$$

■ **Four-Wire System.** Consider the system of a two-wire pair shown in Fig. 11.7, where conductors **1'** and **2'** are the return paths of the signals carried by conductors **1** and **2**, respectively.

11.2 Interconnect Modeling

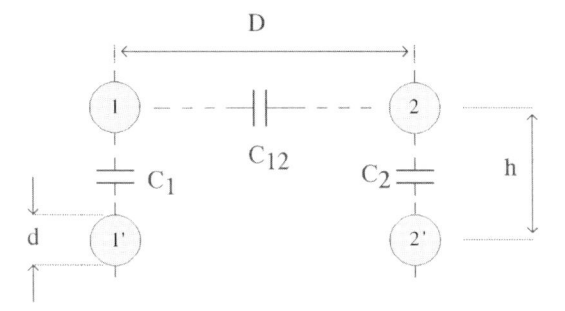

Fig. 11.7. Four-wire system

According to [178], the self- and mutual capacitances per unit length are

$$C_1 = C_2 = \frac{0.121\,\varepsilon_r}{\log\left(\dfrac{h}{d} + \sqrt{\dfrac{h^2}{d^2} - 1}\right)} \quad [\text{pF/cm}] \tag{11.15}$$

$$C_{12} = \frac{C_2 \ln\sqrt{\dfrac{h^2}{D^2} + 1}}{\ln\left(\dfrac{2h}{d} - \dfrac{h}{0.089}\right)} \quad [\text{pF/cm}] \tag{11.16}$$

Note that ε_r denotes the relative permittivity of the dielectric medium between wires, and all dimensions are given in cm. Note also that the self-capacitance of the victim (C_2) plays an important role in coupling.

■ **Capacitance of a Trace (Track).** The usual configuration of a metal trace on a board is shown in Fig. 11.8.

Since the electric field spreads out (fringing) at the edges of any finite thickness conductor, the actual capacitance is greater than that of the parallel-plate model (C_{p1}). Consequently, the total capacitance of the interconnect is [174]

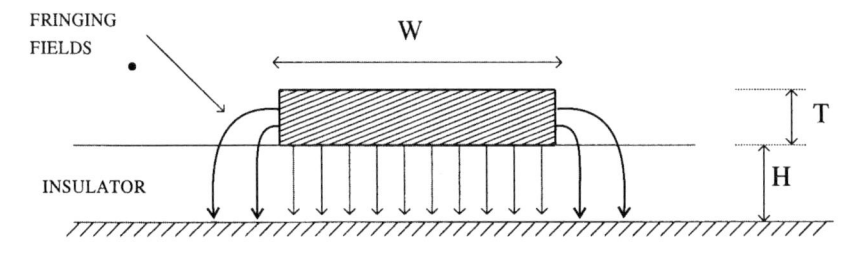

Fig. 11.8. Routing trace (track) on an insulating substrate

11 Interconnection Modeling and Crosstalk

$$C = C_{pl} + 2\, C_{fring} \tag{11.17}$$

with

$$C_{pl} = \varepsilon\, W\, L\, /\, H \tag{11.18}$$

$$C_{fring} \cong \varepsilon\, L \left(\frac{\pi}{\ln\left(1 + \frac{2H}{T}(1 + \sqrt{1 + T/H})\right)} - \frac{T}{4H} \right) \tag{11.19}$$

In the above expressions, L denotes the interconnect length and ε is the permittivity of the insulator.

A much simpler (but less accurate) empirical formula is

$$C \cong \varepsilon\, L \left(\left(\frac{W}{H}\right) + 0.77 + 1.06 \left(\frac{W}{H}\right)^{0.25} + 1.06 \left(\frac{T}{H}\right)^{0.5} \right) \tag{11.20}$$

■ **Multilevel Metal Capacitances.** Modern CMOS processes have at least three metallization layers, at different depths. Modeling of interconnects is usually performed with field simulators, a time-consuming process. For fast estimation, empirical (but less accurate) formula are available.

Consider the typical configuration illustrated in Fig. 11.9.

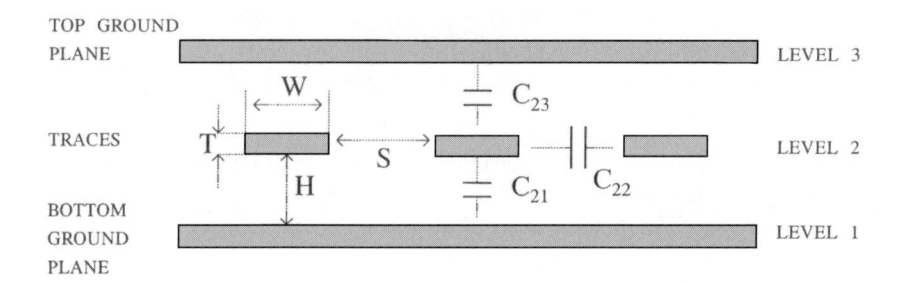

Fig. 11.9. Multilayer structure with three levels

The following expressions are proposed in [174], based on previous work of Chern, Huang et al. [179]:

Trace-to-ground capacitance (one ground plane):

$$\frac{C}{\varepsilon} = \frac{W}{H} + 3.28 \left(\frac{T}{T + 2H}\right)^{0.023} \left(\frac{S}{S + 2H}\right)^{1.16} \tag{11.21}$$

11.2 Interconnect Modeling

Trace-to-ground capacitance (two ground planes):

$$\frac{C}{\varepsilon} = \frac{W}{H} + 1.086\left(1 + 0.685^{-T/1.343S} - 0.9964\exp(-S/1.421\,H)\right)$$

$$\left(\frac{S}{S+2H}\right)^{0.0476}\left(\frac{T}{H}\right)^{0.337} \tag{11.22}$$

Trace-to-trace capacitance (one ground plane):

$$\frac{C}{\varepsilon} = 1.064\left(\frac{T}{S}\right)\left(\frac{T+2H}{T+2H+0.5S}\right)^{0.695}$$

$$+\left(\frac{W}{W+0.8S}\right)^{1.4148}\left(\frac{T+2H}{T+2H+0.5S}\right)^{0.804} \tag{11.23}$$

$$+\,0.831\left(\frac{W}{W+0.8S}\right)^{0.055}\left(\frac{2H}{2H+0.5S}\right)^{3.542}$$

Trace-to-trace capacitance (two ground planes):

$$\frac{C}{\varepsilon} = \frac{T}{S}\left(1 - 1.897\exp\left(\frac{-H}{0.31S} - \frac{-T}{2.474S}\right)\right.$$

$$+\,1.302\exp\left(\frac{-H}{0.082S}\right) - 0.1292\exp\left(\frac{-T}{1.326S}\right)\Bigg) \tag{11.24}$$

$$+\,1.722\left(1 - 0.6548\exp\left(\frac{-W}{0.3477H}\right)\right)\exp\left(\frac{-S}{0.651H}\right)$$

Relations (11.21) – (11.24) relate to geometries satisfying the conditions

$$0.3 \le \frac{W}{H} \le 10 \qquad 0.3 \le \frac{S}{H} \le 10 \qquad 0.3 \le \frac{T}{H} \le 10$$

11.2.4 Inductance Estimation

■ **Discussion.** Although parasitic inductance of interconnections is usually low, it plays an important role in microwave circuits, high-speed I/O buffers, and all VLSI circuits with high-frequency clocks. The major problem is that the actual configurations of interconnects on IC chips or packages are complicated and extremely diverse.

Another problem is more general, and concerns the natural tendency of the magnetic field associated with interconnections to spread out (since no magnetic materials are involved); consequently the values of mutual inductances are slightly lower than those of self-inductances. Note that both categories are independent of the value of the current, and depend only on the wiring geometry.

Therefore, in the following, several simple configurations frequently encountered in practice will be presented [174, 180].

■ **Remark.** A distinction is made between "filament" and "conductor" (or "wire"). *A filament has a negligible cross section, and all lengths are measured with respect to its longitudinal axis.* In contrast, a conductor (wire) has a definite circular or rectangular cross-section, with well-defined dimensions.

■ **Inductance of a Straight Conductor.** The self-inductance of a straight conductor of length l and circular cross section of diameter d is

$$L = 0.002\, l \left(\ln(4l/d) - 0.75 \right) \quad [\mu H] \tag{11.25}$$

where both l and d are given in cm.

A good rule of the thumb for rapid estimation is to adopt the value $7.8\, nH/cm$.

When the conductor material is magnetic and has permeability μ, expression (11.25) becomes

$$L = 0.002\, l \left(\ln(4l/d) - 1 + \mu/4 \right) \quad [\mu H] \tag{11.26}$$

For nonmagnetic materials, a wire of rectangular cross section with sides b and c exhibits a self-inductance

$$L = 0.002\, l \left(\ln \frac{2l}{b+c} + 0.5 - 10^{-5} f(x) \right) \quad [\mu H] \tag{11.27}$$

where f(x) is a function of the aspect ratio $x = b/c$ or c/b (whichever is less than unity), according to Table 11.1.

Table 11.1. Values of f(x) for various aspect ratios

x	0	0.05	0.1	0.2	0.3	0.4	0.5	0.6	0.7	0.8	0.9	1
f(x)	0	146	210	249	244	228	211	197	187	181	178	177

■ **Inductance of Several Parallel Conductors (Common-Mode).** Consider several identical round wires of nonmagnetic material (radius r, length l) connected in parallel. The currents through them all flow in the same direction. All dimensions are given in cm.

- Two parallel straight wires, distance D between their axes, have an inductance

$$L = 0.002\, l \left(\ln \frac{2l}{\sqrt{rD}} - \frac{7}{8} \right) \quad [\mu H] \tag{11.28}$$

- If three mutually parallel round wires are positioned so that their cross sections constitute an equilateral triangle, the inductance is

$$L = 0.002\, l \left(\ln \frac{2l}{\sqrt{r D^2}} - 1 \right) \quad [\mu H] \qquad (11.29)$$

where D is a side of the triangle.

■ **Two Parallel Wires of Equal Length (Differential-Mode).** This is the case of an ordinary transmission line, whose conductors have a length l that is large compared to their separation D. The self-inductance of the circuit is

$$L = 0.004\, l \left(\ln \frac{D}{r} + \frac{1}{4} - \frac{D}{l} \right) \quad [\mu H] \qquad (11.30)$$

It has been assumed that the conductors are nonmagnetic and circular, of radius r; if the wires are made of magnetic material of permeability μ, the term $1/4$ must be replaced by $\mu/4$. Note that the last term is usually negligible.

■ **Mutual Inductance of Two Identical Parallel Straight Filaments.** Consider two parallel filaments of length l, separated by distance D (both expressed in cm). The mutual inductance is

$$M = 0.002\, l \left(\ln \left(\frac{l}{D} + \sqrt{1 + \frac{l^2}{D^2}} \right) - \sqrt{1 + \frac{D^2}{l^2}} + \frac{D}{l} \right) \quad [\mu H] \qquad (11.31)$$

Expression (11.31) is also valid for conductors of circular cross section.

■ **Mutual Inductance of Two Identical Parallel Straight Conductors of Rectangular Cross-Section.** Considering two conductors with identical rectangular cross sections and corresponding sides parallel, the mutual inductance is

$$M = 0.002\, l \left(\ln \frac{2l}{D} - g(x) - 1 + \frac{D}{l} - \frac{1}{4} \frac{D^2}{l^2} \right) \quad [\mu H] \qquad (11.32)$$

Function g(x) yields a correction that depends on the actual dimensions of the rectangles [180]. For traces on a PCB or tracks on IC chips, g(x) can be neglected.

■ **Mutual Inductance of Two Identical Parallel Straight Filaments Above a Ground Plane.** Consider two parallel filaments a distance D apart above a ground plane carrying their return currents; the mutual inductance between the filaments is

$$M = 0.001\, l \left(\ln \left(1 + \left(\frac{2h}{D} \right)^2 \right) \right) \quad [\mu H] \qquad (11.33)$$

11 Interconnection Modeling and Crosstalk

■ **Inductance of a Cylindrical Wire Above a Ground Plane.** A conductor of diameter d at height h above the ground plane carrying its return current has self-inductance

$$L = 0.002\, l \left(\ln \frac{4h}{d} \right) \quad [\mu H] \qquad (11.34a)$$

This equation can be used to estimate the inductance of bonding wires and pins of packages.

For a conductor on a chip of width w and negligible thickness, a rough estimate is [174]

$$L = \frac{\mu}{2\pi} \ln \left(\frac{8h}{w} + \frac{w}{4h} \right) \qquad (11.34b)$$

where h is the height above the substrate (distance to backplane) and μ is the permeability of the material of the wire (typically, $1.257 \cdot 10^{-8}$ H/cm).

■ **Inductance of a Trace Above a Ground Plane.** For the configuration presented at Fig. 11.8 and assuming a narrow trace,

$$L = 0.2 \ln \left(\frac{8H}{W} + \frac{W}{4H} \right) \quad [\mu H/m] \quad \text{if} \quad \frac{W}{H} \leq 1 \qquad (11.35a)$$

while for a large trace

$$L = \frac{1.26}{\frac{W}{H} + 1.393 + 0.667 H \left(\frac{W}{H} + 1.444 \right)} \quad [\mu H/m] \quad \text{if} \quad \frac{W}{H} \geq 1 \qquad (11.35b)$$

(for notation, see Fig. 11.8).

■ **Remark.** In the case of a nonuniform transmission line (conductors A and B with different cross sections), the total inductance is given by the general relation

$$L = L_A + L_B - 2M_{AB}$$

where L_A, L_B, and M_{AB} are each evaluated according to their geometry.

11.2.5 Modeling a Multiconductor Line

■ **Hypothesis.** Let us consider a system of several uniform conductors located in a homogeneous insulator medium. Suppose that for an (n+1)-conductor line, all lines share the same return current path, namely the ground conductor G.

■ **Three Conductor Line.** An elementary section of length Δx is presented in Fig. 11.10, where the transmission lines are constituted by the two pairs A and G (ground) and B and G.

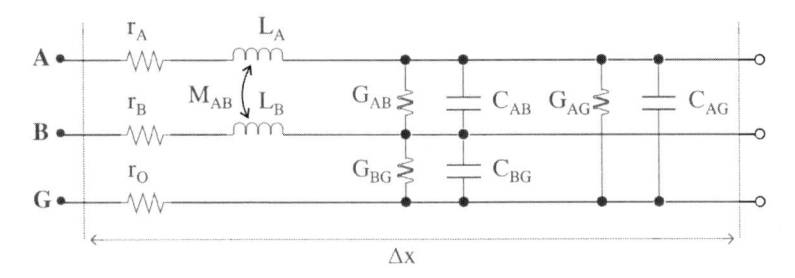

Fig. 11.10. System of three coupled conductors

■ **Notations.** The following notation has been adopted (related to the electrical parameters of each section Δx):

r_A, r_B, r_O — conductor resistances;
L_A, L_B, M_{AB} — self- and mutual-inductances of conductors A and B (if the inductance of the ground return is neglected);
C_{AG}, C_{BG}, C_{AB} — self- and mutual capacitances of the conductors;
G_{AG}, G_{BG}, G_{AB} — leakage conductances between conductors, due to imperfect dielectric medium.

■ **Electrical Parameters.** The elementary section of three conductors is completely characterized by means of the matrices [181]

$$R = \begin{bmatrix} r_A + r_o & r_o \\ r_o & r_B + r_o \end{bmatrix} \qquad L = \begin{bmatrix} L_A & M_{AB} \\ M_{AB} & L_B \end{bmatrix}$$

$$G = \begin{bmatrix} G_{AM} + G_{AB} & -G_{AB} \\ -G_{AB} & G_{BM} + G_{AB} \end{bmatrix} \qquad C = \begin{bmatrix} C_{AM} + C_{AB} & -C_{AB} \\ -C_{AB} & C_{BM} + C_{AB} \end{bmatrix}$$

It can be shown that

$$\boxed{L\,C = C\,L = \mu\varepsilon\, I_2} \qquad (11.36)$$

$$\boxed{L\,G = G\,L = \mu\sigma\, I_2} \qquad (11.37)$$

ε, μ, σ being the permittivity, permeability, and conductivity of the insulator medium; I_2 denotes the second-order unit matrix.

■ **Remarks**

- Matrix R is frequency dependent, because the intrinsic resistance of the conductors is subject to the skin effect.
- Equations (11.36) and (11.37) show that only one matrix must be found, the remaining two being derivable via matrix operations.
- Extending (11.36) and (11.37) to a system of several conductors merely requires augmented matrices.

TRACE ON PCB

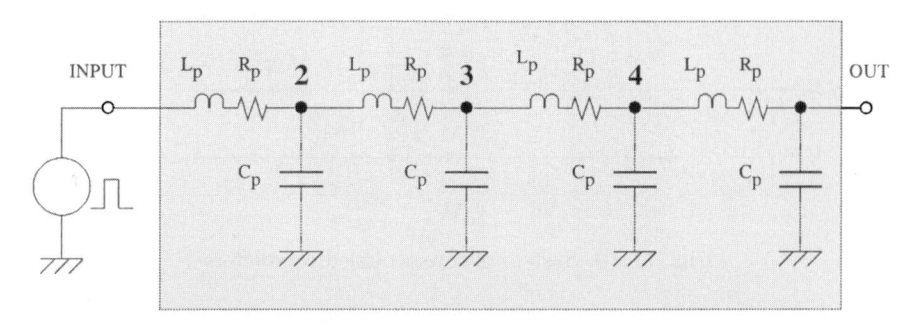

Fig. 11.11. Equivalent circuit of a metal trace on a PCB

11.2.6 Modeling a Trace on a Printed Circuit Board

■ **Comment.** A trace on a PCB (above a ground plane, or parallel to a ground trace) is a distributed RLC structure. However, for analysis with a general purpose simulator (PSPICE), it can be represented as a lumped equivalent circuit containing at least two identical sections. Obviously, increasing the number of sections improves the accuracy of simulation, but the total job time also increases.

■ **Approach.** The global inductance, capacitance, and resistance of the trace, estimated according to Sect. 11.2, are divided by the number of sections in order to obtain the values of parameters per section.

■ **Configuration.** A good compromise between accuracy and computing time is reached with a 4-section lumped RLC circuit (Fig. 11.11). Since the behavior of the trace is critical when it carries fast switching signals, the circuit is driven by a quasi-rectangular pulse generator (to simulate a digital application).

■ **Results.** Simulation with PSPICE has been performed by adopting a pulse period of 10 µs, and a pulse width of 5 µs, switching between a high level of 5 V and a low level of 0 V. The rise- and fall-times are equal; two different values have been selected, namely 1 ns and 100 ns. With component values $L_p = 4\,\mu H$, $C_p = 2\,pF$, and $R_p = 20\,m\Omega$, the voltage at various nodes has been monitored. The following points have been noted:

- Due to the multi-pole, multi-zero configuration, a signal edge rate degradation (ringing) is observed. The amplitude of the damped quasi-sinusoidal oscillation increases with node number (the highest amplitude is perceived at the output). This demonstrates that *a long trace is to be avoided*.
- A "nice" plot of the waveform at node 4 is obtained by conveniently modifying the values of components: $L_p = 4\,\mu H$, $C_p = 20\,pF$, and R_p

11.2 Interconnect Modeling

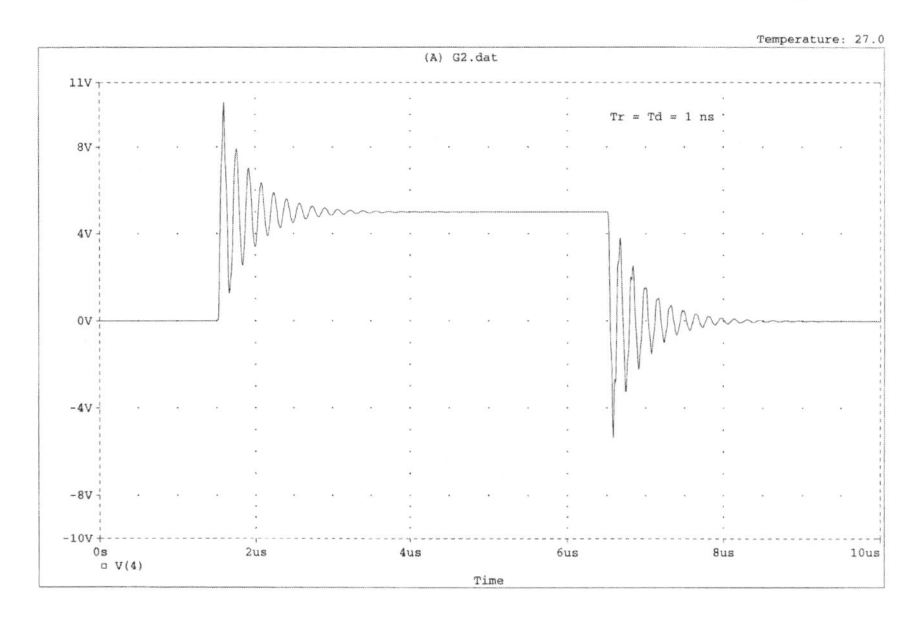

Fig. 11.12. Plot of V(4) versus time for modified component values

$= 20\,\Omega$; this plot is shown in Fig. 11.12. Note that increasing R_p causes *rapid damping of the oscillations, which is beneficial to the reduction of emitted interfering signals* (but a high R_p also attenuates the useful signal transmitted on the trace!).

- Figure 11.13 shows typical waveforms obtained at node 4, together with their corresponding frequency spectra, for $L_p = 4\,\mu H$, $C_p = 2\,pF$, and $R_p = 20\,m\Omega$. Note that for $t_d = t_r = 1\,ns$, significant spectral power is observed up to 20 MHz, while for $t_d = t_r = 100\,ns$, the spectrum is much narrower (up to about 2 MHz). The slower the rise- and fall-times, the simpler the signal spectrum. We conclude that whenever possible, *reduction of the rise- and fall-time of the input signal has a favorable effect on the emission of interfering signals*.
- The characteristic impedance of the trace model shown in Fig. 11.11, at the frequency of the damped oscillations, is roughly $7\,\Omega$. Terminating the trace output with this resistance improves the waveform shape by drastically reducing the amplitude of damped oscillations. Once again, *terminating a trace with its characteristic impedance* is a good policy.

■ **Concluding Remark.** Although this model is specific to a trace on a PCB, the same approach applies to any interconnection above ground, provided that the numerical values of various component are appropriately modified.

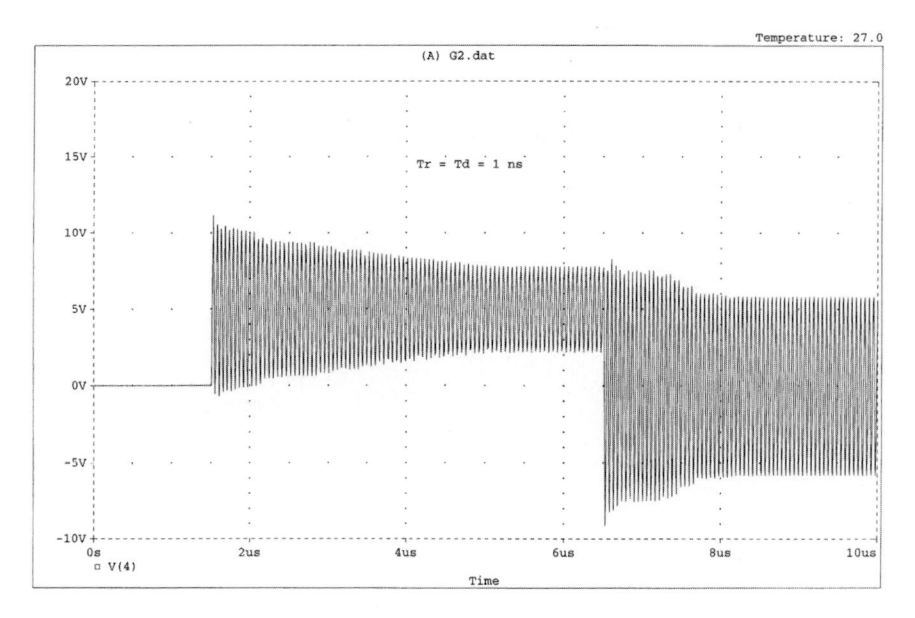

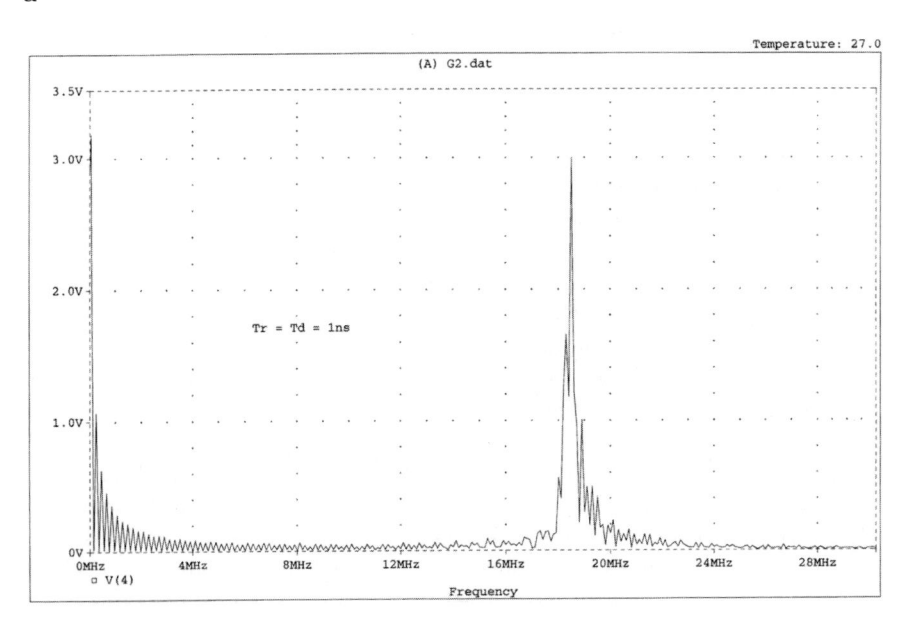

Fig. 11.13. a Waveform obtained with $t_r = t_d = 1$ ns; **b** its frequency spectrum; **c** waveform for $t_r = t_d = 100$ ns; **d** its associated frequency spectrum

11.2 Interconnect Modeling

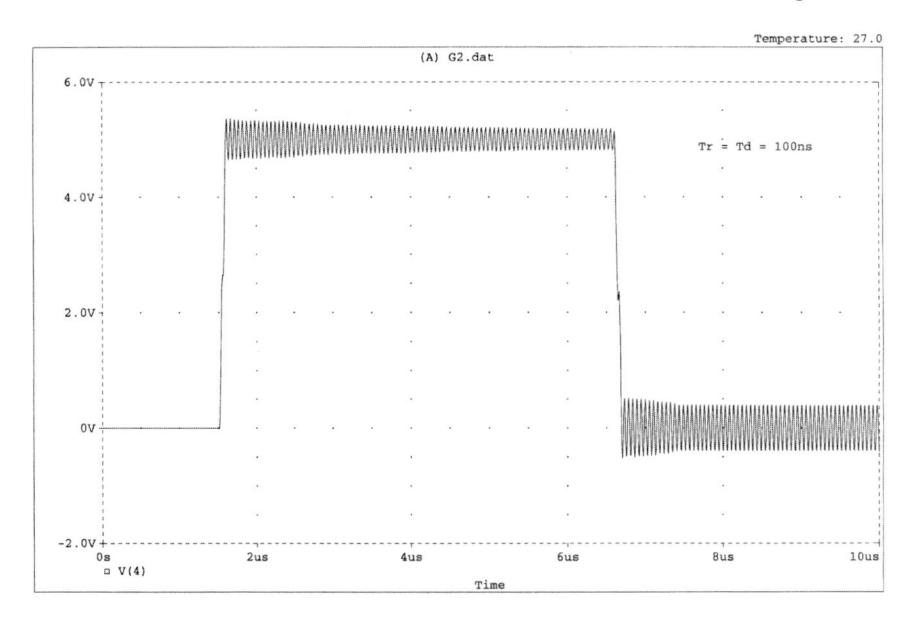

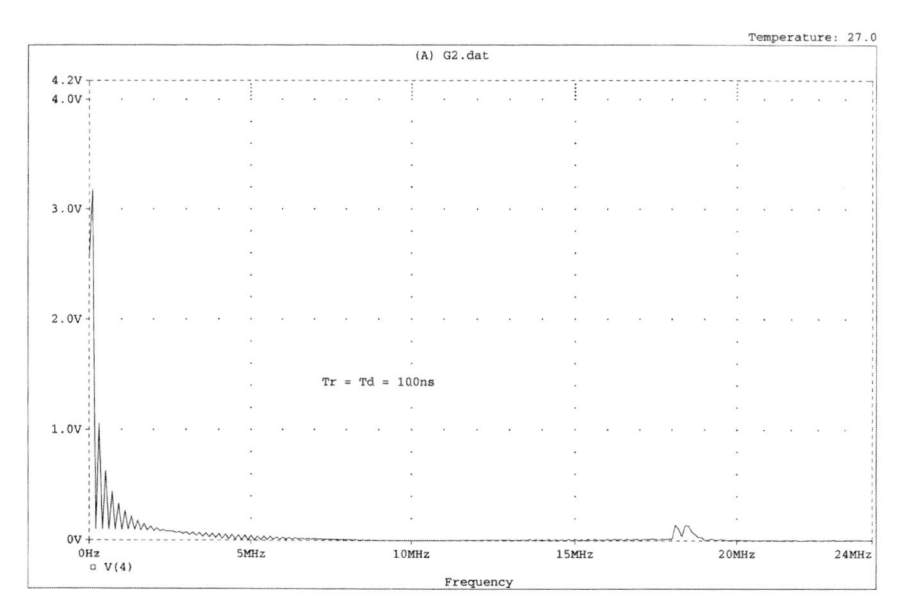

c

d

Fig. 11.13. (continued)

11.3 Crosstalk

11.3.1 Basic Concepts

■ **Definition 1.** *Crosstalk is the penetration of a signal from one channel to another by conduction or radiation.* This definition is pertinent to telecommunication systems, where cross-coupling between speech communication channels leads to intelligible or non-intelligible interference.

■ **Definition 2.** *Crosstalk is the interference caused by stray electromagnetic or electrostatic coupling of energy from one interconnect to an adjacent one.* Due to this coupling, signals (or noise) carried by one line (called the disturbing line) can reach another (called the disturbed line).

■ **Remark.** Crosstalk only applies to circuits in close proximity (near-field).

■ **Classification.** According to the particular circumstances (Fig. 11.14), a distinction is made between near-end crosstalk and far-end crosstalk. The former suggests that the source (on the disturbing line) and the measuring point (on the disturbed line) are at the same end (Fig. 11.11a), while the latter refers to the case in which they are at opposite ends (Fig. 11.11b).

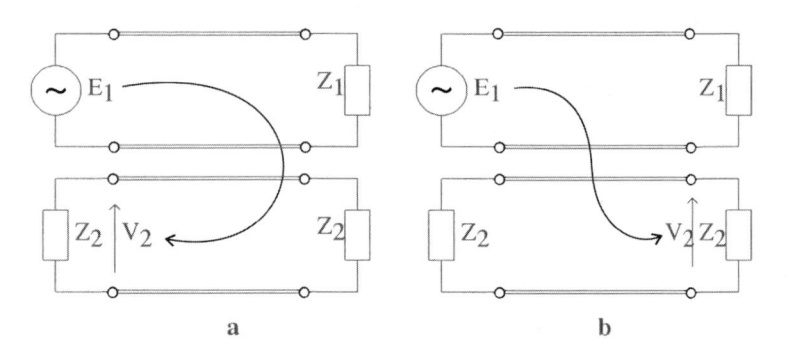

Fig. 11.14. Crosstalk: **a** near-end; **b** far-end

■ **Objective.** The objective of a crosstalk calculation is to predict the undesired signal arriving at the disturbed line, given the geometry of interconnections, medium parameters, and termination characteristics.

■ **Crosstalk Loss.** This is *the ratio, expressed in decibels, of: (1) the power P_1 injected at the input end of the disturbing line, to (2) the undesired power P_2 at the measuring point of the disturbed line, under certain termination conditions*:

$$L_c = 10\ \log\ \frac{P_1}{P_2} = 20\ \log\ \frac{E_1}{V_2} + 20\ \log\ \frac{Z_2}{Z_1} \quad [\text{dB}] \qquad (11.38)$$

■ **Remark.** In practice two situations are encountered:

1) If the length of the conductors is less than $\lambda/4$ (short line), propagation along them can be neglected and the distinction between near-end and far-end crosstalk is meaningless. From a practical point of view, capacitive coupling can be dissociated from its magnetic counterpart, considerably simplifying the analysis.
2) In contrast, when the conductor length exceeds $\lambda/4$, propagation must be taken into account, and the conductors act like long transmission lines. The cross-coupling is then electromagnetic.

11.3.2 Crosstalk Due to Dominant Capacitive Coupling [178, 182]

■ **Overview.** In the context of a rapid increase in both component density and switching rate of digital systems, stray capacitance of interconnections becomes an important limiting factor in the design of integrated circuits. Crosstalk characterization and reduction in very large systems requires knowledge of the capacitances associated with interconnects.

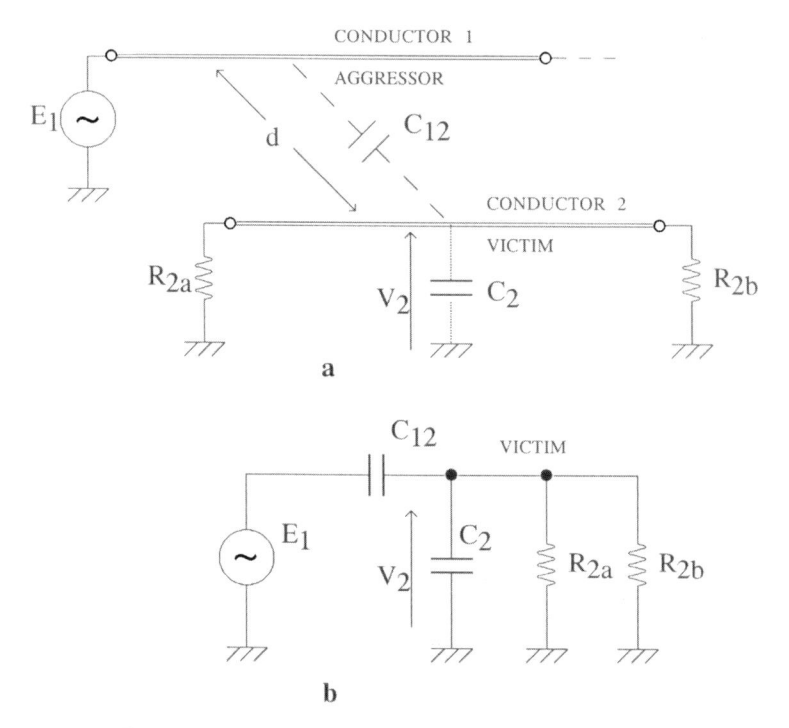

Fig. 11.15. a Capacitive coupling between two parallel wires; b lumped equivalent circuit

11 Interconnection Modeling and Crosstalk

■ **Explanation.** Consider a section of two parallel conductors in close proximity over ground (Fig. 11.15a), whose lengths are smaller than $\lambda/4$. Suppose that conductor 1 is the aggressor and conductor 2 is the victim. Due to the cross-capacitance C_{12}, a fraction of the signal carried by conductor 1 couples into conductor 2 and reaches its terminations R_{2a} and R_{2b}. Since propagation along wires can be neglected, the lumped equivalent circuit shown in Fig. 11.15b can be used successfully.

■ **Calculation.** To find the fraction of the input signal E_1 that contaminates conductor 2 (assuming that resistances R_{2a} and R_{2b} are very high, or even better, open circuits), we may apply the voltage divider formula:

$$V_2 = E_1 \left. \frac{1/j\omega C_2}{1/j\omega C_2 + 1/j\omega C_{12}} \right|_{R_{2a} \to \infty,\, R_{2b} \to \infty} \tag{11.39}$$

which yields

$$\boxed{V_2 = E_1 \left. \frac{C_{12}}{C_2 + C_{12}} \right|_{R_{2a} \to \infty,\, R_{2b} \to \infty}} \tag{11.40}$$

By definition

$$K = \frac{C_{12}}{C_2 + C_{12}} \tag{11.41}$$

is called the *coupling coefficient* [178].

When R_{2a} and R_{2b} are of the same order of magnitude as the reactance of C_2, in (11.39) we must replace the term $1/j\omega C_2$ by the impedance Z_2 of the parallel combination of C_2, R_{2a}, and R_{2b}. For a four-wire system, C_1, C_2, and C_{12} are estimated with (11.15) and (11.16).

■ **Comments.** As simple as they are, (11.40) and (11.41) suggest the following comments:

- The coupling coefficient does not depend upon the self-capacitance of the disturbing line.
- To reduce the coupling coefficient, C_2 must be increased (this can be achieved in practice by locating conductor 2 as close as possible to ground). Another solution is to reduce the mutual capacitance C_{12}, either by introducing shielding or by increasing the separation between conductors 1 and 2.
- The termination resistances can favorably affect crosstalk; as a matter of fact, the fraction of signal contaminating conductor 2 increases when R_{2a} and R_{2b} are larger; one solution is to reduce their values whenever possible.

■ **Conclusion.** *Capacitive crosstalk is dominant in all situations involving lines terminated by high-value impedances and operating at high frequencies.* To minimize the resulting crosstalk, several possibilities exist:

- Shield the disturbing line, the disturbed line, or both. The shield must be connected to ground at both ends.
- Insert a ground trace between the traces 1 and 2. This is certainly less effective than shielding, but it reduces the mutual capacitance.
- Separate the disturbing from the disturbed conductor as much as possible.
- Place the disturbed conductor as close as possible to ground.

11.3.3 Crosstalk Due to Dominant Inductive Coupling [181, 182, 185]

■ **Explanation.** As before, consider an electrically short section of two conductors above ground (Fig. 11.16), where propagation along the lines can be neglected.

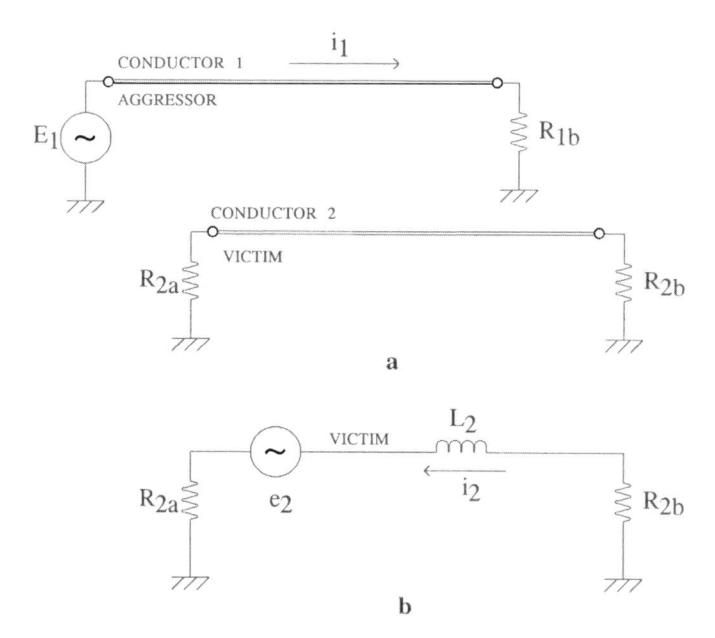

Fig. 11.16. a Inductive coupling between two parallel wires; b equivalent circuit

The signal source E_1 gives rise to a current (i_1) in conductor 1, which in turn produces a time-varying magnetic field around the conductor. The lines of this field cut the loop GROUND, R_{2a}, CONDUCTOR 2, R_{2b}, GROUND. Consequently, a voltage (proportional to the mutual inductance M_{12} and the loop area) is induced in this loop.

■ **Estimation.** According to electromagnetic theory, the induced voltage is

$$e_2 = -j\omega M_{12}\, i_1 \tag{11.42}$$

where the minus sign merely indicates that the induced current is opposed to the current generating it (i_1).

The crosstalk noise voltage at the near end is

$$V_{2a} = e_2 \frac{R_{2a}}{R_{2a} + R_{2b} + j\omega L_2} \qquad (11.43)$$

L_2 being the self-inductance of conductor 2. Similarly, the crosstalk at the far end is

$$V_{2b} = e_2 \frac{R_{2b}}{R_{2a} + R_{2b} + j\omega L_2} \qquad (11.44)$$

In the above expressions, the self- and mutual inductances of conductors are calculated by means of equations given in Sect. 11.2.4.

■ **Remedies.** Relations (11.43) and (11.44) show that the self-inductance of the victim (L_2) is of real help in crosstalk mitigation: a higher L_2 reduces crosstalk noise.

Another possibility is to reduce the mutual inductance, either by increasing the distance separating the conductors or by reducing the height above the ground. Neither is particularly effective, due to the logarithmic function (see (11.33)).

Hence, two solutions are often adopted in practice:

– shielding the victim (conductor 2);
– replacing line 2 by a twisted pair.

□ **Shielding Conductor 2.** This situation is illustrated in Fig. 11.17, where the shield is nonmagnetic and grounded at both ends.

The physical mechanism responsible for crosstalk reduction is the following: current I_1 generates a magnetic field, which induces a current (I_S) in the shield of conductor 2 (the return path for the shield current is provided by the ground plane). The magnetic field produced by I_1 also cuts the unprotected end regions (A – 2 and 2' – B), where interference appears. In turn,

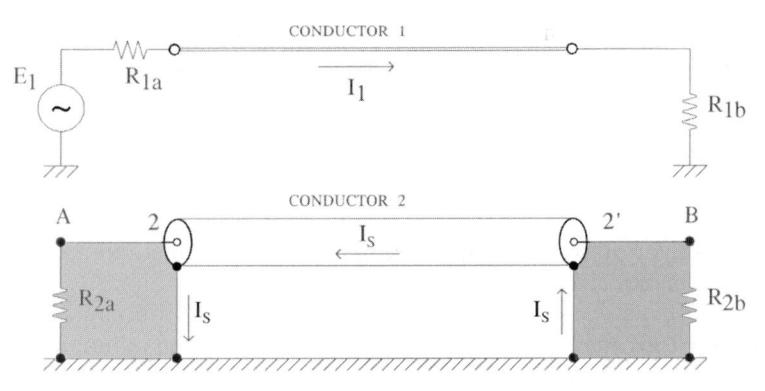

Fig. 11.17. Crosstalk mitigation by shielding the victim wire

the shield current (I_S) generates its own magnetic field, which also cuts the same end regions (A – 2 and 2' – B). Since I_S and I_1 are opposite currents, the two magnetic fields cancel almost completely in the end regions.

Note that grounding the shield at both ends is mandatory; otherwise, no improvement in crosstalk is to be expected.

☐ **Replacing the Victim by a Twisted Pair.** Whenever a twisted pair is employed (Fig. 11.18), the voltages induced in all even loops (B2, B4) are of opposite polarity to those induced in odd loops (B1, B3), since the loop orientation alternates.

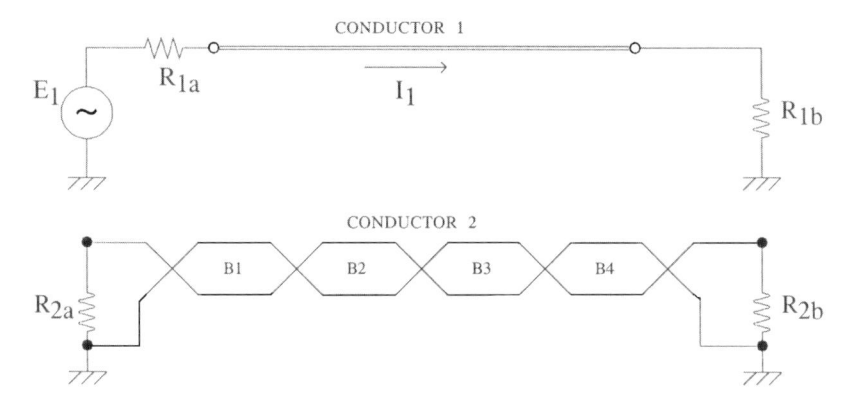

Fig. 11.18. Inductive crosstalk: twisted pair of abrupt loops

Global cancellation is expected, and crosstalk noise should vanish. In practice, significant reduction is observed, but residual crosstalk noise still exists, due to the terminations (which can be regarded as "half-loops"), due to inherent irregularities in the twist, etc.

■ **Remarks**

– Twisted pairs are generally useful at frequencies below 100 kHz [182]; above this limit, losses in the twisted pair increase significantly.
– Since twisting the wires tends to balance the capacitances to ground and to extraneous objects, a balanced circuit using a twisted pair can provide protection against both magnetic and capacitive coupling, without any shield.

11.3.4 Crosstalk Due to Electromagnetic Coupling [178, 185]

■ **Comment.** Whenever an interconnect cannot be ascribed to electrically short lines, magnetic and capacitive coupling can no longer be treated separately. Propagation along conductors must be taken into account by applying classical transmission line theory.

■ **Discussion.** Assume a circuit operating at frequencies such that the corresponding minimum wavelength is much greater than the maximum length of interconnect. Typically, this is the case of electrically short lines, where propagation phenomena can be ignored. However, this is not necessarily true for transients, and perhaps the most unavoidable transient in practice is the turning on of the power supply. Since basically all transients involve short durations, this translates into a broad spectrum extending up to high frequencies, which can greatly exceed the steady-state operating frequency. Therefore, *even in situations when steady-state analysis enables us to calculate the inductive and capacitive crosstalk contributions separately, propagation along conductors must still be considered for fast transients.* As the characteristic impedance of the line plays a decisive role in propagation, attention will be focused on this parameter.

■ **Definitions**

1) The characteristic impedance (Z_o) is defined as the ratio of the complex voltage to the complex current, in the same transverse plane of the transmission line, with the sign so chosen that the real part is positive, when no standing waves are present in the line. For a uniform transmission line, one useful expression is

$$Z_o = \sqrt{\frac{R + j\omega L}{G + j\omega C}} \cong \sqrt{\frac{L}{C}} \qquad (11.45)$$

where R, G, L, C are the line parameters per unit length. Generally, low crosstalk requires low characteristic impedance; practical methods to reduce L and increase C are:
 – minimize loop area;
 – increase surface area for coupling;
 – place line conductors as close to one another as possible.

2) In the case of parallel-coupled lines, *the even mode impedance Z_{oe} is defined as the characteristic impedance of one line to ground when equal currents are flowing in the two lines. The odd mode impedance Z_{oo} is defined as the characteristic impedance of one line to ground when equal and opposite currents are flowing in the two lines.*

These concepts are useful when calculating the crosstalk between a pair of conductors located in close proximity over a ground plane (such as parallel PCB tracks). At microwave frequencies, the characteristic impedance of coupled striplines is

$$Z_o = \sqrt{Z_{oe} Z_{oo}} \qquad (11.46)$$

■ **Remark.** In analog circuits, turning the power supply on is identified with a step current demand; in logic circuits, fast transients are associated with the waveform rise- and fall-time. Assuming that a step current (ΔI) is required when the power is turned on, the corresponding voltage transient (ΔV) across the load depends upon the characteristic impedance of the power supply line:

$$\Delta V = Z_o \, \Delta I$$

Obviously, the smallest the value of Z_o, the weaker the associated voltage transients at the load.

■ **Calculation.** The coupling between a pair of conductors in close proximity, located over a ground plane and terminated in their characteristic impedance, is described by means of the crosstalk coefficient [178]:

$$K = \left(\frac{Z_{oe}}{Z_{oo}} - 1\right) \bigg/ \left(\frac{Z_{oe}}{Z_{oo}} + 1\right) \tag{11.47}$$

□ **Example.** In order to appreciate the crosstalk in embedded, coupled striplines (Fig. 11.19), the coefficient K must be evaluated.

For instance, in a multilayer PCB the signal tracks are sandwiched between a ground plane and a plane connected to the power supply, which corresponds to the structure depicted in Fig. 11.19. Typical dimensions are W = 0.5 mm, S = 0.75 mm, h = 0.86 mm.

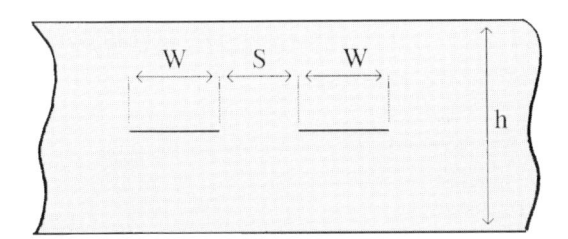

Fig. 11.19. Embedded stripline

Assuming a substrate of epoxy glass ($\varepsilon_r = 4$), this yields $Z_{oe} \cong 210\,\Omega$ and $Z_{oo} \cong 195\,\Omega$ (with the equations given in [183]), which finally yields K = 0.037 (with (11.47)). We may conclude that near-end crosstalk transports 3.7% of the voltage carried by the disturbing track.

■ **Common Configurations.** Table 11.2 represents a collection of some of the most frequently encountered configurations, together with their characteristic impedance [178, 184, 185].

Table 11.2. Characteristic impedance of various lines

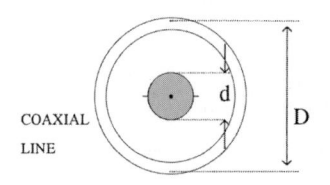

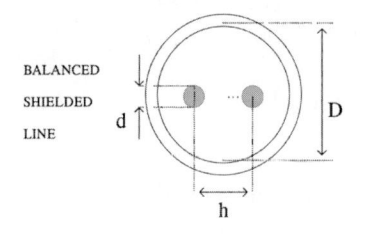

COAXIAL LINE

$Z_o = (60/\sqrt{\varepsilon}) \ln(D/d)$
ε = dielectric constant

BALANCED SHIELDED LINE

For $D \gg d$ and $h \gg d$ with
$v = h/d$ and $\sigma = h/D$

$Z_o = (120/\sqrt{\varepsilon}) \ln\left[2v\left[(1-\sigma^2)/(1+\sigma^2)\right]\right]$

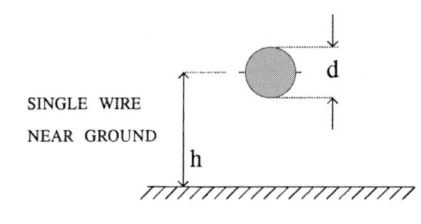

SINGLE WIRE NEAR GROUND

For $h \gg d$

$Z_o = (138/\sqrt{\varepsilon}) \log(4h/d)$

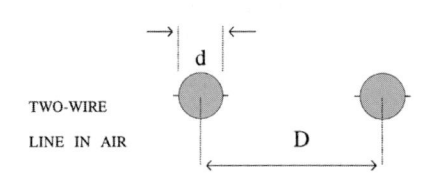

TWO-WIRE LINE IN AIR

$Z_o = 120 \cosh^{-1}(D/d)$
$Z_o \cong 120 \ln(2D/d)$

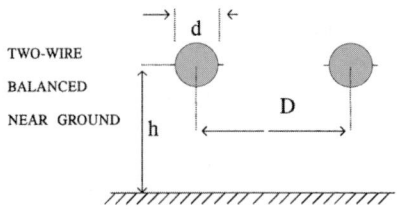

TWO-WIRE BALANCED NEAR GROUND

For $d \ll D, h$ $\quad Z_o = (276/\sqrt{\varepsilon}) \log\left((2D/d)\left(1+(D/2h)^2\right)^{-0.5}\right)$

11.3 Crosstalk

SIDE-BY-SIDE TRACKS

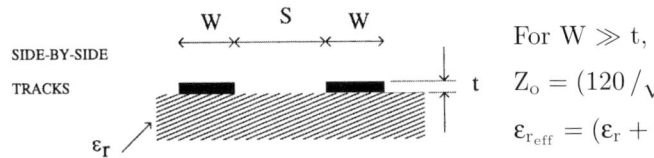

For $W \gg t$, $S \geq W$
$$Z_o = (120/\sqrt{\varepsilon_{r_{eff}}}) \ln\left(\frac{\pi S}{W+t}\right)$$
$$\varepsilon_{r_{eff}} = (\varepsilon_r + 1)/2$$

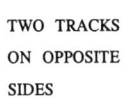

TWO TRACKS ON OPPOSITE SIDES

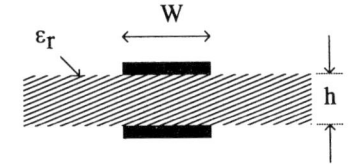

For $h \ll W$
$$Z_o = (120\pi/\sqrt{\varepsilon_r})\left(\frac{h}{W}\right)$$

MICROSTRIP LINE

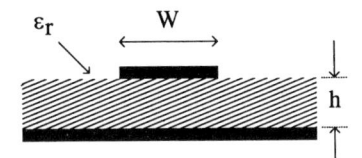

For $W/h \leq 1$, $Z_o = (60/\sqrt{\varepsilon_{r_{eff}}}) \ln\left(\frac{8h}{W} + \frac{W}{4h}\right)$

$$\varepsilon_{r_{eff}} = \frac{\varepsilon_r + 1}{2} + \frac{\varepsilon_r - 1}{2}\left(0.04\left(1 - \frac{W}{h}\right)^2 + 1/\sqrt{1 + 12h/W}\right)$$

For $W/h \geq 1$, $Z_o = (120\pi/\sqrt{\varepsilon_{r_{eff}}})\left(\frac{W}{h} + 1.393 + 0.667 \ln\left(\frac{W}{h} + 1.444\right)\right)^{-1}$

$$\varepsilon_{r_{eff}} = \frac{\varepsilon_r + 1}{2} + \frac{\varepsilon_r - 1}{2}\left(1/\sqrt{1 + 12h/W}\right)$$

MULTILAYER PRINTED CIRCUIT BOARD

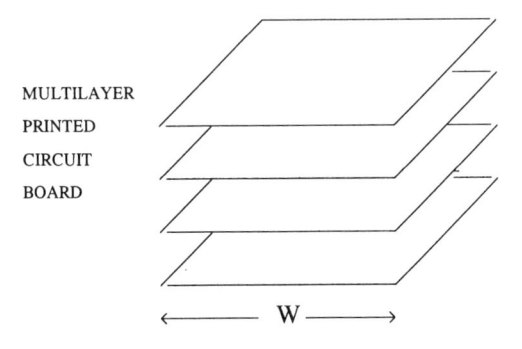

INTERCONNECT
GROUND PLANE
POWER PLANE
INTERCONNECT

$$Z_o = (120\pi/\sqrt{\varepsilon_r})\left(\frac{h}{W}\right)$$

h = height between planes

11.4 Interconnect Optimization

11.4.1 Layout and Printed Circuit Board

■ **Explanation.** Generally, a PCB contains most of the interconnections belonging to a particular circuit or system. If at the circuit level it is rather straightforward to identify the disturbing (aggressor) circuits and the disturbed (victim) circuits, the interconnect may belong simultaneously to both categories. Due to the temporal evolution of various signals, in some "quiet" intervals a particular trace may be a victim, while during "full activity" it may act as an aggressor towards its neighbors.

Globally, the traces of a PCB act as small antennas, able to receive or radiate noise. To mitigate radiated noise, all the arrangements traditionally made to improve the gain of an antenna must be applied in reverse, namely: minimize aperture efficiency, reduce effective area, shield antenna, filter signals, avoid antenna lengths of $\lambda/2$ or $\lambda/4$ at the operating frequency, etc.

■ **Objectives.** A well-designed PCB must satisfy not only the interconnection constraints imposed by the layout, but also several additional requirements:

- high immunity to interfering signals;
- reduced emission of interfering signals;
- low crosstalk between its traces.

■ **Remarks**

- The previous requirements must be taken into consideration early in the design of a PCB. This is not a simple problem, since to the best of the author's knowledge no commercial CAD package is able to predict with enough accuracy the noise emission and susceptibility of the whole product. Of course, electromagnetic computation packages are available, but they don't satisfy the needs of commercial product design. Note that any *post hoc* solution to improve the performance of a PCB is expensive and less efficient.
- Since most problems of electromagnetic compatibility have their origin in the design of PCBs or in improper grounding, emphasis will be placed on these issues.

■ **Approach.** Concerning PCB design, the following actions must be considered [185]:

1. Whenever the size of the circuit is excessive, it is worth *partitioning* the system into a number of sections (each corresponding to a well-defined function). Various partition techniques are presented in [176].

Create a mother board and several daughter boards, if necessary. The benefit is twofold: the lengths of traces are shorter, and the identification of interference sources is much easier on a small board than on a larger one.
Then
- determine which section will be noisy and which not;
- put them in separate areas as far as possible from one another;
- provide interface points to allow optimum common-mode current control.
2. Ensure proper *grounding* by marking all ground points on the circuit schematic diagram. Create a ground map, verify it periodically, and keep evidence of
   - critical grounding bonds (of screens, cabinets, connectors, filters);
   - ground points which may be tied together and those that may not.
3. Once grounding finished, proceed with *PCB design* in the following order:
   a) Always begin by drawing the ground traces; if a CAD package with an automatic routing facility is available and it does not allow drawing the ground traces first, don't hesitate to do it manually, before running the software.
   b) Next, draw the traces carrying currents with strong di/dt variations (clock lines, data bus lines, and the oscillators for switching-mode power supplies), locating them as close as possible to the ground traces (better yet, place the clock trace between two ground traces).
   c) Finally, draw the signal traces and the remaining ones.

11.4.2 Managing PCB Optimization

■ **Objective.** In order to mitigate stray coupling in the near field due to improper design of PCB, the following rules (to be viewed as an expert system) are proposed [182, 185].

■ **Rules**

- Divide the circuit into several parts and allocate separate areas: one for the analog section, another for the digital section, another for I/O, etc. If all must be placed on the same board, provide ground separation for all; whenever required, tied them together at only one point.
- Any PCB contains electrical radiating dipoles (traces) and magnetic radiating current loops. A good policy is to *reduce the lengths of traces subject to high dv/dt, as well as loop areas enclosed by high current*.
- Minimize trace and component lead lengths.
- Use a ground plane (or a grid plane) to minimize inductance of traces.
- Place critical circuits away from the ground plane edges.
- Locate the supply traces, as well as the signal traces, as close as possible to their ground returns.

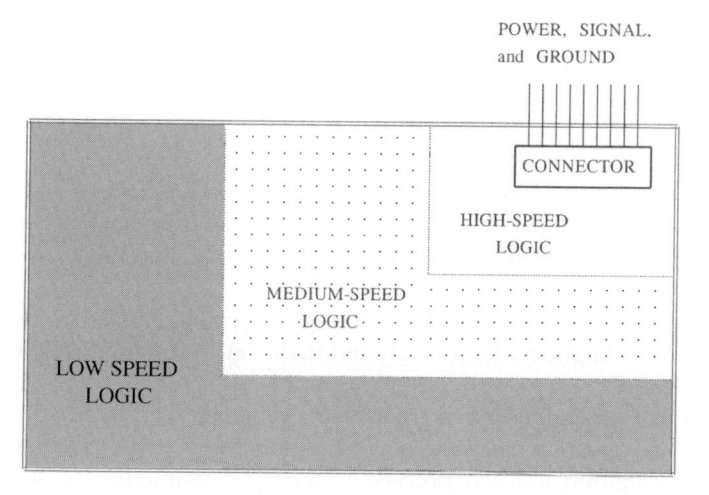

Fig. 11.20. Seperation of logic families by speed on a PCB

- Provide suitable widths for all traces carrying high currents to avoid overheating and electromigration.
- Never extend a digital ground plane over an analog section, nor an analog ground over a digital section.
- Apply segregation by speed (Fig. 11.20), if several logic families are mixed in the design. Place the connector close to the highest-speed logic family, to minimize the lengths of power, signal, and ground traces to it.
- Terminate lines carrying high-frequency analog signals (or any kind of digital signals) by matched impedances.
- Keep trace length smaller than $\lambda/20$ at the operating frequency, to ensure low-efficiency field radiation. If power supply or signal lines are longer, transpose them (but this requires another layer of metal!).
- Use a single-point connection between the digital and the analog ground, only at the system's digital-to-analog converter.
- Transform all unoccupied areas of the PCB into bulk ground regions.
- Remember the 5/5 rule: *when clock frequencies in excess of 5 MHz or when rise times faster than 5 ns are employed, a multilayer board should be used* [176]. Use a plane for power supply ($+V_{CC}$ or $+V_{DD}$) and another for ground in the middle; dedicate top and bottom planes to signal traces. Connections between points belonging to various planes are made by vias.

■ Grounding Concerns for Layout

- *Always separate the analog ground from the digital ground.* If ground planes are used, don't overlap them, because the noise of digital circuits will contaminate the analog circuits.
- *Adopt the largest possible width for ground traces*, to reduce their stray inductance.

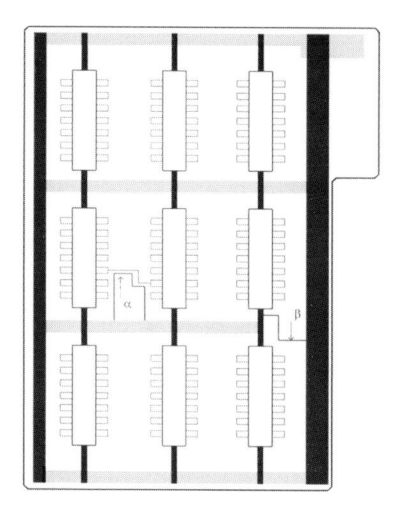

Fig. 11.21. Two-layer PCB with a ground grid system

- Give *priority to ground traces* when creating the PCB.
- If a single-sided PCB is used, encircle the board as much as possible with a large ground trace.
- Minimum trace impedance is obtained with a ground plane, because *a ground plane provides an infinite number of alternate paths, so that the signal can "choose" a return path closest to the signal path.* Whenever this is not possible, employ a ground grid plane instead. With double-sided PCBs, the most frequently encountered technique is to route horizontal traces on the solder side and vertical traces on the circuit side. Plated through-holes are used to interconnect the two sides at the nodes of the grid.

Two special techniques are illustrated in Fig. 11.21:
  - To protect against spurious fields by means of a local return, a ground trace α is run close to a trace carrying a strong disturbing signal (or a sensitive signal).
  - When for different reasons a segment is missing in a grounding grid, it is better to use a narrow trace β to replace the missing segment instead of leaving a gap.

- Place higher-current devices (or circuits) towards ground entry point.
- Select a typical grid size spacing of about 1 cm (but values up to 5 cm can still be accepted). A recommended rule of thumb is to *use a grid size spacing that allows a grid line to exist next to every integrated circuit on the board* [176].
- Split the ground plane into several areas, according to the function of various subsystems. Each such area may be treated as a multiple grounding point configuration. Finally, interconnect different areas, if necessary, using only one point.

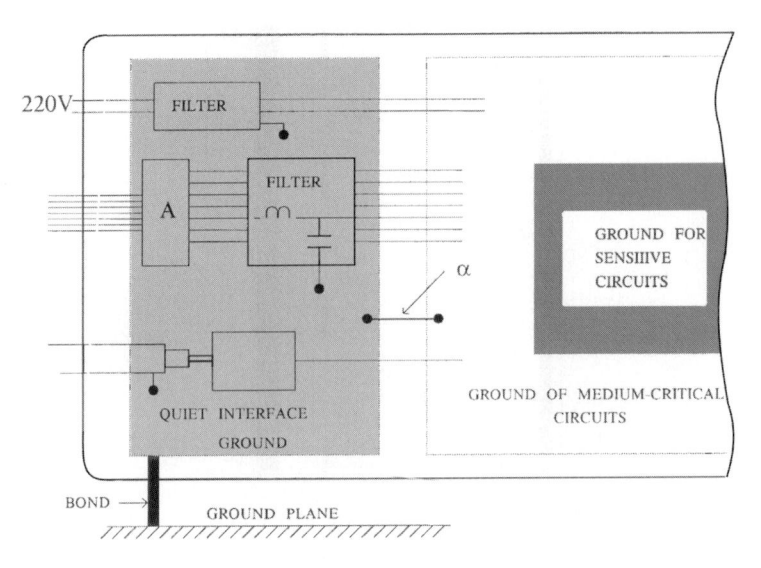

Fig. 11.22. Various grounds on a PCB

– Create "quiet areas" (i.e., sections that are physically isolated from digital systems, analog circuits, power and ground planes). Locate sensitive circuits in such quiet areas.
– Set up an interface ground by grouping all I/O ports in one area, and connect their shields, decoupling filters, and suppressors to an isolated ground plane in this area. Remember that *an effective filtering at high frequency is impossible without a quiet ground*. If required, connect the I/O ground to an internal ground *at only one point* (trace α in Fig. 11.22). For ESD protection, the I/O ground is connected to the earth by means of the chassis (denoted GROUND PLANE on Fig. 11.22); a low-inductance bond is used to connect the interface ground to the chassis.
The ground of medium-critical circuits is the next adjacent area; it houses rectifiers, oscillators, etc. The ground for sensitive circuits refers either to precision amplifiers, low-level analog circuits, etc., or digital circuits.

■ **Supply Routing.** It is important to adopt a proper configuration for supply distribution. When a double-sided PCB is employed, the trace carrying $+V_{CC}$ is located on the component side, while the trace for ground return is placed on the back side. In Fig. 11.23a, the loop area enclosed by the supply current is large, and this is risky.

The alternative routing in Fig. 11.23b is much better, since the loop area is minimized by accurately superposing the ground and supply traces on opposite sides of the board wherever possible. This results in a considerably lower characteristic impedance of the supply line, which is beneficial for interference reduction.

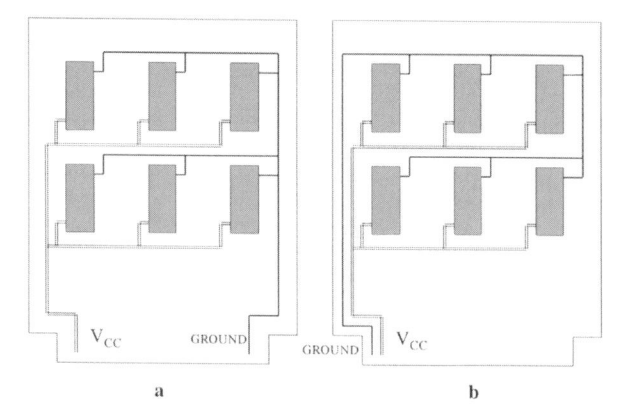

Fig. 11.23. Possible arrangements for supply routing **a** poor solution; **b** good solution

11.4.3 Coupling Effects in VLSI Design

■ **Overview.** Sub-quarter-micron technology is now available for mass production of CMOS analog–digital integrated circuits. With the scaling down of process technology, the geometry of transistors and interconnects becomes smaller, resistance per unit length of the interconnect increases, while capacitance per unit length remains practically constant. However, the coupling capacitance between adjacent parallel tracks can account for over 80% of the total track capacitance (as the distance between adjacent parallel tracks becomes smaller than the distance to the tracks situated above or underneath). Consequently, as transistor delay continues to decrease, the delay introduced by the interconnect becomes dominant. Finally, the quality of a deep submicron VLSI circuit depends more on interconnect performance than on transistor and logic performance.

■ **Paths for Noise Coupling.** Modern demands for portable communication devices (where mixed-signal applications are the rule) led to the integration of high-speed digital circuits with analog subsystems on the same chip (System-on-a-Chip, or SoC). Experience with SoC design shows that in addition to the known path of noise coupling through interconnect, the *substrate* plays an important role in contaminating the relatively quiet analog sections with noise from digital sections. This process is illustrated in Fig. 11.24, where **ND** is the noise generated by the digital circuits which couples to the substrate (since in every CMOS gate, the digital ground is connected to the substrate); **NS** is the noise propagating through the substrate, and **NA** is the noise reaching the sensitive (analog) circuits.

Fluctuations resulting in the substrate voltage due to the current injected into the substrate is called *substrate noise*.

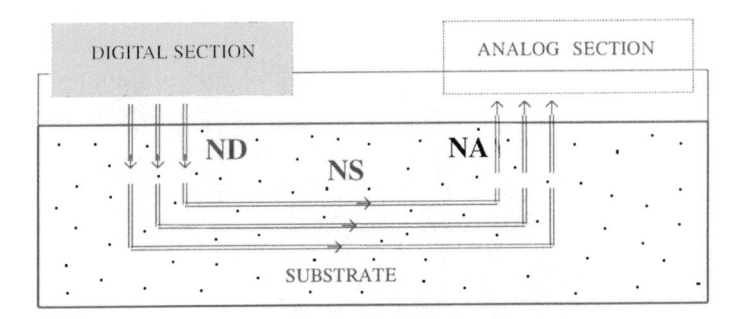

Fig. 11.24. Substrate noise in SoC

■ **Mechanisms.** In CMOS logic circuits, the substrate noise is caused by three mechanisms:

1) Coupling from the digital power supply (inductive noise, resistive voltage drops, and ringing of the power-supply voltage due to on-chip capacitance).
2) Capacitive coupling from switching drain-source terminals
3) Impact ionization in the MOSFET channel (which depends mainly on technology) [186, 187].

The substrate noise level often exceeds the random noise level generated inside integrated circuits.

■ **Effects.** The following effects of substrate noise have been demonstrated [192]:

– it degrades the performance of analog circuits sharing the same substrate with switching digital circuits;
– it affects the dynamic behavior of logic circuits by causing changes in threshold voltage (which leads to erroneous switching and timing faults).

Both effects require remedies at an algorithmic, circuit, and layout levels. Among them, traditional guardbands, Kelvin grounding techniques and reduced supply bounce CMOS logic circuits are discussed in [192].

■ **Crosstalk Reduction.** Several possibilities exist for reducing crosstalk; however, the treatment is different according to the specific type of coupling path.

□ **Substrate.** Whenever the parasitic elements of the substrate are prominent in noise coupling and propagation, several approaches have been proposed to reduce crosstalk: include guard bands to isolate both aggressor and victim, employ SOI (Silicon-On-Insulator) technology or use SIMOX substrates. Significant reduction of crosstalk has been reported with GPSOI

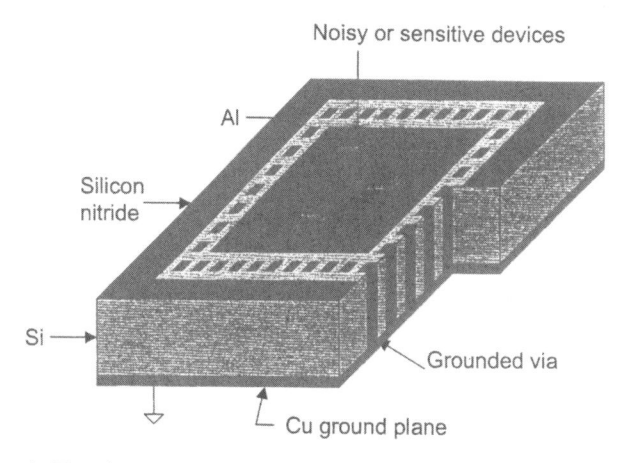

Fig. 11.25. A Faraday cage isolation structure to limit the substrate crosstalk (©2001 IEEE)

(Silicon-On-Insulator substrate with buried Ground Planes) [188]. A Faraday-like cage isolation structure to suppress noise coupling between devices or circuits on the same chip is presented in [189]. This structure consists of a ring of grounded vias that encircles a noisy or sensitive part of the integrated circuit (Fig. 11.25). Each via of the cage has a diameter of 10 µm and is separated by 10 − 70 µm from the next one. The vias are filled with electroplated copper, are shorted together by a ring of metal (Al) at the top, and connected to the Cu grounded backplane at the bottom of the substrate.

Crosstalk attenuation up to 40 dB at 1 GHz has been reported at a transmission distance of 100 µm.

☐ **Interconnect.** To reduce crosstalk associated with interconnect, several issues have to be considered [190, 191], which should be implemented in the design process:

- Define a suitable maximum rise/fall time in the design.
- When routing is performed automatically, limit the size of various blocks.
- Use buffers to isolate critical paths from capacitive loads of noncritical sinks.
- Use specific input and output drivers and one destination only, to limit the number of variables with respect to coupling.
- Insert buffers by effectively dividing a wire into smaller segments and hence reduce the wire resistance (the smaller the wire resistance, the less the coupling effect).
- Separate signals with overlapping switching windows in the layout, by placing the tracks carrying them as far as possible from one another. If two signals don't toggle at the same time, their tracks can be adjacent.

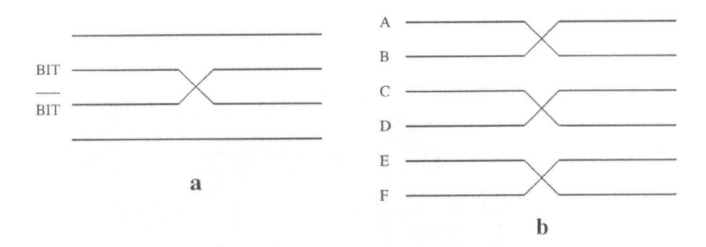

Fig. 11.26. Transposing lines to cancel coupling effects; (a) SRAM pattern; (b) Signal bus pattern

- Increase parallel line spacing (this requires a larger layout area and may not be always practical).
- Insert repeaters on all long tracks, because crosstalk between repeaters is considerably smaller than between lines.
- An effective way of reducing the coupling between lines carrying signals that switch in opposite directions is to transpose the lines in question. This technique is illustrated in Fig. 11.26a, where in the SRAM design the BIT-line and BIT-BAR-line are periodically crossed (note that this implies a technology with at least two different conductive layers). Depending upon the time evolution, signal lines (bus) can also take advantage of transposing (Fig. 11.26b).

Summary

- Crosstalk is caused by mutual electric-field or magnetic-field coupling, in the near field, between two adjacent conductors.
- Capacitive coupling is reduced by shielding or by inserting a ground trace between adjacent traces.
- Inductive coupling is reduced by shielding the conductors and/or by transposing them.
- Printed circuit boards are optimized during layout design. Any later attempt is both inefficient and costly.
- Large circuits should be partitioned for layout according to function: analog, digital, or I/O interface. Grounds follow the same rule.
- In RF and fast logic circuits, minimize the inductance of interconnects. A ground plane (or a grid ground plane) represents a good choice.
- Gridded ground is appropriate when the circuits involved carry similar signals and tolerate a reasonable amount of coupling (logic circuits often benefit from such a ground).
- Any substrate provides a coupling path for crosstalk in VLSI circuits fabricated in deep submicron technologies.

12
Methods of Increasing Immunity to Interfering Signals

"Rest, rest, perturbed spirit"

(Shakespeare: Cymbeline)

Despite careful design and protections to limit the emission of interfering signals by disturbing circuits, some residual noise will still contaminate sensitive circuits by conduction or field coupling. Therefore, the next task is to improve the immunity of perturbed circuits, in order to make them more noise tolerant without sacrificing their electrical performance. Traditional methods include balancing, filtering, and grounding. Although these were discussed in Chap. 10 (in the context of reducing emission from sources of interferences), they will be revisited here with focus on how to use them in order to increase electromagnetic immunity.

12.1 Balancing

■ **Definition.** *A balanced circuit is a two-conductor circuit that is symmetric with respect to a longitudinal axis (ultimately passing through the grounds of its terminal ports).*

The basic condition required is symmetry, which also applies to the transmission path and to the source and load.

■ **Balanced Line.** In a balanced line, the fields around the conductors are symmetric and there is no special ground conductor. For instance, two-wire lines are balanced, but a coaxial line is unbalanced.

■ **Explanation.** The property of symmetry is essential because the ultimate goal of balancing is to make the noise pickup equal in the two longitudinal circuit halves. When this condition is achieved, any external noise is transformed into a common-mode signal, which cancels in a load which has a symmetric configuration with respect to ground.

390   12 Methods of Increasing Immunity to Interfering Signals

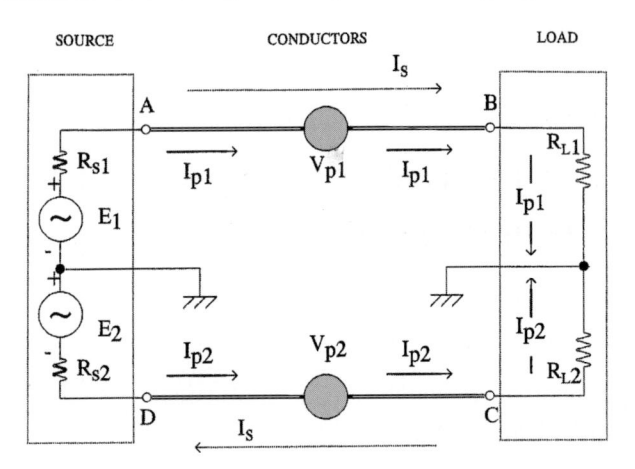

Fig. 12.1. Generic configuration of a balanced circuit

For instance, in Fig. 12.1 suppose that V_{p1}, V_{p2} are the inductive noise voltages picked up by the conductors, which generate common-mode currents I_{p1}, I_{p2}, respectively.

The useful signal is delivered by the symmetric source (E_1, E_2) and causes a current (I_s) through the loop passing through the terminals A–B–C–D; hence, it is a differential-mode current (which is not influenced by load symmetry).

It follows that the total output voltage across the load is

$$V_{AB} = R_{L1}\, I_{p1} - R_{L2}\, I_{p2} + I_S(R_{L1} + R_{L2}) \tag{12.1}$$

The first two terms represent the external noise contributions; they cancel provided that $R_{L1} = R_{L2}$ and $I_{p1} = I_{p2}$. The former condition implies that the load must have a symmetric configuration with respect to ground; the latter requires identical conductors equally exposed to interfering signals (which is a critical condition).

To conclude, perfect balancing requires:

– A signal source that is symmetric with respect to ground ($E_1 = E_2$, $R_{s1} = R_{s2}$).
– A load that is symmetric with respect to ground ($R_{L1} = R_{L2}$).
– Identical conductors, equally exposed to noise.

□ **Example.** To illustrate the difficulties of achieving perfect balancing, consider a system (Fig. 12.2) of two metal traces on a board, denoted (A), (B), coupled to a third one (P), which is the source of interference (perturbing system).

The cross-coupling has two components. Consider the capacitive coupling: the signal (V_p) carried by conductor P is transmitted through C_{PA}, C_{PB} to

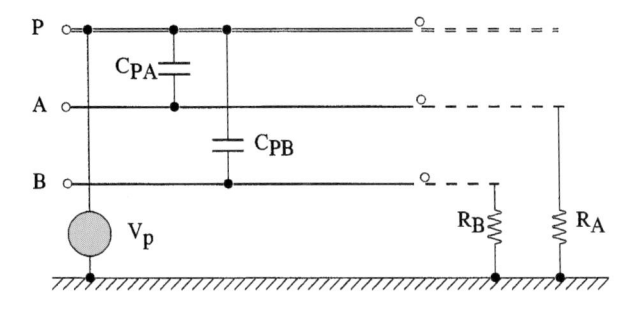

Fig. 12.2. Capacitive coupling in a quasi-balanced system

its neighbors and becomes a perturbation. Let us denote by V_{PA} the fraction of V_P reaching conductor A; thus

$$V_{PA} = V_P \frac{R_A}{R_A + 1/j\omega C_{PA}} \qquad (12.2a)$$

At low frequencies, we may suppose that $R_A \ll |1/j\omega C_{PA}|$, and (12.2a) becomes

$$V_{PA} \cong V_P \, (j\omega C_{PA} R_A) \qquad (12.2b)$$

Similarly

$$V_{PB} \cong V_P \, (j\omega C_{PB} R_B) \qquad (12.2c)$$

When the system is symmetric, $R_A = R_B$, but due to the location of the disturbing conductor (closer to A than to B), $C_{PA} > C_{PB}$, and unequal contamination results.

If inductive coupling is considered, the effect of the magnetic flux generated by the current flowing through conductor P is stronger on conductor A than on conductor B (due to different separation distances). Note that this effect, rather than cancelling the unbalance of capacitive coupling, adds to it instead.

This shows that even when the circuit has perfectly symmetric structure with respect to ground, *it is not perfectly balanced* until the coupling to the source of interference is the same throughout. Possible solutions are:

- modify location of traces to obtain, as much as possible, equal exposure to interfering signals;
- when wire conductors are used instead of traces, twisting represents an effective way to reduce unbalance.

■ **Balance Ratio.** This concept is employed in communication systems, which are designed for a minimum of stray pickup from the outside. The degree to which a transmission device deviates from an ideal balanced condition is expressed by the *balance ratio* or simply *balance* [193]. Note that this

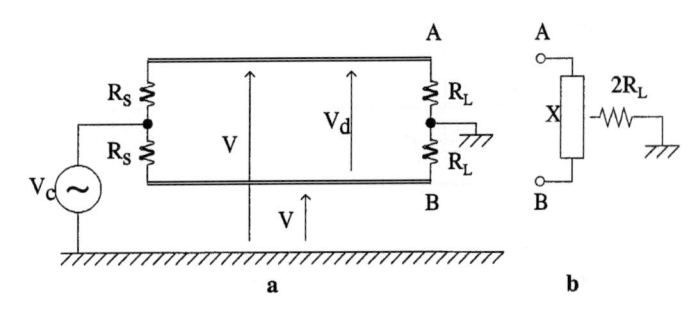

Fig. 12.3. a Measuring the unbalance of a transmission system; **b** equivalent load for well-balanced systems

corresponds to the *common-mode rejection ratio* (CMRR) in the theory of operational amplifiers.

In practice, this parameter can be measured by using the circuit in Fig. 12.3, where we apply a common-mode voltage V_c and measure the resulting differential-mode voltage V_d on the load. For a perfectly balanced configuration, V_d should be zero. The greater the value of V_d, the more unbalanced the circuit.

Often expressed in decibels, the balance ratio (BR) is

$$BR = 20 \log \frac{V_c}{V_d} \quad [dB] \tag{12.3}$$

Whenever the source internal resistance is low, V_c is roughly equal to the voltage (V) measured between each conductor and ground; definition (12.3) can be expressed as

$$BR = 20 \log \frac{V}{V_d} \quad [dB] \tag{12.4}$$

Equation (12.4) is better suited when the distance separating the source and the load is significant, since both measurements are made at the same end. For circuits with high BR, it is advisable to replace the load with the equivalent circuit proposed in Fig. 12.3b, where X is a highly balanced reactor (for instance, a transformer).

■ **Concluding Remarks**

– A well-balanced circuit exhibits a BR factor of the order of 60 to 80 dB.
– In operational amplifiers, a high BR guarantees a high rejection of common-mode interfering signals and therefore increased immunity.
– As a general rule, balancing is employed when shielding has been implemented but fails to sufficiently increase circuit immunity.

12.2 Filtering

As shown in Chap. 9, filters can be employed either to suppress the conducted emission of perturbations from noisy circuits or to improve immunity of critical circuits. The former aspect has been addressed in Chap. 10; now, we shall focus on the latter.

■ **Overview.** Depending on the particular type of action, filters are used in two situations:

1) When the DC power supply is distributed by lines (planes) having a non-negligible internal resistance and inductance, abrupt current demand due to a large number of simultaneous switching events can give rise to considerable inductive noise. As a result, a transient voltage drop in the power supply results, which may introduce logic failures and/or degrade the drive capability of transistors. This noise, propagating along the power and ground planes, may contaminate critical circuits sharing the same power/ground system. In this case, *decoupling filters* must be provided at all sensitive circuit power pins.
2) In electronic equipment, shields are employed to protect critical circuits; however, the cables and wires passing through the shield can transport not only useful signals, but also picked-up interfering signals. To avoid contamination of the protected circuits, careful *filtering* of wires and grounding of interconnect screens must be provided.

12.2.1 Decoupling Filters

■ **Decoupling the DC Supply.** This is mandatory whenever the user has no control over the design of the DC supply and its associated distribution system. In practice, two kinds of filters are used to decouple circuits: RC filters and LC filters. Both are presented in Fig. 12.4, where *filter ground should be connected to the ground of the protected circuit.*

Essentially, they are low-pass pi-section configurations intended to suppress voltage fluctuations due to abrupt transient currents in the DC supply system.

To be effective, the HF series impedance of the filter must be as high as possible, but the DC series impedance must be negligible (to avoid losses in the DC supply). This compromise is rather difficult to obtain with RC filters, since the value of R is not frequency-dependent. From this point of view, LC filters are better suited, but they have the following drawbacks:

- their inductance is sensitive to spurious magnetic fields;
- they present a self-resonant frequency. Provisions have to be made to ensure that the normal operation of the filter will not be affected by it.

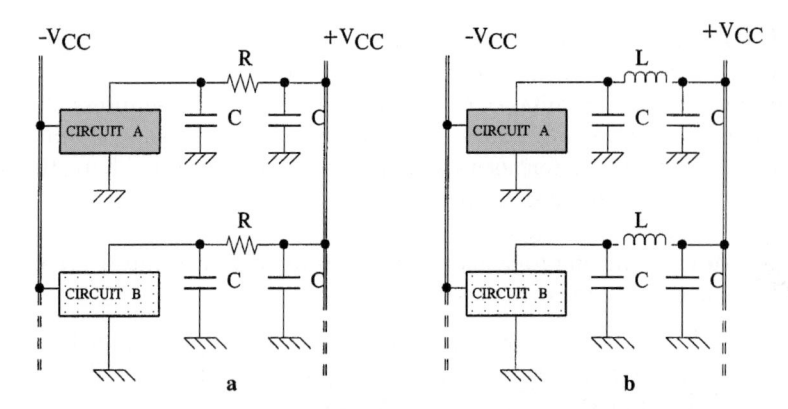

394   12 Methods of Increasing Immunity to Interfering Signals

Fig. 12.4. Decoupling the DC power supply input of every protected circuit with: **a** RC-filters; **b** LC-filters

All in all, RC filters are preferred, mainly because they are less susceptible to electromagnetic interference. Contrary to common wisdom, the value adopted for capacitor C should not be taken much higher than the calculated value. As a general rule, *adopt the smallest decoupling capacitor* that will do the job (since the parasitic series inductance of any capacitor increases substantially with its value, and this will impair filter performance).

■ **Amplifier Decoupling.** To illustrate the necessity of decoupling, consider the single stage amplifier of Fig. 12.5a, where the parasitic inductance of the $+V_{CC}$ trace (between nodes 4 and 5) is denoted by L_p.

It is common practice to draw the AC equivalent circuit of the amplifier by considering node 4 to be at ground potential (since any ideal voltage source has zero internal AC resistance, and the trace inductance is neglected).

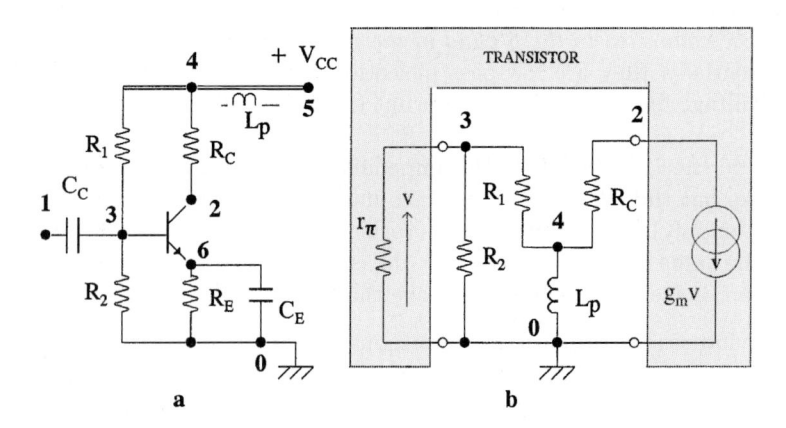

Fig. 12.5. **a** Common-emitter stage; **b** AC equivalent circuit

However, if the trace parasitic inductance (L_p) is taken into account, the equivalent circuit becomes like that shown in Fig. 12.5b. In this context, L_p is responsible for undesired coupling between the output loop (0–4–2–0) and the input loop (0–4–3–0). This kind of coupling is called "common-impedance coupling," because current from at least two circuits flows through the same impedance. This common impedance causes feedback, which, as a general rule, is not anticipated in the design.

A practical solution to this problem is to insert a capacitor between node 4 and ground, to bypass the trace parasitic inductance. This is called a *decoupling capacitor*.

☐ **Decoupling Capacitors.** The type of decoupling capacitor should be carefully selected according to the frequency range of the particular application.

For instance, in analog circuits mica and ceramic capacitors with small values are practical in the frequency range of up to 200 MHz, while metallized paper capacitors can be used only at low frequencies. If broadband operation is required, often two capacitors are connected in parallel (for instance, a higher-value wound aluminum foil capacitor for low-frequency bypass and a ceramic capacitor of lower value, for high-frequency bypass).

In digital applications, remember that *the decoupling capacitor must be effective at frequencies up to 100 times the primary clock frequency of the logic system*. Hence, the parasitic inductance of the capacitor itself, as well as the inductance of its connecting leads, are of prime concern. With the modern trend toward continuously increasing clock frequency (which is already in the GHz range for commercial personal computers), traditional discrete capacitors are obviously no longer suitable for decoupling. Several solutions are proposed instead:

- *Buried capacitors* incorporated in a multilayer printed circuit board have been designed using standard PCB processes [201] for portable and handheld communication products. They consist of two parallel conductor plates (typically situated in the internal power and ground planes) with dielectric material between them and connected by plated via holes. Three different sizes of buried capacitors were reported as being embedded in 530 mm×575 mm panels, corresponding to 9 nF, 21 nF, and 95 nF (the latter when the entire panel area is used as a single capacitor). The small number of required decoupling capacitors for the entire board allows one to clear space on the board and reduce the overall cost of the PCB. Furthermore, a reduction by 80% of the parasitic inductance associated with buried capacitors has been reported, relative to their discrete counterparts
- *On-chip decoupling capacitors*, employed in the design of high-performance CMOS microprocessors. The "white space" available on the chip is used to create MOS decoupling capacitors in the power/ground planes. In this way, decoupling capacitors are somehow placed blindly, with no guaran-

tee that they are located in the right places, where the inductive noise of the power supply is maximum and must be suppressed. Hence, optimization of decoupling capacitors deployment is required, and can be treated either as a post-floorplan step or as an integral part of the floor-planning [202]. This ensures that the decoupling capacitors are eventually optimally placed, very close to the clusters of high-switching activity.

☐ **Multistage Amplifiers.** In this case, the parasitic impedance of traces distributing the same supply ($+V_{CC}$) to various stages can cause problems. The multiple common-mode impedance couplings provide paths to return a fraction of the output signal of a certain stage to the input of a previous stage; if the feedback is positive, instability can ensue.

To decrease the risk of oscillations, we must decouple the overall amplifier, as well as every stage (Fig. 12.6). Note that filters A, B, C should be placed as close as possible to the supply leads (pins) of stages 1, 2, 3, respectively. Note also that good-quality RF grounds should be used for ground connection of all stages.

In practice, whenever the gain of the first stage is much greater than the gain of the following stages, enough protection is obtained by decoupling only the first stage.

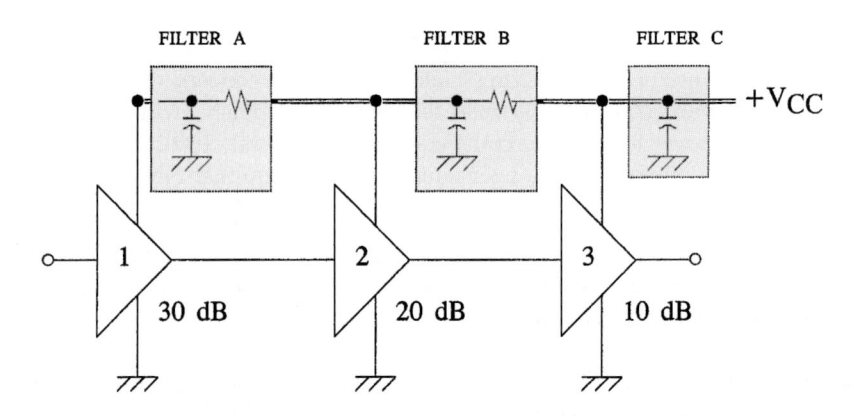

Fig. 12.6. Multistage amplifier feedback decoupling

12.2.2 Filtering of Wires and Cables

■ **Filtering of Wires.** It is important to filter each conductor entering a shielded enclosure, because it can collect spurious signals from the outside and transport them inside. Coaxial feed-through capacitors (C_T) should be used in input filters (Fig. 12.7), and an additional internal shield (IS) is sometimes necessary to confine the radiation of this filter.

12.2 Filtering   397

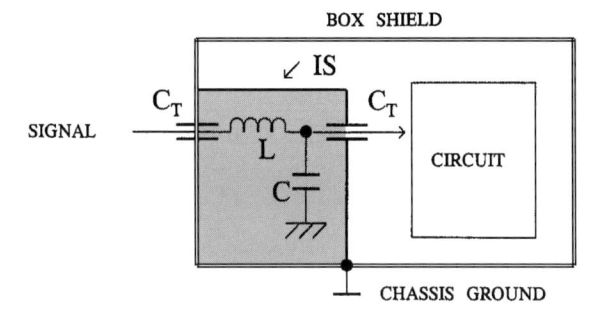

Fig. 12.7. Filtering of conductors entering a shielded enclosure

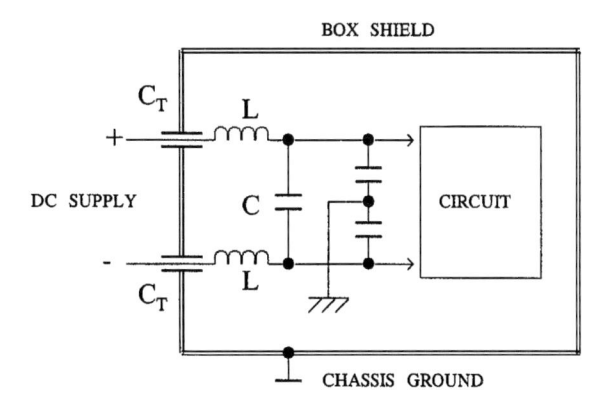

Fig. 12.8. Filtering a DC supply line entering a shield

The filtering of a balanced line (or a DC supply line) entering the shielded enclosure is illustrated in Fig. 12.8, where C_T denotes feedthrough capacitors.

As for general advice when filters are not purchased, but "home-made": it is essential to keep component leads as short as possible, especially capacitor leads and ground connections.

■ **Shielded Cables.** Generally, shielded cables are decoupled by properly grounding the cable shield. Two categories of shielded cables exist:

- *Coaxial cables*; these are unbalanced lines where the outer conductor is employed as a shield, but also as a return path for the inner conductor. It is advisable to ground the shield at the generator end for low-frequency applications. Multipoint grounding of the shield is preferred for high-frequency applications.
- *Shielded twisted pairs*; in this case the screen protects both conductors and is not used as a return path. These cables are employed to connect two distant systems (or circuits), as illustrated in Fig. 12.9.

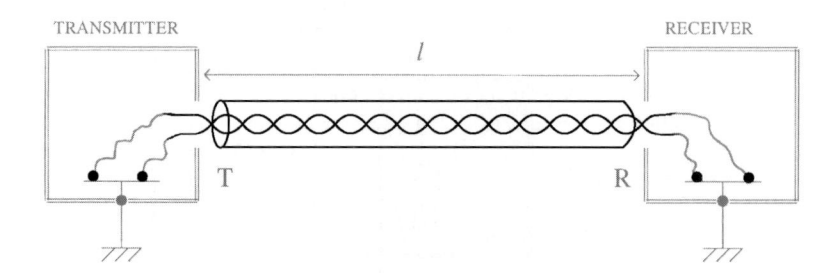

398   12 Methods of Increasing Immunity to Interfering Signals

Fig. 12.9. Connecting two systems with a shielded twisted pair

Table 12.1. Grounding a shielded twisted pair

Objective	$l/\lambda \leq 0.1$	$l/\lambda > 0.1$
Reduce emission	Ground at T	Grounds at T and R
Increase immunity	Ground at R	Grounds at T and R

Depending on the electrical length of the cable and which equipment has priority in protection, optimal grounding of the cable shield is indicated in Table 12.1 [196].

12.3 Grounding

■ **Explanation.** When properly used, grounding is a powerful technique to increase the electromagnetic immunity of a system.

During design, one of the traditional goals is to minimize the size of ground loops in order to reduce the electromagnetic susceptibility to radiated interfering signals. Whenever large ground loops still exist, they are merely the result of particular constraints imposing to interconnect several ground points, which are not in close proximity. Problems can appear when the interconnected grounds have slightly different potentials, since a current will flow in the ground system as a result of the voltage offset between ground points. If the ground also provides the return path for the useful signal, this current contaminates the useful signal.

In practice this happens either when interfacing two different systems (Fig. 12.10), or when the system drives a distant load. Note that any monitoring device inserted between two systems can be considered an interface; the useful signal is transmitted from DEVICE 1 to DEVICE 2 through conductor 1–2 and the return is provided by 2′–1′, which connects together the two grounds.

In many practical situations, although the two grounds are not expected to have the same potential, no current is allowed to flow through the interface.

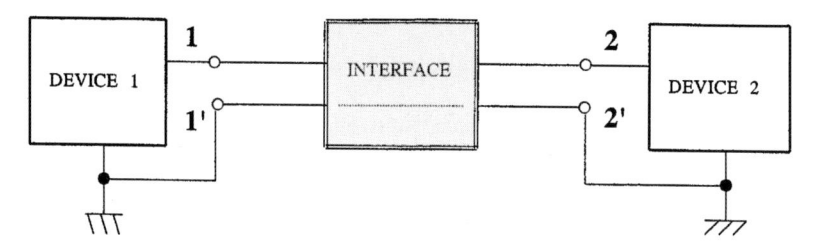

Fig. 12.10. Connecting two devices through an interface

This problem is particularly critical in medical equipment, where it may happen that an electrode applied to the human body must be connected to a particular ground whose potential might differ from the earth potential by as much as several hundreds volts!

■ **Breaking the Ground Loop.** To avoid these potentially harmful situations, the ground loop can be broken by inserting a transformer, an opto-

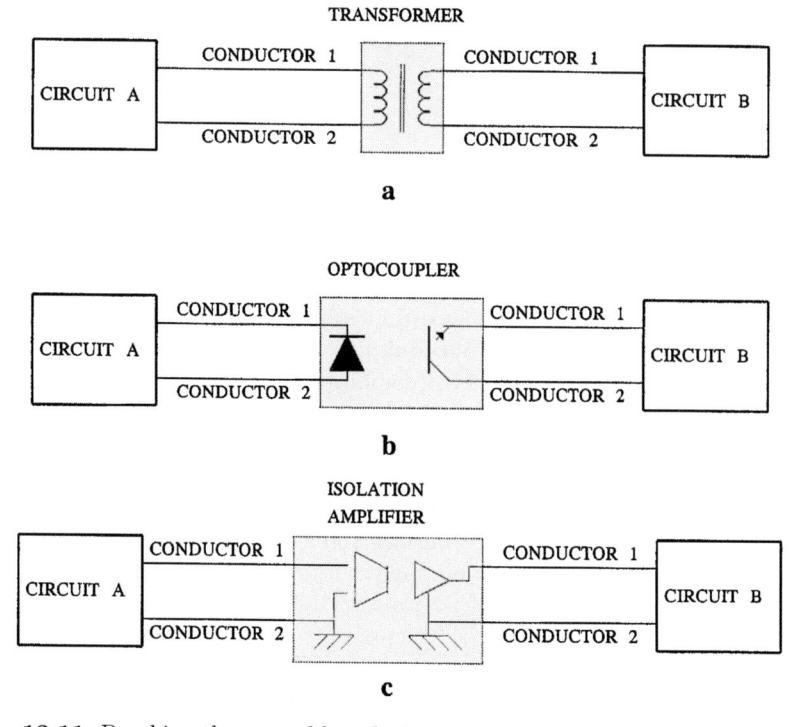

Fig. 12.11. Breaking the ground loop by inserting: **a** a transformer; **b** an optocoupler; **c** an isolation amplifier. Note that in all cases circuit A is galvanically isolated from circuit B

400   12 Methods of Increasing Immunity to Interfering Signals

coupler, or an isolation amplifier (Fig. 12.11). The choice depends upon the frequency of the transmitted signals, but other considerations will be detailed.

☐ **Inserting an Isolation Transformer.** For high-frequency analog or pulse signals, a transformer is mostly recommended (Fig. 12.11a). They have ferrite-type cores with high permeability, requiring a few turns of copper windings (hence, low mutual capacitive coupling between coils). When intended for low-frequency applications, this option is not economical, since it requires bulky transformers. In this case, beyond additional weight and size, the external shielding of the transformer (to avoid pickup of spurious magnetic fields), as well as the internal shielding between the coils, dramatically increases cost, and contribute to degradation of performance.

Figure 12.12 shows two examples of connecting isolation transformers: for an unbalanced line (Fig. 12.12a) and for long-distance communication of digital data over wire lines (Fig. 12.12b), where a transformer is needed at each end [197].

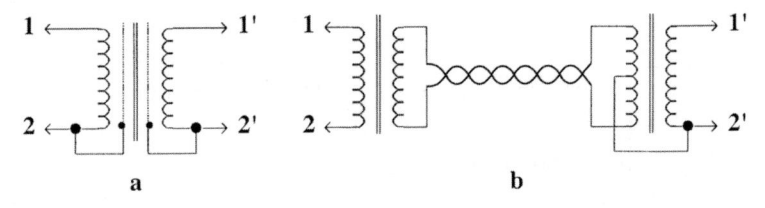

Fig. 12.12. Examples of connecting isolation transformers

☐ **Inserting an Optocoupler.** The main drawback of isolation transformers is the residual parasitic capacitance that still exists between the circuits to be isolated. When this may be harmful, an optocoupler must be adopted.

An optical coupler (Fig. 12.11b) combines a light-emitting device with a light-sensitive device in one package. A light- emitting diode (LED) transforms the electrical signal applied to its input into light, which is sensed by a detector (photodiode) and amplified by a transistor or a Darlington pair (Fig. 12.13). This double conversion of energy is the key to near-perfect isolation between circuits, and as a result electromagnetic susceptibility is greatly improved. Typical performance parameters for an optocoupler are: maximum isolation voltage 2.5 kV, isolation resistance 1 TΩ and residual coupling capacitance of order 1 pF. Globally, it represents an especially good choice for digital applications (linking computers and control devices); for analog applications, its linearity is not satisfactory.

☐ **Inserting an Isolation Amplifier.** For all analog applications where linearity is essential, isolation amplifiers represent a good solution, provided that the transmitted signal does not saturate the amplifier. Usually, isolation

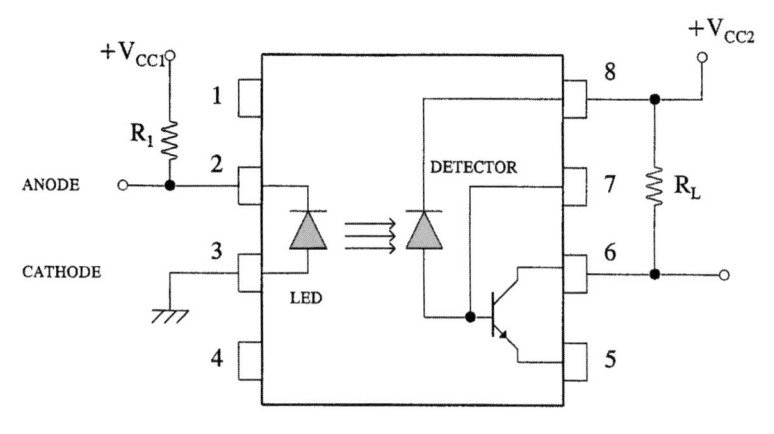

Fig. 12.13. Optocoupler

amplifiers are purchased. A block diagram of a typical isolation amplifier is presented in Fig. 12.14, where the potentials of the grounds on each side of the separation line can be completely different.

Based on the transmission modality of the signal inside the amplifier (across the separation line), these devices can be grouped into three categories:

1) Amplifiers with internal coupling transformer, which suppress any DC component in the transmitted signal (this is the case of isolation amplifiers fabricated by Analog Devices).

   A high-frequency carrier is frequency-modulated (or pulse- width-modulated) by the transmitted signal, whose bandwidth does not exceed 10 kHz. Isolation up to 3.5 kV between input and output is achieved; note that

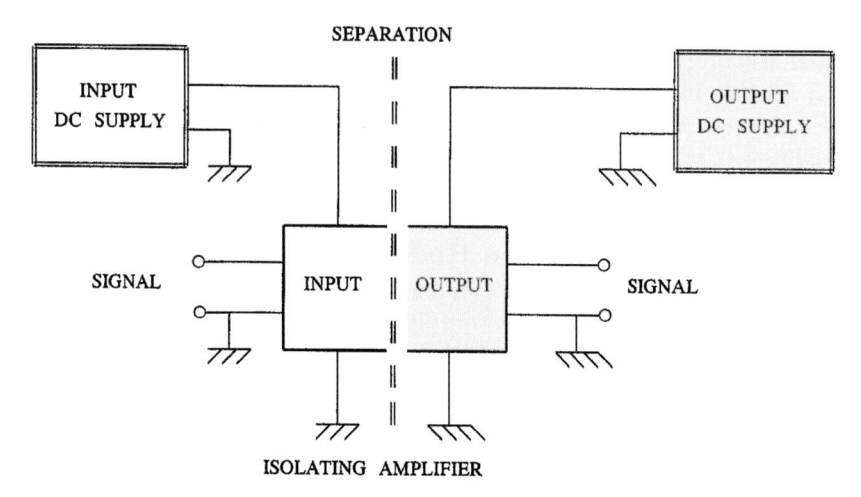

Fig. 12.14. Breaking the ground loop with an isolation amplifier

only one DC supply is needed, since the second one is a converter whose coil shares the same core as the first.

2) Amplifiers with internal input/output optical coupling (such as the IS0100 Burr–Brown isolation amplifier). According to the data sheets, the maximum isolation voltage is about 750 V for IS0100; to improve linearity, a second photodiode is excited by the light emitted by the input LED, and the resulting signal is employed to cancel the nonlinearity of the main conversion process.

3) Amplifiers with internal capacitive coupling, where the transmitted signal is frequency-modulated (like the Burr–Brown IS0106 amplifier, for which a maximum isolation voltage of 3.5 kV and a bandwidth of 70 kHz are specified).

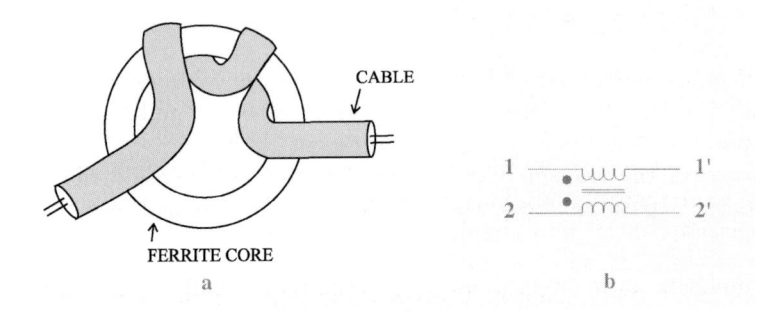

Fig. 12.15. **a** Partial isolation technique; **b** common-mode choke

■ **Remark.** Whenever galvanic isolation between input and output circuits is not required (or is impossible to achieve), a cheaper technique consists in inserting high series impedances in both conductors. In high-frequency circuits this is usually implemented by winding the cable (or the pair of conductors) several times through a toroidal ferrite core (Fig. 12.15a). As a result, two common-mode coupled inductors (Fig. 12.15b) appear in series with the transmission line, providing some isolation between its terminal ports.

12.4 Practical Advice on Reducing Noise and Interference at the Circuit Level

■ **Comment.** In this section we discuss several design guidelines used to minimize interference and noise. It should be emphasized that focusing the effort solely on interference reduction does not guarantee a noiseless system, since intrinsic noise still contaminates the signals. Therefore, fighting interference must be correlated with fighting intrinsic noise (by adopting minimum noise design techniques), and these two activities cannot be dissociated in practice. Bear in mind that all users desire high-quality, noise-free equipment, not explanations in case of failure that one of the two aspects of noise was underestimated during design!

Considering the important difference between the external and intrinsic noise amplitudes, it is obvious that *intrinsic noise becomes a problem only when interfering signals are no longer a problem!* For this reason, in the following, a collection of principles and techniques aimed at the global objective of noise reduction is presented. Although they represent good practice, these rules should be carefully applied, depending on the particular conditions inherent in every design. Very often, more than a single technique is required to satisfy requirements, hence many of them must be simultaneously implemented.

12.4.1 Interference Control

■ **Overview.** The reduction of electromagnetic interference is best achieved when trying to solve this problem during the design process and not latter. The basic rules involve circuit partitioning, selection of components (with criteria for both intrinsic noise and interfering signals in mind), PCB layout, selection of cables, grounding, shielding, and filtering (if the system is fabricated as an integrated circuit, substitute "layout" for "PCB layout" and "routing the interconnects" for "selection of cables").

Any interference problem involves an aggressor circuit and a victim. Often the *aggressor* is a well-designed circuit that satisfies all the electrical requirements, but the electromagnetic compatibility problem has been neglected. Furthermore, the situation is aggravated because testing is conducted on each module separately (or on the module in question connected to the modules with which it must normally operate), but very seldom on the system as a whole. We thus dispose of very limited means of detecting the aggressiveness of a particular circuit, early in the process.

The *victim* is also a well-designed electronic circuit, dedicated to low-level signal processing. Hence, it must be sensitive to weak signals, which is of course its main vulnerability: susceptibility to interfering signals. Therefore, protection is needed to avoid (or, at least, to limit) penetration of interfering signals. Taking into account that the same circuit (or interconnection) can act as aggressor during some time intervals and victim during others, it follows that an exact classification is impossible (and perhaps pointless in any event).

From another stand point, in most practical situations the circuit designer has no control over equipment design, since this task is usually performed by a different team. Ideally, the electromagnetic compatibility problem should be solved at the circuit design level, where the solutions adopted to control interference are easier to implement, less expensive, and above all, more efficient. In the following section, a non-exhaustive list of rules is proposed, to be considered during the *circuit design* and/or *prototyping* step.

12.4.2 Guidelines for Circuit Design

Although this section appears at the end of Chap. 12, it should also be considered as a final conclusion to part II of this book, dedicated to interfering signals. Since controlling conducted and radiated interference is only one side of fighting noise, minimum-noise design principles (referring to intrinsic noise) should be simultaneously considered. Consequently, rules and principles aimed at these both major objectives [194, 197–200] are proposed in the following sequence:

■ **Partition** the circuit to be designed into critical and non-critical sections:

- identify sections prone to generate interfering signals (aggressors). They must be isolated from the rest of the circuit both electrically and physically. *Electrical isolation is achieved by proper design*, but *physical isolation is achieved by layout and shielding;*
- identify noise-sensitive sections (victims);
- locate victims away from aggressors (for instance, on different layers of a multilayer PCB);
- choose proper I/O points.

■ **Select the Circuit Components of Digital Sections** with the goal of reducing emission of interfering signals:

- use slow logic families in regions where there is no impact on operation;
- select high-immunity logic families in noise-sensitive sections;
- group logic families according to functionality and apply segregation by speed.

■ **Select a Circuit Topology** that minimizes both interference and intrinsic noise by paying attention to the following points:

- restrict bandwidth to the minimum required to transmit only useful signals;
- minimize the signal level of aggressor circuits (while remaining consistent with the required S/N ratio);
- maximize the signal level of noise-sensitive circuits;
- adopt impedance values consistent with both performance and minimum crosstalk requirements;
- in RF, microwave, or digital applications, consider matching line terminations.

   For *analog circuits*:

- select the input-stage configuration, active devices, and bias points for minimum noise;
- use resistors with low noise index in all sensitive areas.

12.4 Practical Advice on Reducing Noise and Interference

For *logic circuits*:

- reduce fan-out on clock circuits by using buffers;
- insert a watchdog circuit on every microprocessor.

For *both* categories:

- select power supplies of good quality (noiseless).

■ **PCB Layout.** In general, the *differential-mode emission can be controlled by circuit layout*. The key to achieving reduction of differential-mode radiation is to *minimize loop areas*. Therefore:

- minimize distance between circuit components;
- minimize trace and component lead lengths;
- ensure proper signal returns by using one or more ground planes;
- minimize loop areas subject to large di/dt (for instance, all clock paths must have adjacent ground returns);
- minimize node areas subject to large dv/dt;
- provide a separate power plane.

Another objective of layout is to *avoid crosstalk*; this can be achieved by observing the following rules:

- increase separation between traces;
- avoid parallel runs of traces or minimize parallel trace lengths;
- when parallel traces are unavoidable, increase separation between traces, or better, insert a ground trace between them;
- in multilayer PCB, route adjacent layers orthogonally;
- minimize ground inductance by using a ground plane or ground grid;
- provide matched terminations on traces carrying fast-switching signals or microwave signals;
- reduce signal drive level;
- place clock interconnects away from I/O regions;
- locate all I/O circuitry in one area of the PCB and provide a separate ground to this area;
- place I/O drivers next to the connector.

Miscellaneous:

- minimize loops associated with noise-sensitive circuits;
- avoid sharp angles or 90° angles in traces;
- don't leave unused inputs of logic gates floating: connect them either to ground or to the power supply;
- where signal and power traces must cross, make the crossing so that the traces are orthogonal;
- provide a separate ground for sensitive circuits, and place them away from ground plane edges;

- place decoupling capacitors next to each IC;
- don't leave any metallic area on the PCB floating.

Of the various techniques employed to control interference, note that in practice they are used in the following order: first proper grounding is achieved, then shielding is eventually taken into account, and finally filtering is added. The same order is adopted in the following:

■ **Grounding.** The ground system of any electronic equipment should be created during the design step, then updated and checked during layout, prototyping, and fabrication. The main objectives are to ensure *normal operation* of the system, guarantee *safety conditions*, and provide *protection against hazards* (like ESD, lightning, etc.). Here is a non-exhaustive list of suggestions:

- for circuit dimensions less than $0.03\,\lambda$, use single-point grounding;
- for circuit dimensions larger than $0.15\,\lambda$, use multipoint grounding;
- for circuit dimensions between $0.03\,\lambda$ and $0.15\,\lambda$, provide hybrid grounding (especially for broadband circuits);
- use floating ground only in low-frequency applications;
- create separate ground systems for signal returns, signal shield returns, power supply returns, and hardware (chassis, racks, and cabinets) ground. These returns can eventually be tied together at a single ground point.
- isolate the ground of logic circuits from the ground of analog circuits;
- provide a "clean" ground for decoupling all interfaces;
- provide quality bonding of screens, connectors, cabinets, and filters, and ensure that it is not damaged by operation under adverse conditions;
- minimize the length of ground interconnects and give them adequate geometries;
- avoid common-ground impedances;
- use a ground grid or plane on logic boards to minimize common-mode ground interference;
- break ground loops if problems appear;
- use balanced differential configurations to minimize common-mode ground interference.

■ **Shielding.** This technique is used to control near- and far-field coupling. Here are several basic rules:

- Since shielding is an expensive technique, determine whether a shield is necessary or not. If it is, select the type of shielding required for the frequency range of the particular application.
- Use copper or aluminum for electric field shields. A material thick enough to support itself usually provides good protection.
- Use iron or high-permeability alloys for magnetic shields.

12.4 Practical Advice on Reducing Noise and Interference

- Enclose the most sensitive circuits or noisy systems in additional internal shielding.
- For plane-wave shielding, use electromagnetic gaskets to seal apertures in the metal construction.
- Seams should be welded or overlapped.
- Avoid resonant metallic enclosures (dimensions of $\lambda/2$ are prohibited).
- Avoid large apertures in the shield. Whenever possible, replace a large opening with several small apertures.

■ **Filtering.** Since filters control the spectral content of signal paths, *they represent an efficient technique for eliminating conducted interference.* The following points should be considered:

- filtering is most effective when applied to the interference source (switches, motors, etc.). *All transient interfering signals should be treated at the source.*
- select the proper mains filter for each particular power supply;
- filter all I/O lines with common-mode chokes or capacitors;
- decouple the DC supply of each board (module) with a separate pi-section filter;
- adopt the smallest-value decoupling capacitor that will do the job;
- ensure a good ground to each filter;
- minimize lead lengths and associated wiring of filters.

■ **Cables.** The cabling system requires particular attention since *the key to minimizing common-mode radiation is to reduce the common-mode current on all cables of a given set of equipment.* Thus, all cables entering or leaving the equipment require treatment to control common-mode emission, and the following rules should be considered:

- select the proper type of cable for every application;
- minimize cable lengths;
- for differential-mode input configurations, select equal cable lengths and run them adjacently;
- provide enough separation between cables carrying weak signals and cables transmitting high-level signals. If they must cross, make the crossing orthogonal;
- avoid running signal and power cables in parallel;
- avoid resonant cable lengths;
- run cables close to metallic grounded walls;
- pass cables away from windows or apertures in shielding;
- use signal cables and connectors with adequate shielding;
- use chassis ground or the I/O ground for cable bypassing (never bypass cables to signal ground);
- for balanced configurations, adopt twisted pairs;

- for RF signals, match all cable terminations;
- avoid pigtails when connecting cable shields to ground;
- provide cable damping with ferrite suppressors;
- provide ground termination of cable shields via a 360° solid contact with the shield enclosure.

■ **Connectors and Mechanical Switches.** Metal-to-metal junctions in coaxial connectors, waveguide joints, and mechanical switches can cause poor electrical contact due to corrosion, oxidation, dust, and impurities on surfaces. They are mainly responsible for *contact noise and intermodulation interference (rusty bolt effect)*. To prevent them:

- avoid deposit of dust on the surfaces involved in mechanical contact;
- clean them periodically with special chemical products;
- don't hesitate to replace them when wear makes operation unsafe.

12.5 Increasing System Immunity to Interference: Bluetooth Approach

Until now we have been concerned with techniques to reduce interference at the *circuit* level. A quite different approach is employed to reduce interference at the *system* level in Bluetooth wireless networks.

■ **Definition.** *Bluetooth is a short-range, low-power radio link (10–50 m) between several devices operating in the unlicensed 2.4 GHz industrial, scientific, and medical (ISM) frequency band* [203].

■ **Explanation.** This wireless communication network is intended for home and office applications [204] such as cordless headphones; cordless keyboards; telemetry of physiological signals to health support systems; sharing voice, data, and video among computers, digital cameras and camcorders, TV sets, printers; and so on.

Since the 2.4 GHz band is considerably crowded by other consumer appliances (such as microwave owens, baby monitors, etc.), interference is expected to be large enough to adversely affect the security of transmission.

■ **Approach.** Data transmission is performed using GFSK (Gaussian frequency-shift keying) modulation where a positive frequency deviation corresponds to 1 and a negative frequency deviation means 0.

To increase immunity to interference, Bluetooth uses a pseudo-random hopping sequence over a large number of 1-MHz frequency channels. To support this frequency hopping, the ISM band is splitted into 79 1-MHz channels in the United States and most european countries, and 23 1-MHz channels in France, Japan and Spain (to accomodate smaller ISM frequency bands).

12.5 Increasing System Immunity to Interference: Bluetooth Approach   409

In a point-to-point link (or point-to-multipoing link) the master unit selects the hopping sequence and imposes it to the slave unit. Each channel involved in the hopping sequence is active only during a time slot of 625 µs, then the next on the list channel is selected for another time slot of 625 µs, and so on.

■ **Conclusion.** Even if data can be corrupted by interference when transmitted over a particular channel, global damage is limited due to the continuous frequency hopping and the timing involved in the hopping. This concept, combined with the general Bluetooth architecture, packet structure, data encryption, and access coding techniques largely improves the security of data transfer between two devices. In this way the transmission quality is maintained despite strong interference potential.

Summary

- Differential-mode emission can be controlled by circuit layout. The key to reducing differential-mode radiation is minimizing loop areas.
- Common-mode interference is controlled by balancing the topology of analog circuits. The key to minimizing common-mode radiation is reducing the common-mode current on all equipment cables.
- In logic circuits, use a ground grid or plane to minimize common-mode ground interference.
- Balancing represents an efficient way to pick up equal interfering signals on both conductors of the line. Then, with terminal differential configurations, common-mode interference is eliminated.
- Filters control the spectral content of signal paths and provide an efficient way to eliminate conducted interference.
- All transient interfering signals should be filtered at the source.
- Apply decoupling filters at the DC power supply input on every board, or even on every circuit of a board, to avoid coupling through the source and/or interconnect impedance.
- Adopt the smallest-value decoupling capacitor that will do the job.
- The main objectives of grounding are to ensure *normal operation* of the system, guarantee *safety conditions*, and provide *protection against hazards*.
- Provide a "clean" ground for decoupling all interfaces.
- Bypass all interconnects (cables, wires, etc.) not to the signal ground, but to chassis ground or the I/O ground.
- Whenever it is required to interconnect several ground points which are not necessarily at the same electrical potential, break the ground loop for DC but ensure unimpeded transmission of AC signals.
- Galvanic isolation is achieved by inserting transformers, optocouplers, and isolation amplifiers, or by using capacitive coupling.

Part III

Case Studies

Part III

Case Studies

13
Low-Noise Circuit Design

> "Grau, teurer Freund, ist alle Theorie,
> Und grün des Lebens goldner Baum"
>
> (All theory, dear friend, is grey,
> but the golden tree of actual life
> springs ever green)
>
> (Goethe)

13.1 Introduction

In part I of this book we introduced noise mechanisms, paying attention to noise modeling in passive and active devices. In the second part, emphasis was given to various techniques to reduce interference caused by signals arriving from outside the investigated circuit.

We now attempt to use these topics in the context of low-noise circuit design. It must be remembered that, traditionally, this field is concerned with principles and techniques to reduce the *intrinsic* noise level only. It is assumed that extrinsic noise is not a threat, since circuit immunity to interfering signals has been increased as much as possible. Therefore, a typical formulation of the low-noise design problem is the following [205]: "Given a signal source with known electrical characteristics (impedance, bandwidth, noise, etc.), how do we optimize the amplifier design in order to achieve the minimum noise of the system source-amplifier?"

■ **Methodology.** The design of any electronic circuit is an iterative process. When an amplifier is designed for a specific application, many requirements concerning the DC power supply, gain, bandwidth, input and output impedances, feedback, stability, noise, and cost must be met. To begin, the designer makes strategic decisions, including circuit topology, the technology adopted to fabricate the amplifier, the type of active devices (bipolar or field-effect devices), their bias points, etc. Next, based on previous experience, the designer proposes an initial solution, which is submitted to a simulator package in order to analyze the circuit and check on whether the required specifications are met. Depending on how much the performance departs from the specified values, some element values might be adjusted, or the topology of the circuit might be modified. The analysis is run again, and performance is progressively improved until a final solution is achieved.

■ **Noise Problem.** During the design process, noise can be treated in two different ways [205, 206]:

1) The designer can focus on signal properties, trying to meet specifications on gain, bandwidth, stability, etc., and only at the end will the total noise of the amplifier be evaluated. If the resulting intrinsic noise exceeds the specified level, few things can be done to improve it! Solutions, if any, are expensive to implement at this late stage, and the amplifier must often be redesigned. This is called the *crisis approach.*

2) From the very beginning, the designer anticipates the noise problem. As overall noise is largely determined by the first stage, the designer starts by properly selecting the input stage configuration, the active devices, and their bias points in order to meet the minimum noise requirement. A trade-off against electrical performance is sometimes necessary. For instance, for a resistive sensor, the preamplifier input resistance should be matched for minimum noise, instead of for maximum gain. Coupling networks may be considered, if they are necessary. Once the design of the first stage is finished, the following stages are designed. At the end, the resulting noise is checked to be sure that the specifications are met. As the overall feedback does not affect noise, this can be used to compensate for the bandwidth, modify the input impedance, set the gain, or improve the stability.

This approach is highly recommended, since, as a general rule, *it is less expensive and more efficient to solve the noise problem in the early steps of the design process than later!*

■ **Objective.** The aim of this chapter is to discuss principles and techniques needed in designing low-noise sensor–preamplifier systems. It is assumed that the sensor (or transducer) is available and all its electrical parameters are known; therefore the main optimization effort will be devoted to preamplifier design.

■ **Selecting the Right Parameter.** We first ask what noise parameter(s) must be selected in optimization. Many people propose to minimize the noise factor, but this can lead to erroneous solutions, as we shall see later.

Another possibility is to optimize the signal-to-noise ratio (S/N ratio), which must reach a maximum. From a practical point of view, this translates into a search for minimum input noise. This second approach is much more advisable.

13.2 Low-Noise Design Techniques for Low-Frequency Circuits

■ **Problem Formulation.** Consider a typical application consisting of a preamplifier that processes the signal issued from a particular sensor (transducer). After amplification, the signal is further processed by subsequent stages (Fig. 13.1). Our interest is to reduce the noise of the sensor–preamplifier ensemble (called *the system of interest*) by as much as possible.

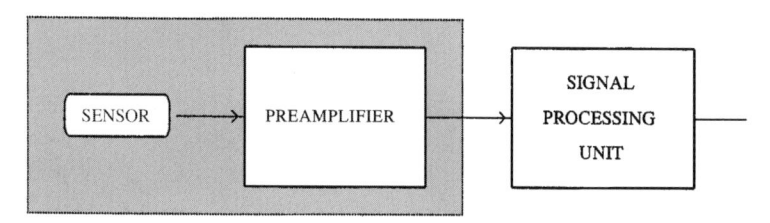

Fig. 13.1. Simplified diagram of a generic low-frequency application

■ **Comment.** No universal method for low-noise circuit design exists. Instead, general guidelines and principles are applied as required; these will be discussed next.

■ **Noise Matching.** *The process of reconciling the noise characteristics of the amplifier and the impedance of the signal source is called noise matching* [212]. This can be achieved in several ways:
- inserting a reactance between the signal source and the amplifier input terminals, in series or in parallel;
- connecting a transformer at the amplifier input, to match the source resistance to the amplifier noise resistance;
- designing the input stage of the amplifier to meet the noise requirements imposed by the source;
- modifying the source impedance; this is rarely possible (for instance, by connecting several smaller sensors in series instead of using a single sensor).

13.2.1 Rules of Low-Noise Design

■ **Rule 1.** *If several uncorrelated noise sources are simultaneously present in the same circuit, the designer must identify the dominant source and focus efforts on reducing it.*

416   13 Low-Noise Circuit Design

In order to illustrate this statement, consider a circuit containing two noise sources $v_1 = 1\,\text{mV}$ and $v_2 = 2\,\text{mV}$ (as rms values). If the sources are correlated, the total noise is:

$$\overline{(v_1 + v_2)^2} = \overline{v_1^2} + \overline{2v_1 v_2} + \overline{v_2^2} = \overline{v_1^2} + 2\rho\,\overline{v_1^2}\,\overline{v_2^2} + \overline{v_2^2} \qquad (13.1)$$

where ρ denotes the correlation coefficient $(0 < \rho < 1)$. For uncorrelated sources $(\rho = 0)$, the total noise is equal to $\sqrt{5} \cong 2.24\,\text{mV}$, which represents barely a 10% increase with respect to the dominant source $(2\,\text{mV})$. Hence, efforts to reduce secondary noise sources are counterproductive and this demonstrates the validity of the formulated statement.

Consider now the case of completely correlated sources $(\rho = 1)$. Certainly this represents the worst case and the total noise is $\sqrt{13} \cong 3.6\,\text{mV}$. For partially correlated sources, some value between $2.24\,\text{mV}$ and $3.6\,\text{mV}$ is expected. It is obvious that for weak correlation Rule 1 still applies, but for strongly correlated sources, both sources need attention.

■ **Rule 2.** *The noise of the preamplifier can be neglected if it represents less than about 1/3 of the source noise* [213].

As the noise of the sensor is not correlated with that of the preamplifier, by applying (13.1) we obtain the total noise $\sqrt{1^2 + 0.3^2} \cong 1.044$, which confirms the statement. From a practical point of view, it makes no sense to try to reduce the preamplifier noise below 1/3 of the sensor noise, since the improvement requires a disproportionate effort.

■ **Rule 3.** One of the most effective ways to reduce the noise of a system is to *restrict the bandwidth to the minimum value required to transmit only the useful signal spectrum.*

Given the continual improvement in device performance and frequency response, with lower and lower costs, the designer tends to select active devices with a much greater bandwidth than required by his particular application. The problem is that both thermal noise power and shot noise power are proportional to the bandwidth. Consequently, a very efficient way to reduce the power of white noise is to narrow the bandwidth to the minimum value required to transmit the useful signal. Using amplifiers with a frequency response wider than the signal spectrum adversely affects noise performance.

■ **Rule 4.** The best strategy to reduce noise in an electronic system is to *diminish the level of noise sources situated at the front-end (if possible, reduce the noise of the signal source).*

In practice, the signal source is either a transducer or an antenna. As they are situated at the front-end, any noise generated by them is necessarily amplified by subsequent stages. Consequently, before considering complicated and expensive solutions, acquire a less noisy transducer or a more directional antenna.

13.2 Low-Noise Design Techniques for Low-Frequency Circuits

Moreover, attention should be paid to the protection of sensitive devices against interference, for instance by shielding the transducer and its preamplifier.

■ **Rule 5.** *Noise can be significantly reduced by cooling the system.*

This is the most expensive solution, which is employed only when all other methods have failed. Generally, such failure is due to the thermal noise contribution, which sets the noise floor in any system. After narrowing the bandwidth, the only way to further reduce thermal noise is to lower the temperature. Therefore, by placing the sensor)preamplifier system in a cryostat, often at liquid helium temperature (4.2 K), the thermal noise contribution can be substantially reduced.

This technique is used in radio astronomy and related fields, where the signals reaching the receiving antennas are often so weak that they cannot be processed in any other way.

■ **Rule 6.** *The size of the sensitive part of the system must be kept to a minimum to reduce the field-coupled noise.*

This translates into placing the preamplifier as close as possible to the transducer (or antenna), and avoiding interconnections (or using the shortest possible cable). Besides the field-coupling effect, the attenuation of the cable causes noise performance degradation. As the loss depends on the cable length, a shorter cable is preferred to a longer one. Of course, the best solution is to use no cable at all.

On the other hand, the preamplifier should have sufficient gain and low enough internal noise, that the noise contribution of the following stages can be neglected.

13.2.2 Noise Performance of Amplifiers

■ **The Noise Figure Fallacy.** The noise factor of a two-port can be predicted with (5.29), which is recalled here:

$$F = F_o + \frac{R_n}{G_s}\left((G_s - G_o)^2 + (B_s - B_o)^2\right)$$

This equation suggests that the minimum noise factor, denoted by F_o, can be attained provided that the signal source admittance is matched to the input admittance of the two-port (i.e., $G_s = G_o$ and $B_s = B_o$). There are two ways to achieve the noise matching condition:

a) Design an amplifier whose parameters G_o and B_o are closely matched to the real and imaginary part of the transducer internal admittance. This is a valuable approach, because we simultaneously attain the minimum noise factor and maximum S/N ratio.

b) If the amplifier is already available and the real part of the sensor admittance is unfortunately far from the optimal value imposed by the amplifier, the designer can attempt to minimize the noise factor by adding resistors either in series or in parallel with the sensor, to appropriately modify its internal resistance. Although the noise factor is effectively decreased, *the total output noise power is augmented* and consequently the S/N ratio is reduced!

To prove this statement, reconsider North's definition of the noise factor (Sect. 4.5.6):

$$F = \frac{Total\ output\ noise\ power\ due\ to\ both\ R_s\ and\ two\text{-}port}{Total\ output\ noise\ power\ due\ only\ to\ R_s}$$

In spirit, *the noise factor is a parameter that merely compares the noise added by the two-port to that produced by the signal source*. It is obvious that by progressively increasing R_s, the source contribution in the numerator becomes more and more significant, until finally the noise of the two-port can be neglected. As a consequence, the noise factor continuously decreases toward unity and the two-port seems to be less noisy. Nevertheless, the total noise of the system amplifier and its signal source has increased!

One may think that when adding a parallel resistor at the amplifier input to lower the input resistance, nothing can go wrong. This is not true, since any resistor inserted between the source and the two-port adds its own thermal noise, effectively reducing the system S/N ratio. Moreover, this resistor attenuates the useful signal and this translates into a further dwindling of the S/N ratio. This represents what is called the noise factor fallacy [207]. To avoid it, *amplifiers should be designed to maximize the S/N ratio, instead of minimizing the noise factor*.

■ **The E_n–I_n Amplifier Noise Model.** The approach based on S/N ratio optimization requires simultaneous knowledge of the levels of noise and signal at a particular location. Naturally, the best-suited location is the input of the two-port, since the source is connected there and the signal level is already known. This explains interest in computing the equivalent input noise.

The E_n–I_n noise model represents the most appropriate tool to evaluate the input S/N ratio, because it reflects all internal noise sources to the input. Figure 13.2 shows this model; here E_{ns} denotes the thermal noise of the source resistance and A is the voltage gain of the amplifier. Note that usually both E_n and I_n are assumed to be uncorrelated white-noise sources [208].

According to [209], the equivalent noise generator E_n (called "equivalent short-circuit input noise voltage") represents the noise voltage expected at the input of the noiseless amplifier when $R_s = 0$. Its value is *frequency dependent* and, at a specified frequency, it is expressed in units of $nV/\sqrt{Hz}$ (or $\mu V/\sqrt{Hz}$, over a given frequency band). The recommended procedure to measure E_n is as follows: 1) the input of the amplifier is short-circuited; 2) the output rms

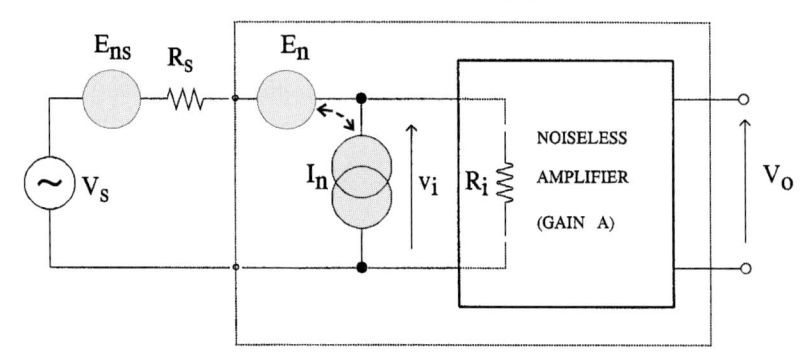

13.2 Low-Noise Design Techniques for Low-Frequency Circuits 419

Fig. 13.2. Amplifier noise representation

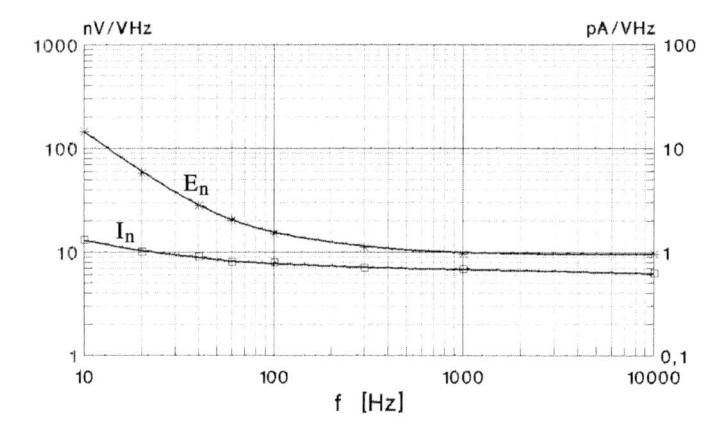

Fig. 13.3. Frequency dependence of E_n and I_n

noise voltage is measured; 3) this value is referred to the input by dividing it by the voltage gain A (this yields the equivalent input voltage); 4) the result is divided by $\sqrt{B}$, B being the bandwidth of the bandpass filter inserted between the output and the rms-meter. In this way, the spectral density of E_n is obtained.

In any amplifier, E_n increases toward the lower end of the bandwidth (see Fig. 13.3) due to the flicker noise contribution.

The equivalent noise generator I_n (called "equivalent open-circuit noise current") is the noise expected to appear at the input of the noiseless amplifier due only to *noise currents*. It is expressed in units of $pA/\sqrt{Hz}$ at a specified frequency, or $nA/\sqrt{Hz}$, over a given frequency band. In order to measure it, a resistor R or capacitor C is connected across the input terminals, so that the noise current gives rise to a voltage drop across it. The output rms noise voltage is measured, then referenced to the input (dividing it by the amplifier voltage gain) and the contribution of R_s and E_n is appropriately subtracted. Finally, the result is divided by $\sqrt{B}$, B being the bandwidth of the bandpass filter, to reach the spectral density of I_n in $pA/\sqrt{Hz}$. Typically, the frequency characteristic of I_n increases at the lower end for operational amplifiers and bipolar transistors, but for field-effect transistors it increases at higher frequencies.

■ **Limitation.** The E_n–I_n amplifier noise model is based on the assumption that the noise sources are *uncorrelated*. Actually, this is not true; a correlation must exist between E_n and I_n, because both lump the effects of the same internal noise sources of the amplifier. It is clear that they cannot be independent! This explains the dependence of the S/N ratio on the correlation coefficient between E_n and the time derivative of I_n. Nevertheless, as stated in [210], high precision in noise calculation is not entirely necessary, because of the manufacturing spread of the parameters, which causes even greater errors (and therefore we are rarely able to determine an accurate value for the correlation coefficient).

■ **Numerical Values.** The numerical values of E_n, I_n for a particular application can be found in one of the following ways:

- By reading the manufacturer's data sheet, they can often be found, together with plots of E_n and I_n versus frequency (this being the case for operational amplifiers and field-effect transistors).
- From measurements, as previously suggested.
- Numerically, using the expressions given in Sect. 7.6.5.

■ **Equivalent Input Noise.** The noise spectral density of the equivalent input voltage (neglecting correlation) is

$$S(E_{ni}) = 4kTR_s + \overline{E_n^2} + \overline{I_n^2}\, R_s^2 \qquad (13.2a)$$

Its rms value is very important in the design process, because it establishes the limit of the minimum signal that can be amplified.

If the equivalent input noise voltage is required in a given frequency band Δf, we proceed by dividing the band into n elementary sections (B_i), calculating the average noise in each section, multiplying by section bandwidth, summing all sections, and finally taking the square root of the sum. Hence

$$E_{ni} = \sqrt{4kTR_s\,\Delta f + \sum_{1}^{n}\left(\overline{E_n^2} + \overline{I_n^2}\, R_s^2\right) B_i} \qquad (13.2b)$$

An interesting property of the equivalent input noise is that it is *independent of the gain of amplifier and its input impedance*. Thus, this parameter is particularly attractive to compare the noise characteristics of various amplifiers or devices.

13.2 Low-Noise Design Techniques for Low-Frequency Circuits

■ **Signal-to-Noise Ratio.** Let us denote by $(S/N)_i$ the input signal-to-noise ratio, expressed in decibels. Thus

$$(S/N)_i = 20 \log \left(\frac{V_s}{E_{ni}} \right) = 10 \log \frac{V_s^2}{\left(4kTR_s + \overline{E_n^2} + \overline{I_n^2} R_s^2 \right) \Delta f} \quad [\text{dB}] \quad (13.3)$$

■ **Noise Factor.** Applying Friis' definition of the noise factor, we obtain

$$\boxed{F = 1 + \frac{\overline{E_n^2} + \overline{I_n^2} R_s^2}{4kTR_s}} \quad (13.4a)$$

and the corresponding noise figure is

$$F_{dB} = 10 \log \left(1 + \frac{\overline{E_n^2} + \overline{I_n^2} R_s^2}{4kTR_s} \right) \quad (13.4b)$$

The minimum noise factor is deduced by differentiating expression (13.4a) with respect to the source resistance R_s

$$\boxed{F_o = 1 + \frac{E_n I_n}{2kT}} \quad (13.5a)$$

This value is attained for an optimal source resistance, which is

$$\boxed{R_o = \frac{E_n}{I_n}} \quad (13.5b)$$

Using (13.4a), (13.5a) and (13.5b), we find that

$$\boxed{F = 1 + \frac{F_o - 1}{2} \left(\frac{R_s}{R_o} + \frac{R_o}{R_s} \right)} \quad (13.6)$$

■ **Faulkner's Contribution** [211,212]. Trying to identify the amplifier noise with a formula developed by Van der Ziel, Faulkner proposed expressing the amplifier noise in terms of two noise resistances, R_E in series with the input and R_I in parallel with the input:

$$R_E = \frac{\overline{E_n^2}}{4kT \, \Delta f} \quad \text{and} \quad R_I = \frac{4kT \, \Delta f}{\overline{I_n^2}} \quad (13.7)$$

The noise figure becomes

$$F = 10 \, \log \left(1 + \frac{R_E}{R_s} + \frac{R_s}{R_I} \right) \quad (13.8)$$

422   13 Low-Noise Circuit Design

With this approach, the optimum source resistance is the geometric mean of R_E and R_I. The ratio R_I/R_E can be interpreted as a "measure" of the low-noise capability of the amplifier.

■ **Note**

1) For a given amplifier driven from a voltage source, (13.3) shows that the S/N ratio is maximum when $R_s = 0$. As R_s increases, the S/N ratio decreases. In all practical applications, R_s has a fixed value; hence, once again, *any resistor inserted in series between the signal source and the amplifier reduces the S/N ratio.*
2) It can be shown that if a resistor R_p is inserted between the source and amplifier, across the amplifier input terminals, the resulting S/N ratio increases as $R_p \to \infty$. Therefore, *any resistor inserted in parallel between the signal source and the amplifier reduces the S/N ratio.*
3) From a physical point of view, it is clear that any resistor inserted between the signal source and the amplifier (in series or in parallel) decreases the S/N ratio because:
   – it attenuates the signal
   – it adds its own thermal noise
4) The noise factor (as opposed to the S/N ratio) is handier for measurement and computation, because there is no need to know the signal amplitude. Furthermore, its expression leads to the optimal value of the signal source resistance (see (13.5b)), while for the S/N ratio the optimal value is zero. These additional arguments explain why many people still consider noise factor reduction the main objective of low-noise design, despite evidence that in any communication link or sensor–amplifier system, it is the S/N ratio, rather than the noise factor, that matters.

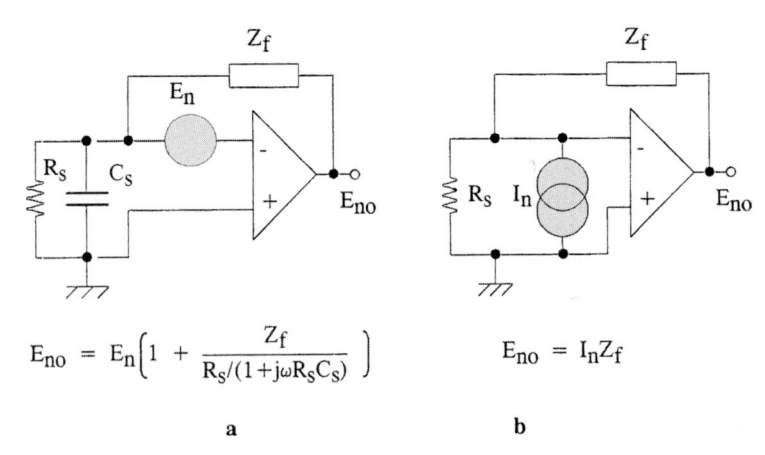

$$E_{no} = E_n\left(1 + \frac{Z_f}{R_s/(1+j\omega R_s C_s)}\right) \qquad E_{no} = I_n Z_f$$

a \qquad\qquad b

Fig. 13.4. Output noise evaluation: **a** when E_n is dominant; **b** when I_n is dominant

13.2 Low-Noise Design Techniques for Low-Frequency Circuits

■ **Case of an Inverting Amplifier.** Consider the inverting amplifier configuration detailed in Fig. 13.4, where R_s denotes the resistance of the signal source and C_s the capacitance of the connecting cable. The expressions for the output noise voltage are given beneath the diagrams. The unexpected conclusion is that the cable capacitance C_s (which, as for any capacitor, must be noiseless) *contributes to the output noise*. This contribution is proportional to the operating frequency, provided that Z_f is purely resistive (case **a**).

■ **Miscellaneous**

☐ **Example 1.** Find the spectral density of the equivalent input noise voltage of an amplifier driven from a signal source having $R_s = 10\,\text{k}\Omega$, which delivers a 200-Hz signal. Suppose that the amplifier noise is described by the plots in Fig. 13.3 and assume $T = 290\,\text{K}$.

Solution

a) $E_{ns} = \sqrt{4kTR_s} = 12.65\,\text{nV}/\sqrt{\text{Hz}}$
b) At $f = 200\,\text{Hz}$, the plots of Fig. 13.3 give $E_n = 12\,\text{nV}/\sqrt{\text{Hz}}$ and $I_n = 0.72\,\text{pA}/\sqrt{\text{Hz}}$.
c) It follows that $E_{ni} = \sqrt{12.65^2 + 12^2 + (0.72 \cdot 10)^2} = 18.86\,\text{nV}/\sqrt{\text{Hz}}$.

Thus, if the signal amplitude does not exceed 18.86 nV, the signal will be flooded by noise and operation will be hampered.

☐ **Example 2.** Find the rms value of the equivalent input noise voltage in the frequency range 20 Hz – 10 kHz, for an amplifier driven from a signal source having $R_s = 3\,\text{k}\Omega$. Assume that the amplifier noise is described by the plots in Fig. 13.3 and $T = 300\,\text{K}$. What is the minimum signal that can be amplified if normal operation requires that the peak-to-peak signal must be at least 5 times the rms value of the noise?

Solution

a) Break the proposed frequency range into 5 segments, about one octave each (where variations are strong) or greater (if in the flat region), according to Table 13.1.
b) For each segment, read the median value of E_n and I_n from the plots, square them, compute the terms required by (13.2b), find the contribution of the segment and enter the result into the table.
For the source resistance:

$$\overline{E_{ns}^2} = 4kTR_s \cong 49.7\,\text{nV}^2/\text{Hz} \quad \text{and} \quad \overline{E_{ns}^2}\,B = 49.7(9980) \cong 495806\,\text{nV}^2$$

c) Applying (13.2b), we find $E_{ni} \cong 0.854\,\mu\text{V}$. Under the condition imposed for proper operation of the amplifier, the minimum signal that can be handled is obviously $5(0.854) = 4.27\,\mu\text{V}$ peak-to-peak.

424   13 Low-Noise Circuit Design

Table 13.1. Noise calculation for each section

B_i [Hz]	Δf [Hz]	$\overline{E_n^2}$ [nV^2/Hz]	$\overline{I_n^2}\, R_s^2$ [nV^2/Hz]	(Sum)Δf [nV^2]
20–40	20	$40^2 = 1600$	$(.95 \cdot 3\,\mathrm{K})^2 = 8.12$	$(1608.12)20 = 32162.4$
40–80	40	$20^2 \cong 400$	$(.80 \cdot 3\,\mathrm{K})^2 \cong 5.76$	$(405.76)40 = 16230.4$
80–200	120	$13^2 = 169$	$(.75 \cdot 3\,\mathrm{K})^2 \cong 5.06$	$(174.06)120 = 20887.2$
200–1000	800	$10^2 = 100$	$(.68 \cdot 3\,\mathrm{K})^2 \cong 4.16$	$(104.16)800 = 83328$
1 K–10 K	9000	$9.3^2 \cong 86.5$	$(.63 \cdot 3\,\mathrm{K})^2 \cong 3.57$	$(90.07)9000 = 81054$

☐ **Example 3.** Estimate the S/N ratio for the amplifier of Example 2 if the signal delivered to its input port has an rms value of 2 mV.

Solution

a) We have already found $E_{ni} = 0.854\,\mu\mathrm{V}$ (rms value).
b) The S/N ratio (at the input terminals) is computed as:
$\mathrm{S/N} = 20\log(2\,\mathrm{mV}/0.854\,\mu\mathrm{V}) \cong 67.4\,\mathrm{dB}$

Note that because the two-port (amplifier) following the equivalent noise generators in Fig. 13.2 is noiseless, the same value is maintained at the output port.

13.2.3 Noise Matching with a Coupling Transformer

■ **Introduction.** This method improves the noise performance of the source-amplifier system by inserting a coupling transformer, which changes the appearance of the source resistance as seen by the amplifier input.

To illustrate the benefit, assume a transformer of winding ratio 1:n. At the amplifier input, the signal voltage is multiplied by n, while the noise remains at the same level (provided the transformer is noiseless). This explains the possible improvement in S/N ratio.

Alternatively, we may regard this technique as a modification of the equivalent noise generators of the amplifier, which when referred to the signal source become nI_n and E_n/n.

■ **Configuration.** The circuit is shown in Fig. 13.5, where r_1 and r_2 correspond to ohmic resistances of the transformer windings.

■ **Analysis.** The amplifier noise is described by its E_n–I_n model. The transformer–amplifier system may be regarded as a new amplifier of voltage gain nA and equivalent input noise generators E_n/n and nI_n. The correlation coefficient between them is the same as that between the generators E_n and I_n. It can be shown [214] that for an ideal transformer and uncorrelated generators, the square of the input S/N ratio is

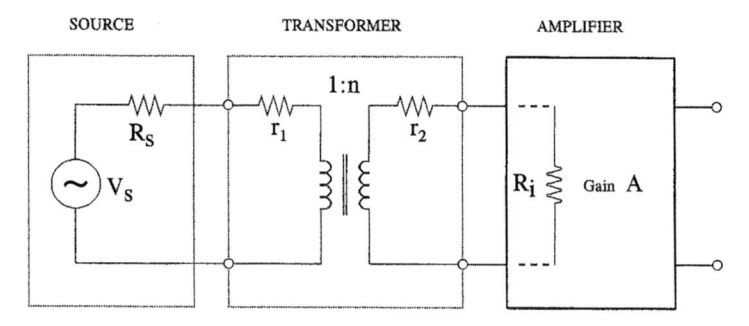

13.2 Low-Noise Design Techniques for Low-Frequency Circuits

Fig. 13.5. Noise matching with a coupling transformer

$$(S/N)_i^2 = \frac{V_s^2}{E_{ni}^2} = \frac{V_s^2}{4kTR_s + \overline{E_n^2}/n^2 + n^2\overline{I_n^2}R_s^2} \quad (13.9)$$

Differentiating this with respect to n^2 and setting the expression equal to zero, the optimal winding ratio is

$$\boxed{n^2 = \sqrt{\overline{E_n^2} / \overline{I_n^2}R_s^2}} \quad (13.10)$$

for which

$$\left(\overline{E_{ni}^2}\right)_{min} = R_s \left(4kT + 2E_n I_n\right) \quad (13.11a)$$

and

$$(S/N)_{Max} = \frac{V_s^2}{R_s(4kT + 2E_n I_n)} \quad (13.11b)$$

This shows that *for a given source resistance and temperature, the product $E_n I_n$ characterizes the inherent performance of the amplifier by establishing the maximum obtainable S/N ratio.*

The effective source resistance seen by the first stage is $n^2 R_s = E_n/I_n$, which is just the optimum resistance given by (13.5b), which yields the minimum noise factor. Consequently, this approach simultaneously ensures the minimum noise factor and maximum S/N ratio.

If the winding resistances r_1 and r_2 are also taken into account, the search for the minimum of E_{ni} is more difficult, since r_2/r_1 is proportional to the ratio n.

■ Limitations

- When R_s is low, the thermal noise of resistances r_2 and r_1 is no longer negligible. In this case, using a coupling transformer may *reduce* the S/N ratio.
- Coupling by means of a transformer is often incompatible with monolithic solid-state technology.

- The coupling transformer drastically reduces the amplifier bandwidth.
- The coupling transformer, due to its imperfections, represents a multiple noisy source: Barkhausen noise (a succession of abrupt changes in the magnetic domains of the core material when a signal is applied), mechanical stress and vibrations, and susceptibility to spurious magnetic fields. The latter property is particularly harmful at power–supply frequency and its harmonics, and requires heavy shielding. Finally, without proper care, the S/N ratio is more likely to be reduced than increased.

This technique is well-suited to RF applications, where excellent quality transformers are available with less core and windings. Nevertheless, it remains the favorite approach in low-frequency circuits driven by sources with very low internal resistance (3 or 4 orders of magnitude lower than the amplifier input resistance).

■ **Illustrative Example.** Consider a typical preamplifier with $E_n = 10\,\text{nV}/\sqrt{\text{Hz}}$ and $I_n = 0.1\,\text{pA}/\sqrt{\text{Hz}}$ (in the intended frequency band), driven by a transducer with $R_s = 10\,\Omega$. Find the optimum winding ratio of a coupling transformer and estimate the improvement in S/N ratio when using the transformer.

Solution

Applying (13.5b), the optimal source resistance is $R_o = E_n/I_n = 10^5\,\Omega$. Then, with (13.10), the optimal winding ratio is $n = \sqrt{E_n/I_n R_s} = 10^2$. Hence, we need a coupling transformer with a winding ratio of 1:100.

The S/N ratio improvement (SNI) is usually defined to be [208]

$$\text{SNI} = \frac{\text{S/N ratio with transformer}}{\text{S/N ratio without transformer}}$$

Therefore, applying (13.3) and (13.11b),

$$\text{SNI} = \frac{4kTR_s + E_n^2 + (R_s I_n)^2}{2R_s(2kT + E_n I_n)} \cong 538.7$$

Note that this result provides an asymptotic value; due to transformer imperfections, the actual improvement is expected to be much more modest.

13.2.4 Noise Matching by Paralleling Input Devices

■ **Introduction.** This technique consists in replacing the active device of the first stage with several identical active devices connected in parallel. It is recognized that this approach is effective whenever the noise originating from E_n is dominant.

■ **Configuration.** The block diagram of a low-noise input stage employing this approach is shown in Fig. 13.6.

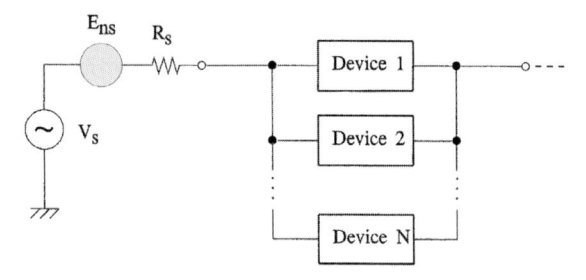

13.2 Low-Noise Design Techniques for Low-Frequency Circuits

Fig. 13.6. Noise matching by paralleling input active devices

■ **Analysis.** Each active device is replaced by its E_n–I_n noise model (at the input) and an equivalent Norton electrical circuit (at the output). The output short-circuit current is the sum of all individual currents. From the equation describing the input, the equivalent squared input noise voltage is

$$\overline{E_{ni}^2} = 4kTR_s + \frac{1}{N}\overline{E_n^2} + N\overline{I_n^2}R_s^2 \tag{13.12}$$

When $R_s = 0$,
$$E_{ni} = E_n / \sqrt{N} \tag{13.13}$$

Equation (13.13) shows that the noise level can be considerably reduced, if N is large enough. However, in the case $R_s \neq 0$, the optimum number of active devices ensuring a minimum of E_{ni} is

$$\boxed{N = \frac{E_n}{I_n R_s}} \tag{13.14}$$

This expression is important, since it shows that N decreases when R_s increases. Hence, for transducers with relatively high internal resistance, this approach is not effective.

■ **Comments**

– The correlation coefficient between the E_n and I_n of the ensemble is N times smaller than that of a single input device.
– Equation (13.13) suggests that this method of noise matching is equivalent to using an input coupling transformer, of winding ratio $1:\sqrt{n}$.
– In practice, when several discrete transistors are connected in parallel, problems can result from non-identical parameters. Although the transistors are initially carefully selected to have almost identical parameters, unbalance may occur with aging, and one of them may drive more current than the others. As a result, the device in question will dissipate additional power and its temperature will rise beyond that of its neighbors. Due to extra heating, its current gain will further increase, as will the dissipation, and so on. Ultimately, that transistor will fail, precluding proper operation

of the ensemble. This is why paralleling active devices does not ensure safe operation, except in solid-state circuits, where it is possible to fabricate almost identical transistors.
- Using several devices in parallel actually increases overall device size. In RF integrated circuits, by properly selecting the device size, it is possible to set R_o to $50\,\Omega$ (or any other value) and simultaneously achieve noise and signal matching [215].

■ **Conclusion.** Paralleling several input devices improves the noise performance of the amplifier, provided that:

- the source resistance is low,
- the contribution of E_n to the overall noise is dominant.

13.2.5 Selection of Active Devices

■ **Principle.** The simplest way to perform noise matching is to properly select the front-end active device with respect to the internal transducer resistance.

■ **Analysis.** Starting with the assumption that *exact* noise matching is of little use, the following approximate expressions are given for bipolar transistors to evaluate the spectral densities of the equivalent input noise generators:

$$S(I_n) \cong 2q\,I_B = 2q\,I_C/\beta \qquad (13.15a)$$

$$S(E_n) \cong 4kT\left(r_{bb'} + \frac{1}{2g_m}\right) \qquad (13.15b)$$

(these are deduced from (7.55) and (7.54) at mid-frequencies). The optimum collector current depends on the source resistance:

$$I_{Copt} \cong \frac{25\sqrt{\beta}}{R_s + r_{bb'}} \quad [mA] \qquad (13.16)$$

Equations (13.15) show that *the quietest bipolar transistors are those with low $r_{bb'}$ and high β* (low $r_{bb'}$ can be achieved by paralleling several transistors). However, when the actual transistor is biased at I_{Copt}, little improvement in noise level is to be expected by using higher-β transistors.

For junction field-effect transistors operating above the $1/f$ noise region, the spectral densities are

$$S(E_n) \cong 0.7(4kT/g_m) \qquad (13.17a)$$

$$S(I_n) \cong 2q\,I_G + 0.7(4kT/g_m)(\omega C_{gs'})^2 \qquad (13.17b)$$

13.2 Low-Noise Design Techniques for Low-Frequency Circuits

where $C_{gs'}$ denotes the internal capacity (about 2/3 of the total value of C_{gs}).

As the leakage gate current is several orders of magnitude lower than the base current of a bipolar transistor, I_n is usually much lower for a JFET than for a bipolar transistor (at least at low- and mid-frequencies).

■ **Practical Advice.** In the design of low-noise amplifiers, the following points must be observed:

- For low values of source resistance (say, between several hundreds ohms and 5 kΩ), the term E_n^2 dominates in (13.2b), (13.3) and (13.4). In this case, it is advisable to select a bipolar transistor for the input stage, since its E_n is lower than that of a field-effect transistor. A proper choice of collector current using (13.16) guarantees a low noise level; consequently, the noise factor can fall to a fraction of decibel (for $R_s \cong 1\,\text{k}\Omega$). When the source resistance is less than 100 Ω, the value of the noise generator E_n becomes so important that we are forced to adopt either a coupling transformer (for discrete circuits) or paralleling several devices (for integrated circuits).
- When the source resistance is high (more than 100 kΩ), it is I_n which dominates, and the best choice for the input stage is a field-effect transistor, with its low I_n.
- Whenever the specified frequency band includes frequencies lower than 10 Hz, it is highly advisable to select field-effect transistors, since the 1/f noise of bipolar transistors is considerably greater. For this reason, the first stage of instrumentation amplifiers is almost always based on field-effect transistors.
- When a monolithic integrated amplifier is selected as the preamplifier (despite their poorer noise performance compared to that of discrete devices), it is useful to remember that a differential amplifier has a noise level higher than that of a single-ended amplifier by about 3 dB. However, this drawback is compensated by the symmetry of the circuit, which rejects common-mode signals. Another point to note is that the feedback resistors used to establish the gain of an operational amplifier can significantly impair overall noise performance.
- Feedback loops over several stages does not modify the equivalent input noise, except for the extra noise added by the feedback resistors.

■ **Conclusion.** Bipolar transistors have a clear advantage when operating with low-to-moderate source impedances; for low source resistances, bipolar transistors with low $r_{bb'}$ must be selected. With high source resistances, field-effect transistors (with their low I_n) are the best choice. Note that these are general suggestions and no clear separation exists between the domains where BJTs or FETs are best-suited. Sometimes, selecting a low-noise device of high quality is as important as making the right choice between FETs and BJTs.

13.2.6 Feedback

■ **Introduction.** One of the basic things students learn is that negative feedback represents an important and elegant means of controlling the amplifier performance (stability, gain, bandwidth, input and output impedances, and even non-linear distortions). In all classical books, it is demonstrated that all of these performances are improved by the reduction factor of the closed-loop gain with respect to the open-loop gain (i.e., $(1 + \beta A)$).

It is natural to wonder whether the same observation applies to noise. In order to answer this question, let us consider the following examples.

■ **Example 1.** Consider a two-stage amplifier and a feedback loop, as depicted in Fig. 13.7. Suppose that a noise signal is injected at the middle summation point. This noise can either be an interfering signal from outside, or produced by the power supply (for instance, A_2 can be a power stage, fed by a supply with a considerable amount of ripple, while A_1 is supplied through a high-quality filter). Assuming that amplifier A_1 has negligible noise, find the output S/N ratio.

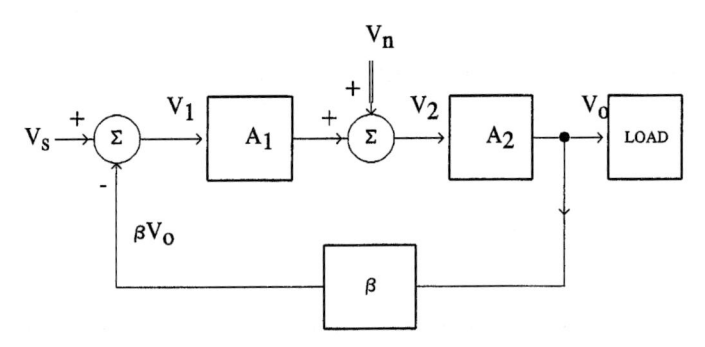

Fig. 13.7. Block diagram of a feedback amplifier

Solution

It is easy to see that the following relations apply:

$$V_1 = V_s - \beta V_o \tag{13.18}$$

$$V_2 = V_n + A_1 V_1 \tag{13.19}$$

$$V_o = A_2 V_2 \tag{13.20}$$

Upon substituting (13.18) into (13.19), and the result into (13.20), we have

$$V_o = V_s \frac{A_1 A_2}{1 + \beta A_1 A_2} + V_n \frac{A_2}{1 + \beta A_1 A_2} \tag{13.21}$$

The output S/N ratio can be evaluated from (13.21) where the contributions of signal and noise are separated. Recall that the S/N ratio is expressed

13.2 Low-Noise Design Techniques for Low-Frequency Circuits

as a power ratio (the squared value of the rms signal to the mean square value of the fluctuation, per unit bandwidth):

$$\frac{S_o}{N_o} = \frac{V_s^2 \left(\frac{A_1 A_2}{1+\beta A_1 A_2}\right)^2}{\overline{V_n^2} \left(\frac{A_2}{1+\beta A_1 A_2}\right)^2} \tag{13.22}$$

hence

$$\boxed{\frac{S_o}{N_o} = \frac{V_s^2}{\overline{V_n^2}} A_1^2} \tag{13.23}$$

☐ **Discussion.** Compared to an open-loop amplifier (Fig. 13.8), the output S/N ratio predicted by expression (13.23) is obviously A_1^2 times greater. However, this improvement is not due to feedback (note that β is not even present in (13.23)), but instead to the inclusion of the preamplifier A_1, which constitutes the difference between the two block diagrams of Fig. 13.7 and Fig. 13.8. The preamplifier (A_1) merely amplifies the signal *before* contaminating it with noise, and under the assumption that the noise of A_1 is negligible, of course the resulting S/N ratio is higher. Note that if $A_1 = 1$, the two situations are identical. Once again, this application is representative of the benefits of using a low-noise preamplifier before processing the signal.

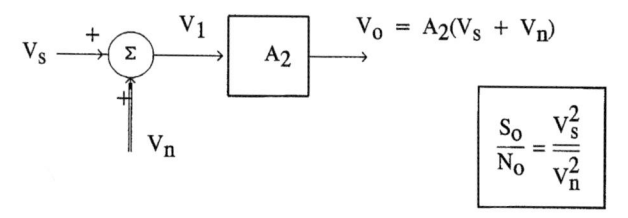

Fig. 13.8. Amplifier without feedback

■ **Example 2.** Consider now the more general configuration shown in Fig. 13.9, in which noise can contaminate the signal path not just at a single point, but everywhere. In order to appreciate the influence of negative feedback, we compare the output signal (V_o) with that derived from an equivalent open-loop system delivering the same output voltage.

Solution

The following equations can be written by inspection:

$$V_1 = (V_s + V_{n1}) - \beta V_o \tag{13.24}$$
$$V_2 = A_1 V_1 + V_{n2} \tag{13.25}$$
$$V_o = A_2 V_2 + V_{n3} \tag{13.26}$$

432   13 Low-Noise Circuit Design

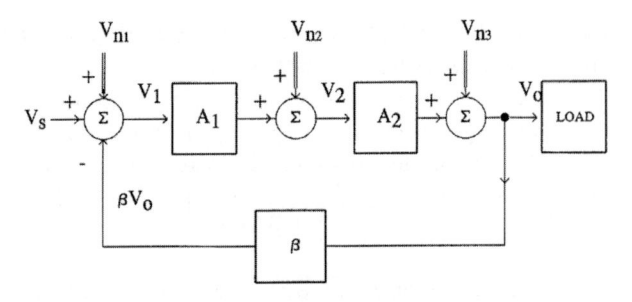

Fig. 13.9. Block diagram of a two-stage feedback amplifier

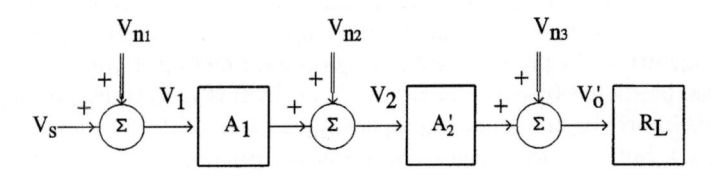

Fig. 13.10. Open-loop two-stage amplifier

Upon substituting (13.24) into (13.25) and the result into (13.26), one obtains

$$V_o = \frac{A_1 A_2}{1 + \beta A_1 A_2}(V_s + V_{n1}) + \frac{A_2}{1 + \beta A_1 A_2} V_{n2} + \frac{1}{1 + \beta A_1 A_2} V_{n3} \tag{13.27}$$

We must compare this with the output voltage of a similar configuration, but without feedback (Fig. 13.10). This time, the following equations apply:

$$V'_o = V_{n3} + V_2 A'_2 \tag{13.28}$$

$$V_2 = V_{n2} + V_1 A_1 \tag{13.29}$$

$$V_1 = V_{n1} + V_s \tag{13.30}$$

With the same approach, we obtain

$$V'_o = A_1 A'_2 (V_s + V_{n1}) + A'_2 V_{n2} + V_{n3} \tag{13.31}$$

Inspecting (13.27) and (13.31), a meaningful comparison requires that

$$A'_2 = \frac{A_2}{1 + \beta A_1 A_2} \tag{13.32}$$

Consequently

$$V'_o = \frac{A_1 A_2}{1 + \beta A_1 A_2}(V_s + V_{n1}) + \frac{A_2}{1 + \beta A_1 A_2} V_{n2} + V_{n3} \tag{13.33}$$

13.2 Low-Noise Design Techniques for Low-Frequency Circuits

Comparing (13.33) and (13.27), it is obvious that *the feedback loop has no effect on the noise injected at the input of either amplifier*. However, the only noise which is attenuated is that contaminating the output (V_{n3}). We may conclude that *negative feedback acts favorably only on that noise which is possibly added at the output* (for instance, by inserting an additional loading device, with its own noise, or when the hum associated with the power supply of the last stage is excessive).

■ **Example 3.** A low-noise amplifier stage employs a MESFET transistor. The available power gain of the stage is $G_a = 8\,\text{dB}$ and its noise factor is $F = 1.5\,\text{dB}$. By applying negative feedback, the input impedance is appropriately modified in order to better match the source resistance ($50\,\Omega$). Laboratory measurements yield a better value of $F_1 = 1.4\,\text{dB}$ but the available power gain has been adversely affected ($G_{a1} = 6.57\,\text{dB}$). Prove that when applying feedback, the noise measure remains unchanged.

Solution

The first step is to transform all quantities expressed in decibels to ratios. Hence

$$F = 1.5\,\text{dB or } 1.41 \qquad F_1 = 1.4\,\text{dB or } 1.38$$
$$G_a = 8\,\text{dB or } 6.31 \qquad G_{a1} = 6.57\,\text{dB or } 4.545$$

Next, applying definition (4.89), the noise measure is evaluated before introducing negative feedback

$$M = \frac{F-1}{1 - 1/G_a} = \frac{1.41 - 1}{1 - 1/6.31} \cong 0.487$$

After applying the feedback, we have

$$M_1 = \frac{F_1 - 1}{1 - 1/G_{a1}} = \frac{1.38 - 1}{1 - 1/4.545} \cong 0.487$$

Hence, *the noise measure is invariant when negative feedback is applied to the amplifier*. Note that the noise factor may eventually change, since feedback modifies the matching of the source impedance, but never the noise measure. Assuming that the feedback loop is noiseless, this shows that

- Negative feedback has no influence on the intrinsic noise of the circuit (despite apparent modification of the noise factor).
- In this context, the noise measure is a more reliable parameter.

■ **Conclusion.** *Thermal and shot noise, which are fundamental in all electronic devices, cannot be reduced by applying negative feedback.* Moreover, feedback resistors *add* their thermal and excess noise to the amplifier's intrinsic noise.

434   13 Low-Noise Circuit Design

Therefore, feedback is useful only to control the critical performance of the amplifier, except for intrinisic noise. Note that feedback resistors must be carefully selected (values as low as possible, to limit thermal noise, and resistors of high quality, to reduce excess noise). For low-noise amplifiers, metal foil or wirewound resistors are particularly recommended in the feedback loop; in contrast, due to their high noise index, carbon resistors must be avoided.

13.2.7 Application 1: Sensor and Its Preamplifier

■ **Comment.** The most often encountered situation is that in which the performance of the sensor (transducer) is well-known and the signal derived from it is processed in electrical form, without energy conversion (to optical, magnetic, or any other form of energy). In this case, the designer has no control over the signal level (once the sensor is selected); hence, to improve the S/N ratio, the philosophy is to *minimize the intrinsic noise of the system by careful design*.

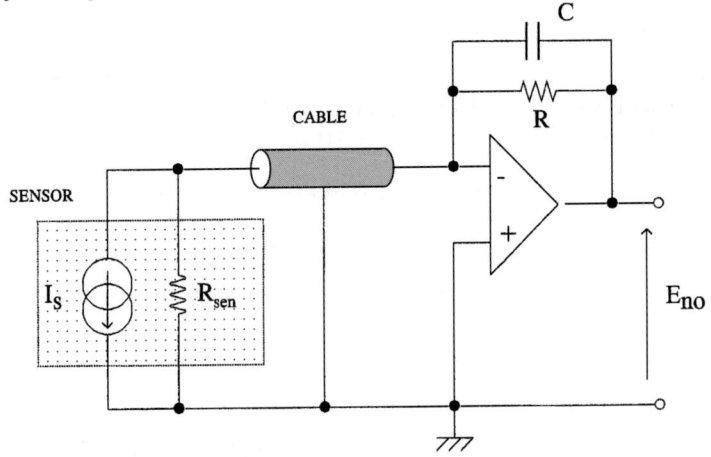

Fig. 13.11. Sensor, cable, and preamplifier

■ **Problem Formulation.** Assume that the sensor is a photomultiplier tube [213], whose anode current consists of a 5 Hz signal with rms value $I_s = 2$ nA (Fig. 13.11), corresponding to light excitation. This signal is added to a DC component $I = 200$ nA (the DC current is due to stray light and/or dark/leakage currents; note that in practice it is removed from the useful signal by performing a background substraction). The internal resistance of the sensor, denoted by R_{sen}, is assumed to exceed 1 MΩ. An operational amplifier connected through a short cable is used as preamplifier, with $e_{nA} = 30$ nV/$\sqrt{Hz}$ and $i_{nA} = 5$ fA/$\sqrt{Hz}$ at 5 Hz. The feedback resistor R is 100 kΩ and C is a stray capacitance, estimated at 2.5 pF. Evaluate the output S/N ratio at T = 290 K and suggest a solution that improves it.

13.2 Low-Noise Design Techniques for Low-Frequency Circuits

☐ **Solution.** The approach comprises the following steps:

Step 1. Find the output contributions of all noise sources, at $f = 5$ Hz:
- The thermal noise associated with the feedback resistor has spectral density

$$e_{nR} = \sqrt{4kTR} = 4 \cdot 10^{-8} \text{ V}/\sqrt{\text{Hz}}$$

and due to the virtual ground on the inverting input, this voltage is transmitted to the output intact.
- The shot noise current spectral density is

$$i_{nc} = \sqrt{2qI} = \sqrt{2(1.6 \cdot 10^{-19})(2 \cdot 10^{-7})} \cong 2.53 \cdot 10^{-13} \text{ A}/\sqrt{\text{Hz}}$$

The corresponding output noise voltage spectral density is

$$e_{nc} = R \, i_{nc} = 10^5 \, (2.53 \cdot 10^{-13}) = 2.53 \cdot 10^{-8} \text{V}/\sqrt{\text{Hz}}$$

- The specified power spectral density of the preamplifier noise, at 5 Hz, is

$$e_{nA} = 30 \text{ nV}/\sqrt{\text{Hz}} \quad \text{and} \quad i_{nA} = 5 \text{ fA}/\sqrt{\text{Hz}}$$

The output contributions of both noise sources are calculated with the expressions in Fig. 13.4. The voltage gain between E_n and the output is roughly unity, so

$$e_{nA1} = e_{nA} = (1)(0.03) = 0.03 \text{ }\mu\text{V}/\sqrt{\text{Hz}}$$

Using the expression in Fig. 13.4b, we obtain

$$e_{nA2} = R i_{nA} = 10^5 (5 \cdot 10^{-15}) = 5 \cdot 10^{-10} \text{ V}/\sqrt{\text{Hz}}$$

All these sources are uncorrelated, since they correspond to different physical mechanisms.

Step 2. Check whether a noise source is dominant. With Rule 1 (Sect. 13.2.1), no source is really dominant, but it is clear that the contribution of e_{nA2} can be neglected. Consequently, the total noise (as a spot value) is calculated by adding all significant contributions:

$$\overline{E_{no}^2} = \overline{e_{nc}^2} + \overline{e_{nR}^2} + \overline{e_{nA1}^2}$$

$$\cong (6.4 + 16 + 9)10^{-16} = 31.4 \cdot 10^{-16} \quad [\text{V}^2/\text{Hz}]$$

Consequently

$$E_{no} \cong 5.6 \cdot 10^{-8} \text{ V}/\sqrt{\text{Hz}}$$

Step 3. Evaluate the output S/N ratio. Due to the stray capacitance C, the amplifier acts as a low-pass filter with a -3 dB cutoff frequency equal to $1/2\pi RC$ and an equivalent noise bandwidth $\Delta f = 1/4RC$. Hence

$$\Delta f = 1/4RC = 1/(4 \cdot 10^5 \cdot 2.5 \cdot 10^{-12}) = 10^6 \text{ Hz}$$

The total output noise voltage in the bandwidth Δf is

$$E_n = E_{no} \sqrt{\Delta f} = 5.6 \cdot 10^{-8} \cdot 10^3 = 5.6 \cdot 10^{-5} \text{ V}$$

The output signal is

$$e_s = RI_s = 10^5 (2 \cdot 10^{-9}) = 2 \cdot 10^{-4} \text{ V}$$

and the ouput S/N ratio can now be calculated

$$S/N = 20 \log \frac{e_s}{E_n} = 20 \log \frac{2 \cdot 10^{-4}}{5.6 \cdot 10^{-5}} \cong \boxed{11 \text{ dB}}$$

Step 4. Improving the S/N ratio. In this case, the only practical way to reduce the sensor noise is to limit the bandwidth as much as possible (Rule 3), to a value that will strictly guarantee signal transmission integrity. If a discrete, high-quality capacitor $C' = 0.2\,\mu\text{F}$ is added across C, then

$$\Delta f' = 1/4RC' = 1/(4 \cdot 10^5 \, 0.2 \cdot 10^{-6}) = 12.5 \text{ Hz}$$

and the total output noise is then

$$E'_n = E_{no} \sqrt{\Delta f'} = (5.6 \cdot 10^{-8}) \sqrt{12.5} \cong 19.8 \cdot 10^{-8} \text{ V}$$

It is necessary to check whether the $-3\,\text{dB}$ cutoff frequency is higher than the signal frequency:

$$f_{3dB} = 1/(2\pi \cdot 10^5 \, 0.2 \cdot 10^{-6}) = 7.96 \text{ Hz}$$

We may conclude that the signal frequency (5 Hz) is not significantly attenuated. The resulting S/N ratio is then

$$(S/N)' = 20 \log \frac{e_s}{E'_n} = 20 \log \frac{2 \cdot 10^{-4}}{19.8 \cdot 10^{-8}} \cong \boxed{60 \text{ dB}}$$

The improvement in S/N ratio is about 50 dB, which shows how powerful the technique of limiting the bandwidth is.

■ Comments

- Attention must be paid when choosing the capacitor C'; it must have low leakage current, so electrolythic capacitors should be avoided. The problem is critical when the specified value of C' is high enough to be impractical for ceramic or mica type capacitors. In this case, paralleling several high-quality capacitors may be a solution.
- The coaxial cable between the photomultiplier and preamplifier has an associated capacitance. The virtual ground input of the operational amplifier has the advantage that with zero volts across it, the cable capacitance cannot be charged and consequently we expect fewer microphonic effects.

13.2.8 Application 2: Dolby Noise Reduction System

■ **Magnetic Tape Recording System.** In all recording systems on magnetic tape, sound is converted to an electrical signal by a transducer (microphone). After amplification, this signal is passed through the head coil of a recording system, which sets up a corresponding magnetic pattern on iron oxide particles suspended on a moving plastic tape. This process corresponds to the *recording phase* (Fig. 13.12). During the *playback phase*, this magnetic pattern is reconverted into electrical signals. They are amplified and processed, and they finally drive a system of speakers (or headphones), that transform the electrical energy back into sound.

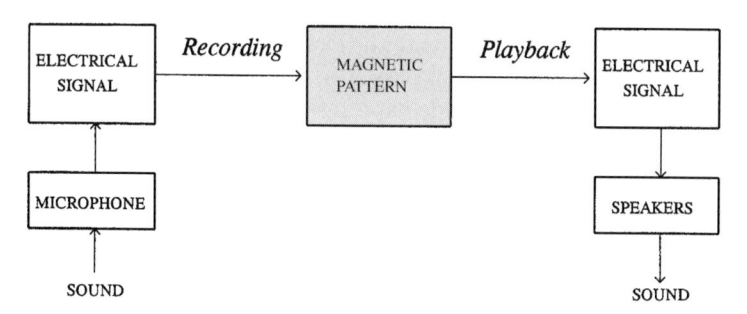

Fig. 13.12. Principle of a tape recording system

■ **Comment.** This application addresses the issue of S/N ratio improvement in a system operating with a double energy conversion. In this case, the magnetic tape is responsible for the dominant noise source, which greatly exceeds the intrinsic noise of the electronic circuits. As a matter of fact, in tape recording, the S/N ratio usually lies between the permissible limit of saturation distortion and the tape's background hiss [219]. In contrast to the application described in Sect. 13.2.7, where the circuit noise was dominant, in this case it is not worthwhile to focus the design effort on further reduction of circuit noise, since the critical process is the storage of information on magnetic medium. This constitutes one of those rare situations in which the designer can control the level of useful signal, but has little (or no) control over the main process responsible for noise.

Therefore, to improve the S/N ratio, a complementary technique to that employed in Sect. 13.2.7 will be discussed, namely *boosting the signal before recording*, to comfortably override the background noise (over which the circuit designer has no control).

■ **Noise in Magnetic Tape Recording Systems.** Frequently the noise heard on audio cassettes is hiss (predominantly containing high frequencies). It can be perceived even when we listen to brand new high-quality tape and is due mainly to the granular structure of the magnetic film deposited on the tape, inherent fabrication inhomogeneities and defects, etc. Progress in technology in recent years has greatly reduced this type of noise, but not eliminated it.

■ **Dolby Technique.** This technique was invented several decades ago by Dr. Ray Dolby; it increases the S/N ratio of the tape recording process [216].

Since the noise of the magnetic tape cannot be eliminated, and moreover, it is quite strong (especially high frequency hiss), the philosophy behind improving the S/N ratio is to *preamplify the useful signal* (at least that part which will be most affected by the noise). Consequently, the Dolby technique raises the volume of the high frequency components of the spectrum by an amount that depends on the signal level. For instance, during the recording of a loud signal, it is no need to activate the Dolby system (this avoids saturation of the amplifier), but during quiet passages, the system boosts high-frequency recorded components (to override tape hiss). During the playback phase, if soft music or speech are present, high frequencies are attenuated to recover the original signal level; however, during loud segments, high frequencies are not attenuated. This process also automatically reduces any noise that was introduced during the recording or playback phases. This technique improves the S/N ratio by about 10 dB.

■ **Definition.** According to [219], *the Dolby system is a noise-reduction system whereby the recording signal is compressed and the replayed signal expanded.*

■ **Implementation.** As high audio frequencies are most affected by the noise of the magnetic tape system, they are isolated from the rest of the spectrum and processed separately. Figure 13.13 presents a block diagram of the pro-

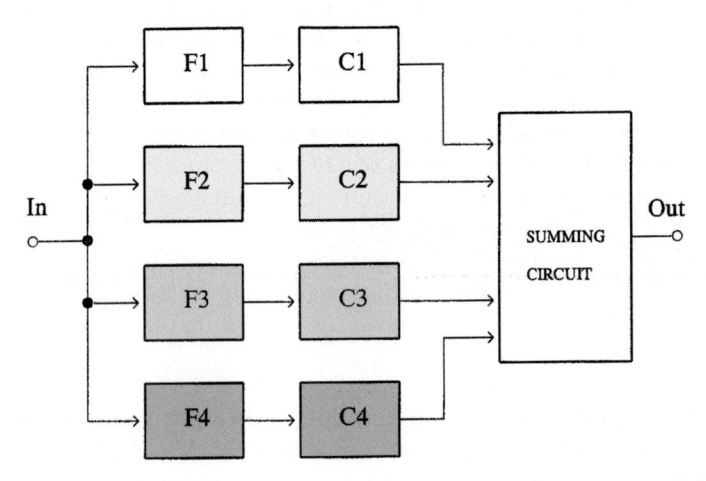

Fig. 13.13. Block diagram of the Dolby A system (recording module)

fessional Dolby A system, which has four independently controlled channels, to increase S/N ratio at low, middle, high, and very high frequencies:

F_1 is a low-pass filter with a cutoff frequency of 80 Hz;
F_2 is a band-pass filter with 80 Hz and 3 kHz cutoff frequencies;
F_3 is a high-pass filter with a cutoff frequency of 3.3 kHz
F_4 is a high-pass filter with a cutoff frequency of 9 kHz.

Each frequency interval is separately compressed by the circuits C1, C2, C3, and C4, in such a way that soft signals (around -40 dB) are amplified by 10 dB. Relatively strong signals (between 0 and -30 dB) are left unchanged. In the playback phase, the reverse operation is performed, and the signals are restorted to their original level.

■ **Remark.** The same kind of processing is applied to each of the four channels; hence, notable quality improvement is expected at every frequency of the audio spectrum.

13.3 Low-Noise Design Techniques for Microwave Circuits

■ **Comment.** One fundamental design concept for microwave circuits is *impedance matching*. In practice, all microwave circuits operate with matched termination (usually 50 Ω), in order to avoid reflected waves. Consequently, all transducers (antennas or any other signal source) for the microwave frequency range have an internal impedance as close as possible to 50 Ω. In this context, since the problem of deliberately increasing or decreasing the source impedance is no longer pertinent, the noise figure fallacy discussed in Sect. 13.2.2 has no reason to exist. We conclude that *for circuits always operating under impedance matching conditions the noise factor minimization is a valuable objective of low-noise design.*

■ **Problem Formulation.** The design of any low-noise microwave amplifier must satisfy the following requirements:

– The specified gain versus frequency profile (which is usually either flat, or has a slope of +6 dB/oct) must be met.
– The maximum specified noise factor in the operating frequency range must not be exceeded.

■ **Constant Noise Circles.** Equation (5.34a), plotted on a Smith chart, represents the locus of all possible source reflection coefficient values that yield a constant noise factor at a particular frequency [217]. The center of this circle is located at

$$C_F = \Gamma_o / (1 + N) \qquad (13.34)$$

and its radius is

$$R_F = \sqrt{N^2 + N(1 - |\Gamma_o|^2)} / (1 + N) \quad (13.35)$$

■ **Constant Available Gain Circles.** Theses circles show the source reflection coefficient values that provide constant available gain (G_A) at a specified frequency. The coordinate of the center is

$$C_a = (g_a C_1^*) / d \quad (13.36)$$

and the corresponding radius is

$$R_a = \sqrt{1 - 2K|S_{12}S_{21}|g_a + |S_{12}S_{21}|^2 g_a^2} / d \quad (13.37)$$

where

$$g_a = G_A / |S_{21}|^2 \quad (13.38)$$

$$d = 1 + g_a(|S_{11}|^2 - |\Delta|^2) \quad (13.39)$$

$$K = \left(1 - |S_{11}|^2 - |S_{22}|^2 + |\Delta|^2\right) / 2|S_{12}S_{21}| \quad (13.40)$$

$$C_1 = S_{11} - \Delta\, S_{22}^* \quad (13.41)$$

■ **Graphical Procedure.** In order to optimize the noise of a microwave amplifier, the following steps are recommended:

1. Consider the first frequency of the specified passband and draw on the Smith chart the constant available gain circle corresponding to the specified power. Next, add several constant noise circles.
2. The best possible noise factor corresponds to the noise circle tangential to the constant available gain circle.
3. If the best noise factor is still too high, change the bias, or even the transistor type. If the noise performance is not sufficiently improved, this simply means that the specified noise level is too optimistic.
4. Repeat the previous steps for each frequency of the specified passband.

■ **Contribution of Link and Gudimetla** [218]. To simplify low-noise microwave amplifier design, the authors proposed using analytic expressions instead of the previous graphical procedure. The expected benefit is improved accuracy, as the algebraic manipulation of equations can be programmed on a computer.

The proposed analytic procedure has two approaches:

- *Find the source reflection coefficient that yields the maximum available gain for a specified noise figure, at a specified frequency.* The condition of having the constant noise circle tangential to the constant available gain circles can be expressed analytically as

13.3 Low-Noise Design Techniques for Microwave Circuits

$$|C_a - C_F| = |R_a \pm R_F| \qquad (13.42)$$

Note that this equation has two solutions; the designer must select the one that ensures a passive source reflection coefficient and results in stable operation. The solutions are given by

$$\boxed{G_{A1,2} = d\,|S_{21}|^2 \left(-A_1 \pm \sqrt{A_1^2 - 4A_0 A_2}\right)/2A_2} \qquad (13.43)$$

where

$$A_0 = -\left(1 - |\Gamma_o|^2\right)^2 / (N+1)^2 \qquad (13.44)$$

$$A_1 = 2\Big((C_F C_1 + C_F^* C_1^*)(|C_F|^2 - 1 - R_F^2) + B_1(1 - |C_F|^2 - R_F^2)\Big) \qquad (13.45)$$

$$A_2 = 4R_F^2 |C_1|^2 - (C_F C_1 + C_F^* C_1^* - B_1^2) \qquad (13.46)$$

$$B_1 = 1 + |S_{11}|^2 - |S_{22}|^2 - |\Delta|^2 \qquad (13.47)$$

An asterisk denotes the complex conjugate.

- *Find the source reflection coefficient that yields the minimum noise figure for a specified available gain.*

By combining (13.34) and (13.35), we obtain

$$R_F^2 = \frac{N}{N+1} - N|C_F^2| \qquad (13.48)$$

Upon substituting (13.48) into (13.42) previously squared, a quadratic equation in terms of the noise parameter N is obtained. Its solutions are

$$\boxed{N = \frac{-D_1 \pm \sqrt{D_1^2 - 4D_o D_2}}{2D_2}} \qquad (13.49)$$

where the coefficients D_o, D_1, and D_2 are

$$D_o = \left(|\Gamma_o - C_p|^2 - R_p^2\right)^2 \qquad (13.50)$$

$$D_1 = 2\left(|C_p|^2 - R_p^2 - 1\right)^2 \left(|\Gamma_o - C_p|^2 - R_p^2\right) + 4R_p^2\left(|\Gamma_o|^2 - 1\right) \qquad (13.51)$$

$$D_2 = \left(|C_p|^2 - R_p^2 - 1\right)^2 - 4R_p^2 \qquad (13.52)$$

The minimum solution of (13.49) is selected; then, with (5.34c) and (5.34a), the minimum noise factor F_o in terms of the required available gain and the amplifier parameters is deduced. Finally, the values of the source reflection coefficient are calculated with the following expressions:

$$\Gamma_s = C_F - \frac{C_a - C_F}{|C_a - C_F|} R_F \quad \text{for} \quad |C_a - C_F| \le R_a \quad \text{and} \quad R_F \le R_a \quad (13.53)$$

or

$$\Gamma_s = C_F + \frac{C_a - C_F}{|C_a - C_F|} R_F \quad \text{otherwise} \quad (13.54)$$

■ **Conclusion.** The derived expressions are of great help in the design of low-noise microwave amplifiers. They provide improved accuracy and considerably simplify the design process.

Summary

- In the low-noise design of electronic circuits, it is important to find the dominant noise source. All efforts must then be directed toward reducing it.
- A very effective way to reduce noise is to limit the bandwidth (by adding filters) to the value required for signal transmission.
- The main objective of the low-noise design process is to maximize the S/N ratio.
- In all circuits operating with matched termination (usually $50\,\Omega$), the equivalent objective of the low-noise design is to minimize the noise factor.
- Noise matching in discrete circuits can be achieved by using a coupling transformer. This is particularly recommended when the source resistance is very low and the equivalent noise generator E_n is simultaneously dominant. The coupling transformer must be shielded.
- Noise matching in integrated circuits is performed by scaling the input active device, or by paralleling several active devices. Both techniques yield satisfactory results when the equivalent noise generator E_n is dominant.
- A very efficient noise matching technique consists in properly selecting the active input device. For source impedance lower than $5\,k\Omega$, BJTs are the best choice. If the source impedance exceeds $100\,k\Omega$, JFETs are preferred.
- Feedback is employed only to control amplifier performance (gain, bandwidth, input/output impedances); it has no (or little) effect on noise.
- In microwave circuits, the best noise performance related to a certain transistor corresponds to the constant noise circle that is tangential to the constant available gain circle.

14
Noise Performance Measurement

"Measurement began our might"

(W.B. Yeats)

■ **Overview.** Low-noise design of electronic circuits requires, among other things, reliable noise models. Many models have been detailed in Chap. 7 for various devices such as bipolar transistors, field-effect transistors, operational amplifiers, etc. Their common feature is a need to find proper numerical values for some noise coefficients, adjusted according to measured data. On the other hand, noise performance measurement is of paramount importance to the manufacturers of semiconductor devices, integrated circuits, resistors, and passive components, in order to provide useful information to the designer.

However, measuring noise signals is not a simple task, because in contrast to conventional signals, intrinsic noise has very low amplitude and power levels. Nowadays, many automated measuring systems exist, offering many capabilities. Unfortunately, they are expensive, and not all electronics laboratories are able to purchase them.

The aim of this chapter is to review some classical noise measurement techniques, that can be successfully employed with the typical equipment available in any conventional laboratory.

14.1 Noise Sources

14.1.1 Introduction

■ **Definition.** *A noise source is a device delivering a signal with a random amplitude* [220].

444   14 Noise Performance Measurement

■ **Classification.** There are two categories of noise sources:

1) *Primary standard*, which delivers a noise signal resulting from a known physical process, whose power can be accurately predicted. The classic example is a resistor at temperature T, producing an available noise power calculated with (3.10), i.e.

$$P_n = \frac{hf\,\Delta f}{\exp(hf/kT) - 1}$$

2) *Secondary standard*, which must be calibrated against a primary standard. A typical example is a Zener diode or a gas-discharge tube. In both cases, the noise cannot be accurately described by mathematical expressions, due to the complexity of the physical processes involved.

For example, consider a gas-discharge tube. It contains a low-pressure gas that conducts current whenever sufficient voltage is applied. Noise is generated at each discharge (in the plasma), with a spectrum covering a broad frequency range. This noise is similar to the thermal noise produced by a resistor at high temperatures (above 10^4 K). Nevertheless, this noise source is not a primary standard, because while the noise power depends mainly on the temperature, it also depends on the gas pressure, plasma cross section, nature of the gas, etc. None of these dependencies can be easily described by mathematical expressions.

■ **Stability.** Stability over various time periods and repeatability measurements performed on noise sources forms an important topic, which is amply treated in reference [230].

■ **White Noise Sources.** These sources are useful in the laboratory to measure the noise factor or the noise equivalent temperature. However, due to circuit constraints, a so-called white source does not have a flat spectrum out to infinity. The term is instead employed to indicate broadband operation. Next section describes several such sources.

14.1.2 Case Studies

CASE STUDY 14.1 [221, 224, 226]

Explain the operation of the diode noise source shown in Fig. 14.1 and comment on the role of various elements. Suggest how one might adjust the noise output power delivered to R_L.

Solution

■ **Explanation.** A vacuum (or thermionic) diode consists of a cathode **K** (usually made of nickel alloy coated with a barium compound) and an anode **A** (a metallic plate) situated at some distance from the cathode, both

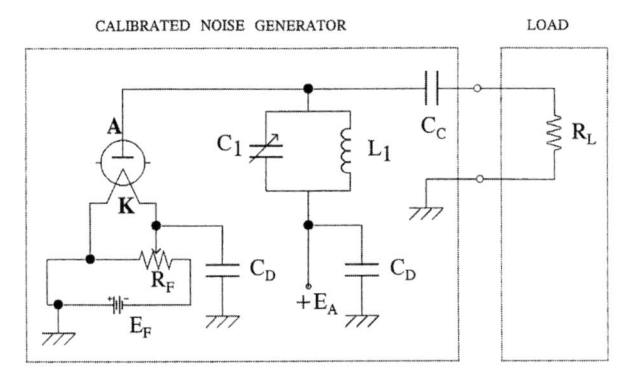

Fig. 14.1. A thermionic diode noise source

in a vacuum enclosure. When the cathode is heated to several hundred degrees Celsius, a large number of electrons is emitted from its surface. The anode is connected to a positive supply voltage (E_A), that attracts all emitted electrons. Therefore, a current I_o (which depends only on the cathode temperature) is established through the device, always flowing in only one direction (from cathode to anode). The diode is temperature limited (no space charge), since all emitted electrons are collected by the anode. The electrons in vacuum have ballistic trajectories; hence, their displacement generates shot noise, whose spectral density is given by (3.20)

$$S(I_n) = 2qI_o$$

■ **Characteristics.** A typical vacuum diode delivers a noise power of about 5 dB, over the frequency range 10 to 600 MHz (above 600 MHz, the transit time of electrons between cathode and anode seriously affects device operation). In order to control the noise level, the best idea is to modify the value of the potentiometer R_F. In this way, the heating current changes, as does the cathode temperature, and this modifies the thermionic emission of the cathode and hence the current I_o.

■ **Equivalent Circuit.** The capacitors denoted by C_D are bypass capacitors, while capacitor C_C is introduced to block the DC component.

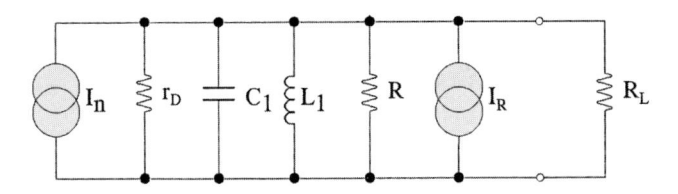

Fig. 14.2. Equivalent circuit of the noise generator

446   14 Noise Performance Measurement

The resonant circuit L_1, C_1 represents the load of the noise generator I_n (L_1 acts as an RF choke; C_1 is chosen large enough to neglect stray capacitance due to interconnections and the thermionic diode). The equivalent circuit is presented on Fig. 14.2, where r_D denotes the diode dynamic resistance.

Note that the equivalent resistance R of the resonant circuit (originating from the ohmic resistance of the inductor and losses in the capacitor) produces a thermal noise (I_R). The total output noise is given by:

$$\overline{I_{tot}^2} = \overline{I_n^2} + \overline{I_R^2}$$

To build a calibrated noise source, we must have $\overline{I_R^2} \ll \overline{I_n^2}$. This means that the resistance R must have a high value, i.e., the resonant circuit must be very selective (high Q). Under this assumption, and also assuming that the thermionic diode operates without charge accumulation (i.e., the high voltage E_A collects all emitted electrons at the anode), the temperature-limited diode becomes a primary noise standard.

■ **Concluding Remark.** Note that due to the resonant circuit, the noise power spectrum is not flat. One may ask why we don't use a simple resistor, instead of the resonant circuit, to avoid this? The reason is twofold:

– The thermal noise of this resistor must be taken into account.
– Stray capacitance (especially the interelectrode anode–cathode capacitance) would introduce a noise power roll-off at high frequencies, which is not easy to control.

CASE STUDY 14.2 [221, 224]

The noise source depicted in Fig. 14.3 is of particular interest at low frequencies. Explain its operation, propose a noise equivalent circuit, and suggest a way to adjust the output noise power.

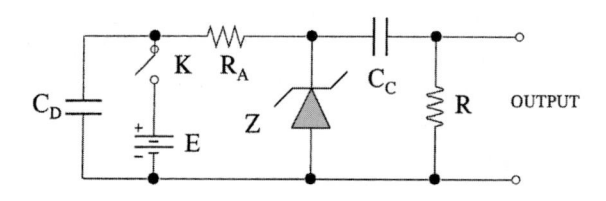

Fig. 14.3. Zener diode noise source

Solution

■ **Explanation.** According to Sect. 7.4.3, the best choice is a low-voltage Zener diode (an avalanche diode adds too much 1/f noise). If possible, select the diode with the lowest corner frequency f_c (we recall that the corner frequency is defined as the frequency where the power of the 1/f noise is equal to the power of the flat noise). Ideally, f_c must be around several Hz. C_D acts as a bypass capacitor (it filters the inevitable spurious signals of the power supply E), while C_C blocks the DC component. Resistor R is used to set the output resistance at the desired level (provided that $R \ll r_Z$ and $R \ll R_A$). The supply voltage E should be much more higher than the Zener voltage of the diode; the benefit is that R_A will wind up with a large value (otherwise, the noise generated by the diode will be internally short-circuited).

■ **Equivalent Circuit.** The noise equivalent circuit is presented in Fig. 14.4, where I_Z is the noise generated by the Zener diode, r_Z is its dynamic resistance, and I_{RA}, I_R represent the thermal noise of R_A and R, respectively.

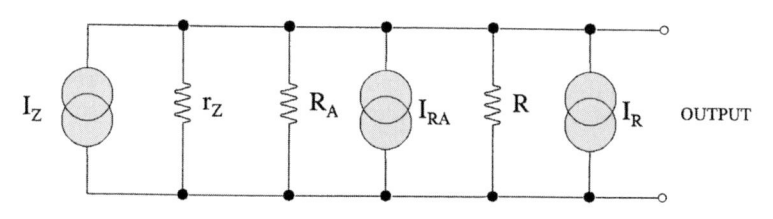

Fig. 14.4. Noise equivalent circuit

Since the noise current generators are not correlated, the total output noise can be written

$$\overline{I_{tot}^2} = \overline{I_Z^2} + \overline{I_{RA}^2} + \overline{I_R^2}$$

Obviously, if

$$\overline{I_{RA}^2} \ll \overline{I_Z^2} \quad \text{and} \quad \overline{I_R^2} \ll \overline{I_Z^2}$$

the noise source is calibrated against I_Z. The first inequality explains why it is desirable to use a high-value R_A. As the dynamic resistance of the diode is generally low, the only way to satisfy the second inequality is to replace R with an LC-resonant circuit (as in Case Study 14.1).

To adjust the output noise level, adjust the value of resistor R_A. As a consequence, the bias of the diode will be modified, as will the noise current I_Z.

CASE STUDY 14.3 [221–223]

Explain the operation of the avalanche diode noise source shown in Fig. 14.5, and suggest how to adjust the noise power delivered to the load (R_L).

Solution

■ **Explanation.** In practice, an avalanche diode is basically a PIN structure specifically designed for noise generators. Usually, it is reverse-biased at V

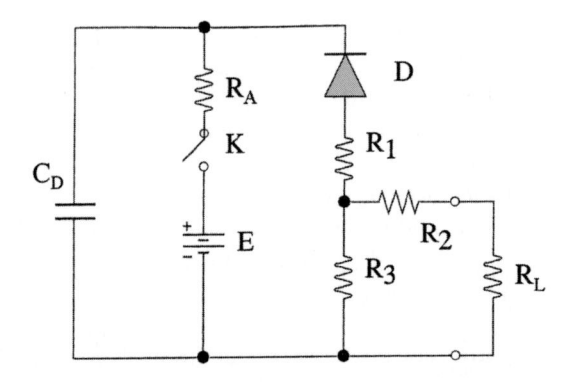

Fig. 14.5. Avalanche diode noise source

$= 13\,\text{V}$ and $I = 50\,\text{mA}$. Using a single device, it is possible to cover the frequency range between 10 MHz and 18 GHz. The power spectrum is flat (with a typical ripple of $\pm 0.3\,\text{dB}$), and the noise power delivered is much greater than that of a thermionic diode.

■ **Circuit.** In the circuit of Fig. 14.5, C_D is a bypass capacitor, used mainly to filter the internal noise of the power supply (E). An easy way to adjust the noise delivered to the load is to adjust resistor R_A, which modifies the current through the device.

The main problem (which applies to all noise sources) is that the output resistance seen by the load changes every time the switch K is open or closed ("cold" or "hot" noise source). When K is closed, the diode internal impedance is about $20\,\Omega$, while with K open (diode turned off) it is around $400\,\Omega$. Hence, a separation stage is necessary to provide a constant output resistance, which guarantees impedance matching in microwave circuits. As stated in [222], this feature is of paramount importance in all applications where the noise source is permanently coupled to the system, as in the measurement of radar noise figure. In Fig. 14.5, the circuit performing this function is the padding attenuator R_1, R_2, and R_3. With the values indicated in Fig. 14.6, the terminal impedance seen by the load takes a value between $47.9\,\Omega$ and $51.8\,\Omega$ (depending on whether the source is on or off).

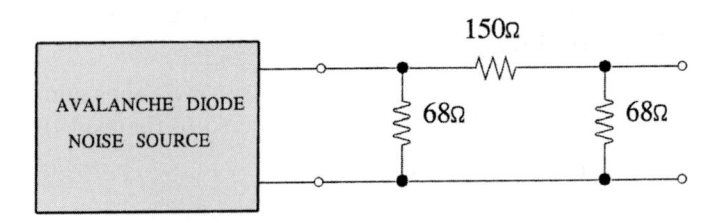

Fig. 14.6. Noise source with an output attenuator

14.1 Noise Sources

One might wonder whether the attenuator will reduce the output noise power to an unacceptable level. At least in this case, the output noise power is always maintained above 15 dB.

CASE STUDY 14.4 [223]

The main characteristic of white noise is that its power is uniformly distributed over the entire frequency range: it exhibits a flat power spectrum. What is traditionally called "pink noise" corresponds to a power spectrum which contains constant power per octave, so, it looks like $1/f$ noise. Propose a circuit to generate pink noise.

Solution

■ **Explanation.** The starting point in this application is an interesting white noise source, which is quite different from these previously discussed, because it is based on logic techniques. The block diagram is presented in Fig. 14.7.

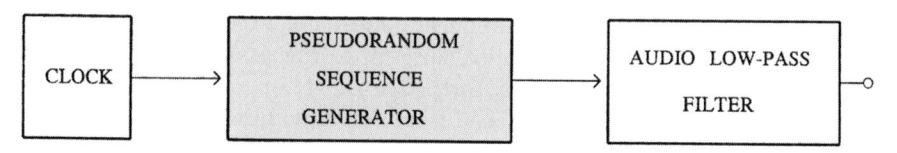

Fig. 14.7. White noise source based on logic circuits

In principle, it consists of a clocked generator that is able to deliver a random sequence of 0 and 1, followed by a low-pass filter. At the output, an analog signal with a white spectrum (at least up to the filter cutoff frequency) and Gaussian distribution is obtained.

However, in practice we are limited by the fact that the so-called random sequence is really a *pseudorandom* sequence. The generator is usually a long shift register, with its input derived from a modulo-2 addition of several of its last bits. Hence, the "random" sequence repeats itself after a time interval that depends on the register length. Nevertheless, time intervals of the order of years can be achieved, for instance with a 50-bit register shifted at 10 MHz (yielding white noise, up to 100 kHz). For most practical applications, a time interval of the order of seconds is quite satisfactory.

■ **Proposed Circuit.** The present goal is to produce a pink noise generator. Since the power spectral density of pink noise drops off at 3 dB/oct, while a traditional RC filter drops at 6 dB/oct, it is clear that we must properly modify the output filter in order to achieve the desired result. Very likely, several filters (with different cutoff frequencies) must be cascaded. A possi-

450   14 Noise Performance Measurement

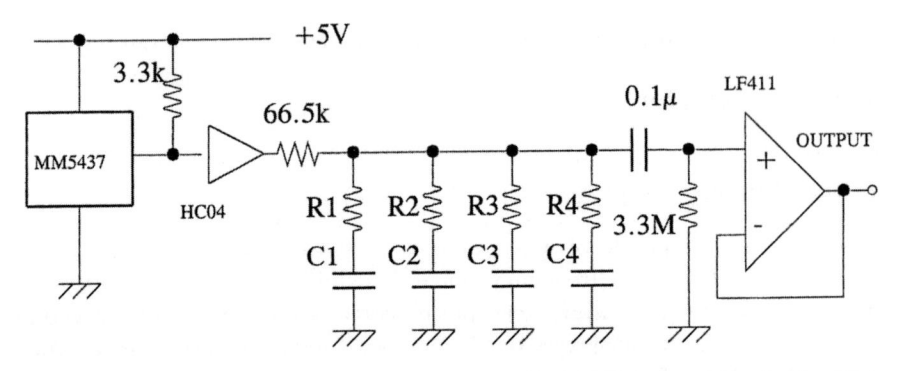

Fig. 14.8. Pink noise source (from 10 Hz to 40 kHz) (Courtesy of Cambridge University Press)

ble realization is proposed in Fig. 14.8, where the MM5437 circuit contains several shift registers [223].

In the present design, a pseudorandom sequence of 0 and 1 is generated; after filtering, white noise is obtained up to the first cutoff frequency (established by R1 and C1). This frequency must not exceed 1% of the clock frequency. Then we have also white noise up to the second cutoff frequency (established by R2 and C2), and so forth. The LF411 operational amplifier is a buffer that isolates the unknown load from the audio filter. In reference [223], values of various filter resistors and capacitors are given (see Table 14.1, where all resistances are in kΩ and the capacitances are in nF).

Table 14.1. Values of multifilter elements

R1	C1	R2	C2	R3	C3	R4	C4
33.2	100.0	10.0	30.0	2.49	10.0	1.0	2.9

■ **Conclusion.** In conclusion, comparing with analog noise sources, their digital counterparts have several advantages:

- the noise frequency range is simply controlled by changing the clock frequency;
- they are more robust with respect to electromagnetic perturbations (since they process high-amplitude logic signals)

Their main limitation is the output noise power, which is not easy to adjust and so rarely can they be used as calibrated noise sources.

14.2 Noise Power Measurement

14.2.1 Introduction

■ **Comment.** Any noise measurement system requires a noise power meter. Usually, this is connected to the output of the device under test (DUT), since the power level is lower at its input and it is therefore difficult to perform accurate measurements.

■ **Accuracy Requirement.** Except in special situations, errors in noise performance measurement up to 10% are quite acceptable, and often a measurement within a factor of 2 is not harmful [221]. As the normalized noise power is proportional to the mean square value of the noise current or voltage, we may deduce the former from the measured value of the latter. Assume δ_E to be the error in measuring the noise voltage E_n, and δ_P the error in the resulting power. Then

$$E_n^2 (1 + \delta_P) = \left(E_n(1 + \delta_E)\right)^2$$

and

$$\boxed{\delta_P = 2\delta_E + \delta_E^2}$$

As a consequence, whenever δ_E is small, *the error in the calculated power is about twice the measurement error of the corresponding rms voltage.*

14.2.2 Case Studies

CASE STUDY 14.5 [221, 224]

When high accuracy is not required, one of the least expensive ways to measure white noise level is to use an oscilloscope. Indicate the constraints of the measurement method.

Solution

If we apply (2.17) to a white noise voltage, we obtain

$$P(v_n) = \frac{1}{\sqrt{2\pi \, \overline{E_n^2}}} \exp\left(-\frac{v_n^2}{2 \, \overline{E_n^2}}\right)$$

where $P(v_n)$ denotes the probability that the fluctuation reaches the value v_n, and $\overline{E_n^2}$ is the mean square value (4kTRΔf for thermal noise). The plot of $P(v_n)$ corresponding to the above equation is given in Fig. 14.9.

It is obvious that small amplitudes are more likely to occur than large ones. Since thermal noise results from an ergodic process, averaging over an

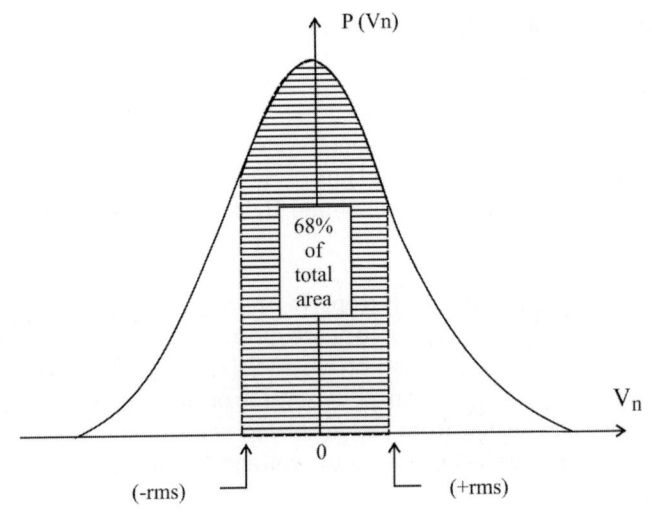

Fig. 14.9. The probability density function of white noise (case of thermal noise)

ensemble yields identical results to averaging over time. This allows us to present the information contained in the plot in an equivalent form, i.e., as the most likely time interval within which the fluctuation exceeds the specified peak-to-peak values (Table 14.2).

Table 14.2. Probability of exceeding peak-to-peak values (normal distribution)

Peak-to-peak	2(rms)	3(rms)	4(rms)	5(rms)	6(rms)	7(rms)	8(rms)
Time	32%	13%	4.6%	1.2%	0.27%	0.046%	0.006%

According to Table 14.2, it is likely that the fluctuation amplitude will exceed twice the rms value during 32% of the monitoring time interval; in other words, during 68% of the monitoring interval the amplitude is inside the range (+rms) and (−rms).

The recommended procedure to measure the noise rms value consists in *excluding one or two of the highest peaks of the displayed waveform, then evaluating the peak-to-peak amplitude, and finally dividing it by 6*. The reason to do so is twofold:

1) Table 14.2 shows that on average, during 99.73% of the monitoring time, the amplitude of the fluctuation remains inside a domain defined by the levels +3(rms) and −3(rms).
2) The displayed waveform may occasionally present high peaks (located between 3(rms) and 8(rms)), if enough monitoring time is spent.

Even when a meter is available, the oscilloscope is very useful as an additional display, because we can control the character of the noise signal and identify when undesired pickup (or another type of noise) is superposed on the white noise.

CASE STUDY 14.6 [224, 226]

In order to measure noise power, a meter may represent a good solution. As various types of meters exist, discuss the merits and limitations of each one.

Solution

■ **Conditions.** In order to accurately measure Gaussian noise, the meter must satisfy several conditions:

- *It must provide an rms indication proportional to the noise power.* A classical meter is designed to respond to the average value of a periodic signal, but its scale is calibrated to read rms values. However, in the case of fluctuations, the amplitude is not constant and the waveform is not sinusoidal. It follows that an unacceptable error may result when trying to measure noise with a classical meter.
- *The peak factor must be higher than 3.* In practice, Table 14.2 shows that more than 99% of the time, the Gaussian noise has a peak factor less than or equal to 3 (see the definition of the peak factor in Sect. 2.2.4). It follows that the meter must have a peak factor at least equal to 3 to avoid saturation when the fluctuation exceeds 3 times the full-scale reading. Some authors [225] prefer a minimum value of 4.
- *The meter bandwidth must be at least 10 times the noise bandwidth of the system delivering the fluctuation.* Assuming that the noisy system and the meter have both a single dominant pole (denoted by f_s and f_m, respectively), it can be shown that the relative meter reading is

$$\sqrt{\frac{f_m}{f_m + f_s}}.$$

Table 14.3 presents the evolution of the relative reading and the relative error of a meter in terms of the ratio f_m/f_s.

It is easy to see that the indication of a meter with the same bandwidth as the measured fluctuation is subject to about 30% error, while the error for another instrument with 10 times the noise bandwidth is only 4.65%.

Table 14.3. Evolution of the relative reading and error

f_m/f_s	Relative reading	Percentage error [%]
1	0.707	−29.28
2	0.816	−18.35
3	0.866	−13.39
4	0.894	−10.55
5	0.913	−8.71
10	0.953	−4.65

454   14 Noise Performance Measurement

■ **Discussion.** In practice, the following categories of meters are employed:

☐ **True rms Meters.** These are able to indicate the rms value of an arbitrary waveform (sinusoid, rectangular, exponential, etc.). They can be successfully employed to measure fluctuations, provided their peak factor is greater than 3 (or 4).

Two common types of true rms meters are encountered:

- *Quadratic device*, whose response is proportional to the square value of the fluctuation. Squaring of the instantaneous value is performed either with a circuit, or by using a Schottky diode operating in the square region of its i-v characteristic. The latter solution is valid only for weak signals (from $-70\,\text{dBm}$ up to $-20\,\text{dBm}$).
- *Devices that respond to heat.* Thermocouple instruments are the best illustration of the classical rms definition given in Sect. 2.2.2. They convert the electrical power of the fluctuation into thermal power, which heats a thermojunction. In turn, this yields a DC current proportional to the heat, and therefore to the squared input current (Fig. 14.10).

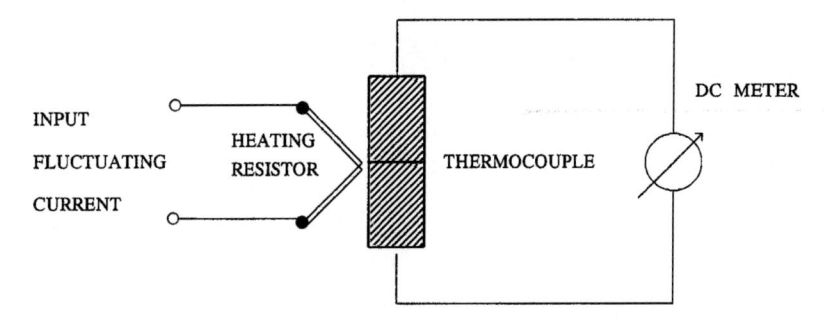

Fig. 14.10. Principle of a thermocouple instrument

In this way, for any waveform applied to the input, the output meter indicates a value proportional to the power of the input signal. For noise measurement, as fluctuations are weak, a low-noise amplifier must be inserted. As the amplifier will add noise, the scale of the DC meter must be properly calibrated. The main shortcomings of thermocouple meters are the risk of burnout (with overload) and also their inertia (slow response).

☐ **Average Response Meters.** Most AC meters and spectrum analyzers fall into this category. Basically they have a half- or full-wave rectifying circuit followed by a DC meter. To indicate the rms value, corrections are needed. For instance, if a sinusoid is measured with a meter having a full-wave rectifier, the average is 0.636 times the peak value, and since the rms value is 0.707 times the peak value, it follows that to calculate the rms value of the sinusoid we must multiply its average by $0.707/0.636 = 1.11$.

When white noise power must be measured, the correction factor becomes $1.11(0.798) = 0.886$ (recall that the average value of a Gaussian noise submitted to a full-wave rectifier is obtained by multiplying its rms value by 0.798).

Finally, *when an averaging meter is employed to measure white noise, the reading of the meter must be multiplied by 1.128 (or, 1 dB must be added)*. Otherwise, we may discover with astonishment that the thermal noise of a known resistor is lower than its theoretical value!

There are several remarks concerning an average meter:

- it is important to check its reading against a calibrated noise source;
- most averaging meters saturate with signals having peak factors around 1.5. As noted above for white noise the peak factor must be at least 3, so a 6 dB attenuator must be inserted between the noise source and the meter, otherwise measurements must be performed on the lower half of the scale to avoid clipping the peaks of the white noise.

☐ **Peak Responding Meters.** These instruments indicate the peak or peak-to-peak value of the signal. They operate correctly provided the peak value is constant. Since this is not the case with noise, their reading depends instead on the charge and discharge time constants of the instrument. Therefore, they should not be used to measure fluctuations.

■ **Conclusion.** A meter represents a reasonable choice, provided that

- its reading is proportional to the noise power;
- the peak factor of the instrument is greater than 3;
- the meter bandwidth is at least 10 times the noise bandwidth of the investigated circuit.

The best choice is a thermocouple instrument; an averaging meter is still acceptable if a correction factor of 1.13 is employed, and a peak responding meter is inappropriate.

CASE STUDY 14.7 [224]

Since the fluctuations to be measured have very low amplitudes, a meter with unusually high sensitivity is required. A practical solution is to insert a low-noise amplifier (LNA) at the front-end of the meter to enhance sensitivity. Propose an LNA circuit built around the AD745 chip and comment on it.

Solution

When designing an LNA, several requirements must be met:

- it should have a low noise equivalent generator E_n (this guarantees proper operation with low resistance sources);
- its 1/f noise must be negligible, if it is intended to amplify DC or very low-frequency signals;

- its gain-bandwidth product must be large, if it is intended to amplify high-frequency signals.

From the data sheet, we learn that the AD745 circuit is a monolithic BIFET operational amplifier with high input impedance and low noise voltage; its gain-bandwidth product is equal to 20 MHz. The proposed circuit of the LNA [224] in Fig. 14.11 cascades two stages, each providing a gain of 33 (the overall resulting gain is therefore around 1000). The bandwidth is about 500 kHz, but can be increased by properly adjusting the feedback resistances of each stage to reduce the gain. Note that a 10 MΩ resistor is added at the input, in order to control the input resistance of the LNA. The equivalent input noise voltage is $2\,\mathrm{nV}/\sqrt{\mathrm{Hz}}$, and the equivalent input current is about $10\,\mathrm{fA}/\sqrt{\mathrm{Hz}}$.

Special attention must be paid when building the LNA, in order not to impair its low-noise performance with a poor implementation. Bear in mind that the proposed configuration guarantees low intrinsic noise, not immunity to interfering signals! Here are several suggestions:

- If possible, for the 9 V supply use batteries (to avoid power supply noise, ground loops, and stray signal pickup through long cables and wires).
- Build the LNA as compactly as possible. A large size is comfortable for eventual adjustments or maintenance, but it increases circuit susceptibility to pickup; a small size reduces pickup, but later intervention may become impossible. A trade-off must be found between these issues.
- Use a printed circuit board with a ground plane.
- Shield the circuit against interfering signals.
- Use shielded cables to drive the input and output.

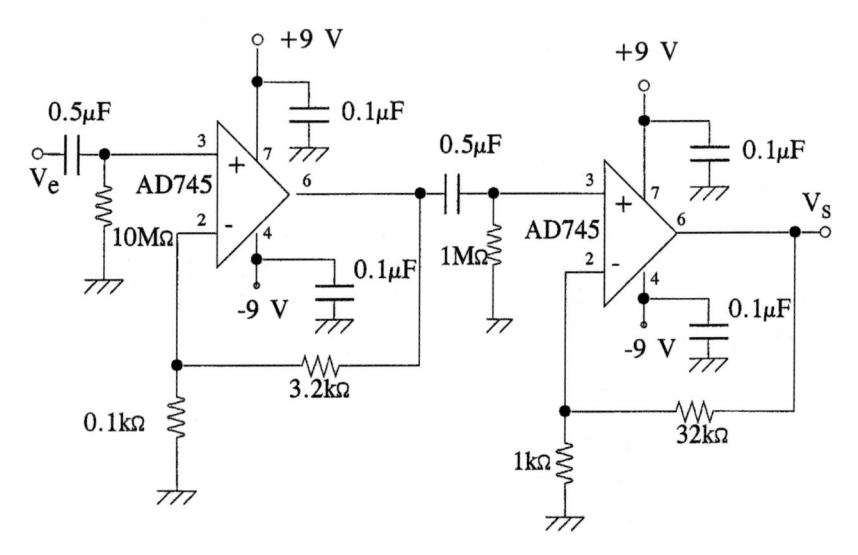

Fig. 14.11. Laboratory low-noise amplifier (Courtesy of John Wiley and Sons)

14.3 Two-Port Noise Performance Measurement

14.3.1 Introduction

The principle of noise temperature measurement was introduced in Sect. 4.5.3.

Now emphasis is given to the main concepts of two-port noise performance measurement, which can be implemented with the standard equipment of any electronics laboratory. Despite their obvious simplicity, these techniques are helpful in at least understanding the operation of more sophisticated equipment that lies beyond the scope of this book.

14.3.2 Case Studies

CASE STUDY 14.8 [224, 226]

Discuss various procedures to measure the equivalent input noise voltage of a two-port.

Solution

Whenever a two-port must be characterized by its S/N ratio, the noise and signal level must be determined at a single location. In a sensor–amplifier system, it is important to select as that special location the sensor itself, because there at least the signal is already known. Therefore, effort must be focused on the equivalent input noise measurement. In practice, noise measurements are made at the amplifier output, and the input noise is deduced by dividing the output noise by the circuit gain.

Two widely used techniques for input noise measurement are the sine-wave and noise generator methods.

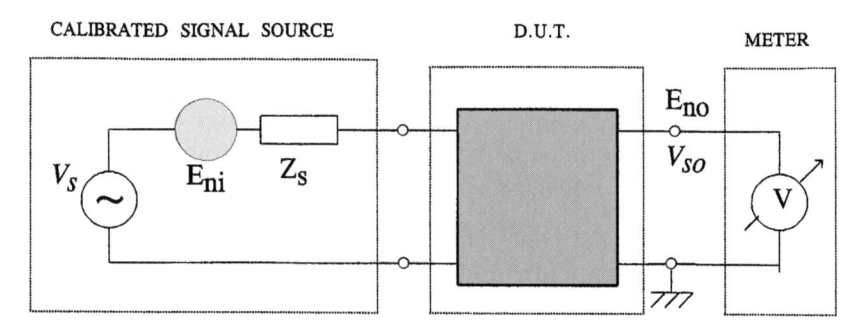

Fig. 14.12. General layout for equivalent input noise measurement with the sine-wave method

■ **The Sine-Wave Method.** Consider the configuration in Fig. 14.12, where V_s is a sine-wave generator simulating the sensor, Z_s is the sensor impedance, and E_{ni} lumps the amplifier noise and thermal noise of Z_s.

For a spot frequency measurement, the procedure is the following:

1. Measure the output signal voltage V_{so} and deduce the voltage gain referred to V_s, i.e., $K = V_{so}/V_s$. Attention must be paid to properly setting the sine-wave generator level, in order to avoid overloading the device under test (amplifier). It is advisable to halve (and then double) the signal level, and check that K is not modified.
2. Measure the output power noise with the output of the sine-wave generator set to zero, and deduce the output noise voltage E_{no} at the operating frequency.
3. Divide E_{no} by K to find the equivalent input noise E_{ni}. Let V_{sen} be the signal delivered by the sensor; hence

$$\frac{S_i}{N_i} = \frac{V_{sen}}{E_{ni}}$$

A simplified setup is proposed in Fig. 14.13, where the two-port (device under test) to be measured is denoted by D.U.T..

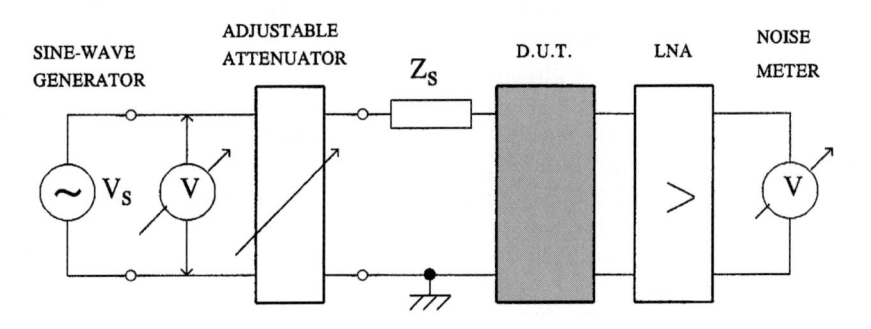

Fig. 14.13. Setup for the sine-wave generator method

Note that it is more convenient to measure the voltage of the sine-wave generator before the attenuator, because it has a larger value. For RF measurements, the reference plane is situated between the attenuator and Z_s.

■ **The Noise Generator Method.** This method requires a calibrated white noise source E_{ng} and a noise meter at the output; a general configuration is proposed in Fig. 14.14. The unknown amplifier noise is measured by comparing it against the noise level of the source. The output noise is monitored throughout the procedure.

The most commonly used procedure is a technique of doubling the output noise:

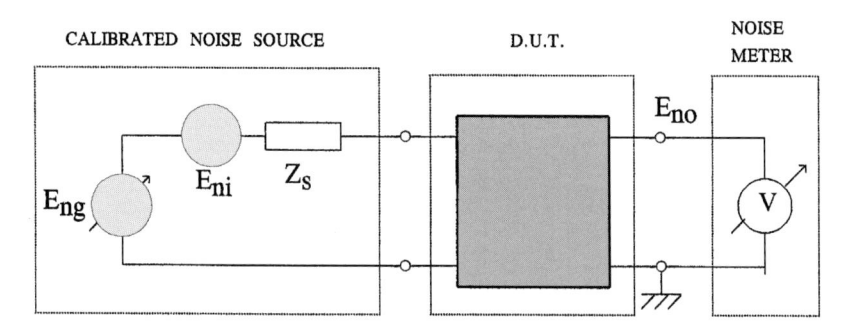

14.3 Two-Port Noise Performance Measurement 459

Fig. 14.14. General layout for the noise generator method

1. Replace the noise generator E_{ng} with a short circuit and measure the output noise power $\overline{E_{no1}^2}$.
2. Insert the noise generator E_{ng} and progressively increase its level up to $\overline{E_{noG}^2}$, where the measured output noise power is $2\overline{E_{no1}^2}$ (or, equivalently, the output noise power is increased by 3 dB).
3. Here the noise generator signal is equal to the amplifier's equivalent input noise, i.e.,
$$\overline{E_{ni}^2} = \overline{E_{noG}^2}$$

Note that the accuracy of this method is primarily determined by the calibration of the noise generator.

■ **Comparison.** The noise generator method is straightforward because we add noise to the amplifier input until the resulting output noise doubles. It is a broadband technique, and consequently we need a broadband noise source. This is not necessarily an advantage, due to the 1/f noise that is added below a few hundred Hz, and the consequent risk of modifying the calibration. The sine-wave generator method needs only standard equipment (no calibrated noise source), but requires two or more measurements. As a general rule, use the sine-wave generator method for low-frequency applications, and the noise generator method for high-frequency applications.

CASE STUDY 14.9 [221, 222, 224]

Suggest several simple techniques to determine the noise figure of a two-port.

Solution

Provided that the source resistance is known, there are several approaches to determining the noise factor:

- Measure the equivalent input temperature (according to Sect. 4.5.3) and apply the equation
$$F = 1 + (T_e/T_o)$$

- Measure the equivalent input noise voltage (see Case Study 14.8) and use the expression

$$F = 10 \log \left(\overline{E_{ni}^2} / \overline{E_{ns}^2} \right)$$

where E_{ns} is the thermal noise of the source resistance.

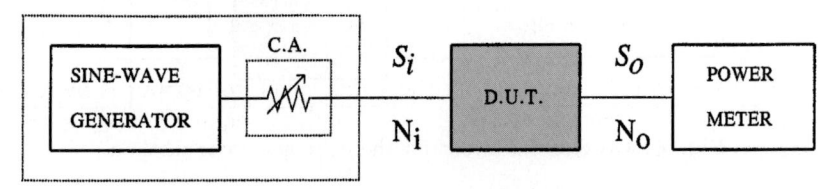

Fig. 14.15. Measuring the noise factor with the sine-wave generator method

- Use the sine-wave generator method. A simplified general layout is depicted in Fig. 14.15, where C.A. denotes the calibrated attenuator of the signal source and D.U.T. is the device under test. S_i, S_o denote the input and output sine-wave power; N_i and N_o represent the input and output noise power. For a spot frequency measurement, the procedure is as follows:

1. With the output of the sine-wave generator set to zero, measure the output noise power N_o. This noise originates from the D.U.T. and the generator output resistance, at the reference temperature T_o.
2. Tune the sine-wave generator at the specified frequency and set the signal power at level S_i by adjusting the calibrated attenuator. The output total power is now N_{ot}, which is

$$N_{ot} = N_o + S_o$$

Consequently

$$S_o = N_{ot} - N_o$$

3. As soon as S_o is found, the power gain (from the signal generator to the D.U.T. output) can be evaluated

$$G_p = \frac{S_o}{S_i} = \frac{N_o}{N_i}$$

Note that the power gain is the same for signal and noise. The power noise referred to the input of the D.U.T. is

$$N_i = \frac{N_o}{G_p} = \frac{S_i}{S_o/N_o}$$

Using (4.70), the noise factor becomes

$$F = \frac{N_i}{4kTR_s \, \Delta f} = \frac{S_i}{4kTR_s \, \Delta f} \frac{1}{S_o/N_o}$$

or equivalently

$$F = \frac{S_i}{4kTR_s \, \Delta f} \frac{1}{(N_{ot}/N_o) - 1}$$

Additional arrangements can then be made to simplify the noise factor calculation:

- In the second step, the signal generator output is set to a value of N_{ot}/N_o equal to 2.
- Alternatively, in the second step, the signal generator output is increased until its power exceeds the noise power, and consequently unity may be neglected with respect to N_{ot}/N_o.

The output power doubling method has the advantage of being straightforward but for narrow-band measurements the integration time becomes excessively long. The second possibility avoids this, but increasing the signal generator power can saturate the D.U.T., which can then become nonlinear.

Finally, note that this method requires that one previously determine the noise bandwidth of the D.U.T. This implies that the voltage gain variation versus frequency must be measured and then, by means of the definition (4.41), Δf is calculated.

CASE STUDY 14.10 [226]

Propose a method to measure the equivalent noise resistance of a two-port.

Solution

■ **Background.** According to Sect. 4.5.1, *the noise resistance is the value of a resistor which when applied to the input of a hypothetical noiseless (but identical) two-port would produce the same amount of output noise.* It is assumed that this resistor is maintained at the reference temperature $T_o = 290$ K.

In practice, it is impossible to find a noiseless (but otherwise identical) two-port, and consequently the definition cannot be applied as is. Therefore, as usual, the procedure of doubling the output noise power is to be preferred, together with the configuration proposed in Fig. 14.16. Here D.U.T. represents the device under test (two-port) and L.N.A. denotes a low-noise amplifier, inserted to improve measurement accuracy.

■ **Procedure.** The output noise is measured with a true rms meter (thermocouple instrument):

1. Connect the input of the D.U.T. to ground by setting the switch K to position **b**. Since the bias must be preserved, a large capacitor C is required to block the DC component and at the same time bypass the AC component to ground. The output noise is measured (E_{nb}); it originates only

462 14 Noise Performance Measurement

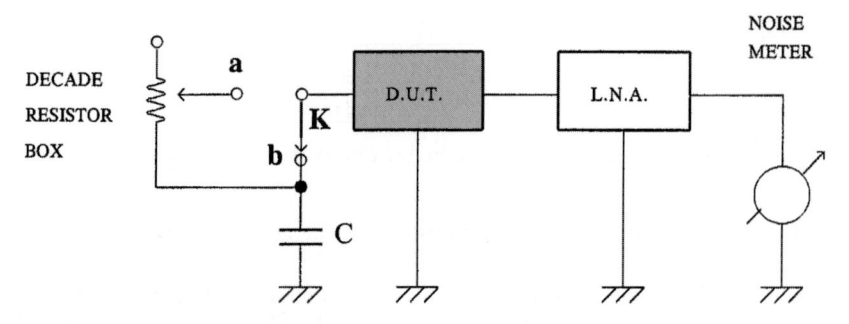

Fig. 14.16. Measuring the equivalent input resistance

from the two-port (i.e., from its equivalent noise resistance R_n), assuming that the noise of the L.N.A. block can be neglected or deduced.

2. With the switch K in position **a**, modify the decade resistor box until significant noise is added to the output. Let R_a be the value of the decade box for which the output noise voltage becomes $E_{na} > E_{nb}$.
3. Calculate the equivalent noise resistance related to coefficient M:

$$E_{na} = M\, E_{nb}$$

But

$$\overline{E_{nb}^2} = \left(4kT_o\,\Delta f\, R_n\right) A_v^2$$

where A_v is the voltage gain between the D.U.T. input and the meter. On the other hand,

$$\overline{E_{na}^2} = \left(4kT_o\,\Delta f\, (R_n + R_a)\right) A_v^2$$

Taking the ratio of these two equations, we find

$$\overline{E_{na}^2}\, \overline{E_{nb}^2} = M^2 = \frac{R_n + R_a}{R_n}$$

so

$$\boxed{R_n = \frac{R_a}{M^2 - 1}}$$

■ **Remarks**

– When $M = \sqrt{2}$ (the output noise voltage has been increased by a factor of 1.41, or the output power increases with 3 dB), $R_n = R_a$.
– Good accuracy requires a high-quality decade resistor box, with excellent mechanical contacts (as low noise as possible). Since the resistors are usually wirewound, they are sensitive to magnetic field pickup, and therefore protection (shielding) must be provided.

CASE STUDY 14.11 [226]

The signal-to-noise ratio (S/N ratio) is the most significant noise quantity employed to describe the performance of a mismatched two-port. Suggest a general layout together with a procedure to measure the S/N ratio.

Solution

The critical point here is that noise always combines with the signal, and consequently any time we measure the signal, we are also measuring noise. This is not important if the S/N ratio is greater than 10. However, for low values of the S/N ratio, accuracy can be seriously affected. To avoid this, a narrow-band voltmeter (N.B.V.) is used for signal measurement. Its main advantage is not that it eliminates the noise, but that it reduces its power to that corresponding to a small bandwidth. A true rms meter is employed for noise measurement, as detailed in Fig. 14.17; TEST POINT means the point located inside the two-port where the S/N ratio is to be measured.

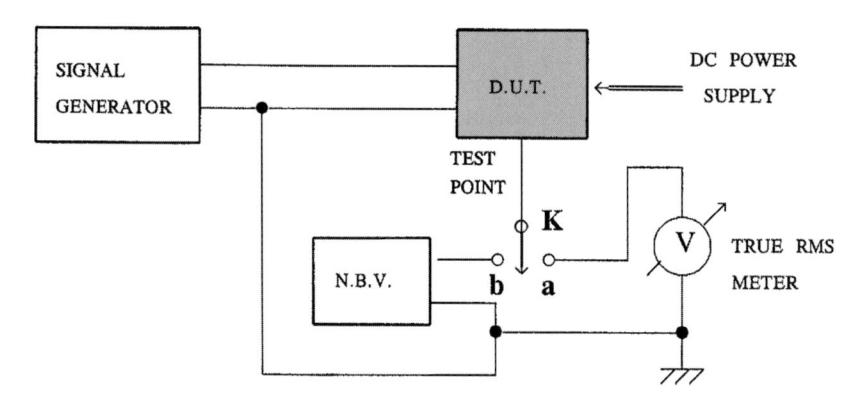

Fig. 14.17. Simplified setup for measuring S/N ratio

The following steps are suggested:

1. Turn on the signal generator; tune its frequency to the specified value and adjust its output to the desired level. Measure the signal voltage V_s at the test point with a N.B.V. (switch **K** at position **b**). The N.B.V. must be tuned to the same frequency as the signal generator.
2. Turn off the signal generator (but leave it connected to the D.U.T.). With a thermocouple instrument, measure the noise voltage V_n at the test point (switch **K** at position **a**).
3. Calculate the S/N ratio:

$$\frac{S}{N} = 10 \log \left(\overline{V_s^2}/\overline{V_n^2}\right) \quad [\text{dB}]$$

464   14 Noise Performance Measurement

4. There are several situations in which a correction is needed (for instance, when the S/N ratio is expected to be low, or when the bandwidth of the N.B.V. is not so small, or even worse, we use the same voltmeter for both measurements). The common element of all these situations is that the measured signal is actually

$$\overline{V_s^2} = \overline{V_{ss}^2} + \overline{V_n^2}$$

with $\overline{V_{ss}^2}$ being the signal component and $\overline{V_n^2}$ the noise component, both at the test point. In practice, we calculate

$$\frac{\overline{V_s^2}}{\overline{V_n^2}} = \frac{\overline{V_{ss}^2} + \overline{V_n^2}}{\overline{V_n^2}} = 1 + \frac{\overline{V_{ss}^2}}{\overline{V_n^2}} = 1 + \frac{S}{N}$$

From the latter expression, it is easy to see that the required correction is

$$\boxed{\frac{S}{N} = \frac{\overline{V_s^2}}{\overline{V_n^2}} - 1}$$

CASE STUDY 14.12 [221, 222, 226]

To measure the input noise temperature (spot value), a hot/cold noise source is required (or one hot source and one cold source). Suppose that the cold source has an equivalent temperature of 290 K; when it is connected to the input of the two-port, a true rms meter at the output indicates 24 µV. If a hot source (whose equivalent temperature is 1940 K) is connected at the input, the meter indicates 42 µV. Suggest a general setup and calculate the input noise temperature.

Solution

■ **Background.** A general setup is suggested in Fig. 14.18, where the L.N.A. is inserted to improve measurement accuracy.

The procedure is the same as that described in Sect. 4.5.3, hence T_e is determined by (4.63), i.e.

$$T_e = \frac{T_h - YT_c}{Y - 1}$$

with the Y-factor

$$Y = N_h / N_c = \overline{E_{nh}^2} / \overline{E_{nc}^2} = 42^2 / 24^2 = 3.0625$$

E_{nh}, E_{nc} denote the noise voltages of the hot and cold source. Therefore

$$T_e = \frac{1940 - (3.0625)(290)}{3.0625 - 1} = 510 \text{ K}$$

14.3 Two-Port Noise Performance Measurement

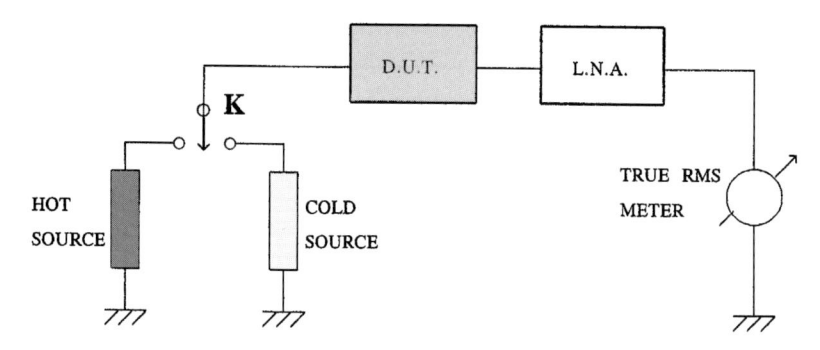

Fig. 14.18. General setup to measure T_e

■ **Correction.** It has been assumed that the noise of the L.N.A. is negligible. Since the resulting value of T_e is significant, it is very likely that this assumption is correct. However, if the resulting value of T_e is low enough, the cascade formula (4.66) must be applied (to the cascade D.U.T.–L.N.A.), in order to subtract the noise contribution of the L.N.A.

■ **Concluding Remark.** In practice, this method is used for RF receivers and microwave circuits. In both cases the two-port must be matched to 50 Ω, and under this condition the noise is measured. Hence, the cold source is an appropriate resistor which, when immersed in liquid nitrogen, presents a 50 Ω resistance; the hot source is another resistor, which when heated (or maintained at the ambient temperature) also presents a 50 Ω resistance. To connect them to the D.U.T. input, low-noise coaxial cables must be employed.

CASE STUDY 14.13 [221, 224, 231]

Suggest a setup together with a procedure to measure the excess noise of resistors.

Solution

■ **Background.** According to Sect. 7.1, excess noise is generated in a resistor when a DC current flows through it. The resulting power spectral density has the form

$$S(P_{ex}) = K_e/f^\alpha$$

where α has a value in the range 0.8–1.2 (usually taken equal to 1) and K_e is a constant that depends on the resistance, the DC current, and the fabrication technology.

■ **Setup.** The proposed setup is shown in Fig. 14.19. It requires a DC constant-current generator I_{DC} (whose current injected into any external cir-

14 Noise Performance Measurement

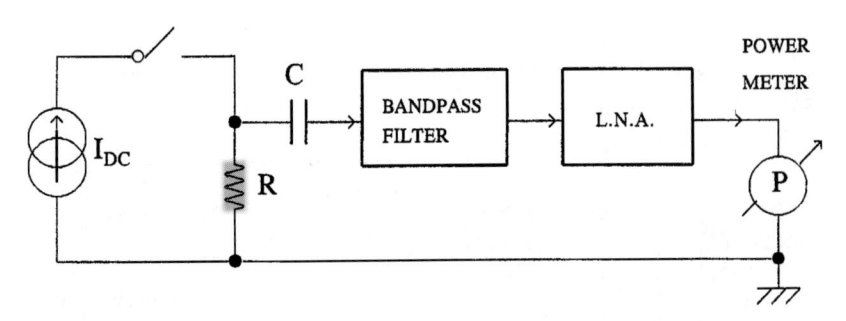

Fig. 14.19. Setup for resistor excess noise measurement

cuit does not depend on the load), a near rectangular bandpass filter, an L.N.A. to increase measurement sensitivity, and a noise meter (for instance, a true rms voltmeter).

The procedure has the following steps:

1. With the switch open, measure the output noise power $\overline{E_{n1}^2}$ (which corresponds to the thermal noise of the resistor plus the noise of the setup).
2. With the switch closed, DC flows through R; measure the output noise power $\overline{E_{n2}^2}$. In addition to the thermal noise and setup noise, this now includes excess noise. Its contribution can be calculated by subtraction:

$$\overline{v_{nex}^2} = \overline{E_{n2}^2} - \overline{E_{n1}^2}$$

Since the excess noise has a 1/f spectrum,

$$\overline{v_{nex}^2} = \int_{f_1}^{f_2} \frac{K_e}{f} G(f) \, df$$

with G(f) being the power gain (defined between R and the meter) and K_e a constant. If G_o is the mid-band gain of the filter (assumed to be ideal, with a rectangular frequency characteristic), and f_1, f_2 are the band-edge frequencies, then

$$\boxed{K_e = \frac{\overline{E_{n2}^2} - \overline{E_{n1}^2}}{G_o \, \ln(f_2/f_1)}}$$

By means of (7.3) we deduce that $K_e = C \, I_{DC}^2 \, R^2$. The DC current is known, as is the value of R. So the constant C can be easily deduced, and therefore the excess noise voltage density can be determined.

■ **Comment.** This method yields good results, provided that the noise of the constant current generator is low. If this is not the case, it would be possible to subtract its contribution from the measured excess noise, but subtracting

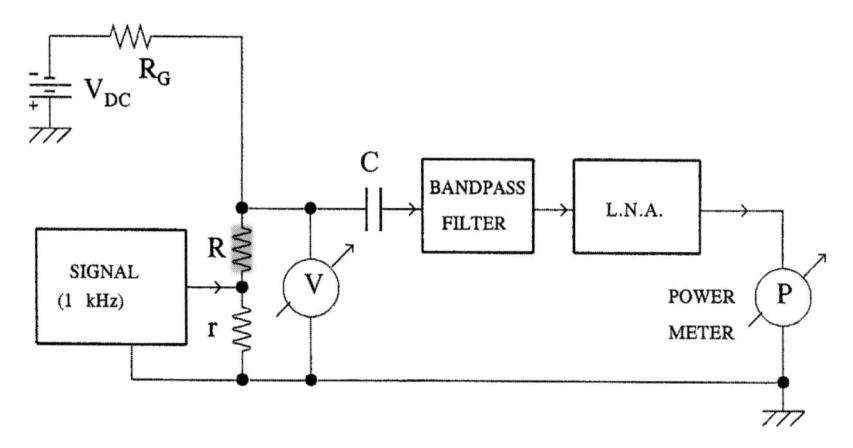

Fig. 14.20. Setup for measuring excess noise in resistors using calibration signal

near-equal random quantities does not improve the accuracy! Hence, another setup, slightly modified, is most appropriate (Fig. 14.20) [221].

The constant-current generator uses a voltage supply V_{DC} and an isolation resistor $R_G \gg R$. A calibration signal of 1 kHz is generated across a resistor $r \ll R$; the voltmeter V indicates the DC voltage drop across the series combination $R + r \simeq R$. The steps to be considered are:

1. With the supply V_{DC} set to zero (not open-circuited), the gain of the setup is measured with a 1-kHz signal having a convenient amplitude. The indication of the voltmeter V must be zero.
2. Turn off the signal generator and measure the output noise, which corresponds to the thermal noise of the resistor combination $(R + r) \parallel R_G \cong R$, to which the noise of the setup is inherently added.
3. Adjust the supply V_{DC} until the voltmeter V indicates the desired value. Measure the output power noise, which this time incorporates the excess noise contribution of R.

□ **Note**

- The excess noise of R_G must be as low as possible (select a wirewound resistor or better, a metal foil type).
- Select a tantalum capacitor C with low losses.
- The role of the resistor r is to guarantee a fixed impedance (R) seen at the filter input at every step.
- To avoid power supply noise and pickup of 50 Hz and its harmonics, use batteries for V_{DC}.
- When selecting the bandwidth (f_2-f_1), consider that: 1) a large bandwidth reduces the measurement time, for a given accuracy; 2) because thermal noise power increases more rapidly with the bandwidth than 1/f noise

power, select a moderately large bandwidth. According to [221], a good compromise is to measure in a 1-kHz bandwidth, in the range 0.5 kHz–1.5 kHz. An interesting investigation of the case when $\alpha \neq 1$ is also discussed in [221].

14.4 Miscellaneous

14.4.1 Passive Circuits

CASE STUDY 14.14 [232]

The white noise produced by a resistor is characterized by a large number of maxima and minima during the monitoring time interval. *The expected number of maxima per second* can be evaluated with the expression [232]

$$\sqrt{\frac{3\left(f_2^5 - f_1^5\right)}{5\left(f_2^3 - f_1^3\right)}}$$

when measuring is performed with an ideal bandpass filter whose bands extends from f_1 to f_2. Also, *the expected number of zeros per second* is [232]

$$2\sqrt{\frac{f_2^3 - f_1^3}{3(f_2 - f_1)}}$$

Find the expected number of maxima and zeros, when the white noise is displayed on the screen of an oscilloscope, in a 10-Hz bandwidth, in two situations:

1) The monitored bandwidth starts at 1 kHz.
2) The monitored bandwidth starts at 100 Hz.

Solution

1) As stated, in this case $f_1 = 1\,\text{kHz}$ and $f_2 = 1.01\,\text{kHz}$. Using the given expressions, we get for the expected number of maxima per second

$$\sqrt{\frac{3}{5}\frac{(1.01^5 - 1)\cdot 10^{15}}{(1.01^3 - 1)\cdot 10^9}} \cong 10^3 \sqrt{\frac{3}{5}\frac{0.051}{0.03}} \cong 1.02 \cdot 10^3$$

The expected number of zeros per second is:

$$2\sqrt{\frac{(1.01^3 - 1)\cdot 10^9}{3(1.01 - 1)\cdot 10^3}} \cong 2 \cdot 10^3$$

Hence, a good quality oscilloscope displaying white noise will very likely show, during each second, 1000 maxima and 2000 zeros, if the narrow bandpass filter of 10-Hz bandwidth is centered around a frequency of about 1 kHz.

2) In the second case, only the scale factor modifies the number of maxima and zeros. Since instead of 10^3 we must use 10^2, the expected waveform will exhibit roughly 100 maxima per second and 200 zeros.

CASE STUDY 14.15

A true rms instrument is used to measure the rms noise voltage of a 10-kΩ resistor, in a 10-MHz bandwidth. Consider the following cases:

- The resistor is at the reference temperature (T = 290 K).
- The resistor is immersed in liquid nitrogen ($T_1 = 77$ K).

1) Supposing that the resistor generates only thermal noise, what is the measured value in each case?
2) What value should a resistor have which, at the reference temperature, would produce the same amount of thermal noise as the cooled resistor (second case)?

Solution

1) Using (3.4), it is possible to calculate the indication of the voltmeter in the first case):

$$\overline{v_n^2} = 4\text{kTR } \Delta f = 4(1.38 \cdot 10^{-23})(290)(10^4)(10^7) \cong 16 \cdot 10^{-10} \text{ V}^2$$

Consequently

$$V_n = 40 \text{ μV}$$

For a cooled resistor (second case)

$$\overline{v_n^2} = 4\text{kT}_1\text{R } \Delta f = 4(1.38 \cdot 10^{-23})(77)(10^4)(10^7) \cong 4.25 \cdot 10^{-10} \text{ V}^2$$

and

$$V_n \cong 20.6 \text{ μV}$$

which is half the previous value.

It is easy to prove that *any resistor, cooled by immersion in liquid nitrogen, generates roughly the half the thermal noise voltage delivered at room temperature*, provided that the bandwidth of the measurement system is the same.

2) The value of a hypothetical resistor which at the reference temperature would produce the same noise as the cooled resistor must satisfy the relation $T_1 R = T R_{eq}$; hence

$$R_{eq} = R\,\frac{T_1}{T} = 10\,\frac{77}{290} \cong 2.65 \text{ k}\Omega$$

470   14 Noise Performance Measurement

CASE STUDY 14.16 [228]

Consider the circuit shown in Fig. 14.21a, where the resistive load $R = 1\,\text{M}\Omega$ is in series with a reverse-biased diode, both maintained at $T = 300\,\text{K}$. The current flowing into the circuit is $I_o = 50\,\text{nA}$.

a) What is the noise voltage across the resistor in a bandwidth of 1 kHz?
b) What is the noise voltage measured between the points M and N of the circuit in Fig. 14.21b? It is assumed that the corresponding elements of both branches are identical.
c) If the points M and N are connected together, what is the noise voltage of these points relative to ground?

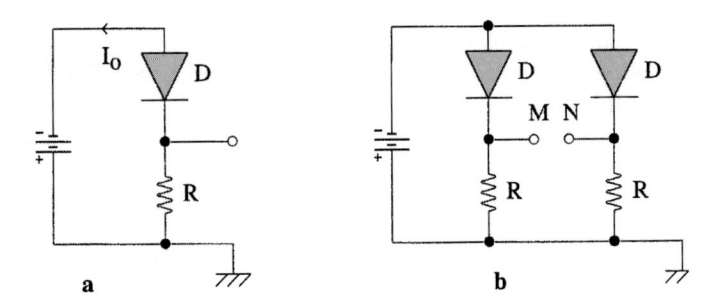

Fig. 14.21. a Reverse-biased diode with series resistor; b circuit with two diodes

Solution

a) Considering the circuit of Fig. 14.21a, three noise contributions are present across R:
  – the thermal noise of R;
  – the shot noise current of D, which produces a voltage drop across R;
  – the excess noise of R, since a DC current flows through it.

As these components arise from different physical processes, they are uncorrelated and their mean-square values can be added; on the other hand, the excess noise of the resistor cannot be evaluated, due to the lack of information concerning its type and characteristics. Consequently, the predicted value of the total noise will certainly be lower than the actual value.

The thermal noise of the resistor is

$$v_n = \sqrt{4kTR\,\Delta f} = \sqrt{4\,(1.38 \cdot 10^{-23})\,300 \cdot 10^6 \cdot 10^3} \cong 4.07\,\mu\text{V}$$

The shot noise of the diode is given by:

$$i_n = \sqrt{2qI_o\,\Delta f} = \sqrt{2\,(1.6 \cdot 10^{-19})(50 \cdot 10^{-9})10^3} \cong 4 \cdot 10^{-12}\,\text{A}$$

The voltage drop produced by i_n across R is

$$Ri_n = 10^6\,(4 \cdot 10^{-12}) = 4\,\mu\text{V}$$

According to expression (2.39), the total noise is

$$v_{nt} = \sqrt{(Ri_n)^2 + v_n^2} \cong 5.7 \text{ μV}$$

As stated before, this is a lower bound, since the excess noise of the resistor is not considered.

b) The previous estimate shows that the noise voltage of point M (or N, since the elements are identical) relative to ground is 5.7 μV. Since the branches are independent, the overall noise voltage measured between M and N is

$$v_{MN} = \sqrt{v_M^2 + v_N^2} = \sqrt{2}\,(5.7) \cdot 10^{-6} \cong 8.05 \text{ μV}$$

c) Upon connecting M and N, the diodes as well as the resistors are in parallel. Each branch involved in a parallel configuration acts as an independent noise source with respect to the other. Consequently, the total shot noise is

$$I_n = \sqrt{2q(I_o + I_o)\,\Delta f} = \sqrt{2}\,4 \cdot 10^{-12} \cong 5.65 \cdot 10^{-12} \text{ A}$$

It generates a voltage drop across the equivalent resistance R_{eq} equal to

$$R_{eq}I_n = 5 \cdot 10^5\,(5.65 \cdot 10^{-12}) \cong 2.82 \text{ μV}$$

The thermal noise of the equivalent resistance is

$$v_{ne} = 4.07/\sqrt{2} \cong 2.88 \text{ μV}$$

Finally, the total noise voltage is

$$V_{nt} = \sqrt{2.82^2 + 2.88^2} \cong 4.07 \text{ μV}$$

Comparing with the result obtained in question a), the total noise is diminished, mainly due to the load, which has been reduced by the parallel connection. We can conclude that *high-value loads are inadvisable in low-noise circuits*.

CASE STUDY 14.17 [227]

An ideal white noise generator with a spectral density of N_G [W/Hz] is connected to the input terminals of the RC filter shown in Fig. 14.22. Estimate the influence of the time constant $\tau = RC$ on the output noise.

Solution

The transfer function $H(\omega)$ of the RC filter is calculated with the voltage divider formula:

14 Noise Performance Measurement

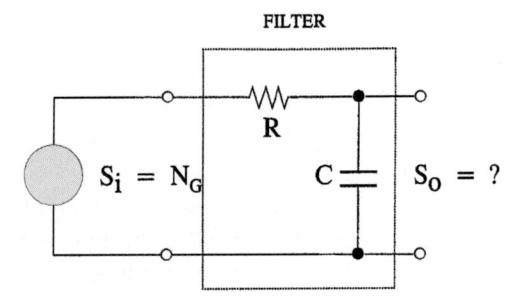

Fig. 14.22. Low-pass RC filter driven by a white noise generator

$$H(\omega) = \frac{1/j\omega C}{R + 1/j\omega C} = \frac{1}{1 + j\omega CR}$$

hence

$$|H(\omega)|^2 = \frac{1}{1 + (\omega CR)^2}$$

Applying (2.41), the output spectral density is

$$S_o = S_i |H(\omega)|^2 = \frac{N_G}{1 + (\omega CR)^2}$$

Note that this equality holds for spectral power density both in terms of f and in terms of ω.

It is easy to see that when $\omega CR = 1$, the output spectral density is half the input value. Traditionally, when the output power is reduced to half its low-frequency value, we have the cutoff frequency $\omega_0 = 1/RC$. We conclude that the higher the cutoff frequency ω_0, the greater the noise power delivered to the output. Hence, *to reduce the total output noise power, the cutoff frequency must be decreased.* This result agrees with Rule 3 of Sect. 13.2.1.

With the IEEE definition of the spectral power density, the total output power can be estimated by integrating the spectral power density over all ω, i.e.

$$P_o = \frac{1}{2\pi} \int_{-\infty}^{+\infty} S_i(\omega)\, d\omega = \frac{N_G}{2\pi} \int_{-\infty}^{+\infty} \frac{d\omega}{1 + (\omega CR)^2}$$

which yields

$$\boxed{P_o = \frac{N_G}{2RC}}$$

Once again, *to significantly reduce the total output power of noise, a RC-filter with a long time-constant must be selected.*

CASE STUDY 14.18 [228]

Consider a reverse-biased diode with a depletion layer width of d = 1 μm and suppose that the minority charge carriers pass through the depletion layer at a nearly constant velocity v = 10^5 m/s. By measurement, it is found that the junction capacitance of the reverse-biased diode is C = 4 pF and its series resistance is R = 5 Ω. What is the frequency at which the power spectrum of the shot noise current rolls off to half its low-frequency value?

Solution

The diode generates shot noise, as predicted by (3.20); its spectrum is flat (provided that transit time is neglected). Two additional physical aspects must be considered:

– The shot noise traverses the equivalent RC circuit of the diode and consequently, at terminals A and K its spectrum is modified (see Case study 14.17). It is implicitly assumed that the external circuit does not significantly load the diode, and thus it has no further effect on the noise spectrum.
– The transit time T_t of the charge carriers through the depleted region must be taken into account.

These two aspects are considered independently. The former needs the previous calculation of the current transfer function of the RC circuit presented in Fig. 14.23 (its expression, as well as the expression of its magnitude, are both given in Fig. 14.23).

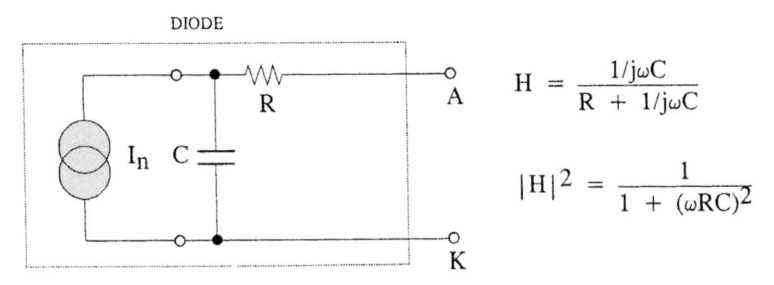

$$H = \frac{1/j\omega C}{R + 1/j\omega C}$$

$$|H|^2 = \frac{1}{1 + (\omega RC)^2}$$

Fig. 14.23. Diode equivalent circuit (A = anode, K = cathode)

Thus, the current spectral density of the shot noise at terminals A and K is

$$S(AK) = S(I_n)\,|H|^2 = 2qI_o \frac{1}{1 + (\omega RC)^2}$$

When ωRC = 1, the spectral density decreases to half its low-frequency value; from this last equality, we deduce the cutoff frequency $f_1 \cong 7.96$ GHz. For the transit time T_t, a crude estimate yields

$$T_t = \frac{d}{v} = \frac{1 \cdot 10^{-6}}{10^5} = 10^{-11} \text{ s}$$

474   14 Noise Performance Measurement

According to Sect. 3.2.3, the cutoff frequency associated with the transit time can be evaluated with the expression

$$f_2 = \frac{1}{2\pi} \frac{3.5}{T_t} \cong 55.76 \text{ GHz}$$

Although f_1 and f_2 are not separated by a factor of 10, it may still be considered that in this case f_1 is the dominant cutoff frequency; hence, the height of the spectrum is halved by around 7.96 GHz.

To conclude, the effect of transit time can largely be neglected with respect to the effect of the junction capacitance and series resistance, at least for the proposed numerical values.

CASE STUDY 14.19

A 150-mH inductor has an impedance of 2.4 kΩ at 1 kHz. A true rms voltmeter is used to measure the thermal noise voltage per unit bandwidth. What will the voltmeter indication be, at the reference temperature? What conditions must be fulfilled to get an accurate reading?

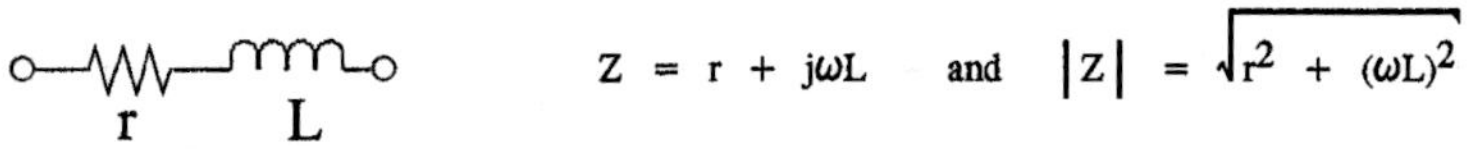

$$Z = r + j\omega L \quad \text{and} \quad |Z| = \sqrt{r^2 + (\omega L)^2}$$

Fig. 14.24. Equivalent circuit of the inductor

Solution

Traditionally, the inductor is fabricated by winding a copper wire around a cylindrical support; therefore, an associated resistance r always exists, i.e., the ohmic wire resistance. Hence, the inductor can be represented by the equivalent circuit depicted in Fig. 14.24.

The proposed impedance value corresponds to the magnitude of the impedance. Hence

$$r = \sqrt{|Z|^2 - (\omega L)^2} = \sqrt{(2.4 \cdot 10^3)^2 - (2\pi \cdot 10^3 \cdot 0.15)^2} \cong 2.2 \text{ k}\Omega$$

Applying (3.1), at T = 290 K we obtain

$$v_n = \sqrt{4KTr} = \sqrt{4(1.38 \cdot 10^{-23})(290)(2.2 \cdot 10^3)} \cong 5.93 \text{ nV}/\sqrt{\text{Hz}}$$

If no special care is taken during measurement, the voltmeter reading can exceed this value. If this is the case, it means that spurious fields have perturbed the measurement. Since any inductor is very sensitive to magnetic fields, to avoid pickup during measurement of the thermal noise voltage the inductor must be placed in a high-μ shielded enclosure.

CASE STUDY 14.20 [221]

Consider the circuit in Fig. 14.25, where $R_1 = 5\,k\Omega$, $R_2 = 1\,k\Omega$, and the inductor is supposed to be ideal, with $L = 15\,mH$. The current source supplies DC at $1\,mA$. According to the manufacturer's data sheet, the excess noise of the resistors is $2\,\mu V/V_{dc}$ (frequency decade)$^{-1/2}$. What is the spectral density of the noise power measured across the current generator, at a frequency of $10\,kHz$ and at the reference temperature?

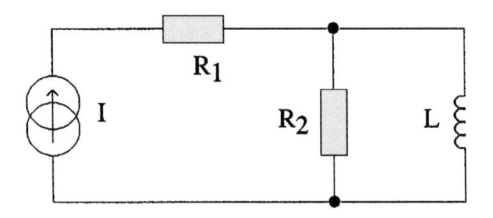

Fig. 14.25. Passive network

Solution

■ **Explanation.** Since the inductor is assumed to be ideal, its ohmic resistance is zero. It follows that the DC voltage across it is also zero, so no DC current can flow through resistor R_2; hence, no excess noise is generated by R_2. The noise sources associated with the circuit of Fig. 14.25 are:

- thermal noise of R_1;
- thermal noise of R_2;
- excess noise of R_1.

In practice, the pickup of the inductance represents a serious noise source, but in this application we neglect it (since the pickup is very difficult to estimate, as it depends on spurious field strength, relative position of equipment, eventual shielding, etc.).

Since all quoted noise sources correspond to independent physical processes, the fluctuations are uncorrelated and their mean square values may be summed.

□ **Thermal Noise.** We apply (3.7) to get the total spectral power of the thermal noise:

$$S(v_t) = 4kT\,\mathcal{R}e\{Z\}$$

with

$$Z = R_1 + \frac{(j\omega L)R_2}{j\omega L + R_2} = R_1 + \frac{(j\omega L)R_2(R_2 - j\omega L)}{(\omega L)^2 + (R_2)^2}$$

Considering that $\omega L \cong 942\,\Omega$ at $f = 10\,kHz$, the real part of Z is

$$\mathcal{R}e\{Z\} = R_1 + \frac{(\omega L)^2\,R_2}{(\omega L)^2 + (R_2)^2} \cong 5\cdot 10^3 + \frac{(942)^2\cdot 10^3}{(942)^2 + 10^6} \cong 5.47\,k\Omega$$

476   14 Noise Performance Measurement

As a result:

$$S(v_t) = 4(1.38 \cdot 10^{-23})(290)(5470) \cong 0.875 \cdot 10^{-16} \text{ V}^2/\text{Hz}$$

☐ **Excess Noise.** Only R_1 exhibits excess noise, which must be calculated at a fixed frequency (f = 10 kHz). Expression (7.3), recalled here, is the most appropriate:

$$S(v_{ex}) = \frac{C\, I_{dc}^2\, R_1^2}{f}$$

The goal is to find the value of the constant C, according to the data sheet; (7.5) yields

$$\sqrt{C\, I_{dc}^2\, R_1^2\, \ln(f_2/f_1)} = 2\, \frac{\mu V}{V_{dc}}\, \sqrt{\text{frequency decade}}$$

The simplest way is to select one frequency decade, i.e., $f_2/f_1 = 10$; since $I_{dc} = 1\,\text{mA}$ and $R_1 = 5\,\text{k}\Omega$, we have

$$C = 0.069 \cdot 10^{-12}.$$

Then

$$S(v_{ex}) = \frac{(0.069 \cdot 10^{-12})(25 \cdot 10^6)(10^{-6})}{1 \cdot 10^4} = 1.737 \cdot 10^{-16} \text{ V}^2/\text{Hz}$$

☐ **Total Spectral Noise Density at 10 kHz**

$$S(v_n) = S(v_t) + S(v_{ex}) \cong 2.61 \cdot 10^{-16} \text{ V}^2/\text{Hz}$$

We may conclude that the excess noise dominates the thermal noise (this is often the case in practical situations).

14.4.2 Impedances at Unequal Temperatures

CASE STUDY 14.21

A circuit is made up of two series impedances (Fig. 14.26) which have unequal temperatures. What is the effective noise temperature of the circuit?

Solution

When two resistors at unequal temperatures are connected in series or in parallel, it makes no sense to apply Thévenin's theorem to find the equivalent resistance, unless an *effective temperature* representing the weighted contributions of the individual resistors is calculated. This effective temperature, when assigned to the Thévenin equivalent resistance, will yield the same amount of noise as that delivered by the actual network at its terminals.

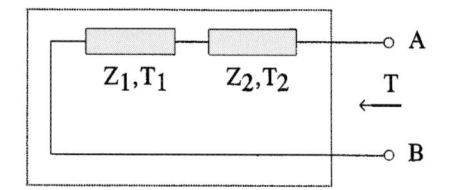

Fig. 14.26. Two series impedances at unequal temperatures

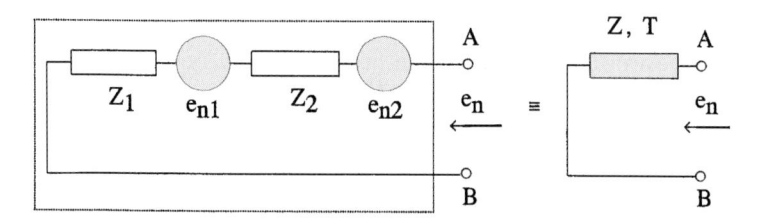

Fig. 14.27. Noise equivalent circuit

The noise equivalent circuit is represented in Fig. 14.27. The goal is to find an equivalent one-port impedance Z and its temperature T, such that the noise power delivered at its terminals is equal to the output noise power of the actual network.

Since the fluctuations generated by Z_1 and Z_2 are uncorrelated, we may calculate the contribution at terminals A and B due to each of them and then add the individual contributions.

With $T_2 = 0$, the contribution of Z_1 is

$$\overline{e_{n1}^2} = 4kT_1 \, \Delta f \, \Re e[Z_1]$$

Similarly, with $T_1 = 0$, the contribution of Z_2 is

$$\overline{e_{n2}^2} = 4kT_2 \, \Delta f \, \Re e[Z_2]$$

The total output noise is:

$$\overline{e_n^2} = \overline{e_{n1}^2} + \overline{e_{n2}^2} = 4kT_1 \, \Delta f \, \Re e[Z_1] + 4kT_2 \, \Delta f \, \Re e[Z_2]$$

Comparing with the noise power delivered by the equivalent impedance, which is

$$\overline{e_n^2} = 4kT \, \Delta f \, \Re e[Z_1 + Z_2]$$

we deduce

$$\boxed{T = \frac{1}{\Re e[Z_1 + Z_2]} \left(T_1 \, \Re e[Z_1] + T_2 \, \Re e[Z_2] \right)}$$

Note that when $Z_1 = R_1$ and $Z_2 = R_2$, we recover (4.53).

From another standpoint, using the expression $T = a_1 T_1 + a_2 T_2$ (Pierce's rule, Sect. 5.2.2), the coefficients a_1 and a_2 become

$$\begin{cases} a_1 = \dfrac{\Re e[Z_1]}{\Re e[Z_1 + Z_2]} \\[2ex] a_2 = \dfrac{\Re e[Z_2]}{\Re e[Z_1 + Z_2]} \end{cases} \quad \text{which verifies} \quad \boxed{a_1 + a_2 = 1}$$

CASE STUDY 14.22

Consider a network with two parallel impedances Z_1 and Z_2, each at a different temperature. What is the effective temperature of the ensemble?

Solution

The noise equivalent circuit is presented in Fig. 14.28, where Z denotes the equivalent impedance of the ensemble.

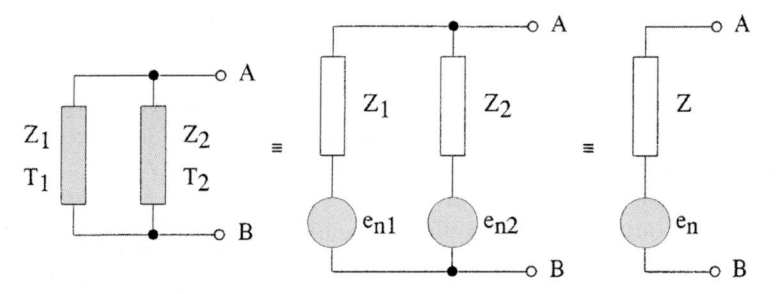

Fig. 14.28. Two parallel impedances at unequal temperatures

The approach is similar to that employed in Case Study 14.21. According to Nyquist's theorem

$$\overline{e_{n1}^2} = 4kT_1 \, \Delta f \, \Re e[Z_1]$$

$$\overline{e_{n2}^2} = 4kT_2 \, \Delta f \, \Re e[Z_2]$$

Each contribution at the output terminals is weighted by its voltage transfer function, namely

$$H_1(p) = \frac{Z_2}{Z_1 + Z_2} \quad (T_2 = 0)$$

and

$$H_2(p) = \frac{Z_1}{Z_1 + Z_2} \quad (T_1 = 0)$$

When computing the mean square value of a complex quantity, remember that taking the average is equivalent to taking its magnitude. Hence:

$$\overline{e_n^2} = \overline{e_{n1}^2} \, |H_1(p)|^2 + \overline{e_{n2}^2} \, |H_2(p)|^2 = 4kT \, \Delta f \, \Re e[Z]$$

where $Z = Z_1 Z_2/(Z_1 + Z_2)$. Upon substituting the expressions for the voltage transfer functions, we obtain

$$T = a_1 T_1 + a_2 T_2$$

with

$$\begin{cases} a_1 = \dfrac{\mathcal{R}e[Z_1]|Z_2|^2}{|Z_1 + Z_2|^2 \, \mathcal{R}e[Z_1 Z_2/(Z_1 + Z_2)]} \\[2ex] a_2 = \dfrac{\mathcal{R}e[Z_2]|Z_1|^2}{|Z_1 + Z_2|^2 \, \mathcal{R}e[Z_1 Z_2/(Z_1 + Z_2)]} \end{cases}$$

When $Z_1 = R_1$ and $Z_2 = R_2$, (5.11b) is recovered, with

$$\begin{cases} a_1 = \dfrac{R_2}{R_1 + R_2} \\[2ex] a_2 = \dfrac{R_1}{R_1 + R_2} \end{cases} \quad \text{which verifies} \quad \boxed{a_1 + a_2 = 1}$$

CASE STUDY 14.23 [222, 229]

This case study is devoted to matched attenuator pads. Consider a matched attenuator pad with loss $L = 10\,\text{dB}$ (Fig. 14.29), having at its input a signal source (typically, an antenna) with resistance R_o at temperature $T_1 = 100\,\text{K}$. Calculate:

1) The effective output noise temperature, if the attenuator is in thermal equilibrium at $T_2 = 300\,\text{K}$.
2) The equivalent input noise temperature.
3) The output noise power spectral density.
4) Compare the output noise power to the noise power applied at the input of the matched pad; consider all possible cases.

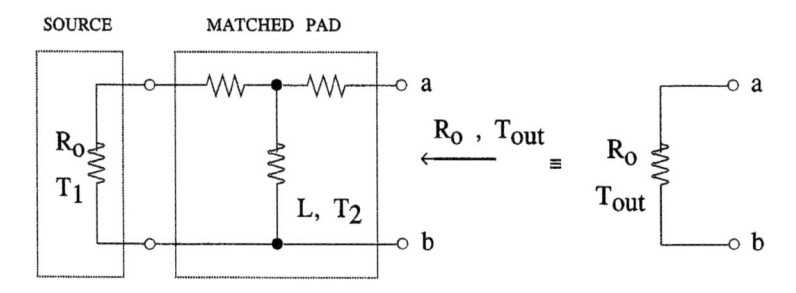

Fig. 14.29. Matched attenuator pad

14 Noise Performance Measurement

Solution

■ **Approach.** R_o represents the matching resistance of the attenuator (i.e., when a resistor R_o is connected to the input, the output resistance seen at terminals a and b is equal to R_o).

The noise power generated by R_o is added to the noise power delivered by the attenuator (since they are uncorrelated).

1) The output noise temperature (T_{out}) is that temperature which assigned to the matching resistance R_o would give rise to the same amount of noise as the actual circuit. To calculate it, we apply Pierce's rule (assuming that unit power is injected at terminals a and b). Let a_1 be the fraction of power dissipated in R_o, and a_2 the fraction of power dissipated by the attenuator. Thus

$$T_{out} = a_1 T_1 + a_2 T_2 \quad \text{with} \quad a_1 + a_2 = 1$$

As the available power gain of the attenuator is $1/L$, and $L = 10\,\text{dB}$ corresponds to a ratio of 10, we deduce that

$$a_1 = \frac{1}{L} = 0.1 \quad \text{hence} \quad a_2 = \left(1 - \frac{1}{L}\right) = 0.9$$

With (5.12), the effective output noise temperature of the attenuator is

$$T_{out} = \frac{1}{L} T_1 + \left(1 - \frac{1}{L}\right) T_2 = 0.1(100) + 0.9(300) = 280\,\text{K}$$

2) The equivalent input noise temperature is established by (5.13):

$$T_e = (L - 1)\, T_2 = (10 - 1)\, 300 = 2700\,\text{K}$$

3) The output noise power spectral density is easily obtained from the effective output noise temperature:

$$S_p(N_o) = k T_{out} = (1.38 \cdot 10^{-23})\, 280 \cong 3.86 \cdot 10^{-21}\,\text{W/Hz}$$

4) The previous expression for T_{out}, both sides of which are premultiplied by $(k\,\Delta f)$, leads to the output noise power

$$N_o = \frac{k T_1\,\Delta f}{L} + (1 - \frac{1}{L}) k T_2\,\Delta f$$

which can be rewritten in the equivalent form

$$N_o = k\,\Delta f \left(\frac{T_1}{L} - \frac{T_2}{L} + T_2\right)$$

14.4 Miscellaneous

In the latter expression, several situations can arise in practice:

- If the attenuator has low loss (L ≅ 1), *the output noise power is merely the input noise power divided by the loss* (noise is transmitted just like any other signal):
$$N_o \cong N_i / L$$
with $N_i = kT_1 \Delta f$. This means that the contribution of the pad to the output noise is negligible.

- When $(T_1/L) \gg T_2$ (i.e., *the input noise is much stronger than the noise generated by the attenuator*), the first term in the sum is dominant and we recover the previous expression:
$$N_o \cong N_i / L$$

- When the attenuator is very noisy relative to the noise power arriving at its input, the last term dominates and *the level of the output noise is almost exclusively determined by the contribution of the attenuator*:
$$N_o \cong k \, \Delta f \, T_2$$

- One might wonder whether the output noise power could ever equal the input noise power. To investigate this situation, in the expression for the output noise power
$$N_o = \frac{N_i}{L} + \left(1 - \frac{1}{L}\right) kT_2 \, \Delta f$$
we put $N_o = N_i$, which yields
$$N_o \left(1 - \frac{1}{L}\right) = \left(1 - \frac{1}{L}\right) kT_2 \, \Delta f$$

Finally
$$N_o = N_i = kT_2 \, \Delta f = \text{attenuator noise power}$$

This means that *the noise level is not affected by an attenuating pad, provided that*:
- The pad is matched at both ports.
- The physical temperature of the pad is equal to the (noise) temperature of the signal source.

■ **Concluding Remark.** A general rule states that when noise is transmitted through an attenuating pad, the pad attenuates the noise (like any other signal), *but always adds its own noise*. The previous discussion is consistent with this rule, and the last result is not surprising: it bears upon the case in which the loss exactly compensates the added noise.

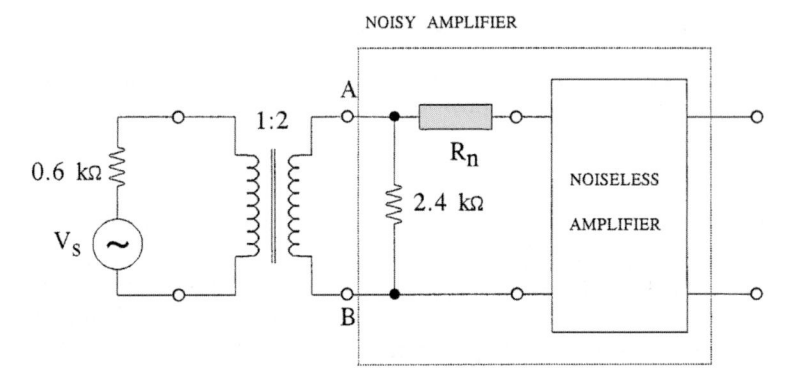

Fig. 14.30. Transformer-coupled amplifier

14.4.3 Low-Frequency Amplifier

CASE STUDY 14.24 [227]

A signal source with 600-Ω internal resistance drives a high-gain amplifier whose input impedance is resistive and equal to 2.4 kΩ over the operating frequency range. A transformer is employed to match the source resistance to the amplifier input resistance. With the signal source set to zero, the noise power is measured at the output terminals of the amplifier. It is observed that when the secondary terminals of the transformer are short-circuited, the output noise power decreases by 4 dB. Calculate (without the short-circuit condition):

1) The noise equivalent resistance of the amplifier.
2) The noise figure.

Solution

It is important to note that the input transformer is *not used to achieve noise matching*, but signal matching. Since the source resistance is 600 Ω and the amplifier input resistance is 2.4 kΩ, the transformer turns ratio must be equal to:
$$1 : \sqrt{2.4/0.6} = 1 : 2$$

The circuit is shown in Fig. 14.30, where R_n denotes the equivalent noise resistance of the amplifier.

The equivalent circuit seen by the secondary of the transformer is presented in Fig. 14.31.

1) Let P_1 be the output noise power measured under normal conditions, and let P_2 be the same quantity measured with the secondary short-circuited. Then
$$10 \log \frac{P_1}{P_2} = 4$$

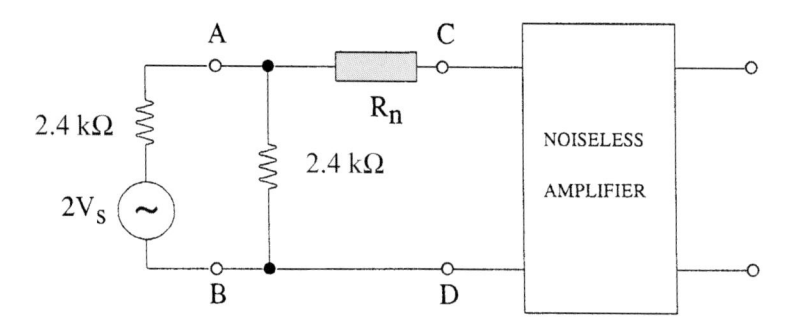

Fig. 14.31. The equivalent circuit seen by the secondary

or
$$P_1/P_2 = 2.5$$

Both powers are proportional to the mean-square value of the output noise voltages; we can write

$$\overline{v_1^2} = 4kT\,\Delta f \left(\frac{2400}{2} + R_n\right)$$

$$\overline{v_2^2} = 4kT\,\Delta f\, R_n$$

and consequently

$$\overline{v_1^2}/\overline{v_2^2} = \frac{4kT\,\Delta f\,(1200 + R_n)}{4kT\,\Delta f\,R_n} = 2.5$$

Solving the equation for R_n, we obtain

$$R_n = 800\,\Omega$$

2) For the noise factor, we apply the definition of Friis:

$$F = \frac{S_i/N_i}{S_o/N_o} = \frac{(2V_s)^2/2400kT\,\Delta f}{(V_s)^2/(1200 + 800)kT\,\Delta f}$$

S_i/N_i can be estimated at terminals A–B (open-circuited) while S_o/N_o is evaluated at terminals C–D (because the amplifier is assumed to be noiseless). We find

$$F \cong 3.333 \quad \text{or} \quad NF = 5.228\text{ dB}$$

15

Noise in Sensing Circuits

"Then at the balance let's be mute,
We never can adjust it;
What's done we partly may compute,
But know not what's resisted"

(John Burns)

15.1 Preamplifiers

15.1.1 Underlying Principles

■ **Definition.** *A preamplifier is an amplifier that is used at the front-end of a transmission system to raise the output of a low level source so that the signal might be further processed by the main part of the system without appreciable degradation in the S/N ratio.*

■ **Note**

1) A preamplifier may also include provision for equalization and/or mixing.
2) A preamplifier is essential when the signal to be processed is so weak that its amplitude becomes comparable to that of spurious signals or intrinsic noise. Typically, the receiving antenna of a radio communication system or a sensor in a noisy industrial environment are low-level sources that both require preamplifiers.

■ **Theory.** We have already seen in Sect. 4.5.6 that for several cascaded two-ports, the overall noise factor can be computed with the expression (4.78):

$$F = F_1 + \frac{F_2 - 1}{G_1} + \frac{F_3 - 1}{G_1 G_2} + \ldots$$

This equation states that if the gain of the first stage is greater than unity ($G_1 > 1$), then all fractions are divided by G_1 and consequently their noise contribution to F is considerably reduced. The greater the gain of the first stage, the less significant the contribution of the following stages.

In contrast, when $G_1 < 1$, the contribution of the following stages is artificially augmented and the overall noise factor is seriously degraded. This simply means that *we must absolutely avoid inserting a passive two-port (such as a connecting cable or waveguide) at the front-end (first position) of a transmission system.*

Instead, a high-gain, low-noise preamplifier must be placed at the front-end, just after the antenna or sensor. Block diagrams illustrating this idea are shown in Fig. 15.1, where the benefit of inserting the preamplifier is suggested by the resulting waveforms (it is assumed that the noise added by the preamplifier is negligible relative to that added by the cable).

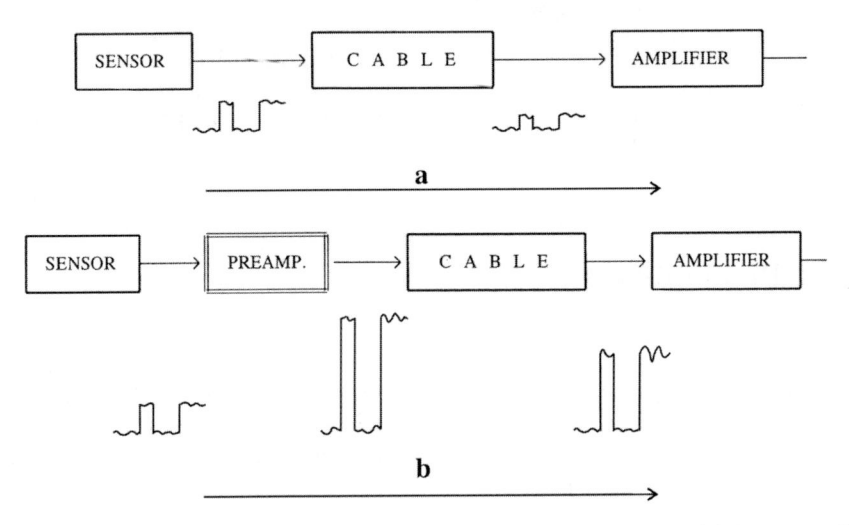

Fig. 15.1. (a) Sensor without preamplifier, S/N ratio deteriorated; (b) Sensor with preamplifier, S/N ratio improved

■ **Conditions.** Any preamplifier must satisfy several requirements:
- Its internal noise must be as low as possible and its power gain as high as possible.
- The physical size of the preamplifier must be reduced, in order to diminish the area of all loops and hence limit the pickup of radiated parasitics. All protection techniques (shielding, filtering, etc.) are eligible.
- The preamplifier must be located as close as possible to the sensor (or antenna) to reduce the length of interconnections and consequently avoid additional pickup.
- When the amplifying chain is not located close to the sensor/antenna (and this is often the case), the preamplifier must incorporate its own power supply (batteries) to avoid long supply wires and thereby limit pickup.

15.1.2 Case Studies

CASE STUDY 15.1 [233]

A sensor provides a 0.18 µV (rms) signal in a 10-Hz bandwidth centered around the operating frequency of 3 kHz. Its internal resistance is $R_s = 1\,k\Omega$. A preamplifier is to be designed employing a 2N930 transistor in common-emitter (CE) configuration. According to the manufacturer's data sheet, the hybrid parameters at $I_C = 1\,mA$ are

$h_{ib} = (25 - 32)\,\Omega$ $h_{ob} = (0 - 1)\,\mu S$
$h_{rb} = (0 - 6 \cdot 10^{-4})$ $h_{fe} = (150 - 600)$

The noise equivalent generators, at $f = 3\,kHz$ and $I_C = 1\,mA$, are $E_n = 3\,nV/\sqrt{Hz}$ and $I_n = 5\,pA/\sqrt{Hz}$.

1) What is the best input S/N ratio that can be expected with this transistor at $I_C = 1\,mA$?
2) Determine the minimum noise factor and the optimal source resistance. Assume $T = 290\,K$.

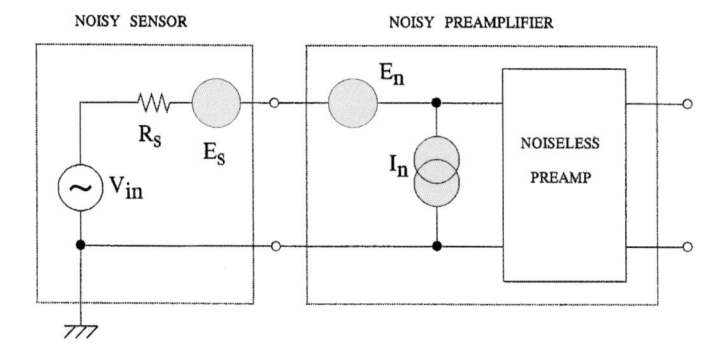

Fig. 15.2. The sensor-preamplifier noise equivalent circuit

Solution

The noise equivalent circuit is shown in Fig. 15.2, where it is assumed that the noise of the sensor is strictly thermal and that all noise sources are uncorrelated. Consequently, the spectral density of the equivalent input noise voltage (referred to the sensor terminals) is obtained by adding the individual contributions.

$$S(E_{ni}) = S(E_s) + S(E_n) + S(R_s I_n)$$

Therefore

$$\overline{E_{ni}^2} = \left(4kTR_s + \overline{E_n^2} + \overline{(R_s I_n)^2}\right) \Delta f$$
$$= \left(16 \cdot 10^{-21}(10^3) + (3 \cdot 10^{-9})^2 + (10^3 \cdot 5 \cdot 10^{-12})^2\right) 10$$
$$= 5 \cdot 10^{-16}\,V^2$$

1) The expected input S/N ratio is

$$\frac{S_i}{N_i} = 20 \log \frac{V_{in}}{E_{ni}} = 20 \log \frac{180}{22.36} \cong 18.12 \text{ dB}$$

If this value is too low, there are two ways to improve it:
- select another bias point (at a different collector current, where perhaps the noise level is lower);
- select another type of transistor.

In order to understand the limitations in each case, it would be convenient to use the noise model of Motchenbacher (see Sect. 7.6.5). That model, however, presumes that all electrical parameters of the Giacoletto model are known (or they can be recovered from the data sheet). This is a hard task, due to the manufacturing spread; in practice, the data sheet does not provide a single value for each parameter, but a large interval instead. Simply choosing the median values of the specified intervals leads to catastrophic results (for instance, a negative value can appear for $r_{bb'}$). It is clear that the only way is to characterize a specific transistor sample by measurement.

Assume that the following values have been measured for a particular 2N930 transistor sample:
$h_{ib} = 31.5\,\Omega$, $h_{fe} = 300$ at $I_C = 1\,\text{mA}$ and $f = 3\,\text{kHz}$
($h_{re} \cong 0$, $h_{oe} = 0$)

The transconductance is

$$g_m = \frac{I_C}{\eta V_T} = \frac{10^{-3}}{1.2(26 \cdot 10^{-3})} \cong 32 \text{ mA/V}$$

(the coefficient η is roughly 1.2 for silicon transistors) and

$$r_{b'e} = \beta_o/g_m = 300/32 = 9.36 \text{ k}\Omega$$

Also

$$h_{fb} \cong \frac{h_{fe}}{1 + h_{fe}} = -0.99667 \quad \text{and} \quad h_{ie} \cong \frac{h_{ib}}{1 + h_{fb}} \cong 9.48 \text{ k}\Omega$$

Thus,

$$r_{bb'} = h_{ie} - r_{b'e} = 9480 - 9360 = 120 \,\Omega$$

The input noise equivalent generators E_n and I_n can be estimated using (7.54) and (7.55) if the frequency-dependent terms are neglected:

$$S(E_n) = 4kTr_{bb'} + 2q\frac{I_C}{\beta_o^2}(r_{bb'} + r_{b'e})^2$$

$$= 1.92 \cdot 10^{-18} + 31.9 \cdot 10^{-20} \cong 2.24 \cdot 10^{-18} \text{ V}^2/\text{Hz}$$

Hence, $E_n \cong 1.49\,\text{nV}/\sqrt{\text{Hz}}$. Similarly
$S(I_n) = 2qI_B \cong 1.06 \cdot 10^{-24}\,\text{pA}^2/\text{Hz}$ and $I_n \cong 1.03\,\text{pA}\sqrt{\text{Hz}}$

Note that these values are of the same order of magnitude as the given measured values, but lower (which is reasonable, since frequency-dependent terms have been neglected). Nevertheless, this calculation shows that to reduce E_n we must either lower g_m (and thus the collector current I_C), or select another type of bipolar transistor with a smaller value of $r_{bb'}$. Reducing I_C is also favorable to I_n reduction.

2) One might wonder why it is necessary to calculate the minimum noise factor when designing the preamplifier.

The minimum noise factor is useful because we can compare it with the actual noise factor of the preamplifier. If they are close enough, it is pointless to try to further reduce the noise, since the additional effort yields negligible noise improvement.

Equations (7.56) and (7.57) are employed to calculate F_o and R_o:

$$F_o = 1 + \sqrt{\frac{2r_{bb'}}{r_{b'e}} + \frac{1}{\beta_o}} = 1 + \sqrt{\frac{2(120)}{9360} + \frac{1}{300}} \cong 1.17$$

$$R_o = \sqrt{\frac{0.05\beta r_{bb'}}{I_C} + \frac{(0.025)^2 \beta}{I_C^2}}$$

$$= \sqrt{\frac{0.05(300)(120)}{10^{-3}} + \frac{(0.025)^2 300}{(10^{-3})^2}} \cong 1409 \,\Omega$$

Next, the actual noise factor is estimated with (13.6) for the specified sensor resistance:

$$F = 1 + \frac{F_o - 1}{2}\left(\frac{R_s}{R_o} + \frac{R_o}{R_s}\right) = 1 + \frac{1.17 - 1}{2}\left(\frac{1}{1.4} + \frac{1.4}{1}\right) \cong 1.18$$

Comparing with F_o, it is obvious that the noise matching condition is nearly reached. Consequently, almost nothing more can be reasonably done to improve the noise of the preamplifier when using this transistor type. Nevertheless, if the noise level is not yet satisfactory, the only solution is to choose a different, less noisy transistor.

CASE STUDY 15.2 [235]

The noise performance of the 2N4221A FET is specified in the data sheet as follows: $E_n = 16\,\text{nV}/\sqrt{\text{Hz}}$ and $I_n = 8\,\text{fA}/\sqrt{\text{Hz}}$, at 1 kHz. It is assumed that $T = 290\,\text{K}$ and no correlation exists between these two generators.

1) What is the optimum noise figure and at what source resistance does it occur?
2) Assume that the signal source is a sensor with an internal resistance of $100\,\text{k}\Omega$. Determine the noise figure when this FET is used in a preamplifier stage.

Solution

1) The optimal source resistance is calculated with expression (13.5b):

$$R_o = \frac{E_n}{I_n} = \frac{16 \cdot 10^{-9}}{8 \cdot 10^{-15}} = 2 \text{ M}\Omega$$

This high value is consistent with the recommendation to employ FETs at the front-end whenever the signal source has high internal resistance. The optimal noise factor is given by (13.5a):

$$F_o = 1 + \frac{E_n I_n}{2kT} = 1 + \frac{(16 \cdot 10^{-9})(8 \cdot 10^{-15})}{2(1.38 \cdot 10^{-23})(290)} \cong 1.016$$

The corresponding noise figure is

$$NF_o = 10 \, \log(1.016) \cong 0.07 \text{ dB}$$

2) If the source resistance is 100 kΩ, (13.6) yields the noise factor

$$F = 1 + \frac{F_o - 1}{2}\left(\frac{R_s}{R_o} + \frac{R_o}{R_s}\right) = 1 + \frac{0.016}{2}\left(\frac{0.1}{2} + \frac{2}{0.1}\right) = 1.16$$

or

$$NF = 10 \, \log 1.7 \cong 0.64 \text{ dB}$$

Since this value is far from NF_o, we might suspect that the transistor is not optimally employed (from a noise point of view). A possible solution is to modify the operating frequency in such a way that the resulting value of R_o is closer to the internal resistance of the sensor. In the present case, increasing the operating frequency at 10 kHz will reduce E_n while I_n remains unchanged. If the matching is still unsatisfactory, the only solution is to select another FET.

CASE STUDY 15.3 [233]

A certain N-channel MOS with channel width W = 60 μm and length L = 1 μm has a gate leakage current I_o = 5 fA measured at a bias point of I_D = 0.6 mA and V_{DS} = 5 V. Assume the MOS parameters in Table 15.1.

This device is employed to design the front-end of a preamplifier intended to amplify the signals delivered by a piezoelectric transducer (assumed noiseless), with an rms value of 7 μV at an operating frequency of 100 Hz and bandwidth 1 Hz. Determine the input S/N ratio when T = 300 K.

Solution

Since the noise contribution of the transducer is neglected, noise arises only from the MOS transistor. Its noise model is represented in Fig. 15.3. The following noise sources are taken into account:

15.1 Preamplifiers 491

Table 15.1. NMOS parameters

Name	Notation	Value
Threshold voltage	V_T	2 V
Channel lenght modulation parameter	λ	10^{-3}
Process gain factor	K'	1.66 µA/V^2
Insulator capacitance per unit area	C_{ox}	1 fF/µm^2
Flicker noise exponent	A_F	1
Flicker noise coefficient	K_F	10^{-27}

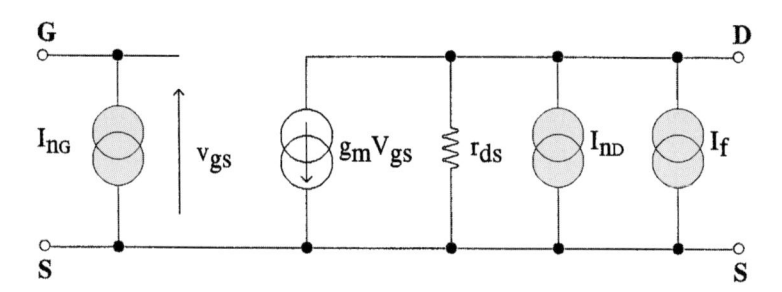

Fig. 15.3. Noise equivalent circuit of the MOS transistor

I_{nG} – shot noise of the leakage current I_o through the SiO$_2$
I_{nD} – drain-source channel noise (Robinson model);
I_f – flicker noise current source

The expressions for the noise currents are

$$\overline{I_{nG}^2} = 2qI_o \, \Delta f$$

$$\overline{I_{nD}^2} = \frac{8}{3} kT \, g_m \, \Delta f$$

$$\overline{I_f^2} = \frac{K_F \, I_D^{A_F}}{f \, C_{ox} \, W \, L} \Delta f$$

To simplify the calculation, we assume that I_{nG} and I_{nD} are uncorrelated. However, we need the value of the transconductance g_m of the transistor, which can be obtained from the expression for I_D. Since the front-end should always be an amplifying stage, the MOS is certainly operating in the saturation region of its current-voltage characteristics. The Shichman-Hodges model (with a correction to include the channel-length modulation effect) yields

$$I_D = \frac{K'}{2} \frac{W}{L} (V_{GS} - V_T)^2 (1 + \lambda V_{DS})$$

Substituting numerical values, we obtain

$$(V_{GS} - V_T)^2 \cong 11.98, \quad \text{hence} \quad V_{GS} - V_T \cong 3.46 \, \text{V}$$

Next, the value of g_m can be found by applying its definition $g_m = \partial I_D/\partial V_{GS}$, i.e.

$$g_m = K'\frac{W}{L}(V_{GS} - V_T)(1 + \lambda V_{DS})$$
$$= 1.66 \cdot 10^{-6}(60)(3.46)(1.005) \cong 346.3 \, \frac{\mu A}{V}$$

It follows that the spectral density of the noise source is

$$S(I_{nD}) = \frac{8}{3}(1.38 \cdot 10^{-23})(300)(3.46 \cdot 10^{-4}) \cong 38.2 \cdot 10^{-25} \, A^2/Hz$$

hence (for $\Delta f = 1 \, Hz$)

$$I_{nD} \cong 1.95 \, pA$$

The spectral density of the gate leakage current shot noise is

$$S(I_{nG}) = 2(1.6 \cdot 10^{-19})(5 \cdot 10^{-15}) = 16 \cdot 10^{-34} \, A^2/Hz$$

which yields (for $\Delta f = 1 \, Hz$)

$$I_{nG} = 0.04 \, fA$$

The spectral density of the flicker noise is

$$S(I_f) = \frac{10^{-27}(0.6 \cdot 10^{-3})}{10^2 \cdot (60)(10^{-15})} = 0.1 \cdot 10^{-18} \, A^2/Hz$$

so, (for $\Delta f = 1 \, Hz$)

$$I_f \cong 0.316 \, nA = 316 \, pA$$

Note that the 1/f noise is dominant; this noise is caused by trapping centers in the gate oxide or at the Si–SiO$_2$ interface. These trap and release charge carriers from the channel in random fashion, producing fluctuations in the drain current. The flicker noise level increases with the density of surface states and temperature.

The mean square value of the total noise current in a 1-Hz bandwidth is

$$\overline{I_{no}^2} = \overline{I_{nD}^2} + \overline{I_f^2} + \overline{I_{nG}^2} \cong (3.8 + 10^5)10^{-24} \cong 10^{-19} \, A^2$$

Next, we need the E_n–I_n noise model of the MOS transistor. I_n is simply identified with I_{nG}, and E_n corresponds to the total noise current (I_{no}) referred to the input gate as an equivalent noise voltage:

$$E_n = I_{no} \, / \, g_m$$

Hence

$$\overline{E_n^2} = \overline{I_{no}^2}/(g_m)^2 = 10^{-19}/(3.46 \cdot 10^{-4})^2 \cong 0.83 \cdot 10^{-12} \, V^2$$

Expression (13.5b) enables us to estimate the optimal source resistance:

$$R_o = \frac{E_n}{I_n} = \frac{0.91 \cdot 10^{-6}}{0.04 \cdot 10^{-15}} \cong 22.85 \cdot 10^{11} \, \Omega$$

This extremely high value suggests that front-end MOS preamplifiers are suited for transducers with huge internal resistance (like piezoelectric sensors).

The equivalent input noise voltage in a 1-Hz bandwidth is

$$\overline{E_{ni}^2} = \overline{E_n^2} + R_s^2\,\overline{I_n^2} + \overline{E_s^2}$$

Since in this application a piezoelectric transducer is assumed, we may reasonably consider that its internal resistance is nondissipative ($E_s = 0$) and has a very high value, such that $R_s \cong R_o$. In this case, $E_{ni} \cong \sqrt{2}\,E_n = 1.29 \cdot 10^{-6}$ V. Finally, the input S/N ratio is

$$\frac{S_i}{N_i} = 20\,\log\frac{7\cdot 10^{-6}}{1.29\cdot 10^{-6}} \cong 14.7\text{ dB}$$

CASE STUDY 15.4 [233]

In designing MOS integrated circuits, the designer can only choose the transistor geometry and bias, since the remaining parameters are fixed by the applicable manufacturing process. The goal of this case study is to explore the influence of the aspect ratio (W/L) on the noise performance of an enhancement type NMOS transistor. Assume that the transistor is fabricated in a process with the following parameters:

$$K' = 24\text{ μA/V}^2 \qquad V_T = 0.75\text{ V} \qquad \gamma = 0.4\,\sqrt{V} \qquad \Phi = 0.6\text{ V}$$

It is desired to design an NMOS with $E_n \leq 1\,\text{nV}/\sqrt{\text{Hz}}$, neglecting the flicker noise contribution. Determine the required aspect ratio and the quiescent drain current (at T = 300 K) to achieve this goal.

Solution

■ **Discussion.** If 1/f noise is neglected, the input noise voltage is due only to drain–source channel noise (I_{nD}) reflected to the input gate (Fig. 15.4). Any voltage fluctuation at the input appears as a current fluctuation at the output proportional to g_m (this being the effect of the voltage-controlled current source). Conversely, any fluctuation in output current can be referred to the input by dividing it by g_m.

According to Robinson's model (also employed by PSPICE)

$$S(I_{nD}) = \overline{I_{nD}^2}/\Delta f = \frac{8}{3}\,kT\,g_m$$

and the noise referred to the input is

$$S(E_n) = \overline{E_n^2}/\Delta f = \frac{S(I_{nD})}{g_m^2} = \frac{8kT}{3g_m}$$

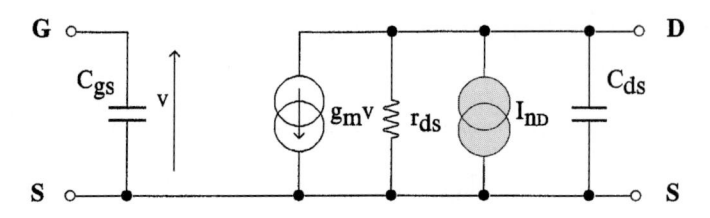

Fig. 15.4. Simplified noise model of a MOS transistor

The transconductance g_m can be calculated from its definition (see Case Study 15.3):

$$g_m = \frac{\partial I_D}{\partial V_{GS}} = K' \frac{W}{L_{\text{eff}}} (V_{GS} - V_T)$$

where W/L_{eff} is called the aspect ratio. Neglecting secondary effects, L_{eff} is identified with the geometrical channel length, and the aspect ratio becomes W/L. Substituting the expression for g_m into the equation for $S(E_n)$, we obtain

$$\boxed{\frac{W}{L} = \frac{8kT}{3K'(V_{GS} - V_T)\, S(E_n)}}$$

Assuming $(V_{GS} - V_T) = 1\,V$ and substituting numerical values, the aspect ratio is found to be

$$\frac{W}{L} = \frac{8(1.38 \cdot 10^{-23})(300)}{3(24 \cdot 10^{-6})(10^{-9})^2} \cong 460$$

■ **Concluding Remarks**

- Since $(V_{GS} - V_T)$ is usually of the order of several volts, the arbitrary adopted value of $1\,V$ yields a good estimate of the order of magnitude of the aspect ratio.
- An aspect ratio of 460 is rather high. To obtain a smaller value, the specified noise performance ($E_n \leq 1\,nV/\sqrt{Hz}$) must be reconsidered. Figure 15.5 shows the aspect ratio as a function of input noise voltage.

This plot shows that if the goal is less ambitious (for instance, $E_n \geq 5\,nV/\sqrt{Hz}$), reasonable values are obtained for the aspect ratio, and consequently the MOS becomes less expensive (a good rule of the thumb is that *the cost of any integrated device is proportional to its occupied area*).

The quiescent drain current is computed with the familiar equation

$$I_D = \frac{K'}{2} \frac{W}{L_{\text{eff}}} (V_{GS} - V_T)^2 = (12 \cdot 10^{-6})(460)(1)^2 = 5.52\,\text{mA}$$

where we adopt $\lambda = 0$, and therefore $L_{\text{eff}} = L$.

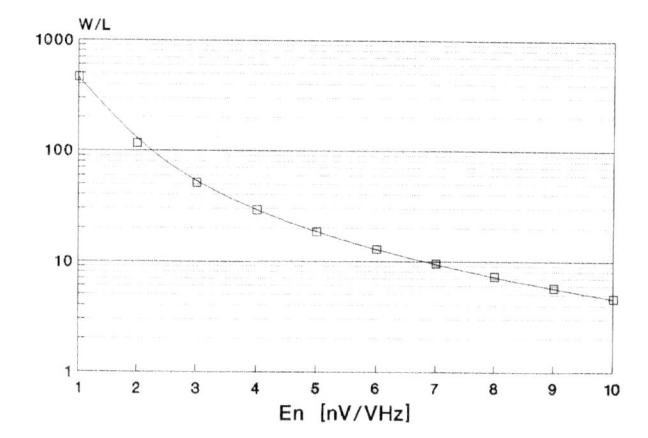

Fig. 15.5. E_{ni} dependence on the transistor aspect ratio

In conclusion, the low-noise condition leads to a high-value g_m; in turn, this requires an unusual large aspect ratio, and consequently larger quiescent drain current than usual.

CASE STUDY 15.5 [233]

In this case study, we emphasize the selection of an appropriate active device when designing a low-noise preamplifier. Consider a given transducer with internal resistance $R_s = 1\,M\Omega$, which delivers a 0.32–µV (rms) signal at an operating frequency of 100 Hz. It is employed in an environment where the ambient temperature is 80°C. To guarantee acceptable operation, the input S/N ratio must be at least 4 dB. Select an appropriate transistor for the preamplifier.

Solution

■ **Discussion.** In this case the goal of the design is to satisfy the specified S/N ratio. To calculate it, the designer needs the spectral density of the input noise voltage. The input noise has its origins in the transducer noise, the front-end preamplifier, and the following stages. The contribution of the following stages is calculated with (4.78). Since in the early design phase we have no information on the following stages, provision must be made on the specified performance. For example, a 50% margin might be considered satisfactory; under this assumption, the input S/N ratio must be at least 6 dB.

■ **Device Selection.** First of all, the designer must decide on the type of transistor (bipolar, FET, or MOS) to be used. As stated in Sect. 13.2.5, since the internal resistance of the transducer is high (1 MΩ), the best choice for the first stage is a field-effect transistor.

Next, an FET must be selected from the data book. At this point, we must remember that no general procedure exists to choose the right FET by inspecting published performance charts and very likely the designer must try different types in succession. Each time, the equivalent input noise voltage must be calculated, followed by estimation of the input S/N ratio. To detail this approach, we shall limit our study to the investigation of a single FET type, assuming that the procedure can be repeated until the goal is achieved.

Let us adopt the N-channel 2N4221A FET. According to its data sheet, I_{DSS} has a value in the range 2 mA to 6 mA, the maximum pinch-off voltage is −6V, the maximum common-source input capacitance C_{iss} is 6 pF, and the gate reverse current is $I_{GSS} = 0.8$ nA (maximum value at 25°C). Since the preamplifier must be located as close as possible to the transducer, it seems reasonable to think that its temperature will also reach 80°C; a rough estimate shows that at that temperature, the gate reverse current will be around 35 nA. With the additional assumption that the transistor will operate in the saturated region, biased at I_{DSS}, we may proceed to a noise calculation.

■ **Noise Estimation.** The FET noise will be described with the Robinson model (Sect. 7.7.4), where only small-signal parameters are required to calculate the noise spectral densities. However, as shown in the data sheet, there are wide variations in I_{DSS} between various samples of 2N4221A FET. On the other hand, for the remaining parameters, only maximum values are listed. It follows that when building the preamplifier, one will very likely encounter discrepancies between the adopted values of electrical parameters (for computational purposes) and the actual ones. To avoid this, any particular FET device must be characterized by measurement, but this is beyond the scope of the present study.

In practice, the adopted value of I_{DSS} is the geometrical mean of the proposed interval, i.e., $\sqrt{12} \cong 3.46$ mA.

The simplified noise equivalent circuit is similar to that given in Fig. 7.18a, and is presented in Fig. 15.6.

According to Sect. 7.7.4, the noise spectral densities are

$$S(I_{nD}) \cong \frac{2}{3} 4kT\, g_m$$

$$S(I_{nG}) = \frac{4}{15} \frac{\omega^2 C^2}{g_m} 4kT + 2qI_o$$

where the shot noise of the reverse gate current (I_o) has been taken into account.

To evaluate the transconductance, we use the familiar expression

$$g_m = \frac{2I_{DSS}}{|V_P|} \sqrt{\frac{I_D}{I_{DSS}}} = \frac{2(3.46)}{6} \sqrt{1} \cong 1.15 \text{ mA/V}$$

C represents the input capacitance; we may approximate it by C_{iss} (which is the common-source input capacitance with the output short-circuited). Actu-

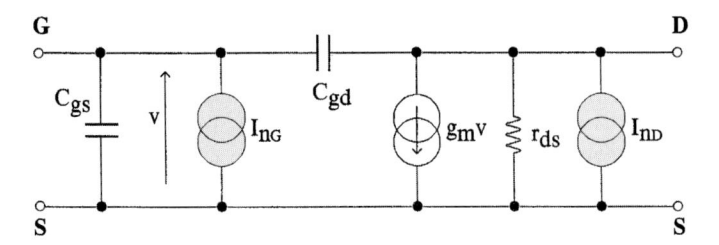

Fig. 15.6. FET simplified noise model

ally, C is greater than C_{iss} due to the Miller effect, but we have no information on the load and therefore no accurate calculation can be performed. Thus

$$S(I_{nD}) \cong \frac{8}{3}(1.38 \cdot 10^{-23})(353)(1.15 \cdot 10^{-3}) \cong 1.49 \cdot 10^{-23} \text{ A}^2/\text{Hz}$$

$$S(I_{nG}) = \frac{16}{15}\frac{(2\pi \cdot 10^2 \, 6 \cdot 10^{-12})^2}{1.15 \cdot 10^{-3}}(1.38 \cdot 10^{-23})(353) + 2(1.6 \cdot 10^{-19})(35 \cdot 10^{-9})$$

$$= 6.42 \cdot 10^{-35} + 1.12 \cdot 10^{-26} \cong 1.12 \cdot 10^{-26} \text{ A}^2/\text{Hz}$$

Note that the dominant contributor to $S(I_{nG})$ is shot noise (second term in the sum). This simply means that neglecting the Miller effect has no effect on the accuracy (even for a two orders-of-magnitude greater input capacitance, the second term of the sum is still dominant).

The noise must be referred to the input, if one is to calculate the input S/N ratio. Equivalence with the noise model of Fig. 7.18b is established via the equation

$$S(E_{nD}) = S(I_{nD})/(g_m)^2$$

$$= (1.49 \cdot 10^{-23})/(1.15 \cdot 10^{-3})^2 \cong 1.13 \cdot 10^{-17} \text{ V}^2/\text{Hz}$$

The resulting FET input noise model is given in Fig. 15.7.

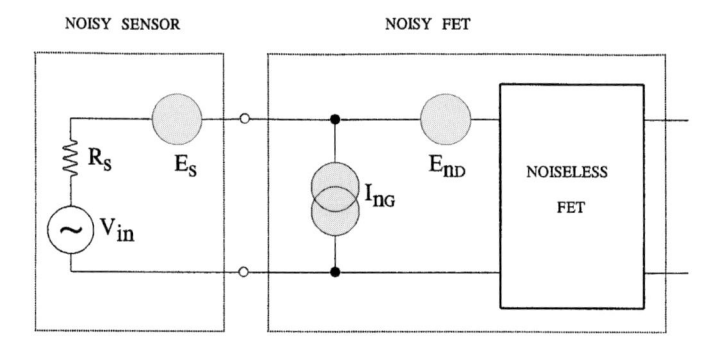

Fig. 15.7. Noise model of sensor-preamplifier system

Under the assumption that I_{nG} and E_{nD} are uncorrelated, the spectral density of the equivalent input noise voltage is obtained by adding the individual contributions of all sources, i.e.,

$$S(E_{ni}) = S(E_{ns}) + S(E_{nD}) + (R_s)^2 S(I_{nG})$$
$$= 4(1.38 \cdot 10^{-23})(353)10^6 + 1.13 \cdot 10^{-17} + (10^{12})(1.12 \cdot 10^{-26})$$
$$= 1.948 \cdot 10^{-14} + 1.13 \cdot 10^{-17} + 1.12 \cdot 10^{-14}$$
$$\cong 3.07 \cdot 10^{-14} \, V^2/Hz$$

Note that the largest contribution is due to the thermal noise of the transducer, followed by the shot noise of the gate reverse current.

The equivalent input noise voltage in a 1-Hz bandwidth is

$$E_{ni} = \sqrt{S(E_{ni})\Delta f} \cong 1.75 \cdot 10^{-7} \, V = 0.175 \, \mu V$$

and the input S/N ratio is

$$\frac{S_i}{N_i} = 20 \, \log \, \frac{V_{in}}{E_{ni}} = 20 \, \log \, \frac{0.32}{0.175} \cong 5.2 \, dB$$

■ **Comment.** Clearly, the design goal has not been achieved, since we are below the 6-dB limit. At this point, another FET type must be selected and the same approach repeated.

The advantage now is our acquired appreciation for the major influence of the shot noise of the gate reverse current. Hence, if possible, we would select the next transistor with an I_{GSS} even lower than that of 2N4221A previously used.

■ **Concluding Remark.** It might be interesting to estimate the impact on the solution accuracy of the hypothesis that E_{nD} and I_{nG} are uncorrelated. In practice, the exact expression for the spectral density of E_{ni} is given by (4.58):

$$S(E_{ni}) = S(E_{ns}) + S(E_{nD}) + (R_s)^2 S(I_{nG}) + 2\mathscr{C} R_s S(I_{nG} E_{nD})$$

where $\mathscr{C}$ denotes the correlation coefficient. According to the Robinson model, the cross-spectral density is given by (7.77):

$$S(I_{nG} E_{nD}) = -\frac{1}{6} \frac{j\omega C \, 4kT}{g_m}$$
$$= j \frac{4}{6} \frac{(2\pi \cdot 100)(6 \cdot 10^{-12})(1.38 \cdot 10^{-23})(353)}{1.15 \cdot 10^{-3}}$$
$$\cong j \, 106.4 \cdot 10^{-28}$$

In the worst case ($\mathscr{C} = 1$), the last term of (4.58) becomes

$$2(j \, 106.4 \cdot 10^{-28})(10^6) \cong j \, 2.13 \cdot 10^{-20}$$

This term is at least three orders of magnitude smaller than the remaining terms of S(E_{ni}), and it consequently yields no significant correction. We may conclude that neglecting the correlation between E_{nD} and I_{nG} has no major influence on accuracy, at least in this case.

CASE STUDY 15.6 [236]

Consider the common-emitter amplifier shown in Fig. 15.8, where the transistor has $\beta = 200$. The transducer connected to its input has internal resistance $R_s = 1\,\text{k}\Omega$ and delivers a signal of 1 mV. Assuming that C_1 and C_E are short circuits in the frequency range of interest, determine the output S/N ratio over a bandwidth $\Delta f = 1\,\text{kHz}$ situated in the low-frequency region. Assume $T = 300\,\text{K}$.

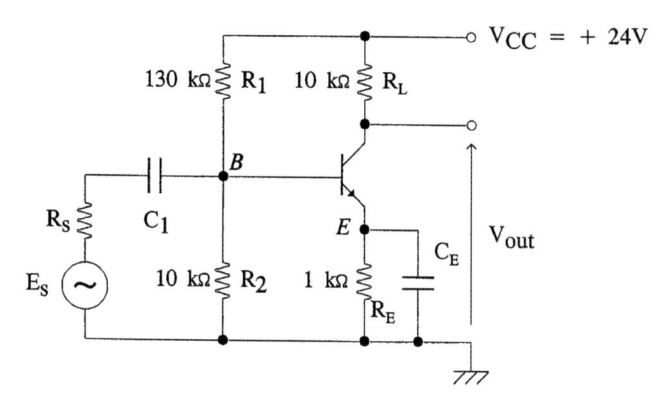

Fig. 15.8. Common-emitter amplifier stage

Solution

■ **Comment.** Since the goal is to calculate the output S/N ratio, we need both the signal power and the noise power at the output. For small-signal AC analysis, the Giacoletto equivalent circuit is most appropriate; however, the electrical parameters of the transistors must first be evaluated. For noise estimation, the model of Motchenbacher will be adopted, since it is based on the same Giacoletto equivalent circuit.

■ **DC Analysis.** If we neglect the base current through R_1, the potential with respect to ground of point B is roughly

$$V_B \cong \frac{R_2}{R_1 + R_2} V_{CC} = \frac{10}{130 + 10} 24 \cong 1.7\,\text{V}$$

Assuming a silicon transistor, $V_{BE} \cong 0.7\,\text{V}$, and the emitter-to-ground voltage is $V_E = 1\,\text{V}$. It follows that $I_C \cong I_E = V_E/R_E = 1\,\text{mA}$ and $I_B = I_C/\beta = 5\,\mu\text{A}$.

15 Noise in Sensing Circuits

■ **AC Analysis.** Adopting $r_{bb'} = 100\,\Omega$, we have

$$r_e = kT/qI_E \cong 26\,\Omega; \quad r_{b'e} = \beta r_e = 5.2\,k\Omega; \quad g_m = 1/r_e \cong 38.4\,mA/V$$

In most practical applications, $r_{b'c}$ is greater than $1\,M\Omega$ and can consequently be neglected. Similarly, r_{ce} (which is greater than $100\,k\Omega$) will be neglected with respect to the load. The resistances of the base voltage-divider network are considerably greater than the sum $(r_{bb'} + r_{b'e})$, and they can be neglected. Also, $C_{b'e}$ and $C_{b'c}$ can be neglected, since it is specified that operation is in the low-frequency range. The simplified equivalent circuit is shown in Fig. 15.9, where the dissipative (noisy) resistors are displayed.

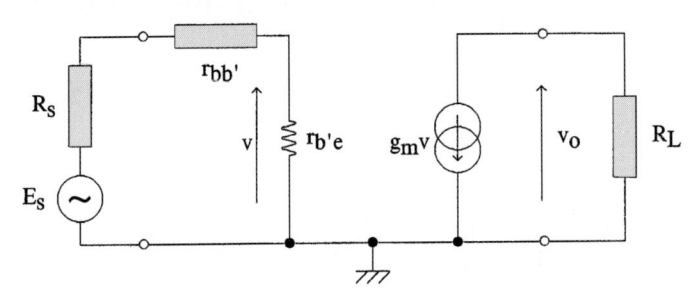

Fig. 15.9. Simplified equivalent circuit of the CE amplifier

Note that $r_{b'e}$ is not noisy, since it corresponds to the flow of charge carriers through the emitter-base junction.

The AC output voltage is

$$v_o = -(g_m v)R_L = -g_m R_L \frac{r_{b'e}}{sum} E_s \quad \text{with} \quad sum = R_s + r_{bb'} + r_{b'e}$$

The output signal power is

$$S_o = v_o^2/R_L = R_L \left(g_m \frac{r_{b'e}}{sum} E_s\right)^2$$

$$= 10^4 \left(38.4 \cdot 10^{-3} \frac{5.2}{6.3} \cdot 10^{-3}\right)^2 \cong 10^{-5}\,W$$

or

$$S_o = -50\,dB$$

■ **Noise Analysis.** The noise sources to be considered here are thermal noise associated with all dissipative resistors, shot noise of the base and collector current and 1/f noise. Since we have no information on the latter, 1/f noise will be ignored. The equivalent noise circuit is shown in Fig. 15.10, where I_{nb}, I_{nc} denote the base and collector shot noise, respectively.

15.1 Preamplifiers

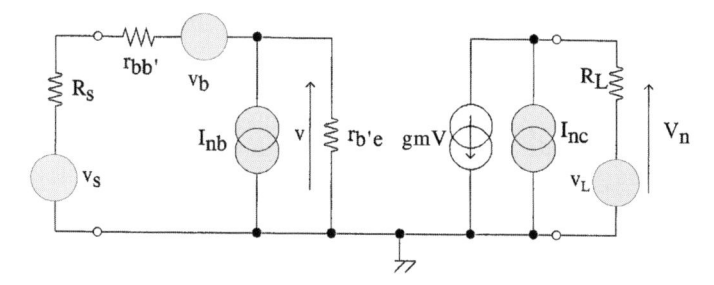

Fig. 15.10. Noise equivalent circuit of the CE stage

These noise sources are assumed to be uncorrelated, so that their mean square contributions add at the output. We have

$$\overline{v_b^2} = 4kT\, r_{bb'}\, \Delta f \qquad \overline{I_{nb}^2} = 2qI_B\, \Delta f$$
$$\overline{v_s^2} = 4kT\, R_s\, \Delta f \qquad \overline{I_{nc}^2} = 2qI_C\, \Delta f$$
$$\overline{v_L^2} = 4kT\, R_L\, \Delta f$$

To proceed, we consider only one source at a time (the remaining ones being set to zero). For each, the voltage drop across $r_{b'e}$ is estimated, and the contribution at the output is calculated, together with its mean square value.

- Contribution of v_b to output (v_{n1})

  The voltage drop across $r_{b'e}$ is

  $$v_1 = v_b \frac{r_{b'e}}{sum}, \quad \text{hence} \quad v_{n1} = -g_m\, v_1\, R_L$$

  $$\overline{v_{n1}^2} = \left(g_m\, R_L \frac{r_{b'e}}{sum}\right)^2 \overline{v_b^2} \qquad \overline{v_b^2} = \left(g_m\, R_L \frac{r_{b'e}}{sum}\right)^2 4kT\, r_{bb'}\, \Delta f$$

  Upon substituting numerical values

  $$\overline{v_{n1}^2} = \left(38.4(10)\frac{5.2}{6.3}\right)^2 4(1.38 \cdot 10^{-23})(300)(100)(10^3)$$
  $$= 16.63 \cdot 10^{-11}\, V^2$$

- Contribution of v_s to output (v_{n2})

  $$v_2 = v_s \frac{r_{b'e}}{sum} \quad \text{and} \quad v_{n2} = -g_m\, v_2\, R_L$$

  $$\overline{v_{n2}^2} = \left(g_m\, R_L \frac{r_{b'e}}{sum}\right)^2 \overline{v_s^2} \qquad \overline{v_s^2} = \left(g_m\, R_L \frac{r_{b'e}}{sum}\right)^2 4kT\, R_s\, \Delta f$$

  Upon substituting numerical values

  $$\overline{v_{n2}^2} = \left(38.4(10)\frac{5.2}{6.3}\right)^2 4(1.38 \cdot 10^{-23})(300)(10^3)(10^3)$$
  $$= 16.63 \cdot 10^{-10}\, V^2$$

15 Noise in Sensing Circuits

- Contribution of v_L to output (v_{n3})

$$\overline{v_{n3}^2} = 4kT\, R_L\, \Delta f = 4(1.38 \cdot 10^{-23})(300)(10^4)(10^3)$$
$$= 16.56 \cdot 10^{-14}\ V^2$$

- Contribution of I_{nb} to output (v_{n4})

$$v_4 = I_{nb}\, \frac{r_{b'e}(r_{bb'} + R_s)}{sum} \quad \text{and} \quad v_{n4} = -g_m\, v_4\, R_L$$

$$\overline{v_{n4}^2} = \left(g_m\, R_L\, \frac{r_{b'e}(r_{bb'} + R_s)}{sum}\right)^2 \overline{I_{nb}^2}$$

$$= \left(g_m\, R_L\, \frac{r_{b'e}(r_{bb'} + R_s)}{sum}\right)^2 2qI_B\, \Delta f$$

$$\overline{v_{n4}^2} = \left(38.4(10^4)\frac{5.2(0.1+1)}{6.3}\right)^2 2\,(1.6 \cdot 10^{-19})(5 \cdot 10^{-6})(10^3)$$

$$= 1.95 \cdot 10^{-10}\ V^2$$

- Contribution of I_{nc} to output (v_{n5})

$$\overline{v_{n5}^2} = 2q\, I_C\, R_L^2\, \Delta f = 2\,(1.6 \cdot 10^{-19})(10^{-3})(10^4)^2(10^3)$$
$$\cong 3.2 \cdot 10^{-11}\ V^2$$

Inspecting the numerical values, the contribution of the collector shot noise and the thermal noise of R_L are clearly both negligible. Hence

$$\overline{V_n^2} \cong \overline{v_{n1}^2} + \overline{v_{n2}^2} + \overline{v_{n4}^2} = 10^{-11}(16.63 + 166.3 + 19.5) \cong 202.43 \cdot 10^{-11}\ V^2$$

or

$$\overline{V_n^2} \cong -87\ \text{dB}$$

■ **S/N Ratio.** The output S/N ratio in a 1-kHz bandwidth is

$$\boxed{S_o/N_o = -50 - (-87) = 37\ \text{dB}}$$

■ **Concluding Remark.** In practice, the S/N ratio will be lower, because 1/f noise has been neglected.

CASE STUDY 15.7

Consider the same common-emitter amplifying stage as in Case Study 15.6. This time, assume that the transition frequency of the transistor is known ($f_T = 300\ \text{MHz}$).

15.1 Preamplifiers

1) Find the E_n–I_n noise model.
2) Determine the optimal noise factor and the optimal source resistance. Assume T = 300 K.

Solution

1) The power spectral densities of noise equivalent generators E_n and I_n can be deduced from their expressions in Sect. 7.6.5, where the 1/f noise contribution is ignored:

$$S(E_n) \cong 4kTr_{bb'} + 2q\frac{I_C}{\beta_o^2}(r_{bb'} + r_{b'e})^2 + 2qI_C\left(\frac{f\, r_{bb'}}{f_T}\right)^2$$

Substituting numerical values

$$S(E_n) = 4(1.38 \cdot 10^{-23})(300)(100) +$$

$$2(1.6 \cdot 10^{-19})(10^{-3})\left(\frac{(100+5200)^2}{200^2} + \frac{(100f)^2}{(3 \cdot 10^8)^2}\right)$$

Finally,

$$\boxed{S(E_n) \cong 18.8 \cdot 10^{-19} + 0.35 \cdot 10^{-34}\, f^2}$$

The power spectral density of I_n is

$$S(I_n) \cong 2qI_B + 2qI_C\left(\frac{f}{f_T}\right)^2$$

$$= 2(1.6 \cdot 10^{-19})\left(5 \cdot 10^{-6} + 10^{-3}\frac{f^2}{(3 \cdot 10^8)^2}\right)$$

yielding

$$\boxed{S(I_n) \cong 1.6 \cdot 10^{-24} + 0.35 \cdot 10^{-38}\, f^2}$$

At mid-band frequencies, the terms in f^2 can be neglected, so

$$S(E_n) \cong 18.8 \cdot 10^{-19}\ \text{V}^2/\text{Hz} \quad \text{and} \quad S(I_n) \cong 1.6 \cdot 10^{-24}\ \text{A}^2/\text{Hz}$$

2) The optimal noise factor is

$$F_o = 1 + \sqrt{\frac{2r_{bb'}}{r_{b'e}} + \frac{1}{\beta_o}} = 1 + \sqrt{\frac{200}{5200} + \frac{1}{200}} \cong 1.208 \quad \text{or} \quad 0.82\ \text{dB}$$

The optimal source resistance is

$$R_o = \sqrt{\frac{0.05\beta r_{bb'}}{I_C} + \frac{(0.025)^2\beta}{I_C^2}}$$

$$= \sqrt{\frac{0.05(200)(100)}{10^{-3}} + \frac{(0.025)^2 200}{10^{-6}}} \cong 1.06\ \text{k}\Omega$$

15.2 Sensing Circuits

15.2.1 Underlying Principles

■ **Background.** By definition, *a sensor (transducer) is any device that converts a nonelectrical quantity, e.g. sound, pressure, humidity, light, etc. into electrical signals, or vice versa.*

The electrical signal delivered from a sensor is eventually amplified, filtered, and so forth. The circuits immediately following the sensor and performing these functions are called *sensing circuits*.

■ **Coupling Networks.** Sensing systems operating over a relatively narrow bandwidth often use a lossless LC-coupling network, inserted between the transducer and the preamplifier. The LC network acts as a sharp passband filter tuned to the operating frequency of the sensor. It greatly attenuates all non-useful signals that fall in the stop band (including noise or interfering signals), and thereby improves the input S/N ratio. Efficiency is considerably degraded when the LC coupling network introduces non-negligible losses (the Q-factor of the filter is low).

■ **Noise Analysis Method.** The most pertinent parameter to describe noise in systems operating with mismatched terminations is the S/N ratio, frequently required at the input. One might ask why we are interested in the *input* S/N ratio. The reason is simple: because once the sensor is available, the magnitude of the signal it delivers is already known (either from measurements or from the manufacturer's data). If the noise is evaluated at the same location (across the sensor), then the input S/N ratio is easily determined. Finally, we need only the equivalent input noise voltage (where *"input" denotes not the amplifier input terminals, but the sensor equivalent signal generator*).

In calculating the input S/N ratio, the following steps are independent of the particular application:

Step 1: Identify all noise sources and draw the noise equivalent circuit.

Step 2: Find the individual contributions of noise sources to the output noise voltage.

Since all noise sources are (or are assumed to be) independent, their individual contributions to the output terminals can be separately calculated. When calculating the contribution of a particular noise source, all remaining noise voltage sources are short-circuited, and all remaining noise current sources are open-circuited. As the sources are assumed to be uncorrelated, their spectral density contributions add at the output. (*Note*: Each spectral density contribution at the output is calculated by means of a transfer function, which must be previously determined. This transfer function is defined between the output terminals of the circuit and the noise source in question.)

Step 3: *Determine the voltage gain K referred to the signal source V_s.* In most cases, the signal source is the sensor; it also has some internal resistance R_s. *Note*: A distinction must be made between the classical voltage gain $A_v = V_o/V_i$ (which refers to the amplifier *input* terminals) and the voltage gain referred to the *signal source* $(K = V_o/V_s)$. The latter is different from the former, for two reasons: a coupling network may be inserted, and $R_s \neq 0$.

Step 4: *Reflect the output noise back to input* by dividing the output noise spectral density by $|K|^2$.

Step 5: *Calculate the input S/N ratio*. Divide the signal voltage by the previously determined equivalent input noise voltage.

■ **Comment.** Steps 2 and 3 of the previous method can also be performed with a circuit simulation package (like PSPICE). The individual contributions to the output of various noise sources, as well as the voltage gain K, are computed by means of harmonic transfer functions. Finally, the output noise results as

$$\overline{v_{no}^2} = |A|^2 \overline{E_1^2} + |B|^2 \overline{E_2^2} + |C|^2 \overline{E_3^2} + |D|^2 \overline{I_1^2} + |E|^2 \overline{I_2^2} + \ldots$$

where A, B, C, D, E, ... are the individual transfer functions defined between the output terminal and the noise source in question. Note that these functions are now evaluated as *numerical quantities, not symbolic expressions*. Insight is lost, but their estimation is painless and considerably faster.

■ **Final Remark.** This general method is employed in all subsequent applications. However, sometimes Steps 2 and 3 (which are independent) can be taken in reverse order.

15.2.2 Case Studies

CASE STUDY 15.8 [233]

Consider the circuit in Fig. 15.11, where $L = 1\,\text{mH}$, $C = 1\,\mu\text{F}$. The transducer has internal resistance $R_s = 1\,\text{k}\Omega$ and delivers a signal $V_{in} = 10\,\mu\text{V}$ (rms value) at $f = 10\,\text{kHz}$. At the same frequency, the noise of the preamplifier is described by two equivalent generators $E_n = 20\,\text{nV}/\sqrt{\text{Hz}}$, $I_n = 3\,\text{nA}/\sqrt{\text{Hz}}$ (which are assumed to be uncorrelated). The input resistance of the preamplifier is $R_i = 10\,\text{k}\Omega$ and its voltage gain is $A_v = 100$.

1) Determine the output noise and the input S/N ratio, per unit bandwidth centered at 10 kHz, with $T = 300\,\text{K}$.
2) Estimate the effect on the S/N ratio of tuning the parallel-resonant circuit to $f = 10\,\text{kHz}$.

506   15 Noise in Sensing Circuits

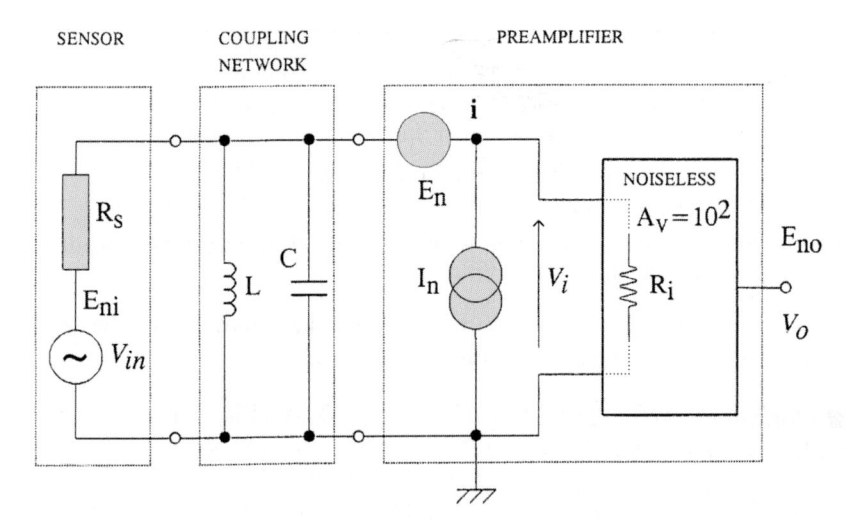

Fig. 15.11. Coupling by means of a parallel-resonant circuit

Solution

■ **Convention.** All voltages referring to signal are denoted by italic fonts (V_{in}, V_i, and V_o). Voltages referring to noise are in standard fonts: E_s (source thermal noise), E_n–I_n (equivalent input noise generators of the preamplifier), and E_i (noise voltage at point **i**).

■ **Noise Equivalent Circuit.** In order to find the equivalent input and output noise voltages, the noise equivalent circuit in Fig. 15.12 must be considered.

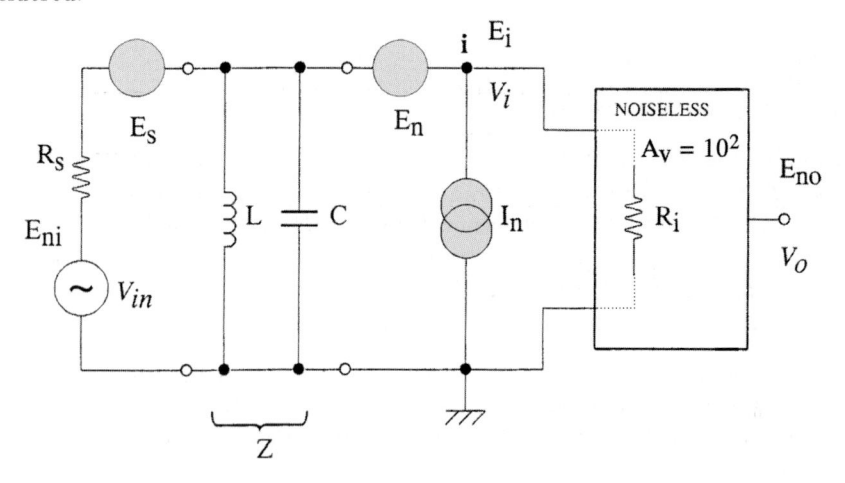

Fig. 15.12. Noise equivalent circuit

■ **AC Analysis.** Clearly, $V_o = A_v V_i$. Thus

$$V_i = V_{in} \frac{R_i \| Z}{R_s + R_i \| Z} \quad \text{and} \quad V_o = A_v V_{in} \frac{R_i \| Z}{R_s + R_i \| Z}$$

where $R_i \| Z$ denotes the parallel combination of R_i and the impedance Z of the LC coupling network.

Consequently, the gain of the circuit referred to the signal generator is

$$K = \frac{V_o}{V_{in}} = A_v \frac{R_i \| Z}{R_s + R_i \| Z}$$

■ **Noise Analysis.** With slight modifications, we can apply the method given in Sect. 15.2.1. There are three noise sources: E_s, E_n, and I_n (note that the input resistance R_i of the preamplifier is considered free of thermal noise, since *all noise generated by the preamplifier is lumped into E_n and I_n*). These sources are assumed to be uncorrelated, so their mean square values add at the output. We may write

$$E_{no} = A_v E_i \quad \text{and} \quad \overline{E_{no}^2} = |A_v|^2 \overline{E_i^2}$$

(in the latter equality, recall that *the mean square value of a complex quantity is calculated by squaring its magnitude*).

The noise voltage at point **i** is obtained by adding all individual contributions per unit bandwidth

$$\overline{E_i^2} = \overline{I_n^2} |Z \| R_s \| R_i|^2 + \overline{E_n^2} \left| \frac{R_i}{Z \| R_s + R_i} \right|^2 + \overline{E_s^2} \left| \frac{R_i \| Z}{R_s + R_i \| Z} \right|^2$$

Hence

$$\overline{E_{no}^2} = |A_v|^2 \left(\overline{I_n^2} |Z \| R_s \| R_i|^2 + \overline{E_n^2} \left| \frac{R_i}{Z \| R_s + R_i} \right|^2 + \overline{E_s^2} \left| \frac{R_i \| Z}{R_s + R_i \| Z} \right|^2 \right)$$

To determine the equivalent input noise (E_{ni}), the output noise must be referred back to the input by dividing it by $|K|^2$, K being the voltage gain referred to the input signal source. After some algebra, we find

$$\overline{E_{ni}^2} = \overline{I_n^2} \left| \frac{R_s + R_i \| Z}{R_i \| Z} \right|^2 |Z \| R_s \| R_i|^2 + \overline{E_n^2} \left| \frac{R_i}{Z \| R_s + R_i} \right|^2 \left| \frac{R_s + R_i \| Z}{R_i \| Z} \right|^2 + \overline{E_s^2}$$

where $\overline{E_s^2} = 4KT R_s$.

At this point it is interesting to evaluate the contribution of each term of the sum. Before doing so, we need to evaluate the following quantities at $f = 10\,\text{kHz}$:

$$Z = \frac{(j\omega L)(1/j\omega C)}{(j\omega L) + (1/j\omega C)} \cong -j\,21.3$$

$$R_i \| Z \cong Z \cong -j\, 21.3 \qquad |R_i \| Z|^2 \cong 454$$

$$R_s + R_i \| Z \cong 1000 - j\, 21.3 \qquad |R_s + R_i \| Z|^2 \cong 10^6$$

$$|Z \| R_s|^2 \cong 454$$

$$|Z \| R_s + R_i|^2 \cong 10^8$$

Then, considering unit bandwidth,

$$\overline{E_{ni}^2} \cong 9 \cdot 10^{-12} + 0.88 \cdot 10^{-12} + 16.54 \cdot 10^{-18} \cong 9.88 \cdot 10^{-12}\ V^2$$

Note that the thermal noise of the signal source resistance is negligible with respect to E_n, which in turn is less than the contribution of I_n (which is the dominant source).

1) The equivalent input noise in 1 Hz bandwidth is

$$E_{ni} \cong 3.14\ \mu V$$

Since at 10 kHz

$$K = 100 \frac{-j\, 21.3}{1000 - j\, 21.3} \quad \text{we have} \quad |K|^2 \cong 4.54 \quad \text{and} \quad |K| = 2.13$$

This yields
$$E_{no} = |K|\, E_{ni} \cong 6.69\ \mu V$$

The input S/N ratio is

$$S_i/N_i = 20\,\log \frac{V_i}{E_{ni}} = 20\,\log \frac{10}{3.14} \cong 10\ dB$$

2) Given the proposed numerical values, the resonant frequency of the LC coupling network is roughly 5 kHz. If the circuit is tuned to the operating frequency of 10 kHz, then $Z \to \infty$ and the gain K reaches a maximum (K = 90.9). The input noise in 1 Hz bandwidth becomes

$$\overline{E_{ni}^2} = \overline{I_n^2}\, R_s^2 + \overline{E_n^2} + \overline{E_s^2}$$

$$\overline{E_{ni}^2} = (3 \cdot 10^{-9})^2 10^6 + (2 \cdot 10^{-8})^2 + 1.656 \cdot 10^{-17}\ V^2$$

$$\overline{E_{ni}^2} = 9 \cdot 10^{-12} + 4 \cdot 10^{-16} + 1.656 \cdot 10^{-17} \cong 9 \cdot 10^{-12}\ V^2$$

yielding $E_{ni} = 3\ \mu V$.
The S/N ratio then becomes

$$S_i/N_i = 20\,\log \frac{10}{3} \cong 10.46\ dB$$

The output equivalent voltage is then: $E_{no} = K\, E_{ni} \cong 272.7\ \mu V$. This high value is mainly due to the spectacular increase in K, not to additional noise.

Table 15.2. Contributions of various noise sources to $\overline{E_{ni}^2}$

Coefficient of	f = 5 kHz	f = 10 kHz (Z → ∞)
$\overline{I_n^2}$	$\left\|\dfrac{R_s + R_i\|\|Z}{R_i\|\|Z}\right\|^2 \|Z\|\|R_s\|\|R_i\|^2$	R_s^2
$\overline{E_n^2}$	$\left\|\dfrac{R_i}{Z\|\|R_s + R_i}\right\|^2 \left\|\dfrac{R_s + R_i\|\|Z}{R_i\|\|Z}\right\|^2$	1
$\overline{E_s^2}$	1	1

■ **Final Discussion.** The expressions of various contributions to $\overline{E_{ni}^2}$ at f = 5 kHz and f = 10 kHz are recalled in Table 15.2.

When both R_i and R_s are considerably higher than $|Z|$ (as is the case at 5 kHz), the coefficient of $\overline{I_n^2}$ asymptotically approaches

$$|R_s + Z|^2$$

The contribution of I_n is slightly higher relative to f = 10 kHz. Similarly, the asymptotic expression of the coefficient of $\overline{E_n^2}$ is:

$$|R_s / Z|^2 \gg 1$$

which proves (Table 15.2) that tuning significantly reduces the contribution of E_n.

Therefore, when both R_i and R_s are considerably higher than $|Z|$ of the LC circuit, the effect of tuning is to reduce the contribution of E_n, while the contribution of I_n remains practically unchanged.

Obviously, under the above conditions, *tuning the input parallel LC-circuit represents an effective technique to improve the input S/N ratio provided that the contribution of E_n is highly dominant.* When both sources have almost identical contributions the improvement is less spectacular.

CASE STUDY 15.9 [233]

A transducer with internal resistance $R_s = 2\,\text{k}\Omega$ generates an rms signal of 190 nV and is coupled to the preamplifier by means of an RL network with L = 10 mH and $R_p = 3\,\text{k}\Omega$ (Fig. 15.13). At the operating frequency of 100 kHz, the input impedance of the preamplifier is 5 kΩ at an angle of −60°. The preamplifier gain is 120, its output resistance is $R_o = 6\,\text{k}\Omega$, and it contributes noise through the uncorrelated generators $E_n = 20\,\text{nV}/\sqrt{\text{Hz}}$ and $I_n = 6\,\text{pA}/\sqrt{\text{Hz}}$. Subsequent stages offer an equivalent load of $R_L = 12\,\text{k}\Omega$ to the preamplifier.

510   15 Noise in Sensing Circuits

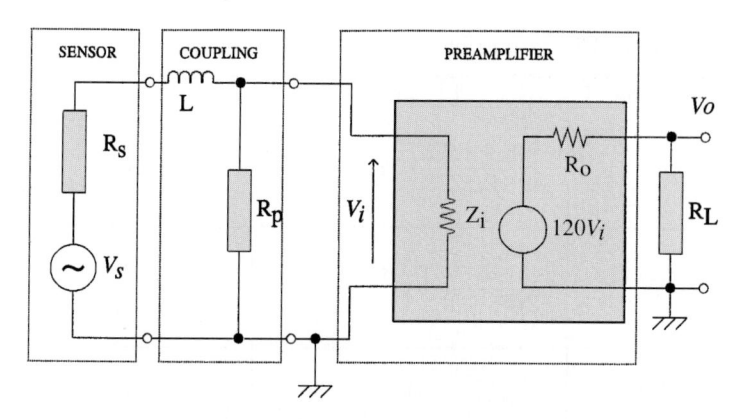

Fig. 15.13. Coupling the transducer with an RL network

1) Assuming T = 300 K, find the output voltage noise spectral density (denoted by $S(E_{no})$).
2) Find the input S/N ratio.

Solution

We apply the method described in Sect. 15.2.1 (see the Final Remark) and the convention of Case Study 15.8.

Step 1: Identify all noise sources and draw the noise equivalent circuit.
The noisy devices (R_s, R_p, R_L, and the preamplifier) shown in Fig. 15.13, must be replaced by their noise models. The resulting noise equivalent circuit is shown in Fig. 15.14. Note that both Z_i and R_o are noiseless, since all preamplifier noise is lumped into E_n and I_n.

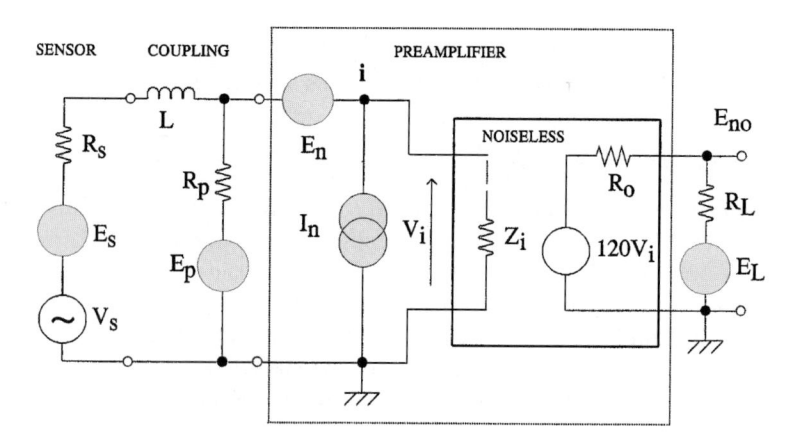

Fig. 15.14. Noise equivalent circuit

15.2 Sensing Circuits   511

Step 3: Determine the voltage gain $K = V_o/V_s$ *referred to the signal source.*
By inspection of the circuit shown in Fig. 15.13, we have

$$V_i = \frac{R_p \| Z_i}{R_s + j\omega L + R_p \| Z_i} V_s$$

(where $R_p \| Z_i$ denotes the parallel combination of R_p and Z_i). But

$$V_o = 120 \, V_i \frac{R_L}{R_o + R_L} = 120 \frac{R_L}{R_o + R_L} \frac{R_p \| Z_i}{R_s + j\omega L + R_p \| Z_i} V_s$$

so

$$K = 120 \frac{R_L}{R_o + R_L} \frac{R_p \| Z_i}{R_s + j\omega L + R_p \| Z_i}.$$

With the indicated numerical values,

$Z_i = 5(\cos 60° + j \sin 60°) = (2.5 - j4.33) \text{ k}\Omega$
$R_p \| Z_i = (1.99 - j0.795) \text{ k}\Omega$

Substituting into the expression for K, we obtain

$$\boxed{|K| \cong 25.27}$$

Step 2: Find the individual contribution of each noise source.
Since it is assumed that no correlation exists between E_n and I_n, the individual contributions to the output noise spectral density can be added. Considering only one noise source to be activated at a time (the remaining ones being set to zero), the contribution to V_i is evaluated in an elementary unit bandwidth $\Delta f = 1 \text{ Hz}$. This contribution is then reflected to the output. Table 15.3 presents all intermediate results.
To illustrate the procedure, the contribution of E_p will be detailed:

$$E_{np} = E_p \frac{(R_s + j\omega L) \| Z_i}{R_p + (R_s + j\omega L) \| Z_i} \left(100 \frac{R_L}{R_o + R_L} \right)$$

$$\overline{E_{np}^2} = \left(120 \frac{R_L}{R_o + R_L} \right)^2 \left| \frac{(R_s + j\omega L) \| Z_i}{R_p + (R_s + j\omega L) \| Z_i} \right|^2 \overline{E_p^2}$$

$$\overline{E_{np}^2} = \left(120 \frac{R_L}{R_o + R_L} \right)^2 \left| \frac{(R_s + j\omega L) \| Z_i}{R_p + (R_s + j\omega L) \| Z_i} \right|^2 (4\text{kTR}_p)$$

$$\overline{E_{np}^2} \cong (80)^2 \frac{45.15}{93.64} \left(4(1.38 \cdot 10^{-23})(300)(3 \cdot 10^3) \right)$$

$$\overline{E_{np}^2} \cong 15.32 \cdot 10^{-14} \text{ V}^2/\text{Hz}$$

The numerical values in Table 15.3 show that E_n is the dominant noise source. Neglecting the contribution of R_L, we obtain

512   15 Noise in Sensing Circuits

Table 15.3. Individual noise contributions of various sources

	Contribution to V_i	Contribution to E_{no}	Contribution to $\overline{E_{no}^2}[V^2/Hz]$
E_s	$\dfrac{R_p \| Z_i}{R_s + j\omega L + R_p \| Z_i} E_s$	KE_s	$2.11 \cdot 10^{-14}$
E_p	$\dfrac{(R_s + j\omega L) \| Z_i}{R_p + (R_s + j\omega L) \| Z_i} E_p$	$120 \dfrac{R_o}{R_o + R_L} \dfrac{(R_s + j\omega L) \| Z_i}{R_p + (R_s + j\omega L) \| Z_i} E_p$	$15.32 \cdot 10^{-14}$
E_n	$\dfrac{Z_i}{Z_i + R_p \| (R_s + j\omega L)} E_n$	$120 \dfrac{R_L}{R_o + R_L} \dfrac{Z_i}{Z_i + R_p \| (R_s + j\omega L)} E_n$	$182.9 \cdot 10^{-14}$
I_n	$\left(Z_i \| R_p \| (R_s + j\omega L) \right) I_n$	$120 \dfrac{R_L}{R_o + R_L} \left(Z_i \| R_p \| (R_s + j\omega L) \right) I_n$	$99.92 \cdot 10^{-14}$
E_L	none	$\dfrac{R_o}{R_o + R_L} E_L$	$22.08 \cdot 10^{-18}$

$$S(E_{no}) = \overline{E_{no}^2}/\Delta f \cong (2.11 + 15.32 + 183 + 99.92)10^{-14} \cong 3 \cdot 10^{-12} \text{ V}^2/Hz$$

from which the output noise voltage for $\Delta f = 1$ Hz is

$$E_{no} \cong 1.73 \text{ μV}$$

Step 4: Reflect the output noise back to the input.
The equivalent input noise voltage is

$$E_{ni} = \frac{E_{no}}{K} = \frac{1.73}{25.27} \cong 68.5 \text{ nV}$$

Step 5: Calculate the input S/N ratio.
The input S/N ratio for $\Delta f = 1$ Hz, is

$$\frac{S_i}{N_i} = 20 \log \frac{V_s}{E_{ni}} = 20 \log \frac{190}{68.5} \cong 8.8 \text{ dB}$$

CASE STUDY 15.10 [235]

A sensor generates a signal of 0.3 μV at f = 300 Hz, and its internal resistance is $R_s = 3$ kΩ. The amplifier designed to amplify the sensor signal has a bandwidth of 100 Hz centered at the operating frequency of 300 Hz, and

15.2 Sensing Circuits

its noise contribution is lumped into two equivalent noise generators, $E_n = 20\,\text{nV}/\sqrt{\text{Hz}}$ and $I_n = 2\,\text{pA}/\sqrt{\text{Hz}}$, which are assumed to be constant over the operating frequency range. In order to achieve input noise matching, a (lossless) coupling transformer is used. This system operates at the ambient temperature of 17°C.

1) Determine the required turns ratio.
2) Find the noise factor of the system, with and without the coupling transformer.
3) Considering the system with the coupling transformer inserted, assume that the sensor is plunged into liquid nitrogen (at T = 77 K). Calculate the improvement in input S/N ratio if the sensor resistance decreases by 20%.

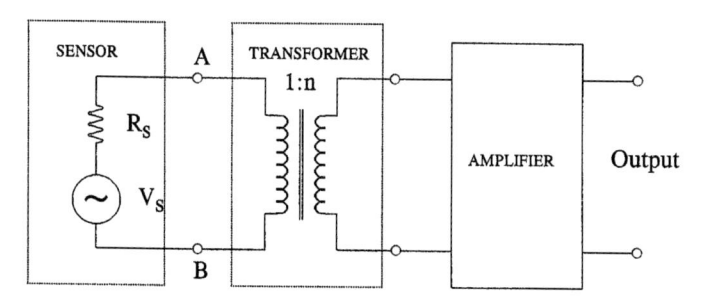

Fig. 15.15. Coupling the transducer with a transformer

Solution

The system is presented in Fig. 15.15.

1) Equation (13.5b) provides the optimal amplifier resistance:

$$R_o = E_n / I_n = (2 \cdot 10^{-8})/(2 \cdot 10^{-12}) = 10\,\text{k}\Omega$$

Hence, the transformer turns ratio for noise matching is

$$n = \sqrt{R_o/R_s} = \sqrt{10^4/(3 \cdot 10^3)} \cong 1.825$$

2) When the transformer is inserted, the noise matching condition is fulfilled. The amplifier exhibits its optimal noise factor, calculated with expression (13.5):

$$F_o = 1 + \frac{E_n I_n}{2kT} = 1 + \frac{2 \cdot 10^{-8}\, 2 \cdot 10^{-12}}{2(1.38 \cdot 10^{-23})290} \cong 6 \quad \text{or} \quad 7.78\,\text{dB}$$

Without the transformer, the noise figure of the amplifier is given by (13.4b):

$$F_{dB} = 10 \log \left(1 + \frac{\overline{E_n^2} + \overline{I_n^2} R_s^2}{4kTR_s}\right)$$

$$= 10 \log \left(1 + \frac{2 \cdot 10^{-16} + 2 \cdot 10^{-24} \, 9 \cdot 10^6}{4(1.38 \cdot 10^{-23})290(3000)}\right)$$

$$\cong 10 \text{ dB}$$

We deduce that by employing the coupling transformer, the noise figure diminishes by about 30% (provided that pickup of spurious signals is avoided by properly shielding and filtering).

3) We must calculate the S/N ratio at terminals A and B when the sensor is at room temperature and then when it is cooled. Remember that with the coupling transformer, the noise equivalent generators E_n, I_n are reflected to the primary as E_n/n and nI_n. The spectral density of the equivalent input noise voltage is

$$S(E_{ni}) = 4kTR_s + \overline{E_n^2}/n^2 + n^2 \overline{I_n^2} R_s^2$$

and the normalized noise power in a bandwidth $\Delta f = 100 \text{ Hz}$ at room temperature is

$$\overline{E_{ni}^2} = \left(4kTR_s + \overline{E_n^2}/n^2 + n^2 \overline{I_n^2} R_s^2\right) \Delta f$$

$$= \left(16 \cdot 10^{-21}(3000) + 4 \cdot 10^{-16}/3.33 + 3.33(4 \cdot 10^{-24})9 \cdot 10^6\right) 100$$

$$\cong 2.88 \cdot 10^{-14} \text{ V}^2$$

At liquid nitrogen temperature ($T_1 = 77 \text{ K}$), we obtain

$$\overline{E_{ni1}^2} = \left(4(1.38 \cdot 10^{-23})77(2400) + 1.2 \cdot 10^{-16} + 1.2 \cdot 10^{-16}\right) 100$$

$$\cong 2.502 \cdot 10^{-14} \text{ V}^2$$

Then the S/N ratios are

$$\frac{S_i}{N_i} = 10 \log \frac{V_s^2}{\overline{E_{ni}^2}} = 10 \log \frac{(0.3 \cdot 10^{-6})^2}{2.88 \cdot 10^{-14}} \cong 4.95 \text{ dB} \quad (T = 290 \text{ K})$$

$$\left(\frac{S_i}{N_i}\right)_1 = 10 \log \frac{V_s^2}{\overline{E_{ni1}^2}} = 10 \log \frac{(0.3 \cdot 10^{-6})^2}{2.502 \cdot 10^{-14}} \cong 5.55 \text{ dB} \quad (T = 77 \text{ K})$$

Ultimately, the improvement is very modest, since it is only $(5.55 - 4.95) = 0.6$ dB. This simply shows that in this case sensor noise is not dominant (in fact, it is the amplifier which establishes the noise floor), and the extra cost of a cooling system for the sensor is not warranted.

In contrast, cooling the amplifier might actually be a viable solution.

15.3 Circuits with Operational Amplifiers

CASE STUDY 15.11 [233]

The operational amplifier of Fig. 15.16 is assumed to be ideal, except for its noise, which is modeled by two uncorrelated noise generators whose values at the operating frequency of 30 Hz are $E_n = 30\,\text{nV}/\sqrt{\text{Hz}}$ and $I_n = 1.2\,\text{pA}/\sqrt{\text{Hz}}$. The circuit is employed to amplify a signal $V_s = 0.15\,\mu\text{V}$ (rms) output by a transducer operating at $T = 300\,\text{K}$. Determine the input S/N ratio, assuming a 1-Hz bandwidth.

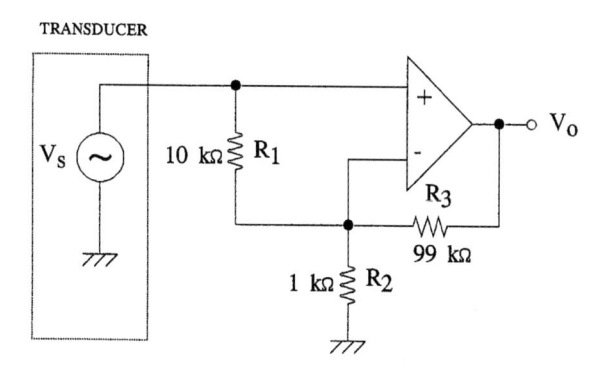

Fig. 15.16. Sensor and its operational amplifier

Solution

We apply the noise analysis method described in Sect. 15.2.1.

Step 1: Identify all noise sources and draw the noise equivalent circuit.
Noise arises in the operational amplifier (E_n and I_n), as well as in the three resistors (thermal noise). The noise equivalent circuit is detailed in Fig. 15.17.

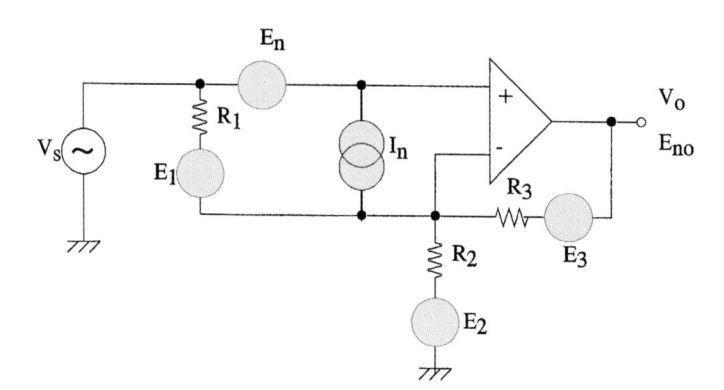

Fig. 15.17. Noise equivalent circuit

15 Noise in Sensing Circuits

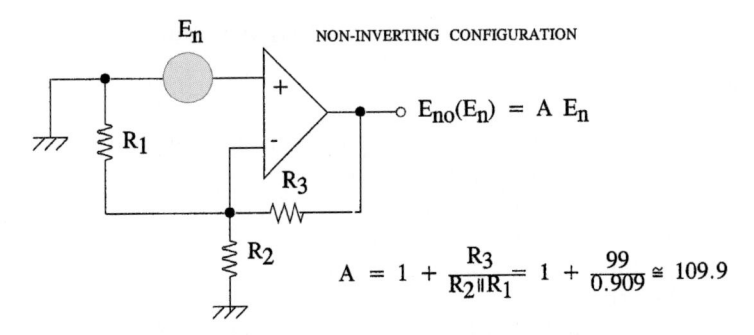

Fig. 15.18. Finding the transfer function A

NON-INVERTING CONFIGURATION

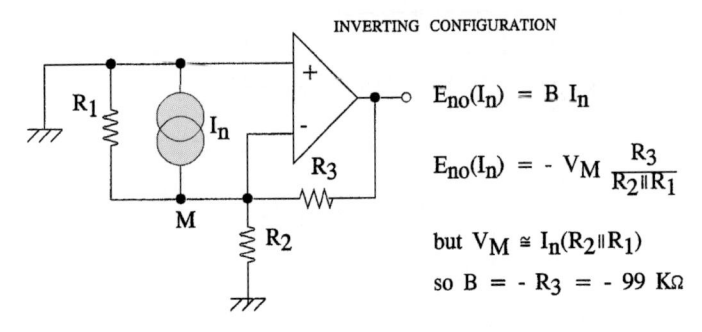

$$E_{no}(E_n) = A\, E_n$$

$$A = 1 + \frac{R_3}{R_2 \| R_1} = 1 + \frac{99}{0.909} \approx 109.9$$

Fig. 15.19. Finding the transfer function B

INVERTING CONFIGURATION

$$E_{no}(I_n) = B\, I_n$$

$$E_{no}(I_n) = -V_M \frac{R_3}{R_2 \| R_1}$$

but $V_M \cong I_n(R_2 \| R_1)$

so $B = -R_3 = -99\ \text{K}\Omega$

Step 2: Find the individual contribution of each noise source.
Letting $E_{no}(E_1)$ be the contribution to the output noise voltage of the source E_1, etc., the individual transfer functions are given in Figs. 15.18 – 15.22.

The spectral density of the output noise voltage is obtained by adding the individual spectral densities, i.e.,

$$S(E_{no}) = \left(|A|^2 E_n^2 + |B|^2 I_n^2 + |C|^2 4kTR_1 + |D|^2 4kTR_3 + |F|^2 4kTR_2\right)$$

The individual contributions in Table 15.4 have been calculated using the indicated numerical values. Clearly, E_n is the dominant source, to which R_2 adds a small correction. Note that the contributions of I_n, R_1, and R_3 are negligible.

Table 15.4. Individual contributions of noise sources

Source	E_n	I_n	E_1	E_2	E_3	Total
Contribution $[10^{-14}\ V^2/Hz]$	1087	1.41	1.623	16.23	0.164	1106.4

15.3 Circuits with Operational Amplifiers   517

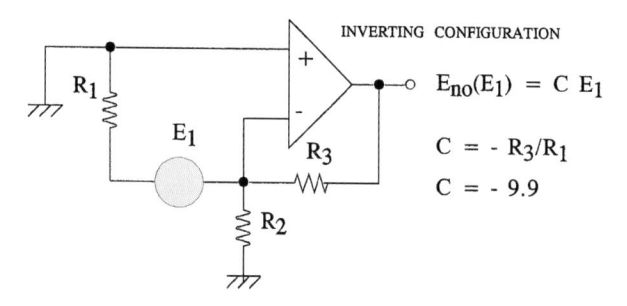

Fig. 15.20. Finding the transfer function C

INVERTING CONFIGURATION

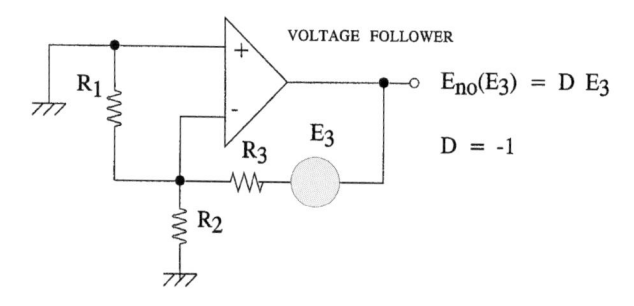

$E_{no}(E_1) = C\, E_1$

$C = -R_3/R_1$

$C = -9.9$

Fig. 15.21. Finding the transfer function D

VOLTAGE FOLLOWER

$E_{no}(E_3) = D\, E_3$

$D = -1$

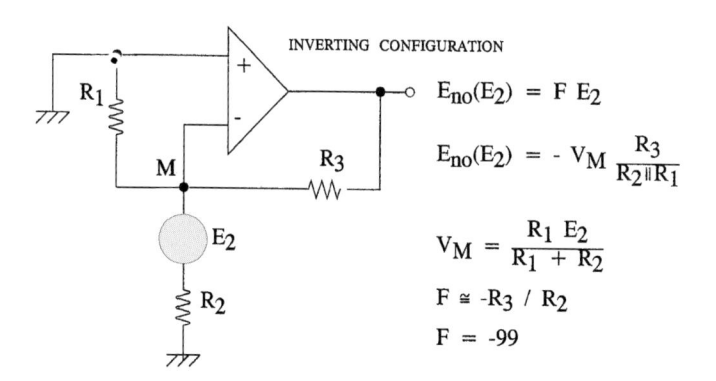

Fig. 15.22. Finding the transfer function F

INVERTING CONFIGURATION

$E_{no}(E_2) = F\, E_2$

$E_{no}(E_2) = -V_M \dfrac{R_3}{R_2 \| R_1}$

$V_M = \dfrac{R_1\, E_2}{R_1 + R_2}$

$F \cong -R_3 / R_2$

$F = -99$

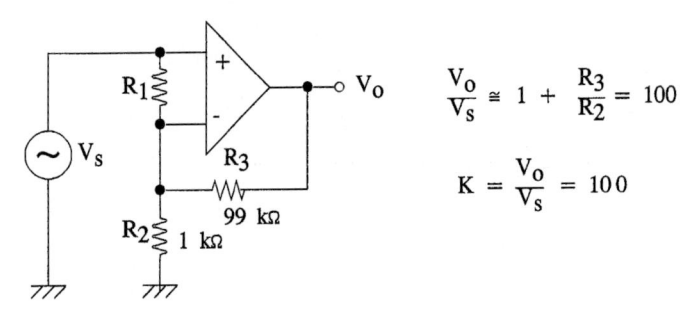

Fig. 15.23. Estimating the voltage gain $K = V_o/V_s$

Step 3: Determine the voltage gain K referred to the signal source V_s.
To calculate the voltage gain referred to the signal source, consider the circuit of Fig. 15.23. The operational amplifier is connected in a non-inverting configuration.

Step 4: Reflect the output noise back to input.
The input spectral density is related to the output spectral density by the expression $S(E_{ni}) = S(E_{no})|K|^2$. In a 1-Hz bandwidth, the equivalent input noise voltage is

$$S(E_{ni}) = (1106.4 \cdot 10^{-14} \cdot 1)/10^4 = 0.11064 \cdot 10^{-14} \text{ V}^2$$

It follows that

$$E_{ni} = \sqrt{0.11064 \cdot 10^{-14}} \cong 0.33 \cdot 10^{-7} \text{ V} = 0.033 \text{ μV}$$

Step 5: Calculate the input S/N ratio.
The input S/N ratio is estimated based on its definition:

$$\frac{S_i}{N_i} = 20 \log \frac{V_s}{E_{ni}} = 20 \log \frac{0.15 \cdot 10^{-6}}{0.033 \cdot 10^{-6}} \cong 13.08 \text{ dB}$$

CASE STUDY 15.12 [233]

Consider the circuit in Fig. 15.24 operating at $f = 1$ kHz, where the operational amplifier is considered ideal except for noise. According to the manufacturer's data, the noise can be simulated by three generators: $E_n = 10 \text{ nV}/\sqrt{\text{Hz}}$ connected between ground and the non-inverting input, and two identical current generators $I_n = 2 \text{ pA}/\sqrt{\text{Hz}}$ (connected between ground and each input). These noise generators are considered to be constant with frequency and uncorrelated. Suppose that a true rms voltmeter is connected to the output and

15.3 Circuits with Operational Amplifiers

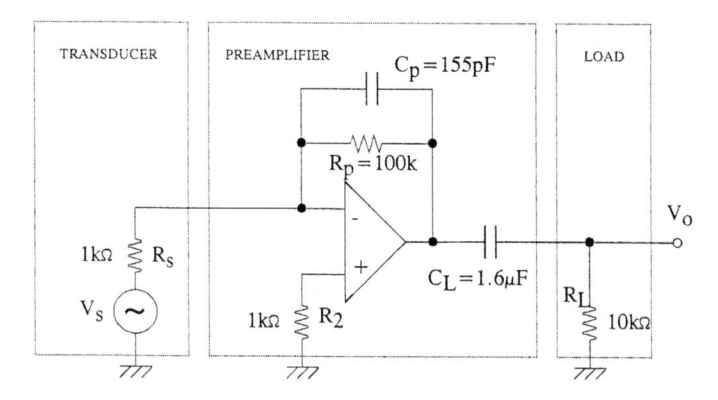

Fig. 15.24. Using an operational amplifier as a preamplifier

that its bandwidth is four times the noise bandwidth of the circuit. What would the total output noise indicated by the voltmeter be, at T = 290 K?

Solution

We follow the first two steps of the method described in Sect. 15.2.1.

Step 1: Identify all noise sources and draw the noise equivalent circuit.
All resistors contribute thermal noise and the operational amplifier contributes its internal noise. The equivalent noise circuit is rather cumbersome, since it contains seven uncorrelated noise sources. However, in order to simplify, we separately calculate the output noise due only to thermal noise, and then that due only to the operational amplifier. Note that at the operating frequency of 1 kHz, the reactance of C_L is close to 100 Ω, and is consequently negligible relative to R_L. On the other hand, the reactance of C_p is 2.56 MΩ, which is much greater than the value of R_p. We conclude that C_L must be considered as a short-circuit at f = 1 kHz while C_p is regarded as an open-circuit. A noise equivalent circuit that takes only the thermal noise into account (the operational amplifier is considered noiseless) is shown in Fig. 15.25.

Step 2: Find the individual contributions to the output noise voltage.
Since all thermal noise sources are uncorrelated, their individual mean square values add. We can write

$$\overline{V_{no}^2} = |A|^2\, \overline{E_S^2} + |B|^2\, \overline{E_2^2} + |C|^2\, \overline{E_p^2} + |D|^2\, \overline{E_L^2}$$

where A, B, C, D are the individual transfer functions defined between the output terminal and the noise source in question. Appropriate expressions are given in Table 15.5.

Concerning the transfer function D, it should be borne in mind that the operational amplifier is assumed to be ideal. Its output resistance is therefore zero, and the contribution of E_L to V_{no} cancels.

520   15 Noise in Sensing Circuits

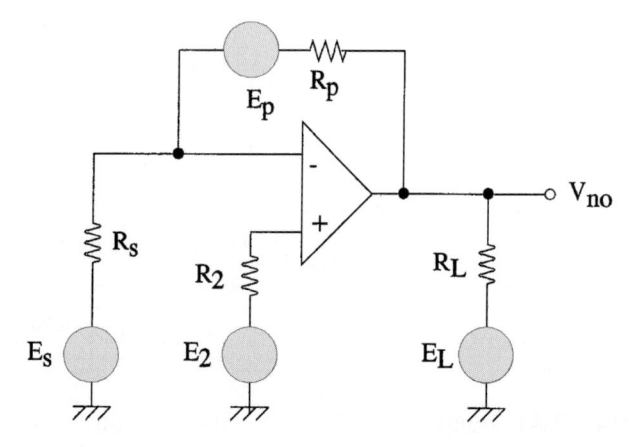

Fig. 15.25. Noise equivalent circuit (thermal noise only)

Table 15.5. Evaluation of transfer functions

Condition	Configuration	Transfer function	Expression
$E_p = E_2 = E_L = 0$	inverting	$A = V_{no}/E_s$	$-R_p/R_s$
$E_s = E_p = E_L = 0$	non-inverting	$B = V_{no}/E_2$	$1 + (R_p/R_s)$
$E_s = E_2 = E_L = 0$	follower	$C = V_{no}/E_p$	1
$E_s = E_2 = E_p = 0$	–	$D = V_{no}/E_L$	0

The output noise spectral density is

$$S(V_{no}) = 4kT\left(R_s(R_p^2/R_s^2) + (1+R_p/R_s)^2 R_2 + R_p\right)$$
$$= 16 \cdot 10^{-21}\left((1)(100)^2 + (1+100)^2(1) + 100\right)10^3$$
$$\cong 32.16 \cdot 10^{-14} \text{ V}^2/\text{Hz}$$

Hence, only the thermal noise is responsible for producing a noise voltage of about $0.567\,\mu\text{V}/\sqrt{\text{Hz}}$ at the output terminals.

Next, we consider only the output contribution of the operational amplifier. All resistors will be assumed to be noiseless; the noise equivalent circuit corresponding to this situation is shown in Fig. 15.26.

As before, the output noise voltage can be written

$$\overline{E_{no}^2} = |F|^2\,\overline{E_n^2} + |G|^2\,\overline{I_{n1}^2} + |H|^2\,\overline{I_{n2}^2}$$

The expressions for various transfer functions are given in Table 15.6. Upon substituting numerical values, we obtain

$$F = 101, \quad G = -10^5\,\Omega, \quad H = 1.01 \cdot 10^5\,\Omega$$

15.3 Circuits with Operational Amplifiers

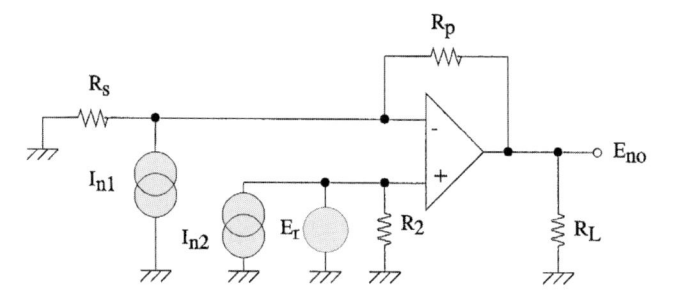

Fig. 15.26. Noise equivalent circuit (amplifier noise only)

Table 15.6. Evaluation of transfer functions

Condition	Configuration	Transfer function	Expression
$I_{n1} = I_{n2} = 0$	non-inverting	$F = E_{no}/E_n$	$1 + (R_p/R_s)$
$E_n = I_{n2} = 0$	inverting	$G = E_{no}/I_{n1}$	$-(R_p/R_s)R_s$
$E_n = I_{n1} = 0$	non-inverting	$H = E_{no}/I_{n2}$	$[1 + (R_p/R_s)]R_2$

Table 15.7. Output contributions of all noise sources

Source	E_n	I_{n1}	I_{n2}	R_s	R_2	R_p	R_L
Contribution 10^{-14} V^2/Hz	102	4	4.08	16	16.32	$16 \cdot 10^{-2}$	0

Table 15.7 collects all individual contributions of all noise sources; clearly, in this example the dominant sources are E_n, R_2, and R_s.

Finally
$$\overline{E_{no}^2} = 142.56 \cdot 10^{-14} \text{ V}^2/\text{Hz}$$

It follows that $E_{no} \cong 1.19\,\mu\text{V}/\sqrt{\text{Hz}}$. In order to find the indication of the true rms voltmeter connected to the output, this value must be multiplied by the instrument bandwidth. It is specified that this is four times the noise bandwidth of the circuit. Hence, we must begin by calculating the circuit noise bandwidth.

■ **Noise Bandwidth.** For a sinusoidal signal, the voltage gain (C_p included) is
$$A_v = -\frac{Z}{R_s} \quad \text{with} \quad Z = \frac{R_p}{1 + j\omega C_p R_p}$$

so
$$A_v = -\frac{R_p}{R_s}\frac{1}{1 + j\omega C_p R_p}, \text{ its maximum being } A_{vo} = -\frac{R_p}{R_s}.$$

Applying definition (4.46) and bearing in mind that the power gain is proportional to the squared voltage gain, the noise bandwidth is

$$\Delta f = \frac{1}{|A_{vo}|^2} \int_0^\infty |A_v|^2 df = \int_0^\infty \frac{1}{1+(\omega C_p R_p)^2} df = \frac{1}{4 R_p C_p}$$

Substituting numerical values, we obtain $\Delta f \cong 16.1$ kHz.

■ **Indication of the Noise Voltmeter.** As stated, the voltmeter bandwidth is four times Δf, i.e., 64.4 kHz. The total output noise voltage in a 64.4-kHz bandwidth is

$$V_{rms} = E_{no} \sqrt{4 \Delta f} = 1.19 \sqrt{(6.44 \cdot 10^4)} \cong 302 \text{ μV}$$

CASE STUDY 15.13 [233]

The bridge amplifier circuit shown in Fig. 15.27 is often employed in low-noise sensing circuits. One of its attractive features is the linear variation of output voltage with respect to sensor resistance, denoted by R_2. Assume that both operational amplifiers are ideal in every respect except for noise, which is described by means of two uncorrelated generators: $E_n = 5$ nV/$\sqrt{\text{Hz}}$ and $I_n = 20$ pA/$\sqrt{\text{Hz}}$. Find the output S/N ratio when the sensor resistance (R_2) is modified by 10% relative to its reference value. Assume that the bridge is balanced for the reference value of R_2, the circuit is operating at T = 290 K, and $V_g = 1$ mV (rms value).

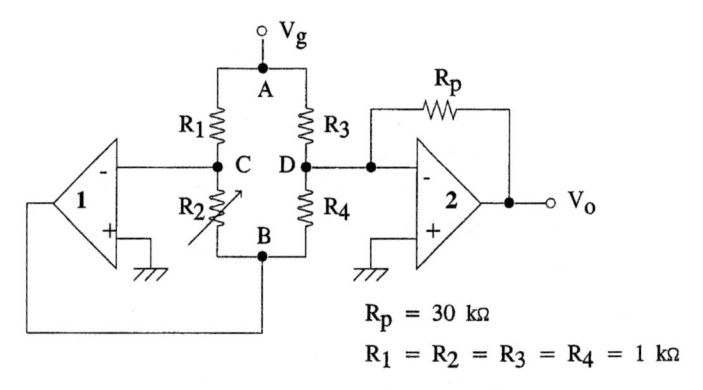

Fig. 15.27. Bridge amplifier

$R_p = 30$ kΩ
$R_1 = R_2 = R_3 = R_4 = 1$ kΩ

15.3 Circuits with Operational Amplifiers

Solution

The goal is to determine the S/N ratio; therefore, we are interested in both signal and noise transmission.

■ **Signal Analysis.** The circuit is rearranged in Fig. 15.28, where it is easier to identify each amplifier configuration.

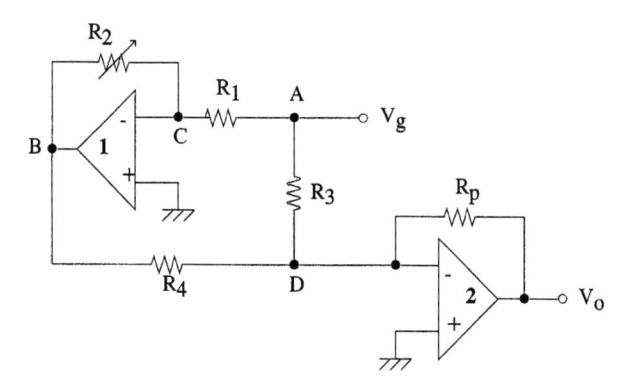

Fig. 15.28. Rearranged circuit of the bridge amplifier

Operational amplifier **1** operates in inverting mode, while **2** is connected as an inverting summing amplifier. Thus

$$V_o = V_A \left(-\frac{R_p}{R_3}\right) + V_B \left(-\frac{R_p}{R_4}\right)$$

But

$$V_B = V_A \left(-\frac{R_2}{R_1}\right) \quad \text{and} \quad V_A = V_g$$

So

$$\boxed{V_o = -V_g R_p \left(\frac{1}{R_3} - \frac{1}{R_4}\frac{R_2}{R_1}\right)}$$

This equation shows that the output voltage is a linear function of R_2 and that the output voltage vanishes under the balance condition ($R_1 = R_2$ and $R_3 = R_4$).

■ **Noise Analysis.** The equivalent noise circuit (Fig. 15.29) is obtained by replacing each element with its noise model. The noise of the operational amplifiers is described according to Fig. 7.37a. Since the noise generators of each amplifier are assumed to be uncorrelated, all noise sources are independent, and their spectral densities add at the output. For noise analysis, if the signal source V_g is ideal, it can be short-circuited.

524   15 Noise in Sensing Circuits

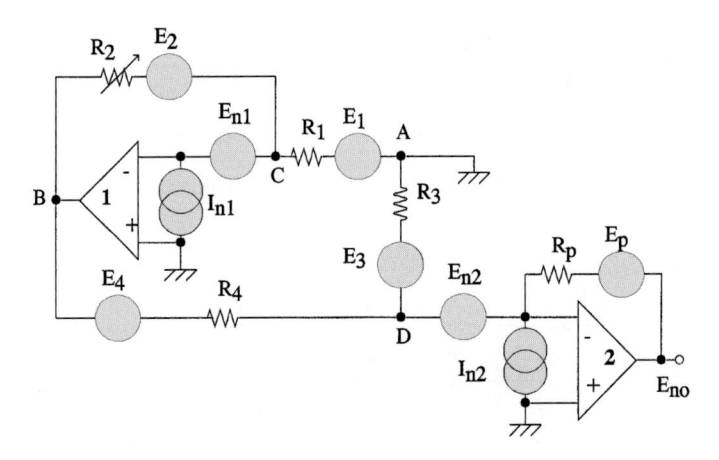

Fig. 15.29. Equivalent noise circuit

The individual contributions of the various noise sources are given in Table 15.8.

■ **S/N Ratio.** We must evaluate the signal variation when the sensor resistance has a drift $\Delta R_2/R_2 = 10\%$. The previously deduced expression for V_o yields by differentiation (and under the condition $R_4 = R_3$)

$$\frac{\partial V_o}{\partial R_2} = V_g \, R_p \, \frac{1}{R_1} \, \frac{1}{R_4}$$

from which

$$\Delta V_o \cong \Delta R_2 \, V_g \, \frac{R_p}{R_1 \, R_4} = 0.1 \, (1) \, \frac{30}{1 \cdot 1} = 3 \text{ mV(rms value)}$$

On the other hand, taking $\Delta f = 1$ Hz, the output noise voltage is $E_{no} = \sqrt{96.108 \cdot 10^{-14}} \cong 0.98 \, \mu V$.

Finally, the output S/N ratio is

$$\frac{S_o}{N_o} = 20 \log \frac{\Delta V_o}{E_{no}} = 20 \log \frac{3 \cdot 10^{-3}}{(0.98 \cdot 10^{-6})} \cong 69.7 \text{ dB}$$

CASE STUDY 15.14 [233]

The operational amplifier in Fig. 15.30 is biased by a single +18 V supply. To set the DC output to 9 V, a Zener diode with reference voltage $V_z = 9$ V and dynamic resistance of 11 Ω is employed. According to the manufacturer's data, this diode generates avalanche noise described by

$$\overline{E_z^2} = (5 \cdot 10^{-20} \, V_z^4)/I_z \quad [V^2/Hz]$$

15.3 Circuits with Operational Amplifiers   525

Table 15.8. Contributions of noise sources in a 1-Hz bandwith

Gen.	Contribution at B	Output contribution	Value V^2/Hz
E_1	$(4kTR_1)(R_2/R_1)^2$	$(4kTR_1)(R_2/R_1)^2(R_p/R_4)^2$	$1.44 \cdot 10^{-14}$
E_2	$4kTR_2$	$(4kTR_2)(R_p/R_4)^2$	$1.44 \cdot 10^{-14}$
E_3	–	$(4kTR_3)(R_p/R_3)^2$	$1.44 \cdot 10^{-14}$
E_4	–	$(4kTR_4)(R_p/R_4)^2$	$1.44 \cdot 10^{-14}$
E_p	–	$4kTR_p$	$4.8 \cdot 10^{-16}$
I_{n1}	$(R_1 I_{n1})^2 (R_2/R_1)^2$	$(R_1 I_{n1})^2 (R_2/R_1)^2 (R_p/R_4)^2$	$3.6 \cdot 10^{-13}$
I_{n2}	–	$(R_3 I_{n2})^2 (R_p/R_3)^2$	$3.6 \cdot 10^{-13}$
E_{n1}	$E_{n1}^2 (1 + R_2/R_1)^2$	$E_{n1}^2 (1 + R_2/R_1)^2 (R_p/R_4)^2$	$9 \cdot 10^{-14}$
E_{n2}		$E_{n2}^2 \left(1 + \dfrac{R_2}{R_3 \| R_4}\right)^2$	$9.3 \cdot 10^{-14}$
Total		$\overline{E_{no}^2} = 96.108 \cdot 10^{-14}$	

The operational amplifier is considered ideal, except for the noise, which is modeled by two uncorrelated, flat spectral density generators $E_n = 38 \,\text{nV}/\sqrt{\text{Hz}}$ and $I_n = 0.4 \,\text{pA}/\sqrt{\text{Hz}}$.

1) Find the total rms noise value that would be measured at the output by a true rms voltmeter whose bandwidth is 10 kHz at T = 290 K.
2) Identify the dominant noise source(s) and suggest a solution to reduce the output noise.

Solution

■ **Circuit Analysis.** At midband frequencies, the coupling capacitor C acts as a short circuit, and the reactance of C_p is much greater than R_p; conse-

526   15 Noise in Sensing Circuits

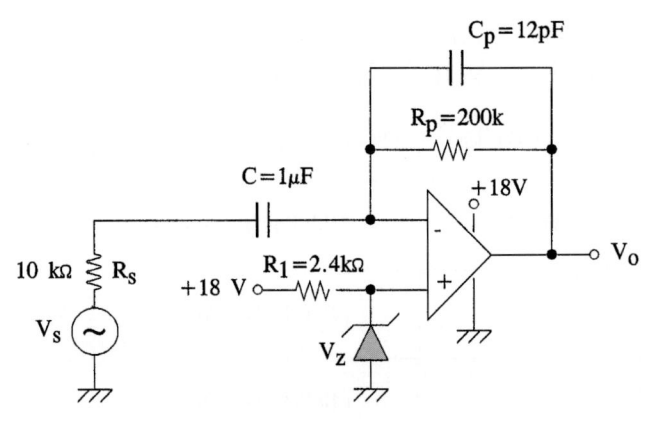

Fig. 15.30. Precision clamping amplifier

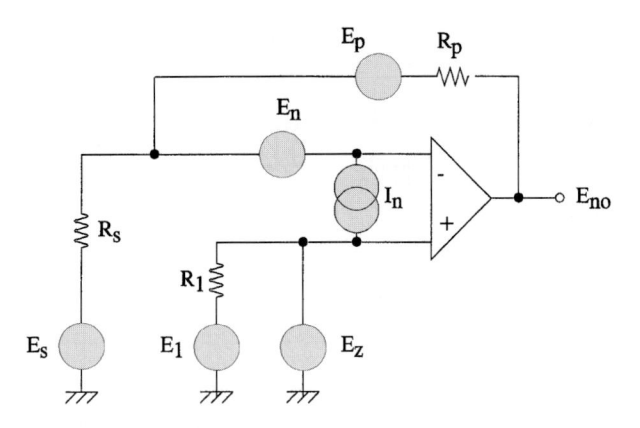

Fig. 15.31. Noise equivalent circuit

quently, C_p can be considered an open circuit. The noise equivalent circuit is detailed in Fig. 15.31.

Since all noise sources are independent, their individual contributions to the output terminals can be calculated separately, and the resulting spectral densities can be added. In calculating the contribution of each noise source, all other noise sources are set to zero. The resulting contributions are given in Table 15.9, sorted in descending order.

The quiescent current through the Zener diode is

$$I_z = \frac{18 - V_z}{R_1} = 3.75 \text{ mA}$$

Applying the proposed expression, the avalanche noise spectral density is $S(E_z) \cong 8.75 \cdot 10^{-14} \text{ V}^2/\text{Hz}$.

Table 15.9. Contributions to the output spectral density

Noise source	Contribution to $S(E_{no})$ ($\Delta f = 1$ Hz)	Value [V^2/Hz]
E_z	$(1 + R_p/R_s)^2 \, E_z^2$	$3858 \cdot 10^{-14}$
E_n	$[-(1 + R_p/R_s)]^2 \, E_n^2$	$63.68 \cdot 10^{-14}$
E_s	$(-R_p/R_s)^2 (4kTR_s)$	$6.40 \cdot 10^{-14}$
E_1	$(1 + R_p/R_s)^2 (4kTR_1)$	$1.69 \cdot 10^{-14}$
I_n	$R_p^2 \left(1 + R_1/(R_s \| R_p)\right)^2 (I_n)^2$	$1.00 \cdot 10^{-14}$
E_p	$4kTR_p$	$0.32 \cdot 10^{-14}$
	Total	$3934.8 \cdot 10^{-14}$

■ **Results**

1) The mean square value of the output noise voltage over a bandwidth of $\Delta f = 10$ kHz is $S(E_{no})\Delta f$. A true rms output voltmeter will indicate

$$E_{no} = \sqrt{S(E_{no}) \, \Delta f} \cong 627.2 \ \mu V$$

2) The results of Table 15.9 show that the Zener diode is the dominant noise source. Clearly, suppressing the Zener diode will improve the value of E_{no}. However, in order to preserve its function (providing a reference voltage of 9 V), we may replace it with a battery. A crude estimate shows that the output noise level is then reduced by a factor of 7. To conclude, *Zener diodes must be avoided in low-noise circuits.*

CASE STUDY 15.15 [233]

The noise figure of an operational amplifier is 6 dB with a source resistance $R_s = 50 \, \Omega$. Assume $T = 290$ K.

1) Find the spectral density of the equivalent input noise voltage.
2) Determine the noise temperature of the amplifier.
3) Find the equivalent noise resistance of the amplifier.
4) Determine the input S/N ratio when a sensor delivering a signal $V_s = 10$ nV (rms value) is connected to the input.

Solution

1) Assuming that all noise is referred to the input, and using the Norton definition, the noise figure is

$$F = 10 \log \frac{\overline{E_{ni}^2}}{\overline{E_{ns}^2}}$$

where E_{ni} denotes the equivalent input noise voltage (including the thermal noise of the source resistance) and E_{ns} is the source noise voltage. Since $T = T_o = 290\,K$, we may write

$$\overline{E_{ni}^2} = \overline{E_{ns}^2}\, 10^{F/10} = 4kT_o\, R_s\, \Delta f\, 10^{F/10}$$

Letting $\Delta f = 1\,Hz$

$$S(E_{ni}) = 4(1.38 \cdot 10^{-23})(290)(50)(10^{0.6})$$

$$\cong 3.184 \cdot 10^{-18}\, V^2/\sqrt{Hz}$$

2) The noise temperature can be calculated with (4.77) after finding the noise factor $F = 3.981$ (which corresponds to a noise figure of 6 dB):

$$T_e = T_o(F - 1) = 290(3.981 - 1) \cong 864.49\,K$$

3) According to Sect. 4.5.1, the equivalent noise resistance is the value of a resistor which, when applied to the input of a hypothetical noiseless (but otherwise identical) amplifier, would produce the same output noise. Consequently, the noise referred to the input must be the same as at point 1, per unit bandwidth; thus

$$4KT_o\, R_n = 4KT_o\, R_s\, 10^{F/10}$$

and

$$R_n = R_s 10^{F/10} = 50 \cdot 10^{F/10} \cong 199.05\,\Omega$$

4) The rms value of the input noise voltage is

$$E_{ni} = \sqrt{\overline{E_{ni}^2}} = \sqrt{3.18 \cdot 10^{-18}} \cong 1.78\,nV$$

The input S/N ratio is therefore

$$S_i/N_i = 20\,\log(10/1.78) \cong 15\,dB$$

CASE STUDY 15.16

Consider the circuit in Fig. 15.32, where the amplifier has a voltage gain K and its input resistance is R_2.

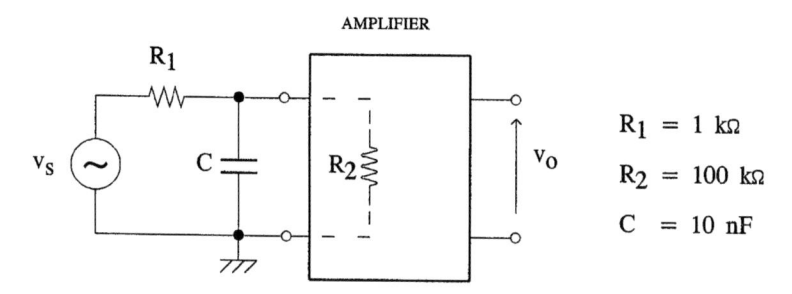

Fig. 15.32. Amplifying circuit

$R_1 = 1 \text{ k}\Omega$
$R_2 = 100 \text{ k}\Omega$
$C = 10 \text{ nF}$

1) Prove that the overall circuit has a noise bandwidth independent of K.
2) Find the noise bandwidth.
3) Estimate the relative error when using the 3-dB bandwidth of the circuit instead of its noise bandwidth, and prove that this error does not depend on any element of the circuit.

Solution

1) The noise equivalent bandwidth is determined with expression (4.41):

$$\Delta f = \frac{1}{|A_{vo}|^2} \int_0^{+\infty} |A_v(f)|^2 \, df$$

where A_{vo} represents the maximum voltage gain referred to the signal source (i.e., the maximum of $A_v = v_o/v_s$). In order to find A_v, we may write

$$v_o = K \, v_s \, \frac{Z}{Z + R_1} \quad \text{with} \quad Z = \frac{R_2(1/j\omega C)}{R_2 + (1/j\omega C)}$$

Consequently

$$A_v = \frac{v_o}{v_s} = \frac{K \, R_2}{R_1 + R_2 + j\omega C R_1 R_2}$$

and

$$|A_v|^2 = \frac{(KR_2)^2}{(R_1 + R_2)^2 + (\omega C R_1 R_2)^2}$$

Clearly the maximum of A_v occurs at $\omega = 0$; hence

$$|A_{vo}|^2 = \frac{(KR_2)^2}{(R_1 + R_2)^2}$$

This expression can be substituted in the definition of the noise equivalent bandwidth; after some algebra, we obtain

$$\Delta f = (R_1 + R_2)^2 \int_0^\infty \frac{df}{(R_1 + R_2)^2 + (2\pi C R_1 R_2)^2 f^2}$$

$$= (R_1 + R_2)^2 \left(\frac{1}{(R_1 + R_2)(2\pi C R_1 R_2)} \arctan \frac{2\pi C R_1 R_2}{R_1 + R_2} \right)_0^\infty$$

$$= \frac{R_1 + R_2}{4 C R_1 R_2} = \frac{1.01 \cdot 10^5}{4 \cdot 10^{-8} \cdot 10^3 \cdot 10^5} \cong 25.25 \text{ kHz}$$

Note that the final expression for Δf does not depend on K; it depends only on the time constant of the equivalent RC circuit seen between the input terminals of the amplifier.

2) According to its definition, the 3-dB circuit bandwidth is

$$B = f_h - 0 = \frac{R_1 + R_2}{2\pi C R_1 R_2} \cong 16 \text{ kHz}$$

f_h being the high-cutoff frequency. Note that the low-cutoff frequency is zero (DC signals are transmitted through the circuit of Fig. 15.32).

If we adopt B instead of Δf (which, unfortunately, often happens in practice), the relative error in bandwidth estimation is found by replacing the previous expressions:

$$\varepsilon = \frac{\Delta f - B}{\Delta f} = 1 - \frac{2}{\pi} \cong 36.33\%$$

To conclude, the noise calculated with B (instead of Δf) will give a more optimistic value, about 36%! This demonstrates the necessity of determining the noise equivalent bandwidth of the circuit, especially when it has several stages.

CASE STUDY 15.17 [235]

The noise of a TL061 operational amplifier is described by two equivalent noise generators, assumed to be uncorrelated, whose values are given in Table 15.10.

1) Assuming that the noise of the first stage determines the frequency behavior, what type of transistors are employed in the input stage of the TL061?
2) Find the optimal source resistance and the minimum noise factor at 10 kHz.
3) This circuit is intended to amplify the signal derived from a sensor with internal resistance $R_s = 400 \text{ k}\Omega$. The useful signal has an rms value $E_s = 1\,\mu V$, and its frequency lies between 9950 Hz and 10050 Hz. In order to

15.3 Circuits with Operational Amplifiers

Table 15.10. Frequency dependence of the noise generators (TL061)

Frequency [Hz]	E_n [nV/$\sqrt{Hz}$]	I_n [fA/$\sqrt{Hz}$]
1...100	43	1
1000	43	8
10000	43	80

reduce the noise, a sharp passband filter (assumed noiseless) is inserted at the output of the operational amplifier to limit the bandwidth to 100 Hz, centered at 10 kHz. Determine the noise factor and the output S/N ratio of this system.

Solution

1) By inspection of the tabulated noise data, we see that the equivalent input noise current increases with frequency, while E_n is constant. Thus, very likely, the input stage employs junction field-effect transistors.
2) Applying (13.5), the optimal noise resistance is

$$R_{opt} = \frac{E_n}{I_n} = \frac{43 \cdot 10^{-9}}{80 \cdot 10^{-15}} = 537.5 \text{ k}\Omega$$

and the minimum noise factor is

$$F_o = 1 + \frac{E_n I_n}{2kT} = 1 + \frac{(43 \cdot 10^{-9})(80 \cdot 10^{-15})}{2(1.38 \cdot 10^{-23}) \cdot 290} = 1.43 \quad \text{or} \quad 1.55 \text{ dB}$$

3) The source resistance is given as 400 kΩ; the noise factor is calculated by applying relation (13.6):

$$F = 1 + \frac{F_o - 1}{2}\left(\frac{R_s}{R_{opt}} + \frac{R_{opt}}{R_s}\right)$$

$$= 1 + \frac{0.43}{2}\left(\frac{400}{537.5} + \frac{537.5}{400}\right) \cong 1.449$$

Note that the noise factor has increased by roughly 1.2% relative to its minimum value, and we may thus expect excellent performance.

To find the output S/N ratio, we can calculate it at the input from the equivalent input noise voltage. We assume that the amplifier is noiseless, and therefore, the input and output S/N ratios are the same. The equivalent input noise voltage is calculated by means of expression (13.2a):

$$\overline{E_{nt}^2} = \left(4kTR_s + \overline{E_n^2} + \overline{I_n^2}R_s^2\right)\Delta f$$

$$= \left(64 \cdot 10^{-16} + (43 \cdot 10^{-9})^2 + (80 \cdot 10^{-15} \cdot 4 \cdot 10^5)^2\right)10^2$$

$$\cong 0.527 \cdot 10^{-12} \text{ V}^2$$

The S/N ratio is estimated according to its definition:

$$\frac{S_o}{N_o} = \frac{S_i}{N_i} = 10 \log \frac{\overline{E_s^2}}{\overline{E_{nt}^2}} = 10 \log \frac{(1 \cdot 10^{-6})^2}{0.527 \cdot 10^{-12}} \cong 2.78 \text{ dB}$$

■ **Concluding Remark.** This case study shows that in practical situations the S/N ratio can be really poor, even if the noise factor is very close to its optimal value. This explains why for *unmatched systems it is recommended to optimize the S/N ratio, rather than the noise factor, which is no longer a significant quantity.*

CASE STUDY 15.18 [237]

Noise of the input stage of an operational amplifier is of primary concern in determining the global performance of the integrated circuit.

It is well known that the equivalent noise voltage of an operational amplifier decreases when the emitter current of its input stage increases. However, increasing I_E will also increase the noise associated with its base current, which begins to dominate. Therefore, an optimum is reached when

$$I_{Eopt} = \frac{kT}{q} \frac{\sqrt{\beta}}{R_s}$$

In order to illustrate this idea, consider the input differential stage of the bipolar transistor operational amplifier shown in Fig. 15.33.

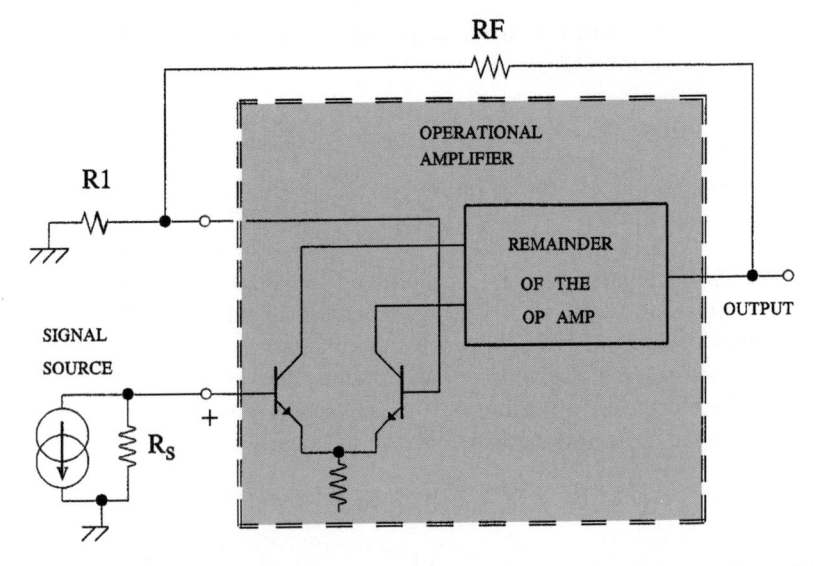

Fig. 15.33. Typical configuration of the input stage of an operational amplifier (Courtesy of National Semiconductor Corp.)

15.3 Circuits with Operational Amplifiers

In this case, with $\beta = 500$ and $R_s = 1\,k\Omega$, I_{Eopt} is roughly $500\,\mu A$, which is quite a high value for the current of the input stage. The problem here is that the resulting high DC base current causes an unacceptable input temperature drift (between 5 and $10\,\mu V/°C$). In order to simultaneously satisfy both noise and drift performance, a properly designed input stage is added before the operational amplifier.

Propose a possible configuration for this stage and comment on the various choices.

Solution

Since we must reconcile the noise and drift performance, it seems reasonable to propose a double input stage, so that each module can fix one of the specified performance. The proposed circuit is detailed in Fig. 15.34.

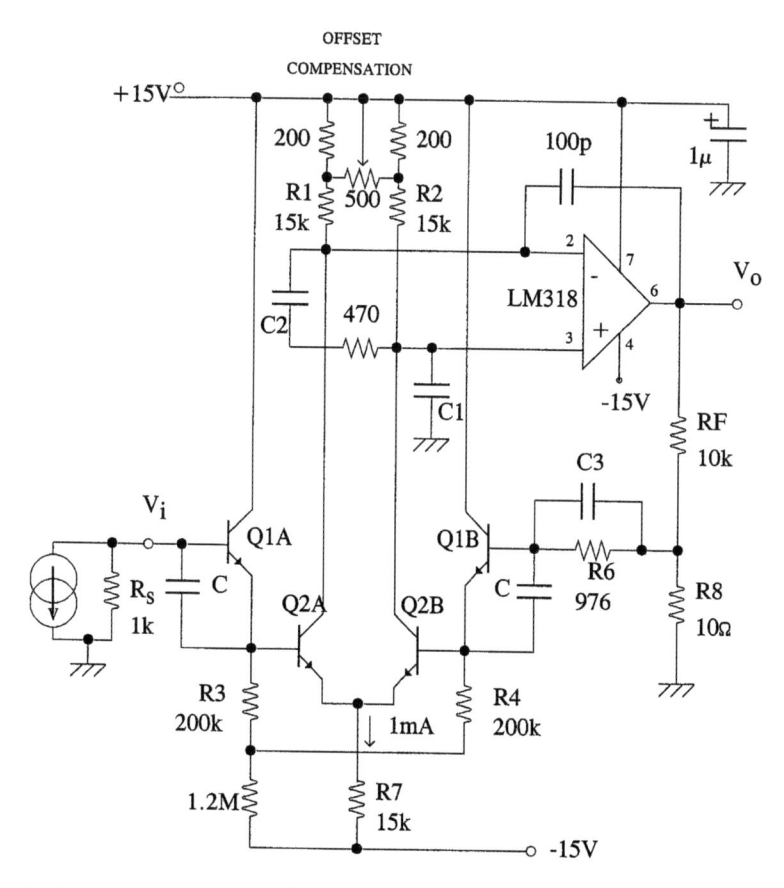

Fig. 15.34. Low-noise, low offset operational amplifier with bipolar transistors (Courtesy of National Semiconductor Corp.)

15 Noise in Sensing Circuits

The additional stage inserted at the operational amplifier input is a differential Darlington pair. The pair of transistors Q1A, Q1B ensures a low offset, while Q2A, Q2B guarantee low noise, as each transistor operates at the optimal emitter current (according to the above expression). Both transistor pairs are of type LM394.

Without special attention, the noise of the first pair can jeopardize overall noise performance (it should be borne in mind that the noise contribution of the first stage of any amplifier critically affects the total output noise). One stratagem to limit the noise level of the first pair is to introduce capacitor C (10 µF) to short-circuit spectral components beyond 10 Hz. According to Sect. 7.2, we must select a low-loss tantalum capacitor, which is almost noiseless. In this way, the equivalent noise voltage of the Darlington stage is reduced to approximately $1.4\,\text{nV}/\sqrt{\text{Hz}}$, which is negligible compared to the thermal noise voltage of $4\,\text{nV}/\sqrt{\text{Hz}}$ generated by the source resistance. The feedback loop includes resistors RF and R1 (which establish the gain of -1000); resistor R6 DC balances the inputs of the differential stage; to avoid an AC contribution, the bypass capacitor C3 = 10 µF is added. Note that this must be also a low-loss (low-noise) tantalum capacitor.

In order to keep the offset voltage as low as possible, R3 and R4 (as well as R1 and R2) are metal-film precision resistors with 0.1% tolerance rating and low temperature coefficient (less than 5 ppm/°C).

We conclude that this amplifier has good noise and offset performance.

16

Noise in Communication Systems

"Noise is so great, that one cannot hear God Thunder"

(Howell)

16.1 Attenuators

16.1.1 Underlying Principles

■ **Definitions.** According to [238]:

– *An attenuator is an adjustable passive network that reduces the power level of a signal without introducing appreciable distortion.*
– *A pad (attenuating pad) is a nonadjustable passive network that reduces the power level of a signal without introducing appreciable distortion.*

■ **Comments**

– Usually pads and attenuators operate under matching conditions, i.e., the input and output impedances of the attenuator (pad) are matched to the impedances loading its terminals.
– Often, the attenuation (or loss) introduced by the attenuator (pad) is denoted by L and expressed in decibels.

16.1.2 Case Studies

CASE STUDY 16.1 [243]

Prove the following theorem: *the noise factor of any reciprocal, passive two-port is numerically equal to the loss of the two-port*, i.e.,

$$\boxed{F = L = \frac{1}{G_a}}$$

where G_a is the available power gain and L the loss. Comment on the conditions under which this equality holds.

Solution

According to (5.13), the input noise temperature of the matched attenuator is

$$T_e = (L-1)T = \left(\frac{1}{G_a} - 1\right)T$$

where T is the attenuator temperature. On the other hand, expression (4.77) establishes an equivalence between the input noise temperature and the noise factor of any two-port:

$$T_e = T_o(F-1)$$

Equating the previous expressions for T_e, we have

$$F = 1 + \frac{T}{T_o}\left(\frac{1}{G_a} - 1\right)$$

At $T = T_o = 290\,\text{K}$, we obtain

$$\boxed{F = \frac{1}{G_a}}$$

Another implicit condition to be satisfied in order to ensure the validity of the theorem is that the *input termination* of the two-port must also have a temperature of 290 K also (see North's definition of the noise factor).

Thus, *the noise factor of any passive, matched two-port is equal to its loss (expressed as a ratio) if and only if the two-port and its input termination are both operating at the reference temperature* ($T_o = 290\,\text{K}$).

CASE STUDY 16.2 [229, 234]

Consider the T-section attenuating pad shown in Fig. 16.1a. Assuming that the pad is driven by a source of negligible internal resistance, determine the noise power delivered at terminals A–B.

Solution 1

The noise equivalent circuit of the attenuator pad is detailed in Fig. 16.1b.

For noise calculation we may assume that the input generator has zero internal resistance, so that the output noise is exclusively due to the attenuator.

Every resistor will be replaced by its noise model, yielding the noise equivalent circuit of Fig. 16.1b. The individual contributions of the various resistors to the output noise are then evaluated (each resistor in turn is considered

16.1 Attenuators

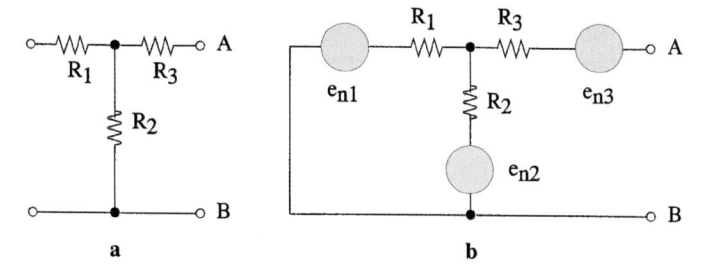

Fig. 16.1. a T section attenuator pad; b noise equivalent circuit (when driven by a source with negligible internal resistance)

noisy, while the remainder are assumed to be noiseless). Since the thermal noise sources are uncorrelated, their output contributions (expressed as mean square values) add.

For example, consider the noise source $e_{n1} = \sqrt{4kTR_1\,\Delta f}$; e_{n2} and e_{n3} are set to zero. Its noise contribution to terminals A–B is denoted by e_1 and can be calculated as follows:

$$e_1 = e_{n1}\frac{R_2}{R_1 + R_2}$$

$$e_1^2 = e_{n1}^2\left(\frac{R_2}{R_1 + R_2}\right)^2$$

$$\overline{e_1^2} = \overline{e_{n1}^2}\left(\frac{R_2}{R_1 + R_2}\right)^2 = 4kTR_1\,\Delta f\left(\frac{R_2}{R_1 + R_2}\right)^2$$

Note that when impedances are considered instead of resistors, the only difference concerns the averaging of the ratio on the right-hand side, which is now a complex quantity. Recall that the mean square value of a complex quantity is calculated by taking the square of its magnitude.

Similarly

$$\overline{e_2^2} = \overline{e_{n2}^2}\left(\frac{R_1}{R_1 + R_2}\right)^2 \quad \text{and} \quad \overline{e_3^2} = \overline{e_{n3}^2}$$

Adding the individual contributions (as mean square values), we obtain

$$\overline{e_n^2} = \overline{e_1^2} + \overline{e_2^2} + \overline{e_3^2} = 4kT\,\Delta f\left(\frac{R_1 R_2}{R_1 + R_2} + R_3\right)$$

which corresponds to Nyquist's formula. Consequently, the total output noise power is the same as that which would be delivered by the equivalent resistance seen at terminals A–B.

Solution 2

When dealing with more complicated configurations, it is advisable to use the voltage transfer functions $H_i(p)$, relating each noise source to the circuit output. These functions are the same for noise and for harmonic signals; therefore, the calculation is straightforward.

In general,
$$e_i = e_{ni} H_i(p) \quad \text{with} \quad i = 1, 2, 3, \ldots$$

so
$$\overline{e_i^2} = \overline{e_{ni}^2} |H_i(p)|^2$$

The output noise becomes
$$\overline{e_n^2} = \sum_{i=1}^{n} \overline{e_{ni}^2} |H_i(p)|^2$$

If the circuit has uniform temperature,
$$\overline{e_n^2} = 4kT\Delta f \sum_{i=1}^{n} R_i |H_i(p)|^2$$

Applying this approach to the circuit in Fig. 16.1b we obtain
$$H_1(p) = \frac{R_2}{R_1 + R_2} \qquad H_2(p) = \frac{R_1}{R_1 + R_2} \qquad H_3(p) = 1$$

which leads to the same result.

CASE STUDY 16.3

Consider the resistive T-section attenuating pad in Fig. 16.2, where $R_1 = R_2 = R_g = 50\,\Omega$. Determine its noise figure at the output terminals.

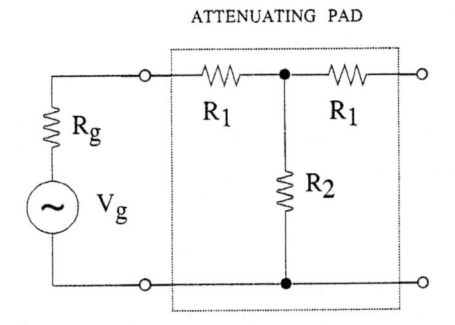

Fig. 16.2. T-section attenuator

Solution

We use the same theorem as Case Study 16.1, namely that the noise factor of a passive two-port is equal to its loss (the loss being defined as the reciprocal of its available power gain). It should be borne in mind that the available

16.1 Attenuators   539

power gain is defined as the ratio of the output available power (P_2) to the input available power (P_1).

According to (4.21b),
$$P_1 = V_g^2 / (4R_g)$$

To find the output available power P_2, the equivalent Thévenin generator and its internal resistance must be calculated at the output:

$$V_e = V_g \frac{R_2}{R_1 + R_2 + R_g} \quad \text{and} \quad R_e = R_1 + \frac{(R_g + R_1)R_2}{R_1 + R_2 + R_g}$$

Then, (4.21b) yields

$$P_2 = V_g^2 \frac{R_2^2}{(R_1 + R_2 + R_g)} \frac{1}{4\big((R_1 + R_2 + R_g)R_1 + R_2(R_1 + R_g)\big)}$$

The available power gain is

$$G_a = \frac{P_2}{P_1} = \frac{R_g R_2^2}{(R_1 + R_2 + R_g)\big((R_1 + R_2 + R_g)R_1 + R_2(R_1 + R_g)\big)}$$

and the noise factor is

$$F = \frac{(R_1 + R_2 + R_g)\big((R_1 + R_2 + R_g)R_1 + R_2(R_1 + R_g)\big)}{R_g R_2^2}$$

Substituting numerical values, the noise factor is

$$F = \frac{3(50)\,[150(50) + 50(100)]}{50^2\,50} = 15$$

and the noise figure becomes

$$F_{dB} = 10 \log F = 10 \log 15 \cong 11.76 \text{ dB}$$

CASE STUDY 16.4

Design a balanced-T matched attenuator for the characteristic impedance $Z_o = 50\,\Omega$, whose attenuation is 3 dB. Find the noise factor of the attenuator.

Solution

The circuit to be designed is presented in Fig. 16.3.

According to [239], the design equations can either be written

$$R_2 = \frac{Z_o}{\sinh \vartheta}, \quad R_1 = Z_o \tanh \frac{\vartheta}{2}, \quad \text{with} \quad \vartheta = \alpha + j\beta$$

or

$$R_1 = Z_o \frac{\sqrt{L} - 1}{\sqrt{L} + 1}, \quad R_2 = \frac{2Z_o \sqrt{L}}{L - 1}, \quad \text{with} \quad L = \alpha \text{ (loss ratio)}$$

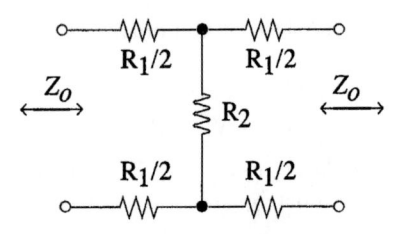

Fig. 16.3. Balanced-T matched attenuator

At the end of the design, it is advisable to check the value of the characteristic impedance against the expression

$$Z_o = R_1 \left(1 + 2\frac{R_2}{R_1}\right)^{1/2}$$

Substituting into the hyperbolic equations with $\vartheta = 2$ (which corresponds to a 3 dB loss), we obtain

$$R_2 = \frac{50}{\sinh(2)} = \frac{50\,(2)}{e^2 - e^{-2}} \cong 13.78\,\Omega$$

and

$$R_1 = 50\,\tanh(1) = 50\,\frac{e^1 - e^{-1}}{e^1 + e^{-1}} \cong 38.08\,\Omega$$

From the algebraic expressions we find

$$R_1 = 50\,\frac{\sqrt{2}-1}{\sqrt{2}+1} \cong 8.58\,\Omega \quad \text{and} \quad R_2 = \frac{2(50)\sqrt{2}}{2-1} = 141.42\,\Omega$$

Note that both solutions yield $Z_o = 50\,\Omega$, with a relative error of less than 0.1% (very likely due to roundoff errors).

The resulting noise factor is $F = \vartheta = 2$, which corresponds to a noise figure of 3 dB.

CASE STUDY 16.5 [229, 243]

Consider the matched attenuator pad in Fig. 16.4, which has a loss $L = 10$ dB. Suppose that a resistor R_o at temperature $T_1 = 100$ K is connected to the input.

1) Find the noise model of the attenuator pad in thermal equilibrium at $T_2 = 300$ K.
2) Determine the output noise spectral density.

16.1 Attenuators   541

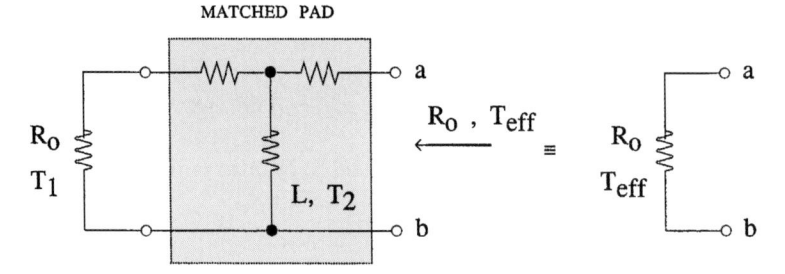

Fig. 16.4. Noise equivalent circuit of the matched attenuator

Solution

1) The effective noise temperature of a passive one-port, denoted by T_{eff}, is the actual temperature which the Thévenin's equivalent resistance of the passive one-port must have in order to generate the same noise power as the actual one-port.

   When the temperature is uniformly distributed within the one-port, T_{eff} is equal to the temperature of the one-port (see Sect. 4.5.5).

   However, in this application the temperature is not uniform, and consequently T_{eff} is derived from the condition that the noise power of the input resistor R_o adds to the noise power delivered by the one-port, since they are uncorrelated.

   Using (5.13), the input noise temperature of the matched attenuator is

$$T_e = (L - 1)\, T_2 = (10 - 1)\, 300 = 2700 \text{ K}$$

The effective noise temperature of the attenuator pad is calculated with expression (5.12):

$$T_{\text{eff}} = \frac{1}{L} T_1 + (1 - \frac{1}{L}) T_2 = 0.1(100) + 0.9(300) = 280 \text{ K}$$

The noise model of the attenuator pad is shown in Fig. 16.5.

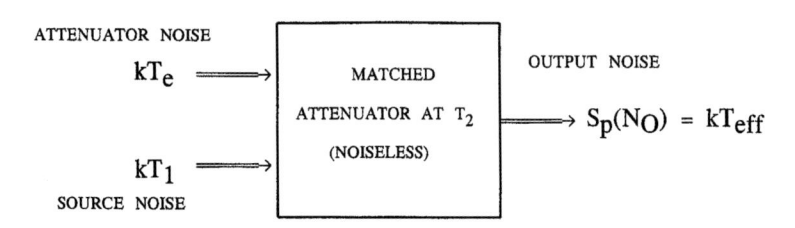

Fig. 16.5. Noise model of the matched attenuator pad

2) The output noise spectral density depends on the effective noise temperature; thus

$$S_{po} = kT_{eff} = (1.38 \cdot 10^{-23}) \, 280 \cong 3.86 \cdot 10^{-21} \quad W/Hz$$

The effect of the attenuator pad on the S/N ratio is investigated in Case Study 16.22.

16.2 Multistage Amplifiers

CASE STUDY 16.6 [244]

Three different amplifiers are available, all matched to the same characteristic impedance. Their electrical performance is detailed in Table 16.1.

In order to amplify a weak signal, these amplifiers must be connected in cascade. Find the optimal order of connecting them and the resulting noise factor.

Table 16.1. Electrical performance of the amplifiers

Amplifier	Power Gain	Noise figure
A	4 dB	1.5 dB
B	8 dB	3.0 dB
C	20 dB	5.0 dB

Solution

Inspecting the expression (4.78), we deduce that the first stage should have the minimum noise, in order not to degrade the overall noise factor; in this case, amplifier A must be placed at the front-end of the amplifying chain. As for the order, we must decide between two combinations: 1) A, B, C, or 2) A, C, B. Since stage B has an intermediate noise level, for comparison purposes it would be useful also to consider the combination 3) B, A, C. The only way to decide which is the most appropriate solution is to determine the resulting overall noise factor for each.

All quantities given in decibels must first be expressed as ratios:
$G_A = 2.51$, $G_B = 6.3$, $G_C = 100$
$F_A = 1.41$, $F_B = 2$, $F_C = 3.16$

1) Combination A, B, C
   Applying (4.78), we obtain

$$F = F_A + \frac{F_B - 1}{G_A} + \frac{F_C - 1}{G_A G_B} = 1.41 + \frac{2-1}{2.51} + \frac{3.16-1}{2.51 \cdot 6.3} \cong 1.945$$

2) Combination A, C, B
   This time we obtain
   $$F = F_A + \frac{F_C - 1}{G_A} + \frac{F_B - 1}{G_A G_C} = 1.41 + \frac{3.16 - 1}{2.51} + \frac{2 - 1}{2.51 \cdot 100} \cong 2.27$$

3) Combination B, A, C
   In this case,
   $$F = F_B + \frac{F_A - 1}{G_B} + \frac{F_C - 1}{G_A G_B} = 2 + \frac{1.41 - 1}{6.3} + \frac{3.16 - 1}{2.51 \cdot 6.3} \cong 2.068$$

Clearly there is no need to explore other combinations, since either amplifier B is in the first position (and consequently F > 2), or F_C is at the front-end (and hence F > 3.16). In any event, we cannot obtain a global noise factor lower than 2, and for this reason combination A, B, C certainly remains the best.

CASE STUDY 16.7 [226]

An amplifier with voltage gain equal to 3 has input resistance $R_i = 5\,k\Omega$ and drives a load $R_L = 10\,k\Omega$. The equivalent noise resistance of the amplifier is $R_n = 1.5\,k\Omega$. Find the equivalent noise resistance of the overall amplifier.

Solution

By definition (Sect. 4.5.1), the equivalent noise resistance *is the value of a resistor which, when connected to the input of a hypothetical noiseless amplifier with the same gain and bandwidth, produces the same output noise.*

Let us denote by G_v the amplifier voltage gain.

In the present application, there are three noise sources: the thermal noise of the input resistance, the thermal noise of the load, and the noise of the amplifier. All these sources are uncorrelated; consequently their contributions add (Table 16.2).

Table 16.2. Noise contributions of various elements

Element	Noise power	Noise power referred to input
R_i	$4kTR_i\,\Delta f$	$4kTR_i\,\Delta f$
R_n	$4kTR_n\,\Delta f$	$4kTR_n\,\Delta f$
R_L	$4kTR_L\,\Delta f$	$(4kTR_L\,\Delta f)/|G_v|^2$

Note that the power gain of the amplifier is proportional to the squared voltage gain and any noise power generated at the output must be divided by the power gain of the amplifier to refer it to the input.

Hence, the total noise power referred to the input is
$$\overline{E_{ni}^2} = 4KT\,\Delta f\,(R_i + R_L/G_v^2 + R_n)$$

544   16 Noise in Communication Systems

The equivalent noise resistance of the system must generate the same equivalent input noise, at the same temperature T:

$$\overline{E_{ni}^2} = 4KT\,\Delta f\,R_{eq}$$

Consequently, the noise equivalent resistance of the overall system is

$$R_{eq} = R_i + R_L/A_v^2 + R_n = 5 + 10/9 + 1.5 \cong 7.61\text{ k}\Omega$$

CASE STUDY 16.8 [247]

An amplifier is driven by an 8-GHz signal delivered by a signal source with internal impedance 50 Ω. The amplifier has the following noise parameters: $R_n = 32\,\Omega$, $G_n = 4\,\text{mS}$, and $Y_{cor} = (8 + j14)\,\text{mS}$.

1) Determine the noise factor of the amplifier.
2) Discuss practical solutions to improve the noise factor.

Solution

1) According to (5.27a), the noise factor is

$$F = 1 + \frac{G_n + R_n\bigl((G_s + G_{cor})^2 + (B_s + B_{cor})^2\bigr)}{G_s}$$

$$F = 1 + \frac{1}{20}\left(4 + 0.032\bigl((20+8)^2 + (0+14)^2\bigr)\right) = 2.768$$

Note that computation is performed with resistance in kΩ and conductance in mS; also, $G_s = 1/50 = 0.02$ S $= 20$ mS.

It is interesting to determine the minimum expected noise factor if perfect noise matching is achieved ($G_s = -G_{cor}$ and $B_s = -B_{cor}$); in this case

$$F_o = 1 + G_n/G_s = 1 + 4/20 = 1.2$$

2) Comparing the value of F_o with the actual noise factor, we see that the mismatch between the signal source and the amplifier is so large, that the amplifier must necessarily be operating suboptimally.

To reduce the noise factor, a coupling network can be inserted in parallel across the input to match the source to the amplifier. According to the discussion of the noise factor fallacy (Sect. 13.2.2), it is unwise to employ a lossy network, since its thermal noise will add to the existing noise (despite an apparent reduction in the noise factor). Consequently, the only possibility is to insert a lossless (purely reactive) network. In this way, all we expect is to compensate the second squared term in the sum of the F numerator (which is not dominant) and hence reach a local minimum.

By inspecting the expression for the noise factor, we deduce that the source admittance $Y_s = 20\,\text{mS}$ must be transformed into $Y_s = (20 - j14)\,\text{mS}$. This is equivalent to introducing a parallel inductance (stub) such that

$$B_{st} = \frac{1}{\omega L} \quad \text{and} \quad L = \frac{1}{\omega B_{st}} = \frac{1}{2\pi(8 \cdot 10^9)14 \cdot 10^{-3}} \cong 1.42\,\text{nH}$$

The new value of the noise factor is

$$F' = 1 + \frac{1}{20}\left(4 + 0.032\left((20+8)^2 + (-14+14)^2\right)\right) = 2.454$$

The noise figure improvement becomes

$$\Delta F = 10\,\log F - 10\,\log F' \cong 0.52\,\text{dB}$$

■ **Remarks**

- The improvement achieved in the noise figure is unimpressive (roughly, 0.5 dB).
- The inserted inductance can modify the gain and the 3-dB bandwidth of the amplifier in an unacceptable way.
- This inductance must be simulated with a lossy transmission line of appropriate length (stub); otherwise, if a discrete inductor has to be added, the risk of picking up spurious magnetic fields must be carefully considered and adequate protections proposed. In any event, a balance between the slight improvement in noise performance and the additional cost must be kept in mind. Ultimately, the best solution may be to change the amplifier.

CASE STUDY 16.9 [240]

The voltage gain A_v of a multistage amplifier is measured at several frequencies. The squared magnitude of the voltage gain is then plotted versus frequency. Finally, a piecewise approximation for $|A_v|^2$ is obtained (Table 16.3). Determine the noise bandwidth of the circuit.

Table 16.3. Piecewise approximation of $|A_v|^2$

#	Interval	$|A_v|^2$
1	$0 \leq f \leq 500\,\text{Hz}$	f^2
2	$500\,\text{Hz} \leq f \leq 30\,\text{kHz}$	$25 \cdot 10^4$
3	$30\,\text{kHz} \leq f \leq 50\,\text{kHz}$	$(-12.5)f + 62.5 \cdot 10^4$
4	$f > 50\,\text{kHz}$	0

546   16 Noise in Communication Systems

Solution

The noise bandwidth is calculated with (4.41):

$$\Delta f = \frac{1}{|A_{vo}|^2} \int_0^{+\infty} |A_v(f)|^2 \, df$$

where A_{vo} is the maximum voltage gain. The calculation is carried out inside each given interval.

■ **Interval 1.** The peak value of $|A_v|^2$ is reached at the end of the first interval (f = 500 Hz), and its value is $25 \cdot 10^4$. Thus

$$\Delta f_1 = \frac{1}{25 \cdot 10^4} \int_0^{500} f^2 \, df = \frac{500}{3} \text{ Hz} \cong 166.6 \text{ Hz}$$

■ **Interval 2.** Since the gain is constant within this interval, we have

$$\Delta f_2 = 30 - 0.5 = 29.5 \text{ kHz}$$

■ **Interval 3.** We first determine the gain at the boundaries: at f = 30 kHz, $|A_v|^2 = 25 \cdot 10^4$ and at f = 50 kHz, $|A_v|^2 = 0$. Between these values the variation is linear; the maximum value is $|A_v|^2 = 25 \cdot 10^4$. Hence

$$\Delta f_3 = \frac{1}{25 \cdot 10^4} \int_{30k}^{50k} \left((-12.5)f + 62.5 \cdot 10^4 \right) df = 10 \text{ kHz}$$

■ **Interval 4.** Here $\Delta f_4 = 0$.

■ **Noise Bandwidth.** The total noise bandwidth is obtained by summing the constituent values:

$$\Delta f = \Delta f_1 + \Delta f_2 + \Delta f_3 = 0.166 + 29.5 + 10 = 39.666 \text{ kHz}$$

CASE STUDY 16.10 [248]

Two different manufacturers present their receiving systems as shown in Fig. 16.6. A potential customer wants to buy one of these receivers for his particular application, where it is known that the input S/N ratio is 12.6 dB. Which receiver will provide the best output S/N ratio?

Solution

For each receiver, the overall noise factor must be calculated with (4.78) and then, applying Friis's definition, the output S/N ratio can be evaluated.

16.2 Multistage Amplifiers

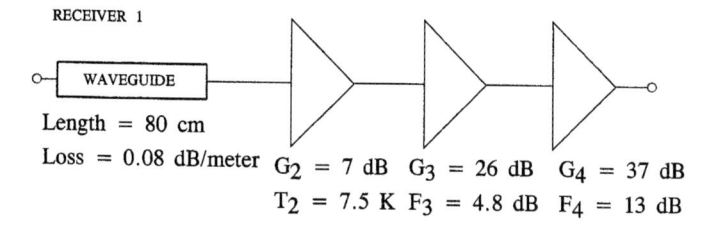

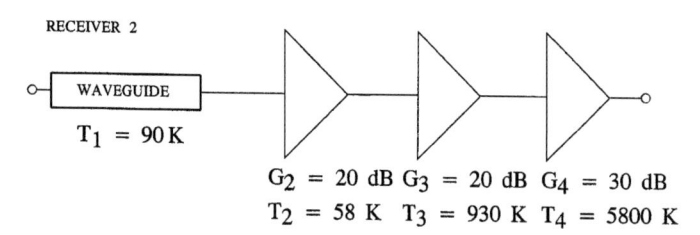

Fig. 16.6. Block diagrams of the proposed receivers

Receiver 1: Waveguide Length = 80 cm, Loss = 0.08 dB/meter; $G_2 = 7$ dB, $T_2 = 7.5$ K; $G_3 = 26$ dB, $F_3 = 4.8$ dB; $G_4 = 37$ dB, $F_4 = 13$ dB.

Receiver 2: Waveguide $T_1 = 90$ K; $G_2 = 20$ dB, $T_2 = 58$ K; $G_3 = 20$ dB, $T_3 = 930$ K; $G_4 = 30$ dB, $T_4 = 5800$ K.

However, in a preliminary step, all individual noise factors must be deduced from the noise figure values or from the noise temperatures.

■ **Receiver 1.** The waveguide has a length of 80 cm, hence its loss is $0.8(0.08) = 0.064$ dB. Expressed as a ratio, this means $L = 1.015$. According to the theorem of Case Study 16.1, the noise factor of the waveguide is $F_1 = L = 1.015$.

The noise temperatures can be converted into noise factors with (4.76); in this way, the block diagram can be redrawn as in Fig. 16.7, where the notation is consistent.

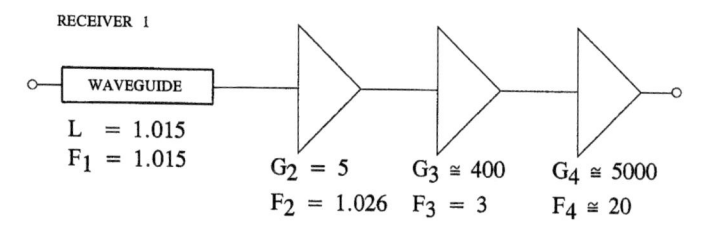

Receiver 1: Waveguide $L = 1.015$, $F_1 = 1.015$; $G_2 = 5$, $F_2 = 1.026$; $G_3 \cong 400$, $F_3 = 3$; $G_4 \cong 5000$, $F_4 \cong 20$.

Fig. 16.7. Equivalent block diagram of Receiver 1

Therefore

$$F_{REC1} = F_1 + \frac{F_2 - 1}{G_1} + \frac{F_3 - 1}{G_1 G_2} + \frac{F_4 - 1}{G_1 G_2 G_3}$$

$$= 1.015 + \frac{1.026 - 1}{1/1.015} + \frac{3 - 1}{(1/1.015)10} + \frac{12.9 - 1}{(1/1.015)(10)(1000)}$$

$$= 1.447 \quad \text{or} \quad 1.6 \, \text{db}.$$

By means of Friis's definition of the noise factor (expressed in decibels), we deduce the output S/N ratio:

$$\frac{S_o}{N_o} = \frac{S_i}{N_i} - F_{REC1} = 12.6 - 1.6 = 11 \text{ dB}$$

Finally, Receiver 1 has an output S/N ratio

$$\left(\frac{S_o}{N_o}\right)_{REC1} = 11 \text{ dB}$$

■ **Receiver 2.** The block diagram, with all parameters expressed in consistent units, is shown in Fig. 16.8. In this case, we can write

$$F_{REC2} = F_1 + \frac{F_2 - 1}{G_1} + \frac{F_3 - 1}{G_1 G_2} + \frac{F_4 - 1}{G_1 G_2 G_3}$$

$$= 1.31 + \frac{1.2 - 1}{0.76} + \frac{4.2 - 1}{(0.76)(100)} + \frac{21 - 1}{(0.76)(100)(100)}$$

This yields:
$$F_{REC2} \cong 1.615 \quad \text{or} \quad 2.08 \text{ dB}$$

and
$$\frac{S_o}{N_o} = \frac{S_i}{N_i} - F_{REC2} = 12.6 - 2.08 = 10.52 \text{ dB}$$

The output S/N ratio of the second receiver is

$$\left(\frac{S_o}{N_o}\right)_{REC2} \cong 10.52 \text{ dB}$$

RECEIVER 2

Waveguide

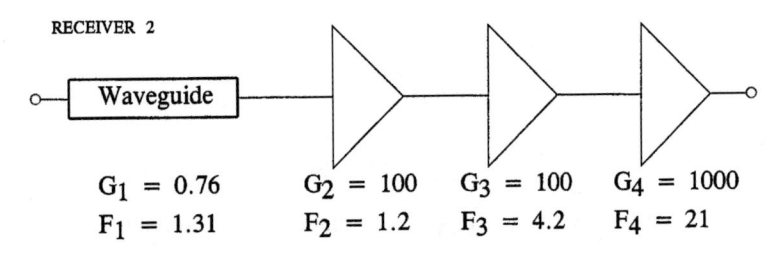

$G_1 = 0.76$ $G_2 = 100$ $G_3 = 100$ $G_4 = 1000$
$F_1 = 1.31$ $F_2 = 1.2$ $F_3 = 4.2$ $F_4 = 21$

Fig. 16.8. Equivalent block diagram of the second receiver

■ **Concluding Remark.** The customer will very likely buy the first receiver, since its output S/N ratio is about 0.5 dB higher than that which would be obtained under equivalent conditions with receiver 2. The result is not surprising, as the loss of the waveguide situated at the front-end of the first receiver is smaller than that of the second receiver waveguide (recall that the noise performance of the front-end stage is of paramount importance).

16.3 Low-Noise Input Stages

CASE STUDY 16.11 [244]

A single-stage amplifier is driven by a signal generator with an internal resistance of 50 Ω by means of a coupling transformer with a winding ratio of 1:4. The equivalent noise resistance of the amplifier stage is 400 Ω. Determine the required rms value of the signal generator that would yield an output S/N ratio of 3 over a 200-kHz bandwidth. Assume T = 300 K.

Solution

The circuit diagram is presented in Fig. 16.9.

The equivalent circuit looking into the secondary of the transformer is given in Fig. 16.10.

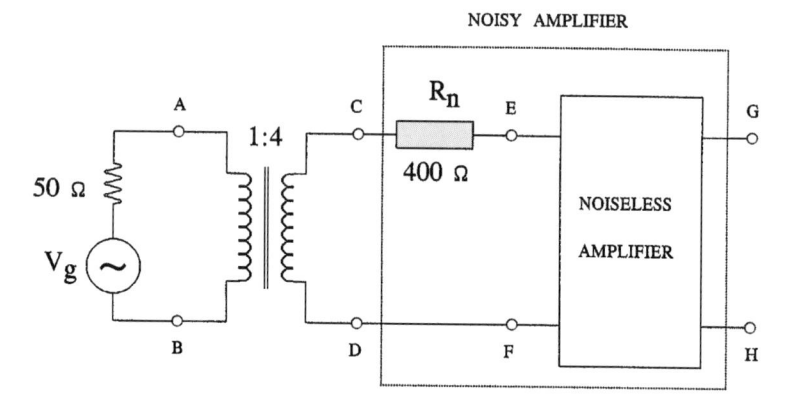

Fig. 16.9. A transformer-coupled noisy amplifier

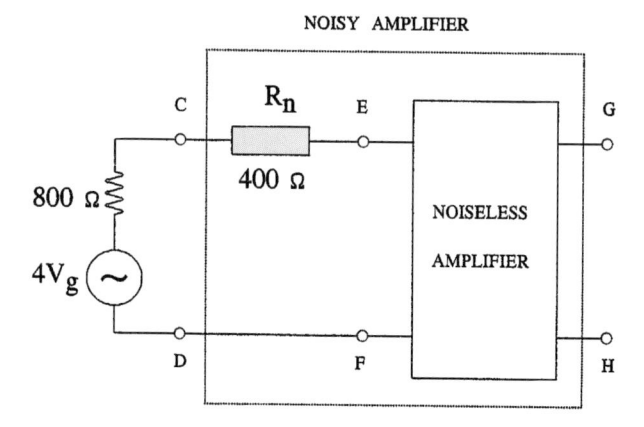

Fig. 16.10. Equivalent circuit looking into the secondary

According to Friis's definition (Sect. 4.5.5), the noise factor is

$$F = \frac{S_i/N_i}{S_o/N_o}$$

where S_i/N_i is the available signal-to-noise ratio at the signal generator terminals A–B and S_o/N_o is the signal-to-noise ratio at output terminals G–H. Note two important points:

1) Since the coupling transformer is considered noiseless, the same signal-to-noise ratio will be observed at terminals A–B or at terminals C–D (the transformer adds no noise and modifies the signal and the noise passing through it in the same way).
2) In this case, it is specified that $S_o/N_o = 3$, and since the two-port of Fig. 16.10 is also noiseless, the same signal-to-noise ratio will appear at terminals E–F.

The first task is to calculate S_i/N_i, ensuring that S_i, N_i are the *available* signal and noise powers, respectively. Therefore

$$S_i = \frac{(4V_g)^2}{4\,(800)} = \frac{V_g^2}{200}$$

With expression (3.3) we have

$$N_i = kT\,\Delta f = 4.14 \cdot 10^{-21}\,(200 \cdot 10^3) = 8.28 \cdot 10^{-16}\text{ W}$$

Now, the noise factor value can be deduced from North's definition. Suppose that N_o' represents the output noise power when the amplifier is assumed to be ideal ($R_n = 0$) and N_o is the output noise power when the amplifier is noisy ($R_n = 400\,\Omega$). Hence

$$F = N_o/N_o' = \frac{800 + 400}{800} \cong 1.5$$

Friis's definition gives

$$F = \frac{S_i/N_i}{S_o/N_o} = \frac{S_i/N_i}{3}$$

so

$$\frac{V_g^2}{200} = 3\,F\,N_i = 3\,(1.5)\,(8.28 \cdot 10^{-16})$$

It follows that in order to ensure the specified S/N ratio, the signal source must deliver a voltage

$$\boxed{V_g = 0.863\ \mu V}$$

16.3 Low-Noise Input Stages

CASE STUDY 16.12 [244]

1) A generator has an electromotive force E and internal resistance R_g. Deduce the general expression for its available signal-to-noise ratio.
2) A microphone is assumed to have an internal impedance of 100 Ω (purely resistive); it generates an open-circuit signal voltage of 60 μV. It is matched to the input of an amplifier (whose minimum noise figure is 1.76 dB) by means of a step-up transformer. If the noise figure of the amplifier is 7 dB and the system bandwidth is 15 kHz, determine:
   a) The output S/N ratio.
   b) The output S/N ratio when the transformer is removed and the microphone is directly connected to the amplifier input.

   Comment on any assumption you make.

Solution

1) The *available* S/N ratio is to be evaluated, the available power of the signal generator is calculated with expression (4.21b):

$$S_a = E^2 / 4R_g$$

Assuming that *the only noise generated by the signal source is thermal* (*Assumption 1*), the available noise power is

$$N_a = kT\, \Delta f$$

Hence, the available S/N ratio of a generator with internal resistance R_g and electromotive force E is

$$\boxed{S_a/N_a = \frac{E^2}{4kTR_g\, \Delta f}}$$

Here, the implicit condition is that *the signal generator operates with a matched load (Assumption 2)*.

2) This study is typical of many practical cases, where the only specifications concerning the amplifier are its minimum noise figure and the noise figure when the amplifier is inserted in the particular circuit (without any additional information on source resistance).
   a) The adopted approach calculates the S/N ratio of the microphone and then, by means of Friis's definition of the noise factor, deduces the output S/N ratio. However, in the rather fuzzy context of this application, several assumptions must be taken into account.

   The circuit detailing the microphone–coupling transformer–amplifier is shown in Fig. 16.11.

   The equivalent circuit looking into the secondary of the matching transformer is given in Fig. 16.12.

16 Noise in Communication Systems

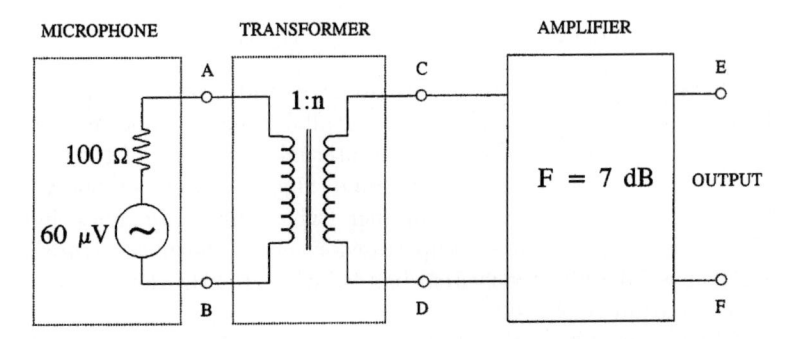

Fig. 16.11. Matching the microphone to the amplifier by means of a step-up transformer

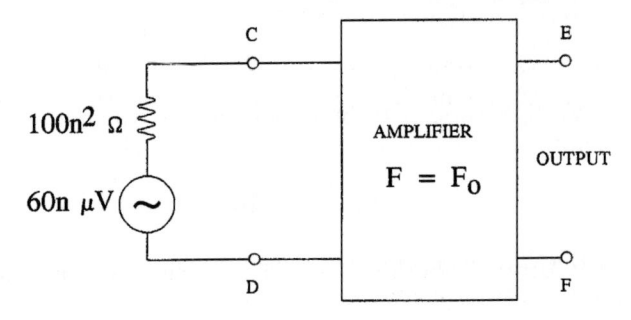

Fig. 16.12. Equivalent circuit looking into the secondary

Applying the expression for the available S/N ratio of a generator (deduced at point 1), we obtain

$$(S/N)_{A-B} = \frac{(60n \cdot 10^{-6})^2}{4\,(1.38 \cdot 10^{-23})\,290\,(100n^2)(15 \cdot 10^3)} \cong 1.5 \cdot 10^5$$

or, expressed in decibels,

$$(S/N)_{A-B} = 10\,\log\,(1.5 \cdot 10^5) \cong 51.76 \text{ dB}$$

Note that *the temperature of the system has been assumed to be 290 K (Assumption 3)*. From another point of view,

$$(S/N)_{C-D} = (S/N)_{A-B} \cong 51.76 \text{ dB}$$

Since the S/N ratio at terminals C–D has been taken equal to the S/N ratio of the microphone, another implicit assumption is that *the transformer generates no noise* (neither thermal, due to its winding resistance, nor pickup of spurious fields). This is *Assumption 4*.

Another hypothesis to be considered is that the step-up transformer achieves *perfect noise matching of the microphone to the input of* the amplifier (*Assumption 5*). Consequently, the equivalent source impedance seen by the amplifier is equal to its optimum impedance, and the amplifier operates at its minimum noise factor (F_o). Therefore, with Friis's definition,

$$F_o = \frac{(S/N)_{C-D}}{(S/N)_{E-F}}$$

or, if all quantities are expressed in decibels,

$$\left(F_o\right)_{dB} = \left((S/N)_{C-D}\right)_{dB} - \left((S/N)_{E-F}\right)_{dB}$$

Hence

$$(S/N)_{E-F} = 51.76 - 1.76 = 50 \text{ dB}$$

b) When the transformer is removed, the noise matching is no longer achieved and the amplifier operates at its prescribed noise figure of 7 dB. Thus

$$(S/N)_{E-F} = (S/N)_{C-D} - F = (S/N)_{A-B} - F = 51.76 - 7 = 44.76 \text{ dB}$$

■ **Concluding Remark.** In this case, noise matching with a step-up transformer improves the output S/N ratio by about 5 dB. In practice, since *Assumption 4* cannot actually be achieved, the improvement must be less than 5 dB (without considering the risk of picking up stray signals by the transformer, which will further degrade the output S/N ratio).

16.4 Receivers

16.4.1 Background

A receiver is the part of a communications system that converts electromagnetic waves into visible or audible form. The quality of a receiver is assessed in terms of the following parameters:

■ **Receiver Sensitivity (Noise Floor).** This is *the lower limit of the input signal available power (S_i) required to achieve the minimum output S/N ratio in normal operation.* In other words, it is the ability of the receiver to bring in weak signals.

Sensitivity is always specified for a given S/N ratio.

In order to estimate the noise floor, consider a two-port with a flat noise bandwidth between its –3 dB cutoff frequencies. Its input is driven by a signal generator with internal resistance R_g. The available thermal noise power of the signal generator is

$$N_i = kT \, \Delta f.$$

16 Noise in Communication Systems

With Friis' definition of noise factor, we obtain

$$S_i = (kT \, \Delta f) \, F \, \frac{S_o}{N_o}. \tag{16.1}$$

Usually, this relationship is written in a different but equivalent form:

$$10 \, \log(S_i) = F_{dB} - 144 + 10 \, \log(\Delta f) + 10 \, \log \frac{S_o}{N_o} \, [\text{dBm}] \tag{16.2}$$

where S_i is expressed in mW and Δf in kHz. The term -144 is related to $-174\,\text{dBm}$, which is the relative power level of kT (at $T = 290\,\text{K}$), adjusted by $+30\,\text{dB}$ originating from F, which is usually given in dB.

■ **Equivalence.** It should be borne in mind that for any quantity, the relationship between dB and dBm is:

$$\boxed{(\text{LEVEL})_{dBm} = (\text{LEVEL})_{dB} + 30}$$

We recall that dBm is the abbreviation for decibels above (or below) one milliwatt [246]. Actually, *it is a quantity of power expressed in terms of its ratio to 1 mW* (0 dBm is equal to 1 mW across a 50-Ω load).

■ **Selectivity.** In the crowded communication bands, good sensitivity must be combined with selectivity, which is *the ability of a receiver to distinguish between signals separated by only a small frequency difference.*

■ **Minimum Discernible Signal (MDS).** By definition [245, 246], this is *the smallest value of input power that produces a discernible signal at the output.* The smaller the input power required, the more sensitive the receiver.

■ **Minimum Detectable Signal.** This is *the input signal voltage which corresponds to the available power S_i* (as stated for the receiver sensitivity). From the definition of S_i, we obtain

$$E_i = \sqrt{4 \, S_i \, R_g} \tag{16.3}$$

The ultimate limit of a receiver's ability to detect weak signals is the thermal noise generated in the input circuit (assuming that a noiseless active device is available and used throughout the receiver).

16.4.2 Case Studies

CASE STUDY 16.13 [248]

To improve the noise of receivers used in communication systems, they often include a narrow bandpass filter that limits the frequency response. For instance, consider a receiver with a noise figure of 5 dB and bandwidth 3 kHz, connected to an antenna matched to its input impedance $R_o = 50\,\Omega$. The system operates at $T_o = 290\,\text{K}$.

1) Find the receiver sensitivity and the minimum detectable signal if the minimum required output S/N ratio is 10 dB. Assume that the antenna is noiseless.
2) Repeat the previous point, assuming that the antenna has a noise figure of 24 dB.
3) Suppose that the receiver is replaced with another unit, which is cheaper and has the same features as the previous one, except that its noise figure is now 20 dB. Connecting the same antenna as in point 2, determine the receiver sensitivity and minimum detectable signal.

Solution

1) We apply (16.2):

$$S_i = 5 - 144 + 10 \log(3) + 10 = -124.228 \text{ dBm}$$

Consequently, the receiver sensitivity is:

$$S_i = 3.77 \cdot 10^{-16} \text{ W}$$

The minimum detectable signal is:

$$E_i = \sqrt{4(50)(3.77 \cdot 10^{-16})} \cong 0.274 \mu V$$

2) For any linear two-port, the available output noise power referred to the input can be evaluated with expression (4.73):

$$\frac{N_o}{G_a} = F \, k \, T_o \, \Delta f$$

If the noise contribution of the source ($kT_o\Delta f$) is subtracted, we obtain the noise available power added by the two-port itself (referred to the input), which is

$$N_Q = (F - 1)kT_o \, \Delta f$$

In the present case, it is convenient to consider the block diagram of Fig. 16.13, where S_i/N_i denotes the signal-to-noise ratio of the waves reaching the antenna and S_o/N_o the signal-to-noise ratio at the receiver output.

Considering only the receiver, we may write

$$N_R = (F_R - 1) \, kT_o \, \Delta f$$

and for the antenna

$$N_A = (F_A - 1) \, kT_o \, \Delta f$$

At the system input, the total noise equivalent power is

$$N_A + N_R + N_t = (F_A + F_R - 1) \, kT_o \, \Delta f = FN_i$$

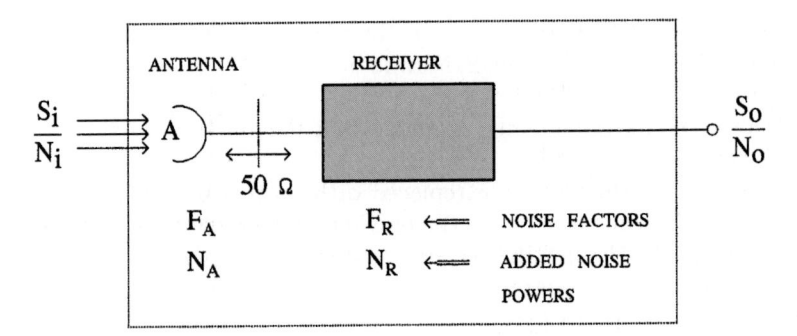

Fig. 16.13. Block diagram of the receiver and its antenna

where N_t is the available thermal noise power of the matching resistance $R_o = 50\,\Omega$ and F is the global noise factor of the system (antenna and receiver). The latter equality makes use of Norton's definition of the noise factor.

By means of Friis' definition, one obtains

$$S_i = \frac{S_o}{N_o}(N_i F) = \frac{S_o}{N_o}(F_A + F_R - 1)kT_o\,\Delta f$$

Upon substituting the given numerical values (all quantities expressed as ratios), we obtain the new value of the receiver sensitivity:

$$S_i = 10(316.22 + 3.16 - 1)(1.38 \cdot 10^{-23})(290)(3 \cdot 10^3)$$
$$S_i = 3.816 \cdot 10^{-14}\,\text{W}$$

It follows that the minimum detectable signal is

$$E_i = \sqrt{4(50)(3.816 \cdot 10^{-14})} \cong 2.76\,\mu\text{V}$$

It is interesting to note that this value is at least 10 times the minimum detectable signal obtained when the noise of antenna was ignored!

3) Now $F_R = 20\,\text{dB}$ (or 100, as a ratio). Adopting the same approach as in the previous case, we obtain

$$S_i = 10(316.2 + 100 - 1)(1.38 \cdot 10^{-23})(290)(3 \cdot 10^3) \cong 5 \cdot 10^{-14}\,\text{W}$$

and the noise floor is now

$$E_i = \sqrt{4(50)(5 \cdot 10^{-14})} \cong 3.16\,\mu\text{V}$$

This result is important, since comparing it to the values obtained at point 2, it is obvious that a drastic reduction in the noise figure of the receiver (from 20 dB to 5 dB) yields an improvement of the minimum detectable signal of only 12%. We may conclude that *whenever the noise* at the antenna is dominant, it is not reasonable to buy a high-quality (less noisy) receiver, because the improvement in overall performance is rather insignificant (this statement is consistent with Rule 4 in Sect. 13.2.1). For communication systems operating up to 30 MHz (LW, MW, and SW bands), a receiver with a noise figure in the range 8–20 dB is quite appropriate (given the usual values of the noise at the antenna). However, above 30 MHz (where antenna noise drops off), a less noisy receiver considerably improves system global performance.

CASE STUDY 16.14 [244]

A receiver has a preamplifier connected to the amplifier by a cable whose length is 24 m. The performance of these components is given in Table 16.4. Determine the minimum gain required in the preamplifier if the global noise figure of the system is not to exceed 6 dB.

Table 16.4. Performance of various components

Noise figure of the preamplifier	Cable loss	Noise figure of the amplifier
2.5 dB	0.2 dB/m	10 dB

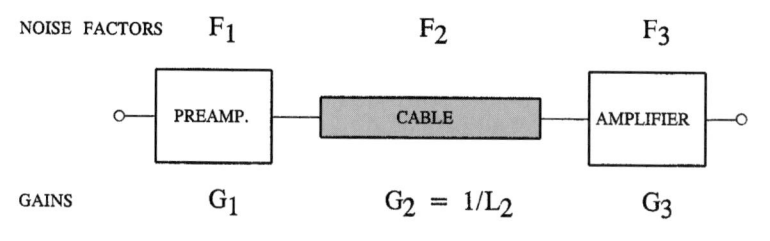

Fig. 16.14. Block diagram of the receiver

Solution

The noise figure (NF) of all components must first be expressed as a noise factor (ratio):

$NF_1 = 2.5$ dB so that $F_1 = 1.778$
$L_2 = 4.8$ dB or 3.02 (as ratio); thus, $G_2 = 1/L_2 = 1/3.02$
$NF_3 = 10$ dB so that $F_3 = 100$
$NF = 6$ dB so that $F \cong 4$.

A block diagram of this receiving system is shown in Fig. 16.14.
Making use of the theorem in Case Study 16.1, the noise factor of the cable is equal to its attenuation (expressed as a ratio). Therefore, $F_2 = 3.02$.

558   16 Noise in Communication Systems

Expression (4.78) then yields the overall noise factor of several cascaded two-ports, which in this case can be written

$$F = F_1 + \frac{F_2 - 1}{G_1} + \frac{F_3 - 1}{G_1 G_2}$$

From this, the preamplifier gain G_1 (required to achieve an overall noise factor of 7.94) becomes

$$G_1 = \frac{1}{F - F_1}\left((F_2 - 1) + \frac{F_3 - 1}{G_2}\right)$$

$$= \frac{1}{4 - 1.1778}\left((3.02 - 1) + \frac{100 - 1}{1/3.02}\right) \cong 135.46$$

Finally, the noise figure of the system does not exceed 6 dB, provided that

$$\boxed{G_1 \geq 10\,\log(135.46) = 21.3\,\text{dB}}$$

CASE STUDY 16.15 [244]

The input matching stage of a receiver has a noise temperature of 100 K and a loss of 4 dB. It is followed by three identical IF stages, each with 10 dB gain, a 6-MHz bandwidth, and a noise figure of 3 dB. With the system matched throughout and operating at T = 290 K, determine

a) the overall noise figure;
b) the equivalent noise temperature of the receiver when connected to an antenna with a noise temperature $T_a = 150\,\text{K}$;
c) the receiver sensitivity, if for high-quality communication an output signal-to-noise ratio of at least 10 dB is required.

Solution

a) Applying (4.76), the noise factor of the input stage is

$$F_1 = 1 + \frac{T_e}{T_o} = 1 + \frac{100}{290} = 1.345$$

A block diagram is shown in Fig. 16.15, where the parameters are expressed as ratios.

The overall noise factor of the cascade is calculated by means of expression (4.78):

$$F = F_1 + \frac{F_2 - 1}{G_1} + \frac{F_3 - 1}{G_1 G_2} + \frac{F_4 - 1}{G_1 G_2 G_3}$$

$$= 1.345 + \frac{2 - 1}{1/2.51} + \frac{2 - 1}{10/2.51} + \frac{2 - 1}{100/2.51}$$

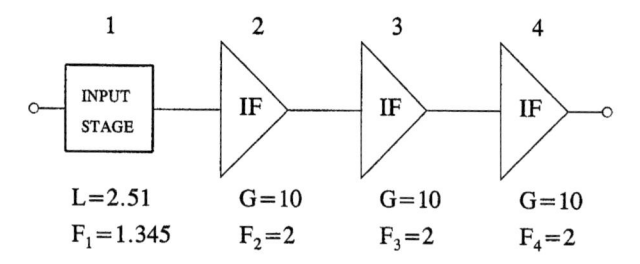

```
    1              2            3            4
┌─────────┐
│ INPUT   │─────▷─────────▷──────────▷──────
│ STAGE   │      IF           IF          IF
└─────────┘
  L=2.51       G=10         G=10         G=10
  F₁=1.345     F₂=2         F₃=2         F₄=2
```

Fig. 16.15. Block diagram of the receiver

Consequently

$$\boxed{F \cong 4.13 \quad \text{or} \quad 6.16 \text{ dB}}$$

b) The equivalent noise temperature of the receiver is calculated with (4.83):

$$T_r = T_o(F-1) = 290(4.13-1) = 907.7 \text{ K}$$

Consequently, the equivalent noise temperature of the system receiver and antenna is obtained by making use of the additive property of the input temperatures:

$$T_s = T_a + T_r = 150 + 907.7 = 1057.7 \text{ K}$$

c) The equivalent noise temperature of the system is useful, since it allows us to estimate the available input noise power spectral density with expression (3.3):

$$S(N_{ai}) = kT_s = (1.38 \cdot 10^{-23})(1057.7) \cong 1.46 \cdot 10^{-20} \text{ W/Hz}$$

As our interest is in calculating the total available input noise power, the noise bandwidth of the IF stages must be estimated. For the useful signal, the overall bandwidth (denoted by B_n) of n identical cascaded stages, depends on the bandwidth B of each stage, as given by

$$B_n = B\sqrt{2^{1/n}-1}, \quad \text{hence} \quad B_n = B\sqrt{2^{1/3}-1} \cong 3.06 \text{ MHz}$$

Assuming that each stage has only one dominant pole, Table 4.1 gives the noise equivalent bandwidth as

$$\Delta f_n = 1.155 B_n \cong 3.53 \text{ MHz}$$

Thus, the available noise power at the input is:

$$N_{ai} = S(N_{ai})\Delta f_n = (1.46 \cdot 10^{-20})(3.53 \cdot 10^6) \cong 5.15 \cdot 10^{-14} \text{ W}$$

Since all the noise has been referred to the input, the receiver is now noiseless and the output S/N ratio is equal to the input S/N ratio. In normal operation, it is specified that the output S/N ratio should be at least 10 dB (or 10, as a ratio). It follows that the input S/N ratio must also be equal to 10. Thus, the receiver sensitivity, equal to the minimum input available power of the signal, is equal to $5.15 \cdot 10^{-13}$ W, which corresponds to -122.87 dB.

Note that the same result would be obtained by directly applying formula (16.2).

CASE STUDY 16.16 [244]

The communication recognition system of a receiver is considered to operate correctly for an output S/N ratio better than 40 dB. At its front-end the receiver has a matching input stage with a noise figure of 10 dB and a loss of 6 dB. The following amplifier has very high gain and a noise figure of 2 dB. At the particular site where it operates, the input noise temperature is 260 K and the input S/N power ratio is 10^5. Determine whether the receiver operates satisfactorily under these conditions, or a preamplifier is needed. In the latter case, discuss the criteria needed to select the preamplifier.

Assume that all receiver stages are linear, matched throughout, and have the same bandwidth. All noise factors are defined with reference to $T_o = 290$ K.

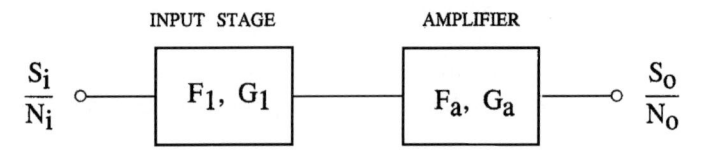

Fig. 16.16. Block diagram of the receiving system

Solution

Before proceeding to calculation, all quantities given in dB must be expressed as ratios. Hence, according to the notation of Fig. 16.16

$F_1 = 10$ dB or 10 (as ratio);
$F_a = 2$ dB or 1.58 (as ratio);
$L_1 = 6$ dB or 4 (as ratio);
$G_1 = 1/L_1 = 1/4$ ($L_1 =$ first stage loss).

The overall noise factor relative to 290 K is calculated with expression (4.78):

$$F_{290} = F_1 + \frac{F_a - 1}{G_1} = 10 + \frac{1.58 - 1}{1/4} = 12.32$$

The relationship between the noise factor at 290 K and that defined at 260 K is deduced by applying (4.77) at the two temperatures:

$$\begin{cases} (F_{260} - 1)T_o = 260 \\ (F_{290} - 1)T_o = 290 \end{cases}$$

Taking the ratio of these equations, we have

$$F_{260} = 1 + (F_{290} - 1)\frac{260}{290} = 1 + 11.32\,\frac{260}{290} \cong 11.15$$

Now, the output S/N ratio can be deduced from Friis' definition of the noise factor:

$$\frac{S_o}{N_o} = \frac{S_i}{N_i}\frac{1}{F_{260}} = \frac{10^5}{11.15} \cong 8.97 \cdot 10^3 < 10^4 \quad \text{(or 40 dB)}$$

We conclude that the output S/N ratio is not satisfactory and a preamplifier is required. The new receiver block diagram is presented in Fig. 16.17.

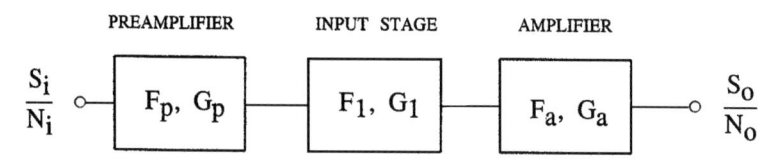

```
         PREAMPLIFIER      INPUT STAGE       AMPLIFIER
Si/Ni ──┤  Fp, Gp  ├──────┤  F1, G1  ├──────┤  Fa, Ga  ├── So/No
```

Fig. 16.17. Employing a preamplifier at the front-end

This time

$$F_{290} = F_p + \frac{F_1 - 1}{G_p} + \frac{F_a - 1}{G_p G_1}$$

and

$$F_{260} = 1 + (F_{290} - 1)\frac{260}{290} = 1 + \frac{26}{29}\left(F_p - 1 + \frac{F_1 - 1}{G_p} + \frac{F_a - 1}{G_p G_1}\right)$$

With Friis' definition, and specifying an output S/N ratio of at least 10^4, we obtain

$$\frac{S_i}{N_i}\frac{1}{F_{260}} \geq 10^4 \quad \text{which yields} \quad F_{260} \leq 10$$

Upon substituting with the proposed numerical values, the following inequality is obtained:

$$F_p + \frac{11.32}{G_p} \leq 11.04$$

This inequality contains two unknowns (F_p and G_p). When selecting the appropriate preamplifier for this receiver, either the noise factor or the gain must be known; the other can be obtained from the above inequality. For instance, assume that a preamplifier with a minimum power gain of 10 (as a ratio) is available. We then find that the noise factor must be better than 9 (as a ratio). Hence, a possible solution is a preamplifier with

$$F_p \leq 9 \quad \text{and} \quad G_p \geq 10$$

CASE STUDY 16.17 [247]

A receiver can operate either with an internal ferrite-rod antenna or with an external antenna. The circuit shown in Fig. 16.18 is employed to isolate the external antenna from the ferrite rod. Its elements are: $R_1 = 600\,\Omega$, $R_2 = 150\,\Omega$, and $R_3 = 90\,\Omega$. Assuming that all resistors are at the standard reference temperature, determine the influence of the isolating circuit on the S/N ratio in two different situations:

a) The external antenna has a characteristic impedance $Z_a = 600\,\Omega$, and its equivalent noise temperature is $T_a = 3000\,K$.
b) The external antenna has the same impedance, but this time its equivalent noise temperature is $T'_a = 3 \cdot 10^5\,K$ (as for an actual wide band antenna like that employed for LW or MW frequency bands).

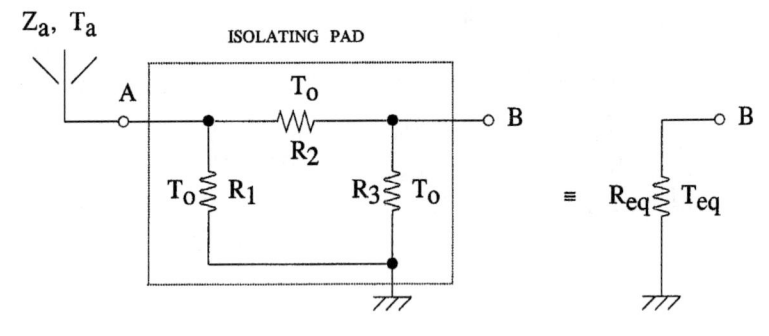

Fig. 16.18. The antenna with its isolating pad, viewed as an equivalent one-port

Solution

■ **Approach.** The isolating circuit with the external antenna is shown in Fig. 16.18.

When the S/N ratio is required, we need to evaluate the noise and the signal power, both at the same location (at point B). Consequently, the equivalent noise power at point B must be calculated, and then the signal power at the same point.

■ **Noise Calculation.** In order to find the equivalent resistance R_{eq} and its noise temperature T_{eq}, we progress from the antenna to the receiver (point B), calculating for each series (or parallel) combination of two resistors the equivalent resistance and its noise temperature (Figs. 16.19a, b and c).

Therefore, the combination of the antenna plus isolating pad is equivalent to a 75-Ω resistor having an equivalent noise temperature of 440.5 K.

■ **Signal Transmission.** Consider the equivalent circuit of Fig. 16.19d.

16.4 Receivers   563

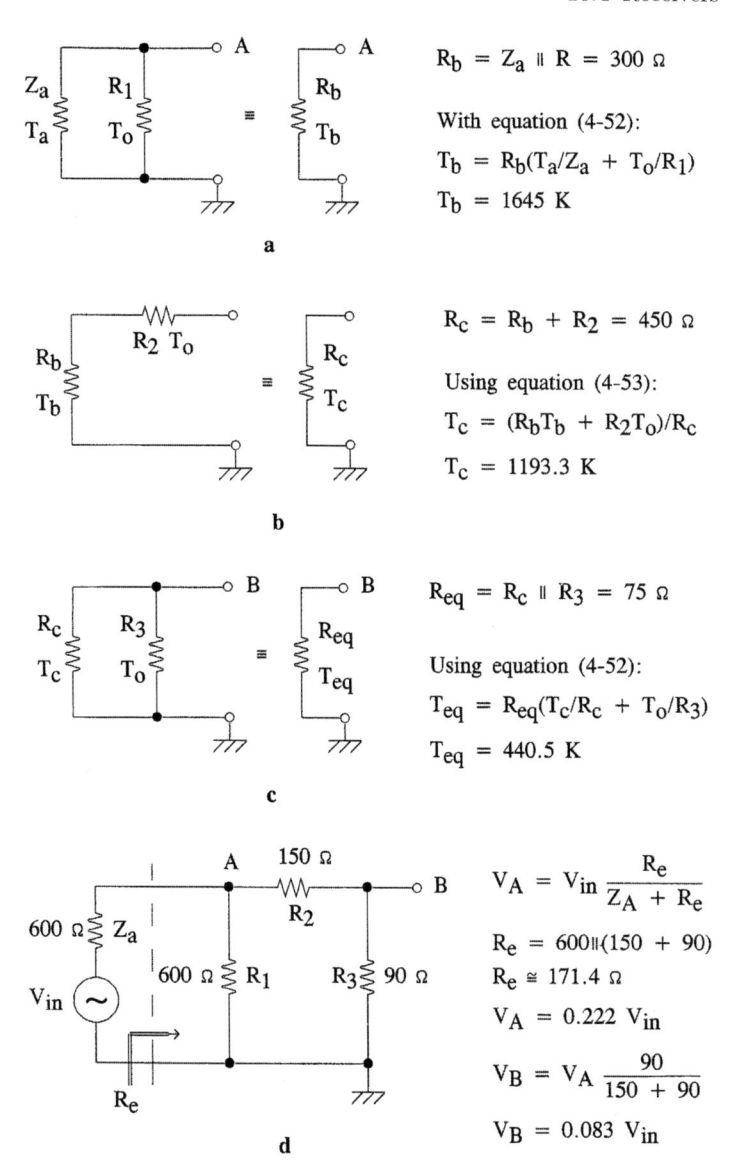

$R_b = Z_a \parallel R = 300 \, \Omega$

With equation (4-52):

$T_b = R_b(T_a/Z_a + T_0/R_1)$

$T_b = 1645 \, K$

$R_c = R_b + R_2 = 450 \, \Omega$

Using equation (4-53):

$T_c = (R_b T_b + R_2 T_0)/R_c$

$T_c = 1193.3 \, K$

$R_{eq} = R_c \parallel R_3 = 75 \, \Omega$

Using equation (4-52):

$T_{eq} = R_{eq}(T_c/R_c + T_0/R_3)$

$T_{eq} = 440.5 \, K$

$V_A = V_{in} \dfrac{R_e}{Z_A + R_e}$

$R_e = 600 \parallel (150 + 90)$

$R_e \cong 171.4 \, \Omega$

$V_A = 0.222 \, V_{in}$

$V_B = V_A \dfrac{90}{150 + 90}$

$V_B = 0.083 \, V_{in}$

Fig. 16.19. a, b, c Step by step calculation of equivalent resistances and their noise temperatures; **d** Calculating the signal attenuation

Clearly, the voltage at node B is roughly 12 times less than the signal voltage delivered by the antenna.

Since the objective is to estimate S/N ratio degradation due to the insertion of an isolating pad, we must calculate the available signal power at point B and the available generator (antenna) power (at point A). The latter is deduced with expression (4.21b):

$$P_a^A = \frac{V_{in}^2}{4(600)}$$

and at point B

$$P_a^B = \frac{V_B^2}{4(75)}$$

■ **Signal-to-Noise Ratio.** In case a, the degradation of the S/N ratio between points A and B is

$$D = 10 \log \frac{S_A/N_A}{S_B/N_B} = 10 \log \left(\frac{P_a^A}{P_a^B} \frac{kT_{eq}}{kT_a}\right)$$
$$= 10 \log \left((18.03) \frac{440.5}{3000}\right) \cong 4.23 \text{ dB}$$

In case b, the noise temperature of the antenna is considerably higher ($T_a' = 3 \cdot 10^5$ K). Following the same method we obtain $T_{eq}' \cong 1.69 \cdot 10^4$ K and this time the S/N ratio degradation is better than 0.07 dB.

■ **Discussion.** We conclude that if the antenna is very selective (i.e., narrow bandwidth, the bandwidth of the antenna being defined as the frequency range over which the gain and impedance are substantially constant), noise pickup is limited and consequently the noise temperature is not excessive. In this situation, the proposed isolating pad is responsible for a S/N degradation of about 4.23 dB (which is the counterpart of its isolation).

However, if the antenna is not selective (as is the case for LW or MW band antennas), its noise temperature is very high, essentially equal to the sky temperature seen by its lobe(s) over a very large bandwidth. S/N ratio degradation is then not at all important, and the pad can be successfully used to isolate the receiver from the unknown impedance of the antenna.

CASE STUDY 16.18 [241]

A TV set with noise figure 14 dB and input bandwidth 4 MHz is connected by means of a long section of coaxial cable to its antenna (Fig. 16.20). The cable characteristic impedance is $Z_o = 75\,\Omega$ and its loss is L = 4 dB. Determine the minimum detectable signal at the antenna terminal if under normal operation conditions a S/N ratio of 40 dB is required at the receiver input.

Solution 1

A block diagram of the antenna-cable-receiver system is presented in Fig. 16.21, where the loss in the cable has been expressed as a ratio.

Our approach is to refer all noise sources to the input receiver (point R). With North's definition of the noise figure we can write

$$\text{NF} = 10 \log \left(\overline{E_{nR}^2} / \overline{E_g^2}\right) \text{ [dB]}$$

16.4 Receivers

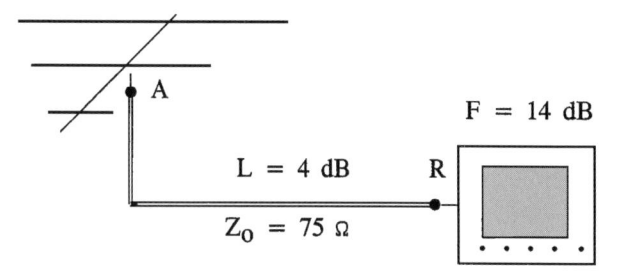

Fig. 16.20. TV receiver connected to its antenna

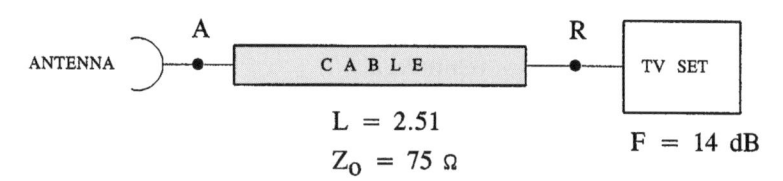

Fig. 16.21. Simplified block diagram

where $\overline{E_{nR}^2}$ is the mean square total noise voltage at point R (including the thermal noise of the signal source) and $\overline{E_g^2}$ represents the thermal noise of the source (in this case, assuming that the cable characteristic impedance is purely resistive, $4kTZ_o\Delta f$). It follows that

$$\overline{E_{nR}^2} = 4kTZ_o\,\Delta f\,10^{F/10} = 4(1.38 \cdot 10^{-23})(290)(75)(4 \cdot 10^6)10^{1.4}$$

which yields

$$\overline{E_{nR}^2} \cong 1.206 \cdot 10^{-10}\ V^2$$

Since the S/N ratio at point R is specified to be 40 dB (or 10^4, as a ratio), the required signal value is

$$S_R^2 = 10^4\,\overline{E_{nR}^2} = 1.206 \cdot 10^{-6}\ V^2$$

or:

$$S_R = 1.098 \cdot 10^{-3}\ V = 1.098\ mV$$

The minimum signal at point A is deduced from the signal at point R multiplied by the cable loss:

$$E_A = E_R\,L = (1.098)(2.511) \cong \boxed{2.758\ mV}$$

Solution 2

Another approach is possible, based on the input noise temperature. This is depicted in Fig. 16.22, where the noise factor of the receiver is marked instead of its noise figure.

566 16 Noise in Communication Systems

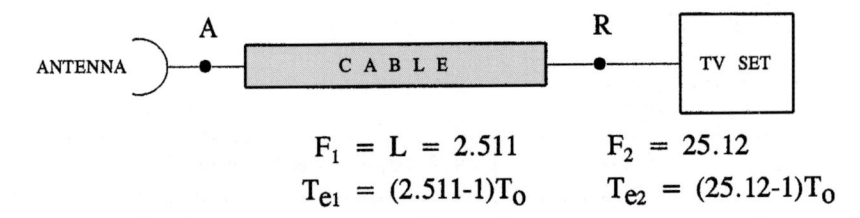

```
              A                          R
 ANTENNA )----•----[  C A B L E  ]----•----[ TV SET ]
```

$$F_1 = L = 2.511 \qquad F_2 = 25.12$$
$$T_{e1} = (2.511-1)T_o \qquad T_{e2} = (25.12-1)T_o$$

Fig. 16.22. Utilization of input noise temperatures

The input noise temperatures have been estimated with formula (4.77). Applying (4.66), the overall noise temperature at point A is obtained with respect to the standard noise temperature T_o:

$$T_{eA} = T_{e1} + \frac{T_{e2}}{1/L} \cong 1.51 T_o + 60.56 T_o \cong 62.07 T_o$$

The total available noise power at point A has two components: 1) the equivalent noise of the cable-receiver system; 2) the thermal noise of the cable characteristic impedance. Hence

$$N_A = k T_{eA}\, \Delta f + k T_o\, \Delta f = 63.07 k T_o\, \Delta f$$
$$= (63.07)(290)(1.38 \cdot 10^{-23})(4 \cdot 10^6)$$

Finally:
$$N_A \cong 101 \cdot 10^{-14}\ \text{W}$$

The available signal power is

$$S_A = N_A \left(\tfrac{S}{N}\right)_A = N_A \left(\tfrac{S}{N}\right)_R F_1$$
$$= (101 \cdot 10^{-14})(10^4)(2.51) = 2.535 \cdot 10^{-8}\ \text{W}$$

Starting from its definition, the available signal power is

$$S_A = E_A^2\, /\, 4 Z_o$$

which yields

$$E_A = \sqrt{4 Z_o S_A} = 2\sqrt{(75)(2.535 \cdot 10^{-8})} = (13.788)10^{-4} = \boxed{2.757\ \text{mV}}$$

16.5 Space Communication Systems

16.5.1 Background

■ **Introduction.** In communications involving geostationary satellites, two situations occur, which are presented in a simplified manner in Fig. 16.23.

Figure 16.23a diagrams a microwave communication system, employing a geostationary relay satellite to provide a link between two ground stations (or one ground station and a ship). The satellite receives the signals emitted by ground station 1, and after amplifying them, sends them back to ground station 2. The signals emitted by ground station 2 are received by the satellite, amplified, and retransmitted to ground station 1.

A similar situation is depicted in Fig. 16.23b, where the communication link involves ground station 1 and a low-orbit user 3 (for instance, a military satellite or an aircraft).

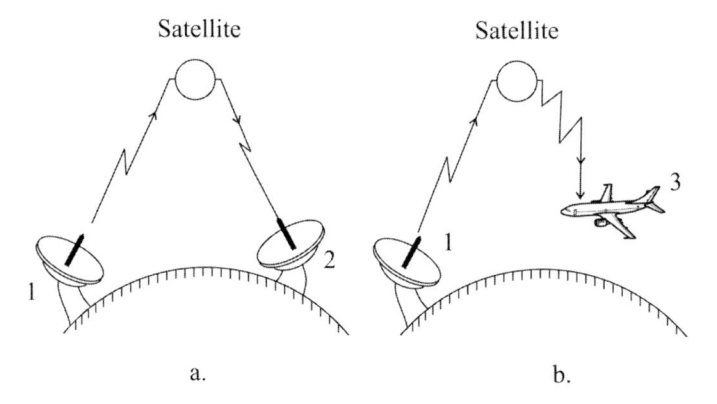

Fig. 16.23. Satellite communications

In both situations, the communication link has two paths:

1) The *uplink*, which is *the radio-communications path from a fixed transmitting site to a moving receiving device*. Usually the fixed transmitting site is a ground station (or ship station), which being on earth (or at sea) has copious energy resources. Along this path the S/N ratio is comfortable, since there are no problems amplifying the transmitted signals to the desired level.
2) The *downlink*, defined as *the radio-communications path from a moving transmitting device to a fixed receiving site*. Here, the transmitter (together with its antenna), being on the satellite, is drastically restricted in power by a series of factors like limited supply power, size, weight, and so on. Consequently, the downlink path is critical from a S/N ratio standpoint, since the useful signals cannot be amplified as much as in the uplink path. Clearly, the global performance of the communications system is constrained by the downlink path.

568   16 Noise in Communication Systems

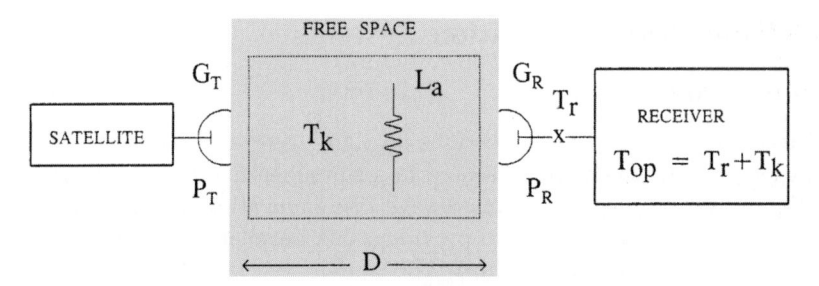

Fig. 16.24. Simplified diagram of the downlink path

■ **Downlink Path.** Figure 16.24 shows the downlink path of a space communications system, which is to be optimized.

The following notation has been employed:

P_T – transmitter output power at the antenna input;
P_R – receiver antenna power output;
G_T – transmitter antenna power gain over an isotropic radiation pattern (it is assumed that the satellite has a directional antenna);
G_R – receiver antenna power gain over an isotropic radiation pattern (it is assumed that the receiver has a directional antenna);
D – distance separating the transmitter and receiver antennas;
T_k – equivalent noise temperature of the sky;
T_r – input noise temperature of the receiver;
L_a – miscellaneous losses affecting signal transmission (except for free-space propagation loss, which is separately treated). The term "miscellaneous losses" refers to additional losses such as atmospheric absorption due to rain (about 2 dB), antenna pointing errors (about 2 dB), transmitter aging (about 1 dB), etc.

■ **Basic Terms.** Before proceeding to calculations, we recall some fundamental definitions:

ERP (Effective Radiated Power). *In a given direction, the directional power gain of a transmitting antenna multiplied by the net power accepted by the antenna from the connected transmitter* [238]. *In other words, it is the amount of power radiated by an antenna, which may be more or less than the power absorbed by it from the transmitter* [246].

EIRP (Effective Isotropic Radiated Power). *The effective radiated power (ERP) of an antenna compared to that of an antenna that radiates equally in all directions* [245].

Free-Space Loss. *The loss between two isotropic antennas in free space, expressed as a power ratio. Note*: *The free-space loss is usually expressed in decibels and is given by the formula* $20\log(4\pi D/\lambda)$, *where D is the separation of the two antennas and λ is the wavelength* [238].

16.5 Space Communication Systems

G/T Ratio (Antennas). *The ratio of the maximum power gain to the noise temperature of an antenna* [238]. It represents a figure of merit that subsumes both the antenna quality and the total noise of the receiver. It is expressed in dB/K.

■ **Downlink Power Budget Calculations.** Assuming an isotropic antenna (i.e., one that radiates uniformly in all directions), the power density at a distance D from the satellite transmitter antenna is

$$p_t = \frac{P_T}{4\pi D^2} \quad [W/m^2] \tag{16.4}$$

When the antenna has directivity, this quantity is multiplied by the gain G_T of the transmitting antenna. The power intercepted by the receiving antenna, at distance D, is then

$$P_R = p_t A_R = \frac{P_T G_T}{4\pi D^2} A_R \tag{16.5}$$

where A_R represents the receiving aperture area of the antenna, which is related to its maximum gain (which depends on the wavelength λ) by the equation $G_R = 4\pi A_R / \lambda^2$. Thus

$$A_R = \frac{G_R \lambda^2}{4\pi} \tag{16.6}$$

Substituting (16.6) into (16.5) we obtain

$$P_R = \frac{P_T G_T G_R \lambda^2}{(4\pi D)^2} \tag{16.7}$$

Taking other losses (L_a) into account, the intercepted power of the receiving antenna becomes

$$P_R = \frac{P_T G_T G_R \lambda^2}{(4\pi D)^2 L_a} = \left(\frac{\lambda}{4\pi D}\right)^2 \frac{P_T G_T G_R}{L_a} \tag{16.8}$$

where the factor $(\lambda/4\pi D)^2$ is the free-space loss.

Recalling that noise input power spectral density is kT_{op} (T_{op} being the effective noise input temperature, as defined in Sect. 4.5.4), the S/N ratio calculation is straightforward:

$$\frac{S_i}{N_i} = \left(\frac{\lambda}{4\pi D}\right)^2 \frac{P_T G_T}{kL_a} \frac{G_R}{T_{op}} \tag{16.9}$$

Expressed in decibels, this equation is sometimes called *the power budget equation*:

$$\boxed{\frac{S_i}{N_i} = 10\log(P_T G_T) + 20\log\frac{\lambda}{4\pi D} + 10\log\frac{G_R}{T_{op}} - 10\log(kL_a) \ [dB]} \tag{16.10}$$

570   16 Noise in Communication Systems

■ **Meaning.** Some terms in (16.10) have special meaning. For instance,

$10 \log(P_T G_T)$ – is the Effective Isotropic Radiated Power (EIRP);
$20 \log(\lambda/4\pi D)$ – is the free-space path loss in decibels;
$10 \log(G_R/T_{op})$ – is the G/T ratio of the receiving earth station; it is obvious that both G_R and T_{op} must be determined at the same reference plane. Nevertheless, the G/T ratio does not depend on the choice of reference plane.

■ **Example.** Consider the Intelsat system [242], which employs a geostationary satellite. It supports point-to-point international communications or TV channels in the frequency band around 6 GHz for the uplink path and 4 GHz for the downlink path. The ground station uses a steerable antenna (an antenna the major lobe of which can be easily shifted in direction). The dish of this antenna is 30 m in diameter, with a G/T ratio of 40.7 dB/K. A simplified block diagram of the system is shown in Fig. 16.25.

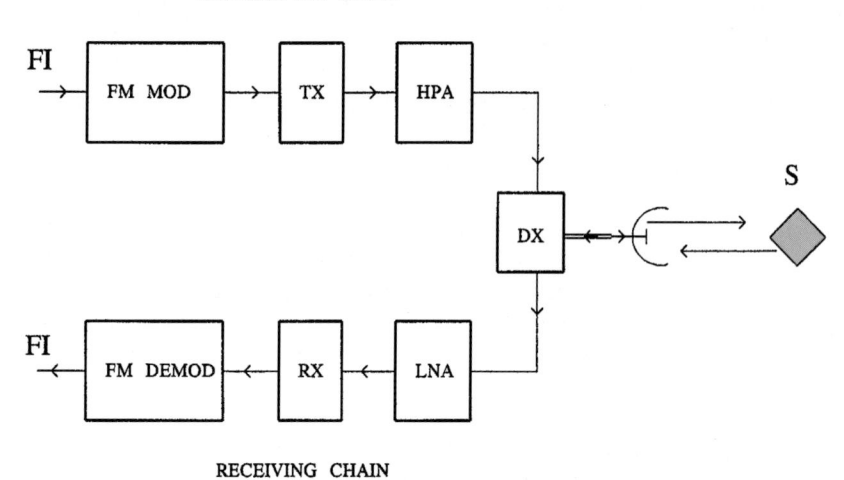

Fig. 16.25. Simplified diagram of Intelsat communications system

The transmitting chain is isolated from the receiving chain by means of a diplexer (DX) – a coupling device that enables the transmitter to operate on the same antenna as the receiver. The intermediate frequency signal reaching the input of the uplink has a frequency of 70 MHz. This signal is up- converted to 6 GHz by the FM modulator, before driving the FM transmitter marked TX. HPA is a high-power amplifier, which employs klystrons, TWT (traveling-wave tubes), or solid-state devices. At the relay satellite, the received signal is amplified and then down-converted to 4 GHz before being transmitted to the ground station.

The receiving chain amplifies the received signals with a low-noise amplifier (LNA), which is usually a parametric amplifier cooled to 15 K. The signals are then sent to the FM receiver (RX) and the FM demodulator, which converts them back to 70 MHz.

The critical parameter of the uplink is the EIRP of the ground station, since the free-space loss is 201 dB at 6 GHz, and the miscellaneous losses are usually estimated at 6 dB.

The critical parameters of the downlink are the EIRP of the relay satellite, the free space loss (197 dB at 4 GHz), the miscellaneous losses (9 dB), and the G/T ratio of the ground station.

If frequency modulation is employed, the S/N ratio at the demodulator input must be better than 10 dB for a signal bandwidth of 500 MHz.

■ **Noise Temperature of the Antenna.** This parameter depends on two components: 1) the noise associated with the radiation resistance of the antenna; 2) the received noise from outside sources (the sky noise). In most situations, the latter is dominant.

For example, a directional antenna (whose main lobe is aligned with the transmitter, while the sidelobes are directed downward) is less susceptible to the galactic noise, atmospheric noise, and solar noise sources outside its main lobe. In this case, the noise temperature of the antenna is almost equal to the average sky temperature in the direction of its main lobe, and is therefore considerably lower (somewhere in the range of tens to hundreds of kelvins). However, if the sidelobes point upward, the received noise power becomes significant, and the noise temperature of the antenna (evaluated with Pierce's rule) is considerably higher than the sky temperature within the main lobe.

In contrast, a broad-band antenna (typically, an antenna for LW or MW frequency bands) collects noise both in the transmitter direction and from all sky noise sources. Consequently, its noise temperature is much more higher (it can reach 10^5 K).

■ **Receiving System.** Traditionally, this term describes the ensemble comprising an antenna, its feeder (transmission line, cable, or waveguide connecting the antenna to the preamplifier), the preamplifier, and the receiver itself.

Satellite (or deep-space) communications systems are characterized by the *system noise temperature*, denoted by T_{sys}.

If the antenna is directly connected to the receiver (whose input noise temperature is T_r), we have

$$\boxed{T_{sys} = T_a + T_r} \qquad (16.11)$$

T_a being the noise temperature of the antenna.

Considering a generic block diagram of a receiving system (Fig. 16.26), T_{sys} is defined at a point situated immediately after the antenna.

572   16 Noise in Communication Systems

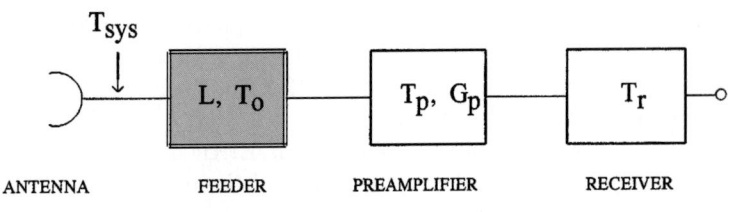

Fig. 16.26. Block diagram of a receiving system

Applying (4.66), we have

$$T_{sys} = T_a + (L-1)T_o + LT_p + LT_r/G_p \qquad (16.12)$$

where

- T_a – noise temperature of the antenna
- L – feed-line loss
- T_o – standard reference temperature (assumed equal to the ambient temperature)
- T_p – input noise temperature of the preamplifier
- G_p – preamplifier gain
- T_r – input noise temperature of the receiver

16.5.2 Case Studies

CASE STUDY 16.19 [242]

A communication downlink between a relay satellite and a low-orbit aircraft (denoted in the following by "user"), illustrated in Fig. 16.23b, has the following parameters:

Satellite transmitter output power	$P_T = 40\,\text{W}$
Satellite transmitter antenna power gain	$G_T = 28\,\text{dB}$
Transmit frequency	$f = 400\,\text{MHz}$
Receiver noise temperature of the user (antenna included)	$T_{op} = 1000\,\text{K}$
User receiver antenna power gain	$G_R = 1\,\text{dB}$
Miscellaneous losses	$L_a = 5\,\text{dB}$
System bandwidth	$B = 2000\,\text{Hz}$
Satellite-user distance	$D = 45000\,\text{Km}$

Miscellaneous losses account for various margins like transmitter aging (practical values lie between 1 and 2 dB), antenna pointing errors (about 2 dB), rain, snow, or fog attenuation (up to 2 dB), etc.

16.5 Space Communication Systems

Determine:

1) The user received signal level (in decibels)
2) The S/N ratio at the user receiver output in a 2000 Hz bandwidth.

Solution

1) First, the attenuation of the signal (or the free space loss) during propagation between the relay satellite and the user must be estimated:

$$20 \log \frac{\lambda}{4\pi D} = 20 \log \frac{0.75}{4\pi(45 \cdot 10^6)} = -177.54 \text{ dB}$$

The next task is to calculate the user received signal level; a power budget estimate yields

Free space loss	-177.54 dB
Effective radiated power (ERP)	44 dB
User receiver antenna power gain	1 dB
Miscellaneous losses (L_a)	-5 dB
Total	-137.54 dB

Therefore, the level of the signal received (by the user) is -137.54 dB.

2) We use (16.10) to estimate the S/N in decibels (per unit bandwidth) at the user's receiver output. Note that in the following expression, P_T, G_T, L_a, and G_R are all expressed in decibels.

$$\frac{S_i}{N_i} = P_T + G_T + 20 \log \frac{\lambda}{4\pi D} + G_R + 10 \log \frac{1}{T_{op}} - 10 \log k - L_a$$

$$= 16 + 28 - 177.54 + 1 - 30 - 5 + 230 - 1.4$$

$$= 61.06 \text{ dB}$$

Since the receiving system bandwidth is B = 2000 Hz, a correction is needed. Note that the signal power is not modified; only the noise power increases with bandwidth. Hence, the S/N ratio at the receiver output over bandwidth Δf is

$$\left(\frac{S_i}{N_i}\right)_{\Delta f} = \frac{S_i}{N_i \, \Delta f}$$

or, in decibels

$$\left(\frac{S_i}{N_i}\right)_{\Delta f} = \frac{S_i}{N_i} - 10 \log (\Delta f)$$

$$= 61.06 - 10 \log (2 \cdot 10^3) = 61.06 - 33 = 28.06 \text{ dB}$$

This parameter is important, since being equal to the S/N ratio at the demodulator input, it provides good insight into what modulation system to

574   16 Noise in Communication Systems

select in order to ensure the best detected S/N ratio. In the present case, pulse code modulation (PCM) would be the best choice, but frequency modulation (FM) might also be employed.

CASE STUDY 16.20 [244]

A receiving antenna with noise temperature 57 K must be connected to the preamplifier by a feeder. The user must decide among the following technical solutions:

a) cheap preamplifier (power gain 14 dB and noise temperature 85 K) that can be located near the antenna, so that a short feeder is required (with a loss of 1 dB);
b) high-quality preamplifier (power gain 30 dB and noise temperature 25 K), which cannot be located nearby, so that a longer feeder (2 dB loss) is needed.

In both cases, the receiver input noise temperature is $T_r = 290$ K. Which choice offers the best noise performance?

Assume an ambient temperature of 290 K.

Solution

The general configuration of the receiving system is shown in Fig. 16.27.

We first transform decibels to ratios. Then, all numerical data are collected in the Table 16.5.

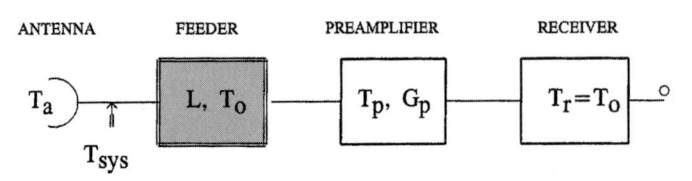

Fig. 16.27. A simplified receiving system diagram

Table 16.5. Numerical values of various system parameters

Parameter	Notation	Case a	Case b
Feeder loss	L	1.26	1.584
Preamplifier gain	G_p	25.12	1000
Preamplifier noise temperature	T_p	85 K	25 K

The next task is to define a figure of merit for the noise performance of the receiving system. As in all low-noise applications, we may reasonably select the system input noise temperature, which can be calculated with (16.12):

$$T_{sys} = T_a + (L-1)T_o + LT_p + LT_r/G_p$$

Let us evaluate it in case a:

$$T_{sys,a} = 57 + (1.26 - 1)290 + (1.26)85 + (1.26)290/25.12$$
$$T_{sys,a} \cong 254 \text{ K}$$

For case b, we obtain

$$T_{sys,b} = 57 + (1.584 - 1)290 + (1.584)25 + (1.584)290/1000$$
$$T_{sys,b} \cong 266.4 \text{ K}$$

The first important observation is that the best choice is case a, despite the considerably better performance of the second preamplifier. Thus, *it makes sense to shorten the feeder and locate the preamplifier as close as possible to the antenna*, even if in the process, a less performant preamplifier must be adopted.

The second remark is that a slight rise in the noise temperature of the preamplifier employed in case a (for instance, T_p equal to 95 K instead of 85 K) restores the balance between the two proposed solutions.

CASE STUDY 16.21 [243]

A receiving system for space communications (Fig. 16.28) has a bandwidth of 100 MHz, employs a narrow-beam antenna, and is matched throughout. The following notation has been adopted: T_k is the average sky temperature within the main lobe of the antenna, T_1 and T_3 are the physical temperatures of the attenuating two-ports, T_{e2} and T_{e4} are the input noise temperatures of blocks 2 and 4, respectively, and G_2, G_4, are the available gains of blocks 2 and 4.

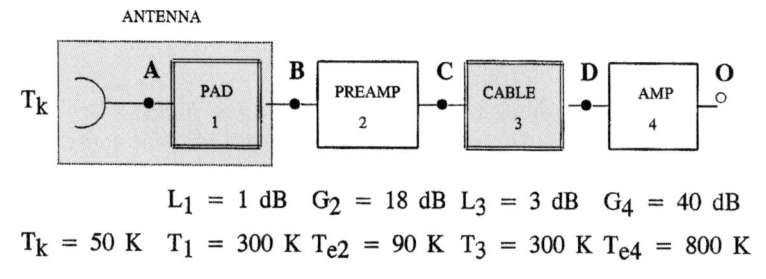

$L_1 = 1$ dB $G_2 = 18$ dB $L_3 = 3$ dB $G_4 = 40$ dB
$T_k = 50$ K $T_1 = 300$ K $T_{e2} = 90$ K $T_3 = 300$ K $T_{e4} = 800$ K

Fig. 16.28. Receiving system for space communications

Note that block 1 does not correspond to a physical attenuator pad, but merely lumps together the losses within the antenna feeder, pedestal, and dome.

a) Determine the output noise power.
b) Find the noise floor, if for normal operation an output S/N ratio better than 10 dB is required.

For simplicity, assume that all noise temperatures are constant with frequency.

Solution

a) Given that a narrow-beam antenna is employed, we may identify T_k with the antenna temperature; hence, $T_a = T_k$.
We first transform decibels to ratios:

$$L_1 = 1\,\text{dB} \rightarrowtail 1.259 \qquad L_3 = 3\,\text{dB} \rightarrowtail 2$$
$$G_2 = 18\,\text{dB} \rightarrowtail 63.1 \qquad G_4 = 40\,\text{dB} \rightarrowtail 10^4$$

Using formula (5.13), the input noise temperatures of attenuating blocks are calculated:

$$T_{e1} = (L_1 - 1)T_1 = (1.259 - 1)300 = 77.7\,\text{K}$$
$$T_{e3} = (L_3 - 1)T_3 = (2 - 1)300 = 300\,\text{K}$$

Since for each block the noise is expressed in terms of its input noise temperature, the overall input noise temperature of the cascade can be obtained with (4.66). *Caution:* the input of the cascade is considered point B, because PAD 1 is not a physical two-port and cannot be dissociated from the antenna. Therefore, all noise contributions of blocks 2, 3, and 4 are referred to point B:

$$T_{eB} = T_{e2} + \frac{T_{e3}}{G_2} + \frac{T_{e4}}{G_2 G_3} = 90 + \frac{300}{63.1} + \frac{700}{63.1(1/2)} \cong 120.1\,\text{K}$$

The next task is to evaluate the noise contribution of the antenna to the same point B, applying (5.12):

$$T_B = \frac{1}{1.259}\,50 + \left(1 - \frac{1}{1.259}\right)300 \cong 101.4\,\text{K}$$

This is a very surprising result, since despite the highly directional antenna (and consequently low sky temperature), the noise contribution of the antenna is almost equal to the overall contribution of the cascaded stages!
The operating noise temperature at point B is the sum of T_{eB} and T_B. The total noise power at point B is calculated assuming that the noise bandwidth is equal to the $-3\,\text{dB}$ bandwidth of the system (which is true when the number of stages is significant):

$$N_B = k(T_{eb} + T_b)\Delta f = (1.38\ 10^{-23})(120.1 + 101.4)10^8$$
$$N_B \cong 3.06 \cdot 10^{-13}\,\text{W}$$

or expressed in decibels,

$$N_B = 10\,\log\,(3.06 \cdot 10^{-13}) = -125.14\,\text{dB}$$

As all noise contributions have been taken into account, we must refer N_B to the output (point O) by transmission through the cascade. Allowing for various gains and losses, the output noise power is

$$N_o = -125.14 + 18 - 3 + 40 = -70.14 \text{ dB}$$

b) In the worst case, the output S/N ratio must be 10 dB. It follows that the output signal level must be -60.14 dB. This corresponds to an input signal available power (point B):

$$S_B = S_o - (18 - 3 + 40) = -115.14 \text{ dB}$$

which represents the *noise floor*.

CASE STUDY 16.22 [243]

This case study is devoted to the problem of feeders connecting antennas to preamplifiers, and specifically the expected degradation in S/N ratio.

A steerable dish antenna has noise temperature $T_a = 100$ K and collects a useful signal at -100 dB. The feeder connecting the antenna to the preamplifier has loss $L = 3$ dB and its temperature is $T_f = 290$ K. The system noise bandwidth is 10 MHz, and it is assumed that all noise quantities are constant with frequency. Assuming that the system is matched throughout, estimate the degradation in the S/N ratio introduced by the feeder.

Solution

■ **Discussion.** Any communication system ought to be match throughout, meaning that the antenna is matched to its feeder, the feeder to the preamplifier, and so on. However, the antenna is not a simple resistor at a known physical temperature, so it is represented by a source whose internal impedance and temperature equal the characteristic impedance and noise temperature of the antenna.

When investigating the influence of the feeder (henceforth denoted the pad, for generality) on the S/N ratio, two situations may be encountered:

– When the signal generator (antenna) has the same noise temperature as the ambient temperature of the feeder ($T_o = 290$ K), the theorem of Case Study 16.1 and Friis' definition of the noise factor can be successfully employed. This situation is depicted in Fig. 16.29a, where the feeder is represented as an attenuating two-port having at its input the signal available power S_i and the noise available power N_i. After passing through the two-port, the signal is attenuated by L (the loss of the two-port), and the noise, like any signal, is also attenuated by L. However, the two-port simultaneously adds its own thermal noise, which according to Case Study 14.23 exactly compensates the loss L. For this reason, at the output, the noise available power is the same as at the input.

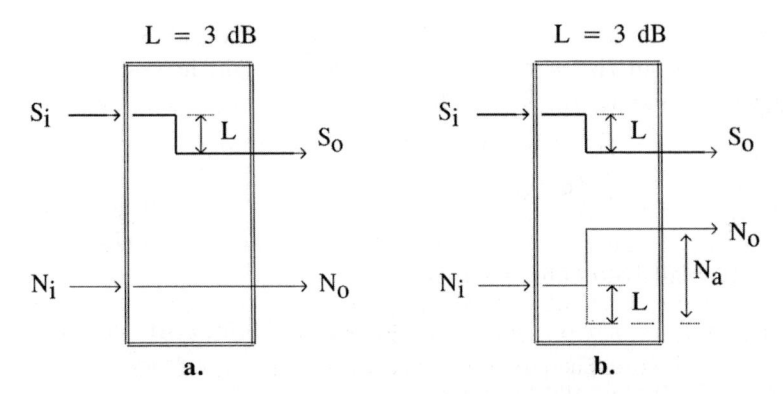

Fig. 16.29. Influence of a lossy two-port (feeder) on S/N ratio; **a** feeder has the same temperature as the antenna; **b** feeder at higher temperature than the antenna

- When the noise temperature of the antenna is lower than the feeder temperature, the situation detailed in Fig. 16.29b must be considered. At the output, the available signal power S_i is attenuated by the two-port loss L. The available noise power is also reduced by L, but this time the noise contribution of the pad (N_a) is much greater than that of the signal source, since its temperature is higher than that of the source. The noise level at the output is obviously then higher than at the input, and significant degradation of the S/N ratio is expected.

■ **Analysis.** The available input noise power in noise bandwidth Δf is

$$N_i = kT_a \Delta f = (1.38 \cdot 10^{-23}) \cdot 100 \cdot 10^7 = 1.38 \cdot 10^{-14} \text{ W} \quad \text{or} \quad -138.6 \text{ dB}$$

The input S/N ratio, expressed in decibels, becomes

$$S_i/N_i = -100 - (-138.6) = 38.6 \text{ dB}$$

The available output noise power derives from the input noise power transmitted to the output (with a loss of 3 dB), to which the intrinsic noise power of the two-port is added. Thus

$$N_o = \frac{N_i}{L} + N_a = \frac{N_i}{L} + \frac{kT}{L}(L-1)\Delta f$$

where use has been made of the expression (5.12). Substituting numerical values and recalling that to a 3 dB loss corresponds a factor of 2,

$$N_o = \frac{1.38 \cdot 10^{-14}}{2} + \frac{1.38 \cdot 10^{-23} \cdot 290 \cdot (2-1) \cdot 10^7}{2} \cong (0.69 + 2)10^{-14} \text{ W}$$

$$N_o = 2.69 \cdot 10^{-14} \text{ W} \quad \text{or} \quad -135.7 \text{ dB}$$

Hence
$$S_o/N_o = (-100 - 3) - (-135.7) = 32.7 \text{ dB}$$

Widespread convention chooses to define the degradation of the S/N ratio as the difference between the two quantities expressed in decibels, i.e.,

$$\text{Degradation} = \left(S_i/N_i\right)_{\text{dB}} - \left(S_o/N_o\right)_{\text{dB}} = 38.6 - 32.7 = 5.9 \text{ dB}$$

■ **Concluding Remark.** If the temperature of the feeder is higher than the noise temperature of the antenna, the feeder is responsible for a S/N ratio degradation that is almost twice its loss. This case study illustrates the impact of cabling on the global noise performance of a system.

CASE STUDY 16.23 [244]

A steerable dish antenna has noise temperature 65 K and is connected by a waveguide to a parametric preamplifier. The loss of the waveguide is 2 dB and the parametric amplifier has a gain of 15 dB and an input noise temperature of 77 K. The receiver has a noise figure of 10 dB. Determine the spectral density of the available input noise power at the reference temperature ($T_o = 290$ K).

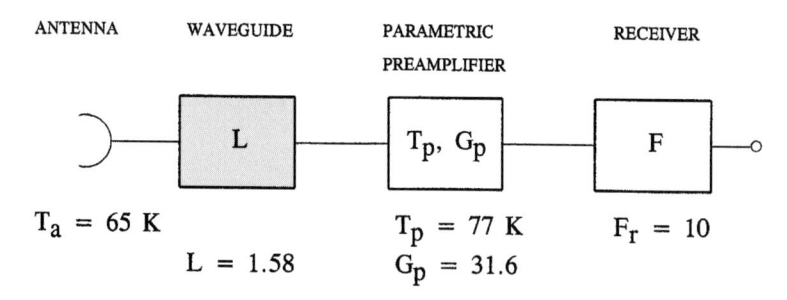

Fig. 16.30. Receiving system for space communications

Solution

The simplified system diagram is shown in Fig. 16.30, where all quantities given in decibels are expressed as ratios.

The available input noise power can be calculated from the system noise temperature (T_{sys}), which lumps all noise contributions at the system input. According to (16.12)

$$T_{\text{sys}} = T_a + (L-1)T_o + LT_p + LT_r/G_p$$

The input noise temperature of the receiver (T_r) can be deduced from its noise factor by means of (4.14):

$$T_r = T_o(F_r - 1) = 290(10 - 1) = 2610 \text{ K}$$

Substituting into the equation for T_{sys}, we obtain

$$T_{sys} = 65 + (1.58 - 1)(290) + (1.58)(77) + (1.58)(2610)/31.6$$
$$\cong 485 \text{ K}$$

Finally, with definition (3.3), the spectral density of the available input noise power is

$$S(p) = kT_{sys} = (1.38 \cdot 10^{-23})(485) \cong 6.69 \cdot 10^{-21} \text{ W/Hz}$$

■ **Concluding Remark.** It is highly desirable to achieve a low T_{sys} that is almost independent of the noise performance of the receiver. In this case, it is important to select a preamplifier with high gain. In this way, the receiver contribution to the overall noise (the term LT_r/G_p) becomes negligible, and eventually a cheaper receiver may be used without any impact on global noise performance.

CASE STUDY 16.24 [247]

A microwave communications system employs a tunnel-diode oscillator. The circuit containing the diode can be represented as a one-port with admittance $G_D = -3 \text{ mS}$ and noise temperature $T_D = -5T_o$, which is in parallel with a resistive load $G_L = 9 \text{ mS}$ (assumed to be at the reference temperature T_o). Determine the equivalent noise temperature of the ensemble.

Solution

The simplified equivalent circuit of the one-port is presented in Fig. 16.31.

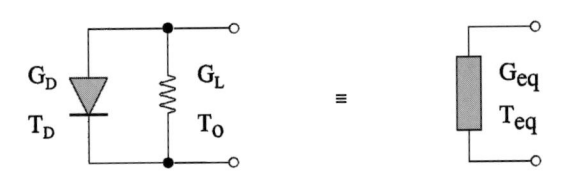

Fig. 16.31. Equivalent circuit of the tunnel diode and its load

This application is typical of many practical situations in which we must extend some traditional definitions of noise quantities. For instance, here we have a tunnel diode with negative admittance; therefore, instead of absorbing power, it generates power. Its noise performance cannot be described with the traditional noise factor, but instead with an *extended noise factor*, based on the *exchangeable power* concept. However, we are not interested here in a noise factor description, but in an equivalent noise temperature. This parameter must also be extended, and in its definition (given in Sect. 4.5.3) the term *available power* must be replaced with the term *exchangeable power*. Consequently, the equivalence between the one-ports of Fig. 16.31 requires that both generate the same noise power per unit bandwidth, i.e.,

$$4kT_D G_D + 4kT_o G_L = 4kT_{eq} G_{eq}$$

which yields

$$\boxed{T_{eq} = \frac{G_D}{G_{eq}} T_D + \frac{G_L}{G_{eq}} T_o}$$

with

$$G_{eq} = G_L + G_D = 9 - 3 = 6 \text{ mS}$$

Substituting numerical values, we have

$$T_{eq} = \frac{-3}{6}(-5T_o) + \frac{9}{6} T_o = 4\, T_o = 1160 \text{ K}$$

17
Computer-Aided Noise Analysis

> "Beware of bugs in the above code: I have only proved it correct, not tried it"
>
> (Donald Knuth)

■ **Overview.** The noise analysis previously performed has been restricted to hand calculation. The resulting advantages are insight into circuit operation, understanding the effects of various noise sources, and how the circuit elements affect the noise parameters. The price to be paid is that calculations are time-consuming and the resulting expressions are often cumbersome. The situation worsens when we are dealing with complex impedances, and when the noise analysis must be carried out at several frequencies.

Therefore, it seems reasonable to use existing electrical simulators, in order to perform fast and accurate noise analysis. However, attention must be paid to the following point: while almost all electrical simulators like SPICE have behind them a history of several decades during which they have been intensely tested and continuously improved, noise simulation is in a quite different position.

In practice, noise analysis is either available as an option in electrical simulators, or it may be performed with the few existing dedicated software packages.

In the former case, the circuit may have any topology and is described using the node-branch concept; a typical example is SPICE or any other member of the family (PSPICE, HSPICE, ISPICE, etc.). As made obvious in Chap. 7, the noise models employed are not necessarily the most efficient ones and up- dating them is a real necessity. Another limitation is the impossibility of directly accessing the four noise parameters (F_o, R_n, G_o, and B_o). Traditionally, the user only obtains information concerning the total noise voltage at a node designated as output or input.

A dedicated noise analysis program (proposed in this book) is NOF, conceived for the simulation of RF and microwave circuits, with specified terminal input/output resistances of $50\,\Omega$. As for SPICE, any topology can be accepted by means of a node-branch description, but, relative to SPICE capabilities, the size of the circuit is limited. The four noise parameters are available, but the electrical simulation is restricted to frequency domain analysis.

As for dedicated software packages, there are several programs concerned with noise analysis, sometimes restricted to a particular category of circuits [250, 251]. Like NOF, they have not been enough used and very likely, the old adage: "Fortunately, the second-to-last bug has just been fixed" still applies.

■ **Classification.** In practice, one of the following two situations may be encountered, depending on the type of circuit subjected to noise analysis:

1) *When the linear circuit operates with unmatched terminal impedances* (as is the case for all sensor–preamplifier systems), *the only meaningful result is the $E_n - I_n$ model* of the circuit. This can be deduced from the equivalent input noise voltage or current, which are directly computed by any simulator of the SPICE family.
2) *When the investigated linear circuit operates with matched 50-Ω terminations* (as is the case for all communication and microwave circuits), *the only significant result is the ensemble of four noise parameters.* In this case, program NOF can be employed successfully.

In the following, a section will be devoted to each category.

17.1 Noise Simulation with PSPICE

17.1.1 SPICE – An Overview

■ **History.** SPICE (Simulation Program with Integrated Circuit Emphasis) was developed at the University of California, Berkeley, in the late 1960s and released in the early 1970s. The Ph.D thesis of Dr. L. Nagel contains the description of the numerical methods and various algorithms implemented in SPICE for DC, AC, and transient circuit analysis. Over the years, SPICE has been continuously upgraded. Perhaps the most significant improvement came with SPICE2, which included advanced simulation methods and models for integrated circuits.

One of the most widely used versions derived from SPICE2 is PSPICE, released by MicroSym Corporation in the 1980s. Two major improvements occurred, one in accelerating convergence and the other in graphics post-processing (module PROBE). Its evaluation form, whose size is limited at most to 10 transistors, is freely distributed, and contributed to rethinking the electrical engineering curriculum at most universities.

In the middle of 1990s new versions under Windows were released, allowing input of the submitted circuit by drawing its schematic.

17.1 Noise Simulation with PSPICE

Since the explanation of noise simulation is better understood in conjunction with the DOS version of PSPICE, this approach will be adopted in the following. It is expected that the reader will have no difficulty performing noise analysis with the later versions of PSPICE, once the basic concepts (common to all versions) have been assimilated.

■ **Structure.** In the DOS version, the circuit is described by means of an input data file; when saved, it receives the suffix ".CIR". When running PSPICE, the following files are automatically generated:
- the output file, which has the same name as the input file, except for the extension, which is ".OUT";
- a file with the same name and extension ".DAT", where all data for the graphic module PROBE are collected;
- two internal backup files, with the same name and the extensions ".CBK" and ".CFG" respectively.

■ **Circuit Description.** When preparing the circuit for analysis with PSPICE, the first task of the user is to identify all nodes of the circuit and to label them. This can be achieved by using positive integers (not necessary in sequential order) or names (any alphanumeric string up to 131 characters). It is advisable (but not mandatory) to let the ground of the circuit be node zero.

■ **Input Data File.** This is the image of the circuit to be simulated obtained when using a node-branch description. The input data file is always created by the user, either with the incorporated text editor (DOS versions), or when drawing the circuit schematics (WINDOWS versions).

In the former case [252], PSPICE disregards upper/lower case; for example, "VOUT" and "vout" are equivalent names.

The typical structure of the input file is the following:

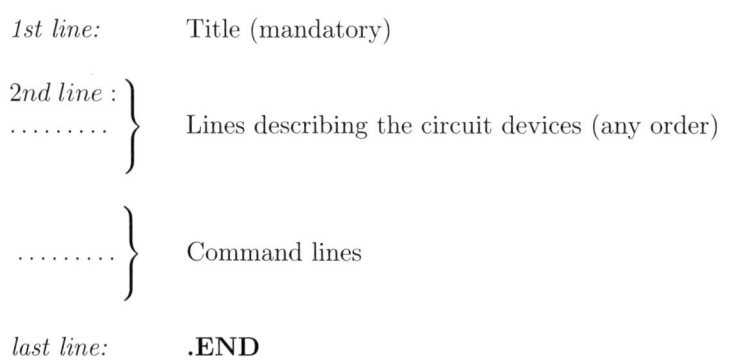

1st line:	Title (mandatory)
2nd line : ⎫ .... ⎬ ⎭	Lines describing the circuit devices (any order)
.... ⎫ ⎬ ⎭	Command lines
last line:	**.END**

Except for the first and last line, as well as lines describing subcircuits, the order of other lines is arbitrary; however, for easy debugging, it is advisable to use a well-structured pattern by grouping lines describing the same category of elements.

Table 17.1. Device description

Let.	Device	Example	Explanation
R	Resistor	R25 5 2 2.2K	2.2-kΩ resistor between nodes 5 and 2
C	Capacitor	CIN 1 0 1U	1-µF capacitor between nodes 1 and 0
L	Inductor	LA 2 3 5M	5-mH inductor between nodes 2 and 3
K	Inductive Coupling	K1 L1 L2 0.9	Mutual coupling of 0.9 between L1 and L2
T	Transmission Line	T1 1 2 3 4 *par*	Transm. line: input (1,2) and output (3,4)
V	Independent Voltage Source	VE 2 4 AC 1V VCC 5 3 DC 12V	AC – altern. current DC – direct current
I	Independent Current Source	I1 2 4 AC 1mA IG 5 3 DC 2A	2 node+ ; 4 node – 5 node+ ; 3 node –
G	Voltage Controlled Current Source	GA 1 3 6 8 5K	1 node + of the source 3 node – of the source
E	Voltage Controlled Voltage Source	EB 1 3 6 8 5K	6 + controlling node 8 – controlling node
F	Current Controlled Current Source	FD 1 3 VSEN 5K	VSEN – current sensor
H	Current Controlled Voltage Source	HC 1 3 VSEN 5K	5K – value of the control coefficient
Q	Bipolar Transistor	Q1 7 8 9 *name*	7 – Collector 8 – Base, 9 – Emitter
D	Diode	DAB 1 2 *name*	1 node +, 2 node – *name* – model name
J	Junction FET	J1 7 8 9 6 *nm*	7 – Drain 8 – Gate 9 – Source 6 – Bulk
M	MOS Transistor	MA 7 8 9 6 *nm*	*nm* – model name
X	Subcircuit	XB 1 2...N *name*	1 2...N subcircuit nodes

17.1 Noise Simulation with PSPICE

□ **Rules**

- Comments lines are marked by "*" in the first column; they may appear anywhere in the file and may contain any text.
- Continuation lines are marked by "+" in the first column.
- The number of blanks between items is not significant; tabs and commas are equivalent to blanks.
- Blank lines may be inserted anywhere in the file.
- Command lines start with "." in the first column and may be inserted anywhere, except for the command **.END**, which must be the last line.

■ **Device Description.** Each device is described with a statement (line) containing the following information: device name, the nodes where it is connected, a model name (not for all devices), and its value (or values of its electrical parameters). The first letter of the device name specifies its type, according to Table 17.1.

The value of the element can be written in the standard floating-point notation or with the scale suffixes given in Table 17.2.

Table 17.2. Scale factors

Suffix	Name	Factor
T	tera	10^{12}
G	giga	10^{9}
MEG	mega	10^{6}
K	kilo	10^{3}
M	milli	10^{-3}
U	micro	10^{-6}
N	nano	10^{-9}
P	pico	10^{-12}
F	femto	10^{-15}

■ **Models.** Almost every device needs a model command line, which starts with **.MODEL**. On the same line are information concerning model type and eventually the values of its parameters. If no values for the parameters are provided by the user, PSPICE loads default values from its internal library. The syntax of this line is:

    .MODEL   *name*   *type name*   *values of parameters*

The type names of various semiconductor devices, as they appear in the internal library of PSPICE, are shown in Table 17.3.

There may be more than one model of the same type in a circuit, although they must have different names.

□ **Example.** Consider the line:

    .MODEL   M1   NPN   BF=250

Table 17.3. Name of models (semiconductor devices)

Type name	Device
D	Diode
NPN	NPN bipolar transistor
PNP	PNP bipolar transistor
LPNP	lateral PNP bipolar transistor
NJF	N-channel junction FET
PJF	P-channel junction FET
NMOS	N-channel MOS
PMOS	P-channel MOS
GASFET	N-channel GaAs MESFET

Here M1 represents the user's name for the bipolar transistor in question, **NPN** is the type name (according to Table 17.3) and **BF** represents the name of parameter β of the transistor, whose value is assigned by the user to be 250, instead of 100 (the default value).

■ **Independent Sources.** These are described either by a constant value, marked on the statement line:

VNAME   N+   N–   AC   12

(for an AC voltage of 12 V connected between nodes N+ and N–), or are associated with one of the following options:

PULSE for a pulse voltage waveform, the syntax is:

VNAME   N+   N-   PULSE(*V1   V2   TD   TR   TF   PW   PER*)

where the parameters of the waveform are: *V1, V2* the initial and final value of the pulse; *TD* is the delay time; *TR* is the rise time of the pulse; *TF* is the fall time of the pulse; *PW* is the pulse width; and *PER* is the period (if no value is assigned to *PER*, PSPICE assumes a single pulse).

SIN for a sinusoidal voltage waveform, the syntax is:

VNAME   N+   N–   SIN(*VO   VA   FREQ   TD   DF   THETA*)

where the parameters of the waveform are: *VO* is the offset value of the sine wave; *VA* is the amplitude; *FREQ* is the frequency in hertz; *TD* is the delay time from zero; *DF* is the damping factor; and *THETA* is the phase.

EXP for an exponential waveform, the syntax is:

VNAME   N+   N–   EXP(*V1   V2   TD1   TC1   TD2   TC2*)

where the parameters of the waveform are: *V1, V2* the initial and peak voltage value; *TD1* is the rise delay; *TC1* is the rise time constant; *TD2* is the fall delay; and *TC2* is the fall time constant.

PWL for a piecewise linear waveform, the syntax is:

VNAME N+ N- PWL(*T1 V1 T2 V2 ... Tn Vn*)

where the parameters of the waveform are: *T1 V1, T2 V2* etc. These represent data pairs (time at corner, voltage at corner). Between the corners, the program performs a linear interpolation.

SFFM for a frequency-modulated waveform, the syntax is:

VNAME N+ N- SFFM(*VOFF VAMPL FC MOD FM*)

where the parameters of the waveform are: *VOFF* is the offset voltage; *VAMPL* is the voltage amplitude; *FC* is the carrier frequency; *MOD* is the modulation index; and *FM* is the modulation frequency.

■ **Circuit Analysis.** Various kinds of analysis can be performed with the commands detailed in Table 17.4.

Table 17.4. Various kind of analyses

Command	Syntax
.DC	.DC LIN START VALUE FINAL VALUE INCREMENT
.OP	.OP
.AC	.AC OPTION POINTS START FREQ END FREQ
.TRAN	.TRAN PRINT STEP FINAL TIME
.TF	.TF OUTPUT VARIABLE INPUT SOURCE NAME
.SENS	.SENS OUTPUT VARIABLE
.FOUR	.FOUR FREQUENCY VALUE OUTPUT VARIABLE
.NOISE	.NOISE VNODE (SOURCE) NAME
.TEMP	.TEMP TEMPERATURE VALUE

☐ **Explanation**

.DC DC sweep analysis. Example: .DC VCC 5 15 0.1 calculates the bias point of the circuit when VCC is swept over the range 5 V to 15 V, with an increment of 0.1 V.

.OP specifies to print detailed information about the bias point, which is calculated whether or not .OP is specified

.AC is used to compute the frequency response of the circuit over a range of frequencies. OPTION is one of the following: LIN, DEC, or OCT, and specify the type of sweep (linear sweep, sweep by decades, or sweep by octaves). Example: .AC DEC 50 1 1MEG indicates frequency analysis over the range 1 Hz to 1 MHz, with 50 points in each decade.

.TRAN specifies a transient analysis of the circuit, with a time step that is internally adjusted as the analysis proceeds (it is *not* under the user's control). Example: .TRAN 1M 100M indicates a time-domain analysis, starting at t = 0, up to 100 ms. 1 ms is the time step used to print (plot) the results.

.TF allows the small-signal transfer function to be calculated, by linearizing the circuit around the bias point. Example: .TF VIN V(8) means that the transfer function defined as V(8)/VIN is evaluated, together with the input and output resistances.

.SENS causes a DC sensitivity analysis to be performed (by linearizing the circuit around the bias point), with respect to all device values. Example: .SENS V(12) I(VCC) is a command to compute the first-order sensitivities of V(12) and I(VCC) with respect to variations in all circuit elements.

.FOUR specifies a Fourier analysis to be performed on the result(s) of a transient analysis. This command requires a .TRAN statement. Example: .FOUR 50 kHz V(3) I(R4) indicates that a Fourier analysis is required on the waveforms of V(3) and I(R4), with respect to the fundamental, which is 50 kHz. Note that the transient analysis must be at least 1/50 kHz long.

.NOISE causes a noise analysis of the circuit to be carried out. This command requires a .AC statement. Examples:
   .NOISE V(5) VIN
   .NOISE V(5) IG
denotes that V(5) is an output voltage and VIN and IGEN are the names of independent voltage or current sources at which the equivalent input noise will be calculated.

.TEMP is used to establish the temperature at which simulation is performed. *Caution:* its default value is 27°C (i.e., 300 K, rather than the standard reference temperature of 290 K). Example: .TEMP−25 +60 means that analysis is performed at −25°C, then repeated at +60°C.

■ **Help on Line.** PSPICE (version 5) provides an on-line user's manual, when stroking the key F3 (when we are in the main menu).

Table 17.5. Internal functions available in PROBE

Name	Explanation	Name	Explanation
ABS(x)	$|x|$	**SIN(x)**	x in radians
SGN(x)	Sign of x; 0 if x=0	**COS(x)**	x in radians
SQRT(x)	$\sqrt{x}$	**TAN(x)**	x in radians
EXP(x)	e^x	**ATAN(x)**	result in radians
LOG(x)	ln(x)	**D(x)**	Derivative of x
LOG10(x)	log(x)	**S(x)**	Integral of x
M(x)	Amplitude (norm)	**DB(x)**	Amplitude in dB
P(x)	Phase (in degrees)	**AVG(x)**	Average of x
R(x)	Real part	**RMS(x)**	Root mean square of x
PWR(x,y)	$|x|^y$		

17.1 Noise Simulation with PSPICE

■ **PROBE.** PROBE is the graphics post-processor of PSPICE. It provides the same information as an ideal oscilloscope connected to any circuit node. The traces can be output variables (available with key F4, from *Add Trace* command of the plot menu) or may be specified by the user, as arithmetic expressions (using the usual operators: "+", "−", "∗", "/", and parentheses). Many internal functions are also available (see Table 17.5).

■ **Example.** Consider the differential amplifier proposed in Fig. 17.1, where the transistors are assumed to be identical, with $\beta = 275$.

The input file which describes this circuit is:

```
Differential amplifier (example)
VCC    1   0    15
VEE    2   0    -9
VIN1   3   0    AC   1M
VIN2   7   0    AC   1.1M
RC1    1   4    10K
RC2    1   6    10K
RE     5   2    4.3K
Q1     4   3    5    MOD1
Q2     6   7    5    MOD1
.MODEL   MOD1   NPN (BF=275)
.AC  DEC 20  1    100MEG
.END
```

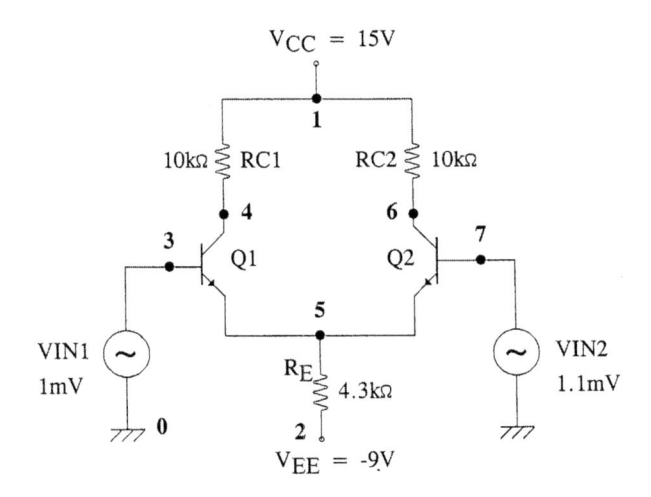

Fig. 17.1. Illustrative circuit for PSPICE simulation

17.1.2 Noise Analysis

■ **Method.** PSPICE assigns thermal noise to each resistor of the circuit, in the form of a noise current generator across the resistor in question (Fig. 3.2c). The noise of every diode and transistor is described as shot noise and 1/f noise, according to the models presented in Sects. 7.4.4, 7.6.7, 7.7.7, and 7.8.2. During analysis, all noise generators are considered to be independent sine generators, whose output contributions are separately evaluated (this is why the frequency response of the circuit must be previously computed). Then, assuming that all noise sources are uncorrelated, their mean square contributions are added at the node designated as output.

It should be remembered that the command for noise analysis is:

.NOISE   V(8)   VIN

where node 8 has been designated as an output (not necessarily the actual output of the circuit) and VIN is the input voltage signal source. In order to reflect the noise to the VIN terminals, PSPICE automatically evaluates the gain between VIN and node 8 (with respect to ground). The total output noise is then divided by this gain and consequently the equivalent input noise is obtained. When the input signal source is of voltage type, the equivalent input noise is expressed in $V/\sqrt{Hz}$; if it is of current type, the units are $A/\sqrt{Hz}$. The output noise is always expressed in $V/\sqrt{Hz}$.

■ **Remarks**

– The results are accurate *provided that all noise sources are uncorrelated*. If at least two noise sources are partially correlated, errors appear because PSPICE cannot take into account the correlation and considers them as being independent.
– It is risky to perform noise analysis when a particular device is replaced by its equivalent circuit, because this circuit may contain resistances free of thermal noise. A typical example is the r_{ce} resistance of Giacoletto's equivalent circuit for a bipolar transistor, which simulates the transport of charge carriers between emitter and collector. This resistance does not correspond to a physical dissipative resistor, and consequently no thermal noise can be associated with it. However, PSPICE will treat it like any resistor, with thermal noise established by Nyquist's formula.

17.1.3 Simulation Techniques

■ **Objective.** The purpose of this section is to recall some background from classical network theory, and to present various techniques of building noise models for advanced simulation.

■ **Independent Sources.** By definition, *an independent source is a source whose terminal voltage and current are determined solely by the source itself and eventually by the load connected to its terminals.*

For instance, the voltage delivered by a battery cell depends exclusively on the device itself (materials and electrolyte employed) and its load.

■ **Controlled Sources.** Also called *dependent sources, these are voltage (or current) sources, whose voltage (or current) depends on the voltage or current in some other part of the circuit*. In practice they are never found as separate devices, but they are useful concepts in the analysis of electronic circuits or filter synthesis. Note that they are *unilateral* elements.

Controlled sources are grouped into four basic categories, shown in Fig. 17.2.

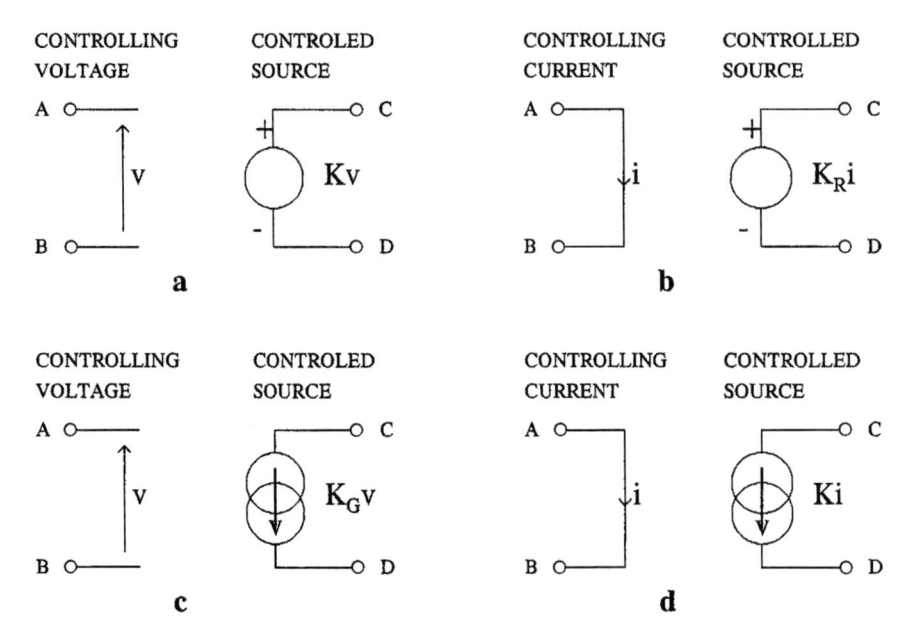

Fig. 17.2. Various types of dependent sources: **a** voltage controlled voltage source (VCVS); **b** current controlled voltage source (CCVS); **c** voltage controlled current source (VCCS); **d** current controlled current source (CCCS)

Note that the control coefficients K are gain factors (dimensionless), but K_G is expressed in units of admittance (S) and K_R has dimensions of resistance (Ω). Typically, a VCVS belongs to the equivalent circuit of any operational amplifier, while a VCCS represents the $i_c = i_c(v_{b'e})$ relationship in Giacoletto's equivalent circuit of bipolar transistors.

For modeling purposes, *dependent sources are the favorite tools to transfer a noise voltage or a noise current from one part of the circuit to another*.

■ **Flat Noise Standard Sources.** Always in noise simulation we need reference (or standard) sources with a flat frequency spectrum. They are built

17 Computer-Aided Noise Analysis

with resistors, whose thermal noise is set to the desired level by appropriately selecting the resistor value, at T = 300 K. The following standards are useful in most applications:

☐ **Noise Standards of $1\,\text{pA}/\sqrt{\text{Hz}}$.** The noise standard of $1\text{pA}/\sqrt{\text{Hz}}$ can be created by using the thermal noise delivered by a short-circuited resistor, whose value must be:

$$R_{ni} = 4kT/\overline{i_n^2} = \frac{4(1.38 \cdot 10^{-23})(300)}{10^{-24}} = 16.56\,\text{k}\Omega$$

The corresponding model and code lines are indicated in Fig. 17.3a; V_{is} is a voltage source set to zero, acting merely as a short-circuit and at the same time as a current sensor, useful to sense the noise to be transferred elsewhere.

☐ **Noise Standards of $1\,\text{nV}/\sqrt{\text{Hz}}$.** The noise standard of $1\,\text{nV}/\text{Hz}$ is obtained with two identical resistors connected in parallel. A noise voltage of $1\,\text{nV}/\text{Hz}$ is delivered at T = 300 K by a resistance:

$$R = \frac{\overline{v_n^2}}{4kT} = \frac{10^{-18}}{4(1.38 \cdot 10^{-23})(300)} = 60.386\,\Omega$$

Since PSPICE requires at least two elements connected to each node, we use two identical resistors connected in parallel, of twice the calculated value (i.e., 2(60.386) = 120.772 Ω). The model, together with its code lines, is shown in Fig. 17.3b. Noise voltage at node A is available to be transferred by means of a controlled source.

MODEL OF $I_N = 1\text{pA}/\sqrt{\text{Hz}}$

R_{ni} 16.56k , V_{is} 0

RNI B 0 16.56K
VIS B 0 DC 0

a

MODEL OF $E_N = 1\text{nV}/\sqrt{\text{Hz}}$

R_{n1} 120.772 , R_{n2} 120.772

RN1 A 0 120.772
RN2 A 0 120.772

b

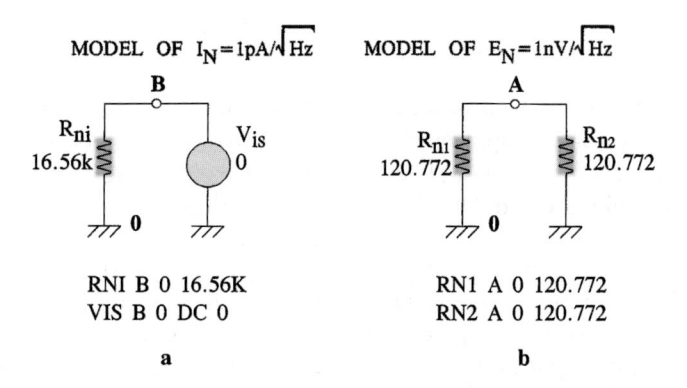

Fig. 17.3. a Current noise reference; b voltage noise reference

☐ **Remark.** The noise standards are created as auxiliary circuits, apart from the main circuit to be simulated. The noise is transferred with controlled sources at the locations where it is needed. Properly selecting the control coefficients (K, K_R or K_G) scales the noise level to the desired value.

17.1 Noise Simulation with PSPICE

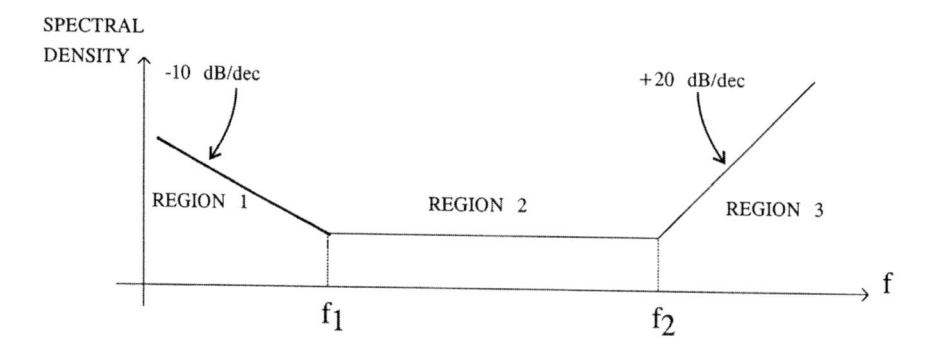

Fig. 17.4. Typical frequency spectrum of the noise of an electronic device (asymptotic representation)

■ **Noise Sources with Frequency-Dependent Spectra.** Seldom does the noise of an electronic device have a flat spectrum. In most situations, its spectrum is frequency-dependent, having three regions: a low-frequency region (dominated by flicker noise, whose slope is $-10\,\text{dB/decade}$), a high-frequency region (where the noise generally rises at $+20\,\text{dB/decade}$), and a flat region extending between them (Fig. 17.4).

□ **1/f Noise Spectrum Modeling.** The only way to simulate the low-frequency region of the spectrum (where 1/f noise dominates) is to use a semiconductor diode, whose noise current spectral density is described in PSPICE by the expression (7.17), recalled here:

$$S(\overline{I_n^2}) = 2q\,I_{DC} + \frac{KF\,(I_{DC})^{AF}}{f} \quad (17.1)$$

The first term of the sum is responsible for the flat region of the spectrum, and the last term corresponds to the excess noise, proportional to 1/f (hence, a slope of $-10\,\text{dB/decade}$). It follows that when adopting a diode as noise generator, regions 1 and 2 are simultaneously simulated.

Two approaches are illustrated in Fig. 17.5: the first [250] consists in injecting a biasing current (IDC) through the diode D1; the resulting noise current flows through the current sensor VS1 (which is an independent voltage source of zero voltage) in series with C1. Capacitor C1 blocks the DC component, but must permit the sensing of all noise components (even those at very low frequencies). Consequently, its value should be as high as possible. In simulation, fortunately, we may adopt an 1-GF capacitor to satisfy this condition.

Another way to eliminate the DC component is suggested in Fig. 17.5b [254]. Two identical (twin) diodes are supplied from two identical current sources I1 and I2. Due to the perfect symmetry imposed in simulation, the common mode is completely rejected and consequently the voltage taken

596   17 Computer-Aided Noise Analysis

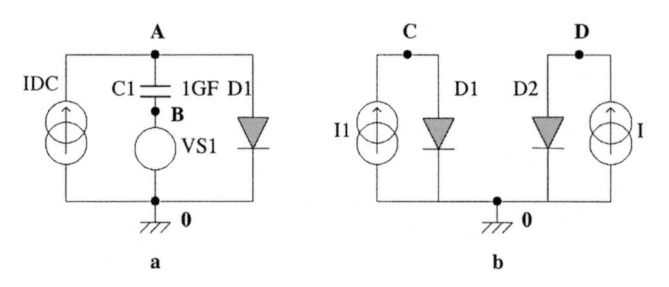

Fig. 17.5. Simulation of noise of regions 1 and 2

between nodes C and D has no DC component, only shot and flicker noise (however, attention should be paid to the spectral density of the noise between C and D, which is double that arising from one diode alone).

The value of coefficient KF is deduced by imposing the condition that the shot noise spectral density be equal to the 1/f noise spectral density at the corner frequency f_1. Coefficient AF is taken equal to unity. Thus,

$$2q\, I_{DC} = \frac{KF\, (I_{DC})}{f_1} \tag{17.2}$$

and consequently:

$$KF = 2qf_1 \tag{17.3}$$

☐ **High-Frequency Noise Modeling.** Region 3 of the spectrum, where noise increases with a slope of +20 dB/decade, can be modeled by noting that the relationship between the current through a capacitor and the voltage between its terminals is

$$I = E\, (j\omega C) \tag{17.4}$$

This expression yields a slope of I versus ω of +20 dB/decade. Consequently, if E is a flat noise source, the resulting current noise will have a spectrum increasing at 20 dB/decade. This is achieved with the circuit (and corresponding code lines) shown in Fig. 17.6.

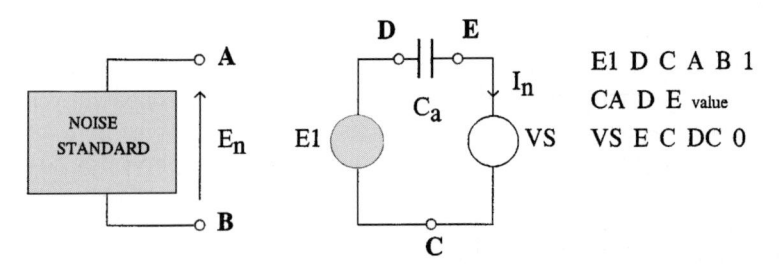

Fig. 17.6. Modeling the high-frequency region of the noise spectrum

17.1 Noise Simulation with PSPICE   597

The value of the required capacitor (symbolically denoted by value) is determined by requiring that the flat noise power density be equal to the high-frequency noise power density at the corner frequency f_2 (details are given in Case Study 17.4).

E1 is a VCVS that transfers the noise of the reference standard to the circuit generating high-frequency noise (after eventually scaling it); VS is a current sensor (its value is set to zero), employed to enable another controlled source (not shown) to insert the resulting noise at the proper location.

17.1.4 Case Studies

CASE STUDY 17.1 [250]

PSPICE automatically adds thermal noise to any resistor; however, in many practical applications there is a need for noiseless resistors (as in the equivalent circuits of active devices, where some resistances are non-dissipative, since they correspond to carrier transport through the semiconductor).

Create the PSPICE model of a 20-kΩ noiseless resistor. Assume that this resistor is connected between nodes 10 and 11 of a particular circuit.

Solution

To avoid the thermal noise associated with every resistor, we must model its resistance as a one-port able to supply the same current into the external circuit as that flowing through the 20-kΩ resistor, when they are both subject to the same voltage.

For instance, a voltage controlled current source can be employed, provided that the control coefficient is appropriately chosen. Ohm's law yields

$$i = \frac{v}{R} = \frac{v}{20 \cdot 10^3} = (5 \cdot 10^{-5}) \, v$$

and this suggests that the control coefficient must equal $(5 \cdot 10^{-5})$. The control voltage is just the voltage across the one-port (in other words, it is self-controlled).

The model, together with the corresponding code line, is presented in Fig. 17.7.

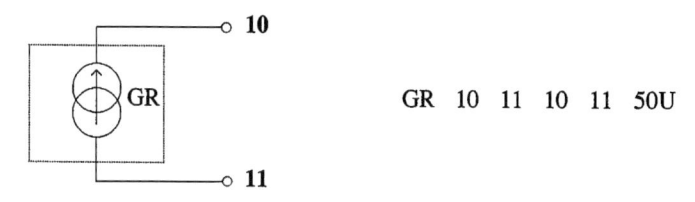

GR 10 11 10 11 50U

Fig. 17.7. Model of a non-dissipative (noiseless) resistor

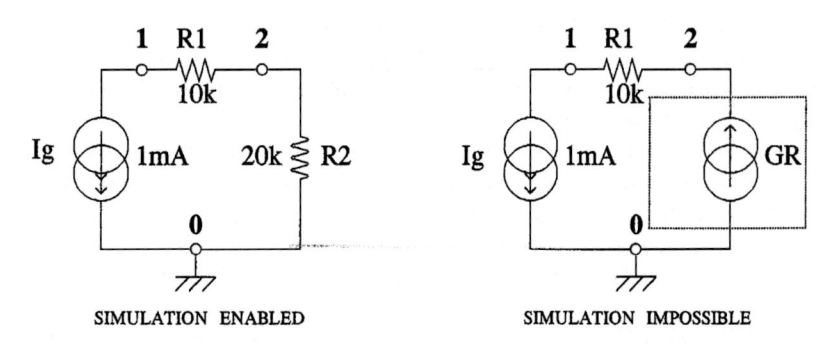

Fig. 17.8. Example of circuit where simulation fails

■ **Beware.** This model acts as a resistance *provided that a voltage drop is applied to its terminals*. This voltage activates the source, which responds by sending a current into the external circuit. If no voltage drop exists between nodes 10 and 11, no resistive behaviour is expected.

A typical situation where this model fails is shown in Fig. 17.8, where the voltage divider is obtained either with two conventional resistors, or with a resistor and the proposed model which replaces the 20-kΩ resistor. In the latter case, no current flows through the circuit and consequently nodes 1 and 2 are floating. PSPICE cannot perform the analysis, since no voltage is applied to terminals 2 and 0. However, the simulation is enabled if instead of driving the voltage divider from an independent current source, an independent voltage source is employed.

■ **Concluding Remark.** This model is particularly useful to simulate the input resistance of a noiseless amplifier. If a resistor of the desired value is introduced between its input terminals, extra thermal noise will be added. Hence, the only solution is to employ the proposed model.

CASE STUDY 17.2 [250, 254]

An amplifier has an input impedance equivalent to a resistance of 500 kΩ in parallel with an input capacitance of 30 pF. Its voltage gain is 750 and its noise is described by means of two uncorrelated noise generators $E_n = 4\,\text{nV}/\sqrt{\text{Hz}}$ and $I_n = 12\,\text{pA}/\sqrt{\text{Hz}}$. Create the macro-model of this amplifier, for electrical and noise simulation with PSPICE.

Solution

A straightforward procedure to model E_n and I_n starts by building two noise standards of appropriate values. Then, by means of two controlled sources their noise is scaled to the specified level and transferred to the nodes of the circuit where it must appear. Note that the type or value of the standard

17.1 Noise Simulation with PSPICE   599

source is not important: it may either be a noise current (of $1\,\mathrm{pA}/\sqrt{\mathrm{Hz}}$) or a node voltage (of $1\,\mathrm{nV}/\sqrt{\mathrm{Hz}}$). What really matters is to appropriately select the type of controlled source and the value of its controlling coefficient, which determines the noise level at the node in question.

Note also that it is imperative to use two independent noise standards, one for each generator; if a single noise standard is used to drive both E_n and I_n, then E_n and I_n will be completely correlated.

■ **Simulating with Noise Standards of $1\,\mathrm{pA}/\sqrt{\mathrm{Hz}}$.** The noise standard of $1\mathrm{pA}/\sqrt{\mathrm{Hz}}$ is created as described in Sect. 17.1.3. To get the value $12\,\mathrm{pA}/\sqrt{\mathrm{Hz}}$, the transfer is made by a CCCS, with a coefficient of 12; to obtain $E_n = 4\,\mathrm{nV}/\sqrt{\mathrm{Hz}}$, the transfer needs a CCVS, whose coefficient is $(4 \cdot 10^{-9})/10^{-12} = 4\,\mathrm{k\Omega}$. The corresponding models and code lines are shown in Fig. 17.9, where V_{es} and V_{is} are voltage sources set to zero, employed solely to sense the noise currents of the standards.

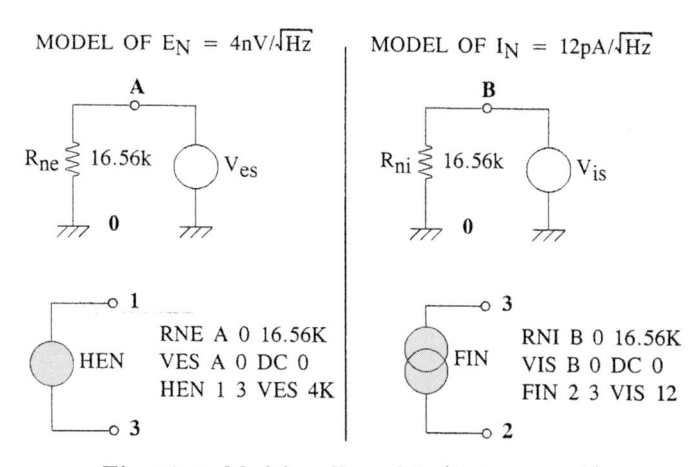

Fig. 17.9. Modeling E_n and I_n (first approach)

■ **Simulating with Noise Standards of $1\,\mathrm{nV}/\sqrt{\mathrm{Hz}}$.** Another possibility is to use noise standards of $1\,\mathrm{nV}/\sqrt{\mathrm{Hz}}$ (as in Sect. 17.1.3). This is shown in Fig. 17.10, where the transfer is made with a VCVS (controlling coefficient equal to 4) and a VCCS (controlling coefficient of $12\,\mathrm{pA}\ /\ 1\,\mathrm{nV} = 12\,\mathrm{mS}$), respectively.

The amplifier macro-model, with the equivalent noise generators modeled according to the first approach, is presented in Fig. 17.11.

Note that the input resistance is modeled as a noiseless resistance, according to Case Study 17.1; otherwise, if a conventional resistor is included, it will add its own thermal noise above that of E_n, I_n (which, in principle, are responsible for *all* the noise of the amplifier).

600   17 Computer-Aided Noise Analysis

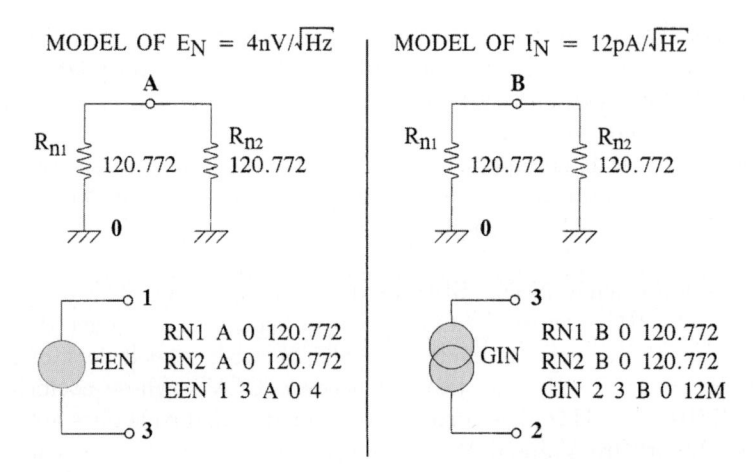

Fig. 17.10. Modeling E_n and I_n (second approach)

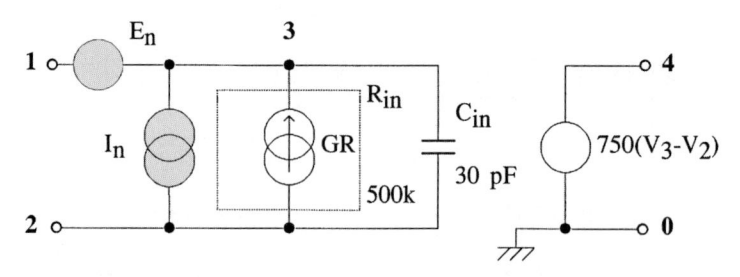

MODEL OF $E_N = 4\text{nV}/\sqrt{\text{Hz}}$

R_{n1} 120.772 R_{n2} 120.772

RN1 A 0 120.772
RN2 A 0 120.772
EEN 1 3 A 0 4

MODEL OF $I_N = 12\text{pA}/\sqrt{\text{Hz}}$

R_{n1} 120.772 R_{n2} 120.772

RN1 B 0 120.772
RN2 B 0 120.772
GIN 2 3 B 0 12M

Fig. 17.11. Amplifier macro-model, noise included

The corresponding code lines that must be included in the input data file are

```
* En generator
RNE  A  0   16.56 K
VES  A  0   DC  0
HEN  1  3   VES 4K
* In generator
RNI  B  0   16.56 K
VIS  B  0   DC  0
FIN  2  3   VIS 12
* Input
CIN  3  2   30P
GR   2  3   2   3   2U
* Output
EA   4  0   3   2   750
```

■ **Concluding Remark.** Note that any hybrid combination between the first- and second-approach is allowed (for instance, E_n modeled with a standard noise voltage and I_n with a standard noise current).

CASE STUDY 17.3 [250, 253]

Often 1/f noise must be included in order to accurately simulate the actual behavior of active devices. Create a model for a noise generator $E_n = 24\,\text{nV}/\sqrt{\text{Hz}}$ with a piecewise linear frequency characteristic and a breakpoint (or corner frequency) situated at $f_e = 100\,\text{Hz}$. Also, model $I_n = 0.8\,\text{pA}/\text{Hz}$ whose characteristic has a breakpoint at $f_i = 300\,\text{Hz}$.

Solution

The *breakpoint* (also called the *corner frequency*) is defined as *the point where the asymptotic frequency characteristic suddenly changes slope* (Fig. 17.12). From a physical standpoint, at the breakpoint the power spectral density of the thermal noise is equal to that of the 1/f noise.

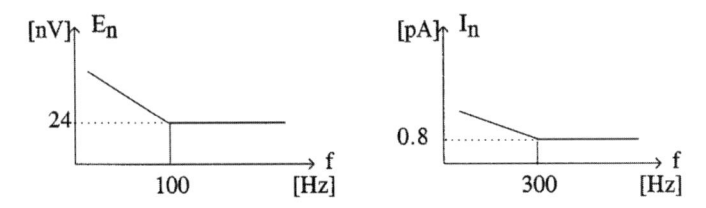

Fig. 17.12. Plots of E_n and I_n versus frequency

■ **Approach.** Instead of using a resistor as the noise source, we employ a semiconductor diode. It should be remembered that its noise spectral density is described by (7.17):

$$S(\overline{I_n^2}) = 2q\,I_{DC} + KF\,(I_{DC})^{AF}/f$$

where f is the frequency and I_{DC} is the DC current through the diode. In conjunction with Sect. 17.1.3, the following steps must be performed:

1. *Create a noise standard of $1\,pA/\sqrt{Hz}$ using shot noise.*
From the first term of the sum, the required bias current is

$$I_{DC} = \frac{\overline{S(I_n^2)}}{2q} = \frac{10^{-24}}{2(1.6 \cdot 10^{-19})} = 3.125\,\mu\text{A}$$

2. *Find the coefficient KF required to model E_n.*
Assuming $AF = 1$, this coefficient is calculated by specifying equality between the power spectral density of the shot noise and 1/f noise at $f_e = 200\,\text{Hz}$. Applying (17.3),

$$KF = 2q\,f_e = 2(1.6 \cdot 10^{-19})(1 \cdot 10^2) = 3.2 \cdot 10^{-17}\,\text{A}$$

3. *Find the coefficient KF required to model I_n.*
As in the previous step,

$$\mathrm{KF} = 2q\,f_i = 2(1.6 \cdot 10^{-19})(3 \cdot 10^2) = 9.6 \cdot 10^{-17}\ \mathrm{A}$$

The noise models of generators E_n and I_n are detailed in Fig. 17.13.

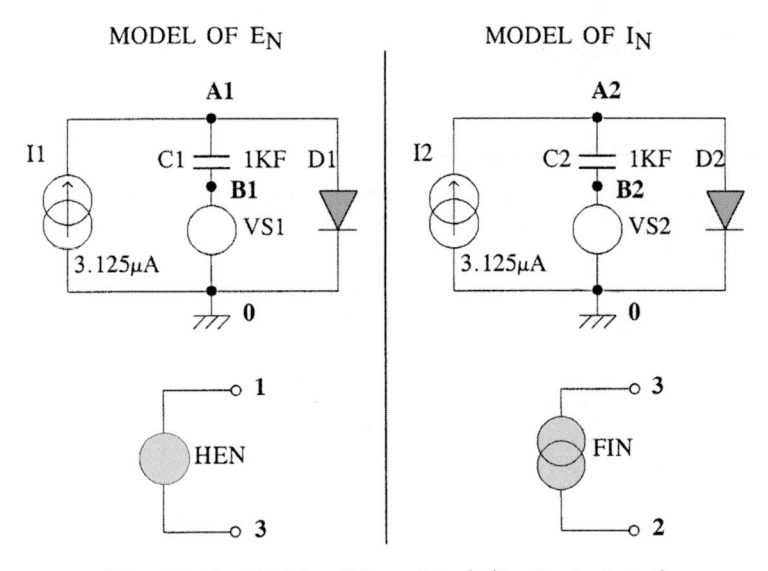

Fig. 17.13. Models of E_n and I_n (1/f noise included)

The role of capacitor C1 (C2) is to prevent the bias current supplied by I1 (I2) from being short-circuited by the voltage source VS1 (VS2), which acts as a current sensor *only for the AC component*. The value of C1 (C2) is impractically large, to avoid bandwidth limitation of the shot noise component. This component must be entirely transferred to the desired node after scaling (by appropriately selecting the control coefficient of the dependent source).

The code lines to be included in the input data file are

```
* En generator
*     Standard Source
      I1 0 A1 DC 3.125U
      C1 A1 B1 1KF
      D1 A1 0 DB1
      VS1 B1 0 DC 0
      .MODEL DB1 D (KF=3.2E-17 AF=1)
*     Transfer
      HEN 1 3 VS1 24K
```

```
* In generator
*    Standard Source
        I2 0 A2 DC 3.125U
        C2 A2 B2 1KF
        D2 A2 0 DB2
        VS2 B2 0 DC 0
        .MODEL DB2 D (KF=9.6E-17 AF=1)
*    Transfer
        FIN 2 3 VS2 0.8
```

CASE STUDY 17.4 [254]

By means of the circuit shown in Fig. 17.14, create a noise generator with $E_n = 10\,\text{nV}/\sqrt{\text{Hz}}$ in the flat region of the spectrum and a corner frequency at $f_c = 1\,\text{kHz}$.

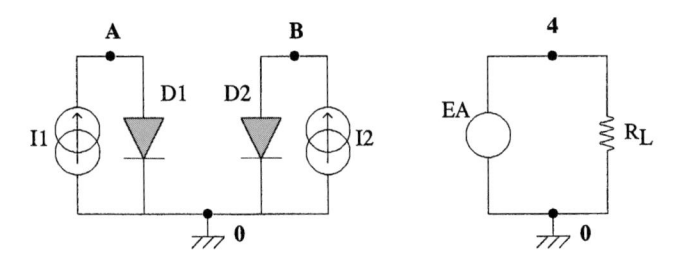

Fig. 17.14. Circuit submitted for simulation

Solution

■ **Approach.** The circuit represents an alternative approach to separate noise from the DC component.

In this case, the voltage employed to control the VCVS, denoted by EA, corresponds to the differential mode (between points A and B). Since in simulation the circuit can be perfectly balanced, V_{AB} is exactly twice the noise voltage V_n appearing across a single diode:

$$\overline{V_n^2} = \left(r_d^2\, \overline{I_n^2}\right) \quad \text{with} \quad r_d = kT/qI_{DC} \tag{17.5}$$

I_{DC} being the bias current of the diode.

The noise level can be modified either by changing the DC currents I1 = I2 through the diodes or by scaling the control coefficients of the VCVS source.

□ **Flat Noise.** The problem which must be solved is to determine the value of the bias current I_{DC}, such as the spectral density of the noise voltage across the diode equals the specified value.

At sufficiently high frequencies, where flicker noise is no longer important, the spectral density of the shot noise is

$$S(I_n) = 2qI_{DC} \tag{17.6}$$

Upon substituting in (17.5), which also holds for spectral densities, we obtain

$$S(E_n) = 2\left(\frac{kT}{qI_{DC}}\right)^2 (2qI_{DC}) \tag{17.7}$$

(the factor 2 accounts for the noise voltage which is taken between points A and B, where the spectral density is twice that of a single diode). From this equation, we find

$$I_{DC} = \frac{4(kT)^2}{q\,S(E_n)} \tag{17.8}$$

or

$$\boxed{I_{DC} = \frac{42.849 \cdot 10^{-23}}{S(E_n)} \quad (\text{at } T = 300 \text{ K})} \tag{17.9}$$

With the specified numerical values, we have

$$I_{DC} = \frac{42.849 \cdot 10^{-23}}{(10^{-8})^2} = 4.2849 \text{ µA}$$

□ **Corner Frequency.** Applying expression (17.3), the coefficient KF of the diode model is

$$KF = 2q\,f_c = 2(1.6 \cdot 10^{-19})10^3 = 3.2 \cdot 10^{-16} \text{ A}$$

■ **Input Data File.** The following code lines must be inserted:

```
Noise generator with 2 regions
I1   0   A   DC 4.2849UA   AC 1
I2   0   B   DC 4.2849UA
D1   A   0   DN
D2   B   0   DN
.MODEL   DN   D   (KF=3.2E-16)
*transfer noise to output without scaling
EA   4   0   A B 1
RL   4   0   5K
*commands
.AC   DEC   50   1   1MEG
.NOISE   V(4)   I1
.END
```

■ **Comment.** See the Remark of Case Study 17.5.

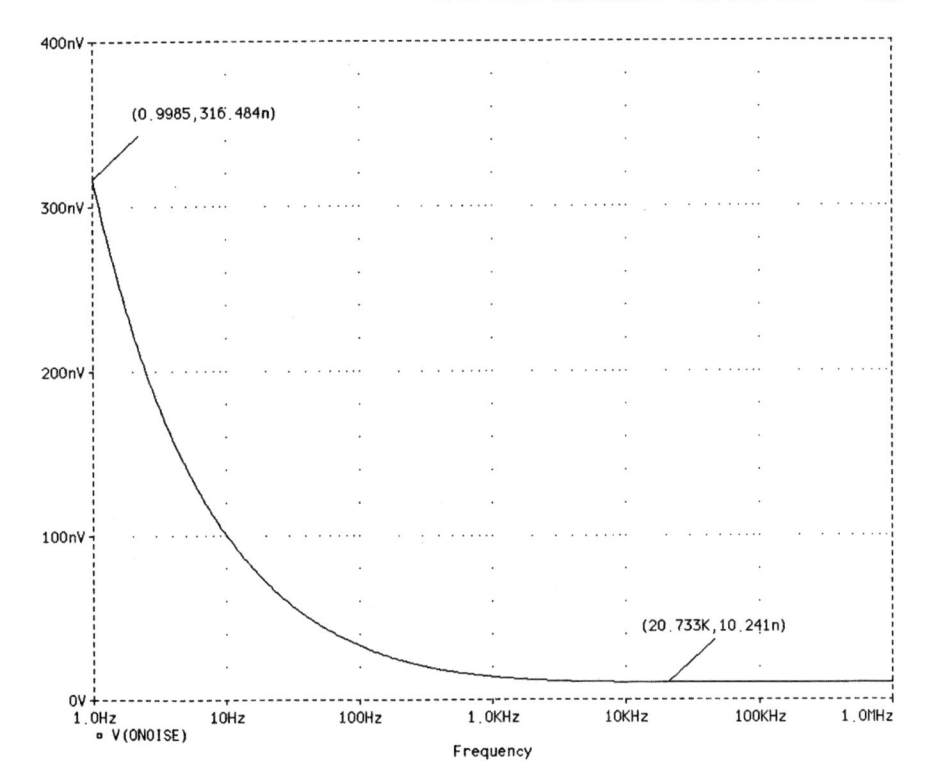

Fig. 17.15. Frequency dependence of E_n delivered at the output

■ **Results.** The plot of output noise (at node 4) is presented in Fig. 17.15.

CASE STUDY 17.5 [254]

Develop the model of a noise generator E_n whose power spectral density depends on frequency as depicted in Fig. 17.16. Region I corresponds to the flicker noise, region II is flat at $10\,\mathrm{nV}/\sqrt{\mathrm{Hz}}$, and region III corresponds to the high-frequency noise (which in practice often has a positive slope of 6 dB/oct).

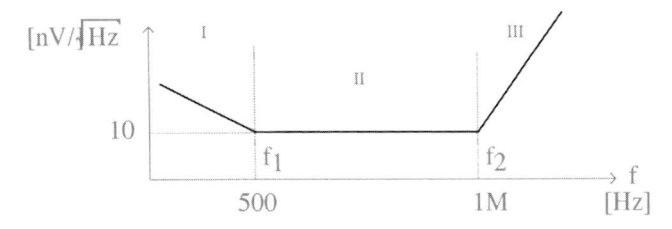

Fig. 17.16. Frequency characteristic of the E_n generator

Solution

▪ **Regions I and II.** The noise corresponding to regions I and II can be simulated as in Case Study 17.4. The starting point is the twin diode noise generator (Fig. 17.13), whose level is controlled by the bias current, while the 1/f corner frequency is established by the value adopted for KF:

$$KF = 2q\,f_1 = 2(1.6 \cdot 10^{-19})(500) = 1.6 \cdot 10^{-16}\text{ A}$$

Since the specified value of E_n is the same as in Case Study 17.4, we adopt the same value for the bias current (4.2849 µA).

The proposed noise model is shown in Fig. 17.17. The transfer of the differential noise voltage $(V_A - V_B)$ is made with the VCVS denoted EEN.

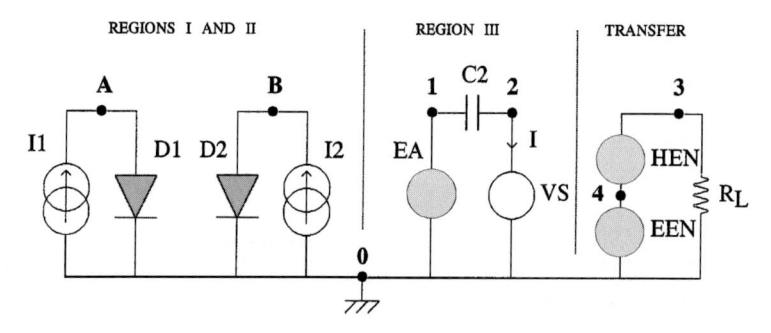

Fig. 17.17. Model of the 3-region noise generator

▪ **Region III.** The high-frequency behavior (region III) is simulated by inserting a capacitor (C_2) in series with the voltage source VS, which being set to zero, acts as a current sensor.

To get the global frequency characteristic (with three regions), the effects must be combined in the load R_L: the flicker noise and the flat noise are transferred by the source EEN, while the high-frequency noise is transferred by HEN (which is a current-controlled voltage source), both in series.

If needed, the 1/f or high-frequency region can be pre-emphasized, by properly selecting the control coefficients of the transfer sources.

The load R_L has been included to avoid letting node 3 float.

The value of C_2 is calculated by requiring that the flat noise spectral density be equal to the HF noise at the corner frequency f_2. Since the HF noise behavior is decided by capacitor C_2, the relationship between the noise current I and the noise voltage E_n is

$$I = (E_n)(j\omega C_2)$$

which shows that current I has a slope of +6 dB/octave (or +20 dB/decade). Imposing equality between the flat noise spectral density and that of region III (at the corner frequency f_2) yields $(2\pi f_2)C_2 = 1$, which gives $C_2 = 0.15915$ µF.

■ **Input Data File.** The file describing the noise generator E_n is

```
3 REGIONS NOISE GENERATOR
* Regions I and II
I1    0   A   DC 3.125U    AC 1MA
D1    A   0   DN
I2    0   B   DC 3.125U
D2    B   0   DN
.MODEL   DN   D   (KF=3.2E-16)
* Region III
EA    1   0   A  B  1
C2    1   2   0.15915UF
VS    2   0   DC   0
*Transfer to output
HEN   4   3   VS    10
EEN   4   0   A   B   0.8
RL    3   0   5K

.AC  DEC  50  1  100MEG
.NOISE  V(3)  I1
.END
```

■ **Remark.** As already explained, noise analysis requires a previous AC analysis of the circuit. The latter asks for an AC independent source in the circuit. To satisfy this requirement, the current source I1 is also provided with an alternating component of 1 mA.

Another way to enable AC analysis is to introduce a dummy AC generator in a separate circuit (Fig. 17.18), which has no node in common (except the ground) with the circuit of interest, in order to avoid perturbing it. However, if the equivalent input noise must be computed, the risk is that the noise will be reflected at the terminals of VAC, instead of the actual input.

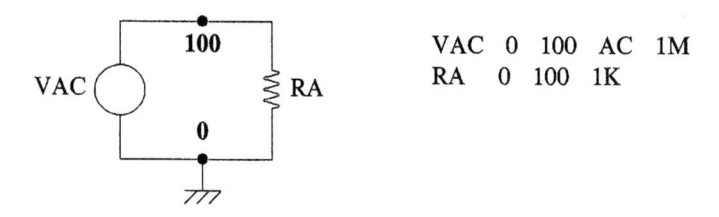

```
VAC  0   100  AC  1M
RA   0   100  1K
```

Fig. 17.18. Dummy source, used only to enable AC analysis

CASE STUDY 17.6 [250]

This case study is dedicated to PSPICE simulation of the excess noise of resistors. Often in simulation attention is exclusively paid to thermal noise, and the excess noise is ignored, despite evidence that the level of the latter

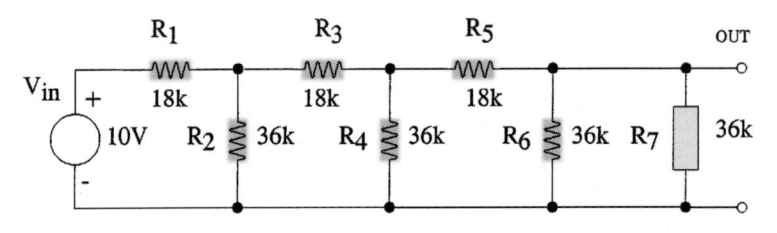

Fig. 17.19. Resistive network with DC input excitation

can greatly exceed that of the former. In order to illustrate the impact of resistor excess noise on circuit performance, consider the passive network shown in Fig. 17.19. Assume that only resistor R_7 has a noise index of 10 dB, and that all other resistors have negligible excess noise (for instance, they are fabricated using a different technology). Establish the noise model of resistor R_7 (excess noise included) and determine the noise voltage at the output of the network (OUT), in two situations:

a) with a poor-quality resistor R_7 (NI = 10 dB);
b) with a high-quality resistor R_7 (excess noise neglected).

Solution

■ **Resistor Noise Model.** It should be remembered that the noise index (NI) was defined in Sect. 7.1 as the ratio of: (1) the rms value (in µV) of the excess noise voltage, to (2) the DC drop across the resistor, in one decade of frequency. According to (7.6), the spectral density of the excess noise voltage is

$$\overline{v_{ex}^2} = \frac{v_{dc}^2 \, 10^{([NI]_{dB}/10)} 10^{-12}}{f \ln 10} \quad [V^2/Hz]$$

where V_{dc} is the DC bias across the resistor. This equation is similar to the first term of the current noise of a diode, recalled here:

$$\overline{I_n^2} = KF \, (I_d)^{AF}/f + 2q \, I_d$$

KF and AF can be calculated by inspection

$$KF = \frac{10^{NI/10} 10^{-12}}{\ln(10)} = 4.343 \cdot 10^{-12}, \quad AF = 2$$

This suggests that the excess noise can be simulated by means of a diode, under the following conditions:

- The shot noise of the diode must be negligible (for practical values, this is always true when the noise index ranges from -40 dB to $+10$ dB, for typical bias values).
- The coefficients of the controlled sources have to be properly selected when the noise is transferred to the main circuit.

The noise model detailed in Fig. 17.20 is proposed.

17.1 Noise Simulation with PSPICE

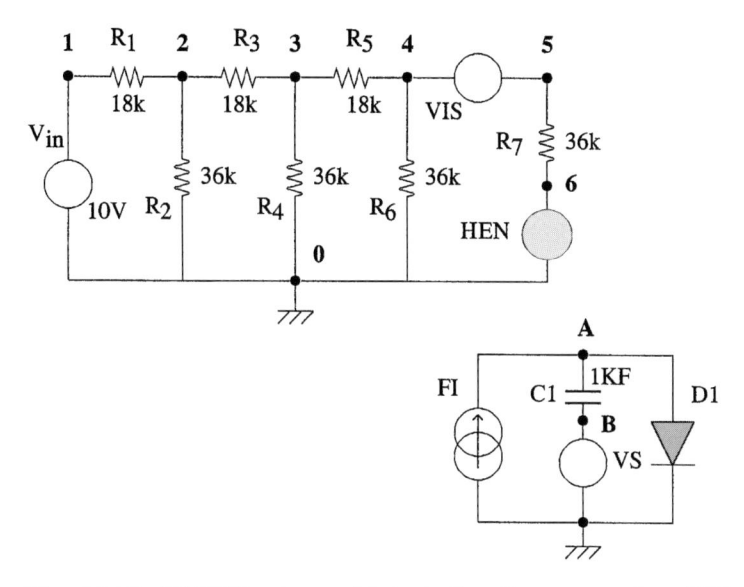

Fig. 17.20. PSPICE noise model, including the excess noise of R7

■ **Explanation.** VIS is a zero-voltage source, employed to sense the DC current through R_7, which in turn controls the current of the DC generator FI. VS also acts as a current sensor only for the noise generated by diode D1, the DC current of FI being blocked by C1. The huge value adopted for C1 avoids unduly affecting the flat spectrum noise. The control coefficient of FI is equal to the value of R_7; the control coefficient of the source HEN is 1.

■ **Input Data File.** The input data file of the proposed circuit is

```
CIRCUIT WITH EXCESS NOISE RESISTOR
VIN   1  0  DC    10
R1    1  2  18 K
R2    2  0  36 K
R3    2  3  18 K
R4    3  0  36 K
R5    3  4  18 K
R6    4  0  36 K
R7    5  6  36 K
VIS   4  5  DC   0
* Noise model
HEN   6  0  VS   1
FI    A  0  VIS  36 K
C1    A  B  1 KF
VS    B  0  DC   0
D1    A  0  DEX
.MODEL  DEX  D  (KF=4.343E-12 AF=2)
```

```
* Commands
.AC DEC 50 1 100
.NOISE V(5) VIN
.END
```

Note that in noise analysis, the output is taken at node 5.

■ **Results**

a) PSPICE simulation yields a noise voltage (at node 5) $V_{nex} = 2.605\,\mu V$ (at f = 1 Hz) and $0.8237\,\mu V$ (at f = 10 Hz), which are in good agreement with the hand-calculated values.
b) If the excess noise of R_7 is assumed to be negligible, the output noise (node 4, this time) becomes 3.3 nV, and is due solely to the thermal noise of the circuit.

■ **Concluding Remark.** Comparing previous results, the excess noise of carbon resistors is highly significant, being much greater than the thermal noise.

CASE STUDY 17.7 [250]

To bias the base of a common-emitter stage amplifier (Fig. 17.21) the designer can select a 1-MΩ resistor from among several categories, with noise indices of +10 dB, 0 dB, −10 dB, and −20 dB, respectively. Compare the total noise obtained at the output, using a cheap carbon resistor (NI = +10 dB) with that obtained in the other cases. Since the cost of the resistor is inversely proportional to its excess noise, indicate the most appropriate choice.

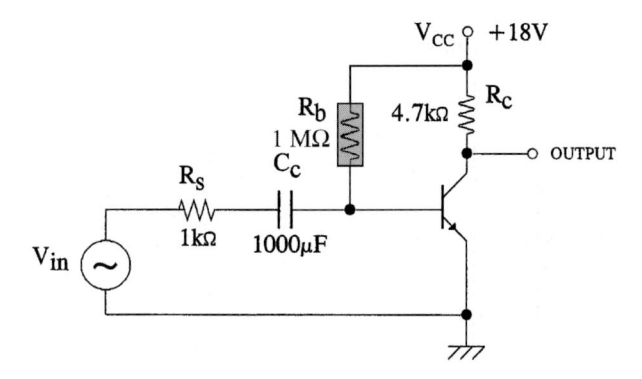

Fig. 17.21. Single-stage amplifier (R_b exhibits strong excess noise)

Solution

■ **Approach.** The model developed in Case Study 17.6 to simulate the excess noise of a resistor is employed again and the circuit submitted for simulation with PSPICE is shown in Fig. 17.22.

VIS is a zero-voltage source sensing the current that flows through R_b. What differs now with respect to Case Study 17.6 is the superposition of an AC component (due to the signal generator V_{in}) on a DC component (due to bias). This means that the current flowing through R_b will also have an AC component and so will the controlled source FI. The problem is how to eliminate it from the current injected into the diode D1. As shown in Fig. 17.22, this is achieved by inserting a low-pass filter, whose huge element values, which are totally impractical but possible in simulation, are appropriate to ideally filter the signal. Note that, as stated in Case Study 17.6, the gain of the source FI is equal to the value of the resistance in question, i.e., 10^6.

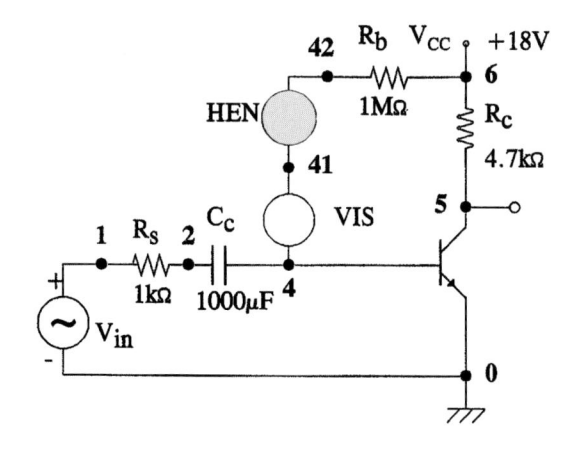

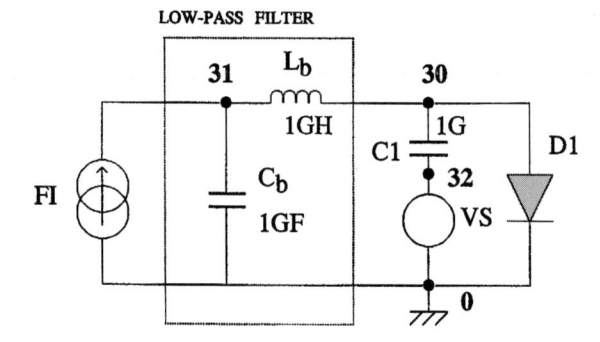

Fig. 17.22. Model of the CE amplifier

17 Computer-Aided Noise Analysis

Following the same approach, the coefficient KF is calculated in terms of the noise index of R_b (which will be taken in the range $+10\,\text{dB}$ to $-20\,\text{dB}$, according to the resistor quality). AF is taken equal to 2.

The noise current generated by D1 is sensed by the voltage source VS (set to zero) and, in turn, this controls the source HEN by means of a coefficient of value 1.

■ **Input Data File.** For simulation with PSPICE, the input data file is

```
Case study 17.7 Rb excess noise NI = +10 dB
VIN   1    0    AC    1
RS    1    2    1K
CIN   2    4    1MF
Q1    5    4    0     NPN1
.MODEL NPN1 NPN
VCC   6    0    DC    18
RB    6    42   1MEG
RC    5    6    4.7K
VIS   41   4    DC    0
HEN   41   42   VS    1
F1    31   0    VIS   1MEG
CB    31   0    1GF
LB    31   30   1GH
CS    30   32   1GF
VS    32   0    DC    0
D1    30   0    DN
.MODEL DN D KF=4.343E-12 AF=2
.AC DEC 100 .1 100 K
.NOISE V(5) VIN
.END
```

■ **Results.** Table 17.6 presents the truncated values of the noise voltage obtained at node 5 when R_7 has different noise indices.

Table 17.6. Noise voltage at node 5 (simulated results)

IB [dB]	+10	0	-10	-20
KF [A]	$4.343 \cdot 10^{-12}$	$4.343 \cdot 10^{-13}$	$4.343 \cdot 10^{-14}$	$4.343 \cdot 10^{-15}$
V(5) at 1 Hz	6.86 µV	2.32 µV	1.02 µV	871 nV
V(5) at 10 Hz	2.30 µV	1.02 µV	871 nV	855 nV
V(5) at 100 Hz	1.11 µV	871 nV	855 nV	853.3 nV

■ **Concluding Remark.** Inspecting the results, it is obvious that for NI = −20 dB, the thermal noise floor is roughly attained for frequencies as low as several Hz. Depending on the lowest frequency specified, the designer should choose between resistors with noise indices between 0 dB and −10 dB, keeping in mind that a good-quality resistor is always more expensive. Selecting an even higher-quality resistor (with NI = −20 dB or better) makes no sense, except when ultra-low frequencies must be processed.

CASE STUDY 17.8

Determine by simulation with PSPICE the noise bandwidth of the Sallen-Key active bandpass filter shown in Fig. 17.23, in two different cases:

1) the operational amplifier is assumed to be ideal;
2) a µA741 operational amplifier is employed.

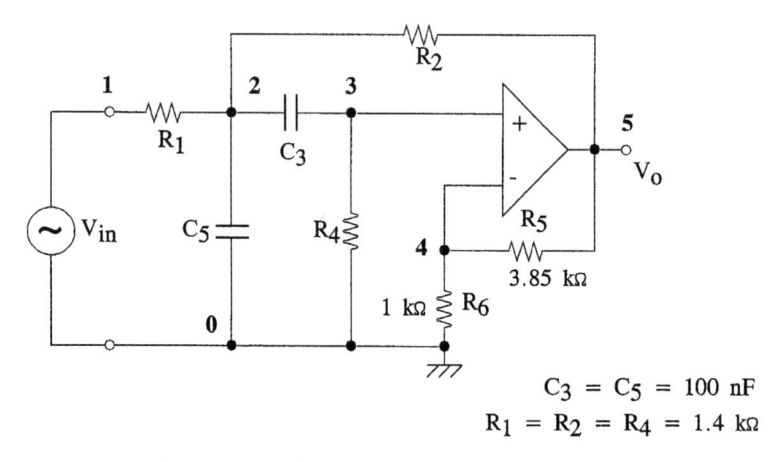

Fig. 17.23. Sallen-Key active bandpass filter

$C_3 = C_5 = 100$ nF
$R_1 = R_2 = R_4 = 1.4$ kΩ

Solution

■ **Input Data File.** To run PSPICE, the following input file is needed:

```
Sallen & Key active filter
VIN   1  0   AC 1
R1    1  2   1.4 K
R2    2  5   1.4 K
R4    3  0   1.4 K
R5    4  5   3.85 K
R6    4  0   1 K
C5    2  0   100 N
C3    2  3   100 N
```

```
* Using μA741
X1  3  4  11  12  5  UA741
V+  11  0  DC  15
V−  12  0  DC  −15
.LIB  EVAL.LIB
* Ideal Op-Amp
*RIN  3  4  1MEG
*EAM  5  0  3  4  1MEG
.AC  DEC  50  1  1MEG
.NOISE  V(5)  VIN
.END
```

As the lines describing the ideal operational amplifier are marked with "*" in the first column, only noise analysis for case 2 is enabled.

■ **Analysis.** Expression (4.41) defines the noise equivalent bandwidth as

$$\Delta f = \frac{1}{A_{vo}^2} \int_0^\infty |A_v(f)^2|\, df$$

where A_{vo} represents the peak value obtained in the plot of V(5), with a unit-value signal generator. We may use the PROBE capabilities to compute Δf by integrating the squared magnitude of A_v. Operating under the "Add trace" option in the PROBE menu, instead of pressing F4 to select a variable from the list, the following expression must be entered:

(1/(5.7082*5.7082))*S(VM(5)*VM(5))

(**5.7082** corresponds to the peak value of output voltage V(5), which must be previously determined).

According to Table 17.5, function S performs integration over the x-range (as specified in the .AC statement, i.e., from 1 Hz to 1 MHz). VM(5) denotes the module of V(5).

■ **Results.** The plots of output noise voltage and input noise voltage are given in Fig. 17.24. The voltage gain versus frequency and the asymptotic plot of the noise bandwidth are shown in Fig. 17.25.

Not surprisingly, when an ideal operational amplifier is used in the circuit, the noise level is lower.

The equivalent input noise increases (in both cases) in both the low- and high-frequency regions, due to the fact that the output noise is divided by the voltage gain, which in these areas is vanishing.

The noise bandwidth is not much modified when the ideal operational amplifier is replaced with a μA741; the results are

$\Delta f = 1.5208\,\text{kHz}$ (ideal amplifier)
$\Delta f = 1.4707\,\text{kHz}$ (μA741)

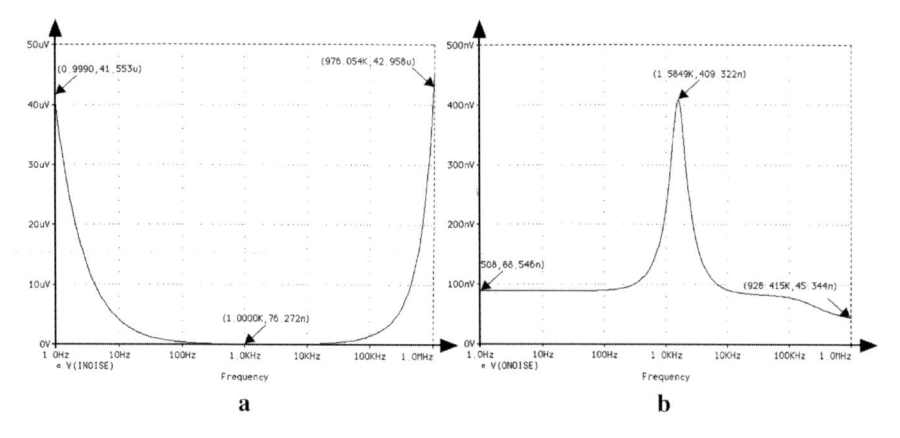

Fig. 17.24. Input and output noise voltage versus frequency

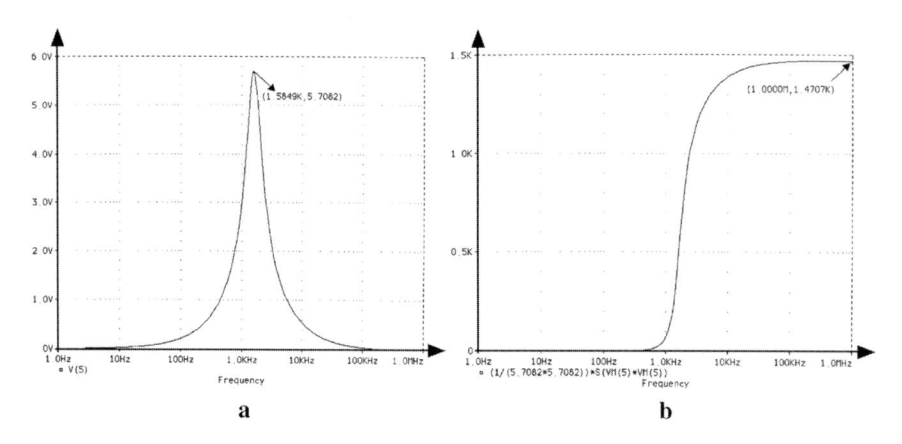

Fig. 17.25. Frequency response and the resulting asymptotic noise bandwidth

CASE STUDY 17.9 [256]

Consider the MMIC (Monolithic Microwave Integrated Circuit) 1.5-GHz amplifier shown in Fig. 17.26. Determine the noise factor of the circuit by simulation with PSPICE.

Solution

■ **Explanation.** Transistor Q2 is employed in the bias network to compensate the V_{BE} drift of Q1 with temperature.

The series negative feedback (L_1 and R_4) is used to match the input (for signal and noise). The parallel negative feedback (resistor R_2) improves the stability of the bias point of Q1 with respect to temperature.

616   17 Computer-Aided Noise Analysis

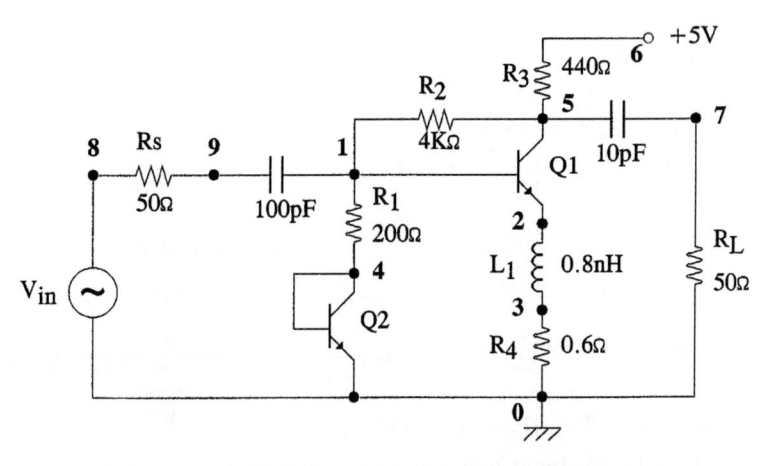

Fig. 17.26. Monolithic 1.5-GHz amplifier

■ **Input Data File.** Adopting default models for transistors, the input file is

```
Case study 17.9
VIN    8  0  AC    1
VCC    6  0  DC    5
RS     8  9  50
C1     9  1  100 P
R1     1  4  200
R2     1  5  4 K
R3     5  6  440
C2     5  7  10 P
RL     7  0  50
L1     2  3  0.8 N
R4     3  0  0.6
Q1     5  1  2    NPN1
Q2     4  4  0    NPN1
.MODEL  NPN1  NPN  BF=80
.TEMP   17
.OP
.AC  LIN  10  1.5 G  1.6 G
.NOISE  V(7)  VIN
.END
```

Note the inclusion of statement **.TEMP 17**, which is required to allow for the standard reference temperature of 290 K (instead of the PSPICE default value 300 K).

■ **Relationship.** The noise factor of the circuit must be related to quantities that are available in simulation. The starting point is the macro-model of the circuit in Fig. 17.27, where both 50-Ω terminations have been added.

17.1 Noise Simulation with PSPICE

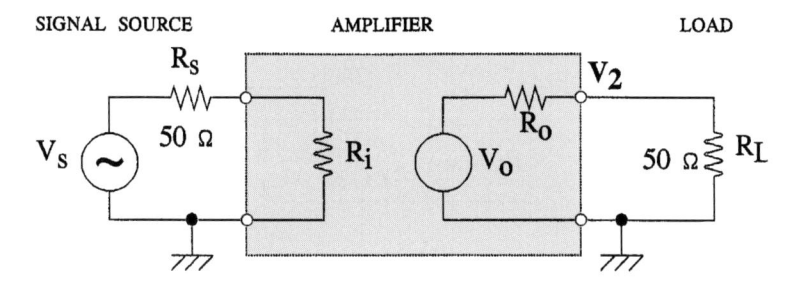

Fig. 17.27. 1.5-GHz amplifier, with its signal source and load

According to (4.73), the noise factor is

$$F = \frac{1}{G_a} \frac{N_o}{kT_o} \qquad (17.10)$$

where G_a is the available power gain, N_o is the available output noise power and $T_o = 290\,K$ (already established).

The available power gain is

$$G_a = \frac{V_o^2 / 4R_o}{V_s^2 / 4R_s} = \frac{V_o^2}{V_s^2} \frac{R_s}{R_o} \qquad (17.11)$$

R_o being the output resistance. However, instead of (V_o/V_s), with PSPICE we can access the voltage gain:

$$A_v = \frac{V_2}{V_s} \qquad (17.12)$$

Since

$$V_2 = V_o \frac{R_L}{R_L + R_o} \qquad (17.13)$$

we deduce

$$\frac{V_o}{V_s} = A_v \frac{R_L + R_o}{R_L} \qquad (17.14)$$

On the other hand, the available output noise power is

$$N_o = \frac{V_{nao}^2}{4R_o} \qquad (17.15)$$

Note that the output noise voltage is related to V_{nao} exactly in the same way as the signal output voltage:

$$V_{no} = V_{nao} \frac{R_L}{R_L + R_o} \qquad (17.16)$$

Substituting (17.14) into (17.11), we obtain the available signal power; substituting (17.16) into (17.15) we obtain the available noise power. Finally, both are substituted into (17.10) to get [256]:

$$F_{SPICE} = \frac{V_{no}^2}{4kT_oR_s A_v^2} \qquad (17.17)$$

☐ **Another Approach.** Applying North's definition

$$F = \frac{P_{no}}{P_{no} \text{ due only to the source}}$$

we let $P_{no} = V_{no}^2/R_L$. The output power noise due solely to the thermal noise of the source resistance is $(4kT_oR_s)A_v^2/R_L$. This yields the same expression for the noise factor.

■ **Results.** The first task is to evaluate the factor $4kT_oR_s \cong 8 \cdot 10^{-19}$ [V^2]. Hence

$$F_{SPICE} = \frac{(V_{(ONOISE)})^2}{8 \cdot 10^{-19} A_v^2}$$

Performing noise analysis with PSPICE (at 1.5 GHz), we get

$$V_{(ONOISE)} = 3.9867 \text{ nV}/\sqrt{Hz} \quad \text{and} \quad A_v = 3.7549.$$

Consequently, the noise factor is $F = 1.409$, which yields a noise figure of 1.489 dB.

CASE STUDY 17.10

Consider the sensor-coupling network-amplifier system discussed in Case Study 15.9, redrawn in Fig. 17.28.

The sensor has internal resistance $R_s = 2\,k\Omega$ and is connected to the amplifier input by means of a coupling network ($L = 10\,mH$ and $R_p = 3\,k\Omega$). At the operating frequency of 100 kHz, the magnitude of the amplifier input impedance is $5\,k\Omega$ and its phase angle is $-60°$. The voltage gain is 120 and the output resistance is $6\,k\Omega$. The amplifier noise can be described by two equivalent input generators $E_n = 20\,nV/\sqrt{Hz}$ and $I_n = 6\,pA/\sqrt{Hz}$, assumed to be uncorrelated. For a load $R_L = 12\,k\Omega$, determine at $T = 300\,K$:

1) The output noise voltage spectral density, $S(E_{no})$.
2) The input noise voltage spectral density, $S(E_{ni})$.

17.1 Noise Simulation with PSPICE 619

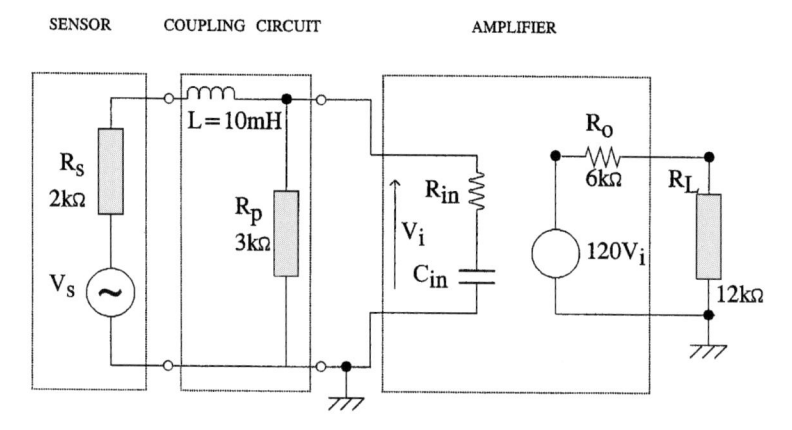

Fig. 17.28. Sensor with its coupling network and amplifier

Solution

■ **Explanation.** The input impedance at f = 100 kHz can be expressed as

$$Z_{in} = 5\,k\,\angle 60° = (2.5 - j4.33)\,k\Omega$$

from which we deduce that it corresponds to an equivalent circuit containing $R_{in} = 2.5\,k\Omega$ and $C_{in} \cong 367\,pF$ connected in series.

The noise model of E_n, I_n requires two noise standards of $1\,pA/\sqrt{Hz}$ each, built with 16.56-kΩ resistors (see Sect. 17.1.3). The noise model for simulation is shown in Fig. 17.29.

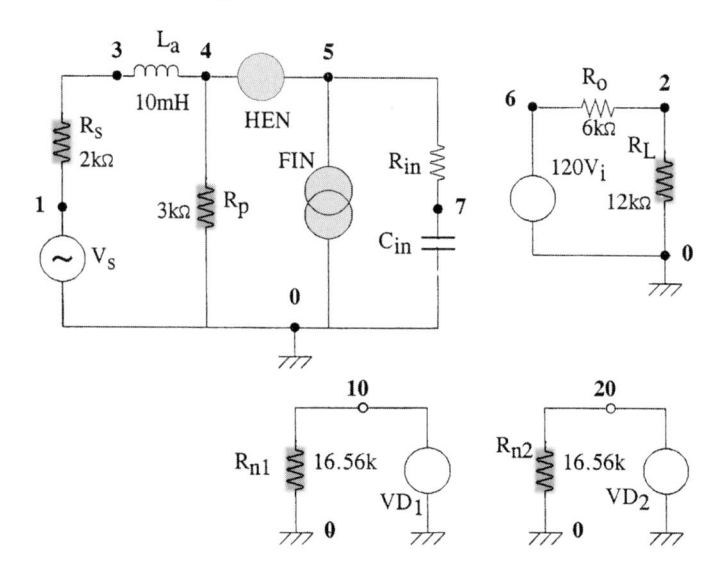

Fig. 17.29. PSPICE noise model (R_{in} and R_o are not detailed)

■ **Input Data File.** The input file for PSPICE is

```
Case Study 17.10
* source En 20 nV/RHz
RN1   10  0   16.56 K
VD1   10  0   DC 0
HEN   4   5   VD1    20 K
* source In 6 pA/RHz
RN2   20  0   16.56 K
VD2   20  0   DC 0
FIN   5   0   VD2    6
* noiseless resistors
GRIN  5  7  5  7  0.4 m
GRO   6  2  6  2  0.1666 m
CIN   7  0   367 pF
LIN   7  0   100 MEG
VS    1  0   AC   1
RS    1  3   2 K
LA    3  4   10 mH
RP    4  0   3 K
RL    2  0   12 K
EA    6  0  5  0   120
.AC   LIN  100  100 k   101 k
.NOISE  V(2)  VS
.END
```

■ **Remarks**

- The input and output resistances (R_{in}, R_o) are modeled as noiseless resistors (see Case Study 17.1). Otherwise, their thermal noise adds to the noise of the amplifier, beyond the level imposed by E_n and I_n. This would be nonsense, since E_n and I_n are responsible for *all amplifier noise*.
- Inductor LIN has been included to provide a DC path between node 7 and the remaining part of the circuit (otherwise node 7 is floating and the DC analysis, which always precedes the AC analysis, will fail and stop the program). As a huge value is allocated to LIN, its reactance is enormous, so it acts as an open circuit and does not perturb the original circuit.

■ **Results.** Analysis performed at 100 kHz yields the following results:

- The output noise voltage spectral density is

$$S(E_{no}) = 1.734 \, \mu V \sqrt{Hz}$$

- The input noise voltage spectral density is

$$S(E_{ni}) = 68.619 \, nV \sqrt{Hz}$$

These numerical values are identical to those obtained in Case Study 15.9.

17.2 Noise Simulation with NOF

17.2.1 NOF – An Overview

■ **History.** The first version of program NOF (acronym for NOise Factor) was developed by the author in the late 1980s. This program was acquired by Technische Fachhochschule Aachen, and due to the valuable comments of Professor Bex, an improved configuration was elaborated in the early 1990s. Finally, a friendly user interface was added in 1997.

■ **Generalities.** The initial objective was to compute the noise parameters of linear microwave circuits with arbitrary topology according to the original method described in Sect. 6.3. Primarily intended for research applications in a university environment, one of the main constraints imposed on NOF was to be easily implemented on any personal computer, independent of the operating system. Hence, it was developed under DOS, which at that time was the "common denominator" of most existing PCs.

The experience acquired during the first years after release showed that the main limitation was its unfriendly constraints imposed on the structure of the input data file. Consequently, a more friendly interface (PPNOF) was added, containing a full-page editor with a pop-up menu.

■ **Running the Program.** To create the input file for simulation, the user must run PPNOF and select the option "File" in the pop-up menu, then "New file". After typing the title of the input file and defining the frequency range, clicking on the "Edit" option gives access to the following sub-menu:

- "Add" a line describing a circuit element; in this case an input window is opened, with all input information;
- "Modify" the information on an existing line;
- "Suppress" a line.

When the input file has been entered, click in the "File" menu and choose the "Save" option (or "Save As" option). Beware: *if the user forgets to save the file, the program has no automatic backup facility and the work is lost.*

Once the input file is saved, formatting in the required syntax of NOF is performed by selecting the option "Execute", then "Create input file". The input file for NOF is automatically named ENTRE001.IN. After running NOF, the output data file automatically receives the name SORTIE001.OUT.

To display the output file on the screen, select the "Load output file" option from the "Execute" sub-menu. A graphical post-processor (not as powerful as PROBE) is available to trace plots as a function of frequency.

■ **Circuit Description.** The circuit is modeled following a branch–node description (close to that employed by SPICE). The user's first task is to identify and number the circuit nodes, with *sequentially increasing integers, starting from unity (node zero is always reserved for ground)*. With respect to PSPICE, note the following differences:

- A input signal generator, with internal 50-Ω resistance is automatically added by NOF. The user's task is only to specify the input as a voltage port (type "E", according to the NOF conventions). Similarly, a 50-Ω load is under the control of NOF, and the user has only to declare the output port as a voltage port (type "E").
- It is *mandatory* to declare the branches in the following order:
  1) the branches describing voltage ports (input and output)
  2) the branches describing two-ports (active devices)
  3) the remaining branches.

  If this rule is not observed, erroneous results are obtained.
- Transistors are described with their Y- or S-parameters (obtained either by measurement or from the transistor data sheet). Each transistor is modeled as a two-port, a line statement corresponding to each port. *The line corresponding to the input of the two-port must be immediately followed by the line corresponding to the output of the two-port.* At the end of the file, the four noise parameters of each active device are required, namely F_o (expressed in decibels), R_n in ohms, and the magnitude and argument of Γ_o. Eventually, the user could describe active devices by their circuit models, provided that only VCCS are used for this purpose.
- Noiseless resistors are included as a distinct device in the list of elements accepted by NOF. This represents a capability beyond PSPICE, where a noiseless resistor must be separately modeled.

■ **Discussion.** During simulation, it may happen that NOF displays the following message:

NO Y-MATRIX – EXECUTION CANCELLED !!

This means that when it computes the Y-matrix of the global passive multiport (resulting after de-embedding all active devices), it finds a singular matrix (see Sect. 6.3) and the program is stopped.

To remedy, the user has to inspect the passive multiport of the circuit and identify loops containing only voltage sources. It should be remembered that any loop containing only independent voltage source is rubbish, since only by chance can Kirchhoff's voltage law then be satisfied. Consequently, each such loop must be broken by inserting a resistor of negligible value (in order not to perturb the normal operation of the circuit). This value can range from 10^{-2} Ω to 10^{-5} Ω, which is essentially a short circuit, but this small value suffices to enable analysis.

We recall the main points of the procedure to create the passive multiport of the circuit:

- the circuit input and output ports are treated as voltage ports;
- all active devices are de-embedded and two additional voltage ports are created for each extracted device, at its input and output, respectively.

Whenever possible, it is advisable to draw on paper the resulting passive multiport, to check *before* simulation that no loop containing only voltage sources exists, and to eventually fix the problem by breaking the loop in question.

17.2.2 Case Studies

CASE STUDY 17.11

Figure 17.30 presents a 8-GHz amplifier, matched at both input and output to 50 Ω. The transmission line has a characteristic impedance $Z_o = 90\,\Omega$ and $\vartheta = 45°$. The relative dielectric constant of the substrate is $\varepsilon_r = 1$ and the attenuation of the line is neglected.

At the operating frequency, the measured MESFET parameters are

$y_{11} = 18.3 + j\,21.22$ [mS] $\qquad F_o = 2.535\,\text{dB}$
$y_{12} = 1.893 - j\,2.431$ [mS] $\qquad R_n = 116\,\Omega$
$y_{21} = 31.46 - j\,60.38$ [mS] $\qquad |\Gamma_o| = 0.23065$
$y_{22} = 3.71 + j\,6.912$ [mS] $\qquad \arg\Gamma_o = 127°$

Find by simulation the noise parameters of the amplifier.

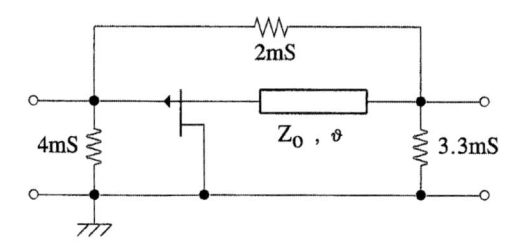

Fig. 17.30. 8-GHz microwave amplifier

Solution

■ **Modeling.** The first task is to develop the passive multiport obtained when the transistor is de-embedded (Fig. 17.31).

By inspection, a loop containing only voltage sources is detected between nodes 1 and 0; due to it, the Y-matrix of the circuit is singular. We must insert a small resistance between the circuit input and the MESFET input (Fig. 17.32).

The next step is to run PPNOF and enter branch information; the branches of the model are declared in the following order:

– Two E-type branches (corresponding to the input and output of the circuit)
– two E-type branches (corresponding to the input and output of the MES-FET)
– the remaining branches, in any order.

■ **Input Data File.** An overview of the circuit branches is given in Table 17.7.

624   17 Computer-Aided Noise Analysis

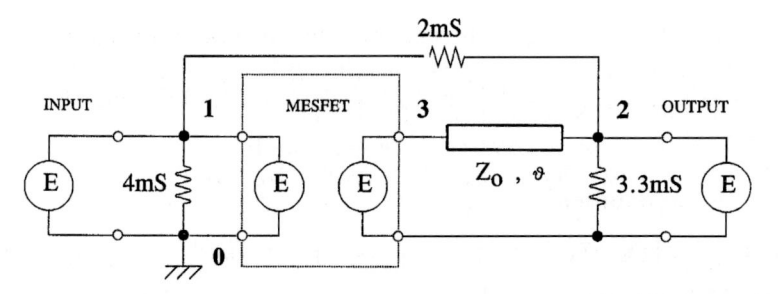

Fig. 17.31. Passive multiport of the circuit

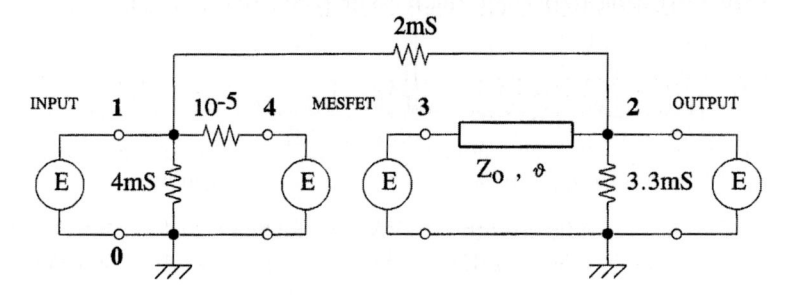

Fig. 17.32. Circuit model for simulation with NOF

Table 17.7. Circuit branches

Line number	Branch type	From node	To node
1	E	1	0
2	E	2	0
3	E	4	0
4	E	3	0
5	G	1	0
6	G	1	2
7	G	2	0
8	LL	3	2

■ **Results.** After running NOF, the following results are displayed:

VOLTAGE GAIN (complex) = .19446E+01+j .26270E+01
VOLTAGE GAIN (mag.) = .3268464E+01
VOLTAGE GAIN (in dB) = .1028687E+02 dB
TRANSDUCER POWER GAIN = .5302576E+01
TRANSDUCER POWER GAIN = .7244869E+01 dB

NOISE CONTRIBUTION OF PASSIVE NETWORK = .2902101E+02+j −.1769696E+02
NOISE INJECTED BY ACTIVE DEVICES = .1043819E+03+j −.8285039E-14

SPOT NOISE FIGURE OF THE CIRCUIT (lin) F = .3012651E+01
SPOT NOISE FIGURE OF THE CIRCUIT (dB) F = .4789488E+01 dB

CORRELATION MATRIX OF THE EQUIVALENT NOISE SOURCES
****
(the circuit is regarded as a noiseless two-port, having an equivalent noise current source connected to each port)

.7172238E+00+ j .0000000E+00   −.2383595E+01+ j.5605849E+00
−.2383595E+01+ j −.5605849E+00   .9024432E+01+ j.0000000E+00

THE EQUIVALENT NOISE PARAMETERS OF THE CIRCUIT
****
Noise resistance Rn = 122.4997961 Ohm
Correlation conduct. GCOR = −15.2131738 mS
Correlation suscept. BCOR = 9.4895100 mS
Noise conductance Gn = 26.4148899 mS
Noise figure (ratio) F = 3.0126508
Noise figure (in dB) F = 4.7894880 dB
Optimum source conduct. G0 = 21.1440951 mS
Optimum source suscept. B0 = −9.4895093 mS
Min. noise fig. (ratio) F0 = 2.4530733
Min. noise fig. (in dB) F0 = 3.8971053 dB
Opt. source reflection coef = .2263682 (mag) 109.8622716 (deg)

Equivalent S Parameters of the circuit
****
	Magnitude	Arg. (degrees)
S11	.578143	−136.438554
S12	.107851	17.264815
S21	2.302732	19.051807
S22	.263793	162.661887

Stop − Program terminated.

17 Computer-Aided Noise Analysis

■ **Conclusion.** Noise analysis with NOF yields the S- and noise parameters of the circuit, regarded as a two-port between its input and output. This suggests a possible method to simulate large circuits, provided that they can be properly partitioned.

CASE STUDY 17.12

A two-stage 11-GHz amplifier is presented in Fig. 17.33. The electrical parameters of all transmission lines are given in Table 17.8. The amplifier employs two MESFETs of type AT10600, whose S-parameters, according to the data sheet, are shown in Table 17.9.

The transistor noise parameters, at f = 11 GHz, are

$$F_o = 0.75\,\text{dB} \quad R_n = 15\,\Omega \quad \Gamma_o = 0.615 \ \angle 65.0$$

Find the noise parameters of the global circuit.

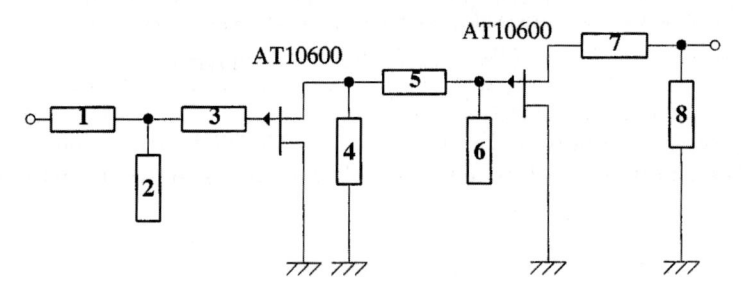

Fig. 17.33. Two-stage, 11-GHz amplifier

Table 17.8. Electrical parameters of transmission lines

Line number	1	2	3	4	5	6	7	8
Z_c [Ω]	18.90	90.50	39.30	185.4	29.20	37.90	95.00	20.20
ϑ [deg]	69.80	100.1	15.50	106.8	58.10	123.3	72.30	95.30
Length [mm]	5.288	7.583	1.174	8.090	4.401	9.341	5.477	7.220

Table 17.9. MESFET S-parameters

f [GHz]	S_{11} Mag.	S_{11} Arg.	S_{21} Mag.	S_{21} Arg.	S_{12} Mag.	S_{12} Arg.	S_{22} Mag.	S_{22} Arg.
10	0.58	$-134°$	2.42	81°	0.11	48°	0.50	$-15°$
11	0.57	$-153°$	2.36	70°	0.11	43°	0.43	$-18°$
12	0.57	$-174°$	2.21	57°	0.11	36°	0.36	$-23°$

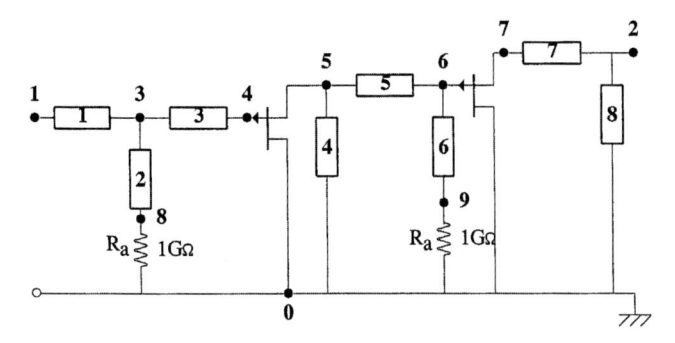

Fig. 17.34. Circuit model for simulation with NOF

Solution

■ **Circuit Model.** In order to simulate with NOF, the model of the circuit is presented in Fig. 17.34.

The role of resistances $R_a = 1\,\text{G}\Omega$ (declared as "noiseless") is to avoid having nodes 8 and 9 float, since in their absence, there is only one element connected to the nodes in question. As the stubs 2 and 6 must have an open circuit at their end, selecting a very large value ($1\,\text{G}\Omega$) is not expected to change this condition.

The transmission lines are modeled as homogeneous lines, without loss and with substrate permittivity $\varepsilon_r = 1$.

■ **Results.** NOF yields the following results:

VOLTAGE GAIN (complex) = .96358E+01+j −.52252E+01
VOLTAGE GAIN (mag.) = .1096135E+02
VOLTAGE GAIN (in dB) = .2079728E+02 dB

TRANSDUCER POWER GAIN =   .9885603E+02
TRANSDUCER POWER GAIN =   .1995003E+02 dB

NOISE CONTRIBUTION OF PASSIVE NETWORK = −.9002521E-13+j −.4294864E+04
NOISE INJECTED BY ACTIVE DEVICES = .1296638E+04+j .4565966E-13

SPOT NOISE FIGURE OF THE CIRCUIT (lin) F = .2049314E+01
SPOT NOISE FIGURE OF THE CIRCUIT (dB) F = .3116085E+01 dB

CORRELATION MATRIX OF THE EQUIVALENT NOISE SOURCES
****
(the circuit is regarded as a noiseless two-port, having an equivalent noise current source connected to each port)

.6320187E-01+ j .0000000E+00   −.5258537E+00+ j .2819616E+00
−.5258537E+00+ j -.2819616E+00   .5959340E+01+ j .0000000E+00

THE EQUIVALENT NOISE PARAMETERS OF THE CIRCUIT
****

Noise resistance Rn = 14.3190911 Ohm
Correlation conduct. GCOR = 1.7185964 mS
Correlation suscept. BCOR = 29.5485000 mS
Noise conductance Gn = 1.7298017 mS
Noise figure (ratio) F = 2.0493139
Noise figure (in dB) F = 3.1160849 dB
Optimum source conduct. G0 = 11.1246324 mS
Optimum source suscept. B0 = −29.5484997 mS
Min. noise fig. (ratio) F0 = 1.3678067
Min. noise fig. (in dB) F0 = 1.3602474 dB
Opt. source reflection coef. = .7188935 (mag.) 116.7934487 (deg.)

Equivalent S Parameters of the circuit
****
	Magnitude	Arg. (degrees)
S11	.128449	138.949685
S12	.021600	-77.133713
S21	9.942637	-23.133714
S22	.126965	-171.429786

Stop − Program terminated.

■ **Concluding Remark.** This case study reveals the advantage of macro-modeling: NOF employs macro-models for both MESFETs and transmission lines, and the input data file is compact and easy to understand. If the same circuit were simulated with PSPICE, besides difficulties reaching the four noise parameters, the description must be made by replacing each MESFET by with equivalent circuit, extracted from its S-parameters. Consequently, the complexity of the model increases considerably.

18
Protection Against Interfering Signals

> "Nothing is as inevitable as a mistake whose time is come"
>
> (Tussman's Law)

18.1 Techniques to Reduce Interference

18.1.1 Shielding

CASE STUDY 18.1 [200, 260]

To protect a harness containing several wire pairs against radiated interfering signals, a tin-plated braided shield is used, made of copper-clad steel. The mass proportion of various metals in the braid are steel 60%, copper 37%, and tin 3%. The thickness of the braid is t = 1 mm. The braid is 80% solid material and 20% air (by volume). When slipped over the harness, it has an average radius of 10 mm. Estimate the magnetic shielding effectiveness at 50 Hz and 100 Hz.

Solution

■ **Calculation.** Since we are dealing with low-frequency magnetic fields, the shielding effectiveness corresponds to the equation of case 1b of Table 9.3, which is recalled here:

$$SE = \left| 131.43\, t\sqrt{f\mu_r \sigma_r} + 74.6 - 10 \log \frac{\mu_r}{f\sigma_r r^2} \right| \text{ [dB]}$$

The problem here is to evaluate the equivalent conductivity and permeability of the braid, which is an inhomogeneous material.

The equivalent relative permeability of the braid (with respect to copper) is calculated taking into account the proportions of various materials:

$$\mu_{req} = 0.8\left(0.6\mu_{Fe} + 0.37\mu_{Cu} + 0.03\mu_{Sn}\right) + 0.2\left(\mu_{air}\right)$$

As the permeability of air is negligible, the last term of the sum can be discarded:
$$\mu_{req} \cong 0.8\left(0.6\mu_{Fe} + 0.37\mu_{Cu} + 0.03\mu_{Sn}\right)$$

Table 9.1 gives $\mu_{Cu} = \mu_{Sn} = 1$, $\mu_{Fe} = 1000$ and upon substituting these values in the latter expression, we find
$$\mu_{req} = 480.32$$

In a similar way, the equivalent relative conductivity (with respect to copper) is
$$\sigma_{req} \cong 0.8\left(0.6\sigma_{Fe} + 0.37\sigma_{Cu} + 0.03\sigma_{Sn}\right)$$

According to Table 9.1 $\sigma_{Fe} = 0.1$, $\sigma_{Cu} = 1$, and $\sigma_{Sn} = 0.15$, so
$$\sigma_{req} = 0.3476$$

Using μ_{req} and σ_{req}, the magnetic shielding effectiveness at $f = 50\,\text{Hz}$ is
$$SE = \left|(131.43)(1)\sqrt{(50 \cdot 10^{-6})(480.32)(0.3476)} + 74.6 \right.$$
$$\left. -10 \log \frac{480.32}{0.3476(50 \cdot 10^{-6})(10^{-4})}\right|$$

Finally
$$\boxed{SE_{50\text{Hz}} \cong 27.8\,\text{dB}}$$

and at 100 Hz we obtain
$$\boxed{SE_{100\text{Hz}} \cong 19.83\,\text{dB}}$$

■ **Remark.** Sometimes the following expression is proposed to evaluate the magnetic shielding effectiveness at low frequency:
$$SE_m \cong 20 \log \left(1 + \frac{t\mu_{req}}{2r}\right)$$

Note that this formula does not depend on frequency; the numerical calculation yields 27.8 dB, the same value as that obtained at 50 Hz.

■ **Conclusion.** This application shows that magnetic shielding effectiveness is rather poor (despite 60% iron in the braid) and decreases with increasing frequency. A possible explanation is that the braid is inhomogeneous, its volume contains 20% air, as well as a significant amount of non-magnetic materials. This example shows the difficulty of achieving good protection against magnetic fields at low frequency.

CASE STUDY 18.2 [151, 260]

A circuit is protected against interference by a double metallic enclosure made of

a) aluminum (thickness $d_1 = 0.8$ mm)
b) steel (thickness $d_2 = 0.8$ mm)

The overall dimensions of the box are length $L = 20$ cm, width $l = 10$ cm, and height $h = 5$ cm.

1) Determine the resonant frequencies of the box and explain how they modify the shielding effectiveness (SE).
2) What is the required thickness of the metallic wall to optimally protect against magnetic fields at 50 Hz?

Solution

1) In any metallic enclosure, there are two kinds of resonance:
   - Resonance related to the wall thickness
   - Cavity resonance

■ **Resonance Related to the Wall Thickness.** The propagation constant of an electromagnetic wave travelling inside a metal is

$$\gamma = \alpha + j\beta = \sqrt{j\omega\mu(\sigma + j\omega\varepsilon)} \cong \sqrt{j\omega\mu\sigma} = (1+j)\sqrt{\pi f \mu \sigma}$$

where it is assumed that $\sigma \gg \omega\varepsilon$. Consequently, $\alpha = \beta = \sqrt{\pi f \mu \sigma}$.

The mechanism of this kind of resonance is the following: any metallic enclosure inherently has leakages, due to apertures and imperfect seams. A small fraction of the outside radiation penetrates inside (through air gaps), and a fraction of the wave incident upon the box is transmitted inside through the metallic walls. Whenever the path length through the metal is an odd multiple of $\lambda/2$ greater than through leakages, a resonant frequency appears. In this case the phase shift between the considered paths is $\beta t = \pi$ (i.e., 180° path delay, t being the wall thickness) and the resulting wave is attenuated at the receiving point. At this frequency, all happens as if the shielding effectiveness has been suddenly increased.

Therefore, the first resonant frequency associated with wall thickness can be deduced from the condition $\beta t = \pi$:

$$f_r = \frac{\pi}{t^2 \mu \sigma} \quad \text{or} \quad f_r = \frac{4.31 \cdot 10^4}{t^2 \mu_r \sigma_r}$$

In the latter equation, t is expressed in mm, and μ_r, σ_r are selected from Table 9.1. In this case, for aluminum $\sigma_r = 0.61$ and $\mu_r = 1$, so that

$$f_{r(Al)} = \frac{4.31 \cdot 10^4}{0.8^2 \, (0.61)(1)} \cong 110.4 \text{ kHz}$$

For the steel wall, $\sigma_r = 0.1$ and $\mu_r = 1000$, which yields

$$f_{r(Fe)} = \frac{4.31 \cdot 10^4}{0.8^2 \, (10^3)(0.1)} \cong 673 \text{ Hz}$$

It is obvious that subsequent resonances are found at all subsequent multiples of these values.

■ **Cavity Resonance.** The mechanism of this kind of resonance is tied to the standing waves within the enclosure. As a consequence, it is expected that the SE of the box will exhibit large variations inside, depending on the location of the receiving point (which may fall at a maximum or minimum). According to (9.14), the lowest frequencies associated with the TE_{011}, TE_{101}, and TE_{110} modes are

$$TE_{011} \text{ mode} \qquad f_r = \frac{212}{L} = \frac{212}{0.2} = 1060 \text{ MHz}$$

$$TE_{101} \text{ mode} \qquad f_r = \frac{212}{l} = \frac{212}{0.1} = 2120 \text{ MHz}$$

$$TE_{110} \text{ mode} \qquad f_r = \frac{212}{h} = \frac{212}{0.05} = 4240 \text{ MHz}$$

■ **Discussion.** The metallic box behavior depends on the operating frequency in the following way:

– up to 673 Hz, SE is close to the calculated value;
– just before 673 Hz and exactly at 673 Hz, SE increases;
– above 673 Hz, SE is less than that defined by calculation.

The same statement applies to the resonant frequency of 110.4 kHz, as well as to multiples of 673 Hz and 110.4 kHz.

At all frequencies equal to or multiples of 1.06 GHz, the distribution of both electric and magnetic fields inside the box becomes highly nonuniform. Consequently, SE exhibits spatial variation with respect to the computed value, which is undesirable.

2) To obtain a high SE value for magnetic fields at 50 Hz, the thickness of steel coverage has to be selected such that the first resonant frequency associated with the metallic wall be situated precisely at 50 Hz. Hence:

$$t = \sqrt{\frac{4.31 \cdot 10^4}{50 \, (1000)(0.1)}} \cong 2.94 \text{ mm}.$$

18.1 Techniques to Reduce Interference

CASE STUDY 18.3 [260, 266]

Often the top cover plates of shielded enclosures are perforated to allow heat to escape from the circuits inside. Determine the shielding effectiveness (SE) of a perforated cover plate having 4 holes per cm^2, each hole being a square of dimension x, with a metallic separation (y) between two adjacent holes. This plate is made of commercial iron, 0.6 mm thick, and is subjected to an incident magnetic field created by a 10-kHz source situated at distance r = 20 cm. Compare the SE of two different panels, having

a) x = 1 mm and y = 4 mm
b) x = 2 mm and y = 3 mm

■ **Background.** According to [260], as long as the operating frequency does not exceed the cutoff frequency of the apertures, the SE (expressed in decibels) of a perforated metal plate is

$$SE = A_a + R_a + B_a + K_1 + K_2 + K_3 \tag{18.1}$$

where the following notation has been introduced:

- A_a is the absorption expressed in decibels:

$$A_a = \begin{cases} 27.3 \, D/W & \text{for rectangular or square holes} \\ 32 \, D/W & \text{for circular holes} \end{cases} \tag{18.2}$$

where D represents the depth of the aperture (in this case, the thickness of the metal plate) and W is the width of the aperture (if square or rectangular, measured normal to the E-field vector), or the diameter of the circular hole.

- R_a represents the aperture reflection loss in decibels, due to mismatch between wave impedance and barrier impedance at the air-to-metal interface:

$$R_a = 20 \, \log \frac{(1+K)^2}{4K} \tag{18.3a}$$

or

$$R_a \cong 20 \, \log \frac{K}{4} \quad \text{if } K \gg 10 \tag{18.3b}$$

K is the ratio of wave impedance to aperture characteristic impedance. It depends on the geometry of the aperture:

$$K = \begin{cases} \pi r/W & \text{square or rectangular holes, near-H field} \\ 3.68r/W & \text{circular holes, near-H field} \\ \dfrac{0.72 \cdot 10^4}{rWf^2} & \text{square or rectangular holes, near-E field} \\ 150/jWf & \text{square or rectangular holes, far field} \\ 173/jWf & \text{for circular holes, far field} \end{cases} \quad (18.4)$$

where
f – frequency in MHz
r – distance from source to panel, in meters
W – in meters.

- B_a is a correction term for aperture reflection that is negligible when $A_a \geq 10\,\text{dB}$. Otherwise

$$B_a = 20\,\log\left(1 - \frac{(K-1)^2}{(K+1)^2}\,10^{-A_a/10}\right) \quad (18.5)$$

- K_1 is a correction factor that depends on the number of holes per unit square. K_1 is always a negative quantity, or equal to $0\,\text{dB}$ (for a single hole):

$$K_1 = 10\,\log\,(1 - an) \quad (18.6)$$

a – area of each hole
n – number of holes per unit square (same units as a)

- K_2 is a correction factor for metal penetration, which has significant values when the metal separation (y) between holes is comparable to the skin depth δ:

$$K_2 = -20\,\log\left(1 + \frac{35}{z^{2.3}}\right) \quad (18.7)$$

where

$$z = \frac{y}{\delta} = \frac{y\sqrt{f\sigma_r\mu_r}}{0.067} \quad (18.8)$$

y being expressed in mm, f in MHz and σ_r, μ_r relative to copper (Table 9.1).

- K_3 is a correction factor for coupling between closely spaced holes:

$$K_3 = \begin{cases} 20\,\log\,\dfrac{1}{\tanh(A_a/8.7)} & \text{for } A_a < 10\,\text{dB} \\ < 1 \text{ (negligible)} & \text{for } A_a > 10\,\text{dB} \\ 20\,\log\,\dfrac{1}{\tanh(3.4D/W)} & \text{for square or circular holes} \end{cases} \quad (18.9)$$

18.1 Techniques to Reduce Interference

Solution

For commercial iron, according to Table 9.1, $\sigma_r = 0.17$ and $\mu_r = 200$.

Applying (18.1) to (18.9), the results presented in Table 18.1 have been found (all expressed in decibels, except for K and z which are dimensionless).

Table 18.1. Numerical values of various quantities

Case	A_a	K	R_a	B_a	K_1	z	K_2	K_3	SE
a)	16.38	628.3	43.92	$\cong 0$	−0.18	34.81	−0.086	0.29	60.9
b)	8.19	314.15	37.9	−1.43	−0.76	26.11	−0.166	2.27	46

To conclude, the perforated panel proposed in case b offers better heat evacuation than that of case a, but since the size of the apertures increases, it has lower shielding effectiveness (near 25% degradation).

This demonstrates that a trade-off must be considered between heat evacuation and shielding effectiveness.

CASE STUDY 18.4 [239, 261]

Most man-made noise sources predominantly generate an electric field. As a consequence, to improve the signal-to-noise ratio of a receiving loop antenna in an industrial area, it would be beneficial to reduce its sensitivity to the electric component of the electromagnetic fields, while maintaining its sensitivity to the magnetic component. Suggest a practical solution to achieve this and comment on the solution.

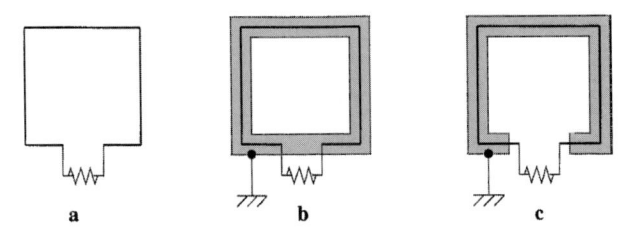

Fig. 18.1. a Loop antenna; b shielded loop antenna; c split shield

Solution

Consider a rectangular loop antenna (Fig. 18.1a), where the resistor closing the loop represents the equivalent load. The voltage induced in the loop by a magnetic field of angular frequency ω is given by the familiar equation

$$V_m = j\omega BA \cos \vartheta$$

where A is the loop area, B is the magnetic induction, and ϑ is the angle between the magnetic field lines and the normal to the plane of the loop.

636   18 Protection Against Interfering Signals

In order to suppress the influence of the electric field, a possible solution is to completely shield the antenna (Fig. 18.1b). However, the shield will also attenuate the magnetic component of the field, due to the flow of induced current through the shield loop (see Sect. 11.3.3). To prevent this, the shield must be split (Fig. 18.1c). In this way, the resulting antenna is sensitive only to the magnetic field component of the electromagnetic wave.

Concerning the place where the shield should be split, this largely depends on the mechanical realization of the antenna. From the above discussion, the only concern is to locate it where the split can be made as small as possible. For instance, at the bottom (as in Fig. 18.1c) it can take advantage of the opening already present for the feeder of the antenna. Another possibility is to split at the top, provided that the continuous shield at the bottom still allows the feeder to emerge without complications.

Possible applications of such antennas are radio direction finders, receiving antennas for broadband receivers, and so on.

CASE STUDY 18.5 [272]

Whenever operational amplifiers are employed in circuits where their input currents are of the order of several pA, a guard shield must be provided to reduce leakage currents (which can rise to the same order of magnitude as the input currents). The guard shield is often placed on the printed circuit board, protecting the amplifier input connections (Fig. 18.2). However, as stated in Sect. 9.1.2, effective shielding only results when the absolute potential of the guard is stabilized with respect to the input signal. Depending on the particular configuration of the amplifying circuit, indicate how the guard should be connected.

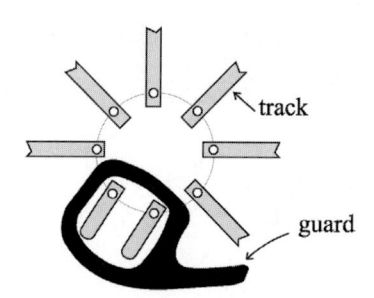

Fig. 18.2. Guarded layout for inputs to a TO–5 case amplifier

Solution

■ **Discussion.** The most frequently encountered linear applications involving operational amplifiers are the non-inverting amplifier, the voltage follower, and the inverting amplifier. Hence, the following two statements apply.

18.1 Techniques to Reduce Interference

- The guard should be connected to a point whose potential is the same as the inputs, with a low enough impedance to ground to absorb leakage currents without introducing excessive offset.
- For any ideal operational amplifier, the inverting input has essentially the same potential as the non-inverting input.

Keeping in mind the above conditions, the configurations shown in Fig. 18.3 are proposed. To reduce input offset, the equation beside each circuit has to be satisfied (R_g denotes the internal resistance of the signal generator connected to the input).

☐ **Configuration of Fig. 18.3a.** According to statement 2, the input potential is nearly equal to the potential of the inverting input, which in turn is equal to the potential at point 0, provided that the current through R_3 is zero. In practice, the input current is low, but never zero (this is called the

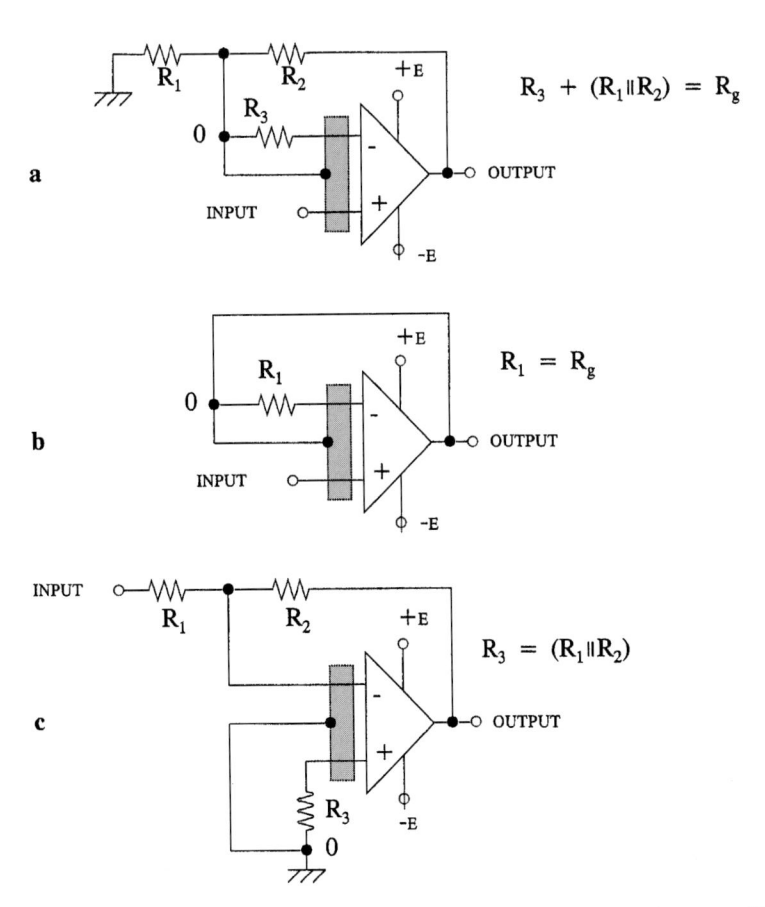

Fig. 18.3. a Non-inverting amplifier; **b** voltage follower; **c** inverting amplifier

Equations shown:
- a: $R_3 + (R_1 \| R_2) = R_g$
- b: $R_1 = R_g$
- c: $R_3 = (R_1 \| R_2)$

offset input current, and it depends on the particular sample; usually it is of the order of several tens of nA). Since the objective is to build a low-noise amplifier, the value of R_3 must clearly not be too high (in order to limit its thermal noise); values of several kΩ are quite practical. Therefore the voltage drop of the input offset current through R_3 usually does not exceed several tens of µV. Finally, the potential difference between point 0 and the INPUT terminal is negligible, and connecting the guard to point 0 is a good choice.

☐ **Configuration of Fig. 18.3b.** In this case point 0 is directly connected to OUTPUT; since the voltage gain is unity, the potential of point 0 is equal to the INPUT potential (second statement), provided that the voltage drop across R_1 produced by the input offset current is negligible.

Once again, connecting the guard shield to point 0 is a good choice.

☐ **Configuration of Fig. 18.3c.** Provided that the voltage drops across R_3 and R_1 are both negligible (or cancel), the potential of point 0 is nearly equal to the potential of INPUT terminal. This shows that connecting the guard to 0 is convenient.

■ **Concluding Remarks.** In all three cases, connecting the guard shield to point 0 represents a good solution, provided that

– the operational amplifier is selected keeping in mind that the input offset currents must be as low as possible
– low-value resistors R_3 and/or R_1 are adopted.

However, additional care must be exercised:

- To limit leakage, the entire circuit board must be carefully cleaned with trichlorethylene or alcohol to remove solder flux, then blown dry with compressed air. Boards should be coated with epoxy or silicone rubber to prevent contamination by dust, water vapor, or any other impurity whose effect is to provide unwanted conduction paths between components.
- The integrated circuit case must be placed as close as possible to the board (to shorten its leads).

CASE STUDY 18.6 [213, 261, 263]

1) Consider a capacitive transducer connected by means of a shielded twin cable to the input of an amplifier (Fig. 18.4). Discuss the problems arising in this circuit.
2) To address those problems, it is suggested to use signal guarding. Propose a configuration illustrating this technique and comment on it.

18.1 Techniques to Reduce Interference

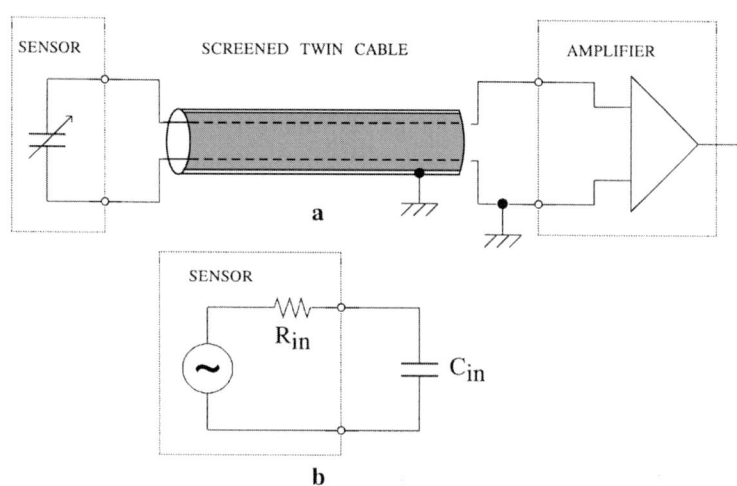

Fig. 18.4. **a** Capacitive transducer connected to an amplifier; **b** equivalent low-pass filter

Solution

1) One of the most critical problems appearing here is the severe bandwidth limitation. Suppose that the input capacitance (C_{in}) of the cable–amplifier system is several tens of picofarads, and the transducer internal resistance (R_{in}) is roughly the insulator resistance (about $1\,G\Omega$). An equivalent low-pass filter appears at the input, with roll-offs situated as low as several Hz. To avoid this, it is necessary to reduce the input capacitance.

   Another problem is the effect of the insulating resistance in the cable. If the amplifier operates with an input bias current in the order of picoamperes, the leakage current can degrade the performance of the amplifier, since the two currents are of the same order of magnitude.

   The solution to both difficulties is *signal guarding*.

2) To reduce the effects of leakage and input capacitance, the circuit shown in Fig. 18.5 is proposed [263].

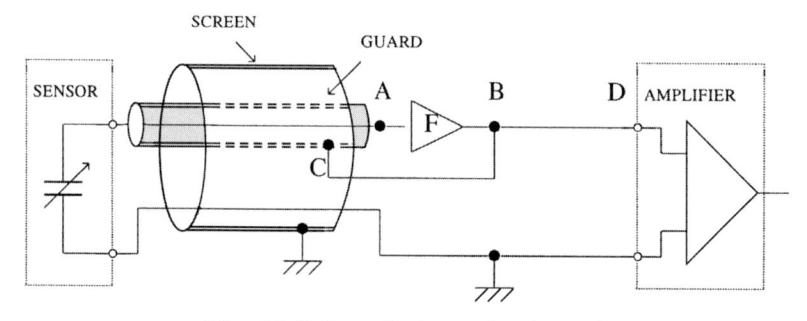

Fig. 18.5. Introducing a signal guard

F represents a voltage follower (unity gain), so that the potential at point A equals the potential at point B. Connecting points B and C ensures zero voltage difference between the signal and its shield; hence the leakage problem disappears. Since the follower offers a low output impedance, it drives the input capacitance of the system (amplifier–conductor BD), avoiding band limitation. Consequently, the follower must be placed as close as possible to the sensor, to reduce the length of cable driven by a high-resistance source.

18.1.2 Filtering

CASE STUDY 18.7 [261, 266]

Draw some simple passive filter configurations, intended to suppress conducted RF interfering signals. Discuss how the efficiency of filtering depends on the terminal impedances.

Solution

■ **Explanation.** Interfering signals often have a considerably higher frequency than the useful signal carried by the various conductors. This explains why the filters used in suppressing RF interference are of *low-pass type*.

A passive low-pass filter (together with its output load) is basically a voltage divider whose ratio depends on frequency in such a way that *spurious signals are attenuated at the load while maintaining the level of the useful signal almost constant* (this is the filtering condition).

To illustrate the principle of operation, consider one of the simplest filter configurations, consisting of a series impedance (Fig. 18.6a). At any instant,

$$E = V(Z_S) + V(Z_F) + V(Z_L) \qquad (18.10)$$

where $V(Z_S)$, $V(Z_F)$, $V(Z_L)$ denote the voltage drop at the source, filter, and load impedance, respectively.

For useful signals, the voltage drop across the filter has to be negligible, i.e.,

$$V(Z_F) \ll V(Z_S) + V(Z_L) \qquad (18.11a)$$

while for interfering signals,

$$V(Z_F) \gg V(Z_S) + V(Z_L) \qquad (18.11b)$$

Since the current through all series impedances is the same, constraints (18.11) translate into the following sufficient conditions:

18.1 Techniques to Reduce Interference

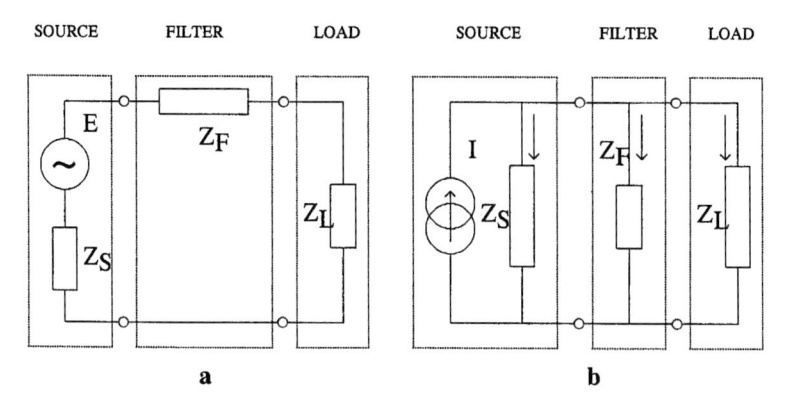

Fig. 18.6. a Filtering with a series impedance; b dual configuration (the size of impedance symbols are correlated with their magnitudes at RF)

at low frequency	at high frequency
$|Z_F| \ll |Z_S|$	$|Z_F| \gg |Z_S|$
and	and
$|Z_F| \ll |Z_L|$	$|Z_F| \gg |Z_L|$

This means that $|Z_F|$ should increase with frequency, so an inductor can be employed. Note also that efficient filtering requires low load and source impedances in this case (in RF and microwave circuits $Z_L = Z_S = 50\,\Omega$).

The dual situation is illustrated in Fig. 18.6b, where at low frequency (useful signal), the current through Z_F must be negligible with respect to those through Z_S and Z_L. At high frequency (for spurious signals), the filtering condition requires that the currents through Z_S and Z_L must be negligible with respect to the current flowing through Z_F. Summarizing,

at low frequency	at high frequency
$|Z_F| \gg |Z_S|$	$|Z_F| \ll |Z_S|$
and	and
$|Z_F| \gg |Z_L|$	$|Z_F| \ll |Z_L|$

Since the magnitude of the filter impedance must decrease with increasing frequency, a capacitor should be adopted.

■ **Basic Configurations.** Figures 18.7a and b present the configurations corresponding to those previously discussed. The conditions imposed on the magnitudes of the source and load impedances are marked beside the components.

More elaborate configurations [266] (which offer better filtering) are given in Figs. 18.7c, d, e, and f. Concerning the level of terminal impedances, a

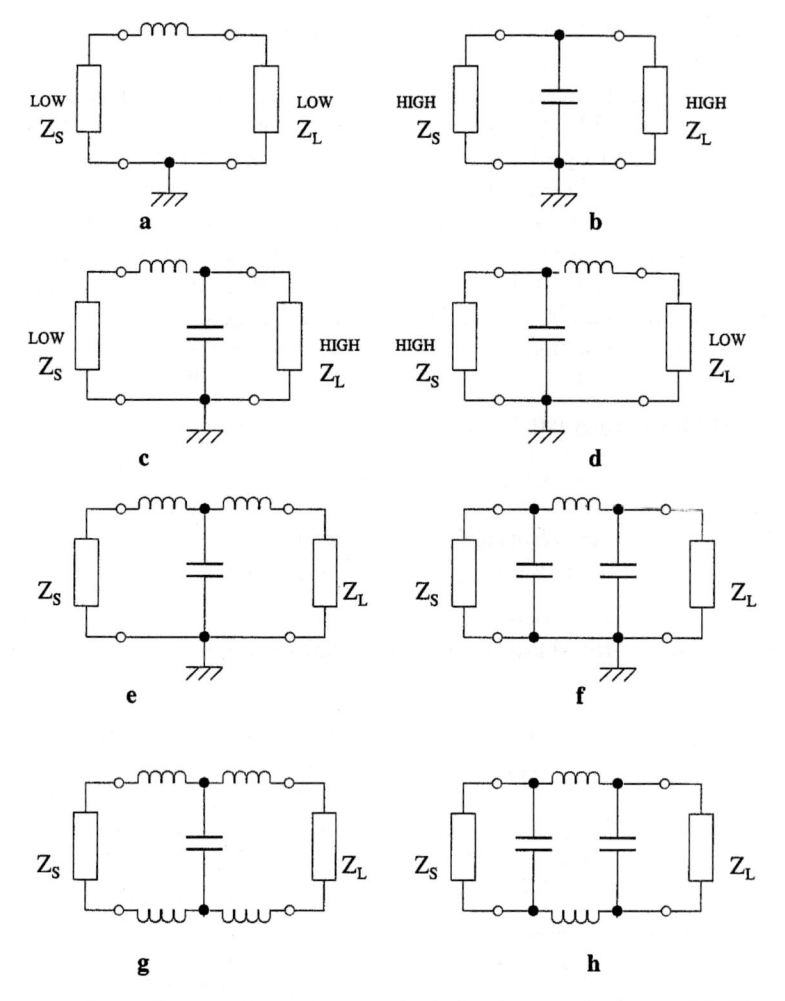

Fig. 18.7. Basic filter configurations: **a**, **b** single element filters; **c**, **d** half-section filters; **e** low-to-medium terminal impedances full-section; **f** high-to-medium terminal impedances full-section; **g**, **h** balanced filters

good rule of the thumb is that an inductor must "see" a low impedance at the nearest port, while a capacitor must "see" a high impedance.

■ **Concluding Remarks**

– Filters intended for RF and microwave applications operate on real terminal impedances of 50 Ω at each end, because this is a universally accepted standard. Note that in practice the terminal impedances are never real (although this is highly desirable) – they still have a reactive component. The problem is the unpredictable behavior of this reactive part at the frequencies of interfering signals, where unwanted resonances can appear and the insertion loss can turn into insertion gain.

- Although it might appear that filtering out a 1-MHz interfering signal from low-frequency useful signals is an easy task, in reality this is true *only if the pertinent circuits* (especially the input stages of amplifiers) *are linear*. If the circuit contains nonlinear elements (like PN junctions), they may rectify the interfering signal and transform it into DC or a very low-frequency voltage. Consequently, the spectrum of the high-frequency interfering signals shifts into a low-frequency region, making the spurious signals indistinguishable from the useful signals. In such situations, filtering is ineffective, and the only remedy is shielding.

CASE STUDY 18.8 [261]

A passive low-pass filter (Fig. 18.8a) intended to filter stray RF signals uses tubular film capacitors, whose band indicates the end connected to the outer foil. The inductors are also tubular (with open magnetic core). A possible layout on a single-sided PCB is proposed in Fig. 18.8b.

1) Comment on the proposed layout.
2) Propose an improved layout, using a single-sided, and then a double-sided PCB.

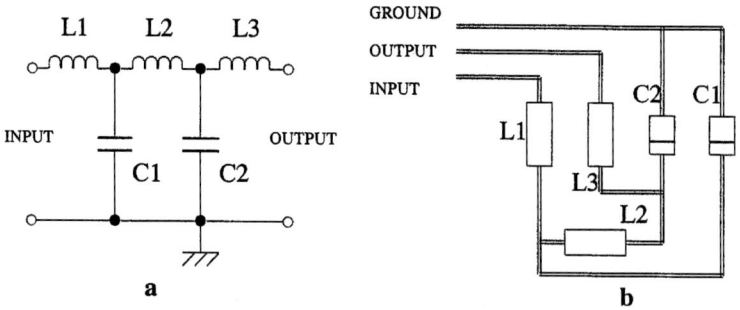

Fig. 18.8. a Circuit topology; b physical layout (reproduced with kind permission of John Wiley & Sons)

Solution

1)

■ **Analysis of the Proposed Layout.** Several drawbacks can be pointed out in the layout of Fig. 18.8b:

a) A general rule states that *the end of a film capacitor connected to the outer foil must be grounded*. This is not done in the proposed layout.
b) Since the inductors are of open-core type, magnetic field leakage is important. To prevent coupling between inductors, parallel placement must be avoided. In the proposed layout, L1 and L3 are placed close together

and parallel. This is a bad choice, yielding maximum unwanted coupling between the filter input and output.
c) Another unwanted coupling between filter input and output results from the position of their respective traces, which are parallel and close; hence, electric and magnetic coupling appears between traces.
d) The ground trace is common to input and output; thus, coupling through a common-mode impedance (in this case, the ground trace impedance) cannot be avoided.
e) The trace between the input and L1 is too long. Since it carries the conducted interfering signals as well as the useful signal, it radiates noise.
f) The trace connecting L1 to C1 is also too long. Its associated stray inductance increases the series inductance of C1, reducing the capacitor self-resonant frequency.
g) The trace connecting L3 to C2 is parallel and close to inductor L2. Unwanted magnetic coupling between them will result.

2)

■ **Improved Solution (Single-Sided PCB).** The improved solution proposed in [261] is shown in Fig. 18.9a. The disadvantages a, b, e, and f have been suppressed. For point c, although the input and the output traces are still parallel, physical separation between them has been considerably increased. Concerning point d, two different ground traces are provided, one for the input, the other for the output, separating the return currents. Located between the input and output traces, they also provide local "shielding".

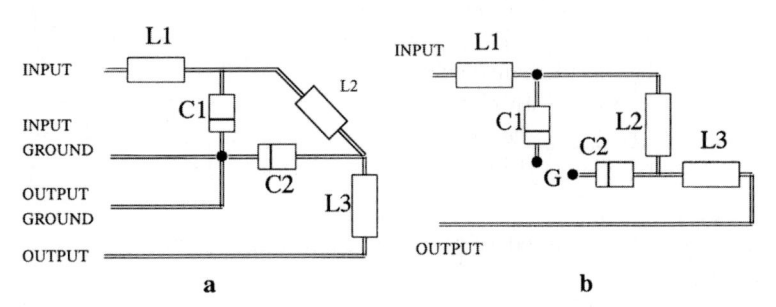

Fig. 18.9. Improved layout: **a** single-sided PCB (with kind permission of John Wiley & Sons); **b** using a ground plane

□ **Another Solution (Using a Ground Plane).** When using a double-sided PCB, a possible solution is shown in Fig. 18.9b, where points G denote vertical connection to the ground plane (situated on the opposite side). The input and output grounds (not shown) should be placed as close as possible to the INPUT and OUTPUT terminals to minimize ground loop area. This improves both susceptibility and noise radiation.

Even if inductors L1 and L3 are still parallel, they are no longer face to face, and coupling is avoided.

CASE STUDY 18.9 [153, 261]

A ferrite bead is made of a magnetic material characterized by a relative permeability whose magnitude is 800 at 100 MHz. It is used to prevent high-frequency interfering signals being conducted out of a switching-mode power supply with internal resistance $0.5\,\Omega$, connected to a load of $10\,\Omega$. The ferrite bead has length $\ell = 4$ mm, inner diameter $a = 1.2$ mm, and outer diameter $b = 3.5$ mm; its tangent loss is equal to 10 (at 100 MHz).

Determine the expected current attenuation of interfering signals at 100 MHz.

Solution

■ **Background.** Ferrites are low-density ceramic materials with composition Fe_2O_3XO, where X is a different metal (cobalt, nickel, zinc, manganese, etc.). Almost all manufacturers have developed their own composition. These magnetic materials have reduced electrical conductivity and therefore very low eddy currents. Cores containing them are useful in high-frequency circuits.

There are several types of ferrite beads: single or multihole beads, tubular or split, spherical, etc. They can be slipped over wires and cables (when split), or wires and component leads can be passed through the beads.

Due to their relatively low impedance ($30\,\Omega$ to $600\,\Omega$) at frequencies above 1 MHz, ferrite beads are most effective in low-impedance circuits (power supplies, series resonant circuits, SCR switching circuits, etc.). Note that multihole beads present a higher impedance than their single-hole counterparts. Attention should be paid to the DC current in the circuit, which must not exceed the limit imposed by the manufacturer to avoid saturation of the ferrite and consequently performance degradation.

Since they represent an elegant and inexpensive way to introduce high-frequency loss (without affecting DC or low-frequency signal transmission), ferrite beads are employed in various applications, including

– RF filtering;
– suppression of conducted interference entering or leaving a circuit or system;
– common-mode chokes, to reduce the effects of common-mode radiation or reception;
– limiting the initial peak in loads with high inrush currents;
– damping high-frequency oscillations generated by fast switching transients or parasitic resonances.

☐ **Explanation.** In ferrites, the magnetic material presents a complex relative permeability μ_r:

$$\mu_r = \mu_r' - j\mu_r'' \quad (18.12)$$

Its real part, denoted by μ_r', is related to the ability of the material to concentrate magnetic flux. The imaginary part, μ_r'', is related to the dissipa-

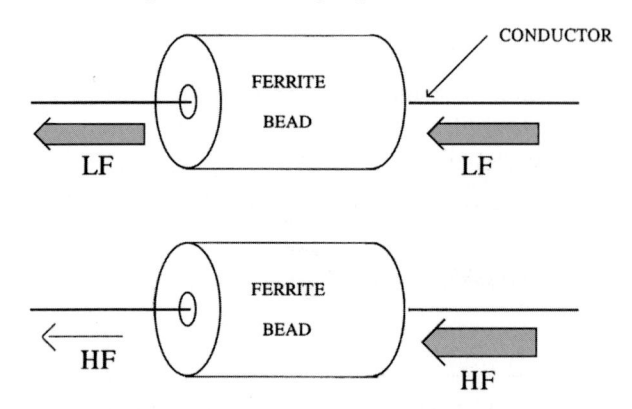

646   18 Protection Against Interfering Signals

Fig. 18.10. Illustrating the effect of selective energy absorption in a ferrite bead

tion of magnetic energy, as it flows through the material. The *loss tangent*, denoted by $\tan(\delta)$, is a function of frequency, defined as

$$\tan(\delta) = \frac{\mu_r''}{\mu_r'} \tag{18.13}$$

As a general rule, ferrites intended for use in interference control exhibit large loss tangent at high frequencies. This means that a large amount of the magnetic energy is transformed into heat inside the ferrite bead. Consequently, *ferrite beads are among the very few devices able to transform noise energy into heat.*

Consider a ferrite bead surrounding a wire transporting broadband signals (Fig. 18.10). The low-frequency (LF) components of the current are not attenuated, since losses at LF are negligible.

However, the magnetic energy of the high-frequency (HF) components is absorbed by the ferrite and internally transformed into heat. The level of the HF output current is considerably reduced relative to the level of the input current. The widths of arrows suggest this mechanism of selective current loss.

■ **Equivalent Circuit.** Consider a cylindrical ferrite bead installed on a conductor (Fig. 18.11). Denote the inner diameter by a and the outer diameter by b.

The inductance per unit length of a wire with a bead around it is

$$L = 2 \cdot 10^{-7} \, (\mu_r' - j\mu_r'') \, \ln \frac{b}{a} \quad [\mu H/m] \tag{18.14}$$

Consequently, the equivalent impedance is

$$Z_e = j\omega L = (j\omega \mu_r' + \omega \mu_r'') \, A \quad \text{where} \quad A = 2 \cdot 10^{-7} \, (\ell) \, \ln \frac{b}{a}$$

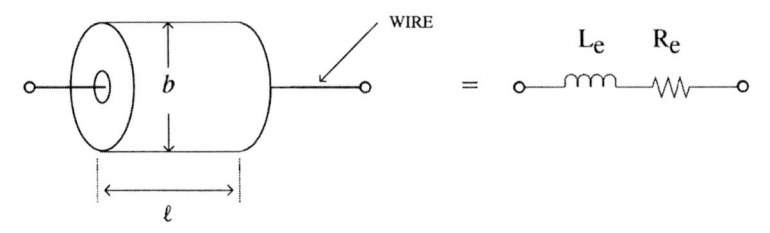

Fig. 18.11. Equivalent circuit of a ferrite bead

from which the equivalent reactance and resistance are

$$\boxed{\omega L_e = A\,\omega\mu_r'} \quad \text{and} \quad \boxed{R_e = A\,\omega\mu_r''} \tag{18.15}$$

Note that the equivalent resistance increases with frequency and, as expected, vanishes at DC.

■ **Calculation.** Considering the indicated numerical values, the values of μ_r' and μ_r'' must first be determined from the following system of equations (corresponding to the specified loss tangent and magnitude of μ_r):

$$\begin{cases} \mu_r''/\mu_r' = 10 \\ (\mu_r')^2 + (\mu_r'')^2 = (800)^2 \end{cases}$$

Solving it, we obtain $\mu_r' \cong 79.6$ and $\mu_r'' \cong 796$. So

$$A = 2 \cdot 10^{-7}(4 \cdot 10^{-3})\ln\frac{3.5}{1.2} = 0.856 \text{ nH}$$

$$R_e = 0.856 \cdot 10^{-9}(2\pi \cdot 10^8)(796) \cong 42.81\,\Omega$$

Since ωL_e must be ten times lower (i.e., $4.28\,\Omega$), it can be neglected.

To appreciate how much RF interfering signals are attenuated with respect to DC, compare the equivalent circuits in Fig. 18.12.

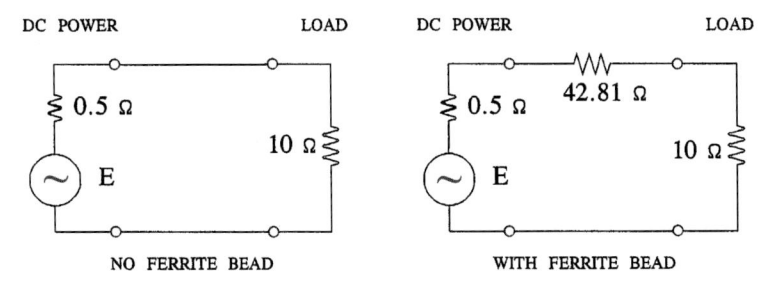

Fig. 18.12. Equivalent circuits at 100 MHz

Without the ferrite bead, the current through the bold wire segment is

$$\frac{E}{0.5 + 10}$$

while when using a ferrite bead the same current becomes

$$\frac{E}{0.5 + 10 + 42.81}$$

Therefore, the current loss in decibels is

$$\text{Loss} = 20 \, \log \left(\frac{E}{10.5} \, \frac{53.31}{E} \right) \cong 14.1 \text{ dB}$$

18.1.3 Grounding

CASE STUDY 18.10 [261, 262]

Consider the power and ground plane system in Fig. 18.13. The current carried by the planes switches at $150 \, \text{mA/ns}$, and the dimensions of the planes are length $l = 2 \, \text{cm}$, width $w = 2 \, \text{cm}$ and separation height $h = 0.2 \, \text{mm}$. Determine the resulting inductive noise in each plane.

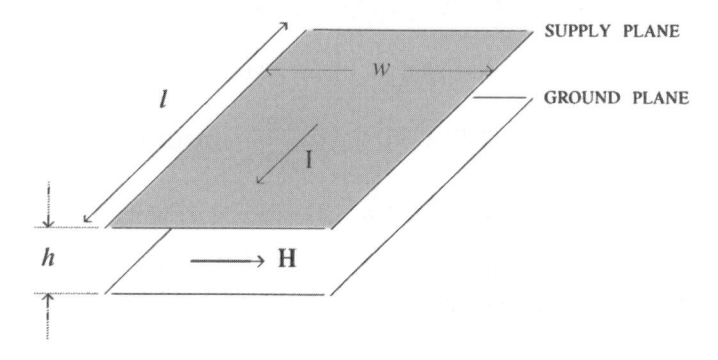

Fig. 18.13. A supply plane and a ground plane in a multilayer system

Solution

■ **Background.** For the configuration in Fig. 18.13, assume that both planes carry equal current I in opposite directions (for simplicity, only the current through the supply plane is shown).

The magnetic field is concentrated in the space between the planes (where it is approximately uniform), while the magnetic field outside the planes can be neglected due to field cancellation.

Under these assumptions, the self-inductance of each plane can be calculated by means of Ampère's law:

18.1 Techniques to Reduce Interference

$$\oint \overline{H} \, \overline{ds} = I \tag{18.16}$$

In this particular case, this means

$$H \, 2w = I \tag{18.17}$$

Since

$$B = \mu H = \mu \frac{I}{2w} \tag{18.18}$$

the inductance of each plane can be calculated from its general definition:

$$L = \frac{N\Phi}{I} = \frac{B \, l \, h}{I} = \frac{\mu \, l \, h}{2w} \tag{18.19}$$

The insulator separating the planes has $\mu = \mu_o = 4\pi \cdot 10^{-7}$ [H/m]; if the height between planes is expressed in cm, we have

$$L = 2\pi \frac{l}{w} h \quad [\text{nH}] \tag{18.20}$$

■ **Calculation.** As soon as the self-inductance of each plane is found with (18.20), the inductive noise voltage can be calculated:

$$v_n = L \frac{dI}{dt} = 2\pi \frac{l}{w} h \frac{dI}{dt} \tag{18.21}$$

With the indicated numerical values, we obtain

$$v_n = 2\pi \frac{2}{2}(0.02) \frac{150 \cdot 10^{-3}}{10^{-9}} 10^{-9} \cong 18.85 \text{ mV}$$

Note that the duration of this inductive noise voltage is approximately equal to the signal rise time [262].

CASE STUDY 18.11 [260, 261, 263]

Indicate the optimum grounding arrangement for two circuits situated at different locations and interconnected by a shielded cable. Consider several possibilities, depending on the transmitted signal type and cable length.

Solution

When grounding between instruments, several situations can be encountered in practice. However, the common objective in all of them is the suppression of common-mode interference produced by the inherent offset voltage between local grounds that have been interconnected.

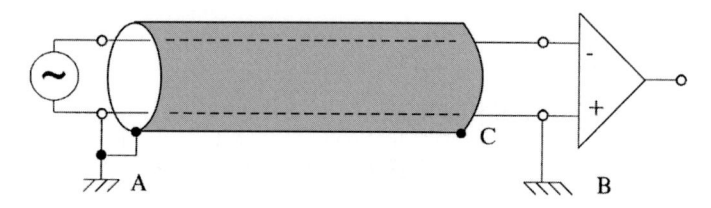

650   18 Protection Against Interfering Signals

Fig. 18.14. Single-point ground connection of the shield (LF)

■ **Low-Frequency Signals (or Cable Lengths Less Than $\lambda/20$).** In the context of the present case study, low-frequency signals are signals whose frequency spectrum extends up to 1 MHz.

Ground point A (Fig. 18.14) denotes the ground of the signal source (which may be a transducer), while ground point B is the ground of the amplifier. For low-frequency signals (or electrically short cables), the cable shield must be tied to ground at a single point (at the driving end, to reduce noise emission by the source, or at the load end, to increase immunity of the amplifier).

■ **Logic Signals (or High-Level Analog Signals, f > 1 MHz).** According to Sect. 9.3.1, HF operation requires multipoint grounding of the shield to ensure a constant potential along it. Since the amplitudes of analog signals (or large logic swings) are several volts, the offset voltage between various grounding points poses no problems. Figure 18.15 shows a cable with the shield grounded at both driving and receiving ends. Assuming that points C and B are not tied together, when the frequency increases the stray capacitance will gradually connect point C to ground; hence, it is better to firmly connect them.

According to [261], the skin effect helps to separate signal and noise currents through the shield: the noise current (due to 50 Hz and its harmonics) flows in the shield bulk, while the signal current (f > 1 MHz) will travel on the shield surface. In this way, the coupling between these two currents has a natural tendency to decrease.

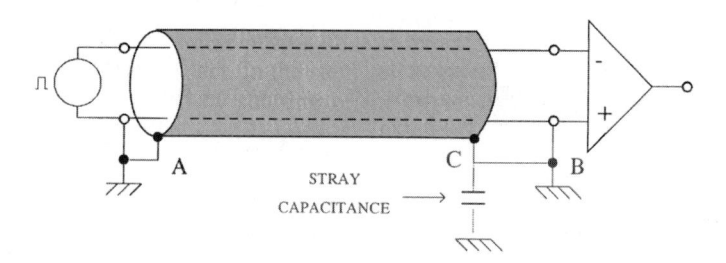

Fig. 18.15. Logic signal transmission requires two grounding points

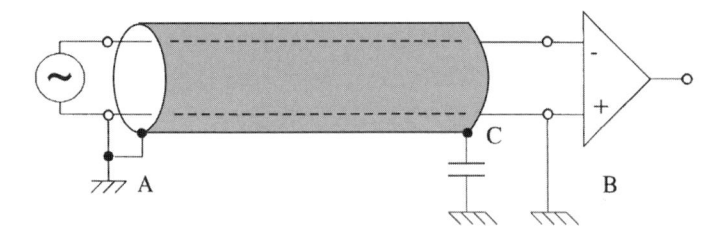

Fig. 18.16. Hybrid ground connection

■ **High-Level Broadband Analog Signals.** The previous discussion suggests that for broadband analog signals, point C can be connected through a capacitor to ground (Fig. 18.16).

In this way, for LF signals the circuit turns into a single-point ground configuration, while for HF signals the grounding arrangement of Fig. 18.15 is obtained. The selection of grounding capacitor is very important, one of the main criteria being its associated *parasitic inductance*, which must be as low as possible (hence, avoid film capacitors, using mica or ceramic capacitors instead). Also, minimize the length of capacitor leads when mounting them, since any wire has a parasitic inductance.

This grounding configuration is useful for circuits operating over a wide frequency range.

■ **Low-Level, Analog Signals and Long Cables.** When we are dealing with weak analog signals (whose amplitude is less than that of interfering signals), the offset voltage between local grounds becomes a problem. A possible solution is indicated in Fig. 18.17 [263].

The shield of the coaxial cable is connected to the case and to the earth ground at the driving end, but is free at the load end (amplifier A2). However, to limit the swing of the signal ground and protect the input of the differential amplifier, a small resistor and a bypass capacitor are introduced at the load end.

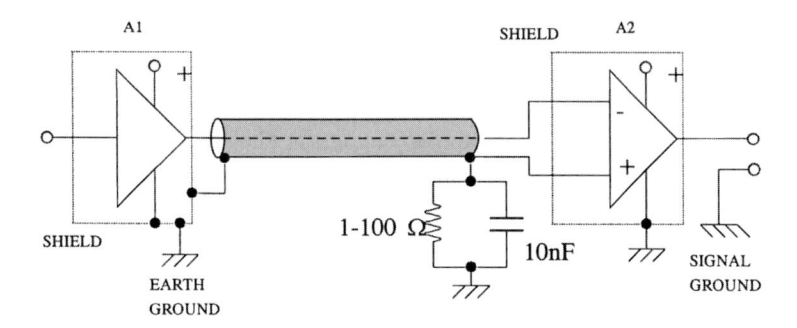

Fig. 18.17. Using a differential amplifier to buffer the signal

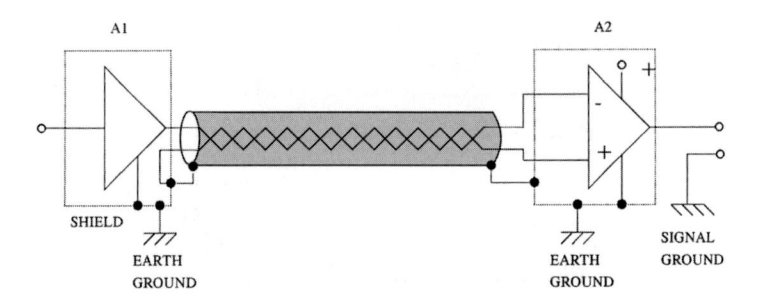

Fig. 18.18. Using a shielded twisted pair for transmission

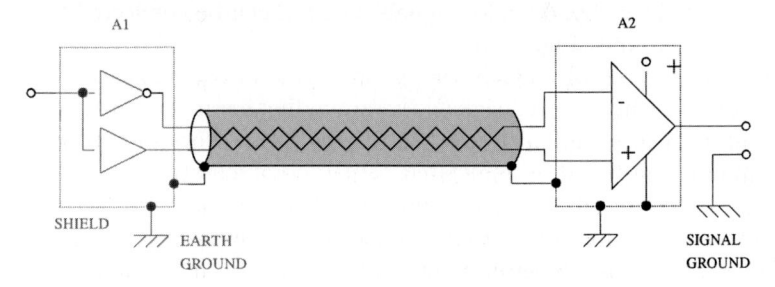

Fig. 18.19. Recommended configuration for logic signals

A better solution is proposed in Fig. 18.18, where a shielded twisted pair is used to transmit the signals between distant sites. Since the cable screen is no longer used as a path for the return current, it can be connected to the earth ground at both ends, via the metallic enclosures of amplifiers A1 and A2.

Note that for both proposed circuits, to avoid common-mode interference the signal ground must be isolated from the cable shield.

To avoid complications with the DC supply of the differential amplifier (at the load end), an easy solution is to power it with batteries (placed inside the shielded enclosure) instead of the AC mains. Otherwise, use a mains transformer with a shield separating the primary and secondary coils.

Double shielding of the differential amplifier A2 can also be considered, with one being tied to the signal ground.

When the configuration in Fig. 18.18 is employed to transmit logic signals, the source end can be balanced as in Fig. 18.19 by means of two complementary logic outputs (or two differential outputs). The source balancing exerts a beneficial effect by increasing system immunity to noise.

CASE STUDY 18.12 [261]

The goal of this case study is to show that when transmitting low-level signals, grounding the cable screen at both ends (as in Fig. 18.15) is counterproductive.

18.1 Techniques to Reduce Interference

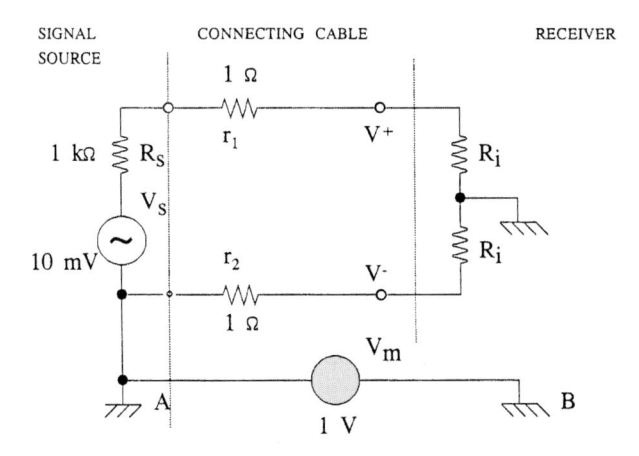

Fig. 18.20. Configuration with two grounding points

Consider the circuit in Fig. 18.20, where different ground symbols have been used for the local ground at the source end (denoted A) and the local ground at the receiving end (denoted B). Assume that interference is due only to the offset voltage between grounds A and B, denoted V_m.

What value must the input resistance R_i have in order to limit the contribution of V_m to less than 0.1% of the signal voltage (V_s) at the receiving end?

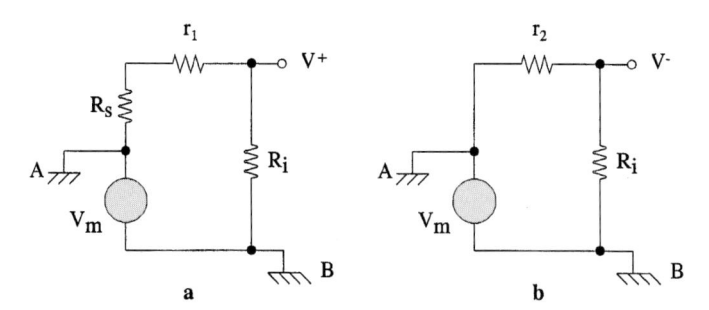

Fig. 18.21. Equivalent circuits

Solution

To estimate the voltage drops generated by V_m between each of the input terminals V^+ and V^- and the local ground B, consider the equivalent circuits in Fig. 18.21.

Recall that *the input voltage of any operational amplifier is always defined between the input terminal and the ground of the signal source (ground A)*. Hence, for the circuit of Fig. 18.21a,

$$V^+ = V_m \frac{R_s + r_1}{R_i + R_s + r_1}$$

while in the circuit of Fig. 18.21b, one can write

$$V^- = V_m \frac{r_2}{R_i + r_2}$$

The differential noise voltage is

$$V_n = V^+ - V^- = V_m \left(\frac{R_s + r_1}{R_i + R_s + r_1} - \frac{r_2}{R_i + r_2} \right) \leq 10^{-3} \, V_s$$

The value of R_i can be found from the latter inequality, noting that with the proposed numerical values $r_1 \ll R_s$ and equally $r_2 \ll R_i$. Thus

$$R_i \geq 10^3 \, \frac{V_m}{V_s} \, (R_s - r_2) \quad \text{from which} \quad \boxed{R_i \geq 100 \, \text{M}\Omega}$$

Such a high value for the input resistance is clearly not practical. Furthermore, if stray capacitances are taken into account, their effect is to take the magnitude of R_i well below this limit, even at relatively low frequencies.

For this reason, *grounding at both ends must be avoided when the level of transmitted signals is at most of the same order of magnitude as the offset voltage between the local grounds.*

However, when transmitting high-level signals, grounding at both ends can be accepted.

18.2 Interconnect Modeling

18.2.1 Evaluation of Stray Elements Associated with Interconnects

CASE STUDY 18.13 [264]

Calculate the mutual inductance of two parallel, unequal straight filaments (or traces on a PCB), positioned as in Fig. 18.22a. Assume $p = 40$ mm, $n = 8$ mm, $m = 20$ mm, and $d = 2$ mm.

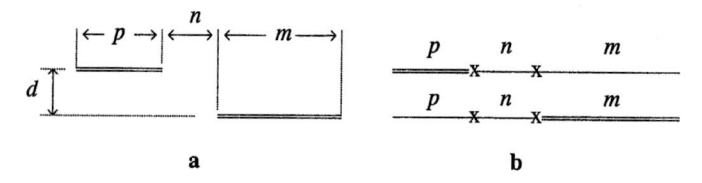

Fig. 18.22. a Proposed configuration; b special case of equal parallel filaments

18.2 Interconnect Modeling

Solution

■ **Method.** The starting point is the case of equal, parallel filaments of length l, spaced a distance d apart, for which the mutual inductance is given by (11.31):

$$M = 0.002\, l \left(\ln\left(\frac{l}{d} + \sqrt{1 + \frac{l^2}{d^2}} - \sqrt{1 + \frac{d^2}{l^2}} + \frac{d}{l} \right) \right) \quad [\mu\text{H}]$$

In order to recover the equal, parallel conductor configuration, the filaments of Fig. 18.22a must be extended by $(n+m)$ and $(p+n)$, respectively (Fig. 18.22b).

For these three-section parallel wires, the total mutual inductance is

$$M_{p+n+m} = M_{pp} + M_{pn} + M_{pm} + M_{np} + M_{nn} + M_{nm} + M_{mp} + M_{mn} + M_{mm}$$

where M_{pp} denotes the mutual inductance between segment p of one filament and segment p of the other, M_{pn} is the mutual inductance between segment p of one filament and segment n of the other, etc.

Note that we always have $M_{ij} = M_{ji}$.

Since the mutual inductance $M_{pm} = M_{mp}$ is to be calculated, consider the identity

$$\boxed{M_{pm} \equiv \frac{1}{2}\left(M_{p+n+m} + M_{nn} - M_{p+n} - M_{n+m} \right)} \qquad (18.22)$$

which can be proved starting with the equations

$$M_{p+n} = M_{pp} + M_{pn} + M_{np} + M_{nn}$$

$$M_{n+m} = M_{nm} + M_{nn} + M_{mn} + M_{mm}$$

■ **Calculation.** All lengths are expressed in centimeters. The following steps must be carried out.

1. Calculate the mutual inductance M_{p+n+m} with (11.31).
   Let $l = p+n+m = 4+0.8+2 = 6.8$ cm. It follows that $l/d = 6.8/0.2 = 34$, so

$$M_{p+n+m} = 0.002(6.8)\left(\ln\left(34 + \sqrt{1 + (34)^2} \right) - \sqrt{1 + \frac{1}{34^2}} + \frac{1}{34} \right)$$

$$\cong 44.18 \text{ nH}$$

2. Calculate the mutual inductance M_{p+n}.
   This time, $l = p+n = 4.8$ cm, $l/d = 4.8/0.2 = 24$. Hence

$$M_{p+n} = 0.002(4.8)\left(\ln\left(24 + \sqrt{1 + (24)^2} \right) - \sqrt{1 + \frac{1}{24^2}} + \frac{1}{24} \right)$$

$$\cong 27.94 \text{ nH}$$

656   18 Protection Against Interfering Signals

3. Calculate the mutual inductance M_{n+m}.
   Here $l = n + m = 2.8 \text{ cm}$, $l/d = 2.8/0.2 = 14$, so

$$M_{n+m} = 0.002(2.8)\left(\ln\left(14 + \sqrt{1 + (14)^2}\right) - \sqrt{1 + \frac{1}{14^2}} + \frac{1}{14}\right)$$
$$\cong 13.46 \text{ nH}$$

4. Calculate the mutual inductance M_{nn}.
   In this case, $l = n = 0.8 \text{ cm}$, $l/d = 4$, so

$$M_{nn} = 0.002(0.8)\left(\ln\left(6 + \sqrt{1 + (6)^2}\right) - \sqrt{1 + \frac{1}{36}} + \frac{1}{6}\right)$$
$$\cong 2.10 \text{ nH}$$

5. Calculate the mutual inductance M_{pm} by applying relation (18.22):

$$M_{pm} = \frac{1}{2}\left(44.18 + 2.63 - 27.94 - 13.46\right) \cong 2.44 \text{ nH}$$

CASE STUDY 18.14 [264]

Calculate the mutual inductance of two unequal parallel traces, partially face to face, as in Fig. 18.23a. Given that $p = 40 \text{ mm}$, $n = 8 \text{ mm}$, $m = 20 \text{ mm}$, and $d = 2 \text{ mm}$.

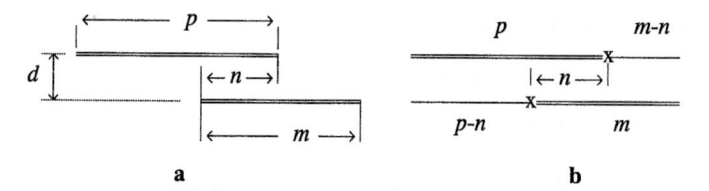

Fig. 18.23. **a** Proposed configuration; **b** case of equal, parallel, face-to-face traces

Solution

■ **Comment.** In this application the configuration involves *traces* on PCB; however, (11.31) refers to *filaments*. As a general rule, *any pair of traces that are long compared to the distance between their axes can be treated as filaments*.

■ **Method.** The procedure is similar to that used in Case Study 18.13: the traces are extended by appropriate lengths to generate two parallel, equal, and face-to-face traces, for which (11.31) holds. This is shown in Fig. 18.23b, where the total length is
$$l = p + m - n$$
resulting from three sections of lengths $(p - n)$, n, $(m - n)$.

18.2 Interconnect Modeling   657

In terms of these three sections, the mutual inductance between the two filaments can be written

$$M_{p+m-n} = M_{(p-n)(p-n)} + M_{(p-n)n} + M_{(p-n)(m-n)}$$
$$+ M_{n(p-n)} + M_{nn} + M_{n(m-n)} \quad (18.23)$$
$$+ M_{(m-n)(p-n)} + M_{(m-n)n} + M_{(m-n)(m-n)}$$

From Fig. 18.23a, we find

$$M_{pm} = M_{(p-n)n} + M_{(p-n)(m-n)} + M_{nn} + M_{n(m-n)} \quad (18.24)$$

Combining (18.23) with (18.24), we have

$$\boxed{M_{pm} = \frac{1}{2}\left(M_{p+m-n} + M_{nn} - M_{(p-n)(p-n)} - M_{(m-n)(m-n)}\right)} \quad (18.25)$$

■ **Calculation.** All lengths are expressed in centimeters. The following steps must be carried out.

1. Calculate the mutual inductance M_{p+m-n}.
   With $l = p + m - n = 4 + 2 - 0.8 = 5.2 \text{ cm}$ and $l/d = 5.2/0.2 = 26$, (11.31) yields

   $$M_{p+m-n} = 0.002(5.2)\left(\ln\left(26 + \sqrt{1 + (26)^2}\right) - \sqrt{1 + \frac{1}{26^2}} + \frac{1}{26}\right)$$
   $$\cong 31.09 \text{ nH}$$

2. Calculate the mutual inductance M_{p-n}.
   In this case $l = p - n = 4 - 0.8 = 3.2 \text{ cm}$ and $l/d = 16$, so

   $$M_{p-n} = 0.002(3.2)\left(\ln\left(16 + \sqrt{1 + (16)^2}\right) - \sqrt{1 + \frac{1}{16^2}} + \frac{1}{16}\right)$$
   $$\cong 16.17 \text{ nH}$$

3. Calculate the mutual inductance M_{m-n}.
   Now, $l = m - n = 2 - 0.8 = 1.2 \text{ cm}$ and $l/d = 6$, so that

   $$M_{m-n} = 0.002(1.2)\left(\ln\left(6 + \sqrt{1 + (6)^2}\right) - \sqrt{1 + \frac{1}{6^2}} + \frac{1}{6}\right)$$
   $$\cong 3.95 \text{ nH}$$

4. Calculate the mutual inductance M_{nn}.
   Here $l = n = 0.8 \text{ cm}$, $l/d = 4$, hence

   $$M_{nn} = 0.002(0.8)\left(\ln\left(4 + \sqrt{1 + (4)^2}\right) - \sqrt{1 + \frac{1}{16}} + \frac{1}{4}\right)$$
   $$\cong 2.10 \text{ nH}$$

5. Calculate the mutual inductance M_{pm}, by substituting the previous values into (18.25):

$$M_{pm} = \frac{1}{2}\left(31.09 + 2.1 - 16.17 - 3.95\right) \cong 6.53 \text{ nH}$$

CASE STUDY 18.15 [264]

Calculate the mutual inductance of two unequal parallel traces, with one end of each adjacent to the other (Fig. 18.24a). Assume $p = 40$ mm, $m = 20$ mm, and $d = 2$ mm.

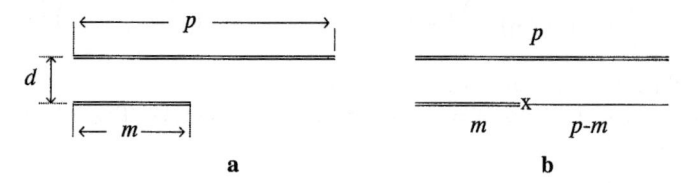

Fig. 18.24. a Proposed configuration; b equal, parallel traces

Solution

■ **Remark.** See comment in Case Study 18.14.

■ **Method.** The procedure is similar to the previous examples: the shorter trace is extended with an appropriate length to generate a configuration of two parallel, equal, and face-to-face traces, for which (11.31) holds. This is shown in Fig. 18.24b, where the length has been extended by $(p - m)$.

We can then write

$$M_{pp} = M_{mm} + M_{m(p-m)} + M_{(p-m)m} + M_{(p-m)(p-m)} \qquad (18.26)$$

Note that
$$M_{mp} = M_{mm} + M_{m(p-m)} \qquad (18.27)$$

Adding the same quantity (M_{mm}) to each side of (18.26) and using (18.27), we find after some algebra that

$$\boxed{M_{mp} = \frac{1}{2}\left(M_{pp} + M_{mm} - M_{(p-m)(p-m)}\right)} \qquad (18.28)$$

■ **Calculation.** All lengths are expressed in centimeters. The following steps must be carried out.

1. Calculate the mutual inductance M_{pp}.
   In this case $l = p = 4$ cm. It follows that $l/d = 4/0.2 = 20$, so

$$M_{pp} = 0.002(4) \left(\ln\left(20 + \sqrt{1 + (20)^2}\right) - \sqrt{1 + \frac{1}{20^2}} + \frac{1}{20} \right)$$
$$\cong 21.9 \text{ nH}$$

2. Calculate the mutual inductance M_{mm}.
   Let be $l = m = 2$ cm. Since $l/d = 2/0.2 = 10$, we obtain

$$M_{mm} = 0.002(2) \left(\ln\left(10 + \sqrt{1 + (10)^2}\right) - \sqrt{1 + \frac{1}{10^2}} + \frac{1}{10} \right)$$
$$\cong 8.37 \text{ nH}$$

3. Calculate the mutual inductance $M_{(p-m)(p-m)}$.
   Now $l = p - m = 4 - 2 = 2$ cm. Thus $l/d = 2/0.2 = 10$, and

$$M_{(p-m)(p-m)} = 0.002(2) \left(\ln\left(10 + \sqrt{1 + (10)^2}\right) - \sqrt{1 + \frac{1}{10^2}} + \frac{1}{10} \right)$$
$$\cong 8.37 \text{ nH}$$

4. Finally, calculate the mutual inductance M_{pm}.
   Applying (18.25), we obtain

$$M_{pm} = \frac{1}{2}\left(21.9 + 8.37 - 8.37\right) \cong 10.95 \text{ nH}$$

CASE STUDY 18.16 [264]

Calculate the mutual inductance of two unequal parallel traces, with their opposite ends lined up along the common perpendicular (Fig. 18.25a). Assume $p = 40$ mm, $m = 20$ mm, and $d = 2$ mm.

Solution

■ **Remark.** See comment in Case Study 18.14.

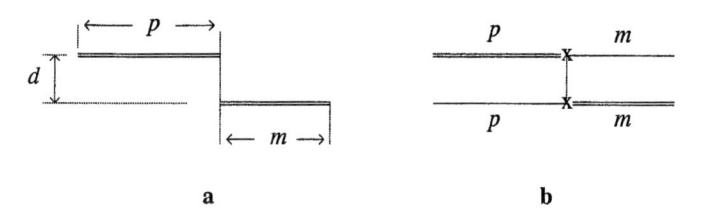

Fig. 18.25. a Proposed configuration; b extension to obtain equal, parallel traces

660   18 Protection Against Interfering Signals

■ **Method.** The procedure is similar to the previous examples: the traces are extended with appropriate lengths to generate the classical configuration of two parallel, equal, and face-to-face traces, for which (11.31) holds. This is shown in Fig. 18.25b, where the lengths have been extended by m and p, respectively.

We can then write

$$M_{(p+m)(p+m)} = M_{pm} + M_{pp} + M_{mp} + M_{mm} \qquad (18.29)$$

Since $M_{pm} = M_{mp}$, we have

$$\boxed{M_{pm} = \frac{1}{2}\left(M_{(p+m)(p+m)} - M_{pp} - M_{mm}\right)} \qquad (18.30)$$

■ **Calculation.** All lengths are expressed in centimeters. The following steps must be carried out.

1. Calculate the mutual inductance M_{pp}.
   In this case, $l = p = 4\,\text{cm}$ and $l/d = 4/0.2 = 20$. Hence

   $$M_{pp} = 0.002(4)\left(\ln\left(20 + \sqrt{1+(20)^2}\right) - \sqrt{1+\frac{1}{20^2}} + \frac{1}{20}\right)$$
   $$\cong 21.9\,\text{nH}$$

2. Calculate the mutual inductance M_{mm}.
   Here $l = m = 2\,\text{cm}$ and $l/d = 2/0.2 = 10$. Then

   $$M_{mm} = 0.002(2)\left(\ln\left(10 + \sqrt{1+(10)^2}\right) - \sqrt{1+\frac{1}{10^2}} + \frac{1}{10}\right)$$
   $$\cong 8.37\,\text{nH}$$

3. Calculate the mutual inductance $M_{(p+m)(p+m)}$.
   Now $l = p + m = 4 + 2 = 6\,\text{cm}$ and $l/d = 6/0.2 = 30$, so

   $$M_{(p+m)(p+m)} = 0.002(6)\left(\ln\left(30 + \sqrt{1+(30)^2}\right) - \sqrt{1+\frac{1}{30^2}} + \frac{1}{30}\right)$$
   $$\cong 37.53\,\text{nH}$$

4. Calculate the required mutual inductance $M_{(p+m)(p+m)}$.
   The application of (18.30) yields

   $$M_{pm} = \frac{1}{2}\left(37.53 - 21.9 - 8.37\right) \cong 3.63\,\text{nH}$$

CASE STUDY 18.17 [264]

Calculate the mutual inductance of two equal traces meeting at a point (Fig. 18.26). This is the most important case for practical purposes. Assume $p = 40\,\text{mm}$, $\alpha = 30°$.

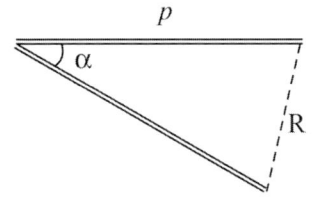

Fig. 18.26. Two equal traces meeting at a point

Solution

■ **Background.** According to [264], the mutual inductance of equal filaments inclined at an angle α to each other is

$$M = 0.004\,p\,(\cos\alpha)\,\frac{1}{2}\,\ln\frac{2p+R}{R}\quad[\mu H] \tag{18.31}$$

where all lengths are expressed in centimeters.

In practice, to describe the proposed configuration, either the distance R between the filaments ends or the angle α is given.

From trigonometry,

$$R^2 = 2p^2(1 - \cos\alpha) \tag{18.32}$$

so that

$$\cos\alpha = 1 - \frac{R^2}{2p^2} \tag{18.33}$$

Equation (18.31) shows that the mutual inductance M reaches a maximum when the filaments point in the same direction ($\alpha = 0$). If α increases, M decreases; the minimum (M = 0) is reached when $\alpha = 90°$. *The mutual inductance between two perpendicular filaments is zero* (which explains why traces in adjacent layers should be routed perpendicular to one another).

Note that when α is obtuse, M becomes negative. To understand the meaning of this negative value, imagine that very close to the limit $\alpha = 0$ the two filaments are almost parallel, and they act like a two-wire circuit whose total inductance is

$$L = L_1 + L_2 - 2M_{12}$$

Note that in this case, currents in the filaments flow *in opposite directions*. By extension, for an obtuse angle asymptotically approaching 180°,

$$L = L_1 + L_2 + 2M_{12}$$

Note that currents in the filaments then flow *in the same direction*.

662   18 Protection Against Interfering Signals

To conclude, the minus sign is not associated with some exotic "negative mutual inductance," but merely indicates that the mutual inductance must be added, rather than subtracted when calculating the total inductance of the system of two filaments.

■ **Calculation.** First of all, the distance R between the ends must be estimated:

$$R = \sqrt{2p^2(1 - \cos \alpha)} = \sqrt{2(4^2)(1 - \cos 30°)} \cong 2.07 \text{ cm}$$

Next, the mutual inductance can be calculated using (18.31)

$$M = 0.004 \, (4) \, (\cos 30°) \, \frac{1}{2} \, \ln \frac{8 + 2.07}{2.07} \cong 10.96 \text{ nH}$$

CASE STUDY 18.18 [264]

Calculate the mutual inductance of two unequal traces meeting at a point (Fig. 18.27). Assume $p = 40$ mm, $m = 2$ cm and $R = 7$ cm between their ends.

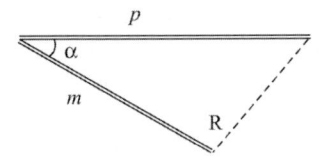

Fig. 18.27. Unequal traces meeting at a point

Solution

■ **Background.** The angle between the traces can be calculated via the generalized Pythagorean theorem:

$$\cos \alpha = \frac{p^2 + m^2 - R^2}{2pm} = \frac{8^2 + 2^2 - 7^2}{2(8)(2)} \cong 0.593$$

According to [264], the mutual inductance can be calculated with the expression

$$M = 0.002(\cos \alpha) \left(\ln \frac{1 + \dfrac{m}{p} + \dfrac{R}{p}}{1 - \dfrac{m}{p} + \dfrac{R}{p}} + \frac{m}{p} \ln \frac{\dfrac{m}{p} + \dfrac{R}{p} + 1}{\dfrac{m}{p} + \dfrac{R}{p} - 1} \right) \quad (18.34)$$

where m denotes the length of the shorter trace ($m/p < 1$).

18.2 Interconnect Modeling

■ **Calculation.** Substituting numerical values into (18.34), we have

$$M = 0.002(0.593)\left(\ln\frac{1+\frac{2}{8}+\frac{7}{8}}{1-\frac{2}{8}+\frac{7}{8}} + \frac{2}{8}\ln\frac{\frac{2}{8}+\frac{7}{8}+1}{\frac{2}{8}+\frac{7}{8}-1}\right) \cong 1.158 \text{ nH}$$

18.2.2 Crosstalk

CASE STUDY 18.19 [261]

Consider a section of two parallel conductors over ground in close proximity (Fig. 18.28a), whose lengths are considerably less than $\lambda/4$. The stray capacitance between conductors is 20 pF, and each conductor has a stray capacitance to ground of 60 pF. Denote by R_2 the parallel combination of R_{2a} and R_{2b}.

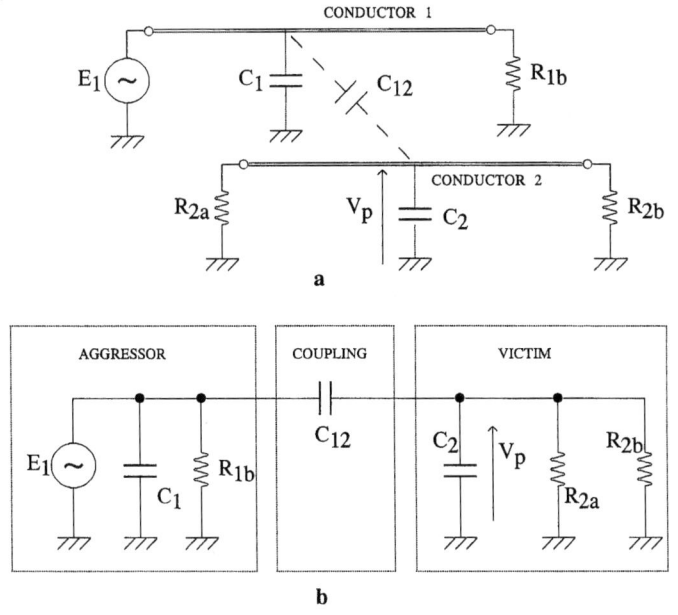

Fig. 18.28. a Two parallel conductors over ground; b equivalent circuit

Suppose that conductor 1 carries an AC signal of 10 V and frequency f = 100 kHz. Neglecting mutual inductance between conductors 1 and 2, determine the crosstalk voltage picked up by conductor 2 via capacitive coupling in the following situations:

1. R_2 has very high values ($R_2 \to \infty$)
2. $R_2 = 600\,\Omega$
3. $R_2 = 50\,\Omega$
4. Discuss the influence of the internal resistance of E_1.
5. What happens if the frequency increases to 500 MHz?

18 Protection Against Interfering Signals

Solution

■ **Comment.** It should be noted that as long as the voltage source E_1 is considered ideal (zero internal resistance), the parallel combination of C_1 and R_{1b} does not modify the fraction of E_1 reaching the victim through C_{12}.

■ **Calculation**

☐ **Case 1 ($R_2 \to \infty$).** When $R_2 \to \infty$ (Fig. 18.28b), the picked-up signal is

$$V_p = E_1 \frac{1/j\omega C_2}{1/j\omega C_2 + 1/j\omega C_{12}} = E_1 \frac{C_{12}}{C_{12} + C_2} \quad (18.35)$$

Substituting numerical values, we obtain

$$V_p = 10 \frac{20}{20 + 60} = 2.5 \text{ V}$$

☐ **Remarks**

– The resulting crosstalk is surprisingly strong.
– With an infinite load, expression (18.35) is not frequency dependent.
– The amount of picked-up voltage through capacitive coupling does not depend on the actual values of C_2 and C_{12}, but only on their ratio. Augmenting C_2 (for instance, by using a ground plane) reduces the crosstalk.

☐ **Case 2 ($R_2 = 600\,\Omega$).** Let Z_2 be the equivalent impedance of R_2 and C_2 in parallel:

$$Z_2 = \frac{(R_2)(1/j\omega C_2)}{(R_2) + (1/j\omega C_2)} = \frac{R_2}{1 + j\omega C_2 R_2}$$

Applying the voltage divider formula and doing the algebra, we have

$$V_p = E_1 \frac{j\omega C_{12} R_2}{(1 + j\omega R_2(C_2 + C_{12}))} \quad (18.36)$$

☐ **Discussion**

- Whenever $\omega R_2(C_2 + C_{12}) \gg 1$ (or, with the given numerical values, $f \gg 66.3\,\text{MHz}$), expression (18.36) reduces to (18.35). This means that regarded from a capacitive crosstalk point of view, *the high-frequency behavior is the same as for an infinite load* ($R_2 \to \infty$).
- If $\omega R_2(C_2 + C_{12}) \ll 1$ (i.e., $f \ll 66.3\,\text{MHz}$), expression (18.36) becomes

$$V_p \cong E_1 (j\omega C_{12} R_2) \quad (18.37)$$

Substituting numerical values,

$$|V_p| \cong 10 \, (2\pi \cdot 10^5) \, (20 \cdot 10^{-12}) \, (600) \cong 75.4 \text{ mV}$$

18.2 Interconnect Modeling

□ **Case 3 ($R_2 = 50\,\Omega$).** Relation (18.37) yields

$$|V_p| \cong 10\,(2\pi \cdot 10^5)\,(20 \cdot 10^{-12})\,(50) \cong 6.28\text{ mV}$$

We conclude that *reducing the line impedance has a beneficial effect on capacitive crosstalk noise.*

□ **Case 4.** Suppose that the source E_1 has internal resistance $R_g = 50\,\Omega$ instead of being ideal; also, assume that $R_{1b} = 50\,\Omega$. In this situation, the voltage reaching C_{12} is half the voltage of Case 3, due to the voltage divider in the aggressor block. Hence, expression (18.37) can be used by replacing E with $E_1/2$, yielding

$$V_p \cong (E_1/2)\,(j\omega C_{12} R_2) = 3.14\text{ mV}$$

Note that the apparent reduction in V_p is due solely to the lower value of the aggressor signal.

□ **Case 5 ($f = 500$ MHz).** Increasing the frequency leads to the same result as in point 1.

■ **Concluding Remark.** Capacitive crosstalk is reduced only when $\omega C_2 R_2 \ll 1$. A sufficient condition is to *provide a low-impedance victim line.*

CASE STUDY 18.20 [261]

A practical method of reducing capacitive crosstalk between conductors is to shield the victim (conductor 2 in Fig. 18.29a). However, residual capacitive coupling still exists at both ends, where the shield is interrupted to allow the inner conductor to connect to the remainder of the circuit. Table 18.2 presents the values of all stray capacitances associated with this system.

Assume that conductor 1 carries an AC signal of 10 V and frequency $f = 100$ kHz. Determine the shielding effectiveness when R_2 (which represents the parallel combination of R_{2a} and R_{2b}) takes the following values:

1) $R_2 \to \infty$
2) $R_2 = 600\,\Omega$
3) $R_2 = 50\,\Omega$

Solution

■ **Comment.** For simplicity, capacitance C_{20} has not been added to the physical representation (Fig. 18.29a), but it appears in the equivalent circuit.

■ **Calculation.** The equivalent circuit (Fig. 18.29b) has the same topology as that of Fig. 18.28b; however, the numerical values are different. Crosstalk is

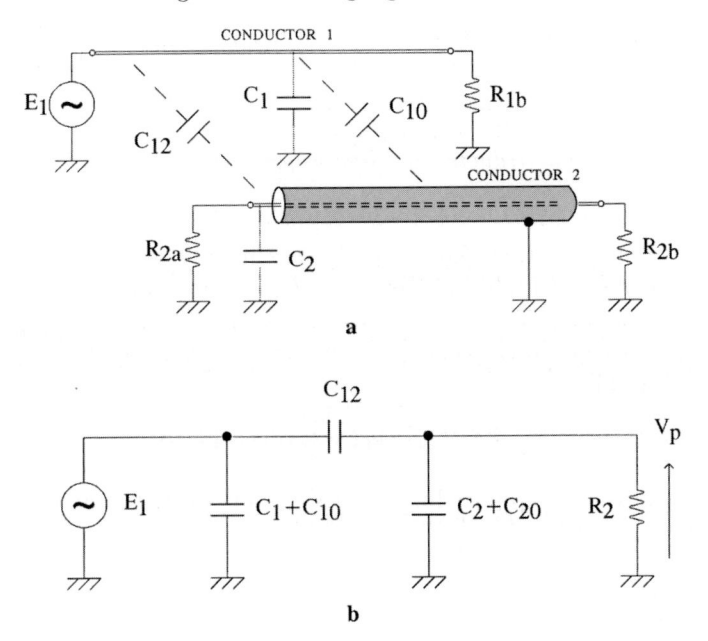

Fig. 18.29. a Physical representation; b equivalent circuit

Table 18.2. Stray capacitances of the system of Fig. 18.29a

Stray capacitance between ...	Notation	Value
Conductor 2 and its grounded shield	C_{20}	100 pF
Conductor 1 and the unshielded end of conductor 2	C_{12}	2 pF
End of conductor 2 and ground	C_2	5 pF
Conductor 1 and shield of conductor 2	C_{10}	50 pF
Conductor 1 and ground	C_1	60 pF

reduced particularly due to the increase in C_{20}, which reduces the equivalent impedance of R_2, C_2 in parallel. As a consequence, V_p decreases.

The expressions deduced in Case Study 18.19 remain valid, provided that we replace C_2 with $(C_2 + C_{20})$. The shielding effectiveness is finally calculated according to its definition given in Chap. 9, expression (9.7).

☐ **Case 1 ($R_2 \to \infty$)**

$$V_p = E_1 \frac{C_{12}}{C_{12} + (C_2 + C_{20})} = 10 \frac{2}{2 + (100 + 5)} \cong 187\,\text{mV}$$

18.2 Interconnect Modeling

The shielding effectiveness is calculated with respect to the previous value of 2.5 V, obtained for a non-screened conductor (Case Study 18.19), i.e.,

$$\mathrm{SE} = 20 \log \frac{2.5}{0.187} \cong 22.5 \text{ dB}$$

☐ **Case 2 ($R_2 = 600\ \Omega$).** This time expression (18.37) can be applied:

$$|V_p| \cong E_1\,(\omega C_{12} R_2) \cong 10\,(2\pi \cdot 10^5)(2 \cdot 10^{-12})600 \cong 7.54 \text{ mV}$$

Without shielding conductor 2 (Case Study 18.19), the picked-up voltage has been found to be 75.4 mV; hence, the shielding effectiveness is

$$\mathrm{SE} = 20 \log \frac{75.4}{7.54} = 20 \text{ dB}$$

☐ **Case 3 ($R_2 = 50\ \Omega$).** In this case we obtain

$$|V_p| = 0.628 \text{ mV}$$

and

$$\mathrm{SE} = 20 \log \frac{6.28}{0.628} = 20 \text{ dB}$$

Note that at low frequency, shielding effectiveness does not depend upon the load.

CASE STUDY 18.21 [261, 264, 266]

1) List the advantages and disadvantages of ribbon cables.
2) Estimate the mutual inductance per unit length between the first and second pair of a ribbon cable, then between the first and third pair, and finally between the first and fourth pair. Assume a ribbon cable whose pairs are uniformly spaced.
3) If the signal transmitted by one pair has the amplitude of 5 V and a frequency of 100 MHz, determine the induced voltage in the second, third, and fourth pair, when the cable terminates in a 500 Ω-load.

Solution

1) Comparing the price of a cable to the price of its connector, the latter is clearly much higher than the former. Therefore, an economical solution is to adopt ribbon cables (whenever possible), where many interconnecting lines share the same connector.

   Another advantage of ribbon cables is the stability of the placement and orientation of their lines with respect to each other, almost the same as for traces on PCB. It follows that two different ribbon cables, when

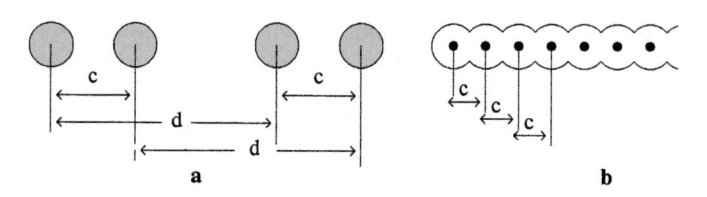

668   18 Protection Against Interfering Signals

Fig. 18.30. a Two pairs in a ribben cable; b uniform ribbon cable

installed under similar conditions, offer almost the same susceptibility to interfering signals.

Their main drawback is performance at high frequency, which is very susceptible to the distribution of the ground returns. This has been detailed in Sect. 9.4. However, if an appropriate grounding configuration is adopted, cable radiation and susceptibility are considerably reduced, and this becomes an advantage.

2) Consider the general case of two pairs of conductors in a ribbon cable (Fig. 18.30a); the separation between pairs is denoted by d, while the separation between two adjacent conductors is denoted by c.

The mutual inductance of two pairs per unit length is

$$M = \frac{\mu}{2\pi} \ln \frac{d^2}{d^2 - c^2} \qquad (18.38)$$

For a uniform ribbon cable (Fig. 18.30b) and two adjacent pairs we have $d = 2c$; adopting $\mu = 4\pi \cdot 10^{-7}$, expression (18.38) yields

$$M_{12} = \frac{4\pi \cdot 10^{-7}}{2\pi} \ln \frac{4c^2}{4c^2 - c^2} = 2 \cdot 10^{-7} \ln \frac{4}{3} \cong 57.5 \text{ nH/m}$$

For the first and third pairs, $d = 4c$:

$$M_{13} = 2 \cdot 10^{-7} \ln \frac{16c^2}{16c^2 - c^2} = 2 \cdot 10^{-7} \ln \frac{16}{15} \cong 12.9 \text{ nH/m}$$

Finally, for the first and fourth pairs $d = 6c$:

$$M_{14} = 2 \cdot 10^{-7} \ln \frac{36c^2}{36c^2 - c^2} = 2 \cdot 10^{-7} \ln \frac{36}{35} \cong 5.6 \text{ nH/m}$$

3) Before calculating the induced voltage, we need the value of the perturbing current:

$$I = \frac{5}{500} = 10 \text{ mA}$$

The induced voltage is $j\omega M_{1k}I$, with $k = 1, 2, 3$. Therefore

$$|V_{12}| = \omega M_{12}I = (2\pi \cdot 10^8)(57.5 \cdot 10^{-9})(10^{-2}) \cong 361.3 \text{ mV/m}$$

Similarly,

$$|V_{13}| = \omega M_{13} I = (2\pi \cdot 10^8)(12.9 \cdot 10^{-9})(10^{-2}) \cong 81 \text{ mV/m}$$

$$|V_{14}| = \omega M_{14} I = (2\pi \cdot 10^8)(5.6 \cdot 10^{-9})(10^{-2}) \cong 35.2 \text{ mV/m}$$

■ **Concluding Remark.** As expected, crosstalk is stronger between adjacent pairs. The only solution is to judiciously select the signals to be carried by adjacent pairs in order to avoid deleterious effects. Otherwise, assign a ground return between the adjacent pairs in question.

18.3 Interfering Signals

18.3.1 Transducers and Associated Circuits

CASE STUDY 18.22 [261]

Consider the monitoring system in Fig. 18.31, where the signal from a remote thermocouple (TC) is transmitted by a two-wire line to amplifiers A1 and A2. A DC motor operating a fan (to exhaust the heat dissipated by the power electronics) is supplied from the same source (denoted V_{CC}). The electric motor can be represented as an inductive load Z_L powered through a periodically closed switch K.

a) Identify the interfering signal sources.
b) List victims and coupling mechanisms.
c) Suggest remedies.

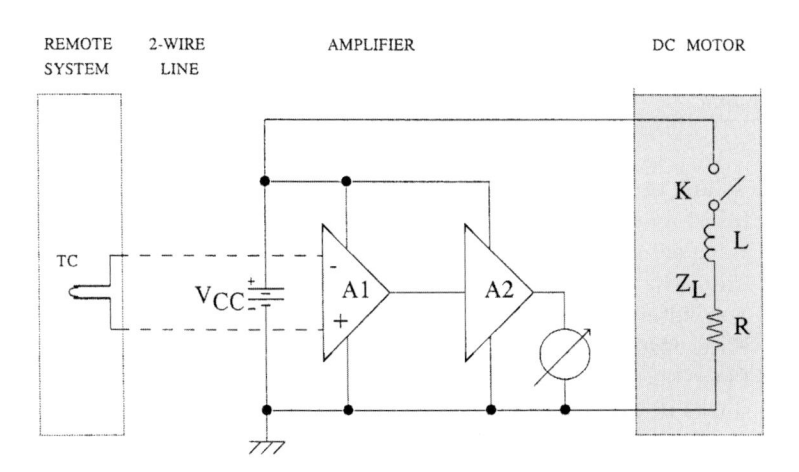

Fig. 18.31. Monitoring the temperature of a remote system

18 Protection Against Interfering Signals

Solution

■ **Comment.** The main objective here is to measure and monitor the signal delivered by the thermocouple, which is proportional to the temperature of the system of interest. Interference is harmful because it can add noise to the weak useful signal from the thermocouple, which is carried by the line and processed by the amplifier. This adversely affects the temperature accuracy of the remote system. Hence, appropriate protections are required to reduce the measurement errors to an acceptable limit.

■ **Answers.** The required answers to a) and b) are presented in a compact way in Table 18.3, where Z_C denotes the common impedance to ground (the impedance of interconnects to ground).

Table 18.3. Interference identification

Noise source	Coupling mechanism	Victims	Protection
Man-made noise	Radiation	2-wire line	6
Arc in switch K	Radiation	TC, A1, A2	1, 6, 7
	Conduction	A1, A2	2
Transient current through motor	Radiation	TC, A1, A2	4, 7
	Coupling through V_{CC}	A1, A2	5
	Coupling through Z_C	A1, A2	7
	Conduction	A1, A2	3, 6 and 8
Steady-state current through motor	Coupling through Z_C	A1, A2	7
Magnetic field of motor	Radiation	TC, A1, A2	4
	Conduction	A1, A2	6 and 8

c) The following remedies are suggested (but others might be considered):
 1) Insert an arc suppressor circuit (in this case, a capacitor of convenient value across the commutator).
 2) Filter the conductor connecting K to the DC supply. To reduce emission, filtering must be applied at the source (i.e., near the motor).
 3) Filter the supply terminals of each amplifier.
 4) Place the motor away from amplifiers and shield it. Connect the shield to ground. Use feedthrough capacitors to pass wires through the shielding.
 5) Decouple the DC power supply.
 6) Shield the sensor (thermocouple). Use a shielded twisted pair to transmit the useful signal, and ground it to improve immunity.

7) Provide a ground plane to reduce Z_C and the size of ground loops.
8) Filter the signal inputs of amplifiers, to suppress high-frequency conducted noise.

CASE STUDY 18.23 [263, 265]

Give a general presentation of interference mechanisms affecting sensors and indicate some ways to limit or eliminate them.

Solution

■ **Explanation.** The most frequently encountered sources of interference affecting electronic equipment are presented in Chapt. 8. *Sensors (or transducers) are devices that convert a nonelectrical parameter (sound, light, pressure, temperature, speed, etc.) into electric signals or vice versa.* Inherently, the efficiency of this conversion cannot be high, and as a result the delivered electrical signals have small amplitudes. Therefore, sensors are potential victims of ambient interfering signals.

Interconnects and packaging make a significant contribution to their inherent low noise immunity. In practice, all interconnects are small receiving antennas, that pick up the perturbing electromagnetic fields and transform them into conducted noise. Depending on how the interfering signals affect the useful signal, two kinds of interference are mostly encountered [265]:

1) *Additive interference*, when noise is added to the sensor signal as a fully independent voltage (or current). This kind of interference is specific to linear or quasilinear sensors and their associated circuits.
2) *Multiplicative interference*, whenever noise affects the sensor in such a way that the amplitude of the signal is modulated by the noise.

■ **Techniques.** The following techniques are widely employed to improve noise immunity.

☐ **Differential Technique.** This method is suited to additive interference. Sensors are fabricated in pairs and installed so that their output signals are subtracted (Fig. 18.32).

Both sensors are subject to the same environmental conditions, but only the main sensor is stimulated by the non-electrical excitation. Assuming that the two sensors are equally exposed to common-mode interference, signal S_o translates the effect of noise on the reference sensor, while S_m is proportional to the common-mode interference *and* the applied excitation.

If a subtracting circuit is used at the output, $(S_m - S_o)$ will represent the electrical image of the excitation only, *provided that the two sensors are identical.* (Note that *identical* means not only sensors with identical parameters, but also placed in close proximity, so that they are equally exposed to interfering signals).

The noise immunity is described by the common-mode rejection ratio (CMRR) [265]:

$$\text{CMRR} = 0.5\,\frac{S_m + S_o}{S_m - S_o}$$

The CMRR depends on the magnitude of the stimulus (it often decreases with increasing excitation).

☐ **Ratiometric Technique.** As before, sensors are fabricated in pairs and connected as in Fig. 18.32. The only difference is that the *ratio* of their output signals is taken. Hence, *multiplicative interference is eliminated,* provided that both sensors are identical and equally exposed to external noise.

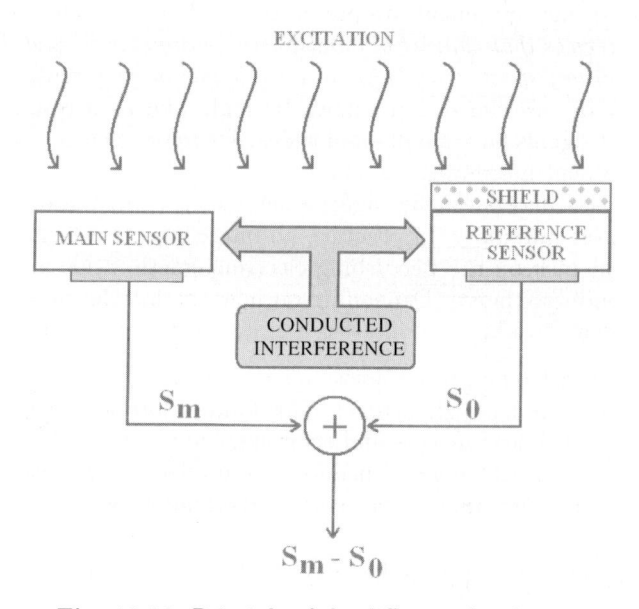

Fig. 18.32. Principle of the differential technique

■ **Electrical Shielding.** Since many sensors and associated circuits contain nonlinear elements (for instance, PN junctions), high-frequency interfering signals can be rectified and appear as DC or very low-frequency voltage, similar to that of the useful signal. In such situations, filtering is ineffective, and the only remedy is to shield both the sensor and its interconnecting line.

Several possibilities are illustrated in Fig. 18.33, where LOAD denotes either a conventional load, or the input of the associated amplifier. Note that the sensor ground is different from the load ground, since they are at different sites.

In Fig. 18.33a the cable shield is connected to the signal reference (at the signal source side). This is a *good solution*.

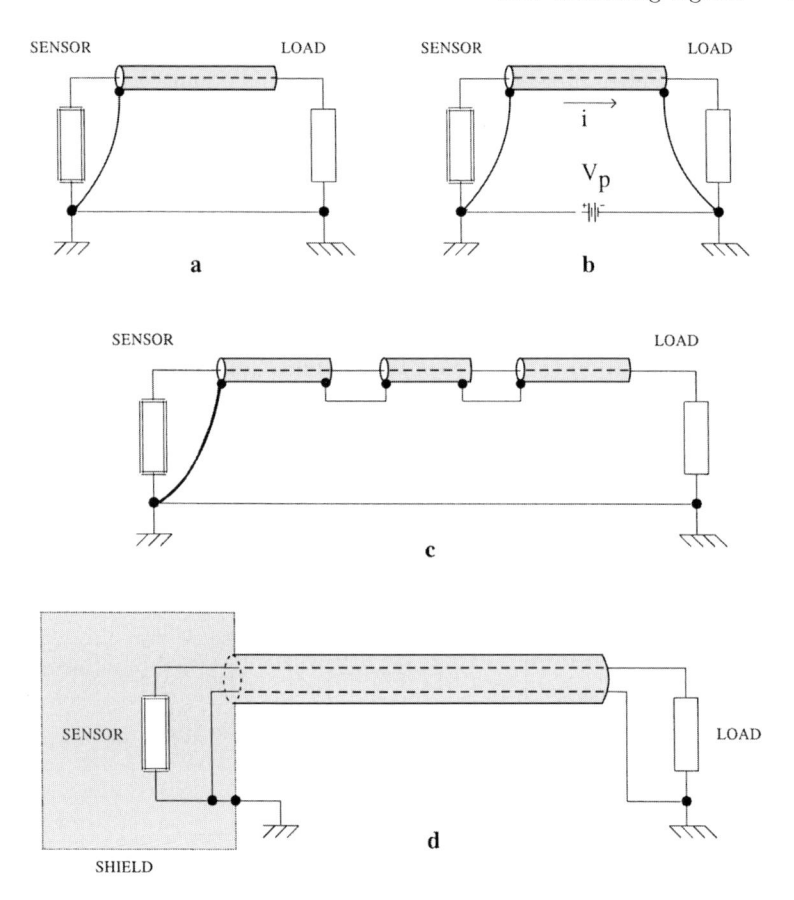

Fig. 18.33. Several shielding possibilities

In Fig. 18.33b the cable shield is grounded at both ends. This is a *bad choice*, since the offset voltage between local grounds (V_p) causes a current (i) to flow through the cable shield. The magnetic field of this current induces a voltage across the inner conductor, degrading the useful signal. Hence, ground loops must be avoided.

In Fig. 18.33c the distance separating the sensor from the load is significant. When connectors are used, several sections of cable are necessary to transmit the useful signal. The shield of each section has to be connected to the shield of the next section; the shield of the first section is tied to the sensor local ground.

The best solution is indicated in Fig. 18.33d, where both signal and return conductor are shielded. In this way, the shield is used only to protect the two-wire line and not as a return path. The sensor is also shielded, and its metallic enclosure should be connected to the cable shield.

■ **Remark.** Whenever an array of sensors is used as a data acquisition system, the number of separate shields required is equal to the number of independent signals to be measured. Each signal must have its own shield, isolated from other shields in the system, but connected to its local ground.

CASE STUDY 18.24 [223, 265]

A temperature sensor uses the Seeback effect to produce its signal voltage. In normal operation, the sensor sensitivity is 50 µV/°C.

The signal is amplified with an operational amplifier connected in a non-inverting configuration (Fig. 18.34). The amplifier draws 5 mA from the DC supply. Determine which of the points A, B, or C is the most appropriate to connect the ground.

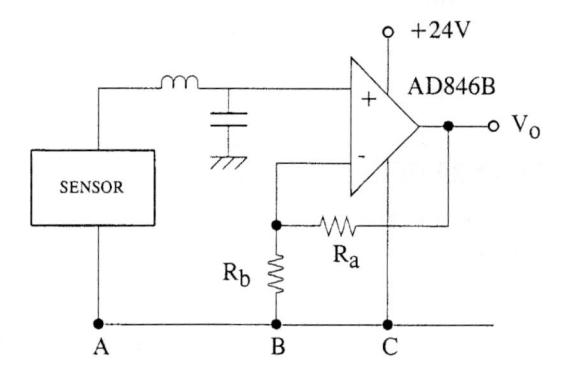

Fig. 18.34. Sensor and its associated amplifier

Solution

■ **Explanation.** In order to evaluate the possible choices, it is useful to redraw the schematic of Fig. 18.34 by inserting the parasitic elements associated only with trace between A and B. Assume that the ground is connected to point A (Fig. 18.35) and that the resistance of the ground trace (r_t) is of the order of 0.08 Ω (for simplicity, l_t is neglected). The flow of the current drawn from the DC supply is indicated by arrows; note that it produces a steady-state voltage drop of about 0.4 mV on the trace connecting points A and B. This voltage is added to that delivered by the sensor and is treated as a useful signal. As a consequence, an error of −5°C is expected, which is unacceptable (actually, the error must be higher, due to the voltage drop added by transients on l_t). Hence, grounding at point A is not a good solution.

Similarly, grounding at point B modifies the feedback voltage applied to the inverting input (due to voltage drop of i on the trace between B and C) and this will also produce large errors.

18.3 Interfering Signals

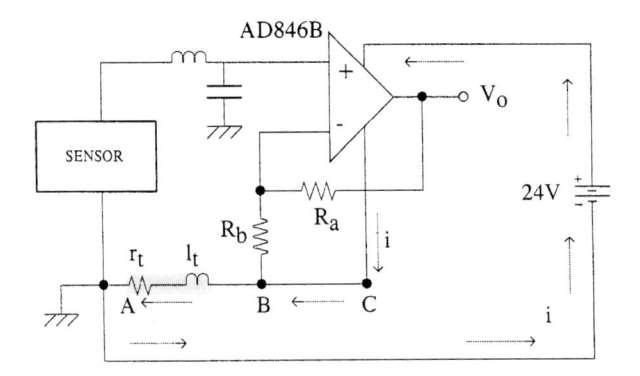

Fig. 18.35. Grounding at point A

Finally, if the ground is connected to point C, the supply current no longer flows through traces A–B and/or B–C and the previous error sources disappear.

To conclude, grounding must be done at point C.

CASE STUDY 18.25 [267]

In order to assess the warping of a wooden beam, a strain gauge measures elongation or flexure by deforming a metal thin-film resistor sticked on the beam. The sensor is connected to the measuring equipment by means of a shielded pair.

The principle of operation consists in sending a constant current through the transducer and sensing the voltage drop across it. Since the sensor and measurement system are not located at the same site, the ground potential of the measurement equipment differs from that of the sensor.

Suggest a way to transmit the useful signal to the measurement system without being affected by the offset voltage between grounds.

Solution

■ **Explanation.** The signal transmission between the sensor and the measuring system, without any accommodations is shown in Fig. 18.36.

Here the offset voltage V_{of} is responsible for a parasitic current (i) through the loop **3 − CS − 2 − 1 − 6 − 5 − 4**, where CS denotes a constant current source and the node numbers are in bold characters.

The constant current source injects a current (I) through the loop **2 − 1 − 6 − 5 − 4 − 3**.

However, the parasitic current i through the sensor adds to the measurement current I, giving rise to unacceptable errors.

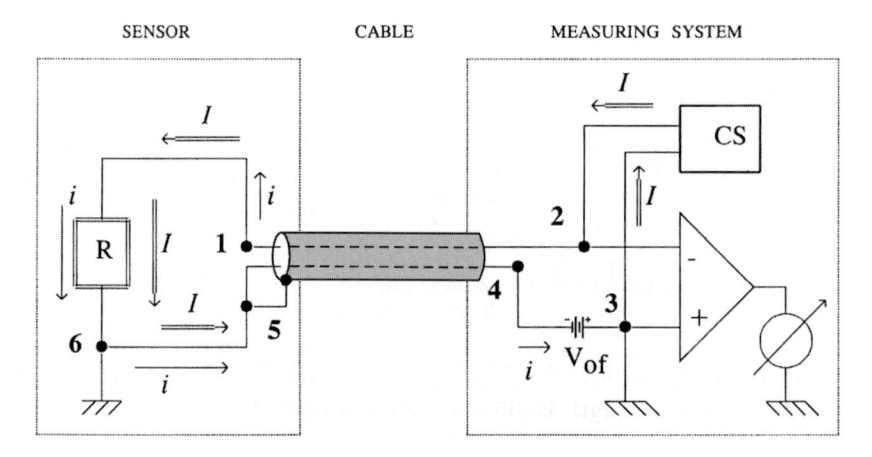

676   18 Protection Against Interfering Signals

Fig. 18.36. Schematics without breaking the ground loop

■ **Proposed Configuration.** To reject the ground potential difference (which is in fact a common-mode voltage), the ground loop must be broken without blocking the flow of current I between the sites.

Since the measurement current I might also be a DC current, breaking the loop by means of a capacitor or an isolation transformer is excluded. The only possibility is to insert an isolation amplifier (Fig. 18.37), provided that the constant current source is floating.

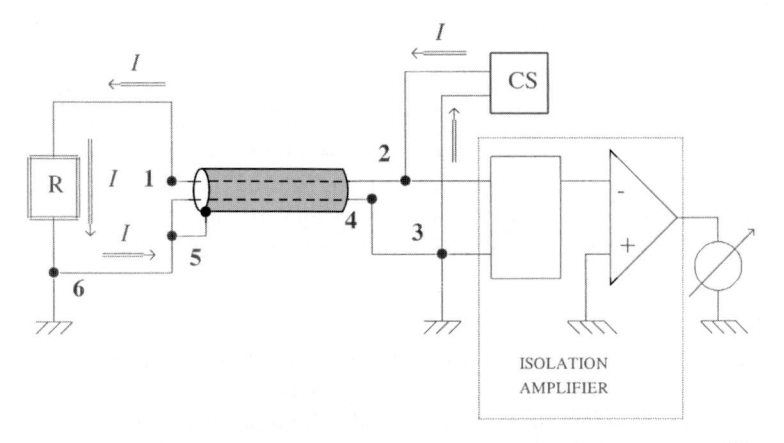

Fig. 18.37. Breaking the ground loop with an isolating amplifier

■ **Concluding Remark.** This technique is limited by the characteristics of the connecting cable [267]. At low frequency, where the cable acts like a simple capacitance, the bandwidth of the system may be seriously restricted. A possible solution is to reduce the sensor resistance R, but the penalty for reduced sensor resistance is a weaker useful signal and consequently a lower S/N ratio.

18.3.2 Logic Circuits

CASE STUDY 18.26

Consider a TTL logic gate supplied at 5 V driving an equivalent load C = 5 pF. The ground trace connecting the gate to ground has a parasitic inductance L = 100 nH and an associated resistance R = 80 mΩ. The slew rate of the output voltage is 1 V/ns.

1) Calculate the peak value of transients in the ground trace when the output of the gate switches from high to low. What is the expected waveform of such transients?
2) Determine the width of the ground trace satisfying the previous conditions if its total length is 26 cm.

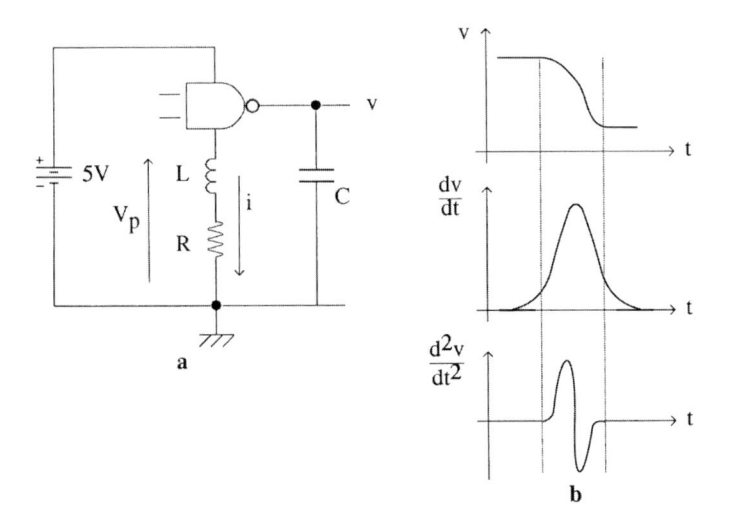

Fig. 18.38. a Schematic diagram; b various waveforms

Solution

The circuit configuration is presented in Fig. 18.38a.

1) When the gate output makes a transition from high to low, the load capacitance discharges through the ground trace; the peak discharge current is

$$i_{MAX} = C \frac{dv}{dt} = 5 \cdot 10^{-12}/10^{-9} = 5 \text{ mA}$$

When flowing through the trace (whose impedance is represented as the series combination of L and R), this current develops a parasitic voltage drop

$$V_{pMAX} = L \frac{di}{dt} + Ri_{MAX} = LC \frac{d^2v}{dt^2} + Ri_{MAX} \quad (18.39)$$

678   18 Protection Against Interfering Signals

As the gate goes to a low state, it is reasonable to think that for a TTL circuit the output voltage excursion is approximately 4 V. With the given slew rate, the corresponding time is 4 ns. For the current waveform, one can assume this time interval is distributed in the following way: 1 ns for rise time, 1 ns for fall time, and 2 ns for the central region.
Applying (18.39), we obtain

$$V_{pMAX} = 100 \cdot 10^{-9} \frac{5 \cdot 10^{-3}}{10^{-9}} + 0.08(5 \cdot 10^{-3}) = (500 + 0.4) \text{ mV} \cong 0.5 \text{ V}$$

One of the merits of this calculation is to show that the second term of the sum (i.e., the contribution of the trace resistance) is negligible. Hence

$$V_{pMAX} \cong L \frac{di}{dt} = LC \frac{d^2v}{dt^2}$$

It follows that *the peak of the parasitic voltage* (inductive noise) *is almost proportional to the second derivative of the output voltage*, and consequently the waveforms of Fig. 18.38 are obtained. Clearly

$$-0.5 \text{ V} < V_p < 0.5 \text{ V}$$

Therefore, the inductive noise is of the same order of magnitude as the gate noise margins. Increasing the noise further, the risk is to unintentionally toggle an identical gate farther down the chain, and hence induce a logic error.

2) The only thing we can control during the design phase is the self-inductance of the ground trace, by means of its geometry. We adopt (11.35b), which gives the self-inductance per unit length. Suppose that the insulator thickness is H = 0.8 mm and the copper trace thickness is T = 40 µm (a typical value).
A first attempt is made by selecting a trace width W = 1 mm:

$$L = \frac{1.26}{\frac{W}{H} + 1.393 + 0.667H\left(\frac{W}{H} + 1.444\right)} \quad [\mu H/m]$$

$$= \frac{1.26}{(1/0.8) + 1.393 + (0.667)(0.8 \cdot 10^{-3})(1.444 + 1/0.8)}$$

$$\cong 4.76 \text{ nH/cm}.$$

Obviously, for a 26-cm trace, the self-inductance exceeds the imposed budget of 100 nH.
Consequently, a second attempt is necessary, by choosing this time a 2-mm width trace:

$$L \cong 3.23 \text{ nH/cm}.$$

This value leads to a total self-inductance of 84.1 nH, which is acceptable. Finally, we must check the resistance of the trace:

$$R = \rho \frac{1}{S} = \rho \frac{1}{T\,W} = (1.72 \cdot 10^{-6}) \frac{26}{0.08 \cdot 10^{-2}} \cong 55.9 \text{ m}\Omega$$

which is better than the given value. Therefore, in order to meet the specifications, the 26-cm ground trace must have a width of 2 mm.

CASE STUDY 18.27 [239]

A personal computer (PC) board contains a 4×5 cm quasirectangular loop, used to distribute the clock signal to all integrated circuits.

The main characteristics of the clock are

- peak current $I = 6$ mA;
- frequency $f = 133$ MHz;
- duty cycle $\delta = 50\%$;
- rise time equal to fall time, $t_r = t_f = 4$ ns.

Determine the radiated spectrum at a distance of 3 m.

Solution

■ **Approach.** For simplicity, we assume that the rise and fall times are defined between amplitude levels of 0% and 100% (instead of the traditional values of 10% and 90%); consequently, the clock waveform corresponds to Fig. 18.39.

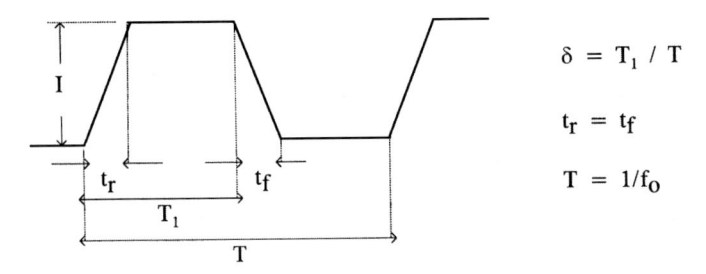

Fig. 18.39. Simplified clock waveform

Under the previous assumptions, the trapezoidal waveform should be decomposed into its Fourier components. For instance, the amplitude of the n-th harmonic is

$$I_n = 2I\delta \,\frac{\sin(n\pi\delta)}{n\pi\delta} \,\frac{\sin(n\pi t_r/T)}{n\pi t_r/T} \tag{18.40}$$

As soon as the spectrum of the trapezoidal waveform is found, the next step is to determine the worst-case radiated electric field:

$$E_n = (2.63 \cdot 10^{-14}) \frac{(nf_o)^2 \, A \, I_n}{d} \qquad (18.41)$$

where A is the loop area. Applying equations (18.40) and (18.41) to various harmonics, we obtain the results in Table 18.4.

Table 18.4. Current amplitude and radiated electric field for various harmonics

n	1	3	5	7	9	11	13	15
I_n [µA]	2270	242	80.1	35.6	17.4	8.47	3.53	0.638
E_n [µV/m]	705	677	621	540	438	318	185	44.5

☐ **Comments**

– The even harmonics of the clock current are zero, due to the function $\sin(n\pi\delta)$ of (18.40), which vanishes at

$$n\pi\delta = k\pi \; (k = 0, 1, 2, 3, \text{etc.})$$

Since $\delta = 0.5$, the previous condition is satisfied for n = 2k (even harmonics).

– Although the amplitude of odd current harmonics substantially decreases with order n, the radiated electric field shows, unexpectedly, a slower variation. To understand why, we substitute the expression for I_n into (18.41):

$$E_n = (2.63 \cdot 10^{-14}) \frac{f_o \, A}{d} (2I) \frac{\sin(n\pi\delta)}{\pi} \frac{\sin(n\pi t_r/T)}{\pi t_r} \qquad (18.42)$$

Equation (18.42) shows that E_n depends on the order n only via the arguments of sinusoidal functions; hence, it is a weak dependence.

■ **Conclusion.** The spectrum of the electric field radiated from the traces carrying the clock signals is much wider than expected. In the present case, for a low clock frequency of 133 MHz, the spectrum extends up to the 15th harmonic (which corresponds to 1.995 GHz). This suggests the necessity of adopting protections like minimizing the loop area (A), slowing the rise and fall times of the clock signal (whenever possible) and shielding the PC board (in practice, this is done by the PC metallic case).

CASE STUDY 18.28 [201, 202, 261]

Consider a logic gate supplied by a 5-V DC voltage source (Fig. 18.40). When the gate switches, the DC supply current undergoes step variations; assume that the corresponding transients have a spectrum extending up to 15 MHz.

18.3 Interfering Signals

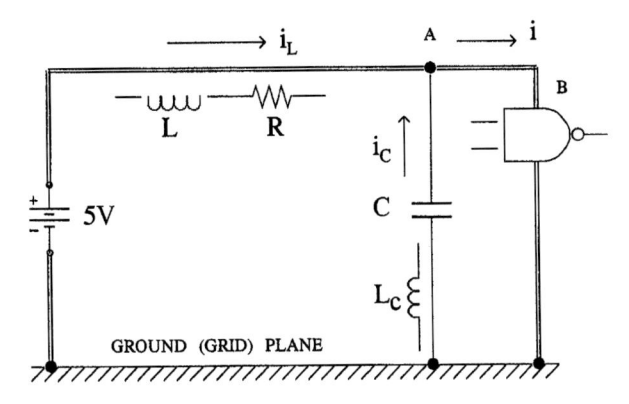

Fig. 18.40. Transient reduction using a bypass capacitor

1) Discuss the necessity of decoupling in logic systems.
2) Explain what benefit can be expected from a bypass (decoupling) capacitor, assuming that it is ideal.
3) What limitations are expected when the capacitor is no longer ideal?
4) Suppose that for a particular application we must choose between two capacitors, one of 10 nF and another of 4.7 nF, both being of the same type (radial ceramic). The manufacturer's data sheet indicates the dimensions of the capacitor leads: 30-mm length, 0.8-mm diameter, and 7.5 mm between the leads. It is reasonable to consider that when installing it, the leads are shortened to 3 mm each. Comment on your choice.
5) Make practical suggestions concerning the installation of the decoupling (bypass) capacitor.

Solution

1) Bypass capacitors are useful to reduce transients when abrupt variations occur in the DC supply current. In logic circuits, these variations appear when several gates or flip-flops switch simultaneously. Due to the physical configuration of the interconnections, resistive, inductive, and capacitive effects are observed. As a global result, they introduce HF *inductive noise* (or ΔI-noise) on the power/ground system, which shows up as a noise voltage at the power pins of the integrated circuits. Among the various consequences, the following are particularly harmful:

- They degrade the drive capabilities of transistors due to the temporal reduction in effective voltage supply seen by the integrated circuits.
- They can induce logic failures and affect the reliability of logic systems, since noise margins are decreased by the temporal reduction in effective power supply.
- They reduce the signal edge rate of the switching currents.

- They can produce radiated and/or conducted interference by coupling to other parts on the board.
- When propagating along power and ground interconnects (or planes), this noise is responsible for exciting resonant frequencies of interconnects. Ringing occurs at every switching edge (see Fig. 11.12).

Consequently, all digital circuits require *decoupling capacitors*. Their positive effects include:
- Providing local sources of charge to switching integrated circuits;
- Bypassing the stray inductances of interconnects, offering the minimum possible impedance between the DC source and the remote location of interest.

To be effective, the bypass capacitor should be installed as close as possible to the leads of the integrated circuit in question.

A schematic diagram is proposed in Fig. 18.40, where the supply traces are drawn with a double line. L and R represent the self-inductance and resistance of the supply trace; L_c includes the self-inductance of the connecting trace plus the stray inductance of the capacitor (the latter has two components, the parasitic inductance of the capacitor itself and the stray inductance of its leads).

As a general rule, *the decoupling system (capacitor and associated parasitics) must be effective at frequencies 10 to 100 times the clock frequency.*

2) To be specific, consider the transition of the logic gate from 0 to 1. In the absence of the capacitor, this will cause a fast, harmonic-rich transition in the DC supply current. When this current pulse flows through the inductance L, a significant amount of inductive noise is generated, as well as a simultaneously radiated magnetic field.

Any reduction in the peak current demand clearly has a beneficial effect on both. Consequently, installing a capacitor in parallel with the gate DC-supply pins must reduce the initial peak current extracted from the 5V-source as the capacitor locally supplies a charge to the gate of concern. Since i_L decreases, so will the radiated magnetic field, and inductive noise. In order to ensure high efficiency of this process, two conditions are required:
   a) the ideal capacitor must have the highest possible value so it can accumulate (and supply) a large charge;
   b) it should be mounted as close as possible to the gate in order to minimize the length of the section AB.

3) In practice, ideal capacitors don't exist. Any real capacitor inherently has an associated parasitic inductance, depending on its fabrication (see Sect. 7.2) and leads configuration. The lead inductance strongly depends on the case shape and type of capacitor. For instance, typical values for lead parasitic inductance range from 6 nH (axial ceramic type) to 2 nH (radial ceramic type) or less than 1 nH (for chip ceramic, surface mounted capacitors).

The problem with the parasitic inductance of any capacitor is that it reduces the so-called self-resonant frequency, which is related to the series resonance of the capacitance with its own parasitic inductance. In the present case, the latter is augmented by the trace parasitic inductance L_c, reducing the self-resonant frequency (f_r). At operating frequencies lower than f_r, the equivalent impedance of the (capacitor + trace) system is capacitive (its magnitude decreases with increasing frequency, which is a quite favorable effect). However, at frequencies higher than f_r, the capacitor looks inductive and its impedance rises with frequency. Right at f_r, the impedance magnitude reaches a minimum and the capacitor provides the most effective bypassing.

It follows that f_r must be placed above the useful frequency band. Note that this requirement conflicts with condition a of point 1, because a large capacitor shifts f_r into the useful band.

4) The trace parasitic inductance must first be evaluated. Since each lead has a final length of 3 mm, (11.25) furnishes the leads parasitic inductance:

$$L = 0.002\, l\left(\ln(4l/d) - 0.75\right)$$
$$= 0.002(0.6)\left(\ln\frac{4(0.6)}{0.08} - 0.75\right) \cong 3.18 \text{ nH}$$

Because information concerning the length of the trace connecting the capacitor to gate is lacking, we make provision for a global length not exceeding 3 cm, with an inductance per unit length of 5 nH/cm. With these assumptions in mind, the total inductance (trace plus leads) becomes

$$L_c \cong 3.18 + (3)(5) = 18.18 \text{ nH}$$

We adopt a final budget of $L_c = 20$ nH. This estimate holds for the mounting of each of the proposed capacitors.

The 10-nF capacitor yields a self-resonant frequency

$$f_{r1} = \frac{1}{2\pi\sqrt{L_c C_1}} = \frac{1}{2\pi\sqrt{(2\cdot 10^{-8})(10^{-8})}} \cong 11.25 \text{ MHz}$$

while the 4.7-nF capacitor yields

$$f_{r2} = \frac{1}{2\pi\sqrt{L_c C_2}} = \frac{1}{2\pi\sqrt{(2\cdot 10^{-8})(0.47\cdot 10^{-8})}} \cong 16.41 \text{ MHz}$$

As the frequency spectrum extends up to 15 MHz, clearly only the 4.7-nF capacitor is suited for this application; the penalty is the smaller amount of charge which can be supplied to the gate during the beginning of a transition, and more inductive noise.

5) Condition b point 2 requires that the decoupling capacitor be placed as close as possible to the integrated circuit of concern. Specific modalities

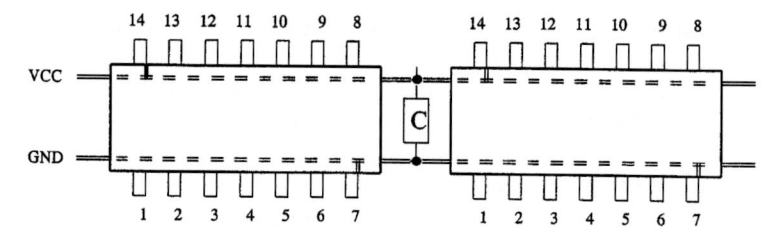

Fig. 18.41. Good placement of the bypass capacitor on a PCB

must be adopted for every type of technology. For instance, if the circuit of Fig. 18.40 is mounted on a PCB with discrete components, one possibility is to place the decoupling capacitor as shown in Fig. 18.41. Here, two different integrated circuits (not simultaneously switching) can share the same decoupling capacitors [266].

Note that if a buried decoupling capacitor is used instead, only plated via holes should be placed close to pins 7 and 14 of each package. This would result in weaker parasitics and better decoupling performance.

If the circuit suggested in Fig. 18.40 must be integrated in a monolithic circuit, then an on-chip decoupling capacitor is the best solution.

■ **Concluding Remark.** Note in Fig. 18.41 that from a parasitics point of view, the traditional assignment of DC-supply to pins 7 and 14 (for 14-pin packages), or to pins 8 and 16 (for 16-pin packages) is not optimal, since the lengths of traces connecting the bypass capacitor cannot be made as short as desired. Obviously, a better choice would be on two adjacent pins, or on two opposite pins (like pairs 7 and 8 or 1 and 14).

18.3.3 Contact Protection

CASE STUDY 18.29 [261, 268]

A TTL logic circuit controls a relay, whose coil is kept energized through the relay auxiliary normally-open bronze contact (denoted K), when the original operating circuit is open. The relay coil has the following characteristics: resistance $R_L = 56\,\Omega$, inductance $L = 120\,\text{mH}$, DC voltage source $E = 5\,\text{V}$. Propose a protection circuit for the switch K.

Solution

■ **Procedure.** Switch protection circuits are useful for two reasons: they suppress interfering signals generated by contact making or breaking, and they reduce material transfer by arcing. With such protection, contacts last longer and operate more safely.

Table 18.5. Selection of protection circuits for various loads

Load	Current drawn	Protection circuit
Noninductive	$< I_m$	no need
Inductive	$< I_m$	R – C
Inductive	$> I_m$	R – C – D
Noninductive	$> I_m$	R – C – D

In all applications of this kind, the following procedure should be observed:

Step 1: Select the proper circuit configuration, according to Table 18.5 [261], where I_m denotes the minimum arcing current (which depends on the contact material, according to Table 10.1).

Since it is specified that the switch has bronze contacts, we have from Table 10.1

$I_m = 310\,\text{mA}$ and $V_m = 13.5\,\text{V}$.

The steady-state current is

$$I_o = \frac{E}{R_L} = \frac{5}{56} = 89.3\,\text{mA}$$

which is less than the minimum arcing value.

Consequently, for an inductive load and a current less than I_m, an RC circuit can be employed, as shown in Fig. 18.42.

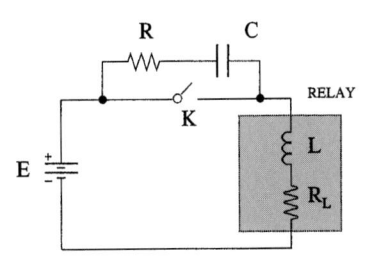

Fig. 18.42. Contact protection circuit across switch

Step 2: Determine the value of capacitor C.

As discussed in Sect. 10.3.2, to avoid discharges two conditions must be fulfilled:

1) preserve the voltage across switch contacts below 300 V;
2) the rate of rise of the voltage across the switch must be kept below 1 V/μs.

To implement the first condition, we may somewhat arbitrarily, neglect resistance R. In this way, the voltage across the switch is equal to the voltage V_c across the capacitor.

18 Protection Against Interfering Signals

Secondly, the role of C is to accumulate the energy liberated by the inductor L when K is opened. Hence

$$\frac{1}{2} L I_o^2 = \frac{1}{2} C V_c^2, \quad \text{from which} \quad V_c = I_o \sqrt{\frac{L}{C}}$$

Requiring $V_c \leq 300\,\text{V}$, one obtains

$$C \geq \left(\frac{I_o}{300}\right)^2 L \qquad (18.43)$$

Substituting numerical values, $C = 10.63\,\text{nF}$.

For the second condition, and assuming that the capacitor is charged by a constant current, we may write

$$C = \frac{\Delta Q}{\Delta V} = \frac{I_o \Delta t}{\Delta V} = \frac{I_o}{\Delta V / \Delta t}$$

from which

$$C \geq \frac{I_o \Delta t}{\Delta V} \qquad (18.44)$$

$$C \geq \frac{0.0893 \cdot 10^{-6}}{1} = 89.3\,\text{nF}$$

The final value of the capacitance must clearly be greater than both limits, i.e.,

$$C = 0.1\,\mu\text{F} \,/\, 300\,\text{V}$$

Step 3: Determine the resistance R.

Resistor R should restrict the discharge current of capacitor C (when closing K) to its maximum allowable value I_m. This establishes the *minimum value* of R.

The *maximum value* of R is deduced from the closing condition on K by taking R equal to the load resistance, to limit the voltage transient developed across K to the DC supply voltage.

Therefore

$$\frac{E}{I_m} < R < R_L$$

With the specified numerical values, we obtain

$$16.13\,\Omega < R < 56\,\Omega$$

For instance, we may choose

$$R = 33\,\Omega$$

CASE STUDY 18.30 [261, 268]

1) Discuss the choice of relay contact materials for practical applications.
2) An electromagnet with $L = 1.5\,\text{H}$ and $R_L = 85\,\Omega$ is operated by closing a relay contact in series with a DC supply $E = 48\,\text{V}$. The relay contacts are tungsten (anode) and silver (cathode). This combination is rated at $I_m = 0.35\,\text{A}$ and $V_m = 13\,\text{V}$. Design the protection circuit for the contacts.

Solution

1) The choice of a particular relay from a manufacturer's list is made according to the electrical performance information and the type of material used for the contacts. These materials have to satisfy various conditions, including high values for minimum arcing current and voltage, low ohmic contact resistance, as well as high contact hardness and melting point. The latter two conditions ensure reduced mechanical deformation of contacts during repetitive operation cycles, and little or no material transfer. Both guarantee a long lifetime.

    For instance, according to [268], microcontacts require gold or gold alloys to give satisfactory reliability in dust-free air. Obviously, dust deposit on the contact surface must be prevented, and a possible solution is to use hermetically sealed packages (as in reed relays).

    Concerning the relays with highly repetitive operation (several operations per second), problems result from resistive contact films and material transfer. To avoid them, a tungsten (or molybdenum) cathode against platinum–iridium or platinum–palladium alloys is recommended. To avoid sticking of contacts after arcing, CdO has been successfully added to these alloys.

    For light-duty relays (such as those used in telecommunication applications), widia (WC) represents a good choice for the contact material, since it considerably reduces material transfer during arcing.

    In integrated circuit applications, silver, silver plated with gold, rhodium, and silver alloys (Ag CdO, Ag Ni 0.15 gold-plated, Ag Pd gold-plated, etc.) are the most frequently encountered materials employed in contact fabrication.

2) The same procedure as in Case Study 18.29 is adopted.

    Step 1: The steady-state current is

    $$I_o = \frac{E}{R_L} = \frac{48}{85} = 0.5647\,\text{A}$$

    Note that this current exceeds the value of I_m. Thus, according to Table 18.4, we need a RCD protection, as shown in Fig. 18.43.

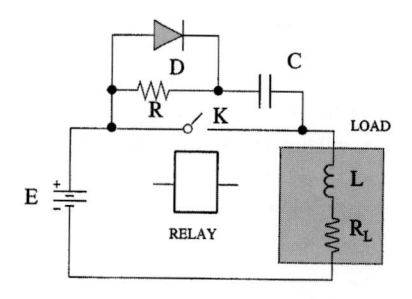

Fig. 18.43. RCD protection circuit across the contacts

Step 2: The capacitor value is found with (18.43) and (18.44) given in Case Study 18.29.

To avoid glow discharge, the voltage across the contacts should always be kept below 300 V, so

$$C \geq \left(\frac{I_o}{300}\right)^2 L = \left(\frac{0.5647}{300}\right)^2 (1.5) \cong 5.31 \text{ µF}$$

To avoid arcing, the voltage rate must not exceed $1\,\text{V}/\text{µs}$; condition (18.44) yields

$$C \geq \frac{I_o \, \Delta t}{\Delta V} = \frac{0.5647 \cdot 10^{-6}}{1} = 0.5647 \text{ µF}$$

Note that when we are dealing with *high* inductive loads, C is established with (18.43), which corresponds to avoiding glow discharge (in contrast with *low* inductive loads, where the condition establishing the value of C is arcing avoidance).

In the present case we adopt a 6.8-µF or 10-µF, 450-V capacitor:

$$C = 10 \text{ µF} \, / \, 450 \text{ V}$$

Step 3: When contact K breaks, the peak of the charging current flows through the small internal resistance of the diode. When K closes, the capacitor discharges through R. The only condition to impose here is to maintain the peak value of the discharge current below a small fraction (say, 10%) of I_m. Then R becomes

$$R \geq \frac{10\,E}{I_m} = \frac{480}{0.35} = 1.37 \text{ k}\Omega$$

A convenient choice is to select a slightly higher value:

$$R = 1.6 \text{ k}\Omega$$

References

1. IEEE(1997): IEEE Standard Dictionary of Electrical and Electronics Terms. Wiley Interscience, New York
2. Pearsall J., Trumble B.(1996): The Oxford English Reference Dictionary. Oxford University Press, Oxford
3. R. F. Graf (ed.) (1977): The Modern Dictionary of Electronics. Howard W. Sams & Co., Indianapolis
4. M. J. Clugston (ed.) (1998): The New Penguin Dictionary of Science. Penguin Books, London
5. Rheinfelder W. A. (1964): Design of Low-Noise Transistor Input Circuits. Iliffe Book Ltd., London
6. Hartmann K. (1976): "Noise characterization of linear circuits". IEEE Trans. on Circ. and Syst., **CAS-23**(10), 581–590
7. Visweswariah C., Haring R. A., and Conn A. R. (2000): "Noise consideration in circuit optimization". IEEE Trans. on Computer-Aided Design of Int. Circ. and Syst., **19**(6), 679–690
8. Shephard K. L., Narayanan V., and Rose R. (1999): "Harmony: static noise analysis of deep submicron digital integrated circuits". IEEE Trans. on Computer-Aided Design of Int. Circ. and Syst., **18**(8), 1132–1150
9. HP Company (1998): "Noise sources in CMOS image sensors".
10. Tian H., Fowler B. and El Gamal A. (2001): "Analysis of temporal noise in CMOS photodiode active pixel sensor". IEEE Journal of Solid-State Electronics, **36**(1), 92–101
11. Balamurugan G., Shanbhag N. R. (2001): "The twin-transistor noise-tolerant dynamic circuit technique". IEEE Journal of Solid-State Circuits, **36**(2), 273–280
12. Jou S-J.,Kuo S-H., Chiu J-T., and Lin T-H. (2001): "Low switching noise and load-adaptive output buffer design techniques". IEEE Journal of Solid-State Circuits, **36**(8), 1239–1249
13. Losee F. (1997): RF Systems, Components, and Circuits Handbook. Artech House, Boston
14. Fraden J. (1997): Handbook of Modern Sensors (Physics, Designs and Applications). AIP Press, Woodbury, New York
15. Bishop C. M. (1996): Neural Networks for Pattern Recognition. Oxford, London

16. Ando B., Graziani S. and Pitrone N. (2001): "Hysteresis shaping in stochastic driven systems". IEEE Transactions on Instrumentation and Measurement, **50**(5),1266–1269
17. Jung P. and Hänggi P. (1991): "Amplification of small signals via stochastic resonance". Phys. Rev. A **44**(12), 8032–8042
18. Sheingold D. H. (Ed.) (1972): Analog-Digital Handbook. Analog Devices Inc., Norwood, Mass.
19. Jay F. (ed.) (1977): IEEE Standard Dictionary of Electrical and Electronics Terms. Wiley, New York
20. Pettai R. (1984): Noise in Receiving Systems. Wiley, New York
21. Ziemer R. E., Tranter W. H. (1976): Principles of Communications: Systems, Modulation, and Noise, Houghton Mifflin, Boston
22. Davenport W.B., Root W. L. (1958): An Introduction in the Theory of Random Signals and Noise. McGraw-Hill, New York
23. Freeman J. J. (1958): Principles of Noise. Wiley, New York
24. Benett W. R. (1960): Electrical Noise. McGraw-Hill, New York
25. Van der Ziel A. (1970) Noise: Sources, Characterization, Measurements. Prentice-Hall, Englewood Cliffs, New Jersey
26. Fish P. J. (1994): Electronic Noise and Low-Noise Design. McGraw-Hill, New York
27. Kleckner K. R. (1965): "Correlation of noise generators". Proc. IEEE **53**(2), 202
28. King R. (1966): Electrical Noise. Chapman and Hall, London
29. Gupta M. (1982): "Thermal noise in nonlinear resistive devices and its circuit representation". Proc. IEEE, **70**(8), 788–804
30. Oliver B. M. (1965): "Thermal and quantum noise". Proc. IEEE, **53**(5), 436–454
31. Twiss R. Q. (1955): "Nyquist's and Thevenin's theorems generalized for non-reciprocal linear networks". J. Appl. Phys., **26**(5), 599–602
32. Agouridis D. C. (1987): "Thermal noise of transmission lines: a generalized solution". IEEE Trans. Instr. Meas., **IM- 36**(1), 132–134
33. Haus H. A. (1961): "Thermal noise in dissipative media". J. Appl. Phys. **32**(3), 493–500
34. Van der Ziel A. (1970): Noise: Sources, Characterization, Measurements. Prentice-Hall, Englewood Cliffs, New Jersey
35. Schottky W. (1918): "Über spontane Stromschwankungen in verschiedenen Elektrizitätsleitern". Ann. Phys., Leipzig, **67**(4), 541–567
36. Buckingham M. J. (1983): Noise in Electronic Devices and Systems. Halstead Press, New York
37. Hooge F. N. (1976): "1/f noise". Physica, **83B**(1), 14–23
38. Hooge F. N. (1994): "1/f noise sources". IEEE Trans. Electron Dev. **41**(11), 1926–1935
39. Bliek L. (1981): "A model for 1/f noise". IEEE Trans. Instr. Meas. **IM-30**(4), 307–309
40. Forbes L.(1995) : "On the theory of 1/f noise in semi-insulating materials". IEEE Trans. Electron Dev. **42**(10), 1866–1868
41. Jaeger R. C., Brodersen A. J. (1970): "Low-frequency noise sources in bipolar junction transistors". IEEE Trans. Electron Dev. **ED-17**(2), 128–134

42. Ambrozy A. (1982): Electronic Noise. McGraw-Hill, New York
43. Vandamme L. K. J. (1994): "Noise as a diagnostic tool for quality and reliability of electronic devices". IEEE Trans. Electron Dev. **ED-41**(11), 2176–2187
44. Laker K. R. and Sansen W. (1994): Design of Analog Integrated Circuits and Systems. McGraw-Hill, New York
45. Jay F. (Ed.) (1997): IEEE Standard Dictionary of Electrical and Electronics Terms. Wiley Interscience, New York
46. Haus H.A., Adler R. (1959): Circuit Theory of Linear Noisy Networks. Wiley, New York
47. Nielsen E. G. (1966): Amplifier Noise. In *Shea R. F. (ed.): Amplifier Handbook*, McGraw-Hill, New York, pp. 7-1–7-56
48. Pettai R. (1984): Noise in Receiving Systems. Wiley, New York
49. Savelli M. (1984): Bruit de fond: caractérisation des composants et des circuits. In: *Techniques de l'ingénieur*, Electronique, vol. 1, Strasbourg, France, pp. E410-1–E411-16
50. Van der Ziel A. (1970): Noise: Sources, Characterization, Measurements. Prentice-Hall, Englewood Cliffs, New Jersey
51. Benett W. R. (1960): Electrical Noise. McGraw-Hill, New York
52. Motchenbacher C.D., Connelly J. A. (1993): Low Noise Electronic Design. Wiley, New York
53. Poitevin J. P. (1967): "Le bruit dans les réseaux liniaires". L'Onde électrique, **47**(1), 43–55
54. Ott H. W. (1988): Noise Reduction Techniques in Electronic Systems. Wiley Interscience, New York
55. Mumford W. W., Scheibe E. H. (1967): Noise Performance Factors in Communication Systems. Horizon House – Microwave Inc., Dedham
56. Haus H. A., Adler R. (1957): "An extension of the noise figure definition". Proc. of the I.R.E., May 1957, 690–691
57. Fukui H. (1966): "Available power gain, noise figure, and noise measure of two-ports and their graphical representation". IEEE Trans. Circ. Theory, **CT-12**(2), 137–141
58. King R. (1966): Electrical Noise. Chapman and Hall Ltd., London
59. Miller C. K. S., Daywitt W.C., Arthur M.G. (1967): "Noise standards, measurements and receiver noise definitions". Proc. IEEE, **55**(6), 865–877
60. Pyati V. P. (1992): "An exact expression for the noise voltage across a resistor shunted by a capacitor". IEEE Trans. Circ. Syst. (I), **39**(12), 1027–1029
61. Van Nie A. G. (1972): "Representation of linear passive noisy 1-port by two correlated noise sources". Proc. IEEE, **60**(6), 751–753
62. Mumford W. W., Scheibe E. H. (1967): Noise Performance Factors in Communication Systems. Horizon House – Microwave Inc., Dedham
63. Pettai R. (1984): Noise in Receiving Systems. John Wiley, New York
64. Montgomery H. C. (1952): "Transistor noise in circuit applications". Proc. I.R.E., **40**(11), 1461–1471
65. Rothe H., Dalke W. (1956): "Theory of noisy fourpoles". Proc. I.R.E., **44**(6), 811–818
66. Lange J. (1967): "Noise characterization of linear two-ports in terms of invariant parameters". IEEE J. Solid State Circ., **SC-2**(6), 37–40
67. I.R.E. Subcommittee 7.9 on Noise (1960): "Representation of noise in linear two-ports". Proc. I.R.E., **48**(1), 69–74

68. Hillbrand H., Russer P. (1976): "An efficient method for computer-aided noise analysis of linear amplifier networks". IEEE Trans. Circ. Syst., **CAS-23**(4), 235–238
69. Haus H.A., Adler R. (1959): Circuit Theory of Linear Noisy Networks. John Wiley, New York
70. Twiss R. Q. (1955): "Nyquist's and Thevenin's theorems generalized for non-reciprocal linear networks". J. Appl. Phys., **26**(5), 599–602
71. Rizzoli V., Lipparini A. (1985): "Computer-aided noise analysis of linear multiport networks of arbitrary topology". IEEE Trans. MTT, **MTT-33**(12), 1507–1512
72. Bosma H. (1967): On the Theory of Linear Noisy Systems. Philips Res. Repts. Suppl., no. 10
73. Dobrowolski J. A. and Ostrowski W. (1996): Computer-Aided Analysis, Modeling, and Design of Microwave Networks. The Wave Approach. Artech House, Boston, London
74. Wedge S. W. and Rutledge D. B. (1992): "Wave techniques for noise modeling and measurement". IEEE Trans. on MTT, **MTT-40**(11), 2004–2012
75. Vasilescu G. (1994): Etude du bruit électrique dans les circuits micro-ondes. Ph. D. dissertation, University Pierre & Marie Curie, Paris, France
76. Vasilescu G. and Alquié G. (1989): "Exact computation of two-port noise parameters". Electronics Letters, **25**(4), 292–293
77. Lane R. Q. (1969): "The determination of device noise parameters". Proc. IEEE, **57**(8), 1461–)1462
78. Kotyczka W., Leupp A. and Strutt M. J. O. (1970): "Computer-aided determination of two-port noise parameters (CADON)". Proc. IEEE, **58**(11), 1850–1851
79. Caruso G. and Sannino M. (1978): "Computer-aided determination of microwave two-port noise parameters". IEEE Trans. MTT, **26**(9), 639–648
80. Mitama M. and Katoh H. (1979): "An improved computational method for noise parameter measurement". IEEE Trans. MTT, **27**(6), 612–615
81. Boudiaf A. and Laporte M. (1993): "An accurate and repeatable technique for noise parameter measurements". IEEE Trans. MTT, **42**(2), 532–537
82. De Dominicis M., Giannini F., Limiti E. and Saggio G (2002): "A novel impedance pattern for fast noise measurements". IEEE Trans. Instr. Meas., **51**(3), 560–564
83. Escotte L., Plana R. and Graffeuil J. (1993): "Evaluation of noise parameter extraction methods". IEEE Trans. MTT, **41**(3), 382–387
84. Hillbrand H., Russer P. (1976): "An efficient method for computer-aided noise analysis of linear amplifier networks". IEEE Trans. Circ. Syst., **CAS-23**(4), 235–238
85. Haus H.A., Adler R. (1959): Circuit Theory of Linear Noisy Networks. John Wiley, New York
86. Rizzoli V., Lipparini A. (1985): "Computer-aided noise analysis of linear multiport networks of arbitrary topology". IEEE Trans. MTT, **MTT-33**(12), 1507–1512
87. Vasilescu G. (1987): "Computer aided noise analysis of linear multiport networks". Electronics Letters **23**(7), 351–353
88. Vasilescu G. (1994): Noise Simulation of Microwave Circuits. Ph. D. dissertation, Université Pierre et Marie Curie, Paris, France

89. Chua L. O., Lin P.M. (1975): Computer-Aided Analysis of Electronic Circuits: Algorithms and Computational techniques. Prentice-Hall, Englewood Cliffs, NJ
90. Wilkinson J. H. (1961): "Error analysis of direct methods of matrix inversion". JACM, **8**, 281–301
91. Motchenbacher C.D., Connelly J. A. (1993): Low Noise Electronic Design. John Wiley, New York
92. Van der Ziel A. (1970): Noise: Sources, Characterization, Measurements. Prentice-Hall, Englewood Cliffs, NJ
93. Buckingham M. J. (1983): Noise in Electronic Devices and Systems. Halstead Press, New York
94. Robinson F.N.H. (1974): Noise and Fluctuations in Electronic Devices and Circuits. Clarendon Press, Oxford
95. Ambrozy A. (1982): Electronic Noise. McGraw-Hill, New York
96. Nielsen E. G. (1957): "Behavior of noise figure in junction transistors". Proc. I.R.E., **45**(7), 957–963
97. Hawkins R. J. (1977): "Limitations of Nielsen's and related noise equations applied to microwave bipolar transistors and a new expression for the frequency and current dependent noise figure". Solid State Electronics, **20**(4), 191–196
98. Pucel R. A., Rohde U. L. (1993): "An exact expression for the noise resistance R_n for the Hawkins bipolar noise model". IEEE Microwave and Guided-Wave Letters, **3**(2), 35–37
99. Motchenbacher C.D., Fitchen F. C. (1973): Low Noise Electronic Design. John Wiley, New York
100. Fukui H. (1966): "The noise performance of microwave transistor". IEEE Trans. Electr. Dev., **ED-13**(8), 329–341
101. Midford T. (1990): FETs: Low-Noise Applications. In *Chang K.(ed.): Handbook of Microwave and Optical Components*, vol. 2: Microwave Solid-State Components, Wiley-Interscience, New York, 550–613
102. Van der Ziel A. (1962): "Thermal noise in field-effect transistors". Proc. I.R.E., **50**(8), 1808–1812
103. Zillmann U., Herzel F. (1996): "An improved SPICE model for high-frequency noise of BJTs and HBTs". IEEE J. Solid State Circ., **31**(9), 1344–1346
104. Bruncke W. C., Van der Ziel A. (1966): "Thermal noise in junction-gate field effect transistors". IEEE Trans. Electr. Dev., **ED-13**(3), 323–329
105. Nicollini G., Pancini D., Pernici S. (1987): "Simulation-oriented noise models for MOS devices". IEEE J. Solid State Circ., **SC-22**(12), 1209–1212
106. Fox R. M. (1993): "Comments on circuit models for MOSFET thermal noise". IEEE J. Solid State Circ., **28**(2), 184–185
107. Wang B., Hellums J. R., Sodini C. G. (1994): "MOSFET thermal noise modeling for analog integrated circuits". IEEE J. Solid State Circ., **29**(7), 833–835
108. Baechtold W. (1971): "Noise behavior of Schottky barrier gate field-effect transistors at microwave frequencies". IEEE Trans. Electr. Dev., **ED-18**(2), 97–104
109. Baechtold W. (1972): "Noise behavior of GaAs field-effect transistors with short gate length". IEEE Trans. Electr. Dev., **ED-19**(5), 674–680

110. Pucel R. A., Haus H. A. and Statz H. (1975): Signal and Noise Properties of Gallium Arsenide Microwave Field-Effect Transistors. In *Marton L.(ed.): Advances in Electronics and Electron Physics*, **38**, Academic Press, New York, 195–265
111. Fukui H. (1979): "Optimal noise figure of microwave GaAs MESFETs". IEEE Trans. Electr. Dev., **ED-36**(7), 1032–1037
112. Fukui H. (1979): "Design of microwave GaAs MESFETs for broadband low-noise amplifiers". IEEE Trans. MTT, **MTT-27**(7), 643–650
113. Podell A. F. (1981): "A functional GaAs FET noise model". IEEE Trans. Electr. Dev., **ED-28**(5), 511–517
114. Gupta M. S., Pitzalis O., Rosenbaum S. and Greiling P. (1987): "Microwave noise characterization of GaAs MESFETs: evaluation by on-wafer low-frequency output noise current measurement", IEEE Trans. MTT, **MTT-35**(12), 1208–1217
115. Gupta M. S., Greiling P. T. (1988): "Microwave noise characterization of GaAs MESFETs: determination of extrinsic noise parameters". IEEE Trans. MTT, **MTT-36**(4), 745–754
116. Heinrich W. (1989): "High-frequency MESFET noise modeling including distributed effects". IEEE Trans. MTT, **MTT-37**(5), 836–842
117. Escotte L. and Mollier J. C. (1990): "Semidistributed model of millimeter-wave FET for S-parameter and noise figure predictions". IEEE Trans. MTT, **MTT-38**(6), 748–753
118. Fjeldly T. A., Ytterdal T., and Shur M. (1998): Introduction to Device Modeling and Circuit Simulation. John Wiley, New York
119. Zimmermann J., Salmer G. (1990): High Electron Mobility Transistor: Principles and Applications. In: *Chang Kai (ed.): Handbook of Microwave and Optical Components*, Wiley-Interscience, New York, 436–470
120. Cappy A., Vanoverschelde A., Schortgen M. et al. (1985): "Noise modeling in submicrometer-gate two-dimensional electron- gas field effect transistors". IEEE Trans. Electr. Dev., **ED-32**(12), 2787–2795
121. Cappy A. (1988): ""Noise modeling and measurements techniques". IEEE Trans. MTT, **MTT-36**(1), 1–10
122. Pospieszalski M. (1989): "Modeling of noise parameters of MESFETs and MODFETs and their frequency and temperature dependence". IEEE Trans. MTT, **MTT-37**(9), 1340–1350
123. Heymann P., Rudolph M., Prinzler H., Doerner R., Klapproth L. and Böck G. (1999): "Experimental evaluation of microwave field-effect transistor noise models". IEEE Trans. MTT, **MTT-47** (2), 156–163
124. Tasker P. J., Schlectweg M., Reinert W. and Braunstein J. (1993): "A novel approach for MODFET noise parameter extraction and its application in mm-wave MMICs". Proc. of MIOP'93, Sindelfingen, Germany, 212–216
125. Hickson M. T., Gardner P. and Paul D. K. (1992): "A semidistributed HEMT model for accurate fitting and extrapolation of S-parameters and noise parameters". IEEE Trans. MTT, **MTT-40**(8), 1709–1712
126. Klepser B-U. H., Bergamaschi C., Schefer M., Diskus, Patrick and Bächtold W. (1995): "Analytical bias dependent noise model for InP HEMTs". IEEE Trans. Electr. Dev., **ED-42**(11), 1882–1890

127. Tedja S., Van der Spiegel J. and Williams H. (1994): "Analytical and experimental studies of thermal noise in MOSFETs". IEEE Trans. Electr. Dev., **ED-41**(11), 2069–2075
128. Ott W. H. (1988): Noise Reduction Techniques in Electronic Systems. John Wiley, New York
129. Trofimenkoff F. N, Onwuachi O. A. (1989): "Noise performance of operational amplifier circuits". IEEE Trans. on Education, **E-32**(1), 12–16
130. Brinson M. E. and Faulkner D. J. (1995): "A SPICE noise macromodel for operational amplifiers". IEEE Trans. Circ. Syst. (I), **CAS 42**(3), 166–168
131. Uchida K., Matsura H., Yakihara T., Kobayashi S., Oka S., Fujita T. and Miura A. (2001): "A series of InGaP/InGaAs HBT oscillators up to D-band". IEEE Trans. MTT, **49**(5), 858–865
132. Bruce S., Vandamme L.K.J. and Rydberg A. (2000): "Temperature dependence and electrical properties of dominant low-frequency noise in SiGe HBT". IEEE Trans. Electr. Dev., **ED-47**(5), 1107–1111
133. Niu G., Zhang S., Cressler J.D., Joseph A.J., Fairbanks J.S., Larson L.E., Webster C.S., Ansley W.E. and Harame D.L. (2000): "Noise modeling and SiGe profile design tradeoffs for RF applications". IEEE Trans. Electr. Dev., **ED-47**(11), 2037–2043.
134. Knoblinger G., Klein P. and Tiebout M. (2001): "A new model for thermal noise of deep-submicron MOSFETs and its application in RF-CMOS design". IEEE Journal of Solid-State Circ., **36**(5), 831–837
135. Voinigescu S., Maliepaard M.C., Showell J., Babcock G., Marchesan D., Schroter M., Schwan P. and Harame D. (1997): "A scalable high-frequency model for bipolar transistors with application to optimal transistor sizing for low-noise amplifier design". IEEE Journal of Solid-State Circ., **32**(9), 1430–1439
136. Brederlow R., Eber W., Dahl C., Schmitt-Landseidel D., and Thewes R. (2001): "Low-frequency noise of integrated polysilicon resistors". IEEE Trans. Electr. Dev., **48**(6), 1180–1187
137. Niu G., Cressler J.D., Zhang S., Joseph A.J., Ansley W.E., Webster C.S., and Harame D.L. (2001): "A unified approach to RF and microwave noise parameter modeling in bipolar transistors". IEEE Trans. Electr. Dev., **ED-48**(11), 2568–2575.
138. Mitama M. and Katoh H. (1979): "An improved computational method for noise parameter measurement". IEEE Trans. MTT, **MTT-27**(6), 612–615
139. Shaeffer D. K. and Lee T. H. (1997): "A 1.5-V, 1.5-GHz CMOS low noise amplifier". IEEE Journal of Solid-State Circ., **32**(5), 745–759
140. Lee T. H. (1998): The Design of CMOS Radio-Frequency Integrated Circuits. Cambridge Univ. Press, Cambridge, UK
141. Westman H. P. (Ed.) (1972): Reference Data for Radio Engineers. Howard Sams & Co., Indianapolis
142. Pettai R. (1984): Noise in Receiving Systems. John Wiley, New York
143. Weston D. A. (1991): Electromagnetic Compatibility: Principles and Applications. Marcel Dekker, New York
144. Fish P. J. (1994): Electronic Noise and Low-Noise Design. McGraw-Hill, New York
145. Jay F. (Ed.) (1997): IEEE Standard Dictionary of Electrical and Electronics Terms. Wiley Interscience, New York

References

146. Graf R. F. (ed.) (1977): The Modern Dictionnary of Electronics. Howard W. Sams & Co., Indianapolis
147. Perez R. (1995): Handbook of Electromagnetic Compatibility. Academic Press, San Diego
148. Ott H. W. (1988): Noise Reduction Techniques in Electronic Systems. Wiley Interscience, New York
149. Williams T. (1996): EMC for Product Designers. Newnes, Oxford
150. White D. R. J. and Mardiguian M. (1974): Electromagnetic Shielding. In *A Handbook Series on Electromagnetic Interference and Compatibility*, vol. 3, Interference Control Technologies Inc., Gainesville, Virginia
151. Schulz R. B. (1995): Electromagnetic Shielding. In *Perez R. (Ed.): Handbook of Electromagnetic Compatibility*, Academic Press, San Diego, 401–441
152. Jay F. (Ed.) (1997): IEEE Standard Dictionary of Electrical and Electronics Terms. Wiley Interscience, New York
153. Weston D. A. (1991): Electromagnetic Compatibility: Principles and Applications. Marcel Dekker, New York
154. Tsaliovich A. (1999): Electromagnetic Shielding Handbook for Wired and Wireless EMC Applications. Kluwer Academic Press, Boston
155. Ott W. H. (1988): Noise Reduction Techniques in Electronic Systems. John Wiley, New York
156. Quine J. P. (1957): "Theoretical formulas for calculating the shielding effectiveness of perforated sheets and wire mesh screens". Proc. of 3rd Conf. on Radio Interference Reduction, Armour Research Foundation, 315–329
157. Wiliams T. (1996): EMC for Product Designers. Newnes, Oxford
158. Graf R. F. (1977): Modern Dictionary of Electronics. Howard W. Sams & Co., Indianapolis
159. Paul R. C. (1992): Introduction to Electromagnetic Compatibility. John Wiley, New York
160. Lewis W. H. (1995): Grounding and Bonding. In *Perez R. (Ed.): Handbook of Electromagnetic Compatibility*. Academic Press, San Diego, 301–399
161. Kodali V. P. (1966): Engineering Electromagnetic Compatibility: Principles, Measurements and Technologies. IEEE Press, New York
162. Kodali V. P. (1996): Engineering Electromagnetic Compatibility: Principles, Measurements, and Technologies. IEEE Press, New York
163. Bogart T. F. (1993): Electronic Devices and Circuits. Merill, New York
164. Morrison R. (1992): Noise and Other Interfering Signals. John Wiley, New York
165. Jay F. (Ed.) (1997): IEEE Standard Dictionary of Electrical and Electronics Terms. Wiley Interscience, New York
166. Horowitz P. and Hill W. (1989): The Art of Electronics. Cambridge University Press, Cambridge
167. Motchenbacher, C.D., Connelly, J. A. (1993): Low Noise Electronic Design. John Wiley, New York
168. Ott, W. H. (1988): Noise Reduction Techniques in Electronic Systems. John Wiley, New York
169. Morrison R. (1977): Grounding and Shielding Techniques in Instrumentation. Wiley Interscience, New York
170. Wilson P. F. (1995): Electrostatic Discharge. In *Perez R. (ed.): Handbook of Electromagnetic Compatibility*. Academic Press, San Diego

171. Holm R. (1967): Electric Contacts: Theory and Application. Springer-Verlag, Berlin/Heidelberg/New York
172. Jay F. (Ed.) (1997): IEEE Standard Dictionary of Electrical and Electronics Terms. Wiley Interscience, New York
173. Glasser L. A. and Dobberpuhl D. W. (1985): The Design and Analysis of VLSI Circuits. Addison-Wesley, Reading, MA
174. Weste N.H.E. and Eshraghian K. (1993): Principles of CMOS VLSI Design – A Systems Perspective. Addison-Wesley, Reading, MA
175. Illingworth V. (1998): The Penguin Dictionary of Electronics. Penguin Books, London
176. Montrose M. I. (1999): EMC and the Printed Circuit Board. IEEE Press, New York
177. Ling D. A. and Ruehli A. E. (1987): Interconnection Modeling. In *Ruehli A. E. (Ed.): Circuit Analysis, Simulation and Design*, vol. 3, part 2, Elsevier, Amsterdam
178. Weston D. A. (1991): Electromagnetic Compatibility: Principles and Applications. Marcel Dekker, New York
179. Chern J.-H., Huang J., Arledge A., Li P.-C. and Yuang P. (1992): "Multilevel metal capacitance models for CAD design synthesis systems". IEEE Electron Device Letters, **13**(1), pp 32–34
180. Grover F. W. (1947): Inductance Calculations. Working Formulas and Tables. Van Nostrand, New York
181. Paul R. C. (1992): Introduction to Electromagnetic Compatibility. John Wiley, New York
182. Ott W. H. (1988): Noise Reduction Techniques in Electronic Systems. John Wiley, New York
183. Matthaei G. L., Young L., Jones E.M.T. (1980): Microwave Filters, Impedance-Matching Networks and Coupling Structures. Artech House, Dedham, MA, pp 174–177
184. Westman H. P. (Ed.) (1972): Reference Data for Radio Engineers. Howard Sams & Co., Indianapolis
185. Williams T. (1996): EMC for Product Designers. Newnes, Oxford
186. Heijningen M., Compiet J., Wamback P., Donnay S., Engels M. and Bolsens I. (2000): "Analysis and experimental verification of digital substrate noise generation for epi-type substrates." IEEE Journal of Solid-State Circ., **35**(7), pp 1002–1008
187. Briaire J., Krisch K. S. (2000): "Principles of substrate crosstalk generation in CMOS circuits." IEEE Trans. on CAD of Int. Circ. & Syst., **19**(6), pp 645–653
188. Hamel J.S., Stefanou S., Bain M., Armstrong B.M. and Gamble H.S. (2000): "Substrate crosstalk suppression capability of silicon-on-insulator substrates with buried ground planes." IEEE Microwave and Guided-Wave Letters, **10**(4), pp 134–135
189. Wu J., Scholvin J., del Alamo J.A. and Jenkins K.A. (2001): "A Faraday cage isolation structure for substrate crosstalk suppression." IEEE Microwave and Guided-Wave Letters, **11**(10), pp 410–413
190. Zhang X. (1996): "Coupling effects on wire delay." IEEE Circuits & Devices, **12**(6), pp 12–18

191. Alpert C. J., Devgan A., Fishburn J. P. and Quay S. T. (2001): "Interconnect synthesis without wire tapering." IEEE Trans. on CAD of Integrated Circ. & Syst., **20**(1), pp 90–104
192. Nagata M., Nagai J., Hijikata K., Morie T., Iwata A. (2001): "Physical design guides for substrate noise reduction in CMOS digital circuits." IEEE Journal of Solid-State Circ., **36**(3), pp 539–549
193. Ott, W. H.(1988): Noise Reduction Techniques in Electronic Systems. John Wiley, New York
194. Williams T. (1996): EMC for Product Designers. Newnes, Oxford
195. Kodali V. P. (1996): Engineering Electromagnetic Compatibility: Principles, Measurements, and Technologies. IEEE Press, New York
196. White D. R. J. and Mardiguian M. (1974): Electromagnetic Shielding. In *A Handbook Series on Electromagnetic Interference and Compatibility*, vol. 3, Interference Control Technologies Inc., Gainesville, VA
197. Fish P. J.(1994): Electronic Noise and Low-Noise Design. McGraw-Hill, New York
198. Horowitz P. and Hill W. (1989): The Art of Electronics. Cambridge University Press, Cambridge
199. Montrose M. I. (1999): EMC and the Printed Circuit Board. IEEE Press, New York
200. Hemming L. H. (1993): Grounding, Shielding, and Filtering. In *Dorf R. C. (Ed.): The Electrical Engineering Handbook*, CRC Press, Boca Raton, pp. 903–919
201. Madou A. and Martens L. (2001): "Electrical behavior of decoupling capacitors embedded in multilayered PCBs". IEEE Trans. on Electromagnetic Compatibility, **43**(4), 549–556
202. Zhao S., Roy K. and Koh C.-K. (2002): "Decoupling capacitance allocation and its application to power-supply noise-aware floorplanning". IEEE Trans. on CAD of Integrated Circuits and Systems, **21**(1), 81–92
203. Wang Wei-Shin (2002): "Bluetooth: a new era of connectivity". IEEE Microwave Magazine, **3**(3), 38–42
204. Fowler P. (2002): "5 GHz goes the distance for home networking". IEEE Microwave Magazine, **3**(3), 49–55
205. Motchenbacher, C.D., Connelly, J. A. (1993): Low Noise Electronic Design. John Wiley, New York
206. Ott, W. H. (1988): Noise Reduction Techniques in Electronic Systems. John Wiley, New York
207. Leach W. M. (1994): "Fundamentals of low-noise analog circuit design". Proc. IEEE, **82**(10), 1514–1538
208. Letzter, S., Webster N. (1970): "Noise in amplifiers". IEEE Spectrum, **7**(8), 67–75
209. Sherwin, J. (1982): "Noise specs confusing?" In National Semiconductor: Linear Applications Databook, Applic. Note 104
210. Netzer Y. (1981): "The design of low-noise amplifiers". Proc. IEEE, **69**(6), 728–741
211. Faulkner, E. A. (1966): "Optimum design of low-noise amplifiers". Electronics Letters, **2**(11), 426–427
212. Faulkner, E. A. (1975): "The principles of impedance optimization and noise matching". Journal of Physics E: Scientific Instr., **8**, 533–540

213. Munroe, D. M. (1982): "Signal-to-noise ratio improvement". In: *Sydenham, P. H. (ed.): Handbook of Measurement Science*. John Wiley, New York, 431–487
214. Netzer, Y. (1974): "A new interpretation of noise reduction by matching". Proc. IEEE, **62**(3), 404–406
215. Shama'a, O., Linscott, I., Tyler, L. (2001): "Frequency scalable SiGe bipolar RF front-end design". IEEE Journal of Solid-State Circ., **36**(6), 888-895
216. Vergers, C. A. (1979): Handbook of Electrical Noise: Measurement & Technology. Tab Books, Blue Ridge Summit, PA
217. Gonzalez G. (1984): Microwave Transistor Amplifiers: Analysis and Design. Prentice-Hall, Englewood Cliffs, NJ
218. Link N. G., Gudimetla Rao V.S. (1995): "Analytical expressions for simplifying the design of broadband low noise microwave transistor amplifiers". IEEE Trans. on MTT, **43**(10), 2498–2501
219. Graf R. F. (1977): Modern Dictionary of Electronics. Howard W. Sams & Co., Indianapolis
220. Van der Ziel A. (1970): Noise: Sources, Characterization, Measurements. Prentice Hall, Englewood Cliffs, New Jersey
221. Fish P. J. (1994): Electronic Noise and Low-Noise Design. McGraw-Hill, New York
222. Pettai R. (1984): Noise in Receiving Systems. John Wiley, New York
223. Horowitz P., Hill W. (1989): The Art of Electronics. Cambridge University Press, Cambridge, New York
224. Motchenbacher C.D., Connelly J. A. (1993): Low-Noise Electronic Design. John Wiley, New York
225. Ott H. W. (1988): Noise Reduction Techniques in Electronic Systems. Wiley Interscience, New York
226. Vergers C. A. (1979): Handbook of Electrical Noise: Measurement & Technology, Tab Books, Blue Ridge Summit, PA
227. Connor F. R. (1982): Noise. Edward Arnold, London
228. Ambrozy A. (1982): Electronic Noise. McGraw-Hill, New York
229. Mumford W. W., Scheibe E. H. (1967): Noise Performance Factors in Communication Systems. Horizon House - Microwave Inc., Dedham, MA
230. Randa J., Dunleawy L.P., Terrell L.A. (2001): "Stability measurements on noise sources". IEEE Trans. on Instr.& Meas., **50**(2), 368–372
231. Conrad G.T., Newman N. and Stansbury A.P. (1960): "A recommended standard resistor-noise test system". IRE Trans. on Component Parts, **CP-7**(3), 71–88
232. Rice S.O. (1944): "Mathematical analysis of random noise". Bell Syst. Tech. Journal, **23**(4), 282–332
233. Motchenbacher C.D., Connelly J. A. (1993): Low Noise Electronic Design. John Wiley, New York
234. Connor F. R. (1982): Noise. Edward Arnold, London
235. Fish P. J. (1994): Electronic Noise and Low-Noise Design. McGraw-Hill, New York
236. Ziemer R. E., Tranter W. H. (1976): Principles of Communications: Systems, Modulation and Noise. Houghton Mifflin, Boston
237. Pease R. A.: "A low-noise precision Op Amp". National Semiconductor Corp., Linear Brief 52

238. Jay F. (Ed.) (1997): IEEE Standard Dictionary of Electrical and Electronics Terms. Wiley Interscience, New York
239. Westman, H. P. (Ed.), (1972): Reference Data for Radio Engineers. Howard W. Sams & Co., Indianapolis
240. Motchenbacher, C.D., Connelly, J. A. (1993): Low Noise Electronic Design. John Wiley, New York
241. Ott, W. H. (1988): Noise Reduction Techniques in Electronic Systems. John Wiley, New York
242. Ziemer R. E., Tranter W. H. (1976): Principles of Communications: Systems, Modulation, and Noise. Houghton Mifflin, Boston
243. Pettai R. (1984): Noise in Receiving Systems. John Wiley, New York
244. Connor F. R. (1982): Noise. Edward Arnold, London
245. Illingworth V. (Ed.), (1998): Dictionary of Electronics. Penguin Books, London, England
246. Graf R. F. (1977): Modern Dictionary of Electronics. Howard W. Sams & Co., Indianapolis
247. Engberg J., Larsen T. (1995): Noise Theory of Linear and Nonlinear Circuits. John Wiley, Chichester
248. Smith J. (1986): Modern Communication Circuits. McGraw-Hill, New York
249. Losee F. (1997): RF Systems, Components, and Circuits Handbook. Artech House, Norwood, MA
250. Motchenbacher C.D., Connelly J. A. (1993): Low Noise Electronic Design. John Wiley, New York
251. Dobrowolski J., Ostrowski W. (1996): Computer-Aided Analysis, Modeling, and Design of Microwave Networks. Artech House, Boston
252. MicroSim Corp. (1987): PSpice. MicroSim Corp., Laguna Hills, CA
253. Meyer R. G., Nagel L., and Lui S. K. (1973): "Computer simulation of 1/f noise performance of electronic circuits." IEEE Journal of Solid-State Circ., **SC-6**(6), pp. 237–240
254. Fish P. J. (1994): Electronic Noise and Low-Noise Design. McGraw-Hill, New York
255. Roberts G. W., Sedra A. S. (1997): SPICE. Oxford University Press, Oxford
256. Ortiz J., Denig C., "Noise figure analysis using Spice". *Microwave Journal*, vol. 35, no. 4, April 1992, 89–94.
257. Rizzoli V., Lipparini A. (1985): "A CAD solution to the generalized problem of noise figure calculation". 1985 IEEE MTT-S Digest, 699-702
258. Vasilescu G. (1994): Noise Simulation of Microwave Circuits. Ph. D. thesis, Université Pierre et Marie Curie, Paris
259. Xia C.-X. (1997): A User's Interface for NOF. Internal report, Université Pierre et Marie Curie, Paris
260. White D. R. J. and Mardiguian M. (1974): Electromagnetic Shielding. In *A Handbook Series on Electromagnetic Interference and Compatibility*, vol. 3, Interference Control Technologies Inc., Gainesville, VA
261. Ott W. H. (1988): Noise Reduction Techniques in Electronic Systems. John Wiley, New York
262. Rainal A. J. (1984): "Computing inductive noise of chip packages". AT& T Bell Lab. Techn. Journal, **63**(1), pp 177–195
263. Horowitz P., Hill W. (1989): The Art of Electronics. Cambridge University Press, Cambridge, New York

264. Grover F. W. (1947): Inductance Calculations. Working Formulas and Tables. Van Nostrand Co., New York
265. Fraden J. (1989): Handbook of Modern Sensors: Physics, Designs, and Applications. AIP Press, Woodbury, New York
266. Wiliams T. (1996): EMC for Product Designers. Newnes, Oxford
267. Morrison R. (1992): Noise and Other Interfering Signals. John Wiley, New York
268. Holm R. (1967): Electric Contacts: Theory and Application. Springer-Verlag, Berlin/Heidelberg/New York
269. Fish P. J. (1994): Electronic Noise and Low-Noise Design. McGraw-Hill, New York
270. Widlar R. (1988): "Working with high impedance Op Amps". National Semiconductor, AN 241
271. Widlar R. (1988): "IC Op Amp beats FETs on input current". National Semiconductor, AN 29
272. Haarmann G. (1989): "Comment réaliser et réparer tous les montages électroniques". Tome III, 4/14.4, 27 complément, Weka Loisiers, Paris, France

Appendix –
Values of General Physical Constants

Constant	Symbol	Value
Electronic charge	q	$1.602 \cdot 10^{-19}$ C
Electronic mass	m	$9.109 \cdot 10^{-31}$ Kg
Boltzmann constant	k	$1.381 \cdot 10^{-23}$ J/K
Planck's constant	h	$6.626 \cdot 10^{-34}$ Js
Stefan-Boltzmann constant	σ	$5.670 \cdot 10^{-8}$ W/m^2(K)^4
Velocity of light	c	$2.998 \cdot 10^{8}$ m/s
Permeability of free space	μ_o	$1.257 \cdot 10^{-6}$ H/m
Permittivity of free space	ε_o	$8.849 \cdot 10^{-12}$ F/m

Index

1/f noise   60

absorber   265
absorption   265, 283
absorption loss   284
active power   72
admittance matrix   122
admittance representation   136
antenna-mode coupling   276
aperiodic signals   17
apertures   291
apparent power   72
arc discharge   328
arc suppression   330
arc welders   260
atmospheric noise   256
attenuating pad   146, 538
attenuation   265
attenuator   535
attenuator pads   479
autocorrelation   32, 36
automotive ignition systems   260
available power   74
available power gain   75, 79
avalanche diode noise source   447
avalanche noise   65, 524
average   26, 29
average input noise temperature   107
average noise factor   107
average normalized power   74
average power   73
average value   17
average values of noise parameters   107

balance ratio   391
balanced circuit   299, 389
balanced line   389
balanced X section   151
balun   265, 299, 300
Barkhausen noise   426
battery noise   177
bilateral power spectrum   35
binomial distribution   27
Bluetooth   408
bolometer   56
bond   265
bonding   265
braided shield   629
bridge amplifier   522
bridge rectifier   321
buried capacitors   395
bus   265
bypass capacitors   681

cable entries   293
cables   309, 407
capacitance estimation   358
capacitance matrix   356
capacitive coupling   271, 371
capacitive crosstalk   665
capacitor noise   169
cascade two-ports   103
cavity resonance   632
central limit theorem   28
chain matrix   122
channel noise   206, 225, 235
characteristic impedance   280, 376
characteristic-noise matrix   141

chokes 298
circuit shielding 286
clock radiation 336
coaxial cables 397
coherent signals 20
collective approach 54
common ground impedance 270
common-emitter amplify 502
common-mode choke 298, 402
common-mode coupling 276
common-mode radiation 407
common-mode rejection ratio 274
complex conjugate power 73
complex power 73
conducted noise 269
connectors 408
contact bounce 328
contact protection 331
contact switching 327
continuous random variable 28
controlled sources 593
corpuscular approach 54
correlation 20, 30, 125
correlation coefficient 22, 31
correlation matrix 39, 41, 134, 145
coupling coefficient 372
coupling modes 273
coupling paths 268
coupling transformer 424
covariance 30
cross-correlation function 37
cross-power spectral density 37
crosstalk 265, 351, 370, 371, 373, 663
crosstalk noise 5
crosstalk reduction 386
cumulative distribution function 25
current spectral density 34

decoupling capacitor 340, 395, 682
decoupling filters 393
deep-submicron noise 4
definitions of noise 1
Delta-I noise 4
dependent source 593
differential mode coupling 275
differential-mode emission 405
diffusion noise 51
diode noise source 444
discrete random variable 27

dish antenna 259, 579
distributed model 220, 230
distributed noise sources 258
Dolby system 437
downlink 567, 572
downlink path 568
downlink power budget 569
drain temperature 242

E_n–I_n model 89, 190, 418
earth grounds 301
effective isotropic radiated power 568
effective noise temperature 97
effective radiated power 568
electric motors 260
electrical noise 2
electrochemical contamination 263
electromagnetic compatibility 265
electromagnetic coupling 273, 375
electromagnetic environment 265
electromagnetic gasket 265
electromagnetic immunity 265
electromagnetic interference 266
electromagnetic leakage 291
electromagnetic perturbation 266
electromagnetic susceptibility 266
electrostatic discharge 343
emission 266
energy spectral density 35
ensemble 24
environmental perturbations 5
equivalent drain temperature 241
equivalent gate temperature 241
equivalent input noise 89, 420
equivalent noise current 85
equivalent noise resistance 83, 91, 461
ergodic process 26
excess noise 164, 465, 475, 607
excess noise power 93
exchangeable power 77
exchangeable power gain 78, 79
extended noise factor 104
extrinsic noise 5, 8

far-field 266, 280
feedback 430
feeder 574, 577
feedthrough capacitor 294
ferrite bead 298, 645

Index   707

ferrite-rod antenna   562
FET   489
FET noise   496
filament   362
filter   296, 317, 326
filter configurations   640
filter installation   319
filtering   407
filtering of wires   396
flicker (1/f) noise   235
floating power supply   324
fluctuating power   72
fluctuation   16, 21, 23
form factor   20
Fourier transform   33
free-space loss   568
full-wave rectifiers   320

G/T ratio   569
galactic noise   257
galvanic action   263
gamma section   149
gas discharge   328
gaskets   289
gate noise   207, 225, 235
gate temperature   243
generation-recombination noise (G-R noise)   58
ground   301
ground (earth)   266
ground bounce   4
ground impedance   301
ground loop   307, 399
ground plane   266, 383, 648
grounding   382, 398, 406
grounding capacitor   651
grounding techniques   303
guard shield   288, 636

Heisenberg principle   56
heterojunction bipolar transistor (HBT)   199
high electron mobility transistor (HEMT)   236
high-voltage transmission lines   260
human body charging   346

ideal differential amplifier   274
impedance matrix   122

impedances unequal temperatures   476
independent sources   592
inductance   361
inductance of a trace   364
induction charging   345
inductive coupling   373
inductive noise   3, 171, 334
inductor noise   171
input noise temperature   92, 93, 109
input noise voltage   457
insertion loss   296
insertion power gain   79
instantaneous power   72
integrated polysilicon resistors   168
interconnect   654
interconnect resistance   353
interconnection   351
interference   2
interference control   403
interfering signal   255, 266
intervalley scattering noise   222
intrinsic noise   4
ISM equipment   261
isolation   266
isolation amplifier   401
isolation transformer   400

joint cumulative distribution function   29
joint probability density function   29
junction field effect transistor (JFET)   202

Kleckner's correlation coefficient   41

linear multiport   108
linear system   38
linear two-port   93, 101
lines   338
loop antenna   635
low-noise design   7, 415
lumped models   220

magnetic coupling   273
matched attenuator   539
mean square value   18
mechanical contacts   264
MESFET transistor   219
metallic enclosure   631

minimum detectable signal   554
minimum discernible signal   554
minimum noise factor   129
model of Rothe and Dahlke   125
MOS   490, 493
MOS transistor   213
multiconductor line   364
multiport representation   140
multistage amplifiers   82, 542
mutual capacitance   354
mutual inductance   354, 363

near-field   266, 280
NOF   158, 621
noise   1
noise bandwidth   80, 529, 613
noise current spectral density   47
noise equivalent power   88
noise factor   421, 535
noise factor (noise figure)   97
noise figure   417, 459
noise floor   553
noise immunity   7
noise in bipolar transistors   178
noise in operational amplifiers   249
noise index   165, 608
noise matching   415
noise matrix   134
noise measure   105
noise of a linear multiport   108
noise parameters   129
noise parameters extraction   132
noise power spectral density   47
noise ratio   87
noise signal   21
noise source   443
noise surface   131
noise temperature   85
noise temperature of the antenna   571
noise voltage spectral density   47
noise wave representation   136
noiseless resistors   597
noisy grounds   306
noisy multiport   140
noisy two-port   124, 127
normal distribution   28
normalized admittance correlation matrix   136

normalized chain correlation matrix   135
normalized energy   34
normalized power   19
North definition   99
Nyquist's theorem   47

on-chip decoupling capacitors   395
one-port at different temperatures   119
one-port at uniform temperature   115
operating noise factor   103
operating noise temperature   96, 110
operational amplifier   515, 532
optocoupler   400

pad   535
Parseval's theorem   34
partial correlation   31
partially correlated   20
PCB design   380
PCB optimization   381
peak factor   20
periodic signals   17
Pierce's rule   119
piezoelectric effect   262
pigtail   310
pink noise   61, 449
pink noise source   450
plane wave   266, 285
Poisson distribution   28
popcorn (burst) noise   64
power spectra   33
power spectral density   34
preamplifier   485, 560
precipitation static   256
probability density function   25

quantum noise   56
quiet ground   384

radiated noise   271
radiated spectrum   679
radiation   267
random process   24, 37
random signal   73, 78
random variable   23
reactive power   72
receiver sensitivity   553, 558
receivers   553

Index

reception loss   285
reflection   267
relationship between S/N ratio and F   131
relative conductivity   267, 283
relative permeability   267, 283
relay   687
resistor noise   163
reverse gamma section   150
ribbon cable   311, 667
ripple   326
root-mean-square value   18
RTS noise   65

sample function   24
Schottky diodes   55
Seebeck effect   264
selectivity   554
self capacitance   357, 358
self- and cross-power spectra   39
sensing circuits   504
sensor   434, 487, 504, 671
sheet resistance   353
shield   267
shield impedance   280
shielded cable   310
shielding   279, 406, 629
shielding effectiveness   267, 288, 289, 633
shielding enclosure   267
shot noise   52
signal   17
signal grounds   301
signal guarding   639
signal-to-noise ratio   88, 91, 421
simultaneous switching noise   4
skin depth   283
skin effect   282
sky noise   258
solder joints   264
spectrum   34
SPICE   584
spot noise parameters   83
spurious signal   2
standard reference temperature   86
standing waves   292
static noise   3
stationary process   26

stochastic resonance   11
stochastic signals   17
substrate noise   385
superposition of fluctuations   36
supply routing   384
surge suppressors   317
switch protection circuits   684
switching-mode power supplies   325

temporal noise   3
thermal noise   46
trace capacitance   359
trace on a PCB   366
trace-to-ground capacitance   360
trace-to-trace capacitance   361
track   351, 359
transducer   509, 669
transducer power gain   79
transformation matrix   145
transformer noise   340
transformer shielding   319
traps   59
triboelectric effect   262, 344
tunnel-diode oscillator   580
twisted pair   375, 397
two-ports cascade   95

unbalanced Π section   150
unbalanced T section   150
unilateral power spectrum   35
unilateral spectrum   81
uplink   567

ventilation slots   291
voltage regulator   322
voltage spectral density   34

wave impedance   280, 281
wave reflection   284
waveguide   291
white noise   468
Wiener-Khintchine theorem   36, 40

x-caps   297

y-caps   297

Zener diode noise source   446

Printing: Krips bv, Meppel
Binding: Litges & Dopf, Heppenheim